"John is the only person I know that can fill the gap between technology and philosophy; and he really did in this book. I recommend this text for students and professional engineers, as well as for non-experienced people who are interested in getting a frank and sometimes humorous assessment of gas turbine technology."

—Alberto Traverso, University of Genoa

"This book is clearly written by an expert with a lot of industry experience in OEMs (original equipment manufacturers) and operation, resulting in a book that shows a very practical approach to design and analysis of turbomachinery that matters in the real world without lacking the theoretical depths that are necessary to understand the topic thoroughly. This very well-written book covers the theoretical basics of thermodynamics as well as components such as the compressor, the combustor, the turbine, the whole engine, and additional topics needed to understand the analysis and design of electrical power generation equipment. It provides a comprehensive overview for everyone interested in this fascinating topic, be it a practitioner with OEMs and utilities, or academics, such as a researcher or a new student of the field.

This long-awaited book closes a gap in the literature between the practitioner's view and a purely theoretical approach."

—Hans-Juergen Kiesow, ABB, Siemens (retired)

"It is rare that one comes across a book that can be considered seminal in the area of gas turbine engineering and that provides an excellent blend of theory and practice of the state of the art of heavy- duty advanced gas turbines. The book provides a detailed, lucid, and insightful treatment of a wide range of gas turbine topics, including history, cycles, components and their interactions, and technology trends. It provides a quantitative and qualitative treatment of the subject matter with usable equations, insights, and rules of thumb that enable quick design checks and calculations. It will be of immense value to designers and users of gas turbines. John's technical leadership over the past two and a half decades has contributed immeasurably to the current understanding of large advanced gas turbines. Much of this expertise has been successfully encapsulated in this book. This book is of archival quality and will endure and enrich gas turbine engineers for decades to come."

—Cyrus B. Meher-Homji, PE, Bechtel Fellow and Turbomachinery Technology Manager, Bechtel Corporation

Gas Turbines for Electric Power Generation

In this essential reference, both students and practitioners in the field will find an accessible discussion of electric power generation with gas turbine power plants using quantitative and qualitative tools. Beginning with a basic discussion of thermodynamics of gas turbine cycles from a second law perspective, the material goes on to provide an in-depth analysis of the translation of the cycle to a final product, facilitating quick estimates.

In order to provide readers with the knowledge they need to design turbines effectively, there are explanations of simple- and combined-cycle design considerations and state-of-the-art performance prediction and optimization techniques, as well as rules of thumb for design and off-design performance and operational flexibility and simplified calculations for myriad design and off-design performance. The text also features an introduction to proper material selection, manufacturing techniques, and the construction, maintenance, and operation of gas turbine power plants.

S. Can Gülen (PhD) PE, Bechtel Fellow, ASME Fellow, has a combined 25 years of mechanical engineering experience covering a wide spectrum of technology, system, and software design, development, assessment, and analysis in the field of steam and gas turbine combined-cycle process and power plant turbomachinery and thermodynamics at Thermoflow, Inc., General Electric, and Bechtel.

He has written numerous technical papers and journal articles on design practices and technical assessment reports. He holds more than 20 US patents on gas turbine performance, cost, optimization, data reconciliation, analysis, and modeling.

Gas Turbines for Electric Power Generation

S. CAN GÜLEN
Bechtel Infrastructure & Power, Inc.

CAMBRIDGE
UNIVERSITY PRESS

University Printing House, Cambridge CB2 8BS, United Kingdom

One Liberty Plaza, 20th Floor, New York, NY 10006, USA

477 Williamstown Road, Port Melbourne, VIC 3207, Australia

314-321, 3rd Floor, Plot 3, Splendor Forum, Jasola District Centre, New Delhi - 110025, India

103 Penang Road, #05-06/07, Visioncrest Commercial, Singapore 238467

Cambridge University Press is part of the University of Cambridge.

It furthers the University's mission by disseminating knowledge in the pursuit of education, learning and research at the highest international levels of excellence.

www.cambridge.org
Information on this title: www.cambridge.org/9781108416658
DOI: 10.1017/9781108241625

First published 2019

A catalogue record for this publication is available from the British Library

Library of Congress Cataloging in Publication data
Names: Gülen, S. Can, 1962– author.
Title: Gas turbines for electric power generation / S. Can Gülen (Bechtel Infrastructure & Power, Inc.).
Description: Cambridge ; New York, NY : Cambridge University Press, 2019. | Includes bibliographical references and index.
Identifiers: LCCN 2018039284 | ISBN 9781108416658 (hardback : alk. paper)
Subjects: LCSH: Gas-turbines. | Electric power production. | Electric generators.
Classification: LCC TJ778 .G8284 2019 | DDC 621.31/2133–dc23
LC record available at https://lccn.loc.gov/2018039284

ISBN 978-1-108-41665-8 Hardback

Contents

Preface

He who defends everything, defends nothing.

Frederick II of Prussia

To paraphrase the great Prussian emperor, to cover everything is to cover nothing. This book is a monograph on a specific class of heat engines, namely the heavy-duty industrial gas turbine for electric power generation. It does not cover gas turbines used for aircraft, marine, or land-based vehicle propulsion (i.e., there is no discussion of turbofan, turbojet, or turboprop engines, except in a historical context). Its focus is fully on land-based (i.e., stationary shaft) power generation and conversion thereof into electric power via alternating current synchronous machines; especially so-called frame machines with outputs of 100 MWe or more, which are also known as heavy-duty industrial gas turbines. In other words, a discussion of "microturbines" is not to be found herein.

This is not a textbook, although it can be used by a student of mechanical engineering with sufficient background in thermodynamics, fluid mechanics, and heat transfer as a useful reference to help complete certain class assignments.

This is not a handbook of generalized information either. In our times, the necessity of handbooks is debatable. A wide range of detailed information on any given subject is only one mouse click away. This book is a compendium of expert knowledge, which is either impossible to find online or, even if found, is either too sketchy or too diluted or too obscure to be of immediate, practical use.

The intended audience is primarily professionals (i.e., engineers and researchers) who are working in the industry or in research organizations on various aspects of electric power generation with gas turbines. Graduate and undergraduate students who are working on projects toward their degree with an ultimate goal of joining the industry can be added to this group as well.

It is an advanced text with little material of an introductory nature (mostly a few paragraphs to start the main narrative). In other words, one will not find the derivation of Navier–Stokes equations from the analysis of an infinitesimally small control volume of fluid in this book. The goal of the author is to provide the reader with specialized knowledge, calculation methods, and tools that can be readily applied to the solution of the day-to-day problems encountered in the design, development, optimization, operation, and maintenance of gas turbine power plants. Said methods and tools comprise specific data (some hard to find – even on the Internet, at least not in a compact and readily usable form), practical formulae, Visual Basic code, charts, and rules of thumbs. Most of the specific methods and tools have been developed and used by the author over the course of more than two decades spent in the industry.

What is the use of such a monograph? After all, at the time of writing (i.e., near the end of the second decade of the twenty-first century), almost all aspects of gas turbine power plant design are dominated by highly sophisticated, extremely expensive (i.e., not available to individuals) computer software with steep learning curves. These "black box" tools incorporate the latest techniques in computational fluid dynamics and finite element analysis fortified with flashy graphical user interfaces and other "digital" accoutrements on the most advanced computing platforms to enable engineers (some fresh out of school) to design, say, advanced airfoils in order to squeeze the last 0.01 percent efficiency from the compressor or the turbine. Most complicated gas turbine combined cycle calculations for tens or hundreds of cases can be done in a matter of seconds by user-friendly heat balance simulation software.

The goal of this monograph, as envisioned by the author, is to provide the junior engineer or researcher using those tools, as well as his or her supervisor with decades of experience under his or her belt, with a single source of reference to put every little detail in its rightful place in the proverbial big picture, which will be expounded upon in the next few pages.

Before moving on, however, there is a simple fact that needs to be stated unequivocally. This book is dedicated to the memory of Mustafa Kemal Atatürk (1881–193∞) and his elite cadre of reformers. Without their vision, sacrifices and groundbreaking work, there would not have been a fertile ground where my parents, teachers, mentors, family and friends could shape me into the author of this book.

1 Introduction

The "big picture" mentioned at the conclusion of the Preface can be best described in an analogy to the Russian nesting doll (*matryoshka*) comprising a set of wooden dolls of decreasing size placed one inside another.

- The biggest, outermost doll is the electric power industry.
- The second doll represents heat engines in electric power generation.
- The third doll represents gas turbine power plants in the realm of heat engines.
- The fourth doll is the gas turbine itself.
- The fifth to seventh dolls are major gas turbine components: compressor, combustor, and expander.
- The remaining dolls represent individual building blocks of each component: vanes, blades, disks (wheels), combustor cans, tie rods, turbine rings, etc.

Consider a design engineer working on the first turbine stage stator vane design using a proprietary 3D computational fluid dynamics code. In case you are wondering, original equipment manufacturers (OEMs) deploy teams of engineers for the stage-by-stage design of an advanced gas turbine's "hot gas path." Each engineer, many of them with PhDs in thermofluids from prestigious universities, specializes in one part among many using highly complex software to solve, say, the full Reynolds-averaged Navier–Stokes equations in three dimensions, which describe the flow of hot combustion gas across rows of vanes and blades in each stage of the turbine. It is easy to imagine the requisite skill and knowledge to perform such tasks adequately.

Yet even more knowledge and skill (and experience) are required to assess the impact of a particular tweak to, say, the stage 1 nozzle vane geometry on the efficiency and output of the gas turbine in question, in a simple or combined cycle, and, in turn, the impact of said gas turbine's operation in the field on the overall generation portfolio's, say, carbon footprint.

This book is not about how to design a particular part; there are tomes written on the underlying theory of governing aerothermodynamic principles (they will be mentioned later in the book). Specialized and highly proprietary knowledge is available in OEM design practices, theory manuals of specific software packages, and, to a certain extent, in archival papers published in academic journals (containing mostly results and few background details). Even if one is armed with all of these tools and resources, nobody sets out to design an entire heavy-duty industrial or aeroderivative gas turbine (or one of its components or one of the parts thereof) from a proverbial "blank sheet" on his or her own anymore.

This book is about the first four "dolls" of the *matryoshka* described above. The fifth to seventh dolls will be covered in sufficient detail for one to gain a good grasp of the impact of particular design decisions on the first four. Finally, the remaining dolls will be touched upon briefly mainly to illustrate where the technology is, where it came from, and how far it can go.

Electric power is the lifeblood of contemporary human civilization. Electricity consumption per capita is shown to have a strong correlation with key social development and economic indices (i.e., human development index, gross domestic product per capita, etc.). This simple fact is both a boon and a bane simultaneously, as evidenced by the great disparity between developed and underdeveloped countries. Consider that China and India, two up-and-coming countries with populations well above one billion each, lag well behind, say, the USA in this key measure: about 4,000 and 600 kWh/person for China and India, respectively, vis-à-vis more than 12,000 for the USA! There is increasing concern that the unstoppable striving of these two giants of the so-called Third World toward a better life standard will push fossil fuel-based electricity production to levels where the global drive to limit carbon dioxide emissions will become an exercise in futility. It is not a stretch to assume that readers of these lines are already well versed in the dangers of global warming and its connection to anthropomorphic greenhouse gas emissions, of which CO_2 is by far the biggest culprit.

This brings us to the subject of electric power (or electricity) generation from burning a particular fossil fuel – namely, natural gas – in advanced gas turbine-based power plants. At the time of writing, natural gas-fired power plants, primarily in the form of combined cycles, are by far the most efficient means of electric power generation via combustion of fossil fuels. Their *rated* performance (i.e., similar to the sticker performance of a passenger car) is more than 60 percent net thermal efficiency at power outputs of 500–1,000 MWe or even more. Best-in-class power plants can reach 56–57 percent thermal efficiency during actual field operation. The average efficiency of all natural gas-fired power plants in the USA in 2015 was about 46 percent, vis-à-vis about 33 percent for all coal-fired plants (as reported by the US Energy Information Administration [EIA]). Furthermore, in terms of specific CO_2 emissions (i.e., pounds of CO_2 emitted per MWh of electricity generated), coal-fired generation is by far the worst culprit, at nearly 150 percent more than that by natural gas-fired generation (again, in 2015 per the EIA).

In 2015, natural gas and coal each accounted for about a third of all US electricity generation, with more than 1,700 power plants utilizing the former and more than 500 power plants utilizing the latter. Almost all natural gas-fired power plants are based on gas turbine technology. Finally, in several months of 2016, US natural gas-fired electricity generation surpassed coal-fired generation. The top two driving factors leading to this trend are low natural gas prices due to the shale gas boom and the much more favorable emissions characteristics of natural gas vis-à-vis coal and all other types of fossil fuels.

Increasing penetration of intermittent renewable technologies (i.e., solar and wind) into the electricity generation portfolio requires readily available backup from fossil fuel-fired generating capacity. In the USA, the contribution of wind and solar to electricity generation in 2015 was about 5 percent (mostly from wind). When wind stops blowing, especially on short notice, the most efficient, clean, and fast-responding power plants to pick up the slack are natural gas-fired gas turbine power plants. In the early years of the second decade of the twenty-first century, this dynamic was thought to be a very important driver of gas turbine-based electricity generation.

In the age of the Internet, it is pointless to cite a lot of statistics, which will be obsolete within the span of a few years (if not sooner), in a print book. All sorts of numbers are readily available online. Barring a "black swan" event of a catastrophic nature (e.g., an accident associated with shale gas extraction via fracking leading to huge loss of life) or of a fortuitous one (e.g., successful culmination of cold fusion research into the large-scale commercialization of it in electric power generation), heavy-duty industrial gas turbines will play a significant role in electric power generation in the USA, Europe, and many other places in the world in the remainder of the twenty-first century.

Nevertheless, it is difficult to guess the magnitude of the aforementioned "significant role." Around 2010, it was foreseen that there would be about 300 large gas turbines sold each year, but in 2013, just 212 were ordered worldwide. In 2017, this number was 122. Major OEMs reported declining sales and, even worse, the most advanced turbines were being sold at no margin or sometimes at a loss. At the time of writing (late 2017), it is difficult to gauge whether the decline in industrial gas turbine sales is just a cyclical problem or whether there is a structural problem in the industry.

The underlying technology is more than a century old and its basic features are not expected to change drastically. There are many excellent treatises of the gas turbine theory out there, going back to the monumental work of Aurel Stodola in the early twentieth century. This book does not intend to retread the path of giants in the field. In fact, they constitute the foundation with which any serious student of the subject matter must be familiar. Herein, core scientific principles of mechanical, electrical, operational, and economic aspects of the land-based gas turbine generator are going to be distilled into easily usable "bites" to help the practitioner and student alike in their daily work.

The book is divided into four parts (plus Appendices):

I. Prerequisites
II. Fundamentals
III. Extras
IV. Special Topics

Part I includes material that the reader must be familiar with in order to derive the maximum benefit from the book and ends with a brief history of the gas turbine (Chapter 4). This chapter will start with three "vignettes" covering three historical gas turbines.

It is interesting to note that the concept of a gas turbine has been around much longer than is commonly realized. In his treatise *Mathematical Magick* of 1648 (a second

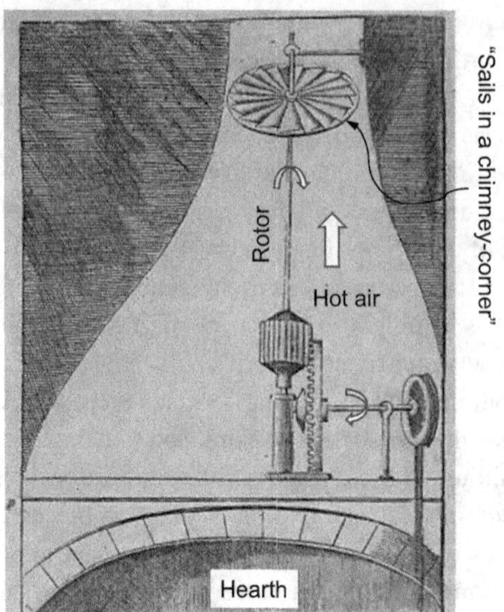

Figure 1.1 "Chimney gas turbine" described by John Wilkins, Lord Bishop of Chester.

edition was published in 1683), English clergyman, natural philosopher, polymath, and author, John Wilkins (1614–1672), Lord Bishop of Chester, described "a better Invention to this purpose, mentioned in Cardan, whereby a Spit may be turned (without the Help of Weights) by the Motion of the Air that ascends the Chimney."[1] The "moving of Sails in a Chimney-corner" (i.e., a "chimney gas turbine"), as depicted in Figure 1.1, is the purpose that Wilkins refers to.[2] Bishop Wilkins indicated that the "invention" was mentioned in the 1559 treatise *De Varietate Rerum* by the Italian polymath, Gerolamo Cardano ("Cardan" as Wilkins calls him), who is well known for his achievements in algebra, but was also an inventor of mechanical devices such as the combination lock. While the contraption in Figure 1.1 is unmistakably a gas turbine, the aforementioned "vignettes" are bona fide examples of turbomachinery, which laid the foundation for the modern-day variant.

Once the historical foundations of the gas turbine are laid down, the current state of the art is going to be assessed in realistic terms. The "class hierarchy" of heavy-duty industrial gas turbines will be established in this chapter as well.

[1] The complete title is *Mathematical Magick, or, The wonders that may by performed by mechanichal geometry: in two books, concerning mechanical powers [and] motions. Being one of the most easie, pleasant, useful (and yet most neglected) part of Mathematicks. Not before treated of in this language* (London: Printed for J. Nicholson, at the Kings-Arms in Little Britain; A. Bell, at the Croft-Keys in Cornhill; B. Tooke, at the Middle-Temple-Gate in Fleetstreet; and R. Smith under the Piazza's of the Royal-Exchange. MDCCVIII).

[2] According to Wilkins, "the Motion of these Sails may likewise be serviceable for sundry other Purposes; for the Chiming of Bells, for the Reeling of Yarn or the Rocking of a Cradle, with diverse the like domestick Occasions."

Part II comprises fundamental material one has to be thoroughly familiar with in order to be conversant in more advanced subject matter related to the design and operation of gas turbines for electric power generation.

In Chapter 5, it will be shown that the basic gas turbine concept logically follows from the second law of thermodynamics (even though, in reality, the evolution of heat engines and the science of thermodynamics went pretty much hand in hand). Once that logical path is established, it is amazing how many practical calculations, which typically require hours of analysis with sophisticated heat balance simulation software, can be distilled into a few simple formulae amenable to simple spreadsheet analysis.

From the thermodynamic cycle to the final end product (Chapter 14), there are several intermediate steps. They separate "concept" from "reality." In a roughly logical order, they comprise aeromechanical design of compression and expansion (turbine) sections to achieve particular cycle conditions (Chapters 10 and 11), combustion with minimal or – even better – no harmful emissions (Chapter 12), and the availability and selection of suitable materials of construction (Chapter 13). All of these individual design and development steps must be undertaken with a keen eye on cost, operability, maintainability, and reliability.

Part III comprises subject matter of a more advanced nature, which one has to master in order to claim comprehensive knowledge of industrial gas turbines for electric power generation. Note that the coverage in the book does *not* deliver the said mastery; that requires many years of active involvement in the field. The goal herein is simply to shed light on the path to be taken toward that goal.

At the end of the day, the gas turbine itself generates mechanical (shaft) power. The synchronous alternating current machine, which transforms the shaft power into electric power, is an integral part of the system (i.e., the gas turbine generator). In the past, when gas turbines were primarily utilized for peaking duties (i.e., a few hundred hours annually) or for base load duties in a combined cycle configuration, treatment of the generator was barely an afterthought (i.e., generator efficiency was 98.9 percent – that's it!). Modern grid requirements such as grid code regulations and reactive power support necessitate a closer look at the alternating current machine and its ramifications for running the gas turbine efficiently and reliably. A whole chapter in Part III will be devoted to this (Chapter 15).

Other chapters in Part III deal with:

- Reliability, availability, and maintainability (Chapter 16)
- Combined cycle power plant (Chapter 17)
- Off-design operation (Chapter 18)
- Transient operation (Chapter 19)
- Economics (Chapter 20)
- The Hall of Fame (Chapter 21)

The last chapter in Part III is a deep dive into the "latest and greatest" in gas turbine technology, as of the end of 2017, via the flagship products of four major OEMs.

Part IV comprises chapters on unique applications of gas turbine technology:

- Closed cycle gas turbines (Chapter 22)
- Aeroderivative gas turbines (Chapter 23)

Finally, now that the reader is equipped with a trove of knowledge on the state of the art, a glimpse into future development possibilities will conclude the treatise. The final, concluding chapter will recap the status of gas turbine technology and look into the future with a keen eye on the past.

With this brief introduction complete and without further ado, it is time to move on to the nitty-gritty details of large, stationary gas turbines for electric power generation.

Part I

Prerequisites

2 The Tool Chest

Let us state the obvious first: this is a book for engineers and engineering students. A solid understanding of the key concepts of thermodynamics, fluid mechanics and heat transfer is more than a prerequisite; it is a *must*. Without that, slogging through the material covered in this book will become a tedious chore (which is definitely *not* the author's intention).

One does not have to hold a degree in advanced mathematics to be a good gas turbine engineer. Nevertheless, a strong foundation in calculus is requisite to understanding the laws of thermodynamics, especially the second one. Furthermore, the foundation of the thermal design of power plants in general and gas turbines in particular consists of the famous "TdS" equations. The reader must be readily conversant in them and in Maxwell relationships (if you draw a blank here, it is time to pick up your thermodynamics textbook and blow the dust off its cover) – simply put, there is no shortcut around it. If your math skills are rusty, the excellent book by Zel'dovich and Yaglon is the perfect resource for getting up to speed [1].

It also helps greatly to have an adequate understanding of basic statistical concepts, which, in turn, require a firm grasp of calculus. This is imperative for understanding the difference between "engineering" and "commercial" performance (i.e., commercial "margins"). It is also very handy in order to prevent being lured into unrealistic expectations by the proverbial "fool's gold." A very useful tool to help you in statistical analysis is the Oracle Crystal Ball, which is a Microsoft Excel-based software for predictive modeling, forecasting, Monte Carlo simulation, and optimization.[1] Monte Carlo simulation in particular is a very valuable tool to differentiate between substantial and insignificant differences when comparing design options. Its use will be demonstrated later in the book (see Chapter 20).

2.1 Computer Software

In order to derive the maximum benefit from this book to impact your daily work, you need to perform calculations. Some of the equations developed in the following chapters are amenable to quick implementation with a pen, a piece of paper, and a $10 drugstore

[1] For more information about Crystal Ball, visit www.oracle.com/crystalball (last accessed by the author in December 2017).

calculator (or the calculator app on your smartphone). Once you go beyond a few "back of the envelope"-type estimates, you will probably need to implement quite a few such equations in an Excel spreadsheet. Nevertheless, at a certain point, you will realize that it is a good idea to distill some of your calculations into compact snippets of computer *code*. This is where the real fun starts.

It is highly recommended that you are well versed in at least one of the following programming languages[2]:

1. Visual Basic (VB; especially VBA in Excel)
2. C or C++
3. FORTRAN
4. Matlab (from MathWorks®)

The last one is extremely popular among the younger generation – students *and* practitioners. Even though the author has not had extensive experience with it, he is aware that it is a very simple and powerful tool for computational purposes. The most beneficial one in this author's opinion, which is also quite easy to learn and implement, is VB. In particular, VBA (Visual Basic for Applications) is the programming language of Excel and other Microsoft Office programs. (It is highly unlikely that the reader of this book does *not* have access to Microsoft Office and Excel through his or her organization or on his or her own.) If you are not already familiar with VBA, you can get started by automating tasks in Excel by using the Macro feature, which essentially records a sequence of your Excel spreadsheet actions in a VB function.

An alternative to Microsoft Office is Apache OpenOffice (AOO), which is an open-source office productivity software suite. It contains a word processor (Writer – equivalent to Microsoft Word), a spreadsheet (Calc – equivalent to Microsoft Excel), and a presentation application (Impress – equivalent to Microsoft PowerPoint), among others. Readers who are more familiar with it can of course substitute the AOO variant for Excel, which is the software program most familiar to the author.

If you are already familiar with C++ and/or FORTRAN and, even better, if you already have old programs that you had developed earlier in your educational and/or professional career, you can easily translate them into VB. It is very intuitive and easy. Another way to capture such legacy code in Excel or similar software programs is via converting them into a *dynamic-link library* (DLL; i.e., an executable file encapsulated in a DLL).

In Appendix C of the book, some examples of Excel VBA code based on the formulae developed for the subject matter hand will be provided. You can use them in your own applications *as is* or as starting points. For the younger generation of readers, these code snippets (or formulae and methods contained therein) can be used as building blocks or starting points to design your own "apps."

However, at some point, for serious work, the calculations requisite for accurate and reliable engineering design analysis require specialty software. This is especially true

[2] There are many new computer languages out there. Whatever works for you is just fine.

for transient (dynamic) analysis of the power plant for operability analysis and control system development. The software used for the latter task is extremely complex, with a steep learning curve (and it is very expensive, of course), and it requires a huge number of man-hours to develop and execute fully functional power plant models. It is no exaggeration to state that some people spend an entire career on such programs and associated tasks. Unfortunately, there are no shortcut calculation methods to replace them (which require the solution of combinations of partial differential equations), and even simplified approaches to estimate their outcomes are quite complicated and require a deep engineering knowledge of the underlying phenomena. This will be touched upon again in Chapter 19.

Software for steady-state performance calculations are commonly known as "heat balance simulation tools." The name derives from the fact that the underlying fundamental principle is the first law of thermodynamics (also known as conservation of energy) along with mass continuity (also known as conservation of mass). These two laws of conservation are applied to the individual pieces of equipment comprising the power generation system in question. In the end, a "balance" is established between the two forms of energy transfer – work and heat – and mass transfer across the entire system (i.e., the "control volume").

There are several steady-state, heat balance simulation tools widely used in the industry:

1. **GateCycle**. Gate, as it is commonly known, is a PC-based simulation software with a graphical user interface that allows the user to build a system by interconnecting individual components, which are available as icons (compressor, combustor, expander, etc.). In 2000, Enter Software was fully acquired by General Electric. Since then, however, marketing and support activities for Gate continuously dropped until 2013 or 2014, when, for practical purposes, GE Enter all but ceased to function. Some readers might be familiar with Gate and have access to its legacy (and functional) copies through their organizations.
2. **Thermoflow Suite**. This is a suite of several software packages, developed by Thermoflow, Inc., since 1987, which are classified into the following two distinct groups:
    a. **Application-specific software**. A special-purpose tool focusing exclusively on a specific type of power plant. The program includes a general model, from which the user selects a subset via a guided, structured procedure. They comprise:
        i. **GT PRO** and **GT MASTER** for gas turbine combined-cycle and cogeneration (i.e., combined heat and power, or CHP as it is commonly known in Europe) design and off-design performance.
        ii. **STEAM PRO** and **STEAM MASTER** for fossil plant (e.g., coal-fired boiler and steam turbine generator) design and off-design performance.
    b. **Fully flexible software**. A general-purpose tool that allows the user to construct any real or conceptual power plant model by connecting

appropriate building blocks in a flexible and unconstrained (up to a certain limit, of course) manner. Thermoflex is Thermoflow's fully flexible software, which is similar in functionality to GateCycle.

3. GateCycle and Thermoflex belong to a specialized family of engineering software commonly known as *flow-sheet simulators*, which are widely used in chemical process industry. Some of these tools (e.g., HYSYS [formerly known as HYSIM], ASPEN [at the time of writing, the company owning it is the owner of HYSYS as well] and PRO/II) are also used by engineers to develop fossil fuel-fired power plant performance modeling (especially for integrated gasification combined-cycle applications, which require simulation of chemical process components). HYSYS and ASPEN also have transient (i.e., dynamic or unsteady-state) simulation capability.
4. Two recent additions to this family of software are Ebsilon® Professional, developed by Steag Energy Services GmbH, and the suite of software developed by the SoftInWay, Inc., which includes cycle heat balance tools (i.e., AxCycle™) as well as stage-by-stage design tools (i.e., AxSTREAM®). More information on these software products can be found online.

Note that this brief introduction is not intended as a commercial for particular software programs. In addition to the aforementioned software tools, *with which the author had hands-on experience*, there are many other commercially available software packages for steady-state and/or transient simulation. Examples of the former are IPSEpro (offered by SimTech in Austria) and PEPSE (now offered by Curtiss-Wright Nuclear Division). There is also Simulink by Mathworks, the developer of Matlab, which can be used for dynamic simulation of gas turbine power plants and controls. Similar to Matlab, Simulink is widely used by the new generation of engineering students and practitioners and might come in handy especially in doing the gas turbine transient calculations described in Chapter 19.

Gas turbine original equipment manufacturers (OEMs) utilize their own proprietary codes for the design of flange-to-flange gas turbines. To a large extent, these are legacy FORTRAN codes going back 50 years or more and incorporate "lessons learned" from a huge amount of laboratory, factory test bed, and field operation data in the form of empirical correlations. As such, they are simply irreplaceable, unknown to the general public, and, even to some of their current users, they are quasi "black boxes." Even their in-house name, "cycle deck," goes back to the times when these codes, compiled in punched-card decks, were run on mainframe computers.

In terms of commercially available software for flange-to-flange gas turbine design, in the author's opinion, the most user-friendly one with sufficient detail to model a hot gas path reasonably accurately is the GASCAN code encapsulated in the "Cooled Turbine Stage" component of Thermoflow's Thermoflex software. This program will be frequently used herein to illustrate concepts and design principles.

Another similar program, with a focus on aircraft propulsion, is GasTurb by Dr. Kurzke. Once a "one-man show" by Dr. Kurzke, in future, this program is going to be developed and maintained by Dr. Jeschke and his team at the Institute for

Jet Propulsion at RWTH Aachen University, Germany. More information on this can be found online. For underlying theory and origins of the computer codes (in FORTRAN), readers with knowledge of German can refer to the somewhat dated (but still very valuable) book by Münzberg and Kurzke [2].

For transient simulation and off-design, unsteady-state performance calculations, in addition to HYSYS, commercially available software products include ProTRAX (formerly known as PC-TRAX) and EASY5. ProTRAX was originally developed as a power plant operator training simulator. Basically, the software creates a "digital twin" of the entire fossil fuel-fired power plant, which is then run in real time with all pertinent controls and operator screens to create real-life scenarios. Thus, operators can be trained in all features and procedures of the power plant they will be working in. (In fact, the first dynamic simulators were developed for safety analysis of nuclear power plants almost 50 years ago.) Such dynamic simulators are indispensable for plant controls design, testing, and commissioning, especially when they are combined with the actual control system hardware and software, including operator consoles (e.g., General Electric's Mark series systems). Examples of other commercially available dynamic simulation software are Simcenter Amesim (now part of Siemens PLM Software), Apros (by Fortum and VTT Technical Research Centre of Finland Ltd. for modeling and dynamic simulation of power plants), ISAAC Dynamics (a dynamic simulation software by the TransientGroup – formerly Struttura Informatica – in Florence, Italy), 3KEYMASTER (modeling and simulation platform by WSC, Inc., in Maryland, USA), and Modelica (developed by the non-profit Modelica Association and is free).

As mentioned earlier, dynamic simulation software is extremely complex and expensive and requires a significant amount of training for an engineer to become well versed in its use. While their steady-state counterparts (especially Thermoflow software) are comparatively easy to use, they are also very expensive (initial purchase *and* annual maintenance fees for upgrades and customer support can run to tens of thousands of dollars) and are unlikely to be affordable to individuals. In any event, all software tools require a certain experience level achievable via a sometimes steep and lengthy learning curve. While they make our lives easier and increase our productivity, they hide the fundamental principles that are at work from us – sometimes with disastrous results (i.e., the famous GIGO: garbage in, garbage out).

The material in this book is intended to make the inner workings of such software tools "transparent" to the reader, who may already be a user of them. (Thermoflow software, especially GT PRO/MASTER and Thermoflex, will be used in some of the coverage.) By the same token, the methodology and fundamental principles, expressed as simple physics-based equations, introduced in the following chapters should provide a companion to or an alternative for carrying out certain *basic* design and analysis tasks.

Before moving on, a few words on computational fluid dynamics (CFD) are in order. In most gas turbine theory books (e.g., by Saravanamuttoo and his coauthors [3]), compressor and turbine design processes are covered via one-dimensional "free vortex" theory fortified by empirical correlations. This was indeed how actual gas turbines were designed and developed by major OEMs in the post-war period until the sixties. Thereafter, improved analytical methods for the solution of two-dimensional inviscid flows made inroads in research and development toward a better understanding of secondary flows and losses. The first successful viscous CFD calculations of turbine secondary flows were reported in the early 1980s. In the 1990s, progress in CFD methods and their deployment in actual turbomachinery design went hand in hand with the rapid increase in computational speed via better hardware and software. A very good, brief history is provided by Horlock and Denton [4]. For the underlying three-dimensional thermofluids, the single most important resource (in the opinion of this author, of course) is the book by the late Dr. Lakshminarayana [5].

Modern turbomachinery design relies fully on three-dimensional viscous CFD. Even the major OEMs use sophisticated commercial tools such as PHOENICS and ANSYS combined with their proprietary in-house codes and post-processing add-ins for design and development. It is interesting to note that, starting in 2015, said OEMs (initially led by General Electric) started a big move toward "digital business" and "industrial software," which may have been a factor in the acquisition of many such independent software developers by hardware suppliers. In fact, in such a move, Siemens acquired CD-adapco, a leading engineering simulation company with a broad portfolio including CFD software, in early 2016.

The following was already mentioned in the Introduction and is worth repeating here as well: there is no point (and no practical way) to encapsulate highly sophisticated numerical computing methods and design practices in a book. Even if it could be done, the reader of such a book would not be able to develop his or her own CFD code. Those programs take many *man-years* to develop – they are simply not intended for "basement tinkerers." The best resource for help in daily work involving using such software is specific theory manuals (they usually come with the software product) and applicable OEM design practices, which cannot be publicly disseminated. Even the most bare-bones approach amenable to implementation in a relatively simple FORTRAN program (as described, say, by the late Dr. Traupel in the second volume of his magnum opus [6]) would be a huge chore for any skilled engineer. There would be no point in such an endeavor either because, once again, these days nobody is asked to design a heavy-duty industrial gas turbine with F, G, H, or J technology from the proverbial "blank sheet."

2.2 Books

Let us assume that you are already equipped with Microsoft Excel (or its AOO variant, Calc) on your PC (desktop or laptop), a scientific calculator (these days probably in your smartphone), a pen, and paper. You also possess sufficient VB skills to develop small snippets of code as needed to facilitate fairly involved calculations. What else is needed? Books, of course. At a minimum, you need a good reference book each for basic thermodynamics and basic gas turbine theory. For engineering thermodynamics, the author's preference is the first edition of the textbook by Moran and Shapiro [7]. In the remainder of this treatise, this particular book will be referred to as "Moran and Shapiro." For basic gas turbine theory, the excellent book by Saravanamuttoo et al., which went through six editions by the time of writing, is the *must-have* reference [3].

While primarily a treatise on aircraft gas turbine engines, its clarity and detail on all relevant aspects of gas turbine theory is simply unmatched. Any serious practitioner must own a copy (and, most likely, does). From here on, it will be referred to as "Saravanamuttoo et al." Finally, a very good resource with practical data, formulae, and methods is the gas turbine performance book by Walsh and Fletcher [8]. This book is highly recommended for those readers who want to develop their own calculation programs and apps.

There are several gas turbine "handbooks" out there written by a single author that provide general information on the subject matter (e.g., by the late Dr. Boyce [9], Soares [10] and others). The author has only the first edition of one of them in his library and has perused several others. If one goes with the definition of "handbook" in the Oxford English Dictionary (OED; i.e., "any book ... giving information such as facts on a particular subject, guidance in some art or occupation, instructions for operating a machine, or information for tourists"), such references are only good in terms of the first part of the definition (or even the fourth judging by the quality of the coverage in some – but certainly not all). They are essentially of no practical use for day-to-day activities of someone working in the industry and only of marginal use for those involved in academic research.

There are, however, two gas turbine handbooks out there containing a large number of chapters each written by an expert or team of experts in the field (mostly from the major OEMs) and guided by an editor or team of editors also active in the industry and/or academia [11,12]. The author has a copy of both of them in his library. The first one is available in English [11] and the other – by far the superior one – is available only in German [12]. In terms of providing information in excellent alignment with the second and third parts of the definition of "handbook" in the OED, the latter is simply unmatched, and is likely to stay so for the foreseeable future. In fact, it is the aim of the current book (in its author's humble opinion) to supplement [12] from the perspective of more in-depth and detailed calculations of the performance and operability of "stationary" gas turbines.

One other handbook deserves to be mentioned in passing: namely, the turbomachinery handbook edited by Logan and Roy [13]. Several chapters of that handbook provide decent coverage of key gas turbine components.

It is a good idea to own a copy of a general reference book for typical engineering calculations. The handbooks recommended by the author are (the reader can, of course, use their own favorites):

1. *Marks' Standard Handbook for Mechanical Engineering* (any fairly recent edition will do).
2. *Perry's Chemical Engineer's Handbook* (any fairly recent edition will do).
3. For those readers who know the German language, *Dubbel – Taschenbuch für den Maschinenbau* (by Springer-Verlag; any fairly recent edition will do) is by far the best handbook known to the author.

The Professional Engineer (PE) exam reference manual (for the USA) by Lindeburg is an excellent resource for many practical formulae in thermofluids, hydraulics and other disciplines of mechanical engineering [14]. It is concise, to the point, and contains many examples. It is highly recommended that you obtain a recent edition if you do not already have one left over from your PE exam days.

You should have access to property calculations for steam (e.g., ASME steam tables) and gas mixtures (e.g., JANAF tables[3]). Once upon a time, they were indeed "tables" on paper, which engineers used, along with their slide rules, to look up enthalpies and entropies (e.g., see the appendix of Moran and Shapiro [7]). Nowadays, many organizations have these tables encapsulated as Microsoft Excel add-ins or DLLs. If this is not the case, you need to acquire (buy) your own (e.g., from the ASME or NIST) or create your own property package (e.g., by programming the tables and/or equations in your thermodynamics textbook or in Walsh and Fletcher [8] in VBA). The latter option might be tedious and time-consuming, but it is pretty straightforward and needs to be done only once.

An indispensable reference book that you *must* have if you are serious about doing your own calculations is *Numerical Recipes in FORTRAN* (or *C*) by William H. Press and his coauthors [15]. The codes in those books (e.g., for finding the roots of a highly non-linear function) are extremely robust, easily translatable into VB, and very useful. A few samples (e.g., for double interpolation) are provided in Appendix C.

2.3 Journals, Standards, and Codes

The best sources for new and emerging technologies are academic journals (requiring subscription and/or membership to professional organizations), conference proceedings (ditto), and trade publications (almost all of them available online at no charge). The reader is encouraged to consult them for up-to-date information on new research and development. The list below, while by no means comprehensive, is provided as a starting point:

1. *ASME Journal of Energy Resources Technology*
2. *ASME Journal of Engineering for Gas Turbines and Power*
3. *ASME Journal of Turbomachinery*
4. *Chemical Engineering* (www.che.com)
5. *Combined Cycle Journal* (www.ccj-online.com)
6. *Electric Power & Light* (www.elp.com)
7. *Energy, The International Journal*, Elsevier
8. *Gas Turbine World* (www.gasturbineworld.com)
9. *Hydrocarbon Processing* (www.HydrocarbonProcessing.com)
10. *Modern Power Systems* (www.modernpowersystems.com)
11. *POWER* (www.powermag.com)

[3] JANAF is an acronym for "Joint Army, Navy, and Air Force." JANAF thermochemical property tables are probably the most widely used and reliable data sets, and they are currently offered by the National Institute of Standards and Technology (NIST).

12. *Power Engineering* (www.power-eng.com)
13. *Turbomachinery International* (www.turbomachinerymag.com)

The two most important sources of technology information are the annual "handbooks" published by *Gas Turbine World* (GTW) and *Turbomachinery International* (TMI). These publications will be referred to frequently in this book (e.g., GTW 2016 for the 2016 handbook).

Plant performance tests are performed to demonstrate that the plant meets the guaranteed performance (primarily power output and efficiency or heat rate) offered by the engineering, procurement, and construction contractor to the plant owner. ASME Power Test Codes (PTC) have been developed to provide guidance on how to conduct power plant performance testing. The codes provide comprehensive, practical information on calculations pertaining to major fossil fuel-fired power plant equipment and, as such, constitute very useful resources for information on the core subject matter of the chapter. The test codes most relevant to the reader of this book are listed below.

1. PTC 22 – Gas Turbines
2. PTC 4 (or 4.1) – Fired Steam Generators
3. PTC 4.4 – HRSGs
4. PTC 46 – Overall Plant Performance
5. PTC 47-2006 – Integrated Gasification Combined Cycle Power Generation Plants
6. PTC 6 – Steam Turbines (Rankine Cycle)
7. PTC 6.2 – Steam Turbines (Combined Cycle)

Obviously, one can add to the list the International Organization for Standardization (ISO) standards (in all likelihood more familiar to readers based in Europe). In particular, ISO 2314:2009 "Gas turbines – Acceptance tests," last reviewed and confirmed in 2013, is particularly recommended for practitioners in the gas turbine industry (along with ASME PTC 22).

First published in 1914, the ASME Boiler and Pressure Vessel Code (BPVC), especially its sections on the boiler and pressure vessel, is an indispensable resource in the USA, as well as in the global electric power generation industry (in addition to other industries). An important, related code is ASME B31.1 Power Piping Code.

The European equivalent of the ASME BPVC is the Pressure Equipment Directive (PED) 2014/68/EU (formerly 97/23/EC) of the European Union (EU), which sets out the standards for the design and fabrication of pressure equipment (e.g., steam boilers, pressure vessels, piping, safety valves, and other components and assemblies subject to pressure loading) generally over one liter in volume and having a maximum pressure of more than 0.5 bar (about 7.3 psi) gauge. The PED has been mandatory throughout the EU since May 30, 2002, with the 2014 revision fully effective as of July 19, 2016.

Finally, VGB PowerTech e.V. is an international (voluntary) technical association for the generation and storage of power and heat based in Germany. The organization brings together companies and supports their generation and internal utilization of electricity, heat, and the by-products resulting therefrom based on the most up-to-date technologies. Their publications and the *VGB PowerTech Journal* constitute a significant resource for engineers and researchers alike. The reader is strongly encouraged to consult their website and the lists of publications therein (many available in both German and English) when looking for hard-to-find technical information (www.vgb.org).

References

References 3, 5, 6, 8, and 12 are "must-have" books on gas turbine technology, which will be referenced frequently in the following chapters of this book.

1. Zel'dovich, Ya. B., Yaglom, I. M., *Higher Math for Beginners (Mostly Physicists and Engineers)* (Englewood Cliffs, NJ: Prentice-Hall, Inc., 1987).
2. Münzberg, H.-G., Kurzke, J. T., *Gasturbinen – Betriebsverhalten und Optimierung* (Berlin and Heidelberg, Germany: Springer-Verlag, 1977).
3. Saravanamuttoo, H. I. H., Rogers, G. F. C., Cohen, H., Straznicky, P. V., *Gas Turbine Theory*, 6th edition (Harlow: Pearson Education, Ltd., 2009).
4. Horlock, J. H., Denton, J. D., A Review of Some Early Design Practice Using Computational Fluid Dynamics and a Current Perspective. *Journal of Turbomachinery*, **127**:1 (2005), 5–13.
5. Lakhshminarayana, B., *Fluid Dynamics and Heat Transfer of Turbomachinery*, 1st edition (New York: Wiley-Interscience, 1995).
6. Traupel, W., *Thermische Turbomaschinen*, 2nd edition (Klassiker der Technik), 2 vols. (Berlin, Germany: Springer, 2000).
7. Moran, M. J., Shapiro, H. N., *Fundamentals of Engineering Thermodynamics* (New York: John Wiley & Sons, Inc., 1988).
8. Walsh, P. P., Fletcher, P., *Gas Turbine Performance* (Fairfield, NJ: Blackwell Science, Ltd., 1998).
9. Boyce, M., *Gas Turbine Engineering Handbook*, 4th edition (Oxford: Butterworth-Heinemann, 2011).
10. Soares, C., *Gas Turbines: A Handbook of Air, Land and Sea Applications*, 2nd edition (Oxford: Butterworth-Heinemann, 2014).
11. Jansohn, P. (Editor), *Modern Gas Turbine Systems: High Efficiency, Low Emission, Fuel Flexible Power Generation (Woodhead Publishing Series in Energy)*, 1st edition (Cambridge: Woodhead Publishing, Ltd., 2013).
12. Lechner, C., Seume, J., *Stationäre Gasturbinen, 2. Neue Bearbeitete Auflage* (Heidelberg, Germany: Springer, 2010).
13. Logan, Jr., E., Roy, R. (Editors), *Handbook of Turbomachinery, Second Edition (Revised and Expanded)* (New York: Marcel Dekker, Inc., 2003).
14. Lindeburg, M. R., *Mechanical Engineering Reference Manual for the PE Exam*, 10th edition (Belmont, CA: Professional Publications, Inc., 1998).
15. Press, W. H., Teukolsky, S. A., Vetterling, W. T., Flannery, B. P., *Numerical Recipes in FORTRAN: The Art of Scientific Computing*, 2nd edition (New York: Cambridge University Press, 1992).

As stated in the Preface, this is not a book for beginners. There is no introductory material including the derivation of key equations from first principles. The reader is expected to have at least a four-year degree in engineering under his or her belt and to be familiar with the fundamental disciplines of thermodynamics, heat transfer, and fluid mechanics (including hydraulics, fluid dynamics, and gasdynamics). In case you got rid of your textbooks a long time ago and need a refresher, here are the author's favorites:

Incropera, F. P., Dewitt, D. P., *Introduction to Heat Transfer*, 4th edition (New York: John Wiley & Sons, Inc., 2002).

Landau, L. D., Lifshitz, E. M., *Fluid Mechanics*, 2nd edition (Oxford: Pergamon Press, 1987).

Liepmann, H. W., Roshko, A., *Elements of Gasdynamics* (New York: John Wiley & Sons, 1957).

Thompson, P. A., *Compressible Fluid Dynamics* (New York: McGraw-Hill, 1988).

White, F. M., *Fluid Mechanics* (New York: McGraw-Hill Book Company, Inc., 1979).

Zemansky, M. W., *Heat and Thermodynamics*, 4th edition (New York: McGraw-Hill Book Company, Inc., 1957).

3 Ground Rules

There are several important items the reader should be aware of while reading this book. Instead of pointing them out over and over in various chapters, they are summarized below for easy reference. It is highly recommended that the reader goes through this chapter before moving on. This "one-stop shop" approach, so to speak, will hopefully ensure a steady reading pace without the need to go back and forth and search for clues. I would also encourage the reader to read the "nomenclature" section in Appendix A before reading the rest of the book.

3.1 Units

This book primarily uses the US Customary System for units. The reason for this is obvious. The author's entire professional career and graduate study took place in the USA. As such, the huge trove of material condensed into this book was exclusively developed using the – admittedly tedious and cumbersome – US customary units. Conversion of a huge list of formulae, charts, codes, etc., would unavoidably lead to errors. Therefore, with apologies upfront, the unenviable task of charting a course through the maze of conversion factors is left to the reader (if he or she is really keen on having everything handy in SI units).[1] In particular:

- Flows are expressed
  - in lb/s (pounds per second) for gas turbines (about 1,000–2,000 lb/s for most "frame" units).
  - in lb/h (pounds per hour) or kpph (1,000 lb/h) for steam turbines (of the order of million lb/h or thousand kpph).
- Temperatures are expressed in °F (degrees Fahrenheit), but used as R (Rankine) in formulae (there may be a few exceptions).
- Pressures are expressed in psi (pounds-force per square inch)
  - either as "absolute" (i.e., psia), or
  - as "gauge" (i.e., psig).

[1] However, this is not exactly a dogmatic stance by any means. For example, turbine inlet temperatures are cited in degrees Celsius due to the simple beauty of nice, round numbers separated by even 100 degrees. Furthermore, some examples are done in SI units just to keep the "mess of numbers" to a minimum.

- Specific enthalpies are expressed in Btu/lb.
- Specific entropies are expressed in Btu/lb-R.

The units for power (rate of mechanical or electrical energy transfer) are Btu/s or kWe (kilowatt *electric*).

The units for heat transfer rate are Btu/s or kWth (kilowatt *thermal*).

One Btu/s is equal to 1.05506 kilowatts.

One thousand kilowatts (kW) are equal to one megawatt (MW).

Although very rarely used in power generation calculations of the type covered in this book, for the sake of completeness, 1,000 MW is 1 gigawatt (GW) and 1,000 GW is 1 terawatt (TW). After that comes peta-, exa-, zetta-, and yottawatt. (Incredulous readers – yottawatt? – can google the terms.)

For conversion to SI units, one can turn on his or her laptop (or smartphone) and google the particular conversion online with immediate results. In case the battery is drained or Wi-Fi access is not available, by far the best compact table of conversions and the most frequently used constants available on paper is in the back of Moran and Shapiro. The author strongly advises the reader to make Xerox copies of those two pages and to keep them handy – as he still does.

Unfortunately, even without conversion to SI units, US customary units present certain difficulties in calculations. As is well known, the beauty of the SI units is in their self-consistency (i.e., when one plugs the parameters into a particular equation in their expected units, the result is automatically obtained in its expected units as well). Consider the total temperature

$$T_0 = T + \frac{V^2}{2c_p}, \tag{3.1}$$

where T_0 is the total (stagnation) temperature in K (degrees Kelvin), T is the static temperature in K, V is the fluid velocity in m/s, and c_p is the fluid-specific heat in J/kg-K. For a 100-m/s fluid flow at T = 373 K, with c_p = 1,000 J/kg-K, the total temperature is

$$T_0 = 373 + 100 \times 100/(2 \times 1{,}000) = 378 \text{ K}$$

In passing, in *all* scientific formulae derived from first principles, temperatures are used in the *absolute* frame of reference (i.e., degree Kelvins [K] or degree Rankines [R]). (Exceptions are occasional empirical relationships such as curve-fits to data, etc.)

In the US Customary System, Equation 3.1 becomes

$$T_0 = T + \frac{V^2}{2c_pJg_c}, \tag{3.2}$$

where g_c = 32.174 ft-lbm/lbf-s^2 (g_c = 1 in SI units; note the differentiation between "pound-mass," lbm, and "pound-force," lbf) and J = 778.16 ft-lbf/Btu (J, of course, is unity in the SI system). Thus,

$$T = 373 \text{ K} \times 1.8 = 671.4 \text{ R}$$

$$V = 100\ m/s \times 3.28084 = 328.1\ ft/s$$

$$c_p = 1{,}000\ J/kg\text{-}K \times 0.000238846 = 0.239\ Btu/lb\text{-}R$$

$$T_0 = 671.4 + 328.1 \times 328.1/(2 \times 0.239 \times 778.16 \times 32.174) = 680.4\ R = 378\ K$$

In this book, even though the base unit system is the US Customary System, the conversion factors g_c and J will be omitted from the formulae so that universally known and recognized forms of standard equations are preserved and excessive clutter is avoided. However, the reader must always be cognizant of the "presence" of g_c and J "lurking" in the background.

Another dangerous spot in the conversion minefield is the difference (or, rather, lack of it) between "mass" and "weight" in the US customary units. In the USA, when someone enquires about your weight, you would answer, say, "180 pounds (lb)." However, to be precise, 180 pounds is your "mass" (i.e., 180/2.2046 = 81.6 kg). (In passing, the proper unit to use is "lbm" for pound-mass, not "lb," but this is usually ignored.)

In SI units, your "weight" is

$$w = 81.6\ kg \times 9.807\ m/s^2 = 800\ N\ (\text{Newton})\ \text{or}$$

$$w = 81.6\ kp\ (\text{kilopond})\ \text{or}\ 81.6\ kgf(\text{kilogram-force})$$

One Newton is equal to the force that would give a mass of one kilogram an acceleration of one meter per second squared. What is the force, then, that would give a mass of one pound an acceleration of one foot per second squared? By analogy, since the gravitational constant in US customary units is 32.174 ft/s^2 and one foot is 0.3048 m, it would be 32.174 × 0.3048 = 9.807 lbm-ft/s^2, which defines one pound-force as

$$1\ lbf = 9.807\ lbm\text{-}ft/s^2.$$

It is easy to see that 1 lbf is the equivalent of 1 kgf (or 1 kp) in metric units. Note that the latter is a *non-standard* unit and is classified in the SI Metric System as a unit that is *unacceptable* for use with SI. (In fact, the last time the author came across 1 kp was in the 1970s while at high school.)

Similarly, by analogy, your "weight" in US customary units is

$$w = 180\ lbm \times 9.807\ ft/s^2 = 1{,}765.3\ lbm\text{-}ft/s^2$$

$$w = 1{,}765.3\ lbm\text{-}ft/s^2/9.807\ lbm\text{-}ft/s^2 = 180\ lbf$$

Thus, when asked about your "weight," the standard answer in US customary system is *not* 1,765.3 lbm-ft/s^2, but "180 pound-force or 180 lbf!" To summarize:

	Mass	Weight
US Customary (English)	1 lbm	1 lbf
SI (Metric)	1 kg	9.807 N
SI (Metric)	0.453592 kg (= 1 lbm)	4.448377 N

In other words, 1 lbf is equal to 4.45 N.

Before concluding this section, especially in case the reader is wondering about the "ink wasted" on this mundane subject, let us look at a cautionary (and true) tale. In 1999, NASA lost its $125 million Mars Climate Orbiter because of a unit conversion error. The navigation team at the Jet Propulsion Laboratory, who used the SI system in their calculations, mistook acceleration readings measured in *pound force-seconds* for *Newton-seconds* (instead of 4.45 N-s!). This happened because Lockheed Martin Astronautics in Denver, which designed and built the spacecraft, provided crucial acceleration data in US customary units.

In layman's terms, then, the spacecraft was lost in translation!

3.2 Important Metrics

Thermal efficiencies are expressed on a *lower heating value* (LHV) basis. In particular, natural gas, which is assumed to be 100 percent CH_4 (methane), has an LHV of 21,515 Btu/lb at 77°F (about 50 MJ/kg at 25°C). This is not a capricious choice. The *higher heating value* (HHV) is the "true" energy content of the fuel, which includes the latent heat of vaporization that is released by the gaseous H_2O in the combustion products when they are cooled to room temperature. In other words, HHV is the value measured by a *calorimeter*. In a real application (e.g., a gas turbine), the combustion products, by the time they reach the exhaust gas stack (e.g., around 180°F for a modern combined cycle power plant), are not cooled to a temperature to facilitate condensation, which, depending on the amount of H_2O vapor in the gas mixture, is around 110°F. Thus, the latent heat of vaporization is *not* recovered and utilized. (This *can* be done by adding a condensing heat exchanger before the heat recovery steam generator [HRSG] stack, but it would be highly uneconomical.)

This is why LHV is the logical measure of fuel energy input that should be used in engineering calculations. There are, however, the following two exceptions:

1. The efficiency of the integrated gasification gas turbine power plant is referenced to the HHV of the gasifier feedstock (typically coal or pet-coke). This goes back to the practice historically adopted by US utilities when coal was the primary fossil fuel used in electric power generation.
2. The price of fuel in economic calculations is always expressed in dollars per million Btu (HHV). This is because when plant operators purchase fuel, the contract is always based on the "true" energy content of the fuel.

Note that LHV is not measurable; it can only be calculated from the laboratory analysis of the fuel by subtracting the latent heat of vaporization. For 100 percent CH_4 (methane), the ratio of HHV to LHV is 1.109. Many handy formulae can be found in handbooks (e.g., in Marks or Dubbel) to estimate the LHV and HHV for various fuels and fuel gas compositions.

Most of the thermodynamic treatment of gas turbines in this book will revolve around the two cardinal performance parameters: (i) generator power output and (ii) thermal efficiency. The reason for the former is obvious; it is the raison d'être of the book you are reading. *Thermal efficiency* is the ratio of power output (net or gross; see below) to *heat consumption*. Its importance cannot be overstated due to its impact on the following:

- Conservation of limited fuel resources
- Minimization of operating costs
- Reduction of criteria pollutants and CO_2 emissions

Heat consumption is the product of the fuel mass flow rate into the combustor of the gas turbine (plus fuel supplied into the duct burners of the HRSG in a combined cycle power plant, if applicable) and fuel LHV. For example, a gas turbine burning 100 percent CH_4 fuel at a rate of 30 lb/s has a heat consumption of 30 × 21,515 = 645,450 Btu/s, which is equal to about 681 MWth. If the gas turbine in question generates 275 MWe of power, its efficiency is 275/681 = 40.4% (again, net or gross; see below).

Another unit commonly used for heat consumption is MMBtu/h (million Btus per hour). For the example above, 645,450 Btu/s is 2,324 MMBtu/h. For a 40 percent gas turbine at different ratings between 275 and 375 MWe, heat consumption ranges between ~2,300 and 3,200 MMBtu/h (see Table 3.1).

In gas turbine performance calculations for electric power applications, the fuel flow rate and heating value are rarely expressed in "per unit volume" terms. Nevertheless, there may be instances when one has to resort to them (e.g., in "gas-to-power" applications with liquefied natural gas). The two common standardized volumes are *standard cubic feet* (scf, commonly pronounced as "scuff") and *normal cubic meters* (Nm^3). A standard cubic foot corresponds to 1 cubic foot of gas at 60°F (15.6°C) and 14.73 psia, and a normal cubic meter of gas corresponds to 1 cubic meter at 15°C at 101.325 kPa. The HHV of natural gas is approximately

- 38 MJ/Nm^3 or
- 0.038 GJ/Nm^3 or
- 1,030 Btu/scf

The HHV of 100 percent CH_4 is 1,012 Btu/scf and its LHV is 911 Btu/scf at 1.013 bara and 20°C. At 15°C, the corresponding values are 1,030 and 927, respectively. Thus, in volumetric terms, the values in Table 3.1 become those given in Table 3.2.

Frequently, thermal efficiency is expressed as a *heat rate*, which is given by 3,412 Btu/kWh divided by thermal efficiency. It is the land-based counterpart of *specific fuel*

Table 3.1 Output – heat (fuel) consumption (40 percent efficient gas turbine)

Output (MWe)	MMBtu/h	MWth/h
275	2,346	687.5
300	2,559	750.0
325	2,772	812.4
350	2,986	875.1
375	3,199	937.5

Table 3.2 Output – heat (fuel) consumption (40 percent efficient gas turbine). MMSCFC = million standard cubic feet per day

Output (MWe)	MMSCFD	Nm^3/day
275	60.8	1.60
300	66.3	1.74
325	71.8	1.89
350	77.3	2.03
375	82.8	2.18

Table 3.3 Heat rate – efficiency conversion (simple cycle)

Efficiency (%)	Btu/kWh	kJ/kWh
36	9,478	10,000
37	9,222	9,730
38	8,979	9,474
39	8,749	9,231
40	8,530	9,000
41	8,322	8,781
42	8,124	8,571

Table 3.4 Heat rate – efficiency conversion (combined cycle)

Efficiency (%)	Btu/kWh	kJ/kWh
56	6,093	6,429
57	5,986	6,316
58	5,883	6,207
59	5,783	6,102
60	5,687	6,000
61	5,594	5,902
62	5,503	5,806

consumption, which is a widely used metric for aircraft engines. For example, the heat rate of a gas turbine with 40 percent efficiency is 3,412/0.4 = 8,530 Btu/kWh. Typically, large "frame" gas turbine efficiencies range between 36 and 42 percent (ISO base load rating; see below for its definition). Thus, their heat rate range is ~8,000–9,500 Btu/kWh (~8,500–10,000 in SI units; see Table 3.3).

Typically, large "frame" gas turbine *combined cycle* efficiencies range between 56 and 62 percent (ISO base load rating; see below). Thus, their heat range is ~5,500–6,000 Btu/kWh (~5,800–6,500 kJ/kWh in SI units; see Table 3.4).

Another important parameter is gas turbine *specific power output*, which is the ratio of generator output to compressor airflow. Gas turbines are large, air-breathing machines. Size and cost considerations dictate the maximum amount of kilowatts squeezed from each lb/s of airflow. As will be demonstrated later via thermodynamic

Table 3.5 Gas turbine specific power outputs

Gas Turbine	Btu/lb	kJ/kg
E Class	140	325
F/G/H Class	190–200	440–465
J/HA Class	210–220	490–510

Table 3.6 Output – fuel oil consumption (40 percent efficient gas turbine). gpm = gallons per minute

Output (MWe)	MMBtu/h	MWth/h	lb/s	ft^3/min	gpm	Nm^3/h
275	2,346	687.5	35.6	39.6	296	67.3
300	2,559	750.0	38.8	43.2	323	73.4
325	2,772	812.4	42.1	46.8	350	79.5
350	2,986	875.1	45.3	50.4	377	85.6
375	3,199	937.5	48.6	54.0	404	91.7

arguments, specific output is a direct function of the "firing temperature" (see below for the definition of the latter). Typical values are shown in Table 3.5.

There was a time when gas turbines only burned all sorts of liquid fuels, from the dirtiest and most difficult to burn to diesel fuel. At the time of writing, except on specific occasions, which will be covered in Chapter 12 on fuel flexibility, the only liquid fuel worth mentioning is number 2 fuel oil, which is used as a "backup fuel" (and rarely at that). As a "quick and dirty" reference, consider that:

- The LHV of number 2 fuel oil is 18,297 Btu/lb.
- Its specific gravity is 0.8654.
- Thus, its density is 865.4 kg/m^3 or 54 lb/ft^3.
- Its LHV is 18,297 Btu/lb × 54 lb/ft^3 = 988,038 Btu/cuft.

In terms of the number 2 fuel oil flow rate, Table 3.1 can be revised and expanded as given in Table 3.6.

It is strongly recommended that the reader becomes familiar with the numbers in Tables 3.1–3.6. This is the best way to acquire a "feel" for the magnitudes involved in modern electric power generation technology with gas turbines and to catch and/or prevent errors in many complex calculations (or performance quotations).

It is of utmost importance to specify whether a quoted efficiency or power output is on a "net" or "gross" basis. This is by far the most common source of confusion, obfuscation, and error (sometimes intentional!). As far as this book is concerned:

- *Gross* refers to the power measured at the low-voltage terminals of the gas turbine (or other prime mover; e.g., steam turbine) generator.
- *Net* refers to the power after subtracting all plant power consumers (pumps, compressors, lighting, *heating, ventilation, and air-conditioning [HVAC]*, etc.) from the gross.

- The difference between the net and gross is commonly known as the *auxiliary load* or auxiliary power consumption.
- Strictly speaking, auxiliary load must include the losses incurred in the step-up transformer because this is the power supplied by the power plant to the electric grid.

For a gas turbine combined cycle power plant, the difference between net and gross power output can be anywhere between 1.6 and more than 3 percent, mainly dictated by the steam turbine heat rejection system. This will be discussed in detail in Chapter 17.

A gas turbine's performance (power output and efficiency or heat rate) with a given type of fuel is strongly dependent on its boundary conditions:

- Site ambient conditions (pressure [altitude], temperature, and relative humidity [RH])
- Load (expressed as a percentage)
- Inlet pressure loss
- Exhaust pressure loss

Typical *rating* performance is quoted at ISO conditions (i.e., 59°F [15°C], 14.7 psia [1 atm], and 60 percent RH) and zero inlet/exhaust losses (unless specified otherwise) at 100 percent load with 100 percent methane (CH_4) gaseous fuel.

Another significant source of confusion, obfuscation, and error is the specification of "base" and "full" load. Strictly speaking, base load refers to full (i.e., 100 percent) load at particular site ambient conditions (e.g., ISO). Thus, the rating performance of the gas turbine at 100 percent load and ISO conditions can be referred to as "base load" performance. In fact, typical "sticker performance" of a gas turbine in trade publications is referred to as "ISO base load performance."

However, a gas turbine can run at full load (i.e., 100 percent load) at any given site ambient conditions (e.g., on a hot day at 95°F and 70 percent RH). In that case, when referenced to its ISO base load, the gas turbine's output will be much less than 100 percent. Thus, specification of full load must be accompanied by the applicable site ambient conditions, and it refers to the gas turbine loading with:

- Inlet guide vanes at their nominal fully open position (i.e., nominal volume flow)
- Fuel flow per the specific control curve (i.e., not over- or under-fired)

Therefore, whenever one comes across a gas turbine performance number quoted at "x percent" load, he or she *must* ascertain what is exactly meant by that "x percent."

Comparison between technology options and optimization cases is the most frequent analysis activity. This requires calculating differences between key performance parameters. This is another source of confusion and error. Strictly speaking:

- Power output, air or fuel flows, and heat rate deltas can be expressed in *relative* (i.e., as a percentage) or absolute terms (i.e., kW/MW, lb/s, or Btu/lb).
- Temperature deltas should ideally be expressed in *absolute* terms (i.e., degrees F/R or C/K).
- Efficiency deltas should ideally be expressed in absolute terms (i.e., *percentage points*).

This last point is a particularly troublesome problem and particular attention *must* be paid to it. In this book 1 percent better efficiency for a base efficiency of 58 percent, say, means 58% + 1% = 59% (i.e., it is a delta in *absolute* terms). If it is misused and applied as a *relative* delta, it would mean 58% × (1 + 1%) = 58.6%. It is indeed very common to come across the latter convention in publications and presentations *without explicitly specifying its definition* and misinterpret the magnitude of efficiency improvement (or deterioration, depending on the context). If you really want to do a "relative" comparison of efficiencies, it is highly recommended to do so on a heat rate basis.

3.3 Cycle Terminology

The cycle of interest herein is the Brayton cycle, which is a theoretical approximation of the gas turbine via fundamental thermodynamics. The reader is referred to his or her favorite thermodynamic textbook for its definition (e.g., Moran and Shapiro). The textbook Brayton cycle is an *ideal* cycle – specifically, *an air-standard* cycle – which is based on the following two assumptions:

1. Working fluid is air modeled as an ideal gas.
2. All four processes (i.e., compression, heat addition, expansion, and heat rejection) are internally reversible.

In other words, there is no combustion and change of working fluid composition. Furthermore, if the specific heat of air is constant throughout the cycle (i.e., air is a *calorically perfect* gas), it is referred to as a "cold" air-standard cycle. For a detailed discussion, please see Appendix A.

Obviously, the air-standard cycle is an *ideal* cycle. But an ideal cycle does *not* have to be an air-standard cycle. An ideal cycle can have air modeled as a "real" gas with combustion replacing cycle heat addition and the working fluid undergoing the expansion process is a gas mixture (i.e., combustion products). In that case, the "idealness" of the cycle stems from 100 percent component efficiencies and zero losses.

In the rest of this book, there will be three variants of the *ideal* Brayton cycle under consideration:

1. Ideal, (cold) air-standard Brayton cycle
2. Ideal Brayton cycle with heat addition
3. Ideal Brayton cycle with combustion

The efficiency of the first variant is a function of the cycle pressure ratio (PR) only. The efficiency of the other two variants is a function of PR and the cycle maximum temperature, T_3. This will be covered in detail in Chapters 5 and 8.

As far as the processes in a "real" gas turbine with hot gas path cooling are concerned, it is futile to define a "real" Brayton cycle. In the first place, there is no "cycle" to speak of. Air is sucked into the machine from one end and still hot combustion gases are expelled from the other. Multiple side streams of cooling air, exiting the main flowpath and then reentering it, make matters even worse. As such, no reference is made to a "real" Brayton cycle in this book. The "reality" resides in the actual hardware.

3.4 Total or Static?

3.4.1 Temperature

The heart of this book is turbomachinery thermodynamics. This requires a full understanding of thermodynamic property calculations, of which the two most important are temperature (T) and pressure (p). In a turbomachine, a distinction must be made between *total* (also known as *stagnation*) and *static* values of T and p, which derive from the application of the *steady-state, steady-flow* (SSSF) version of conservation of energy to the turbomachinery component under study. Thus, for example, for the compressor in Figure 3.1, per unit mass flow rate and ignoring the potential energy terms, the SSSF energy balance reduces to

$$W_{comp} = (h_1 - h_2) + \left(\frac{V_1^2}{2} - \frac{V_2^2}{2}\right). \tag{3.3}$$

For an ideal gas, enthalpy is a function of temperature only. For a calorically perfect gas with constant c_p, Equation 3.3 becomes

$$W_{comp} = c_p(T_1 - T_2) + \left(\frac{V_1^2}{2} - \frac{V_2^2}{2}\right), \tag{3.4}$$

which, using Equation 3.1, is simplified to

$$W_{comp} = c_p(T_{0,1} - T_{0,2}). \tag{3.5}$$

In Equation 3.5, the temperatures in the parentheses on the right-hand side are known as stagnation or total temperatures; i.e.,

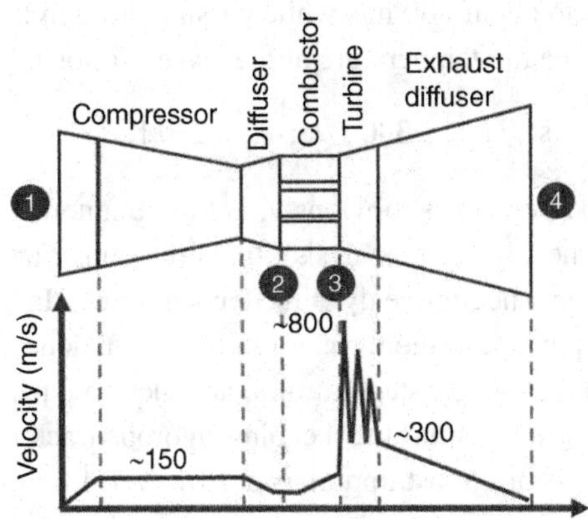

Figure 3.1 Typical land-based industrial gas turbine velocity profile.

$$T_{0,1/2} = T_{1/2} + \frac{V_{1/2}^2}{2c_p}. \tag{3.6}$$

The first part of the total/stagnation temperature is the *static temperature*. In aircraft performance calculations, the second part of the total temperature, $V^2/2c_p$, which is known as the *dynamic temperature*, is of significant importance and must be kept track of rigorously. In land-based (i.e., stationary) industrial gas turbines, one can be a bit more lenient. A typical air/gas velocity profile is shown in Figure 3.1. In particular:

- Axial air velocity across the compressor is fairly constant.
- It is slowed down in the diffuser, prior to entry into the combustor, to maintain stable combustion.
- Hot combustion gas enters the stage 1 rotor at a fairly high speed.
- Expansion and work production slow down the gas.
- Dynamic recovery in the exhaust diffuser brings the gas speed down prior to the stack or HRSG duct inlet.

Using the SI units (to prevent unnecessary conversion complication), ignoring the dynamic temperature at the compressor inlet is worth about

$$150 \times 150\ m^2/s^2/(2 \times 1{,}000\ J/kg\text{-}K) \sim 10\ K.$$

Considering the typical ISO ambient (inlet) temperature of 59°F (15°C or 288K), this is not an insignificant error. However, state-point 1 in Figure 3.1 refers to the ambient (i.e., zero dynamic temperature) and it can be considered as the compressor inlet. In that way, inlet losses are implicit in the stated compressor isentropic efficiency. This simplification is, of course, not recommended when dealing with gas turbine inlet conditioning equipment such as an inlet air filter, evaporative cooler, or electric chiller.

At the turbine inlet (i.e., combustor exit), a rough estimate of the dynamic temperature is

$$150 \times 150\ m^2/s^2/(2 \times 1{,}380\ J/kg\text{-}K) \sim 8\ K.$$

However, the hot gas accelerates to about 800 m/s while passing through the first-stage stator nozzle vanes so that the dynamic temperature at the stage 1 rotor inlet is

$$800 \times 800\ m^2/s^2/(2 \times 1{,}380\ J/kg\text{-}K) \sim 230\ K.$$

A contribution of several hundred degrees is, obviously, not to be ignored. Luckily, in most practical calculations (e.g., heat balance analysis), the turbine inlet and exit state-points in Figure 3.1 will be used (i.e., negligible dynamic temperatures). In other words, everything in between will be lumped into the turbine isentropic efficiency.

In summary, when carrying out stage-by-stage turbine aerothermodynamic calculations, even for a land-based industrial gas turbine, keeping rigorous track of static and total temperatures (and pressures) is of utmost importance. However, for more practical cycle heat balance calculations, at the four important state-points of prime interest, the distinction is immaterial:

1. State-point 1 is the same as ambient (i.e., dynamic temperature is zero).
2. For convenience, the inlet cone/duct can be lumped into the compressor (for simple cycle analysis).
3. At state-points 2 and 3, as shown above, the dynamic temperature contribution is minor and can be ignored.
4. State-point 4 is at the end of the exhaust diffuser where the gas is brought to a very slow speed and, once again, static/dynamic difference can be ignored.

Therefore, in most of the discussion in the later chapters, with the exception of the two chapters on turbine and compressor aerothermodynamics (Chapters 10 and 11, respectively), we will simply refer to "temperature" and omit the subscript "0," which denotes the total (stagnation) value thereof. A few very important points must be kept in mind:

1. Thermocouples in the fluid flow path measure the stagnation (total) temperature (e.g., gas turbine exhaust temperature rakes).
2. In thermo-physical property calculations, however, one must use the static temperature. In other words, the temperature substituted into the equation of state for, say, enthalpy calculation is the static temperature.
3. The distinction between static and total temperatures can only be ignored when flow velocities are low (i.e., Mach number 0.4 or lower).

The first item in the list above requires some elaboration. In order to measure the *true* total temperature, the measuring device (i.e., the thermocouple) has to stop the flowing gas adiabatically *and* convert its kinetic energy to temperature rise. This is the theory. In reality, the gas flowing around a thermocouple immersed in the gas is slowed down, but not stopped and not adiabatically. The former diversion from ideality leads to the *velocity error*, and the latter to the *heat transfer error* (mainly via conduction and radiation). Total measurement error is quantified by a *recovery factor* (RF) defined as

$$\mathrm{RF} = \frac{\mathrm{T_{TC}} - \mathrm{T}}{\mathrm{T_0} - \mathrm{T}}.$$

The RF accounts for the heat transfer and velocity errors. A good thermocouple requires that RF is ~1 so that the measured value, $\mathrm{T_{TC}}$, is very close to $\mathrm{T_0}$. (Commonly, this is achieved by a *shield* around the thermocouple.) ASME Standard PTC 22-2014 *suggests* that, if the average velocity in the area of exhaust temperature measurement exceeds 100 ft/s (30.5 m/s), readings from thermocouples with very low RFs are adjusted via Equation 3.6.

Thus, for state-points 1–4, by ignoring the static-total (stagnation) difference, we facilitate easy calculations of enthalpies and entropies from property tables or equations of state and use the isentropic relation for component performances for quick estimates.

Coming back to the equation defining the total temperature (Equation 3.6), it can be rewritten as a function of the Mach number, Ma = V/a, where a is the local speed of sound given by

$$a = \sqrt{\gamma RT},$$

where R is the gas constant for the fluid in question (e.g., air) and T is the *static* temperature. Thus, dropping the state-point subscripts, Equation 3.6 becomes

$$T_0 = T\left(1 + \frac{\gamma - 1}{2} Ma^2\right). \tag{3.7a}$$

If we apply Equation 3.7a to a *thermocouple* with an RF, we obtain

$$T_{TC} = T\left(1 + RF\frac{\gamma - 1}{2} Ma^2\right). \tag{3.7b}$$

It is very important to understand that the definition of total/stagnation temperature via Equations 3.6 or 3.7 is an *artificial* construction enabled by the assumption of calorically perfect gas. Strictly speaking, there is no such thing as a "total/stagnation temperature"; what there *is* is the "total/stagnation enthalpy." In other words, this is when Equation 3.5 is written as it correctly should have been in the first place,

$$w_{comp} = h_{0,1} - h_{0,2}, \tag{3.8}$$

where

$$h_{0,1/2} = h(p, T) + \frac{V_{1/2}^2}{2}, \tag{3.9}$$

and the first part of the total/stagnation enthalpy definition, h(p,T), is derived from an "equation of state" with *static* values of p and T (or any two independent thermodynamic properties; say, v and s). The steps taking us from the *rigorous* thermodynamic definition in Equation 3.9 to the *approximate* (but still quite accurate for most practical problems) Equations 3.6 and 3.7 are:

1. Equation of state is $p = \rho RT$.
2. Thus, enthalpy is *not* a function of pressure.
3. Specific heat, c_p, is constant.
4. Specific heat ratio, γ, is constant.

Without these simplifying assumptions (again, it is worth repeating that they are quite good for most practical problems), there is no way to back-calculate a "total/stagnation temperature" directly from Equation 3.9.

3.4.2 Pressure

Stagnation or total pressure is given by

$$p_0 = p \cdot \left(1 + \frac{\rho V^2}{2p}\frac{\gamma - 1}{\gamma}\right)^{\frac{\gamma}{\gamma-1}}. \tag{3.10}$$

Using the definition of Mach number (see above), Equation 3.10 becomes

$$p_0 = p \cdot \left(1 + \frac{\gamma - 1}{2} Ma^2\right)^{\frac{\gamma}{\gamma-1}}, \tag{3.11}$$

so that for Mach numbers of 0.4 or lower, we can assume that $p_0 \approx p$. Note that, *incompressible* fluid pressure measured by a Pitot tube inserted into the stream is given by

$$p' = p + \frac{\rho V^2}{2}, \tag{3.12}$$

and should not be confused with the stagnation (total) pressure given by Equation 3.11.

By comparing Equation 3.11 with Equation 3.7, we immediately recognize that it is merely a restatement of the basic isentropic p–T correlation; i.e.,

$$p_0 = p \cdot \left(\frac{T_0}{T}\right)^{\frac{\gamma}{\gamma-1}}. \tag{3.13}$$

3.5 50 or 60 Hertz?

Electric grids in the world are either 60 Hertz (Hz; e.g., in the USA) or 50 Hz (e.g., in Europe). In some countries (e.g., Saudi Arabia and Japan), both are present.

When a prime mover generator (i.e., a gas or steam turbine generator) is "synchronized" to the grid, it runs either at 3,600 rpm (60 Hz grid) or at 3,000 rpm (50 Hz), where rpm denotes "revolutions per minute." In angular terms:

- 3,600 rpm is 120π *radians per second* (rad/s).
- 3,000 rpm is 100π rad/s.

Since connecting a heavy-duty industrial gas turbine rated at, say, 300 MWe and with a speed other than 3,000 or 3,600 rpm to a generator running at those speeds would require a very large, expensive, and parasitic power-consuming gearbox, large "frame" machines are designed to run at 3,000 or 3,600 rpm. There are, however, some notable exceptions. In particular:

1. Smaller aeroderivative gas turbines with self-contained gas generators and power turbines (e.g., some of General Electric's LM-2500 units), which run at 6,100 rpm.
2. Smaller industrial gas turbines such as GE's Frame 6, which runs at 5,100 rpm.
3. Alstom (now owned by GE) GT11N2, which runs at 3,600 rpm for both 50 *and* 60 Hz versions (this is a not-so-small 115-MWe gas turbine).

In general, 50– and 60–Hz gas turbines by a particular original equipment manufacturer (OEM) are "speed-scaled" versions of the same basic design. The goal is to maintain the tangential speed at the blade tips to preclude undesirable shock losses. Accordingly,

$$\dot{m} \propto A = \pi \frac{D^2}{4},$$

$$U = \omega \frac{D}{2} = \pi \frac{ND}{60}$$

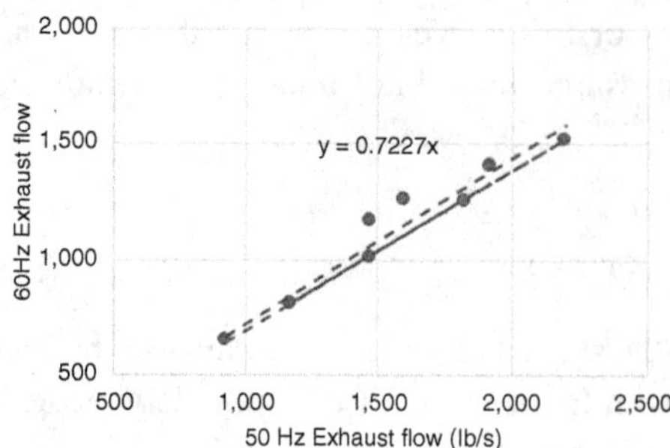

Figure 3.2 Gas turbine exhaust flows for 50 and 60–Hz units. The dashed line is the curve-fit to the data. The solid line is the theory (slope of 0.694).

where N is the rotational speed (rpm) and D is a characteristic engine diameter. Thus,

$$\frac{\dot{m}_{60}}{\dot{m}_{50}} \propto \frac{D_{60}^2}{D_{50}^2} = \left(\frac{3,000}{3,600}\right)^2 = \frac{1}{1.44} = 0.694.$$

In other words, 50–Hz units are roughly 40 percent larger than corresponding 60–Hz units. This is illustrated by the data extracted from the *Gas Turbine World 2014–2015 Handbook* [1] as shown in Figure 3.2. Deviations usually stem from upgrades made by the OEM to certain units over the years. Since such upgrades are typically not done for 50- and 60–Hz machines in lockstep, the aerodynamically "correct" size ratio of 1.44 gets slightly off-kilter.

Very rarely, a 60-Hz gas turbine can be used in a 50-Hz grid or vice versa, driven mainly by the unavailability of the appropriate product at the time for one reason or another. In that case, there are two choices for electrical frequency conversion of the output power:

- A reduction gearbox to change the generator speed from 3,600/3,000 rpm to 3,000/3,600 rpm, or
- A static frequency conversion using a direct current (dc) rectifier.

In general, suitable dc rectifiers are very large and expensive. In such a project in Kuwait, for example, engineers went ahead with a gearbox and requisite system modifications to use a 60-Hz GE 7EA gas turbine in the country's 50-Hz grid.

On the other hand, in the Birr test facility of Ansaldo Energia (in Switzerland, with a 50-Hz grid), an "active generator" is used to connect 60–Hz gas turbines on the test bed to the generator synchronized to the grid. The "active generator" is a four-pole machine at 120–Hz frequency with a static frequency converter between itself and the grid.

In this book, examples are used from either class of gas turbine. Other than its size, whether a gas turbine is a 50-Hz (3,000 rpm) or 60-Hz (3,600 rpm) unit is immaterial to the discussion at hand – unless, of course, the discussion is on compressor or turbine aerothermodynamics. One difference that can be seen in combined cycle applications is due to the larger steam flow generated in the HRSG of a larger 50-Hz gas turbine (roughly the same exhaust temperature, but ~40 percent higher mass flow rate). Since steam turbine efficiency is a function of the volumetric flow rate of steam, combined cycles with 50-Hz gas turbines are slightly more efficient than their 60-Hz counterparts (everything else being the same, of course). By the same token, at a given speed, multi-gas turbine units are slightly more efficient than single-gas turbine units.

3.6 Firing Temperature

A significant source of ambiguity is associated with the definition of highest cycle temperature (yes, there are *many* fuzzy terminology problems in the industry – this is one big reason why this book has been written). Furthermore, an associated terminology jumble pertains to the hot gas path parts. There is a very precise convention used in this book, which is explained below with the help of Figure 3.3, wherein:

- CDT is the compressor discharge temperature and CAC is the cooling air cooler.
- Cooling flows that enter the hot gas path before the stage 1 rotor buckets (S1B) inlet are *nonchargeable* (nch) flows.
- Cooling flows that enter the hot gas path after the S1B inlet are *chargeable* (ch) flows.

Strictly speaking, the highest cycle temperature is the *flame temperature* in the combustor, which is well above 3,000°F for modern industrial gas turbines with premix Dry-Low-NOx combustors. In practical terms, the logical choice is the combustor exit temperature after dilution with combustor liner cooling flow and before the inlet of the turbine section. This is commonly referred to as the *turbine inlet temperature* (TIT). This temperature is used to define the class hierarchy of the gas turbines (see Chapter 4). It is also the ISO-rated firing temperature as defined by API Standard 616 in its §3.16 [2].

For practical purposes, the cycle maximum temperature is the temperature of hot combustion gas at the inlet of the S1B before it starts producing useful turbine work. This temperature is commonly known as the *firing temperature* (TFIRE) and it is several hundred degrees Fahrenheit *lower* than the hot gas temperature at the combustor exit (at the inlet of the stage 1 nozzle vanes [S1N]; i.e., the "true" TIT in essence). The reason for this reduction is due to dilution with nch cooling flow used for cooling S1N and wheel spaces. Thus, another term for this is the *rotor inlet temperature* (RIT), where

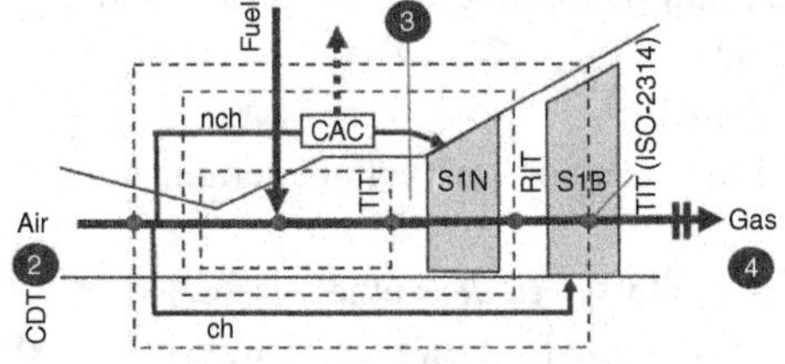

Figure 3.3 Schematic diagram of a gas turbine hot gas path (partial) with temperature nomenclature.

the rotor in question is the stage 1 rotor. The RIT is essentially what is defined in the ANSI Standard B 133.1 (1978) [3], which is referred to as the ISO-rated *cycle* temperature in API Standard 616 in its §3.15.

The TFIRE (or RIT) is about 100°C *higher* than the fictitious temperature as defined in ISO-2314 and adopted by European OEMs [4]. Based on this definition, all compressor airflows, including ch and nch cooling flows, are assumed to enter into reactions with the combustor fuel. Note that ISO-2314 defines *and* outlines a calculation methodology with requisite equations, whereas API 616 and ANSI B 133.1 do *not*.

Even in this day and age, many industry practitioners (inadvertently or not) use the term TIT when they really mean RIT/TFIRE or vice versa. It is thus *absolutely imperative* to understand the distinction between different definitions and to ascertain which one is being used in order to correctly interpret the particular gas turbine technology in question.

3.7 Rating Performance

In this book, quite frequently, rating performances of gas turbines by different OEMs will be used as examples in calculations, discussions, etc. Rating (or rated) performances of gas turbines are annually published in the following two major trade publications:

- *Gas Turbine World* (GTW; www.gasturbineworld.com)
- *Turbomachinery International* (TMI; www.turbomachinerymag.com)

The data in these publications are directly obtained from the marketing departments of the OEMs. As such, they are identical (i.e., for a given year, say, 2016, whether you look up numbers from GTW or TMI is immaterial). Barring typos (which are quite rare), you will get the exact same information.

Gas turbine performance is defined by the following four "cardinal" parameters:

- Output (kWe or MWe)
- Heat rate (Btu/kWh or kJ/kWh)
- Exhaust gas mass flow rate (lb/s or kg/s)
- Exhaust gas temperature (°F or °C)

Rated/rating gas turbine performance is based on the following assumptions:

- ISO conditions (14.7 psia/1 atm [i.e., sea level], 59°F/15°C and 60 percent RH)
- 100 percent (i.e., "full") load
- 100 percent methane (CH_4) fuel
- Gross output as measured at generator low-voltage terminals (i.e., no step-up transformer losses)
- Zero inlet and exhaust pressure losses (*unless* otherwise noted)
- No auxiliaries (*unless* otherwise noted – however, shaft-driven auxiliaries must have been accounted for in OEM-provided data)
- "New and clean" (i.e., no deduction for field deterioration)

For a precise evaluation of the performance, one needs the following information stated clearly:

- Inlet loss (if not stated as zero), how many inches of H_2O (or millibar).
- Exhaust loss (if not stated as zero), how many inches of H_2O.
- Fuel temperature.
- Any overboard air leaks.
- Whether there is an exhaust/aft bearing blower. If yes, the flow rate and delivery temperature are needed.
- Casing heat loss (typically as a fraction of heat consumption).
- Shaft friction losses (typically as a fraction of shaft output).

This, of course, is not the case for ISO base load rating data found in trade publications. Why this information is needed and how it is used will be discussed in detail in Chapter 7. Reasonable assumptions for missing information will be provided in the main body of the text. In other words, missing or "fuzzy" information is not a big impediment to proper evaluation of a particular gas turbine's performance if one is really determined to carry it out with as much accuracy as possible.

Nevertheless, in this book, ISO base load rating data are almost always used "as they are." No attempt has been made to "correct" published numbers by "informed guesses." It is believed that this is the best approach to eliminate any bias and/or error piled on top of other biases and/or errors already inherent in the original data set. What does this exactly mean? For example, let us assume that a gas turbine's exhaust temperature is listed as 1,150°F at ISO base load. No further information is available. Furthermore, let us assume that we want to estimate the exhaust gas *exergy* of this gas turbine in order to estimate the output of a steam turbine generator in a combined cycle arrangement with the said gas turbine. First of all, note that:

- Overall exhaust pressure loss imposed by the HRSG on the gas turbine back end is roughly 15 inches of H_2O (for three-pressure, reheat designs with advanced-class, heavy-duty industrial gas turbines).[2]
- Each 4 inches of H_2O *additional* pressure loss *increases* the exhaust temperature by about 2°F.

Thus, if the quoted 1,150°F number is at zero losses, in a combined cycle arrangement, it would be

$$1,150 + (15/4) \times 2 \sim 1,157^{\circ}\mathrm{F}.$$

If it were quoted at a specified exhaust loss, the error would be somewhere between 0 and 7°F. Rather than hunting for the exact value of the exhaust temperature, which would also be affected by other uncertainties, a decision has been made to evaluate each customer *at their word* and to use their listed exhaust gas parameters as the

[2] These are later versions of the F class (e.g., GE's F.04 and F.05), G/GAC, H, HA, and J/JAC class gas turbines.

technology drivers. Thus, "OEM's word" becomes the common basis of evaluation with absolutely no other bias, error, etc., imposed by a (interested or non-interested) third party.

References

1. *Gas Turbine World 2014–15 Handbook*, Volume 31 (Fairfield, CT: Pequot Publication, Inc.).
2. API Standard 616: Gas Turbines for Refinery Services, 5th edition (Washington, DC: American Petroleum Institute, 2011).
3. ANSI B113.1: *Gas Turbine Terminology* (New York: American Society of Mechanical Engineers, 1978).
4. ISO-2314:2009: *Gas Turbines – Acceptance Tests* (Geneva, Switzerland: International Organization for Standardization, 2013).

4 Past and Present

First, a word of caution: this chapter on the history of gas turbines is *not* written for the casual reader. It includes a thermodynamic *critique* of certain milestone technologies. All thermodynamic concepts and calculations used in this chapter are explored in detail in the chapter on thermodynamics (Chapter 5). If you have difficulty following some aspects of the discussion below, it is suggested that you familiarize yourself with the subject matter of that chapter first. Alternatively, of course, you can ignore the technical subtleties and focus purely on the timeline and historic nuggets.

Famous French philosopher René Descartes (1596–1650) stated in his *Discourse on the Method* that "there is nothing so strange and so unbelievable that it has not been said by one philosopher or another." The Bible predated Descartes by about 16 centuries and was a bit more succinct ("There is nothing new under the sun" – Ecclesiastes 1:9). While one should not take these statements literally (otherwise we would still be living in caves as hunter–gatherers), they contain a considerable amount of truth. (Recall Bishop Wilkins' "chimney gas turbine" in Figure 1.1.) Having a full grasp on the history of technology progress leading to the state of the art in a particular field is sine qua non for getting ready for the next step in the evolution of said field. In this book, a historical perspective will always be provided while discussing a particular aspect or application of gas turbines. In this particular chapter, comprehensive coverage of the relatively recent past of the gas turbine (as we know it today) is provided so that the reader can appreciate the significant gains that have been made in this technology and can form a good idea of what the future might hold.

Gas turbine development history is somewhat convoluted. It does not follow a straight line starting from an early prototype and progressively advancing along a thermodynamically logical path to its current state of the art. Obviously, the ultimate enabling idea – namely, realization of the useful work-generating ability of a fluid at high pressure and temperature – can be traced back to the high-pressure steam engine of Richard Trevithick, who valiantly, and ultimately successfully, defended his concept as the more efficient alternative to James Watt's vacuum-driven atmospheric steam engine (which itself was an improvement on Thomas Newcomen's engine via a separate condenser).[1]

[1] James Watt maintained that high-pressure steam was "criminally insane," although his initial concepts recognized the possibility and superiority of it vis-à-vis the atmospheric variant, which depended on very low pressure (i.e., vacuum) on the other side of the piston. It is not clear whether this was a genuine concern of his or whether he was simply acting to protect his and his partner's (Matthew Boulton) commercial interests.

(Of course, the proper thermodynamic concept of *exergy* [*availability* in US textbooks] would be introduced more than a century later.) Around the same time, in 1791, John Barber filed a patent (UK patent no. 1833 – *A Method of Rising Inflammable Air for the Purposes of Procuring Motion, and Facilitating Metallurgical Operations*), which is generally accepted as containing the basic features of a gas turbine. Nevertheless, the sketch from his patent does not look like anything one would recognize today as a gas turbine (it is widely available online).[2] The next patent, which contained a "compressor" (a multi-stage axial one no less) and a multi-stage reaction "turbine" was filed by Frank Stolze in 1877 (curiously enough, the patent was not awarded). The first experimental gas turbine that actually ran was built by Lemale and Armengaud in France between 1903 and 1906. It was a 400 hp (about 300 kW) unit (rated but not measured), which was tested in 1905 in Société Anonym des Turbomoteurs in Paris with essentially zero net output (i.e., the turbine generated just enough power to drive the compressor with a pressure ratio [PR] of about 4 at 4,250 rpm). It is interesting to note that the combustion chamber of that test unit was cooled with water.

A comprehensive but compact history of the evolution of gas turbine technology can be found in the paper on turbomachinery development by Meher-Homji [1]. The reader can use that paper (available online) and its list of references as a starting point to delve deeper into the historical aspects of gas turbines. In the same vein, the 26-page section titled "A brief history of turbomachinery" in Wilson and Korakianitis also provides a concise but very informative look into the historical development of gas turbines [2]. Another similar source is chapter 1 of the book by Bathie, which covers the history of gas turbines in myriad applications (aviation, land-based, transportation, etc.) until the 1990s [3]. A recent book on industrial gas turbine development by Eckardt, with a focus on the venerable Swiss industry giant Brown Boveri Co. (BBC) and its descendants, is highly recommended as a source of general gas turbine development history [4]. An Internet search would definitely reveal many other sources covering basically the same inventors and machines, but with varying focus (in most cases significantly skewed by the author's nationality and/or organization). However, in this author's opinion, when it comes to industrial (i.e., land-based) gas turbines, it is very difficult to match the attention to detail as well as depth and breadth of the information provided in Eckardt's book. Its closest match, again in this author's opinion, is the technical history of Pratt & Whitney (P&W) engines by Jack Connors [5], which obviously focuses on aircraft jet engines, closely followed by Bill Gunston's book, also on aircraft jet engines [6]. One criticism that can be leveled against the last two sources is their heavy focus on US and British engine development history. The lack of detailed coverage of the considerable jet engine development work in Germany by Connors and Gunston is somewhat balanced by Meher-Homji's excellent papers on the German pioneers von Ohain and Bentele [7–9] and Leiste's paper on the history of Siemens' industrial gas turbine development [10]. For early Russian (Soviet) jet engine development, which has

[2] The April 28, 1972, issue of *Nuneaton Evening* (a local paper in the UK) included a feature with a picture taken at Hannover Trade Fair, which showed an accurate working model of Barber's gas turbine built by the Kraftwerk Union (predecessor of Siemens).

received scant notice in the West, the reader should consult the excellent short book (published by the Rolls-Royce Heritage Trust) by Kotelnikov and Buttler [11]. Finally, it would be amiss if one did not mention a recent book written on jet engine development by the professional historian, H. Giffard, which is a superb history of the inventive and industrial activities simultaneously undertaken by three major industrialized nations in the 1930s and 1940s [12].

Key milestones in the development of the gas turbine (in addition to those briefly touched upon above) can be summarized as follows:

1903 First gas turbine worldwide delivering net power by Ægidius Elling, Norway [13]
1905 Holzwarth's explosion turbine concept
1908 First explosion turbine built by BBC for Dr. Holzwarth
1930 Frank Whittle patent 347,206 "Improvements in Aircraft Propulsion"
1937 First "Houdry" turbo-compressor set (with 900 kW *excess* power)
1937 First test run of Whittle jet engine in a factory in Rugby, England
1939 BBC Neuchâtel 4 MW gas turbine goes commercial
1939 First jet-powered flight (von Ohain's HeS 3b gas turbine engine in Heinkel He 178)
1941 First flight of Whittle jet engine in Gloster Pioneer (E.28/39 at RAF Cranwell)
1943 First flight of Me-262 with Jumo-004B gas turbine engine

Herein, three entries from the list above are going to be discussed as bona fide examples that are of prime significance in laying the foundation for the modern heavy-duty industrial machines commercially available for electric power generation:

1. Holzwarth's "explosion" turbine (1905)
2. The BBC's 1939 Neuchâtel unit
3. The Jumo-004 engine of the World War II German jet fighter Messerschmitt 262 (1943–1945)

The reader might be questioning the choice of the last machine. Why an aircraft engine? If an aircraft engine is necessary, why not Sir Frank Whittle's W.1A or Hans von Ohain's Heinkel-Strahltriebwerk 3 (HeS 3), both of which predated Jumo-004? It is hoped that the reason for this will be made amply clear in Section 4.3. In essence, Jumo-004 was a *serial-production gas turbine* as we know it today (i.e., *not* a "one-and-done" novelty and *not* a prototype) with all the proverbial "bells and whistles," alas sans superalloys.

An Unsung Pioneer – Arkhip Mikhailovich Lyul'ka

The early history of gas turbine development is primarily driven by prime mover applications for fighter aircraft (1930s and 1940s). The two household names, which are widely known to students and the practitioners alike in the West, are Whittle and von Ohain, who are considered "fathers" of jet engines and rightly so. But there is a third engineer whose name is largely forgotten (if it were known to many to begin with), who was a contemporary of Whittle and von Ohain and led the independent jet engine design work in the Soviet Union: Arkhip Lyul'ka.[3] While this is not an aircraft jet engine book, the author would like to use this opportunity to help introduce the pioneering work of Lyul'ka to the readers. The material in this section is largely culled from the book by Kotelnikov and Buttler [11].

Lyul'ka's first design, in 1937, was RTD-1 ("rocket turbojet engine") with a centrifugal compressor and single-stage axial turbine generating 500 kg (1,102 lb) of thrust. In 1939, he started the design work on RD-1 with an axial compressor and 525 kg (1,157 lb) of thrust. In April 1941, Lyul'ka received an "inventor's certificate" (the Soviet equivalent of a patent) for a "bypass engine" (i.e., a turbofan engine) designated TRD. Thus, Lyul'ka may very well be the inventor of the turbofan engine. (Kotelnikov and Buttler remark that Frank Whittle had first "suggested" the bypass engine in 1936.)

When Germans invaded the Soviet Union in June 1941, the work on RD-1 (in Kirov factory in Leningrad) was interrupted and Lyul'ka had to move to Chelyabinsk, where he worked on diesel engines for tanks. Although jet engine and jet-engined aircraft work resumed in 1942, in 1944, the Soviets recognized that the Germans and the Soviet's own allies, the British and the Americans, were well ahead of them in both areas.

Lyul'ka began his design work on the S-18 turbojet in autumn 1944. At the end of the war, when S-18 was compared with the captured German jet engine Jumo-004, it proved to be more fuel efficient with higher thrust, but with inferior reliability. The primary reason for this was the eternal bane of gas turbine development (i.e., the lack of adequate materials by the Soviets, more specifically high-temperature alloys for turbine blades [more on this subject in Chapter 13]). The stopgap solution was to use captured Tinidur blanks while establishing the chemical composition of that alloy and eventually to produce their own, which they did, but these were found to be inferior to the original German alloy.

Lyul'ka kept working on advanced versions of S-18, TR-1, later TR-2, and eventually TR-3 turbojets. In the meantime, the Soviets used the German Jumo-004 and BMW-003 copies – RD-10 and RD-20, respectively – as well as copies of Rolls-Royce Nene, RD-45 (with a centrifugal compressor). Lyul'ka's axial compressor gas turbine jet engines, evolutions of the TR-3, under the Soviet designations of AL-5 (using his initials) and AL-7, eventually found their way into Soviet military aircraft in the 1950s. The latter was the first mass-production Lyul'ka jet engine, which was used in various types of aircraft. Detailed pictures and drawings of early Lyul'ka engines can be found in [11].

4.1 Holzwarth's Explosion Turbine

Stodola credited Hans Holzwarth for having built the first "economically practical" gas turbine [14]. The august "father" of steam and gas turbine engineering was dead on, of course. Holzwarth built several types of his "explosion" turbine (very descriptively

[3] There is also György Jendrassik from Hungary who ran a 100-hp turboprop on a test bench in 1937. See [2] for more on him.

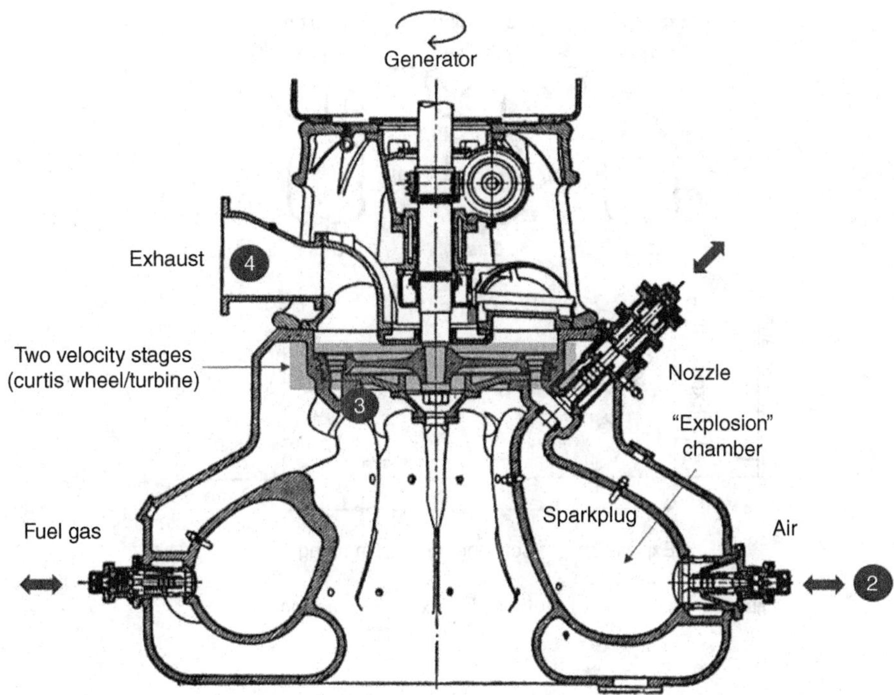

Figure 4.1 Cutout of Holzwarth's 1909–11. Mannheim explosion turbine [14].

called *Verpuffungsturbine mit intermittierender Zündung* in German) in the first quarter of the twentieth century. His gas turbines were "real" (i.e., not on-paper patent ideas), and they operated as commercial units generating shaft power for customers (i.e., they were not laboratory novelties). The turbine was, strictly speaking, a hybrid construction combining the spark-ignition (constant volume) combustion process of an Otto cycle with the axial expansion process of a Brayton cycle. Admittedly, similar to John Barber's strange-looking device, Holzwarth's machine did not look like a gas turbine as we know it today either. A cutout of the 1909–1911 Mannheim turbine (quite similar to the 1908 Hannover turbine, which can be seen in Deutsches Museum in Munich) is shown in Figure 4.1. The positions of the nozzle, air, and fuel gas valves during one cycle and the corresponding variations of the combustion (explosion) chamber pressure are shown in Figure 4.2. (You can close your eyes and imagine the "bang! – pffft! – bang!" sequence of Holzwarth combustion chamber operation – hence the German name *Verpuffungsturbine*!)

Stodola calculated a thermal efficiency of 25.6 percent at the Curtis wheel for the test of a 1,500-rpm experimental turbine in Mülheim-Ruhr in 1919 [14]. Burning a gas with 434 Btu/ft^3 heating value, the said turbine produced about 725 kW. Air and fuel gas were compressed by steam turbine-driven compressors and sequentially injected into the explosion chamber (gas first) at about 30 psia. Stodola reports an average maximum explosion pressure of 160 psia. The combustion products expanded through a two-stage velocity-compounded impulse turbine (i.e., a *Curtis wheel*). Exhaust gas at about 800°F

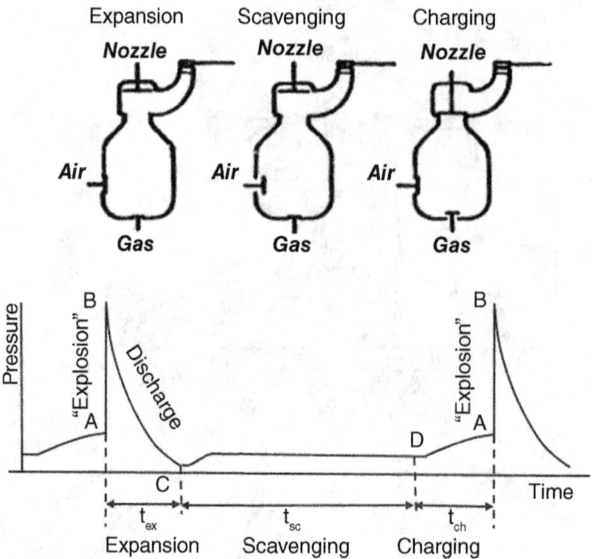

Figure 4.2 One cycle of Holzwarth's explosion turbine.

was recovered in an exhaust heat boiler and drove the steam turbine, which in turn drove the blower (precompressor). In that sense, this unit could be considered the first *combined-cycle* power plant in the world. Holzwarth designed and developed several different variants of his turbine between 1907 and 1928. From 1928 on, BBC took over the development of the Holzwarth turbine, and in 1933 installed a blast furnace gas-fired unit in a German steel mill, which was destroyed during World War II.

Coming back to the experimental turbine in Mülheim-Ruhr, at the combustion (explosion) chamber exit, thermocouple readings indicated 880–1,020°F. Thus, prima facie, a Carnot cycle efficiency of 65 percent is possible with an assumed T_1 of 59°F (15°C). This, however, is incorrect because in an ideal, constant volume explosion, the gas temperature rise in the explosion chamber would be proportional to the pressure ratio, which was about 5 in the Mülheim-Ruhr machine. Based on the test data provided by Stodola, this would correspond to a maximum chamber temperature of ~2,900°F. (Essentially, this is exactly what would happen in the cylinder of an Otto or Diesel cycle piston cylinder [reciprocating] internal combustion engine.) Thus, the implied Carnot efficiency is about 84 percent.

Note that the air-standard cycle describing the operation of the Holzwarth turbine is the *Atkinson* cycle. The real cycle effect with a PR of about 10 (i.e., 2×5, with precompression) and a turbine inlet temperature (TIT) of 2,900°F indicates an ideal (i.e., Carnot-like) cycle efficiency of about 77 percent for the Atkinson cycle. Stodola predicted an overall efficiency of 25 percent by applying "known means to overcome the observed obstacles in the test unit." This translates into a technology factor of $25\% / 77\% \approx 0.32$. Note that the overall efficiency of the Mülheim-Ruhr machine with the exhaust heat recovery steam turbine and blower was only 13 percent. Today's state of the art in heavy-duty industrial gas turbines is equivalent to a technology factor of about 0.7 on average. In other words, if an original equipment manufacturer (OEM) decides to construct and build Holzwarth's Mülheim-Ruhr machine with the same cycle parameters but with modern materials and design tools, the achievable efficiency could be as high as 0.7 × 77% ≈ 54%.

4.2 Brown Boveri's Neuchâtel Unit

The first commercial (utility) stationary gas turbine for electric power generation was erected in Neuchâtel, Switzerland, by the former BBC in 1939, three years before the first flight of the Messerschmitt Me-262 interceptor powered by twin Jumo-004B turbojet engines, which will be covered in detail in Section 4.3. This fuel oil-burning 4-MWe machine, primarily used for standby and peaking duties, was operational for nearly 70 years (see Figure 4.3). The combustion chamber was derived from the turbocharged *Velox* boiler, which itself resulted from the BBC work done on the Holzwarth turbine. The BBC turbine was tested under the supervision of Stodola, who reported an overall thermal efficiency of 17.4 percent, which is less than half of the efficiency of early F–class gas turbines. This machine was designated by the ASME as a *Historic Mechanical Engineering Landmark* in 1988. Today, it can be seen on display in Ansaldo Energia's development and production facility in Birr, Switzerland (inherited from Alstom as a result of EU-enforced divestiture imposed on General Electric (GE) after its acquisition of Alstom in 2015).

For a PR of 4.4 and 1,500°C (2,732°F) combustion temperature, the Neuchâtel gas turbine's Carnot cycle efficiency is 83.5 percent (with 20°C ambient temperature or T_1). Note that this is a bona fide Brayton-cycle engine and, as such, its ideal, air-standard cycle efficiency would be a function of PR only, which results in 34.5 percent for a cycle effect of 0.41. Using the measured 17.4 percent efficiency, this translates into a technology factor of 17.4% / 34.5% ≈ 0.50. The test run of the new gas turbine was conducted in the BBC Test Center on July 19, 1939, under the direction of Aurel

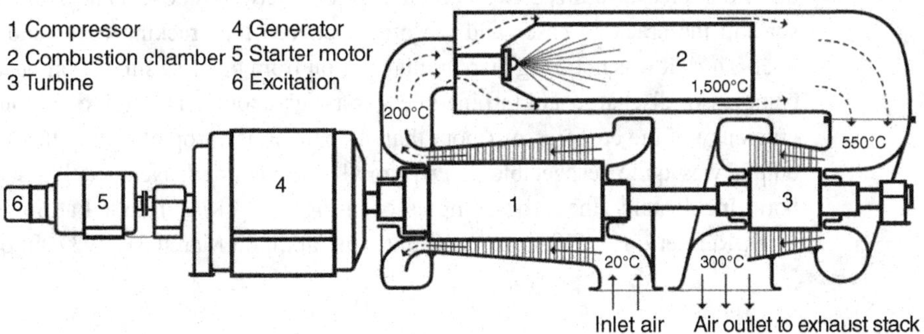

Figure 4.3 Schematic diagram of BBC's 1939 gas turbine [3].

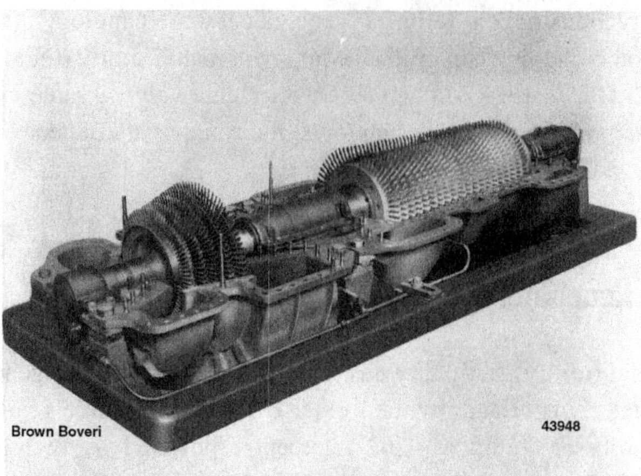

Figure 4.4 BBC turbocharger set for the Houdry process. (Courtesy of Historisches Archiv ABB Schweiz)

Stodola, who was 80 years old at the time, and the unit was presented later that year as a world first at the Swiss National Exhibition in Zurich. The five-stage turbine had an isentropic efficiency of 88.4 percent and generated 15.4 MW, 11.4 of which was consumed by the 15-stage axial-flow compressor with 84.9 percent isentropic efficiency.

The precursor of BBC's Neuchâtel gas turbine was the BBC turbocharger set for compressed hot air generation in connection with the Houdry oil cracking process (see Figure 4.4). This machine was essentially a turbo-compressor to deliver 20 m^3/s air at a PR of 4.2 to the fixed-bed, three-case cracker unit.[4] The axial-flow compressor had 20 stages and was driven by a five-stage turbine. Although it was not designed for electric power generation per se, the output of the turbine (5.3 MW) more than compensated for the power consumed by the compressor (4.4 MW) so that the synchronous AC machine connected to the drivetrain via a gearbox actually *generated* 900 kWe of electric power. The similarity in drivetrain arrangement between the turbo-compressor in Figure 4.4 and the gas turbine in Figure 4.3 is unmistakable. Note that the Houdry turbo-compressor utilized two combustors: one between the discharge of the compressor and the cracking cases and the other between the cracking cases and the turbine inlet. Thus, it was not a big leap in imagination from that to a single combustor between compressor discharge and turbine inlet for a high enough TIT and to exploit the superb efficiency of the compressor (more than 84 percent isentropic) so that the net generator output was quite respectable. Thus, voila! The landmark Neuchâtel gas turbine was born. Incidentally, the turbo-compressor in Figure 4.4 was installed at Sun Oil Marcus Hook Refinery, PA, USA, and went into operation on March 31, 1937. In other words,

[4] At approximately 1.2 kg/m^3, roughly 250 kg/s or 550 lb/s of airflow.

by the time Dr. Stodola oversaw the first run of the Neuchâtel gas turbine, this unit had already demonstrated 2.5 years of constant service.

4.3 Junkers Motoren Jumo-004

The Messerschmitt Me-262 was the world's first operational jet fighter equipped with two *Junkers Motorenwerke* Jumo-004 turbojet engines, whose development started in 1939 under the leadership of Anselm Franz [7]. The first flight of Me-262 was in 1942. Luckily for the free world, mainly due to Hitler's inane interference with its final development, the plane's entry into full combat service was delayed until 1944. On top of that, since the Nazis' utter stupidity surpassed even their depravity (again, lucky for the free world), despite Albert Speer's best efforts to move the production into mountain tunnels for protection from Allied bombing attacks, his inability to make the Nazi leadership understand the value of skilled labor and the pragmatism of humane treatment of "slave labor" hampered production and quality significantly [12]. Even then, almost 6,000 engines were manufactured by the end of the war, of which about 4,750 were delivered to the aircraft factories.

This last statistic represents the main significance of the Jumo-004; it was neither a one-off experimental setup nor a laboratory specimen that had been tested and thrown out. It was a *mass production* gas turbine engine for a mass production fighter aircraft, which proved itself very successful in combat operations in the hands of expert fighter aces like Galland, Nowotny, and Bär, who remains the world's highest-scoring jet fighter ace with 17 aerial combat victories.

The Jumo-004 gas turbine engine reflects a superb compromise between ideal design intent and available materials and production facilities (including a skilled workforce). Furthermore, it was a modern gas turbine incorporating almost all of the key features of a state-of-the-art aircraft propulsion or electric power generation engine (see Figure 4.5):

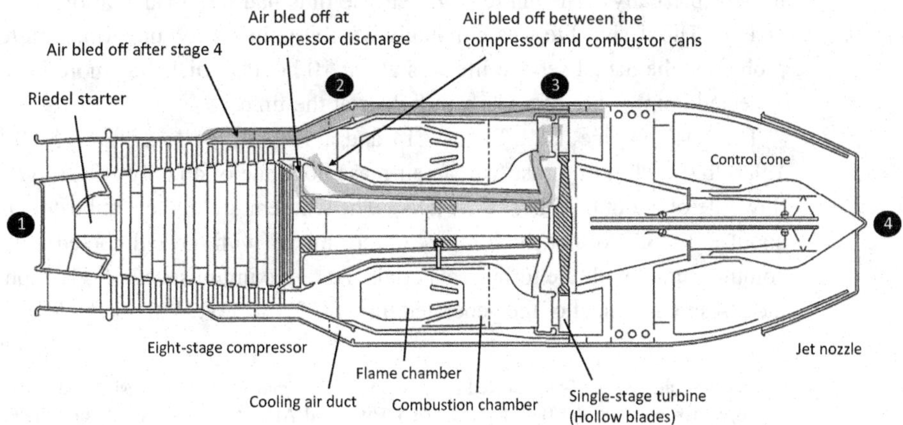

Figure 4.5 Cutout of a Jumo-004 gas turbine engine (turbojet).

1. An eight-stage axial compressor with a PR of about 3 with a stacked-disk rotor construction (single tie bolt and Hirth serration for torque transmission)
2. A single-stage turbine cooled with air bled (about 7%) from the compressor (hollow turbine blades)
3. Six combustor cans arranged circumferentially (i.e., can-annular design)

Rudolf Friedrich, who was responsible for the compressor design, came up with the stacked disk with the single tie-bolt design due to the aforementioned wartime difficulties requiring material savings and relatively easy constructability. For example, frequent scrapping of a part (e.g., a single disk or the tie bolt itself) in the mass production line was practically unavoidable. Multi-part design made replacement of scrapped parts much easier. Furthermore, the lightweight structure, as opposed to a solid shaft forged from a large block, increased resistance to low-cycle fatigue stemming from thermal stresses induced by rapid load changes.

The entire engine took about 700 man-hours to manufacture. Parts were manufactured in different factories and the assembly took place in well-protected "caves" [12]. The single-stage turbine was manufactured in the *AEG Turbinenwerke* in Berlin Huttenstrasse, which later became the turbine manufacturing plant for Siemens gas turbines in Europe. Rudolf Friedrich joined Siemens after the war, and he was instrumental in originating the design concepts of Siemens' heavy-duty industrial gas turbines (designated "VM," *Verbrennungsmaschine* being German for combustion machine), which have been adhered to since then (e.g., the same shaft design as in Jumo-004 is used by modern Siemens H-class heavy-duty industrial gas turbines).

As mentioned above, engine development was overseen by Anselm Franz, who, from the outset, aimed at a realistic and conservative design with the maximum chance of success.[5] The first variant, 004A, proved unsuitable to mass production due to its weight and the scarcity of strategic materials (i.e., nickel, cobalt, and molybdenum) for hot gas path parts. In the production version, 004B, the combustion chamber was made from mild steel and protected against oxidation by aluminum coating. Hollow turbine blades (to facilitate air cooling) were manufactured from "Cromadur" sheet metal by folding it in two (basically). The entire 004B engine thus had less than 5 lb of chromium (zero nickel). The design life was nominally 25 hours; in reality, due to the aforementioned problems, the actual service life was about 10 hours, which was more in line with the expected combat life of a German fighter at the time.

The 004B engine had a PR of 3.14 and a TIT (T_3) of 1,427°F (775°C) to generate 1,980 lb (900 kg) of thrust with an air mass flow rate of 46.6 lb/s (21.1 kg/s) and a fuel flow rate of about 0.75 lb/s (0.34 kg/s). The temperature in the combustor chamber was about 2,000°C, so that the hot gases were mixed with air and cooled down to a TIT commensurate with the turbine materials. At a maximum speed of 541 mph (242 m/s), the power equivalent of the generated thrust is 2,130 kW. With the fuel consumption of

[5] After the war, Anselm Franz moved to the USA. In Lycoming, he led the design and development of the T53, which powered the Bell Aircraft UH-1 Huey and AH-1 Cobra helicopters and the OV-1 Mohawk ground attack aircraft. In the 1960s, he led the design efforts for a new turbine engine for land-based propulsion, which eventually developed into the AGT-1500 used in the M1 Abrams tank.

about 14,500 kWth, the equivalent shaft efficiency of the 004B is 14.7 percent. (Compressor isentropic efficiency was 78 percent and turbine isentropic efficiency was 79.5 percent with combustor efficiency of 95 percent.)

For a Brayton-cycle engine with a combustor temperature of 2,000°C, the Carnot efficiency is 87.3 percent. The ideal, air-standard cycle efficiency is about 28 percent for a cycle effect of 0.32 (via Equation 5.2). Using the shaft power-equivalent 14.7 percent efficiency, this translates into a technology factor of 17.4% / 28% ≈ 0.53 for the Jumo-004B, which is about the same as BBC's Neuchâtel gas turbine.

In her otherwise excellent book, Giffard almost dismisses Jumo-004B as a technically *inferior* engine (inferior to others in development in the UK and in the USA, that is) with poor performance [12]. Since the Allies deployed no jet fighters of their own against the twin-Jumo-powered Me-262, this is a meaningless assertion in and of itself. Furthermore, Jumo-004 was a successful end product of masterful engineering, which combined (i) judicious trade-off in available materials and manufacturing technology with performance and (ii) innovative design features (such as the aforementioned rotor design – currently used in Siemens' H–class gas turbines – and variable-area exhaust nozzle). It is truly unfortunate that the Jumo-004 gas turbine jet engine was designed, developed, and deployed in the service of an evil regime. It is also true that it was only a matter of time before a similar and/or superior gas turbine jet engine would be introduced by the Allies in much larger quantities. (In that sense, there was never a realistic chance that Me-262 could have changed the outcome of the war.) Nevertheless, it is indisputable that Jumo-004 was a superb example of mechanical engineering of the highest ingenuity, which made it into mass production under highly adverse conditions and performed as designed in battle and should be recognized as such.

4.4 Recap: 1900–1950

The major technology characteristics of the three historical gas turbines discussed above and a modern J–class gas turbine are summarized in Table 4.1. The key takeaways from Table 4.1 can be summarized as follows:

1. The limited availability of high alloys for the turbine hot gas path severely limited allowable TIT and caused significant temperature (exergy) loss between the combustor and turbine inlet. This is quantified by the difference in Carnot efficiencies.
2. For the two Brayton-cycle gas turbines, the lack of full understanding of aerothermodynamics and commensurate design tools limited compressor PRs (the main driver of Brayton-cycle efficiency). This is quantified by the low "cycle factor" (less than 0.5).
3. The superiority of constant-volume combustion (CVC) over constant-pressure combustion (CPC) is strikingly illustrated by the high cycle PR (driven by heat input) and high cycle factor.

Table 4.1 Key technology data for the three historic gas turbines (in comparison to a modern J–class gas turbine)

	Holzwarth Explosion Turbine	BBC Neuchatel GT	Jumo-004B Turbojet	2016 J Class
Active Years	1908–1930	1940–2000	1943–1945	2016
Air-Standard Cycle	Atkinson	Brayton	Brayton	Brayton
Combustion	Constant volume	Constant pressure	Constant pressure	Constant pressure
Fuel	Fuel oil, anthracite gas	Fuel oil	Kerosene	Natural gas
Cycle PR	10	4.4	3.14	23.1
Combustion Temp., °F	~2,900	~2,750	~3,500	~3,100
TIT, °F	1,020	1,022	1,427	2,912
Cooled Turbine	No	No	Yes	Yes
Carnot Efficiency (%)	84.4	83.5	87.3	85.4
Carnot Eff. (@ TIT) (%)	64.9	64.4	72.5	84.6
Ideal Cycle Efficiency (%)	77.1	34.5	27.9	59.2
Cycle Factor	0.91	0.41	0.32	0.70
Actual Efficiency (%)	13.0	17.4	14.7	42.2
Technology Factor	0.17	0.50	0.53	0.71

4. At the same time, difficulties associated with the practical design of an actual CVC turbine vis-à-vis the CPC turbine are also dramatically quantified by the difference in technology factors.

It was only a matter time before the Brayton-cycle gas turbine design fulfilled its potential via advances made in materials, coatings, and the 3D aero design of axial compressors (especially) and expanders. Serious design work commenced immediately after the war in the USA and in the UK. In fact, it had already started in the USA during the war. Westinghouse, apparently without any knowledge of British or German efforts, developed the first US jet engine with an axial compressor and an annular combustor [15]. The engine, designated WE19A, was developed in about 15 months upon receiving a contract from the US Navy. An upgraded version, WE19B, with 1,365 lb of thrust (about 70 percent of Jumo-004B) was test flown in January 1944 on a Chance Vought Corsair as a booster engine [15]. Later designated as J30 and upgraded to 1,600 lb of thrust, this engine powered the Navy's McDonnell Douglas XFD-1 (later FH-1) Phantom, the first jet-powered aircraft to operate off a carrier [15]. The next Westinghouse aircraft propulsion product was WE24C (J34 in military designation), a 3,000-lb thrust jet engine, which powered Vought F6U Pirate. Westinghouse J34 was also deployed in the Douglas Skyrocket, a jet engine and rocket-powered research aircraft with swept wings, which was used for data collection in transonic and supersonic flight. Westinghouse eventually exited the aircraft engine business in 1960 after delivering 1,223 engines to the US Navy.

Also in January 1944, GE completed and delivered the first I-40 jet engine to test [16]. Development of this engine was started by GE at the request of Army Air Forces (the predecessor of the US Air Force) in 1943. The I-40 had a centrifugal compressor and a can-annular, reverse-flow combustor arrangement with 14 cans. From the outside, it resembled the British jet engines of the time (i.e., the Whittle Rolls-Royce Welland engine, which first powered the Gloster Meteor twin-engine jet fighter in 1944 when it was deployed against the V1 flying bombs).[6] The I-40 had a compressor PR of 4.126 and a TIT of 1,492°F for a net thrust of 4,000 lb [16]; it was used to power Lockheed's XP-80A Shooting Star jet fighter, which made its first flight with I-40 in June 1944.[7] Thus, near the end of World War II, two US engine manufacturers, contracted by two different branches of the US military with different performance requirements, were independently moving along two different paths. It must be pointed out that, around the same time, GE was also actively developing two axial-compressor jet engines, TG-100 and TG-180, for propeller and jet propulsion, respectively [17]. Interestingly, just as I-40 was a copy of the British Whittle engine with the centrifugal compressor, the TG-180 axial compressor was a copy of the German Jumo-004 compressor (but with a better turbine, as told by Gerhard Neumann in his 1989 AeroAstro Gardner Lecture). TG-180 (military designation J35) was used in the Republic XP-84 Thunderjet. Late in 1947, complete responsibility for the development and production of the engine was transferred to the Allison Division of the General Motors Corporation. The engine was also used in the transonic research aircraft Douglas Skystreak.

P&W entered the jet engine business when the US Navy asked them to manufacture 130 of 261 Westinghouse 19B engines [5,15]. They unwillingly accepted it in order not to alienate the Navy (P&W had previously dismissed gas turbines for aircraft propulsion as "uncompetitive"). This order gave P&W good experience with axial-flow gas turbines. Finally, in 1947, upon the Navy's request to "Americanize" the Rolls-Royce Nene engine for Grumman's F9F-2 at 5,000 lb of thrust, P&W entered the aircraft gas turbine development business [5]. Designated as J42, this engine had a double-entry centrifugal compressor and nine combustion chambers with a single-stage turbine. P&W's long history in aircraft gas turbine development is presented in the excellent book by Jack Connors [5].

The Germans resumed from where they left off almost immediately after the war under the leadership of Rudolf Friedrich at Siemens (see Leiste [10]). Initially, their activities were limited by the Allied Control Council to theoretical work, but eventually, in the mid-1950s, Siemens manufactured its first commercial gas turbine, VM 5.

The Soviets also resumed their own research and development that had first been interrupted then hampered by the war, initially based on captured Jumo-004B and BMW 003A engines. Further development was supported by Rolls-Royce Nene and Derwent engines obtained from the British [11]. As mentioned earlier in this chapter, A.M. Lyul'ka, who led the Soviet jet engine design in the 1930s, continued to work on his own designs (rather than on copies of German and British engines), which

[6] This, of course, was no surprise, because GE started off with the Whittle technology, including a full-size engine and all the drawings. Frank Whittle made a visit to GE in Schenectady in 1942 to advise American engineers and ensure that all was going well [10].

[7] The first XP-80A was powered by a British De Havilland Halford H.1 engine and made its initial flight in January 1944 [10].

eventually made their way into frontline Soviet aircraft. The main focus in all four countries and elsewhere was initially on military aircraft propulsion. For very informative but concise coverage of early British, German, American, and, to a lesser extent, Soviet engine development efforts from a technical perspective, the reader is referred to the book by Gunston [6].

Initially, centrifugal-compressor and even "turbocompound" reciprocating (i.e., piston-cylinder) engines were still very much in the picture (Puffer and Alford in [17]). Eventually, axial compressors with their potential for high PR and high efficiency at large air flow rates (for a given frontal area) overtook their centrifugal counterparts for high-power, high-performance applications. Turbocompound engines offered excellent efficiency (e.g., Napier's Nomad), but eventually fell out of the race due to the unmatchable simplicity and thrust-to-weight ratio of axial-flow jet engines

Land-based (stationary) industrial gas turbines also entered the picture soon after the war. The first gas turbine installed in an electric utility in the USA (Oklahoma Gas & Electric, Belle Isle Station, Oklahoma) was a 3.5-MWe GE Frame 3 unit (about 17 percent efficiency) that entered service in 1949.[8] In addition to generating power, the exhaust gas of this gas turbine was utilized to heat the feed water of a conventional steam plant. In other words, the first US electric utility gas turbine was in a "combined"-cycle configuration [18]. In fact, as mentioned earlier, the combined-cycle concept can be traced back to Holzwarth's experimental turbine in Mülheim-Ruhr in 1919. Another combined-cycle (or, more properly, *binary*-cycle, because the *topping* mercury cycle is a Rankine cycle as well) example from the same time period is Emmet's mercury-vapor process (1925) [19]. In short, by the mid-twentieth century, all requisite pieces for a modern electric power generation station utilizing gas turbines were already in place, except the requisite materials and high-speed computer hardware and software to solve the full Navier–Stokes equations to design efficient axial-flow compressors and expanders. Later cycle-specific developments can, with some justification, be described as "variations on the theme."

GE's Frame 3 was originally developed with the help of the American Locomotive Company (Alco) to power railroad locomotives.[9] Its general design was similar to the TG-180 (see above) with a 15-stage axial-flow compressor (PR of 6) and a firing temperature of 1,400°F. The first Frame 3 was deployed in an Alco gas turbine electric locomotive (GTEL). This unit was eventually given to the Union Pacific Railroad Company for extensive testing (1948–1949). Union Pacific later ordered 25 of these GTEL engines burning cheap but difficult-to-handle Number 6 fuel oil (also known as residual or "bunker C" fuel oil). The fuel treatment included water-washing, centrifuging, and adding magnesium sulfate to inhibit vanadium and slow down hot corrosion

[8] A second unit was installed in 1952. These units were taken out of service in 1980, and the first one, designated an ASME National Historic Landmark, was installed in front of Building 262 in GE's Schenectady works. It was relocated to GE's gas turbine manufacturing plant in Greenville, SC, in 2013.

[9] Based on reminiscences of Ivan G. Rice as told in a Turbomachinery International blog that appeared in March 2017.

to ensure 5,000 hours of hot gas path parts life before replacement. As mentioned earlier, the second Frame 3 ended up in Oklahoma Gas & Electric's Belle Isle Station.

Around the same time frame, Westinghouse developed a 2,000–hp (~1.5–MW) gas turbine generator set, the W21, with a thermal efficiency of 18 percent. The first application of the W21 in an industrial setting was as a gas-compressor drive installed at the Mississippi River Fuel Corp. facility located at Wilmar, Arkansas, in 1948. Westinghouse also built a 4,000–hp gas turbine-driven locomotive with the Baldwin Company (Chester, PA) that used two W21 gas turbines. Initial operation was on the Union Railroad burning distillate fuel oil. Later operation was on the Pittsburgh and Lake Erie Railroad using residual fuel oil. W21 had a 23-stage axial compressor for a cycle PR of 5 (35.5–lb/s airflow), a 12-can combustion system, and an eight-stage turbine. Its firing temperature was 1,250°F (677°C) for a thermal efficiency of 18 percent.

At a cycle PR of about 5 but with a relatively modest firing temperature, W21 already had a cycle factor of 0.53, significantly higher than that for the Jumo-004B and BBC gas turbines, but, not surprisingly, the technology factor was still about the same (i.e., 0.53). It would take about a half century of sustained development for the cycle factor to nearly double, from 0.3–0.4 to 0.70, and for the technology factor to increase by about 50 percent to 0.75.

4.5 Post–World War II

Starting with the Belle Isle Station, the history of the land-based gas turbine and combined-cycle power plants for electric power production can be summarized in the following five generations:

1. Generation 1 (1949–1968)
    a. Smaller than 30–MW gas turbines (GE Frames 3 and 5, Westinghouse W251A/B)
    b. Firing temperatures 1,500–1,800°F (primarily with fuel oil)
    c. For peak power, repowering, and cogeneration
2. Generation 2 (1968–1990)
    a. GE B and E Technology gas turbines (50–120 MW); Westinghouse W251B, W501A/D (50–120 MW)
    b. Firing temperatures ~2,000°F with diffusion combustors
    c. NOx emission control using gas turbine water/steam injection or selective catalytic reduction
    d. Non-reheat steam cycles
3. Generation 3 (1990–1998)
    a. GE and Westinghouse F Technology gas turbines (75–260 MW)
    b. Three or four air-cooled turbine stages, 16–18 axial-compressor stages
    c. Firing temperatures ~2,400°F
    d. Performance fuel heating (365°F)

   e. Dry-Low-NOx combustion system for NOx control
   f. Three-pressure, reheat steam cycles
4. Generation 4 (1998–2010)
   a. Steam-cooled G (Westinghouse, Mitsubishi) and H Technology (GE) gas turbines (~300 MW)
   b. Closed-loop steam cooling of the first two turbine stages (both stator and rotor); four turbine stages (GE's H-System)
   c. Cooling of cooling air for turbine wheel-spaces and subsequent stages via a heat exchanger (a kettle reboiler) that generates intermediate-pressure (IP) steam to be used in the bottoming cycle
   d. Active clearance control (compressor and turbine)
   e. Firing temperatures ~2,600°F
   f. Performance fuel heating (>400°F)
   g. Dry-Low-NOx combustion system for NOx control
   h. Three-pressure reheat steam cycle integrated with the gas turbine Brayton cycle
5. Generation 5 (2010–present)
   a. Air-cooled H technology (Siemens H, GE's HA) with four-stage turbine
   b. Advanced, high-pressure axial compressor with reduced number of stages
   c. Steam- and air-cooled J technology (Mitsubishi Hitachi Power Systems [MHPS])
   d. Firing temperatures ~2,700°F

The first generation of industrial gas turbines was primarily deployed in three capacities:

1. Railroad locomotive propulsion
2. Pipeline service
3. Peak power supply

As mentioned above, Union Pacific ordered twenty-five 2,500-hp Frame 3 engines in 1951, which stayed in service until the early 1960s, when, after over 300,000 hours of operation, they were replaced by diesel engines. In 1958–1961, Union Pacific ordered thirty 8,500-hp GE Frame 5 engines. These engines, named "Big Blows" by Union Pacific, were rated at 6,000 feet elevation and 90°F air inlet (Kennedy, Rees, and Russ, chapter 23 of [18]). At 1,000 feet and 80°F, the engine could produce about 12,000 hp. The Frame 5 was a single-shaft, two-bearing design with a 16-stage axial compressor and a two-stage turbine.[10] Ten combustors were fed by air-atomizing fuel nozzles. The atomizing air was supplied by compressor extraction air boosted by a shaft-driven rotary air compressor. Heated tender cars, custom made from old steam tenders, were provided for each locomotive to keep the fuel from solidifying. These locomotives stayed in operation until the mid-1970s when residual fuel oil, no longer a "waste" product, became too costly for Union Pacific (refineries found a way to crack it to make gasoline and it became an ingredient for making plastics). These gas turbines could move trains

[10] This was the first gas turbine to require compressor bleed to avoid stalling during starting.

at speeds averaging 40 mph and, in the early 1960s, they hauled close to 20 percent of the total freight gross ton-miles on the Union Pacific system.

In late 1940s, large gas transmission operators in the USA recognized the cost-effectiveness of gas turbine-driven centrifugal compressors in pushing natural gas over large distances. Although diesel engines were much more efficient than even recuperative gas turbines, with natural gas at about 20 cents per thousand cubic feet, lower maintenance costs were the dominant economic factor. Absence of vibration and lower installation costs (mainly due to simpler foundations) were other factors in favor of gas turbines.

GE redesigned the Frame 3 into a two-shaft, 7,000-hp recuperative machine, MS3002, with 25 percent thermal efficiency and sold 28 of them to The El Paso Natural Gas Company to pump natural gas from West Texas to California.[11] The high-pressure turbine powered the 15-stage, axial-flow compressor and the low-pressure turbine drove the load compressor from a different manufacturer (e.g., Cooper-Bessemer) at about 6,000 rpm. Later models were rated at 11,000 hp at approximately the same speed. These turbines were used primarily in the gas pipeline and petrochemical industries and some of them still remain in use throughout the world. (The recuperator, which utilized the hot exhaust gas to preheat combustion air, was a large, vertical tube-type heat exchanger – see Figure 14.4.) Later on, more of these gas turbines as well as Frame 5 machines were sold to pipeline companies across the USA. There were also 10 intercooled-recuperated gas turbines, which were installed across the USA for electric power generation (see Section 14.2 for more on these unique gas turbines).

In the 1960s, turbojet aircraft engines were investigated for the possibility of super-fast but also safe and cheap railroad travel. In the USA, two GE J47-19 jet engines (designed for use in the Convair B-36 Peacemaker intercontinental bomber) were mounted on the roof of a diesel railroad locomotive (Budd Rail Diesel Car, RDC-3). The original blunt nose of RDC-3 was modified by adding a streamlined front cowling (many good pictures are available on the Internet). Designated M-497 and nicknamed *Black Beetle*, the train was tested on an "arrow-straight" track between Butler, Indiana, and Stryker, Ohio. On July 23, 1966, the car reached a speed of 183.68 mph (295.6 km/h), which was an American rail speed record still unbroken to date. (One can see a video of M-497 running on YouTube.) A similar effort was undertaken in the Soviet Union in 1970 with a design similar to M-497 (an ER22 diesel locomotive with two turbojets from a Yakovlev Yak-40 airliner mounted on the roof). The locomotive, referred to as SVL (Russian acronym for High-Speed Laboratory Railcar), was tested in 1971 and reportedly got up to about 160 mph. Neither M-497 nor SVL went beyond the experimental stage. High fuel consumption, the deafening noise of the turbojets (imagine one of those pulling into a railway station), and the need for perfectly straight tracks to prevent derailing at stupendous speeds prevented them from being more than gimmicks (*not* safe and *not* cheap).

[11] Ivan G. Rice, ibid., Dave Lucier from PAL Turbine Services, LLC, website.

Starting in the early 1960s, natural gas or liquid petroleum gas-fired gas turbines were also utilized to drive liquid pipeline pumps. Another use for gas turbines was "repressurization," where the natural gas byproduct of oil production was compressed sufficiently to force it back into the oil formations through injection wells (see T. C. Heard in chapter 26 of [18]). By 1954, they had been deployed in Venezuela and Saudi Arabia.

In the 1960s, especially after the Great Northeast Blackout of November 9, 1965, aeroderivative gas turbines played a significant role in electric power generation, especially as peaking and "black start" facilities. The heavy government research and development investment into military aircraft engines for performance (in effect, with little abatement, from the end of World War II to the present) and, especially in the 1970s, the investment by the OEMs pressed by airline customers into commercial airliner engines to improve fuel efficiency as well as reduce noise and emissions rapidly brought aircraft gas turbine technology forward. Many of those technologies (especially superalloys, coatings, and the aerodynamic design of axial compressors) later found their way into land-based industrial gas turbines. (Aeroderivative gas turbines are covered in detail in Chapter 23.)

The second generation of industrial gas turbines went through a rocky patch in the 1970s. The worldwide gas turbine electric power generation capacity had grown quite slowly from 1952 to 1962 to about 4 GWe. Fewer than ten years later, in 1970, the capacity reached 34 GWe with 17 MWe in the USA. A big driver in this impressive jump was, of course, the infamous blackout in 1965, which is demonstrated by the dramatic rise in US gas turbine orders between 1965 and 1970, as shown in Figure 4.6.

Then the tide changed almost overnight. Already weakening in 1971–1972, the US gas turbine market collapsed due to the impact of the 1973 Arab–Israeli war and the following OPEC oil embargo against the USA (in retaliation for the US decision to resupply the Israeli military), which stoked fears of fuel supply instability.

Due to the excessive US dependence on imported oil, the embargo caused a major disruption of the national economy. Consecutive US administrations developed plans to increase domestic production and reduce the use of imported oil. During the Jimmy Carter administration, there was a concerted move in the natural gas industry for deregulation, and a supply shortage of pipeline gas was created to drive their point home.

Eventually, all of the turmoil in the energy supply chain led to Jimmy Carter's *National Energy Plan*, consisting of 10 fundamental principles including reducing demand through conservation and the creation of a Department of Energy. After lengthy debates going on for months, the National Energy Act of 1978 was passed and signed into law by Jimmy Carter. Two of the major provisions of the National Energy Act had profound impacts on the gas turbine industry:

- *The Fuel Use Act* (FUA), which, among other things, prohibited the use of oil and natural gas as fuel for new base load power plants. Only "alternative fuels" – such as coal! – were allowed for that purpose. Peaking and intermediate-load combined-cycle power plants (<3,500 hours of operation per year) were exempt from the prohibitions of FUA, as were "cogeneration facilities."

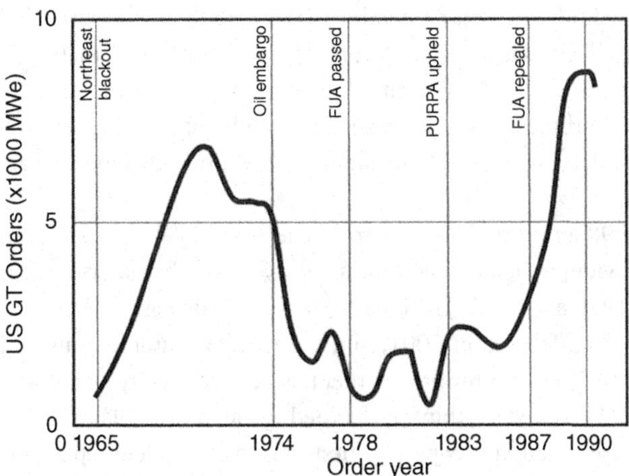

Figure 4.6 US gas turbine (GT) orders 1965–1990. (The chart is adopted from the inset of a chart in the article by Jaeger and Owen [19].) FUA = Fuel Use Act; PURPA = Public Utility Regulatory Policies Act.

- *The Public Utility Regulatory Policies Act* (PURPA), which paved the way for deregulation of the electric utility industry and, among other things, established the rules governing a new *independent power producer* (IPP) industry by requiring electric utilities to purchase power from *non-utility generators* as long as the generating facility also delivered at least a specified minimum amount of thermal energy to an industrial process plant (i.e., that the generating plant "qualified" as a cogeneration facility referred to as a *qualifying facility*).

It is not surprising that, following the passage of FUA and PURPA, there was little new domestic business for gas turbines and the market bottomed out in 1982. The IPP market was very slow to develop since a number of state Public Utility Commissions refused to implement the PURPA regulations. Finally, in 1982, in *FERC* v. *Mississippi*, the Supreme Court of the USA decided in favor of the Federal Energy Regulatory Administration (FERC) and upheld the law, following which the IPP market took off and reached about 10 GW of annual sales in 1990. This was also the beginning of the deregulation of the electricity market. By 1990, most large US utilities had set up independent deregulated subsidiaries, which aimed to expand into other regions and internationally.

Meanwhile, gas turbine orders in Europe also rose in response to environmental concerns and regulations. In Asia, gas turbine power plants, with their lower costs and shorter construction schedules, gained favor at the expense of other options to match the growing power demand. The prime beneficiary of this worldwide boom and the following "bubble" was the third generation of industrial gas turbines, the F class.

In the summer of 1998, the need for controlled blackouts and brownouts in Alberta and Chicago triggered a market panic. Reserve margins throughout North America were at an all-time low. Some power marketers who had sold forward electricity were caught short when supply could not meet demand. In 1997–1998, an enormous bubble started and drove the annual US sales of gas turbines to a record level of 60 GW, whereas worldwide orders peaked at 108 GW (in 2000). The period represented a huge sellers' market for large gas turbines that were in such demand that the suppliers were rationing their shop space and forcing buyers to sign "reservation agreements" and pay non-refundable deposits.

Between June 1998 and June 2001, approximately 330 GW of gas turbine equipment was ordered worldwide, which equaled in just three years the orders total registered in 15 years between 1979 and 1995. Eventually, the wild ride came to an end, which was triggered by the Enron debacle in 2001, and the market bottomed out in 2003. Since then, at least until 2017, the gas turbine market has been relatively stable with cyclic ups and downs. In the USA, gas turbine combined-cycle power plants built during the bubble ran well below their full capacity, first due to the sudden spike in natural gas prices (before the shale gas "boom") and then due to the economic crisis in 2008. Soon thereafter, though, capacity factors started inching steadily up toward the 60 percent level (see Figure 4.11).

4.6 Key Players

Before delving into the details of the "class hierarchy" of heavy-duty industrial gas turbines, a brief description of the key market players (i.e., OEMs) is in order. At the time of writing (in 2017), there are four major OEMs:

1. GE (including former Alstom, acquired at the end of 2015)
2. Siemens (formerly Siemens-Westinghouse, which were formerly separate companies; i.e., Kraftwerk Union [KWU], Siemens Power Generation, and Westinghouse Electric Corporation)
3. MHPS (formerly Mitsubishi Heavy Industries [MHI])
4. Ansaldo Energia

Siemens acquired Westinghouse Electric's power division in 1998 from CBS Corp. (yes, the media company). For the first five years after the acquisition, the new entity was known as Siemens-Westinghouse Power Corporation (SWPC), but in 2003 the Westinghouse name was commercially phased out. The legal entity became Siemens Power Generation, Inc. In 2004, Siemens announced a consolidated naming strategy for their mixed portfolio of power generation products and services. All legacy "V Series" gas turbines (the "V" designation is for *Verbrennung*, which means *combustion* in German) with the KWU heritage and the "W Series" gas turbines with the Westinghouse heritage acquired new, uniform names based on English-language abbreviations. This transition is summarized in Table 4.2 for the heavy-duty industrial gas turbine products. The "SGT" designation is for "Siemens Gas Turbine" followed by 5 or 6 for 50- or 60-Hz units, respectively. For a combined-cycle power plant, the designation becomes "SCC" for "Siemens Combined Cycle." The rest of the sequential model designation including frequency (5 or 6) and class (E, F, G, or H) stays the same.

Table 4.2 Naming convention for Siemens heavy-duty industrial gas turbines after 2004

New Name	Old Name	Frequency	Class	Status	Original OEM
SGT6-1000F	V64.3A	60	F	Deleted	KWU
SGT6-2000E	V84.2	60	E		KWU
SGT6-3000E	W501D5A	60	E	Deleted	Westinghouse
SGT6-4000F	V84.3A2	60	F	Deleted	KWU
SGT6-5000F	W501FD2	60	F		Westinghouse
SGT6-6000G	W501G	60	G	Deleted	Westinghouse
SGT5-1000F	V64.3A	50	F	Deleted	KWU
SGT5-2000E	V94.2	50	E		KWU
SGT5-3000E	V94.2A	50	E	Deleted	KWU
SGT5-4000F	V94.3A2	50	F		KWU

Table 4.3 Siemens KWU "V Series" gas turbines

Turbine	Year	PR	TFIRE °F	TFIRE °C	MAIR (lb/s)	TEXH (°F)	Output (kWe)	Efficiency (%)	Cycle Factor	Technology Factor
V84.2	1987	10.8	1,985	1,085	778	1,018	106,000	33.7	0.62	0.68
V94.2	1981	10.9	1,984	1,085	1,122	1,015	154,000	33.9	0.62	0.69
V64.3	1990	15.7	2,124	1,162	408	990	60,700	35.1	0.67	0.64
V84.3	1992	16.0	2,191	1,200	954	1,022	152,000	36.1	0.67	0.66
V94.3	1993	16.0	2,191	1,200	1,375	1,022	219,000	36.1	0.67	0.66

For a while, both Siemens and Westinghouse gas turbine models were offered in the 60–Hz markets worldwide, while the 50–Hz markets were served by existing Siemens products. The key performance characteristics of Siemens KWU "V Series" gas turbines are summarized in Table 4.3 [20]. After some time, it was decided that the Westinghouse designs would be the primary basis of the Siemens offerings for all 60-Hz markets (e.g., most of the Americas, South Korea, Saudi Arabia), served by the Orlando division (where the old Westinghouse power group was based). The 50-Hz markets (i.e., Europe, Africa, most of Asia, and part of South America) was served by Siemens in Germany. As Siemens developed new gas turbine products with more advanced technology (e.g., the H–class gas turbine, 8000H), the new, optimized offerings incorporated features of both Westinghouse and Siemens technology and design traditions.

At the time of writing, as indicated in Table 4.2, all Westinghouse-heritage machines, with the exception of W501F, as well as some old KWU-heritage ones, have been removed from the Siemens product lineup. (This list also included the smaller industrial gas turbine W251, which was offered for a while as SGT-900.) The evolution of the Westinghouse 501F gas turbine line over the three decades until Siemens acquisition is summarized in Table 4.4 [15].

In passing, note that Mitsubishi (then MHI) was a partner with Westinghouse Electric Corporation (WEC) for many years, and they were licensed to manufacture

Table 4.4 Westinghouse 501F gas turbine evolution

Turbine	Year	PR	TFIRE °F	TFIRE °C	MAIR (lb/s)	TEXH (°F)	Output (kWe)	Efficiency (%)	Cycle Factor	Technology Factor
W501A	1968	**7.5**	1,615	879	548	885	42,000	27.1	0.59	0.62
W501AA	1971	10.5	1,630	888	744	798	60,000	29.4	0.66	0.60
W501B	1973	11.2	1,819	993	746	907	80,000	30.5	0.65	0.61
W501D	1975	12.6	2,005	1,096	781	982	95,000	31.2	0.65	0.61
W501D5	1981	14.0	2,070	1,132	781	987	107,000	33.2	0.66	0.63
W501D5	1995	14.8	2,150	1,177	830	1,006	118,000	34.0	0.67	0.63
W501F	1995	14.6	2,300	1,260	960	1,083	160,000	35.6	0.65	0.66

Table 4.5 Mitsubishi 50- and 60-Hz (701 and 501, respectively) gas turbine evolution

Turbine	Year	PR	TIT °F	TIT °C	MEXH (lb/s)	TEXH (°F)	Output (kWe)	Efficiency (%)	Cycle Factor	Technology Factor
M501D	1981	14.0	2,282	1,250	863	1,006	114,000	34.9	0.65	0.66
M501F	1989	14.0	2,462	1,350	965	1,072	153,000	35.3	0.64	0.67
M501F3	1999	16.0	2,552	1,400	1,025	1,125	185,000	37.0	0.66	0.68
M701D	1984	14.0	2,282	1,250	1,243	1,006	144,000	34.8	0.65	0.66
M701F	1992	16.0	2,462	1,350	1,470	1,020	234,000	36.6	0.67	0.67
M701F3	1998	17.0	2,552	1,400	1,470	1,087	270,000	38.2	0.67	0.69

Westinghouse-designed W501D5 gas turbines in Japan. The license agreement, which had started around 1962, came to an end in 1984. Thereafter, MHI started to independently design and develop their own products while continuing to manufacture gas turbines for WEC as well. In addition, 60-Hz W501F and W501G (with steam-cooled combustor liners and transition pieces) were joint development efforts by MHI and WEC.[12] For the 50-Hz version, 701F, Fiat-Avio joined MHI and WEC in a tri-partnership agreement. The collaboration between MHI and WEC came to an end when WEC was acquired by Siemens and MHI retained the 50-Hz versions of the co-developed F- and G-class products, W701F and W701G. The evolution of MHI's 50- and 60-Hz F-class (air-cooled) gas turbine line over the two decades until Siemens' acquisition of WEC is summarized in Table 4.5.

One interesting aspect of the MHI philosophy is the disciplined approach to product development, with in-house research, design, manufacturing, and testing utilizing only proven concepts. The company culture dictates absolute control over every aspect of the design and development process (no outsourcing of manufacturing

[12] In fact, the 501F program permanently changed the relationship between the two companies, giving each independent and royalty-free manufacturing and marketing rights to the new engine. The prototype 501F engine was built and shop-tested at MHI's turbine factory and development center at Takasago in mid-1989.

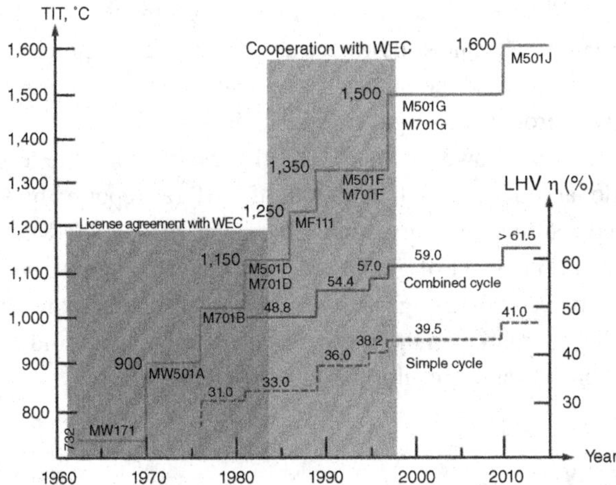

Figure 4.7 MHI product evolution in terms of TIT and thermal efficiency. LHV = lower (net) heating value.

except for blade castings) combined with exhaustive testing and validation prior to commercialization so that each new product is ensured to be a milestone in a very long-term program in the pursuit of the "ultimate" Brayton-cycle machine. This is concisely illustrated by the historical chart of MHI product evolution in Figure 4.7 [21], which is essentially a snapshot of a half-century's worth of progress achieved by the *global* gas turbine industry. Note that this disciplined approach to product development is continuously supported by Japan's Ministry of International Trade and Industry (MITI) via projects like "Moonlight" (for 1,300°C TIT-class gas turbines in the 1980s) and the current "National Project" (for 1,700°C TIT-class gas turbines). Ultimately, gas turbine and power plant designs receive MITI certification after passing a required qualification test.

Roughly 15 percent of the industrial gas turbine market (in terms of GW capacity – 1 GW is 1,000 MW) is aeroderivatives. GE has been the perennial market leader with a ~40 percent share (aeroderivative plus large industrial gas turbines). Siemens was a close second with ~30 percent (mostly large industrial units), with MHPS coming up strong with 15–20 percent. Until recently (i.e., before 2015), Alstom and Ansaldo did not even add up to 10 percent (as bookkept before the recent acquisitions).

After Alstom's acquisition, GE was forced by the EU to divest part of Alstom's gas turbine portfolio to preserve competitiveness. As such, the 50-Hz sequential (reheat) combustion GT26 and the new J–class gas turbine GT36 were sold to Ansaldo Energia, while GE kept the 60–Hz GT24 and the rest of the Alstom lineup.

Supported by its aircraft engine arm, at least until recently, GE was expected to remain a technology leader as well. In fact, other OEMs also tried to emulate GE by acquiring and/or forming strategic partnerships with aircraft engine manufacturers. For example, in 2013, MHPS (then MHI) completed its acquisition of Pratt & Whitney Power Systems, the small- and medium-sized aeroderivative gas turbine business unit of P&W. In 2014, Siemens bought the land-based gas turbine and compressor business of

Rolls-Royce. As part of the deal, under a 25-year licensing agreement, Rolls-Royce also granted Siemens access to relevant Rolls-Royce aeroderivative technology for use in the 4–85–MW power output gas turbine range. Thus, the three major industrial gas turbine OEMs have their own aeroderivative gas turbine divisions.

In a dramatic turnaround, however, in 2017–2018, MHPS took the leadership with *more than half* of global orders in the first quarter of 2018 (as reported by Barclays Plc, a London, UK-based multinational investment bank, on May 3, 2018). According to Barclays, Siemens had 26 percent of global orders in the same period, with GE seeing its share drop to 14 percent. Although Mitsubishi won the most contracts, Barclays noted that Siemens led the market in the number of gas turbine units sold, at 30 percent. GE was next at 27 percent, and Mitsubishi had 14 percent.

4.7 The Class Hierarchy

The "class hierarchy" of heavy-duty industrial gas turbines became quite confusing in the 2000s. Earlier, the terms "H technology" or "*H class*" were reserved by GE and MHI (now MHPS) for *fully* steam-cooled turbines (i.e., the first two turbine stages, nozzles, and buckets) [22]. The term is now used by Siemens for advanced *F–class* (air-cooled) units with TITs of 1,500°C (2,732°F) or higher.[13] These machines are thus capable of reaching firing temperatures of 2,600°F, which were previously possible only with the steam-cooled *H-System* (a registered trademark of GE; MHI eventually dropped its H technology program).

The *J class* (its air-cooled variant is referred to as *JAC*) of MHPS is capable of 1,600°C (2,912°F) TIT (or higher). It is referred to as the *HA class* (i.e., air-cooled H) by GE. In the meantime, GE retired its "fully" steam-cooled H technology (with only six units operating in the world). Note that the J–class gas turbines retain steam cooling for the combustor transition piece and first-stage turbine ring. Along with high firing temperatures and cycle PRs (about 23), J–class gas turbines are characterized by high air flow rates and power ratings (almost 500 MWe for the 50–Hz units). The class hierarchy of heavy duty industrial gas turbines is illustrated in Figure 4.8.

The chart in Figure 4.8 shows the E-, F-, G-, H-, and J–class turbine lineup as a function of TIT in degrees Celsius (preferred over degrees Fahrenheit because it changes in round 100-degree steps). All major OEMs have their own E- and F–class technologies. Mitsubishi owns the steam-cooled G technology and its air-cooled cousin, GAC (they also refer to the E class as "D class," which goes back to their collaboration with Westinghouse in the 1980s and 1990s). Note that in G–class gas turbines, the combustor transition piece and the first-stage turbine ring are cooled with steam extracted from the bottoming cycle.

[13] As a historical sidenote, H class was the designation for a series of super-large battleships intended for the German navy (Kriegsmarine) in the late 1930s (see *German Warships: 1815–1945* by Erich Gröner, Annapolis, MD: Naval Institute Press, 1990). The H class comprised five designs, codenamed H-39 and H-41 to H-44 in increasingly large displacement and main gun calibers. Just like their counterparts in other nations (e.g., US Montana class and Japan's A-150 class), none came to fruition.

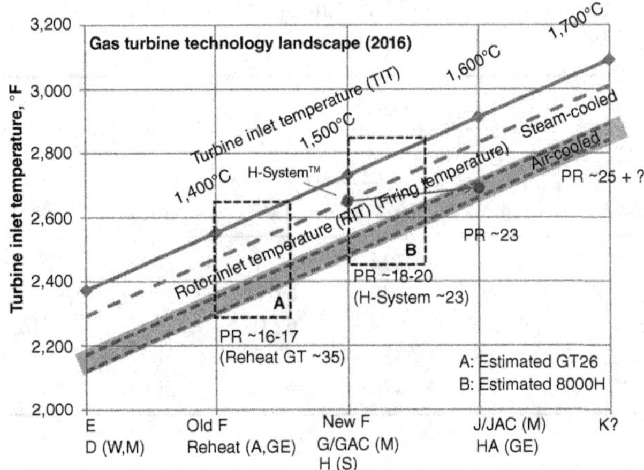

Figure 4.8 Class hierarchy of heavy-duty industrial gas turbines based on TIT (A = ABB/ Alstom/Ansaldo; GE = General Electric; GT = gas turbine; M = Mitsubishi; S = Siemens; W = Westinghouse).

The term H was originally used by GE for its fully steam-cooled gas turbine under the name *H-System* (a GE trademark). In 2011, Siemens introduced its own air-cooled H–class gas turbines. This gas turbine had nominally the same TIT as the H-System (i.e., 1,500°C, 2,732°F), but a lower firing temperature. The reason for this is the smaller temperature drop in a steam-cooled stage 1 nozzle (vane) from TIT to rotor inlet temperature (RIT; about 80°F) vis-à-vis the air-cooled turbine (about 180–200°F). Thus, as shown in Figure 4.8, by 2003 in Baglan Bay in the UK GE achieved a firing temperature (RIT) well above 2,600°F with the first H-System, 109H, a single-shaft, combined-cycle power plant. Today, the H–class Siemens gas turbine has a TIT probably closer to 1,550°C (2,822°F) and a firing temperature (RIT) of >2,600°F, essentially on a par with the now-defunct, steam-cooled H-System of GE.

While the fully steam-cooled H-System was an unqualified success from a technology perspective, it was a commercial failure. The gas turbine was only available in combined-cycle configuration with expensive alloy piping for cooling steam transfer between the topping and bottoming cycles. Single-crystal hot gas path components with advanced thermal barrier coatings added to the cost and complexity with longer-than-typical major maintenance outages. Furthermore, the original claim of "60 percent net efficiency capability" was not demonstrated in the field – initially due to the performance shortfall of the bottoming cycle and later due to unsuitable site ambient conditions. (Note that Siemens broke the 60 percent thermal efficiency barrier in 2011 with an 8000H gas turbine in Irsching, Germany, with a steam turbine condenser drawing cooling water from the Danube to achieve a super-low back pressure of only 0.6 inches of Hg. Combined with a 56-inch long titanium last-stage buckets – which was quite expensive – this was a big facilitator of combined-cycle efficiency [23].)

Altogether, six H-System combined-cycle power plants were built and are in commercial operation. While not a 60 percent performer per se, it is interesting to note that

Table 4.6 2016 OEM data for combined-cycle (CC) performances (1 × 1 and 2 × 1) of 50–Hz heavy-duty industrial gas turbines (GTs). SC = steam cooled; ST = steam turbine

		1 × 1			2 × 1		
	GT	ST	CC		ST	CC	
Turbine	MW	MW	MW	η (%)	MW	MW	η (%)
F+/H 60 Hz	269.1	130.4	394	60.4	263.4	792	60.7
F+/H 50 Hz	337.8	178.7	508	61.1	357.5	1,033	61.4
J 60 Hz	334.7	180.5	509	62.0	364.2	1.021	62.3
J 50 Hz	502.8	280.8	774	62.7	562.0	1,552	62.8
J 60 Hz (SC)	322.0	148.0	470	61.5	298.9	943	61.7

the 60–Hz H-System at the Inland Empire Energy Center has been in top 20 in terms of heat rate three times since 2010. It was also ranked number 1 in NOx emissions in 2011 and 2012, with less than 0.004 lb/MMBtu. Ultimately, the H-System did not fulfill its promise, but it achieved significant design benchmarks in a field-proven, reliable platform. We will come back to the H-System in Chapter 21 due to its connection to GE's flagship heavy-duty industrial gas turbine product line (i.e., the air-cooled HA class), and will look into those achievements in detail.

Aside from the aforementioned cost and complexity factors, the demise of the H-System (which is not commercially offered anymore) is easy to infer from comparison with the new air-cooled J/HA–class gas turbines. Modern air-cooled J- (JAC) and HA–class gas turbines with 1,600°C (2,912°F) can easily achieve ~2,700°F firing temperatures, higher than that of the H-System, with a much simpler architecture.

The current status of the heavy-duty industrial gas turbine technology with advanced F- (F+ or Siemens H) and J–class gas turbines (including GE's HA class) is summarized in Table 4.6 based on data extracted from *Gas Turbine World*'s 2016 Performance Specs (32nd edition), a trade publication issued every year. A more up-to-date performance summary reflecting the state of the art, as announced at the end of 2017, and projecting well into 2020s can be found in Chapter 21.

4.8 The Golden Age: 1980–2015

Gas turbine efficiency evolution over the last four decades is illustrated in Figure 4.9, which is an updated version of the chart published in a paper by Gülen [23]. The data are taken from myriad sources listed in the references of the same paper. The "handbook" data (A–C, E) and field-measured data (D) are highlighted. (No correction is applied to the published numbers.)

As shown in Figure 4.9, gas turbine efficiency evolution over the last 35 years does not display striking technology breakthroughs per se. Nevertheless, the introductions of the F, H, and J classes are quite discernible. It is represented reasonably well by the asymptotic growth formula

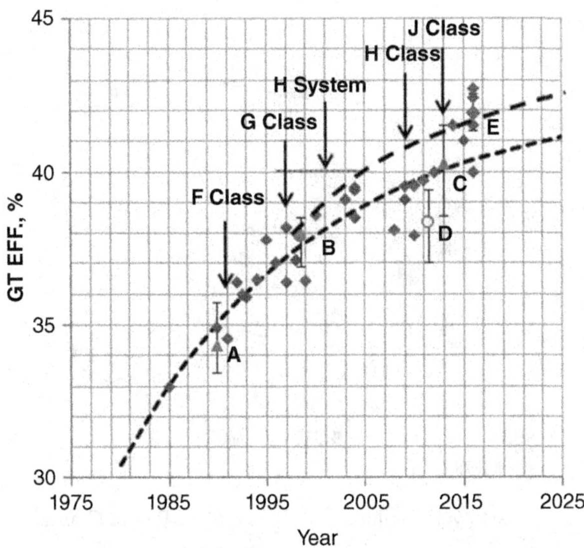

Figure 4.9 Gas turbine evolution, 1985–2016. (A: *Turbomachinery International* 1990 Handbook, B: *Gas Turbine World* (GTW) *1998–1999 Handbook*, C: GTW 2013 Handbook; D from measurements by Ol'khovskii et al. [24], E from GTW 2016 Handbook for J/HA class). Arrows denote the introduction year of particular technologies.

$$\eta = \eta' + (\eta_0 - \eta') \cdot e^{-\frac{\Delta t}{\tau}},$$

with a time constant of $\tau = 20.7$ years, an asymptotic limit of $\eta' = 42.5\%$, and a base value of $\eta_0 = 30.4\%$ at $\Delta t = 0$ (in the year of 1980). Thus, the technology development history of the last 30+ years suggests that to expect more than 43 percent net gas turbine efficiency by 2035 is unrealistic (although perhaps not *impossible* – see Chapter 21, which discusses the late-2017 performance "claims" of major OEMs).

Most gas turbine power plants for electricity generation are in a *combined-cycle* configuration. Gas turbine combined-cycle efficiency evolution over the last 30+ years is shown in Figure 4.10. Once again, similar to the situation depicted in Figure 4.9 for gas turbines, the gas turbine combined-cycle efficiency evolution over the last three decades does not display any discernible technology breakthroughs (i.e., step changes) and is represented very well by the same asymptotic growth formula with a time constant of $\tau = 14.5$ years, asymptotic limit of $\eta' = 62.3\%$, and a base value of $\eta_0 = 42.6\%$ at $\Delta t = 0$ (in the year of 1980). Thus, the technology development history of the last 30 years suggests that to expect more than 63 percent net gas turbine combined-cycle efficiency by 2035 is unrealistic. Then again, claimed (but unproven) efficiencies by major OEMs in late 2017 (see Chapter 21) may make this statement invalid.

At this point, the reader may raise the following, quite reasonable question: "Wait a minute; some OEMs already report almost 63 percent efficiency in their 2016 ratings. Why do you state that 63 percent by 2035 is probably the maximum possible?"

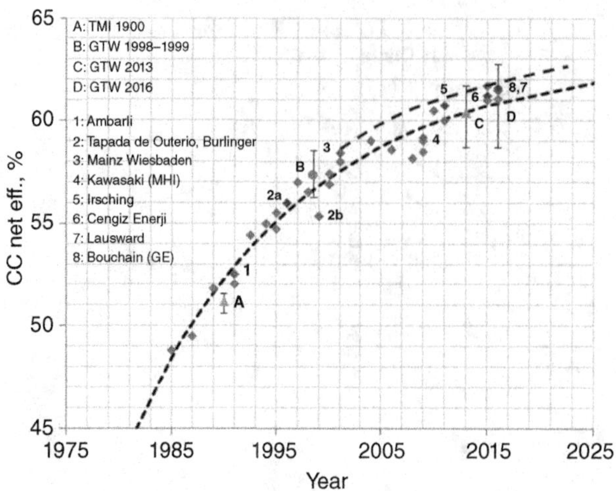

Figure 4.10 Gas turbine combined-cycle evolution, 1985–2016. A, B, and C are the same as in Figure 4.12.

This question will be answered in detail in the following chapters via detailed and rigorous thermodynamic analysis (supported by economic arguments as well). The 63 percent (see Chapter 3) in question is net thermal efficiency, achievable on a real site with existing ambient conditions (on a year-round average basis, in many places in the world it is indeed pretty close to the ISO conditions) and incorporating a cost-effective bottoming cycle and balance of plant design, especially a realistic steam turbine condenser heat rejection system as dictated by environmental and other regulations.

In order to illustrate the severe disconnection between ISO base load rating with the most optimistic (let us be honest, quite unrealistic) assumptions and real plant performance (from EIA-923 forms) when the proverbial rubber meets the road, consider the data in Figure 4.11. The US Energy Information Administration (EIA) collects detailed electric power data – monthly and annually – on electricity generation, fuel consumption, fossil fuel stocks, and receipts at the power plant and prime mover levels via Form EIA-923. The data can be accessed on US EIA website (www.eia.gov) and are available to the public. The results of the analysis of EIA data by Energy Ventures Analysis (VA, USA) are published under the title of "Power Plant Performance Report" in *Power Engineering* magazine and the *Electric Light & Power* magazine website (until 2017).

Figure 4.11 summarizes the performances of "top 20 in heat rate" combined-cycle plants from annual performance reports over the least 10 years. Also shown in Figure 4.11 are the average capacity factors of those power plants.

The picture depicted in Figure 4.11 is unmistakable. Rating performances quoted by the OEMs steadily increase, but they are not matched by a lockstep increase in the year-round average efficiencies recorded by the crème de la crème of the combined-cycle power plants in the USA. The contributing factors are as follows:

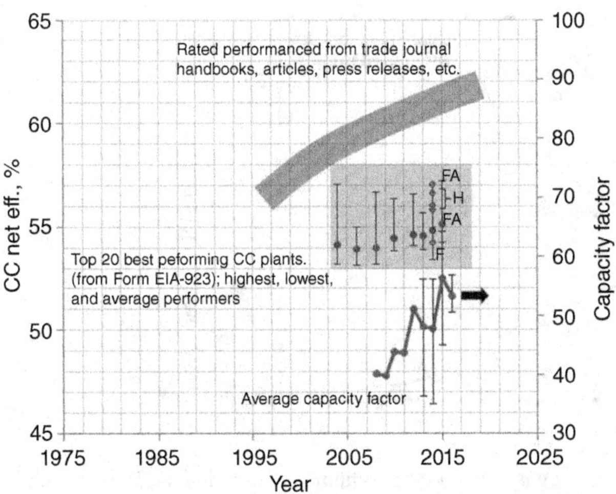

Figure 4.11 Field performance of the top 20 combined-cycle (CC) power plants (in terms of heat rate) in the USA.

1. High plant auxiliary load (air-cooled or mechanical-draft cooling tower heat rejection)
2. Supplementary firing in the heat recovery steam generator (especially in the summer)
3. Frequent start–stop cycles and part load operation in load-following mode

For example, in 2014, the best-performing power plant, with ~57 percent lower (net) heating value efficiency, was based on an advanced F–class gas turbine (FA or F+), ahead of the two then-newly commissioned H–class gas turbine power plants (nominal 60 percent rating). The situation was similar in 2015, with an FA–class gas turbine combined-cycle power plant leading the pack followed by the two H–class power plants.

After >60 percent net combined-cycle efficiency was first achieved by an H–class gas turbine (60.75 percent to be exact – with a steam turbine condenser pressure of 0.6 inches of mercury), several other H–class power plants achieved that performance and, most likely, >61 percent. Finally, in early 2016, an HA–class gas turbine was reported to break the world record with 62.22 percent *net* combined-cycle efficiency. Publicly available information does not clearly enumerate the gross-to-net steps. It does not even indicate whether the performance in question is corrected to ISO conditions or not (see Section 21.2 for more on this particular power plant). Whatever the case may be, there is little doubt that a J/HA–class gas turbine is readily capable of 61 percent combined-cycle efficiency (net ISO) or even a bit higher *with favorable site conditions* (i.e., access to a natural coolant source such as river, lake, ocean, etc., for very low steam condenser pressures) when running at full load.

In conclusion, there is no question that gas turbine technology has progressed quite impressively since the early 1980s. In particular, over a 20-year span covering the late 1970s and through to the 1990s, in Japan, Europe, and the USA, multiple advanced gas

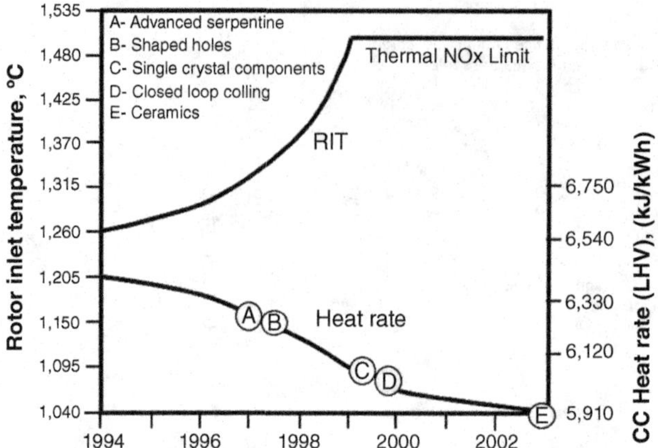

Figure 4.12 Gas turbine hot gas path cooling technology evolution. A, B, C, D and E are the same as in Figures 4.9 and 4.10. CC = combined cycle; LHV = lower (net) heating value.

turbine projects were established with the aim of reaching 60 percent combined-cycle efficiency. Examples are:

1. Japanese MITI's Moonlight Project (1978–1987)
2. European Union's COST-522 Project (launched in 1998)
3. US Department of Energy's (DOE) Advanced Turbine Systems (ATS) Project (1992–2001)

As dictated by fundamental thermodynamics, such programs were cognizant of the need to increase the firing temperature (same as the RIT) to at least 1,450°C (~2,650°F) or, preferably, 1,500°C (~2,750°F) in order to achieve this goal. Two major US OEMs focused on closed-loop steam cooling as a key enabler (i.e., GE and WEC, which was later acquired by Siemens, in a collaborative effort with MHI). One reason for steam cooling was to achieve high firing temperatures (RIT) without raising the flame zone temperature in lockstep beyond the thermal NOx limit (see Figure 4.12). Another reason was to reduce the chargeable cooling flows by fully steam-cooling the first two turbine stages (see Section 9.3 for more on this). Even with single-crystal components in the hot gas path, extreme dilution of the hot gas between the combustor exit and the S1B inlet with open-loop air cooling could not achieve that goal (at the time). As mentioned earlier, these gas turbines were labeled as the *H class*.

While both parties tried to develop H–class machines with fully steam-cooled first- and second-stage components (both stationary *and* rotating parts), ultimately only GE succeeded in bringing the true H–class gas turbines to the market in the early 2000s under the trademark *H-System*. Westinghouse eventually settled at the intermediate *G–class* firing temperature (between the air-cooled F- and fully steam-cooled H-class machines) with closed-loop steam cooling limited to the combustor transition piece (and later to the first-stage turbine blade ring). Today, Westinghouse G–class technology is present in MHI (now MHPS) products. Mitsubishi, on the other hand, pursued the

H technology further, although it never made its way into the Mitsubishi product lineup (e.g., see Figure 4.7).

The ATS program began in 1992 with a target of commercially available industrial (<20 MW) and utility power plant (>400 MW) turbines by the year 2000 with 60 percent efficiency, less than 10 ppm NOx emissions, and 10 percent lower operating costs. The ATS program led to numerous technology breakthroughs and improvements, such as:

- High-efficiency axial compressors with high PRs (20 or even higher)
- Advanced dry-low-NOx combustors, closed-loop steam cooling (the H-System)
- Advanced nickel-based alloys and castings for hot gas path components (i.e., single crystal)
- High-temperature thermal barrier coatings
- 3D aerodynamic design of compressor and hot gas path components via computational fluid dynamics
- Advanced brush seals

Nevertheless, after half a billion dollars being spent by the DOE and participating OEMs, it took another decade for an actual combined-cycle power plant with an *air-cooled* H-class gas turbine to reach (and surpass) the target efficiency [23].

A quite interesting depiction of gas turbine combined-cycle efficiency predictions made by other engineers and scientists going back as far as the early 1980s is summarized in Figure 4.13. Extrapolation of the data from the OEMs straddles the asymptotic curve in Figure 4.10. This is to be expected because, per their coincidence with the ATS program target, data points from OEMs A and B reflect the most optimistic theoretical efficiencies, whereas those from OEM C reflect commercial

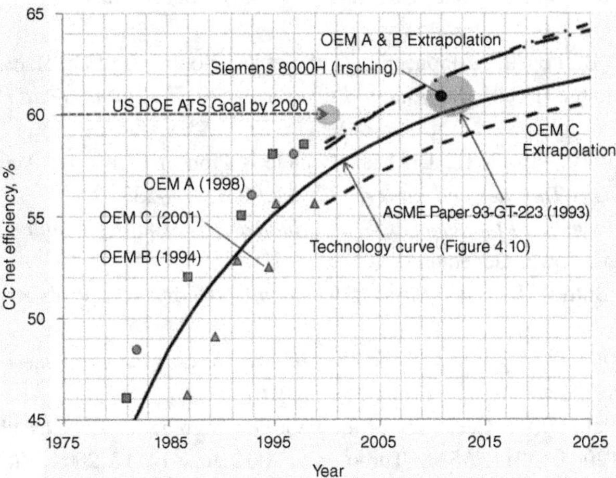

Figure 4.13 Historic gas turbine combined-cycle efficiency data (with extrapolations) from papers published by major OEMs and academia (ISO base load). CC = combined cycle.

performances. On the other hand, a more than 20−year-old paper from academia seems to be right on the money with its prediction of the current state of the art [25].

Since 60.75 percent net efficiency at Irsching (2011), all major OEMs have been quoting efficiency ratings exceeding 60 percent, even 61 percent (e.g., see Table 4.6). Some of the reliable >60 percent net efficiency combined-cycle power plants in commercial operation are identified in Figure 4.10. There may be a few more. (The LNG-fired Dangjin 3 and Andong gas turbine combined-cycle power plants in South Korea, commissioned in 2013 and 2014, respectively, are reported to have achieved 60 percent efficiency. They are similar in configuration to the Irsching power plant.) Regardless of the reality in the field, for the last 10 years or so, the new goalpost seems to have been set at 65 percent net combined-cycle efficiency by 2020 or 2025.

References

1. Meher-Homji, C. B., The Historical Evolution of Turbomachinery. Proceedings of the 29th Turbomachinery Symposium, 281–322.
2. Wilson, D. G., Korakianitis, T., *The Design of High Efficiency Turbomachinery and Gas Turbines*, 2nd edition (Uppersaddle River, NJ: Prentice-Hall, 1998).
3. Bathie, W. W., *Fundamentals of Gas Turbines*, 2nd edition (New York: John Wiley & Sons, Inc., 1996).
4. Eckardt, D., *Gas Turbine Powerhouse* (Munich, Germany: Oldenbourg Verlag, 2014).
5. Connors, J., *The Engines of Pratt & Whitney: A Technical History*, (Reston, VA: AIAA, Inc., 2010).
6. Gunston, B., *The Development of Jet and Turbine Aero Engines*, 4th edition (Newbury Park, CA: Haynes North America, Inc. (Patrick Stephens Ltd.), 2010).
7. Meher-Homji, C. B., The Development of the Junkers Jumo 004B – The World's First Production Turbojet. *Journal of Engineering for Gas Turbines and Power*, **119**:4 (1997), 783–789.
8. Meher-Homji, C. B., Prisell, E., Pioneering Turbojet Developments of Dr. Hans von Ohain – From the HeS 1 to the HeS 011, *Journal of Engineering for Gas Turbines and Power*, **122**: 2 (2000), 191–201.
9. Meher-Homji, C. B., Prisell, E., Dr. Max Bentele – Pioneer of the Jet Age, *Journal of Engineering for Gas Turbines and Power*, **127**: 2 (2005), 231–239.
10. Leiste, V., *Development of the Siemens Gas Turbine and Technology Highlights* (Erlangen, Germany: Siemens Power Generation).
11. Kotelnikov, V., Buttler, T., *Early Russian Jet Engines* (Derby, UK: Rolls-Royce Heritage Trust, 2003).
12. Giffard, H., *Making Jet Engines in World War II: Britain, Germany, and the United States*, (Chicago, IL: The University of Chicago Press, 2016).
13. Bakken, L. E. et al., 2004, "Centenary of the First Gas Turbine to Give Net Plant Output," ASME paper GT2004-53211, ASME Turbo Expo 2004, June 14–17, 2004, Vienna, Austria.
14. Stodola, A., *Steam and Gas Turbines, Authorized Translation from the 6th German Edition by L. C. Löwenstein* (New York: McGraw-Hill Book Company, Inc., 1927).
15. Scalzo, A. J., Bannister, R. L., DeCorso, M., Howard, G. S., Evolution of Heavy-Duty Power Generation znd Industrial Gas Turbines in the United States. ASME paper 94-GT-488,

International Gas Turbine and Aeroengine Congress and Exposition, June 13–16, 1994, The Hague, The Netherlands.

16. Streid, D. D., *Type I-40 Jet Propulsion Gas Turbine, Aircraft Gas Turbine Engineering Conference. Restricted Report No. 171* (West Lynn, MA: General Electric, 1945).
17. Howard, A., *Aircraft Gas Turbines with Axial-Flow Compressors, Aircraft Gas Turbine Engineering Conference, Restricted Report No. 171* (West Lynn, MA: General Electric, 1945).
18. Foster-Pegg, R. W., Utility Applications of Gas Turbines, chapter 25 in *Sawyer's Gas Turbine Engineering Handbook*, 1st edition (Stamford, CT: Gas Turbine Publications, Inc., 1966).
19. Horlock, J. H., Combined Cycle Power Plants – Past, Present, and Future. *Journal of Engineering for Gas Turbines and Power*, **117**:4 (1995), 608–616.
20. Maghon, H., Becker, B., Schulenberg, T., Termühlen, H., Krämer, H., The Advanced V84.3 Gas Turbine. American Power Conference, April 13–15, 1993, Chicago, IL.
21. Ai, T., Masada, J., Ito, E., Development of the High Efficiency and Flexible Gas Turbine M701F5 by Applying "J" Class Gas Turbine Technologies. *Mitsubishi Heavy Industries Technical Review*, **51**:1 (2014).
22. Pritchard, J. E., H System™ Technology Update, GT2003-38711. ASME Turbo Expo 2003, June 16–19, 2003, Atlanta, GA.
23. Gülen, S. C., Étude on Gas Turbine Combined Cycle Power Plant – Next 20 Years. *Journal of Engineering for Gas Turbines and Power,* **138**:5 (2015), 051701.
24. Ol'khovskii, G. G., Radin, Yu. A., Mel'nikov, V. A., Tuz, N. E., Mironenko, A. V., Thermal Tests of the 9FB Gas Turbine Unit Produced by General Electric. *Thermal Engineering*, **60**:9 (2013), 607–612.
25. Chiesa, P., Consonni, S., Lozza, G., Macchi, E., Predicting the Ultimate Performance of Advanced Power Cycles Based on Very High Temperature Gas Turbine Engines, 93-GT-223. ASME 1993 International Gas Turbine and Aeroengine Congress and Exposition, May 24–27, 1993, Cincinnati, OH.

Part II

Fundamentals

5 Thermodynamics

This is a "crash course" on thermodynamics with the focus on "bits and pieces" of the much broader subject matter pertinent to the analysis of gas turbines. A conscious effort has been made to make the coverage of some of the said "bits and pieces" as pedantic as possible in order to make the reader confident that there is a profound scientific argument behind even the simplest empirical equation. Nevertheless, without a solid foundation in the principles of thermodynamics, the chapter will be of a very limited use. If the reader "hated" thermodynamics in school (let us face it, this is true of many students), it is feared that the material below might even fortify that hatred. This is simply unavoidable. As a precaution, you might want to keep your favorite thermodynamics textbook handy while reading this chapter (e.g., Moran and Shapiro cited in the references section of Chapter 2).

While the coverage is fully believed to be scientifically rigorous, quibbles from experts even more pedantic than the author are also unavoidable. Some of those have been anticipated and necessary explanations are provided (lack of curvature in constant-pressure lines in temperature–entropy [T–s] diagrams and ignoring the presence of liquid H_2O at the reference state used for the exergy, etc.). Suffice to say that the intended audience are practitioners in need of "useable tools" and not modern-day "natural philosophers" after the secrets of the universe. In that sense, the author vouches 100 percent for the thermodynamic validity of every single statement herein.

Let us start with definitions and terminology. A gas turbine is a *heat engine* operating in a *power cycle*. The "heat" in question is generated via the following three major practical means:

1. Combustion of a fossil fuel
2. Nuclear reaction
3. Solar radiation

A gas turbine for electric power generation is a combustion heat engine.
There are two types of combustion heat engines:

1. Internal combustion engine
2. External combustion engine

Internal combustion engines (ICEs) operate in *open power cycles*.
External combustion engines operate in *closed power cycles*.

Consequently, a gas turbine is an ICE operating in an *open* cycle. (Yes, there are *closed*-cycle gas turbines proposed for nuclear or solar power applications. The operative word is "proposed." Those types of gas turbines do *not* yet exist. We will come back to that in Chapter 22 when we discuss closed-cycle gas turbines with *external* combustion – they *did* exist as external combustion engines.) In fact, some industry practitioners, especially old-timers with extensive coal-fired steam turbine power plant experience under their belts, refer to the gas turbine as a "combustion turbine" – after all, steam is a gas as well![1] The operation of the gas turbine is described by the *Brayton* cycle. Other similar cycles are the Otto and Diesel cycles for automotive and marine reciprocating (piston-cylinder) engines and the Atkinson cycle (i.e., "explosion" turbine).

The best-known closed power cycle is the *Rankine* cycle for steam turbine power plants. As such, the Rankine cycle is an important part of the gas turbine *combined cycle*. Due to their relative position on the T–s diagram, in a gas turbine combined cycle

1. The gas turbine Brayton cycle is referred to as the *topping cycle*.
2. The steam turbine Rankine cycle is referred to as the *bottoming cycle*.

Gas and steam turbines (heat engines) are also referred to as *prime movers*. This goes back to the fact that, historically, they started off as marine/land (steam turbine) and aircraft (gas turbine) propulsion units. When the gas turbine is the sole prime mover in the power plant, it is referred to as a *simple cycle*.

The *Kelvin–Planck* statement of the second law of thermodynamics dictates that "it is impossible for a heat engine to operate in a power cycle and deliver net work to its surroundings while receiving energy by heat transfer from a single thermal reservoir." The Kelvin–Planck statement is a *negative* statement. As such, according to the rules of formal logic, it cannot be proven. In other words, asking "why so?" is futile. It just is. The only proof for its validity is the absence of experimental proof contradicting it (so far on this planet at least).

The translation of the *qualitative* Kelvin–Planck statement into *quantitative* terms is the *Carnot* cycle. Accordingly, *it is impossible to devise a heat engine operating in a power cycle between two thermal reservoirs and that is more efficient than a Carnot cycle operating between the same two reservoirs*.

Recall that, in Chapter 4, it was stated that the idea of heat engines for useful work generation, whether using steam or air as working fluids, and their translation into actual end products *preceded* the full development of the science of thermodynamics. In this chapter, we will resort to a *gedankenexperiment*. Here is how it goes: let us assume that the gas turbine is not invented yet. However, we are in possession of the full arsenal of engineering thermodynamics; our copy of Moran and Shapiro is on the desk in front of us. We are also familiar with the theory and practice of steam engines and steam

[1] As an interesting sidenote, in 1979, reacting to the prohibition on the use of natural gas under the Fuel Use Act (FUA), Westinghouse Electric Corporation (WEC) decided to rename their gas turbine product as a "combustion turbine." In a similar vein, the Gas Turbines Systems Division of WEC was renamed as Combustion Turbines Systems Division. In that way, WEC attempted to hide the fact that the primary fuel of their gas turbines was indeed natural gas!

turbines (i.e., we are fully aware of the Rankine cycle). Since air is even more plentiful than water and is available for free everywhere (unlike steam, which *must* be generated first), we want to develop a better heat engine with air as the working fluid. We know that the available materials and manufacturing technologies enable us to achieve temperatures as high as 1,600°C (2,912°F) and pressures as high as 25 bara (about 360 psia).

Armed with the second law, we know that the theoretical maximum in efficiency can only be obtained in a Carnot cycle, which is given by

$$\eta = 1 - \frac{T_1}{T_3} = 1 - \frac{59 + 460}{2,912 + 460} = 84.6\%.$$

The first process in a Carnot cycle is *isentropic compression*. We are confident that the state of the art in axial compressor design can bring us very close to that ideal. Thus, from a conceptual design perspective, we are satisfied with this assumption. But there is a problem: the requisite pressure ratio (PR) to achieve 2,912°F (starting from T_1 = 59°F, p_1 = 14.7 psia) from the isentropic correlation with γ = 1.4

$$\frac{T_2}{T_1} = \left(\frac{p_2}{p_1}\right)^{\frac{\gamma-1}{\gamma}}, \tag{5.1}$$

is an astronomically high 699 (i.e., a bona fide "pipe dream"). Even if it could be done, designing a process for *isothermal heat addition* (the second process of a Carnot cycle) would be a big headache and it would result in a significant pressure loss – defeating the purpose of achieving a high PR in the first place. Eschewing heat addition and resorting to expansion immediately would result in a useless engine with net cycle work of zero! In other words, a "true" Carnot cycle and its promise of nearly 85 percent efficiency is out of the question. Thus, we proceed with a compressor PR of 23 and achieve a T_2 of 811°F. With a constant heat capacity of c_p = 0.25 Btu/lb-R, we consume

$$w_{comp} = c_p(T_2 - T_1) = 0.25 \cdot (811 - 59) = 188 \text{ Btu/lb}$$

to drive the compressor. For the heat addition process, we have the following two options:

1. Constant-volume heat addition
2. Constant-pressure heat addition

Our objective is to design the simplest possible heat engine that will be much smaller (and lighter) than a steam turbine and provide the highest possible power density (per unit weight *and* per unit footprint). We know that this requires a fully axial turbomachine and thus we quickly find out that squeezing a constant-volume "explosion" chamber into a steady, axial flow path is a very difficult task. Even if it can be done (as Holzwarth showed in the early 1900s), it will have limited airflow capability because it is hampered by the same practical limit as large diesel engines: cylinder (explosion/combustion chamber) size. Thus, we choose the constant pressure route and add

$$q_{in} = c_p(T_3 - T_2) = 0.25 \cdot (2,912 - 811) = 525 \text{ Btu/lb}$$

of heat into the cycle. From the *isentropic expansion* (the third process in the Carnot cycle) to $T_4 = 917°F$, which we obtain from the same isentropic correlation we used for compression with $p_4 = 14.7$ psia, i.e.,

$$\frac{T_3}{T_4} = \left(\frac{p_3}{p_4}\right)^{\frac{\gamma-1}{\gamma}},$$

we calculate

$$w_{turb} = c_p(T_3 - T_4) = 0.25 \cdot (2{,}912 - 917) = 499 \text{ Btu/lb}$$

of useful work, which, after subtracting 188 Btu/lb spent for compression, gives us a net cycle work of 311 Btu/lb for a cycle efficiency of

$$\eta = \frac{w_{turb} - w_{comp}}{q_{in}} = \frac{311}{499} = 59.2\%.$$

Thus, the best we can do, *even on paper with zero losses of any kind*, is 59.2% / 84.6% = 0.7, or about 70 percent of the ultimate theoretical limit imposed by the second law of thermodynamics. This is the *real cycle effect.*

Note that, in these conceptual calculations, we have used a calorically perfect gas model, pv = RT with constant c_p. With this assumption, it can be shown that the cycle efficiency is a function of the cycle PR only, i.e.,

$$\eta = 1 - \frac{1}{PR^{\frac{\gamma-1}{\gamma}}}. \tag{5.2}$$

Substituting PR = 23 and $\gamma = 1.4$ into Equation 5.2, the result is

$$\eta = 1 - \frac{1}{\left(\frac{p_2}{p_1} = \frac{p_3}{p_4}\right)^{\frac{\gamma-1}{\gamma}}} = 1 - 23^{-0.2857} = 59.2\%.$$

In other words, we did not have to go through each component one by one. This, of course, is a side effect of using the extremely simplistic perfect gas assumption. This is not a big problem in that its key "message" is incorrect; on the contrary, the cycle PR *is* indeed the strongest driver of Brayton-cycle efficiency. But, as will be discussed later on, once we start using more accurate thermodynamic property calculations, we will realize that there is more to the story.

There is, however, another interesting aspect of ideal cycle thermodynamics hidden in Equation 5.2. The efficiency of our conceptual cycle with the calorically perfect gas model is also given by the compressor temperature ratio, i.e.,

$$\eta = 1 - \frac{T_1}{T_2} = 1 - \frac{49 + 460}{811 + 460} = 59.2\%.$$

In essence, there is a Carnot cycle with infinitesimally small heat addition and infinitesimally small work output. This is simply a mathematical limit, which follows from

$$\frac{w_{turb}}{w_{comp}} = \frac{T_3}{T_1} \frac{1}{PR^k},$$

where $k = \frac{\gamma - 1}{\gamma}$ and becomes equal to unity as $T_3 \rightarrow T_2$ (derivation is left to the reader as an exercise).

5.1 Mean Effective Temperatures

Here is the *correct* approach to establishing a *Carnot-equivalent* cycle, which will give us a clearer picture of cycle thermodynamics and its connection to the second law. The heat addition process from state-point 2 to 3 can be described by the first "Tds" equation as follows:

$$\delta q = dh = Tds + vdp.$$

(We will discuss the Tds equations in detail later in the chapter – including the "δ" preceding heat transfer "q." For now, we will treat it as a "fact.") Integrating between state-points 2 and 3 at constant pressure:

$$q_{in} = h_3 - h_2 = c_p(T_3 - T_2) = \int_2^3 Tds.$$

Let us assume that the same amount of heat can be transferred in a hypothetical, isothermal process at (constant) temperature, $\bar{T}$, i.e.,

$$c_p(T_3 - T_2) = \bar{T} \int_2^3 ds = \bar{T}(s_3 - s_2).$$

The entropy delta can be translated into the following form via *Maxwell relations* (more on them later):

$$s_3 - s_2 = c_p \ln\left(\frac{T_3}{T_2}\right) - R \ln\left(\frac{p_3}{p_2}\right).$$

The pressure term becomes zero ($p_3 = p_2$) so that, via substitution, we obtain

$$c_p(T_3 - T_2) = \bar{T} c_p \ln\left(\frac{T_3}{T_2}\right), \tag{5.3a}$$

$$\bar{T} = \text{METH} = \frac{(T_3 - T_2)}{\ln\left(\frac{T_3}{T_2}\right)}. \tag{5.3b}$$

which defines the *mean effective heat addition temperature* (METH from here on – H stands for "high") for the cycle in question. In mathematical terms, METH is the logarithmic mean of T_2 and T_3.

By going through the same logical reasoning, we can also define a *mean effective heat rejection temperature* (METL from here on – L stands for "low") for our cycle, i.e.,

$$\bar{T} = \mathrm{METL} = \frac{(T_4 - T_1)}{\ln\left(\frac{T_4}{T_1}\right)}. \tag{5.4}$$

In mathematical terms, METL is the logarithmic mean of T_1 and T_4.

Thus, our conceptual cycle, which, as you know by now, is a *Brayton* cycle, is *equivalent to a Carnot cycle operating between two thermal reservoirs at METH and METL.* In particular,

$$\mathrm{METH} = \frac{(2{,}912 - 811)}{\ln\left(\frac{2{,}912 + 460}{811 + 460}\right)} = 2{,}154\ \mathrm{R} = 1{,}694°\mathrm{F},$$

$$\mathrm{METL} = \frac{(917 - 59)}{\ln\left(\frac{917 + 460}{59 + 460}\right)} = 879\ \mathrm{R} = 419°\mathrm{F},$$

$$\eta = 1 - \frac{\mathrm{METL}}{\mathrm{METH}} = 1 - \frac{879}{2154} = 59.2\%.$$

This brings us to the "engineering version" of the Kelvin–Planck statement of the second law of thermodynamics when applied to a heat engine operating in a power cycle (Brayton, Rankine, etc.):

The maximum possible efficiency of a heat engine operating in a power cycle is given by the efficiency of a Carnot cycle operating between two thermal reservoirs, one at the METH and the other at the METL of the said engine.

For any ideal heat transfer process between two state-points at constant pressure, the mean effective temperature can be found as

- $\bar{T} = \frac{(h_{out} - h_{in})}{(s_{out} - s_{in})}$ for a **real fluid** with known equation of state (5.5a)
- $\bar{T} = \frac{\int_{in}^{out} c_p(T)dT}{\int_{in}^{out} \frac{c_p(T)}{T}dT}$ for an **ideal gas** (5.5b)
- $\bar{T} = \frac{(T_{out} - T_{in})}{\ln\left(\frac{T_{out}}{T_{in}}\right)}$ for a **calorically perfect gas** (5.5c)

Furthermore, from the first law of thermodynamics,

$$w_{net} = q_{in} - q_{out}.$$

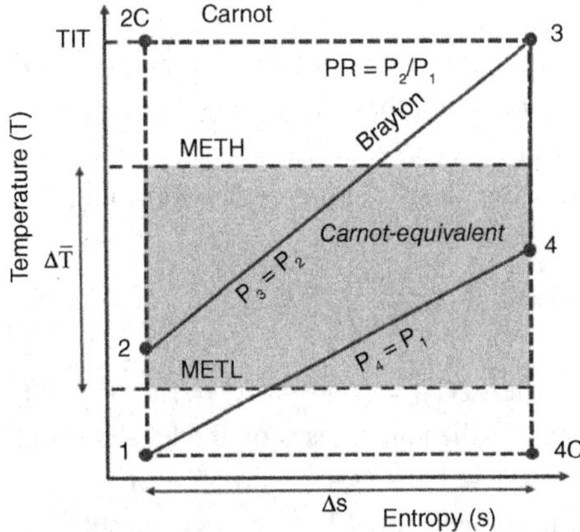

Figure 5.1 Temperature–entropy diagram of a Brayton cycle and its Carnot equivalent.

For our ideal gas turbine Brayton cycle, this leads to

$$w_{net} = \mathrm{METH} \cdot (s_3 - s_2) - \mathrm{METL} \cdot (s_4 - s_1).$$

Since entropy deltas for heat addition and rejection are identical, we obtain

$$w_{net} = (\mathrm{METH} - \mathrm{METL}) \cdot \Delta s = \Delta \bar{T} \cdot \Delta s. \tag{5.6}$$

In other words, *the area encompassed by the Brayton cycle on the T–s surface between the two mean effective temperatures and "clamped" by the two isentropic processes on both sides is equal to the net cycle work per unit mass.* The ideal cycle analysis is summarized in Figure 5.1.

While the T–s diagram in Figure 5.1 may look too simplistic to be of any practical use, the wealth of information hidden in this simple diagram is truly immense and extremely reliable for very accurate estimates. We will come back to that in a short while. Before moving on, let us recognize a few important facts.

1. The triangular area {2–2C–3–2} is the cycle *heat addition irreversibility* and is quantitatively equal to the rectangular area given by

$$i_H = (T_3 - \mathrm{METH}) \cdot \Delta s. \tag{5.7a}$$

2. The triangular area {1–4–4C–1} is the cycle *heat rejection irreversibility* and is quantitatively equal to the rectangular area given by

$$i_L = (\mathrm{METL} - T_1) \cdot \Delta s. \tag{5.7b}$$

The sum of the two heat-exchange irreversibility terms gives us the total cycle irreversibility as

$$i_{Cycle} = \left(1 - \frac{\mathrm{METH} - \mathrm{METL}}{T_3 - T_1}\right) \cdot (T_3 - T_1)\Delta s. \tag{5.7c}$$

However, we note that

$\Delta T_B = \Delta\bar{T}$= METH – METL is the Brayton cycle's thermal "driving source."

$\Delta T_C = T_3 - T_1$ is the Carnot cycle's thermal driving source.

$w_C = (T_3 - T_1)\cdot\Delta s$ is the Carnot cycle work.

Thus, Equation 5.7c quantifies the "lost" Carnot cycle work, i.e.,

$$w_{Lost} = \left(1 - \frac{\Delta T_B}{\Delta T_C}\right)\cdot w_C. \tag{5.8}$$

From the values calculated earlier, ΔT_B = 1694 – 419 = 1,275°F, ΔT_C = 2,912 – 59 = 2,853°F, and thus the term in the parentheses on the left-hand side of the lost work formula, Equation 5.8, is 1 – 1,275/2,853 = 0.553. In other words, going from the Carnot entitlement to the *still-ideal* Brayton cycle obliterates more than half of the cycle work potential. The mechanism behind this severe reduction is the "truncation" of the original driving temperature potential, ΔT_C, down to ΔT_B. This truncation also reduces cycle heat input by almost a third. Thus, the reduction in ideal cycle efficiency potential is less than the reduction in ideal cycle work potential.

3. For a given T_3, there are two ways to increase METH, reduce heat addition irreversibility, and hence increase cycle efficiency:
   a. Increase cycle PR
   b. Reheat (we will come back to this later; for now, if you have to, please consult Moran and Shapiro or your favorite thermodynamics textbook)
4. For a given T_1 (almost always the ambient temperature), there are two ways to reduce METL, reduce heat rejection irreversibility, and hence increase cycle efficiency:
   a. Increase cycle PR
   b. Add a "bottoming" cycle (e.g., steam Rankine cycle)
5. On an ideal cycle basis, increasing T_3 does not change cycle efficiency; METH and METL increase in lockstep. This is not the case for a real cycle; T_3 has a positive effect on Brayton-cycle efficiency. However, more importantly
   a. Increasing T_3 increases net cycle work.
   b. Increasing T_3 increases the "combined"-cycle efficiency.

In a nutshell, the entire *raison d'être* of modern gas turbine power plant engineering is covered by one T–s diagram (in Figure 5.1) and the five facts listed above. For conceptual analysis of gas turbine cycles, the T–s diagram is the most convenient visual tool, which summarizes pages of formulae in one easy-to-decipher graphic rather dramatically.

Note that the Brayton-cycle T–s "loop" {1–2–3–4–1} in Figure 5.1 is not an "exact" depiction of the function f = T(p,s); it is rather a "sketch" showing the relative positions of state-points 1, 2, 3, and 4 with respect to their Carnot counterparts (i.e., 1, 2C, 3, and 4C). For one thing, the constant-pressure lines 2–3 and 4–1 are not exactly straight lines. From the Maxwell correlation (e.g., see equation 11.55 of Moran and Shapiro – we will get to Maxwell relations later in the chapter)

$$\left(\frac{\partial T}{\partial s}\right)_p = \frac{T}{c_p},$$

it is clear that the isobars 2–3 and 4–1 have a curvature (albeit a slight one). Nevertheless, calculating those isobars rigorously from an equation of state for plotting purposes is quite tedious, and depiction of the slight curvature "exactly" does not add anything to the understanding of the subject matter. For "realism," it is sufficient to depict them as diverging lines (*not* parallel, though, which implies *zero* net work).

At this point, we know our Brayton-cycle gas turbine on a conceptual basis. Now, it is time to add realism to it in a step-by-step manner. First, let us get rid of the calorically perfect gas assumption (i.e., enthalpy, entropy, and specific heats are known functions of temperature [*and* pressure for entropy]). For this, we will use Thermoflex software with air at 59°F–60 percent relative humidity with a composition of

O_2 = 20.738%
N_2 = 77.292%
CO_2 = 0.030%
H_2O = 1.009%
Ar = 0.931%

Component efficiencies are set to 100 percent with zero losses and 1 lb/s of mass flow rate is assumed to obtain results in terms of per unit mass. Results from the Thermoflex model for a PR of 23 and a turbine inlet temperature (TIT) of 2,912°F are compared with the perfect gas model is in Table 5.1. Mean effective temperatures are calculated from the state-point data using enthalpy and entropy differences.

Clearly, using a more accurate ideal gas model does not change the numerical results by much. While this is quite easy to set up in a software tool, it introduces significant computational baggage if you have to do it manually. In that sense, it is not worth the extra effort. Let us go one step further and introduce component efficiencies and losses as follows:

Compressor polytropic efficiency = 90%
Turbine polytropic efficiency = 85%
Inlet loss = 1% (about 4 in. H_2O)

Table 5.1 Ideal vs. perfect gas model predictions

	Perfect Gas	Ideal Gas	Delta
w_{comp}	188	190	1.1%
w_{turb}	499	542.3	8.7%
w_{net}	311	352.3	13.3%
q_{in}	525	635.4	21.0%
η	59.2%	55.4%	–3.8% (points)
METH	1,694	1,701	+7°F
METL	419	503	+84°F

Table 5.2 Ideal (with losses) vs. perfect gas model predictions

	Perfect Gas (No Loss)	Ideal Gas w/ Losses	Delta
w_{comp}	188	223.4	18.8%
w_{turb}	499	478	–4.2%
w_{net}	311	254.6	–18.1%
q_{in}	525	602	14.7%
η	59.2%	42.3%	–16.9% (points)
METH	1,694	1,753	+59°F
METL	419	590	+170°F

Exit loss = 1.5% (about 6 in. H_2O)
Heat addition pressure loss = 5%
Cycle heat loss = 0.5%

The results are summarized in Table 5.2.

It is very interesting to note that the efficiency predicted by this simple model with reasonable assumptions for component efficiencies and pressure losses is essentially the same as that for a "real" gas turbine. (From the discussion in Chapter 4, it is hoped that the reader recognizes the cycle parameters herein – they are for a modern J/HA-class heavy-duty industrial gas turbine with an ISO base load rating of 42.2 percent efficiency.) At first glance, this might be a cause for celebration; after all, a very simple model is adequate to describe the performance of the most advanced land-based gas turbine technology available. This, alas, would be a premature declaration of victory, as signified by the big error in gas turbine exhaust temperature (i.e., 1,345°F) vis-à-vis the "true" value for the actual gas turbine (i.e., ~1,150°F). The source of this big error can be directly traced to the dilution effect of the turbine cooling flows. Another error, also tied to the same source, is the error in specific output (i.e., ~255 Btu/lb from the ideal model vis-à-vis about 218 Btu/lb for the actual machine). Obviously, ignoring the presence of secondary (cooling) flows bypassing power-generating turbine stages leads to an overestimation of the turbine power. We certainly know that, without cooling, a turbine flow path made from the most advanced nickel-based alloys would melt within mere seconds when subjected to a hot gas flow of nearly 3,000°F!

Before moving on to the turbine (hot gas path) cooling, there is one more tweak to the simple model we are working with. At this point, we recognize that a realistic heat exchanger designed to heat the working fluid to 2,912°F in a cost-effective manner is simply not possible. The only option available to the designer is to utilize a *combustor* burning 100 percent methane (CH_4) gaseous fuel at 77°F. Note that, in this case, state-points 3 and 4 have a different working fluid composition, i.e.,

O_2 = 9.149%
N_2 = 73.234%
CO_2 = 5.279%

Table 5.3 Ideal (with combustion) vs. perfect gas model predictions

	Perfect Gas	Ideal Gas with Losses	Ideal Gas with Combustor	Delta with Respect to Perfect Gas
w_{comp}	188	223.4	212	12.8%
w_{turb}	499	478	487	–2.4%
w_{net}	311	254.6	275	–11.5%
q_{in}	525	602	663	26.3%
η	59.2%	42.3%	41.5%	–17.7%
METH	1,694	1,753	NA	–
METL	419	590	NA	–

H_2O = 11.457%
Ar = 0.882%

The heat adder in the Thermoflex model is replaced by a combustor and the model is rerun. The results are summarized in Table 5.3.

In Table 5.3, METH and METL for the "combustion turbine" are listed as NA (not applicable) because state-points 2 and 3 as well as state-points 4 and 1 do *not* have the same working fluid. In terms of adding value to the analysis, switching from simple heat addition to combustion does not make a big difference. If anything, errors in specific power output and turbine exhaust temperature increase significantly (e.g., turbine exhaust temperature is above 1,400°F!). It is now time to turn to a very simple cooled turbine model.

There are two types of turbine cooling flows:

1. *Nonchargeable* flows (upstream of stage 1 rotor)
2. *Chargeable* flows (downstream of stage 1 rotor)

The term "chargeable" is shorthand for "chargeable toward net turbine output." Since the working fluid (i.e., hot combustion products) starts generating power at the stage 1 rotor, any cooling flow upstream of that point is not "chargeable." Nonchargeable cooling flows determine the hot gas temperature drop between TIT and rotor inlet temperature (RIT), which is of the order of 200°F. They are roughly of the order of 10 percent of compressor airflow.

Chargeable flows are distributed across the first three turbine stages (in a four-stage turbine). They are anywhere between 10 and 15 percent of compressor airflow depending on component materials, coatings, and particular film cooling techniques.

Calculation of chargeable and nonchargeable cooling flows is the heart of gas turbine calculations. For now, let us make the following assumptions:

- Cycle PR (at the compressor outlet) of 23
- TIT of 2,912°F with 100 percent CH_4 fuel at 77°F
- Same component efficiencies and losses
- Nonchargeable flows 11%
- Chargeable flows 13%
- PR of 2 across pseudo-stage 1

Table 5.4 Ideal (with cooling flows) vs. perfect gas model predictions

	Perfect Gas	Ideal Gas with Combustor	Cooled Turbine	Delta with Respect to Perfect Gas
w_{comp}	188	212	212	−12.6%
w_{turb}	499	487	409	18.1%
w_{net}	311	275	197	36.7%
q_{in}	525	663	497	5.3%
η	59.2%	41.5%	39.6%	−19.6% (points)
METH	1,694	NA	NA	–
METL	419	NA	NA	–

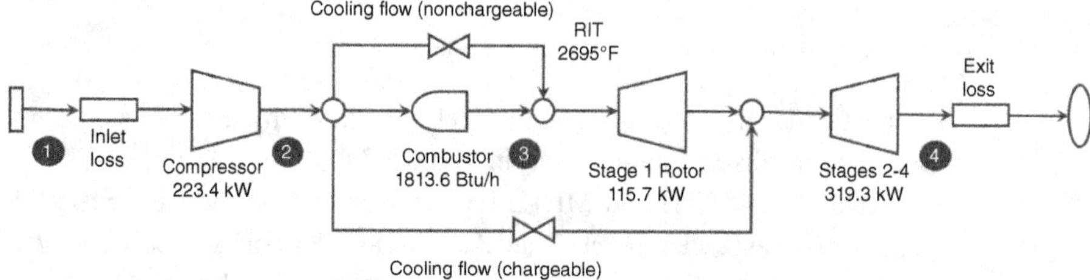

Figure 5.2 Thermoflex Brayton-cycle model (with chargeable and nonchargeable flows).

The Thermoflex Brayton-cycle "combustion" turbine model with cooling flows is shown in Figure 5.2. The firing temperature (RIT) implied by 11 percent nonchargeable flow and 2,912°F TIT is 2,695°F. The results in Table 5.4 are quite close to the aforementioned actual J/HA-class gas turbine performance (i.e., 218 Btu/lb specific output at 42.2 percent efficiency). Obviously, an exact match can be obtained easily by tweaking component efficiencies and other assumptions.

In particular, by repeating the model run with 93 percent compressor polytropic efficiency and 85.5 percent turbine polytropic efficiency, we obtain a reasonably good approximation of the actual gas turbine (see Table 5.5).

Let us pause here and summarize what we have learned so far:

1. We can define our gas turbine by only two parameters, T_3 and PR.
2. The second law of thermodynamics shows us the ultimate limit on efficiency (i.e., the Carnot-cycle efficiency).
3. Realistic cycle analysis (ideal Brayton cycle with no losses) gives us the best we can hope to achieve (i.e., the *Carnot-equivalent* cycle efficiency).
4. The ideal, air-standard Brayton-cycle efficiency is driven by *one and only one* parameter (i.e., cycle PR).
5. Using a very simple model with losses, we can get a very good handle on the "actual" gas turbine efficiency. Key technology enablers are quantified by:
    a. Component polytropic efficiencies.
    b. Chargeable and nonchargeable (secondary) cooling flows.
    c. Pressure and heat losses.

Table 5.5 Calibrated simple Thermoflex model

	7HA.02 (GTW 2016)	Simple Thermoflex Model
TIT	?	2,912
RIT	?	2,695
Fuel	100% CH_4	100% CH_4
Fuel Temperature	?	77
Inlet Loss	?	4 in. H_2O
Exhaust Loss	?	6 in. H_2O
Compressor Efficiency	?	93%
Turbine Efficiency	?	85.5%
Nonchargeable Cooling Flow	?	11%
Chargeable Cooling Flow	?	13%
PR	23.1	23
Specific Output	218.2	211.9
Efficiency	42.2%	41.3%
Exhaust Temperature	1,153	1,153

5.2 Optimum Brayton-Cycle Pressure Ratio

What we need to nail down now is the optimal value of cycle PR for a given T_3. What is "magic" about the value 23.1 for the 7HA.02 gas turbine with TIT (T_3) of 2,912°F? Why not 15? Why not 35? After all, did we not already establish the strong impact of the cycle PR on cycle efficiency? In order to answer these questions, let us establish the following:

1. Imagine a *very high PR* in Figure 5.1 for the same T_3; you will realize that the net cycle work of the cycle will be very small (zero in the limit of PR ≈ 699 – e.g., see the beginning of our *gedankenexperiment*).
2. Imagine a *very low PR* in Figure 5.1 for the same T_3; once again, you will realize that the net cycle work of the cycle will be very small (zero in the limit of PR = 1).
3. Therefore, there *must* be an *optimum* cycle PR between the two very high and very low values for which the net cycle work of the cycle is a maximum. Let us find it.

First, let us derive a non-dimensional version of the cycle net output (per unit mass), i.e.,

$$w = \left(PR^k - 1\right) \cdot \left(\frac{\tau_3}{PR^k} - 1\right), \tag{5.9}$$

$$k = \frac{\gamma - 1}{\gamma}, \quad \tau_3 = \frac{T_3}{T_1}. \tag{5.10}$$

Setting the first derivative of w with respect to PR, we find that the *optimum* PR for *maximum* net specific output is given by

$$PR_{opt} = \sqrt[2k]{\tau_3} = \tau_3^{1/2k}. \tag{5.11}$$

Thus, for T_3 = 2,912°F, T_1 = 59°F, and γ = 1.4 (which gives k = 0.2857), we have

$$PR_{opt} = \left(\frac{2{,}912 + 460}{59 + 460}\right)^{\frac{1}{2\times 0.2857}} = 28.1.$$

This is right in the proverbial "ballpark." (By the way, you can verify that the optimum PR value indeed corresponds to a *maximum* by evaluating the *second* derivative of w given by Equation 5.9 with respect to PR; it is *negative* at PR_{opt}.) One could not expect more from a very simple equation with a constant value of γ. At this point, it is a good idea to find out the existing industry practice. In order to do that, we refer to *Gas Turbine World* (GTW) 2016 Performance Specs and plot the cycle PR values for 60-Hz gas turbines from three original equipment manufacturers (OEMs) as a function of TIT. Since the rating data do not include TIT, we use some judgment based on Figure 4.8 and estimate a "pseudo-TIT" from the GTW 2016 rating values for exhaust temperature and PR as follows:

$$TIT_{est} = (TEXH + 460) \cdot PR^{0.2345} - 460. \tag{5.12}$$

A careful reader will immediately recognize this as the isentropic pressure–temperature (p–T) correlation where the exponent *k* is replaced by a curve-fit parameter. Furthermore, we postulate that the exponent *k* in the optimum PR formula should be a function of TIT as follows

$$k' = c_1 \cdot \left(\frac{TIT}{TIT_{ref}}\right)^{c_2}, \tag{5.13}$$

$$PR_{opt} = \tau_3^{\frac{1}{2k'}}, \tag{5.14}$$

where c_1 and c_2 are curve-fit parameters and TIT_{ref} is a *reference* TIT, which is set to 1,600°C (2,912°F) for the 7HA.02 data point with PR of 23. Temperatures in Equation 5.13 should be in absolute scale. The results are shown in Figure 5.3.

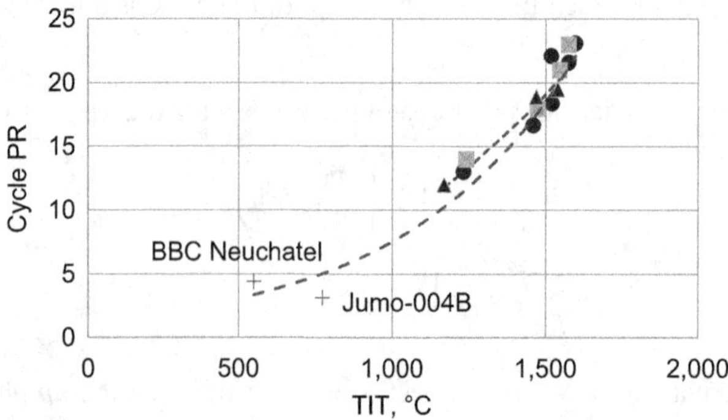

Figure 5.3 Cycle PR of large, industrial gas turbines. (data from GTW 2016)

Also shown in Figure 5.3 are the data points for the two historic gas turbines. The long dashed line is a curve-fit formula calculated by the Microsoft Excel Trend function:

$$\mathrm{PR} = 1.2455 \times \exp\left[0.0018 \times \mathrm{TIT}\left({}^{\circ}\mathrm{C}\right)\right]\left(\mathrm{R}^2 = 0.9215\right). \tag{5.15}$$

The short dashed line in Figure 5.3 is the optimum PR formula from Equation 5.14 with k' from Equation 5.13 and curve-fit constants

$$c_1 = 0.3093$$
$$c_2 = -0.2939,$$

which are determined using the Microsoft Excel Solver function to minimize the sum of squares of errors between the GTW 2016 data and the proposed formula. The fit between the published OEM rating data and our theoretical formula derived from first principles is indeed excellent. There is clearly an unmistakable logic behind the selection of a cycle PR for commercially available large, industrial gas turbines, which is recognized by *all* major OEMs. The selection is based on a *maximum gas turbine-specific power output* for a given TIT.

This, of course, makes perfect sense from an *intuitive* perspective. The major advantage of gas turbines over all other types of internal external combustion engines derives from its tremendous power density. For example, the 346–MWe 7HA.02 60–Hz gas turbine is listed at 602,000 lb of approximate weight (about 275 metric tons!) for a power density of about 1.3 kWe/kg vis-à-vis *less than 0.1 kWe/kg* for a large, gas-fired diesel engine (about 20–MWe rating).

Before continuing, let us look at the plot of specific outputs of modern industrial gas turbines as a function of (estimated) TIT. Figure 5.4 clearly illustrates the translation of the basic thermodynamic principle encapsulated by Equation 5.11 into practice.

There is little doubt that the most cost-effective gas turbine design intuitively leads to maximizing the specific output per unit mass flow rate. However, there is another consideration at play here, which is also implied by the ideal T–s diagram in Figure 5.1. In particular, we pay attention to the triangular area {1–4–4C–1}, which quantifies the cycle *heat rejection irreversibility*. What if we could place another power cycle in there

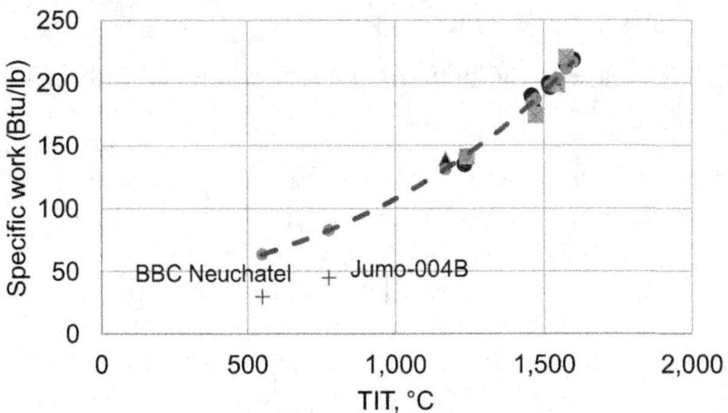

Figure 5.4 Specific output of large, industrial gas turbines. (data from GTW 2016)

that has a mean effective heat addition temperature of METL and a mean effective heat rejection temperature of T_1? The efficiency of that ideal *bottoming* cycle would be given by Equation 5.16:

$$\eta_{BC} = 1 - \frac{T_1}{METL}. \tag{5.16}$$

From the cycle T–s diagram in Figure 5.1, the ideal "combined" efficiency with ideal Brayton (topping) and ideal bottoming cycles can be deduced as depending on the ratio of METH and the ambient temperature T_1, which, after rearranging the terms a bit, leads us to the combined-cycle efficiency:

$$\eta_{CC} = 1 - \frac{T_1}{METH}, \tag{5.17a}$$

$$\eta_{CC} = 1 - \frac{\ln\left(\frac{\tau_3}{PR^k}\right)}{\tau_3 - PR^k}. \tag{5.17b}$$

Ignoring the miscellaneous topping- and bottoming-cycle losses and minor inputs, a simplified version for the combined-cycle efficiency can also be written as

$$\eta_{CC} = \eta_{TC} + (1 - \eta_{TC}) \cdot \eta_{BC}, \tag{5.18}$$

where the subscripts TC and BC denote topping and bottoming cycles, respectively. Taking the derivative of both sides with respect to the PR and setting the combined cycle efficiency derivative to zero to find its maximum, noting that the bottoming and topping cycle efficiencies are approximately the same in magnitude, we find that

$$\frac{\partial \eta_{TC}}{\partial PR} \approx -\frac{\partial \eta_{BC}}{\partial PR}. \tag{5.19}$$

This correlation states that the maximum combined-cycle efficiency occurs at the point where the rate of increase of the topping-cycle efficiency with PR is the same as the rate of decrease of the bottoming-cycle efficiency. The rate of increase of the topping-cycle efficiency with increasing PR can be found from the Brayton-cycle efficiency

$$\frac{\partial \eta_{TC}}{\partial PR} = \frac{k}{PR^{k+1}}. \tag{5.20}$$

Similarly, the rate of decrease of the bottoming cycle efficiency with increasing PR is found as

$$\frac{\partial \eta_{BC}}{\partial PR} = \frac{k}{PR} \cdot \frac{\left(1 - \frac{T_4}{METL}\right)}{\left(\frac{T_4}{T_1} - 1\right)}, \tag{5.21}$$

$$\frac{T_4}{METL} = \frac{T_4}{T_1} \cdot \frac{\ln\left(\frac{T_4}{T_1}\right)}{\left(\frac{T_4}{T_1} - 1\right)}, \tag{5.22}$$

$$\frac{T_4}{T_1} = \frac{\tau_3}{PR^k}. \tag{5.23}$$

At this point, we can go ahead and state that a steam *Rankine* cycle is an ideal candidate for the bottoming cycle because

1. Current gas turbine technology with exhaust temperatures of around 1,150°F or higher constitute a very good heat source for generating high-pressure steam at 1,000–1,100°F in a "waste heat recovery" boiler (the term used in combined-cycle parlance is the *heat recovery steam generator* or HRSG).
2. Heat rejection from the steam turbine condenser at constant pressure and temperature is the perfect fit to the fourth ideal Carnot-cycle process (i.e., *isothermal* cycle heat rejection).

We find the optimum topping (Brayton)-cycle PR for maximum combined-cycle efficiency by substituting Equations 5.20–5.23 into Equation 5.19. We use the exponent k' as defined by Equation 5.13. Note that we substituted the ambient temperature for the topping-cycle T_1 because it is the logical choice. We do not, however, have to use it as the isothermal heat rejection temperature for the bottoming cycle. (Typical condenser steam temperatures are around 90–100°F.) The results for two different selections, 59°F and 95°F, are superimposed on the plot in Figure 5.3, which is reproduced in Figure 5.5.

Let us summarize the knowledge we have acquired up to this point in a very simple, ideal cycle analysis.

1. Our starting goal was to design the best possible heat engine with air as the working fluid.
2. The theoretical maximum set by the second law of thermodynamics, via its Kelvin–Planck statement, is the Carnot cycle (set by maximum and minimum cycle temperatures T_3 and T_1, respectively).
3. Practical considerations led us naturally to a gas turbine operating on the Brayton cycle with cycle PR = p_2/p_1, cycle maximum temperature T_3, and constant-pressure heat addition and heat rejection processes.

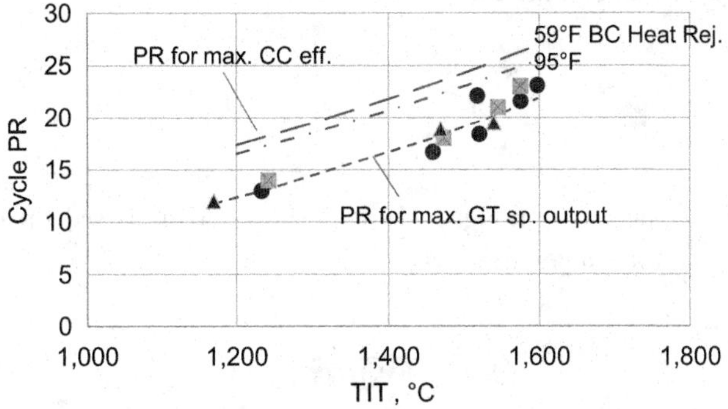

Figure 5.5 Cycle PR of large, industrial gas turbines (data points are from GTW 2016). BC = bottoming cycle; CC = combined cycle; GT = gas turbine.

4. Further analysis revealed that the ideal Brayton cycle is, in fact, *equivalent* to a Carnot cycle, but with mean effective heat addition and heat rejection temperatures, METH and METL, respectively (instead of T_1 and T_3).
5. We then calculated the optimum Brayton-cycle PR for a given T_3, which resulted in the maximum specific cycle output.
6. Comparison with "real" gas turbine rating data listed in a recent trade publication revealed that this was indeed the design basis adopted by the OEMs.
7. Realizing that the Brayton-cycle exhaust, at a mean effective temperature of METL, can be the heat source for another ideal, "bottoming" cycle, we established the ideal performance of a "combined" cycle with cycle maximum temperature T_3, topping cycle PR of p_2/p_1, mean effective heat addition temperature METH, and isothermal heat rejection at T_1.
8. Our analysis led us to an optimal topping-cycle PR for maximum combined-cycle efficiency, which is depicted in Figure 5.5.

As shown in Figure 5.5, the optimum PR for maximum combined-cycle efficiency is higher than that for maximum gas turbine-specific output. As we will show later in the book via more rigorous cooled gas turbine calculations, this is indeed the case, and this has been verified by OEM studies as well [1].

Herein, just a preview of the results should suffice. Using the curves from a paper by Gülen [2] (specifically, figure 8 in that paper, which is qualitatively the same as Figure 8.10 later in the book), general trends can be summarized as follows:

1. The optimum gas turbine Brayton-cycle PR increases with increasing TIT (T_3).
2. The optimum PR for maximum combined-cycle efficiency is higher than that for maximum gas turbine-specific output.
3. The difference in combined-cycle efficiency at optimum PR for gas turbine-specific output and at the "true" optimum is negligible.
4. Since the cost-effective choice leads to a gas turbine design at a lower PR, this is where OEMs set their product designs.

This is where we pause and define the most important thermodynamic property for gas turbine cycle analysis; namely, the *exergy*.

5.3 Exergy

The efficiency of the ideal bottoming cycle was shown to be based on its *Carnot-equivalent* cycle reservoir temperatures as

$$\eta_{BC} = 1 - \frac{T_1}{METL}. \tag{5.24}$$

Thus, the *maximum* output of the ideal bottoming cycle can be found by multiplying the topping (i.e., Brayton) gas turbine cycle's exhaust energy by Equation 5.24 as

$$w_{BC} = c_p \cdot (T_4 - T_1) \cdot \left(1 - \frac{T_1}{METL}\right), \tag{5.25a}$$

$$w_{BC} = c_p \cdot (T_4 - T_1) \cdot \left(1 - \frac{T_1}{(T_4 - T_1)} \ln\left(\frac{T_4}{T_1}\right)\right). \tag{5.25b}$$

Note that in Equation 5.25, the implicit assumptions are as follows:

1. The gas turbine exhaust at T_4 can be cooled down to the ambient temperature T_1.
2. The bottoming-cycle (isothermal) heat rejection takes place at T_1 as well.

Combining the terms on the right-hand side of Equation 5.25, we obtain

$$w_{BC} = c_p \cdot (T_4 - T_1) - c_p \cdot T_1 \cdot \ln\left(\frac{T_4}{T_1}\right). \tag{5.26}$$

We recognize the first term on the right-hand side of Equation 5.26 as the enthalpy change between states 4 and 1. Similarly, the second term is the entropy change between states 4 and 1 (at constant pressure) multiplied by T_1. Thus, we can rewrite Equation 5.26 as

$$w_{BC} = (h_4 - h_1) - T_1 \cdot (s_4 - s_1). \tag{5.27}$$

Since any combination of two or more arbitrary thermodynamic properties (in this case, strictly speaking, T_1, T_4, and $p_1 = p_4$, which lead to h_1, h_4, s_1, and s_4 via the equation of state) defines another thermodynamic property, we realize that *the theoretically possible maximum bottoming-cycle work (per unit mass) for any given bottoming cycle and working fluid is a property of the exhaust stream of the topping cycle.*

The property in question is known as *specific flow availability* or *specific flow exergy* and it is given by (ignoring the kinetic and potential energies)

$$a = (h - h_{DS}) - T_{DS} \cdot (s - s_{DS}), \tag{5.28}$$

where the subscript DS refers to a "dead state," which is defined (by Moran and Shapiro) as the equilibrium state between the system and its environment. Obviously, for our purposes, the logical choice for the dead state is the ambient at T_1 and p_1. (The simplified derivation here assumed a calorically perfect gas. However, the exergy of any given working fluid [i.e., hot combustion gas at the gas turbine exhaust] can be rigorously calculated by a known equation of state with known pressure, temperature, and composition.)

The first term on the right-hand side of the exergy definition, Equation 5.28, is the maximum work production *potential* of a fluid at a given state, p, T, and composition, and the dead state.

The second term on the right-hand side of Equation 5.28 is the portion of that maximum potential, which is unavailable due to the Kelvin–Planck statement of the second law of thermodynamics. It is frequently referred to as "exergy destruction" or "lost work." In order to understand this better, let us start at the gas turbine topping-cycle state-point 4 (i.e., the exhaust). Note that

1. The maximum heat transfer from the hot exhaust at T_4 is via cooling it to the ambient temperature T_1.
2. The maximum work extraction is via an ideal heat engine operating on a Carnot cycle, which receives this heat and rejects a portion of it to the surroundings (Kelvin–Planck statement of the second law) at T_1.

Consequently, the heat transfer from the hot exhaust to the Carnot cycle is

$$q_{in} = c_p \cdot (T_4 - T_1) = h_4 - h_1. \tag{5.29}$$

Similarly, the heat transfer from the Carnot cycle to the surroundings is

$$q_{out} = T_1 \cdot (s_4 - s_1), \tag{5.30}$$

because the entropy change during heat rejection in a Carnot cycle is equal to the entropy change during heat addition (see Figure 5.1). Finally, the net Carnot-cycle work is found by subtracting Equation 5.30 from Equation 5.29, i.e.,

$$w_{net} = q_{in} - q_{out}, \tag{5.31a}$$

$$w_{net} = (h_4 - h_1) - T_1 \cdot (s_4 - s_1), \tag{5.31b}$$

which brings us back to Equation 5.28.

In many gas turbine problems, a common feature is mass and/or energy stream exchange between two systems. Examples are:

1. Gas turbine cooling air cooling
2. Gas turbine closed-loop steam cooling
3. Steam import/export to/from the gasifier (in integrated gasification gas turbines)

A sample system diagram illustrating gas turbine cooling air cooling is shown in Figure 5.6. Hot compressor discharge air is extracted for cooling; say, stage 1 stator nozzle vanes. Especially in gas turbines with high cycle PRs, in order to minimize the

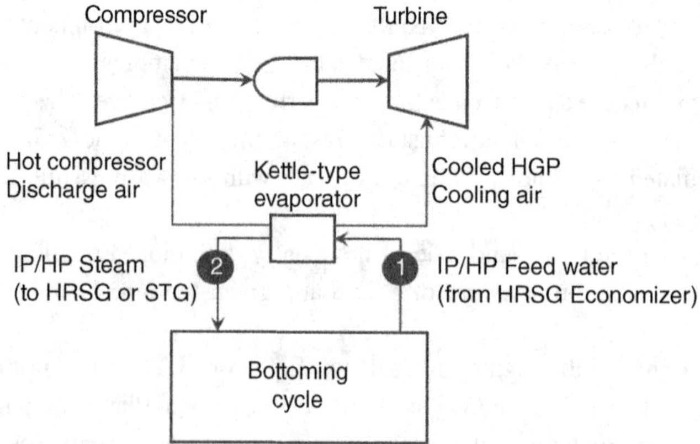

Figure 5.6 Generic gas turbine cooling air cooling scheme. HGP = hot gas path; HP = high-pressure; IP = intermediate-pressure; STG = steam turbine generator.

coolant flow rate, hot air is first cooled in a kettle-type evaporator by generating steam. High- or intermediate-pressure feed water is bled from the HRSG economizer and sent to the evaporator, where it boils by extracting heat from the hot gas turbine air. Steam generated in the evaporator is returned to the bottoming cycle at a suitable location.

We intuitively know that the heat transferred to the bottoming cycle (i.e., $q = h_2 - h_1$) contributes X amount of kilowatts to the steam turbine generator output. We can determine the *entitlement* (i.e., maximum possible) value of said contribution by imagining that the heat transfer in question is utilized in a hypothetical Carnot cycle, which rejects heat to the environment at a temperature equal to that of the "dead state." Consequently, from what we have just learned about exergy, we easily calculate that maximum value as

$$w_{max} = a_2 - a_1 = (h_2 - h_1) - T_{DS}(s_2 - s_1), \tag{5.32a}$$

$$w_{max} = (h_2 - h_1)\left(1 - T_{DS}\frac{(s_2 - s_1)}{(h_2 - h_1)}\right), \tag{5.32b}$$

$$w_{max} = q_{in}\left(1 - \frac{T_{DS}}{METH}\right). \tag{5.32c}$$

Equation 5.32c is the net cycle work for the Carnot cycle operating between two reservoirs at METH and T_{DS} and receiving heat from the former in the amount of q_{in}. Of course, we do not have a Carnot engine in our hands. The question is thus "how much of that maximum can we actually achieve?" The answer depends on the "quality" of the energy in question. In particular,

- If cooling air cooling is used to generate high-pressure (HP) steam from HP feed water, the conversion effectiveness is very high (i.e., about 0.88–0.89).
- If cooling air cooling is used to generate hot reheat steam from intermediate-pressure (IP) feed water, the conversion effectiveness is rather high as well (i.e., about 0.83–0.88).
- If cooling air cooling is used to generate low-pressure (LP) steam from LP feed water, the conversion effectiveness is still quite respectable (i.e., 0.64–0.72).

5.4 Kilowatt "Thermal"

In cogeneration (combined heat and power [CHP] as preferred by the Europeans) applications, there are two products from the system: (1) electricity measured in kWh; and (2) thermal energy (usually in the form of steam or hot water) also measured in kWh. The "kilowatts" in either measure are not equivalent and must be distinguished. In terminology, this is done by referring to them as kilowatt (electric), kWe, and kilowatt (thermal), kWth.

The numerical equivalence is established by the "Carnot multiplier," χ, which is given by

$$\chi = 1 - \frac{T_0}{T_{prod}},$$

where T_{prod} is the temperature of the thermal product stream (steam or hot water) and T_0 is the temperature of the "dead state," which is best represented by the ambient temperature. The multiplier as stated above is incorrect for a real fluid (e.g., H_2O in the form of steam or water). The *energy* of the product stream is represented by its enthalpy, h, which is a function of temperature and pressure via the applicable equation of state (see Section 5.6). The *exergy* of the product stream can also be exactly calculated using Equation 5.28, i.e.,

$$a = (h - h_0) - T_0 \cdot (s - s_0).$$

Thus, the exact Carnot multiplier is found as

$$\chi = \frac{a(p, T)}{h(p, T)}.$$

For practical purposes, if the product is hot water, χ can be set to 0. It is true that you will get a finite, non-zero value from the ASME steam tables, but it is very small. Practically, though, it must be considered to be zero – you cannot generate power from hot water cost-effectively. (Building a water turbine, while certainly doable, would be economically infeasible.)

If the product is steam, in the age of smartphones and apps, it is no big deal to calculate χ exactly using ASME steam tables. Note, however, in almost all cases, steam supplied to the cogeneration host is saturated. In that case, for quick estimates you can use the following correction to the original formula (pressure in psia, temperatures in degrees Rankine)

$$\chi = \left(1 - \frac{T_0}{T_{prod}}\right) p_{prod}^{-0.02813}.$$

(The value of the correction factor is about 0.85 for pressures of 150–450 psia.) In the range of typical pressures encountered in cogeneration applications with saturated steam supply to a host, one kilowatt (thermal) is less than 0.5 kilowatt (electric).

5.5 "Tds" Equations

Perhaps the two most important equations in the entire body of thermodynamics are the two "Tds" equations:

$$du = Tds - pdv, \tag{5.33a}$$

$$dh = Tds + vdp. \tag{5.33b}$$

Equation 5.33a is known as the *Gibbs equation* or the *first* Tds equation. It is the differential form of the first law of thermodynamics for a closed system undergoing an internally reversible process.

Equation 5.33b is known as the *second* Tds equation. It is the differential form of the first law of thermodynamics for an open or steady-state, steady-flow (SSSF) system undergoing an internally reversible process.

The first term on the left-hand side of Equation 5.33b is the internally reversible heat transfer in differential form:

$$\delta q_{rev} = Tds.$$

Note the Greek letter δ denoting the "differential" of heat transfer, q. It signifies that q is a "path function," i.e.,

$$\int_1^2 \delta q \neq q_2 - q_1,$$

whereas thermodynamic properties such as enthalpy are "exact differentials," i.e.,

$$\int_1^2 dh = h_2 - h_1.$$

The second term on the left-hand side of Equation 5.33b is the internally reversible work transfer in differential form:

$$\delta w_{rev} = vdp.$$

Just like heat transfer, work transfer is also a "path function."

If a process is reversible *and* adiabatic, the term Tds becomes zero and we have the isentropic enthalpy change for the control volume in question (in differential form) as

$$dh = \delta w_{rev} = vdp, \tag{5.34a}$$

$$dh = \delta w_{rev} = \frac{1}{\rho} dp. \tag{5.34b}$$

Applying it to a control volume around a pump, for example, and an incompressible fluid (i.e., constant density, ρ), we obtain the *actual* pump work as

$$w = \frac{h_{2s} - h_1}{\eta_s} = \frac{1}{\eta_s \rho}(p_2 - p_1), \tag{5.35}$$

where we make use of the isentropic efficiency of the pump. Note that we identify the *isentropic* end state 2 explicitly by adding an "s" to the state number (i.e., "2s").

For a compressible fluid such as air or combustion products undergoing compression or expansion, we have to integrate Equation 5.34a using the ideal gas equation of state and the isentropic process definition,

$$pv = RT, \tag{5.36}$$

$$pv^{\gamma} = \text{const.} \tag{5.37}$$

Thus,

$$h_{2s} - h_1 = \int_1^{2s} \left(\frac{p_1 v_1^{\gamma}}{p} \right)^{\frac{1}{\gamma}} dp, \tag{5.38}$$

which, after integration and some algebra, brings us to the well-known correlation for an isentropic process (and calorically perfect gas)

$$h_{2s} - h_1 = c_p T_1 \left(\left(\frac{p_2}{p_1} \right)^{\frac{\gamma-1}{\gamma}} - 1 \right). \tag{5.39}$$

Equation 5.39 is, of course, the basis of all simplified compressor and turbine/expander calculations. The isentropic process formula, Equation 5.37, is also derived from the second Tds equation, Equation 5.33b, i.e.,

$$dh = c_p dT = 0 + v dp, \text{ or}$$

$$dh = \frac{1}{\rho} dp.$$

Assuming calorically perfect gas and substituting the definition of c_p and the ideal gas equation, we can derive everything else we need therefrom, i.e.,

$$c_p dT = \frac{RT}{p} dp \quad \rightarrow \quad c_p \int_1^2 \frac{dT}{T} = R \int_1^2 \frac{dp}{p} \quad \rightarrow \quad \frac{\gamma - 1}{\gamma} \int_1^2 \frac{dT}{T} = \int_1^2 \frac{dp}{p}.$$

In essence, the entire set of thermodynamic building blocks of gas turbine theory can be derived from the Tds equations and the ideal gas correlation. The details are left as an exercise for the reader. If help is needed, all the salient details can be found in Moran and Shapiro (or your favorite thermodynamics textbook) or in Saravanamuttoo et al.

Another implication of the isentropic version of Equation 5.33b is important for compressor- and turbine-stage aerothermodynamics. Ignoring the change in density or simply assuming that the flow is incompressible, it is realized that $dh \propto dp$ or $\Delta h \propto \Delta p$. This implies that the stage degree of reaction is a measure of static enthalpy change distribution as well as static pressure change distribution between the stator and the rotor. (This nugget of information will be useful in Chapters 10 and 11.)

If a given process *is* reversible but *not* adiabatic, it is *not* isentropic and we have to retain the reversible heat transfer and work terms in their integral form, i.e.,

$$q_{rev} = \int_1^2 T ds, \tag{5.40}$$

$$w_{rev} = \int_1^2 v dp, \tag{5.41}$$

and by substituting Equations 5.40 and 5.41 into the second Tds equation, we obtain the reversible work as

$$w_{rev} = (h_2 - h_1) - \int_1^2 Tds. \tag{5.42}$$

At this point, we recognize that the property *exergy* is a special case of Equation 5.42 such that

- State 1 is a "dead state" where the fluid is in equilibrium with its surroundings,
- The heat transfer takes place *isothermally* at temperature $T_1 = T_{DS}$,

and Equation 5.42 becomes

$$w_{max} = (h_2 - h_{DS}) - T_{DS}(s_2 - s_{DS}). \tag{5.43}$$

The second law of thermodynamics states that *any arbitrary* process between two states 1 and 2 (i.e., it does *not* have to reversible) must satisfy the following inequality:

$$\int_1^2 Tds \geq q_{rev}. \tag{5.44a}$$

(In passing, if asked to identify a single equation that best explains the second law of thermodynamics from a *concrete* engineering point-of-view [i.e., without resorting to pseudo-philosophical explanations using the "arrow of time," etc.], the answer should point to Equation 5.44a. It tells us that

- In *any arbitrary process* between two equilibrium thermodynamic states 1 and 2 with heat transfer, entropy can increase *or* decrease. In other words, entropy does *not* always increase!
- If the process is reversible *and* adiabatic (i.e., no heat transfer), entropy does *not* change.
- If the process is irreversible *and* adiabatic, entropy *does* increase.

It is the entropy of an *isolated system*, which does not have any work, heat, or mass interaction with its surroundings *by definition*, which must either increase or, at best, stay constant. Since the universe is the ultimate isolated system, again, *by definition* [i.e., what is the "environment" of the universe?], the entropy of the universe must either increase or, at best, stay constant. This is the source of the popular belief that entropy *always* increases, which really pertains to a special case.)

Returning back to the more mundane world of engineering, another way to write the inequality in Equation 5.44a is

$$\int_1^2 Tds = q_{rev} + \sigma, \tag{5.44b}$$

where σ

- Is zero for a reversible process (i.e., Equation 5.44 reverts back to Equation 5.40)
- Is nonzero *and* positive for an *irreversible* process
- Can *never* be negative.

The term σ is the *rate of entropy production*, which, when multiplied by a reference temperature, T_0, i.e.,

$$i = T_0 \cdot \sigma,$$

is known under several different names such as *irreversibility* or *exergy destruction*. One other name, which is of most interest to practicing engineers, is "lost work." How "work is lost" will be obvious when we integrate Equation 5.33b and combine it with Equation 5.44b, i.e.,

$$h_2 - h_1 = \int_1^2 v dp + (q_{rev} + \sigma).$$

Since the first term on the left-hand side is the isentropic work given by

$$h_{2s} - h_1 = \int_1^2 v dp,$$

we end up with the following result:

$$h_2 - h_1 = (h_{2s} - h_1) + (q_{rev} + \sigma) \tag{5.45}$$

The wealth of information carried inside the deceptively simple Equation 5.45 is illustrated in Figure 5.7. The two most common uses of Equation 5.45 in gas turbine thermodynamic analysis are for *adiabatic compression* and *adiabatic turbine expansion*. In either case, Equation 5.45 becomes

$$h_2 - h_1 = (h_{2s} - h_1) + \sigma, \tag{5.46}$$

$$\sigma = (h_1 - h_{2s})\left(1 - \eta_{t,s}\right) \text{ for a turbine,} \tag{5.47a}$$

$$\sigma = (h_{2s} - h_1)\left(\frac{1}{\eta_{c,s}} - 1\right) \text{ for a compressor,} \tag{5.47b}$$

and thus makes the quantification of "lost work" by σ quite clear.

This very brief section is the *entirety* of the thermodynamics you need for gas turbine engineering (well, with the addition of *Maxwell relations*, that is – see Section 5.6). It is, of course, an extremely distilled summary of one-and-a-half centuries' worth of extreme scholarship by giants in the field such as Boltzmann, Helmholtz, Clausius, Lord Kelvin, and others. It is not something to fully grasp in one or two semesters of undergraduate thermodynamics and requires years of absorption in its day-to-day applications to real problems. If the reader feels that he or she is not fully comfortable with the ramifications

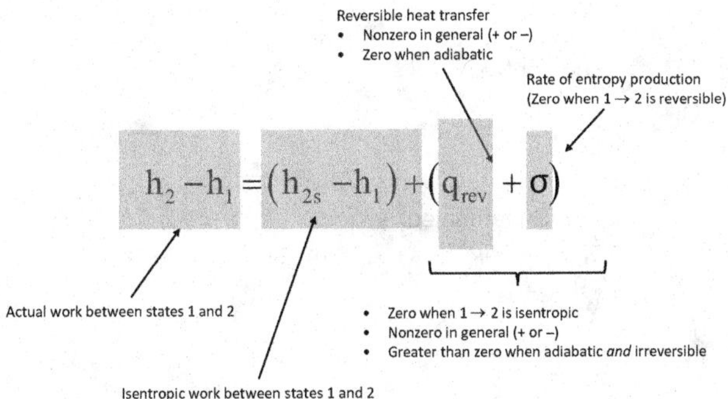

Figure 5.7 The second Tds equation for a general process between two arbitrary states 1 and 2.

of the thermodynamic concepts outlined above, it is suggested that he or she uses this section as a roadmap and goes over the relevant parts of the suggested textbooks.

5.6 Property Calculations

Obviously, the assumption of calorically perfect gas makes life very easy. Calculations done using that assumption can be quite adequate for simple systems and rough estimates. For real work, however, bona fide property evaluation is a must. Even with the simplest form of state equations, closed algebraic formulae are either too long and messy or not obtainable at all. In that case, you have to resort to numerical calculations (e.g., see the Visual Basic [VB] functions COMPP and TURBP in Appendix C). This, of course, requires a good understanding of equations of state and the property calculations using them.

For typical heavy-duty, industrial gas turbine calculations, the ideal gas assumption is adequate. Air is a gas mixture with a critical pressure of 550 psia and a critical temperature of –221°F (see Figure 5.8 and the associated text for the definition of the *critical point* [CP]). Consequently, one does not have to worry about phase change from gas to liquid or vice versa. Furthermore, pressures are not high enough to warrant correction for "real gas" effects. The error incurred by not using real gas properties is less than 0.5 percent in c_p in the pressure–temperature space typical of gas turbine compressor discharge and turbine inlet [3].

For closed-cycle gas turbine calculations with, say, CO_2 as the working fluid, a "real fluid" property package such as *NIST Reference Fluid Thermodynamic and Transport Properties Database* (REFPROP) must be used to account for the real gas effect at very high pressures (e.g., up to 3,000 psia or higher) and the possibility of phase change from gas to liquid and vice versa.

ASME Standard STP-TS-012-1 contains a collection of ideal gas and real gas data for 14 chemical species and air in temperature and pressures ranges typically encountered in gas turbine performance calculations [3]. It is prepared by ASME's Air Properties Committee and provides equation representations for enthalpy, entropy, heat capacity,

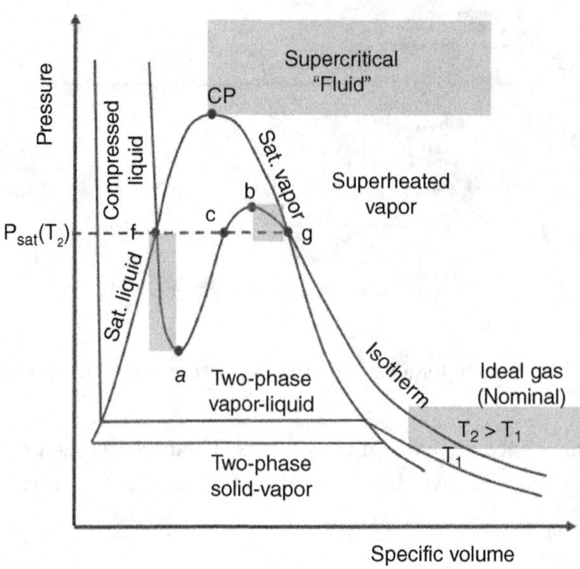

Figure 5.8 Thermodynamic phase diagram on p–v surface for property evaluation. The bell-shaped, vapor–liquid coexistence region is commonly known as the "vapor dome."

and enthalpy of formation as published by NASA and VDI.[2] One can easily program the equations therein in a VBA function for property calculations in an Excel spreadsheet.

Accurate property calculations near the vapor–liquid CP are imperative for supercritical CO_2 cycle calculations. As shown on the pressure–volume (p–v) diagram in Figure 5.8, at the CP, saturated liquid and vapor lines connect (i.e., the distinction between the two phases disappears). Furthermore, at the CP, the first and second derivatives of the isotherm p(v) become equal to zero, which means that the compressibility of the fluid (and its heat capacity) becomes very large. Consequently, fluid behavior near the CP is highly unstable and property calculation becomes susceptible to large errors.

Thermodynamic property calculations require an *equation of state*, which can be derived from the principles of statistical mechanics as an infinite series, i.e.,

$$Z = 1 + \frac{B(T)}{v} + \frac{C(T)}{v^2} + \frac{D(T)}{v^3} + \ldots, \tag{5.48}$$

where Z is the *compressibility factor*, Z = pv/RT, and the coefficients B, C, D, etc., are the first, second, third, etc., virial coefficients, respectively. The equation itself is known as the *virial equation of state*.

The ideal gas equation is obtained from the virial equation by setting all virial coefficients to zero (i.e., compressibility factor of unity) and Z to 1. Strictly speaking,

[2] Verein Deutscher Ingenieure (The Association of German Engineers) is the counterpart of ASME in Germany.

it is only valid at low pressures and low densities (i.e., for a dilute gas), although it has been used successfully well beyond its theoretical limits. Nevertheless, the ideal gas equation has its limits; for example, it cannot differentiate between liquid and vapor phases of a pure fluid. When state-points become too close to the "vapor dome," this becomes a big deficiency.

The next approximation is the *cubic equation of state*, which is found by truncating Equation 5.48 after the second virial coefficient, i.e.,

$$Z = 1 + \frac{B(T)}{v} + \frac{C(T)}{v^2}. \tag{5.49}$$

Thus, it gives a third-order polynomial in specific volume, v, hence its name:

$$\frac{pv^3}{RT} - v^2 - B(T)v - C(T) = 0. \tag{5.50}$$

Cubic equations of state predict liquid and vapor "coexisting states" (f and g, respectively, in Figure 5.8) via Maxwell's "equal area construction" rule. For a given temperature below the critical, at the saturation pressure, there are *three roots* of Equation 5.50, i.e.,

- "Liquid-like" v_f
- "Vapor-like" v_g
- v_c

It can be shown that the area bounded by {f–*a*–c–f} is equal to the area bounded by {c–b–g–c}.

Furthermore, there are two local extrema for a subcritical isotherm: points *a* (a local minimum) and b (a local maximum) on the p–v diagram in Figure 5.8. Point *a* is known as the *liquid spinodal*; point b is known as the *vapor spinodal.*

Fluid states between f and *a* are *metastable states*, which are also known as *superheated liquid states*. In other words, the fluid at a pressure between p_f and p_a has a temperature higher than that corresponding to the saturation (equilibrium) temperature at the same pressure.

Fluid states between v and b are metastable states as well, which are also known as *subcooled vapor* states. In other words, the fluid at a pressure between p_v and p_b has a temperature lower than that corresponding to the saturation (equilibrium) temperature at the same pressure.

Metastable states can be achieved for short durations (they are not stable, equilibrium states); in particular,

1. Superheated liquid states are reached upon sudden expansion of a hot liquid (e.g., loss-of-coolant or LOCA accidents in nuclear power plants) such that there is not enough time to reach equilibrium via slow evaporation (i.e., via homogeneous or heterogeneous nucleation). Evaporation of superheated liquid droplets happens explosively, causing violent shock waves and damage.
2. Subcooled vapor states are reached upon sudden expansion of a hot vapor. The best-known example is expansion of steam in the low-pressure section of a steam turbine. There is not enough time to reach equilibrium via slow condensation

(i.e., via homogeneous or heterogeneous nucleation). Evaporation of subcooled bubbles happens suddenly, causing a *condensation shock* and leading to losses. Where the condensation shock happens is known as the *Wilson line*, which is the practical counterpart of the theoretical vapor *spinodal* (point b).

The simplest cubic equation of state is the *van der Waals* equation of state:

$$p = \frac{RT}{v - b} - \frac{a}{v^2}. \tag{5.51}$$

Here, the term b accounts for the finite volume occupied by the fluid molecules (i.e., $v > b$) and the second term on the right-hand side of Equation 5.51 accounts for the forces of attraction between the molecules. When constants *a* and b are zero, the equation reverts to the ideal gas equation of state. The van der Waals equation, while theoretically correct, does not have enough accuracy for engineering calculations. More advanced cubic equations such as Peng–Robinson, Benedict–Webb–Rubin (BWR), Soave–Redlich–Kwong (SRK), etc., are available in the literature and in chemical process simulators such as HYSYS and ASPEN. They can be used for pure fluids and mixtures when accurate property prediction and phase equilibrium information are required.

For steam-cycle calculations, ASME steam tables must be used. For property estimations in a limited region, specific curve-fits can be used. However, due to the fact that boiling and condensation processes constitute the bulk of the steam-cycle calculations, comprehensive yet easy-to-use transfer functions cannot be developed.

Until the late 1990s, the IFC-67 package was the basis of the ASME steam tables. This formulation can be utilized for pressures up to 14,503 psia (1,000 bar) and temperatures up to 1,472°F (800°C). For calculations involving evaluation of cycles developed in the past, this package must be used. It can be used for cycles with supercritical states, or temperatures beyond 1,200°F (649°C) as well.

IAPWS-IF97 is a newer property formulation published in 1997 by the International Association for the Properties of Water and Steam. This formulation was adopted by ASME in 1999 to replace the IFC-67 formulation. This system can also be utilized for pressures up to 14,503 psia (1,000 bar) and temperatures up to 1,472°F (800°C). It can also be used for cycles with supercritical states, or temperatures beyond 1,200°F (649°C).

Commercially available heat balance simulation software (unless you have a very old version) typically contains both versions of the ASME steam tables.

5.6.1 Critical Point Blues

Recently, there has been an increased interest in closed-cycle gas turbines with supercritical CO_2 as the working fluid. This will be covered in some detail in Chapter 22. One advantage of a closed-loop Brayton cycle where the compressor suction is near the CP of the working fluid (which does not have to be CO_2) stems from the "weird" behavior of the fluid near this singularity. Let us take a closer look at this, albeit within the limits of this brief chapter. As noted above, the CP is the apex of the vapor–liquid equilibrium

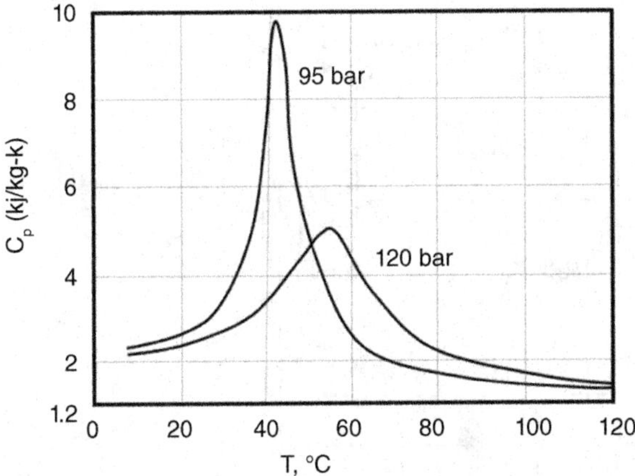

Figure 5.9 Specific heat (at constant pressure) for CO_2.

dome of a substance, where the distinction between the two phases disappears. From an engineering perspective, the important thing to know about the behavior of a fluid in the vicinity of its CP is that the first and second derivatives of pressure with respect to specific volume (or density) while holding temperature constant are nearly zero. (They are exactly zero at the CP itself, which is next to impossible to sustain as a stable, equilibrium state-point and – as such – pointless to dwell on.) Since with two known properties any third thermodynamic property is unambiguously fixed via the equation of state and Maxwell relations, it is a safe bet that when something has a very small (near-zero) value, something else is bound to have a very large value, for instance:

- Specific heat at constant pressure, c_p, becomes very large (see Figure 5.9).
- Isothermal compressibility becomes very large.

Furthermore, as shown in Figure 5.10, near the CP, the compressibility factor, $Z = Pv/RT$, becomes quite small (it is near unity for most gases, and exactly unity for the *ideal* gas). Also, it can be shown that the specific heat ratio, γ, becomes close to unity. When so many physical quantities become 0 or 1, things start going haywire. This is why the best environment for making CP measurements is inside a spacecraft in orbit with zero gravity! This also is why reliable equation of state data near the CP are very difficult to gather and distill into compact formulae suitable to engineering calculations.

The significance of a low compressibility factor shows itself in the compressor work formula, shown below in non-dimensional form:

$$\frac{\dot{w}}{RT_1} = \frac{\gamma Z}{\gamma - 1}\left[\left(\frac{P_2}{P_1}\right)^{\frac{\gamma-1}{\gamma}} - 1\right].$$

Thus, with a compressor operating near the CP, significant reduction in parasitic power consumption is possible.

At present, reliable prediction of supercritical CO_2-cycle performance via detailed heat balance simulation is subject to some uncertainty. The reason for that is the lack of

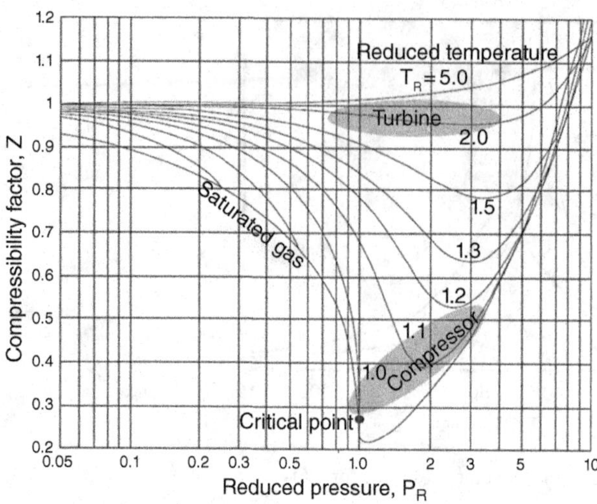

Figure 5.10 Compressibility factor of CO_2.

highly accurate equation of state data that are widely accepted by the industry (similar to the ASME steam tables). The performance of the cycle is highly sensitive to the fluid state at the pump or compressor inlet near the vapor–liquid CP, which is the key driver of the remarkable cycle efficiency (it also presents problems from a system control perspective). Thus, it is very difficult to verify and/or compare myriad performance claims via rigorous simulation models on a widely agreed-upon thermodynamic basis.

5.7 Maxwell Relations

Once an equation of state is known, any other thermodynamic property can be found by using the Maxwell relations. In engineering, the two most common independent properties are pressure and temperature, p and T, respectively, because they are readily amenable to direct measurement. Everything else (i.e., enthalpy, h, entropy, s, etc.) can be derived from the equation of state and the applicable Maxwell relations using known p and T. Calculation of enthalpy will be given as an example. We start with the definition of an "exact differential," i.e.,

$$dh = \left(\frac{\partial h}{\partial T}\right)_p dT + \left(\frac{\partial h}{\partial p}\right)_T dp. \tag{5.52}$$

This formula stems from the fact that any given thermodynamic property is a function of any two other, arbitrary thermodynamic properties; in this case, h = f(p,T). Therefore, we can evaluate "partial derivatives" of h with respect to p and T separately, while holding the other variable constant, respectively. Furthermore,

$$\frac{\partial}{\partial p}\left(\left(\frac{\partial h}{\partial T}\right)_p\right)_T = \frac{\partial}{\partial T}\left(\left(\frac{\partial h}{\partial p}\right)_T\right)_p. \tag{5.53}$$

These two mathematical "facts," combined with several a priori thermodynamic "facts," form the foundation on which four key equalities involving, p, T, v, and s, known as *Maxwell relations*, are arrived at via combination and algebraic manipulation, i.e.,

$$\left(\frac{\partial T}{\partial v}\right)_s = -\left(\frac{\partial p}{\partial s}\right)_v, \left(\frac{\partial T}{\partial p}\right)_s = \left(\frac{\partial v}{\partial s}\right)_p, \left(\frac{\partial p}{\partial T}\right)_v = \left(\frac{\partial s}{\partial v}\right)_T, \left(\frac{\partial v}{\partial T}\right)_p = -\left(\frac{\partial s}{\partial p}\right)_T.$$

From the four equations listed above, an entire library of thermodynamic correlations can be built, and they are indispensable in calculating thermodynamic properties from a known equation of state.

Since c_p is the partial derivative of enthalpy, h, with respect to temperature, T, with pressure, p, held constant, we obtain

$$dh = c_p dT + \left(\frac{\partial h}{\partial p}\right)_T dp. \tag{5.54}$$

Since equations of state are usually formulated in p, v, and T, a suitable equation for the second partial derivative is found by combining several Maxwell relations as

$$\left(\frac{\partial h}{\partial p}\right)_T = v - T\left(\frac{\partial v}{\partial T}\right)_p, \tag{5.55}$$

so that for the exact differential enthalpy, combining Equations 5.54 and 5.55, we have

$$dh = c_p dT + v - T\left(\frac{\partial v}{\partial T}\right)_p dp. \tag{5.56}$$

Armed with a known equation of state (e.g., Equation 5.51), we can evaluate the partial derivative on the right-hand side and carry out the integration to find

$$h_2 - h_1 = \int_1^2 c_p dT + \int_1^2 \left(v - T\left(\frac{\partial v}{\partial T}\right)_p\right) dp. \tag{5.57}$$

Enthalpy cannot be measured directly. In other words, there is no counterpart of a thermometer or barometer (i.e., an "enthalpymeter") that can tell us the enthalpy of a particular substance at an equilibrium thermodynamic state as X Btu/lb or kJ/kg or whatever units one prefers. Thus, strictly speaking, the statement "substance XYZ has an enthalpy of X Btu/lb" has no meaning per se. Only the difference between the enthalpies of two equilibrium states can be measured (indirectly via Equation 5.57 with p, v, and/or T measurements) and has a meaning. Nevertheless, for practical purposes, one needs to assign a value to the enthalpy at a given equilibrium state, and this is done by (arbitrarily) assigning the value 0 Btu/lb to a *reference* state.

In most property tables such as JANAF, enthalpy is listed as a *difference* per Equation 5.57 with respect to a reference temperature (usually 25°C or 77°F). Furthermore, the other assumption in generating property tables is that state 1 is at a very low pressure so that the ideal gas correlations apply. In that case, we know that

$$h_{1,id} = \int_{ref}^{1,id} c_{p,id}(T)dT, \tag{5.58}$$

where "ref" designates a "reference state" at a temperature, T_{ref}, such that enthalpy is assumed to be 0 Btu/lb. Consequently, for the enthalpy at an arbitrary state point with known p and T, we have

$$h(p,T) = h_{id}(T) + \int_{ref}^{p,T} \left(v - T\left(\frac{\partial v}{\partial T}\right)_p \right) dp. \tag{5.59}$$

The first term on the right-hand side of Equation 5.59 is the "ideal gas part of the enthalpy." The second term on the right-hand side of Equation 5.59 is the "real part of the enthalpy." (In some references, it is also referred to as a *departure function* – as in a departure from the ideal gas state.) As one could verify easily via pv = RT, $(\partial v/\partial T)_p$ = R/p and the real part in Equation 5.59 become zero for an ideal gas.

It is easy to appreciate the tedious (albeit straightforward) work involved in evaluating the "real part" via differentiation and integration, even for a very simple equation such as Equation 5.51 (i.e., the van der Waals equation). Unfortunately, equations of engineering interest with sufficient accuracy are much more complex. Luckily for us practitioners, somebody already went through all that hard work and tabulated the results for our convenience. In power plant engineering, the most important and well-known property table package is the ASME steam tables, which are now available as ready-to-use C++, FORTRAN, and many other types of dynamic link libraries, Microsoft Excel add-ins, etc.

Even when you use such libraries, however, you *must* be cognizant of the underlying theory outlined above. This will become apparent when we work through a heat balance exercise for a gas turbine with steam or water injection for NOx control.

5.8 Simple Property Calculation

For quick estimates, one clearly needs an easy-to-use transfer function to calculate exhaust gas enthalpy and exergy as a function of exhaust gas temperature. Since entropy is needed for exergy calculation, with a transfer function for exergy, a separate transfer function for entropy is not necessary. For combustion with natural gas (mostly methane), the composition effect can be ignored with no appreciable loss in fidelity.

Note that gas turbine combustion products typically contain about 11–12 percent H_2O vapor by volume (natural gas fuel with requisite excess air – see Chapter 12). At 59°F and 14.7 psia (i.e., standard ISO reference conditions), part of the H_2O condenses. This is so because the partial pressure of 11–12%(v) H_2O in the gas mixture is about 1.5 psi. At that pressure, the saturation temperature of H_2O is about 115°F, whereas 59°F corresponds to a saturation pressure of 0.25 psi. Therefore, cooling the gas below 115°F until it reaches 59°F will knock out H_2O from the mixture via condensation until its content reduces to 0.25/14.7 $\approx$ 1.7% by volume. Accounting for the presence of liquid H_2O (or not) makes a difference in exergy calculation.

Since the objective in evaluating the exergy of gas turbine exhaust gas is to determine the theoretically possible maximum work output from a bottoming cycle making use of said exhaust gas, we typically ignore the contribution of latent heat of condensation of water vapor. This is the same principle at work when choosing the lower (net) heating value (LHV) of a gas turbine fuel over the higher (gross) heating value; it is not cost-effective to make use of that extra heat content. The difference in calculated exergy is about 5–8 percent.

With that in mind, the exergy of gas turbine exhaust gas burning natural gas fuel with a "dead state" of 59°F and 14.7 psia is given by

$$a_{exh} = 0.1961 \cdot \mathrm{T}_{exh}\,[\mathrm{in}^\circ\mathrm{F}] - 86.918 \qquad \mathrm{Btu/lb.} \tag{5.60a}$$

A zero-entropy reference is defined as 59°F and 14.7 psia (since ideal gas entropy is a function of temperature *and* pressure). With a "dead state" of 77°F, the transfer function becomes

$$a_{exh} = 0.1909 \cdot \mathrm{T}_{exh}\,[\mathrm{in}^\circ\mathrm{F}] - 87.0 \qquad \mathrm{Btu/lb.} \tag{5.60b}$$

The difference in exergy calculation between the two "dead state" (reference point) assumptions is about 5 to 6 Btu/lb. The error introduced by not including the latent heat of condensation of water vapor is 7 to 9 Btu/lb.

Similarly, with a 0 Btu/lb enthalpy reference of 59°F, the enthalpy of the exhaust gas is given by

$$h_{exh} = 0.3003 \cdot \mathrm{T}_{exh}\,[\mathrm{in}^\circ\mathrm{F}] - 55.576 \qquad \mathrm{Btu/lb.} \tag{5.61a}$$

With a 0 Btu/lb enthalpy reference of 77°F, the transfer function becomes

$$h_{exh} = 0.3003 \cdot \mathrm{T}_{exh}\,[\mathrm{in}^\circ\mathrm{F}] - 59.897 \qquad \mathrm{Btu/lb.} \tag{5.61b}$$

The difference in enthalpy calculation between the two reference point assumptions is about 4.3 Btu/lb. The error introduced by not including the latent heat of condensation of water vapor is 35 to 40 Btu/lb.

Equations 5.60 and 5.61 can be used for the temperature range 900–1,200°F with reasonable accuracy. The error resulting from using these simple equations in lieu of bona fide property calculations should not be more than $\pm$1–2 percent. The information implicit in Equation 5.61 is the specific heat of the gas turbine exhaust gas (i.e., c_p = 0.3003 Btu/lb-R [1.2573 kJ/kg-K]), which is approximately, but not exactly, constant in the range of its applicability.

A very useful parameter is the ratio of exergy to energy (enthalpy). Using Equations 5.60 and 5.61, the data in Table 5.6 are generated as a function of gas turbine exhaust temperature. This is a very handy reference. It is essentially a measure of the ultimate bottoming-cycle efficiency (i.e., a Carnot cycle). Note the improvement in maximum attainable efficiency with better exhaust gas energy "quality" (i.e., increasing exhaust gas temperature). A "real" cycle (e.g., a modern steam Rankine cycle) would achieve only a fraction of this (about 0.8, to be exact, per Figure 6.4 in the next chapter).

Table 5.6 Gas turbine exhaust gas exergy/enthalpy ratio

Exhaust Temperature (°F)	Enthalpy (Btu/lb)	Exergy (Btu/lb)	E/Q
1,000	244.7	110.2	0.450
1,050	259.7	120.5	0.464
1,100	274.8	130.7	0.476
1,150	289.8	140.9	0.486
1,200	304.8	151.1	0.496

For combined-cycle calculations, we also need to estimate the enthalpy of the HRSG stack gas, which can be done by using

$$h_{stack} = 0.2443 \cdot T_{stack}\,[\text{in°F}] - 13.571\ \text{Btu/lb}, \tag{5.62a}$$

$$h_{stack} = 0.2443 \cdot T_{stack}\,[\text{in°F}] - 17.892\ \text{Btu/lb}, \tag{5.62b}$$

with a 0 Btu/lb enthalpy reference of 59°F (in Equation 5.62a; 77°F in Equation 5.62b) and H_2O in the mixture in gaseous form. Equation 5.62 is reasonably accurate for the temperature range 150–240°F. The information implicit in Equation 5.62 is the specific heat of the HRSG stack gas (i.e., c_p = 0.2443 Btu/lb-R [1.0228 kJ/kg-K]), which is approximately, but not exactly, constant in the range of its applicability. For the exergy of the HRSG stack gas, use

$$a_{stck} = 0.0519 \cdot T_{stck}\,[\text{in°F}] - 6.1539 \quad \text{Btu/lb}, \tag{5.63a}$$

$$a_{stck} = 0.045 \cdot T_{stck}\,[\text{in°F}] - 5.7748 \quad \text{Btu/lb}, \tag{5.63b}$$

with a dead state temperature of 59°F (in Equation 5.63a; 77°F in Equation 5.63b). For convenience in automated calculations (e.g., in a VB/VBA code), the zero-enthalpy reference can be chosen as absolute zero (i.e., T_{ref} = –459.67°F or 0 R). In that case, for gas turbine exhaust and HRSG stack gas enthalpies, we simply have

$$h_{exh} = 0.3003 \cdot T_{exh}\,[\text{in R}]\ \text{Btu/lb}, \tag{5.61c}$$

$$h_{stack} = 0.2443 \cdot T_{stack}\,[\text{in R}]\ \text{Btu/lb}. \tag{5.62c}$$

Two useful quantities are tabulated below in both US customary and SI units:

1. Condenser pressure and temperature (define the steam Rankine bottoming-cycle METL)
2. Steam Rankine bottoming-cycle METH
   a. For HP steam pressures of 1,500–2,400 psia
   b. Hot reheat (HRH) set at one-sixth of HP pressure
   c. Steam temperatures 1,000–1,112°F

The data in Tables 5.7–5.9 should facilitate easy calculation of ideal Rankine cycle efficiency from

Table 5.7 Typical steam turbine condenser conditions

Condenser Pressure			Saturation Temperature	
in. Hg	psi	mbar	°F	°C
1.0	0.49	33.9	79.0	26.1
1.5	0.74	50.8	91.7	33.2
2.0	0.98	67.7	101.1	38.4
2.5	1.23	84.7	108.7	42.6

Table 5.8 METH for a reheat steam Rankine cycle (US customary units)

	HP/HRH Steam Temperatures (°F)		
HP Pressure (psia)	1,000	1,050	1,112
2,400	553.1	564.9	579.4
2,100	547.2	558.7	573.0
1,800	540.1	551.3	565.2
1,500	531.2	542.2	555.8

Table 5.9 METH for a reheat steam Rankine cycle (SI units)

	HP/HRH Steam Temperatures (°C)		
HP Pressure (bara)	538	566	600
166	289.5	296.1	304.1
145	286.2	292.6	300.6
124	282.3	288.5	296.2
103	277.3	283.4	291.0

$$\eta_{\text{ideal}} = 1 - \frac{\text{METL}}{\text{METH}}. \tag{5.64}$$

Performance fuel gas heating is a feature of all modern gas turbines for power generation (e.g., in a combined cycle configuration with hot boiler feed water bled from the HRSG). For heated fuel delivered to the gas turbine combustor, the total heat input into the gas turbine control volume is the total of LHV (at 77°F) and the sensible heat evaluated by using

$$h_{\text{fuel}} = 0.5832 \cdot \text{T}_{\text{fuel}}\,[\text{in}^{\circ}\text{F}] - 44.9\,\text{Btu/lb}, \tag{5.65a}$$

which is for 100 percent methane at a zero-enthalpy reference temperature of 77°F. Similar to the gas enthalpies discussed above, Equation 5.65a assumes a constant fuel gas-specific heat of 0.5832 Btu/lb-R (2.4417 kJ/kg-K), and for a zero-enthalpy reference at the absolute zero, it becomes

$$h_{fuel} = 0.5832 \cdot T_{fuel}\,[\text{in R}]\ \text{Btu/lb}. \tag{5.65b}$$

For quick estimates, using 0.6 Btu/lb-R for fuel gas-specific heat is sufficient.

Note that while gas turbine *heat input* is

$$\dot{Q}_{in} = \dot{m}_{fuel} \cdot (LHV + h_{fuel}),$$

gas turbine *heat consumption* (HC) is

$$HC = \dot{m}_{fuel} \cdot LHV.$$

In this book (and in many other books), LHV of fuel is at a reference temperature of 77°F. Slight variations in cited values, even at the same reference temperature, are possible. For most calculations herein, a value of 21,515 Btu/lb is used for 100 percent CH_4.

Typical temperatures for performance fuel gas heating used to be 365°F (185°C) for F–class gas turbines by using IP feed water extracted from the IP economizer of the HRSG. In advanced-class gas turbines, heating up to 410–440°F (210–225°C) is common. The increase in heat content supplied to the gas turbine via fuel gas heating can be up to 1 percent higher.

Two more tables are provided with enthalpies for typical water and steam injected into the gas turbine combustor for NOx control. The pressures in Table 5.10 and Table 5.11 are determined by gas turbine PR plus 100 psia, but any combination in the tables can be used. These enthalpies come in handy when doing manual heat balance calculations (see Chapter 7).

Table 5.10 Water injection enthalpies

Gas Turbine (PR)	Water Pressure (psia)	Water Temperature (°F)		
		350	400	450
12	370	322.1	375.2	437.0[a]
15	438	322.2	375.3	430.2
18	505	322.3	375.4	430.3
21	573	322.4	375.5	430.3

[a] Saturated liquid.

Table 5.11 Steam injection enthalpies

Gas Turbine (PR)	Steam Pressure (psia)	Steam Temperature (°F)		
		550	600	650
12	370	1,280.5	1,309.8	1,337.8
15	438	1,273.6	1,304.4	1,333.4
18	505	1,266.4	1,298.7	1,328.7
21	573	1,258.8	1,292.8	1,324.0

Before continuing, let us wrap this section by continuing with the example above. Recall that we identified an F–class gas turbine with exhaust temperature 1,150°F and exhaust mass flow rate of 1,500 lb/s, but did not say anything about the output and efficiency.

Equation 5.12 with known exhaust temperature and cycle PR gives an estimate of cycle TIT (written again below for convenience):

$$\mathrm{TIT_{est}} = (\mathrm{TEXH} + 460) \cdot \mathrm{PR}^{0.2345} - 460.$$

We know that the cycle PR is set to maximize the gas turbine-specific output and is given by Equations 5.13 and 5.14 (using the numerical values of the constants c_1 and c_2 in the former) as

$$k' = 0.3093 \cdot \left(\frac{\mathrm{TIT}}{\mathrm{TIT_{ref}}}\right)^{-0.2939} \quad \text{and}$$

$$\mathrm{PR_{opt}} = \tau_3^{\frac{1}{2k'}}.$$

From these three equations, we determine PR as 20.1 and TIT as 2,792°F. Using the ideal, air-standard Brayton-cycle efficiency correlation (Equation 5.2, written again below for convenience)

$$\eta = 1 - \frac{1}{\mathrm{PR}^{\frac{\gamma-1}{\gamma}}},$$

with $\gamma = 1.4$, the *Carnot-equivalent* efficiency is found as 57.5 percent. Figure 6.1 sets the technology factor at 0.7 (a very solid number!), so our gas turbine has an efficiency of 0.7 × 57.5% = 40.3%, which corresponds to a heat rate of 8,471 Btu/kWh.

From Table 3.5, the F–class-specific power output is 200 Btu/lb, which is the ratio of power output to gas turbine airflow (i.e., roughly exhaust flow minus fuel flow). Using the "Goal Seek" feature of Microsoft Excel, we find the output as 309,400 kWe with a fuel mass flow rate of 33.84 lb/s (100 percent methane with 21,515 Btu/lb LHV) and an airflow of 1,466 lb/s by iterating on

$$\begin{array}{c} \mathrm{HC}\ [\mathrm{Btu/h}] = 8,471\ \mathrm{Btu/kWh} \times 200\ \mathrm{Btu/lb} \times 1.05506 \times (1,500 - \mathrm{MF}) \\ \mathrm{MF}\ [\mathrm{lb/s}] = \mathrm{HC}/21,515\ \mathrm{Btu/lb}/3,600 \end{array},$$

where MF is the fuel flow rate (in lb/s) and HC is heat consumption (in Btu/h).

As a final check, the ratio of the steam turbine output of 145,000 MWe to the gas turbine output is 0.47, which is in accordance with the famous one-to-two rule of thumb (more on this in Chapter 17).

5.9 Enthalpy

As was explicitly stated in the Preface, this is not an introductory book. Including a section on enthalpy might thus seem contradictory at first glance. This is why this section is placed near the end of this chapter on thermodynamics – almost as an "afterthought." Many experienced engineers use enthalpy in compressor or turbine/expander calculations automatically without giving a lot of thought to it. For most practical purposes, this is just fine. Nevertheless, a deeper understanding of enthalpy might come in handy later when discussing stage aerothermodynamics.

Enthalpy is simply flow energy (i.e., the combination of the thermal and mechanical energies of a fluid). It can be found via combining the Tds equations and integrating as

$$h = u + pv \tag{5.66a}$$

On a *specific* (per unit mass; i.e., Btu/lb or kJ/kg) basis or as

$$H = U + pV \tag{5.66b}$$

on a *total* basis (i.e., Btu or kJ). The internal energy, U, is the macroscopic manifestation of the kinetic energy of molecules constituting the substance in question. It can be derived rigorously from the statistical mechanics.

The second term, pV, can be translated into the following form below by substituting F/A for pressure, p, and Ax for volume, V (note that F is force and A is area where x is an arbitrary distance perpendicular to area A)

$$p \cdot V = F \cdot x.$$

Force times distance is total work done by said force over said distance. When you consider a cylindrical mass of fluid flowing in a pipe with velocity C (see Figure 5.11), U is the internal energy of the fluid mass, $m = \rho V = Ax/v$, and $pV = Fx$ is the energy representing the work done by the fluid to "open up" for itself a space of $V = Ax$ by exerting its pressure on its surroundings across the flow cross-sectional area of A. While U represents the "thermal" energy of the fluid mass m, pV represents its "mechanical" energy, which is *different* from the "kinetic" energy of the same as given by $\frac{1}{2}mC^2$.

When we write the generic first law equation for an adiabatic compressor or expander (turbine) with constant mass flow rate as

$$\dot{W}_{c/t} = \dot{m}(h_1 - h_2)$$

(with the implicit assumption that kinetic energy change is negligible), what we really have is

$$\dot{W}_{c/t} = \dot{m}(u_1 - u_2) + \dot{m}(p_1 v_1 - p_2 v_2), \text{or}$$

$$w_{c/t} = (u_1 - u_2) + (p_1 v_1 - p_2 v_2).$$

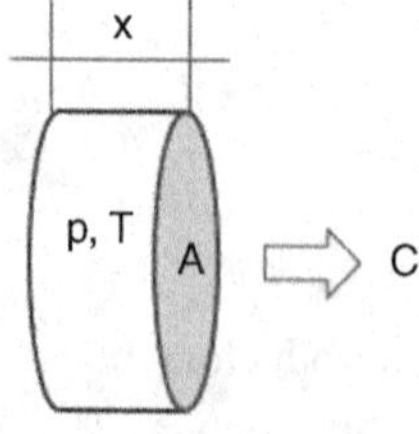

Figure 5.11 Cylindrical body of fluid (mass of m) in flow.

Table 5.12 Examples illustrating h = u + pv

		Steam			Gas		
		HP Throttle	Cold Reheat	Δ	Stator Inlet	Rotor Inlet	Δ
p_{tot}	psia	2,500	485		205	199	
T_{tot}	°F	1,112	672		2,475	2,359	
T_{dyn}	°F	Negligible	Negligible		16	427	
p_{stat}	psia	~same	~same		~same	101.6	
h	Btu/lb	1,531	1,343	188	689	524	166
p·v	Btu/lb	158.5	117.2	41	199.9	163.9	36
C	ft/s	200	200		500	2,591	
KE	Btu/lb	0.8	0.8		5.0	134.1	–129

In other words, work done in the machine, *on* the fluid or *by* the fluid, has two parts: thermal (the first term on the right-hand side) and mechanical (the second term on the right-hand side). The relative contribution of the mechanical term is illustrated via two examples:

1. Typical high-pressure section of a steam turbine.
2. Flow of hot combustion gas across the first stage stator of a gas turbine.

Calculations are shown in Table 5.12.

As was stressed earlier, in thermodynamic analysis, it is the *change* in enthalpy that has a real meaning because the absolute value is dependent on the – somewhat arbitrary – choice of a zero-enthalpy state. With that in mind, for the two examples in Table 5.12, change in p·v is about 20 percent of the enthalpy change, with the remainder coming from the internal energy, u.

Also note the negligible contribution of kinetic energy at the steam turbine inlet and exit in Table 5.12. Thus, there is no distinction between total and static values of pressure and temperature. The same can be said for the hot gas at the stator inlet with little loss in accuracy. However, acceleration of the hot gas while flowing through the stator nozzle vanes results in a high velocity just before entering the rotor. The change in kinetic energy is of the same order as the change in enthalpy (of which nearly 20 percent is due to p·v). The change in *total* flow energy, $h + C^2/2$, is about 5 percent loss mainly due to friction, heat transfer to, and mixing with cooling air. Since no work is done by the gas flowing across the stator, if the nozzle vanes were uncooled and the flow were frictionless, the total flow energy change would be zero. The change in enthalpy (reduction) would be exactly equal to the change in kinetic energy (increase). Without this translation from thermal to kinetic energy in a turbine and vice versa in a compressor, one would not have a turbomachine.

Yet, unless we investigate the stage-by-stage performance of a compressor or a turbine, we tend to ignore this fundamental aspect of turbomachines. As clearly illustrated by the steam turbine example in Table 5.12, steam velocity is totally inconsequential in most calculations (unless, of course, they are for stage-by-stage turbine design). Since this "ignoring" convention is done so frequently and so matter-of-factly, many engineers doing heat balance simulations on a daily basis cannot even cite the typical steam velocity in, say, the main steam pipe of a combined-cycle steam turbine. (In passing, 200 ft/s used in Table 5.12 is on the high side.)

One side effect of this practical simplification of the underlying physics is the mess of terminology prevalent in the technical literature. Therefore, the reader should keep the following facts in mind (even if he or she keeps working with existing – sometimes downright incorrect – technical "shorthand"):

1. *Total* temperature is a useful concept facilitated by the simplifying assumption of constant specific heat.
2. Strictly speaking, there is only the *static* temperature, which, unfortunately, *cannot* be measured directly in a flow system (whereas the total temperature *can* be measured).
3. Using an equation of state in rigorous calculations results in "total flow energy" comprising
    a. Enthalpy
    b. Kinetic energy
4. Some texts use the term "total enthalpy" in lieu of "total flow energy," which is incorrect because
    a. Enthalpy is a property of the fluid in question.
    b. It can be calculated from a known equation of state as a function of
        i. *Static* temperature only (ideal gas).
        ii. *Static* temperature *and static* pressure (real fluid).
    c. Thus, by definition, enthalpy is a static property.
5. Whenever you are using a property package such as ASME steam or JANAF gas tables to calculate enthalpy, entropy, or another thermodynamic property of a fluid in flow,
    a. The temperature and the pressure of the fluid you obtain from respective "transducers" are *total* (stagnation) values.
    b. Therefore, when plugging temperature and pressure values into the property subroutines,
        i. You either make an implicit assumption that the velocity of the fluid in question is low enough to assume that the static and total values are close enough, or
        ii. You already evaluated the contribution of the fluid kinetic energy and *extracted* the *static* values of temperature and/or pressure.
        iii. All calculated properties are, by definition, *static* values.
6. Only changes in enthalpy between two equilibrium states are meaningful quantities.
7. Absolute values of enthalpy can be used to represent the *thermal energy* of a fluid (main steam in a steam turbine power plant, gas turbine exhaust gas at the inlet to the HRSG, etc.).
8. In such cases, you *must* be aware of and/or clearly state the applicable *zero-enthalpy reference state.*

9. The most logical choice for the zero-enthalpy reference is the ambient conditions (i.e., 59°F and 14.7 psia per ISO definition), which is also the "dead state" for exergy calculations. Another useful zero-enthalpy reference for heat balance calculations is 77°F because the fuel heating values are typically quoted at that reference temperature. The difference between the two reference temperatures is 4.3 Btu/lb in enthalpy.

5.10 Maximum Power/Minimum Entropy

Several theoretical physicists have shown that the efficiency of an irreversible heat engine at maximum power output is given by the formula

$$\eta_{\max p} = 1 - \sqrt{\frac{T_L}{T_H}}, \tag{5.67}$$

where T_H and T_L are the thermodynamic temperatures of the high- and low-temperature reservoirs, respectively (e.g., see Curzon and Ahlborn [4]). In a later paper, Bejan (a very prolific mechanical engineering professor who wrote several excellent books on thermodynamics and heat transfer) demonstrated that the same result can be derived by minimizing the rate of entropy generation [5]. In his paper, Bejan made reference to an earlier paper by Landsberg and Leff, who had shown that most practical heat engine cycles, including the Brayton cycle, were special cases of a more universal "generalized" cycle and had derived the same equation [6].

Equation 5.67 is, of course, no surprise to the readers of this chapter. In Section 5.2, it was shown that the *optimum* PR for *maximum* net specific output was given by Equation 5.11:

$$PR_{opt} = \sqrt[2k]{\tau_3},$$

which can be rewritten as

$$PR_{opt} = \sqrt[2k]{\frac{T_3}{T_1}}, \text{ or}$$

$$PR_{opt} = \sqrt[2k]{\frac{T_H}{T_L}}.$$

Substituting into Equation 5.2 (note that $k = 1 - 1/\gamma$), one ends up with Equation 5.67.

In accordance with this fundamental thermodynamic result, one should expect that actual gas turbine efficiencies at ISO base load should be predicted reasonably well by the formula

$$\eta_{act} = 1 - \left(\frac{METL}{METH}\right)^{\beta}, \tag{5.68}$$

where the exponent β has a value close to 0.5. Indeed, examination of published rating data for gas turbines from three major OEMs (see Tables 6.1–6.3) confirmed this expectation with a value of 0.5943 for the exponent β (see Figure 5.12). The values

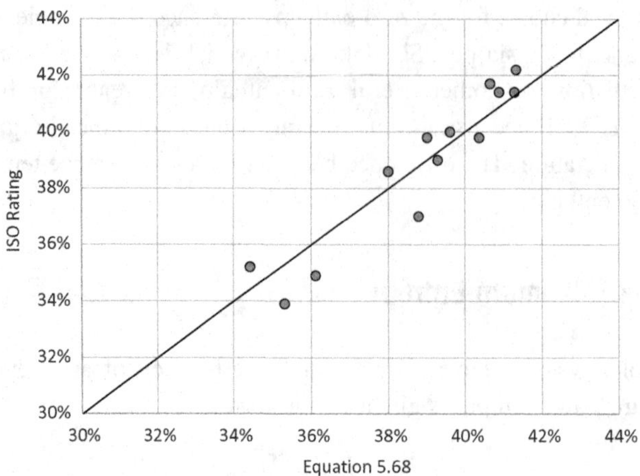

Figure 5.12 Heavy-duty industrial gas turbine efficiencies.

of METH and METL for each gas turbine are determined using the simple logarithmic mean of listed values of turbine exhaust (exit) temperature and turbine ambient temperature (59°F) with TIT and compressor discharge temperature from the cycle PR with k = 0.2857. With a little algebra, it can be shown that the corollary of Equation 5.68 is

$$\eta_{act} = 1 - PR^{-\alpha}, \quad (5.69)$$

where α is 0.1698 (note that $\alpha = k = 0.2857$ in Equation 5.2 with $\gamma = 1.4$). Thus, state-of-the-art, heavy-duty industrial gas turbine efficiency can be predicted rather accurately with ideal, air-standard Brayton-cycle formulae for a hypothetical fluid with $\gamma = 1.205$ with just one known parameter (i.e., the cycle PR).

References

1. Kano, K., Matsuzaki, H., Aoyama, K., Aoki, S., Mandai, S., Development Study of 1500°C Class High Temperature Gas Turbine. ASME Paper 91-GT-297, International Gas Turbine and Aeroengine Congress and Exposition, June 3–6, 1991, Orlando, FL.
2. Gülen, S. C., A Simple Parametric Model for Analysis of Cooled Gas Turbines. *Journal of Engineering for Gas Turbines and Power*, **133** (2011), 011801.
3. ASME Report STP-TS-012-1. Thermophysical Properties of Working Gases Used in Gas Turbine Applications. March 5, 2012, ASME Standards Technology, LLC, Air Properties Committee.
4. Curzon, F. L., Ahlborn, B., Efficiency of a Carnot engine at maximum power output. *American Journal of Physics*, **43** (1975), 22–24.
5. Bejan, A., Models of Power Plants That Generate Minimum Entropy While Operating at Maximum Power. *American Journal of Physics*, **64** (1996), 1054–1059.
6. Landsberg, P. T., Leff, H. S., Thermodynamic cycles with nearly universal maximum-work efficiencies. *Journal of Physics A: Mathematics and General*, **22** (1989), 4019–4026.

6 Technology Factors

These days, very few people, if any, attempt to design perpetual motion machines of the first or second kind (i.e., fantastic machines imagined by Rube Goldberg-type garage inventors violating the first or second laws of thermodynamics, respectively). Nevertheless, marketing hyperbole for bragging rights (i.e., the fight among original equipment manufacturers [OEMs] for the "world record holder" title) is still alive and well. Quite frequently, exaggerated claims of gas turbine power plant performance are made not only in sales-oriented articles in trade publications (essentially product advertisements in disguise), but even in (supposedly) scientific research papers in archival journals.

British Prime Minister Benjamin Disraeli (as attributed to him by Mark Twain) divided lies into three categories, the last of which was *statistics*. Herein, the author would like to add a fourth category: *thermal efficiency*. It is nearly impossible to prove or disprove cycle efficiency claims without delving into the minute details of the heat balance calculations, including myriad hardware design assumptions, thermodynamic property packages, system boundary conditions, etc. However, by using the fundamental principles developed in Chapter 5, it is eminently possible to separate the proverbial wheat from the chaff quite easily, but also rigorously and, most importantly, in an unassailable manner with absolutely zero room for "ifs, ands, or buts" by using just two parameters and a few very simple formulae.

If there is one chapter in this book that you want to pick to read over and over until you memorize it, let this chapter be the one.

If you haven't done so, please read Sections 3.5 and 3.6 prior to reading this chapter. They might clarify some questions you may have while following the key arguments made below.

Although not necessary per se, having gone through Chapter 17 before reading this chapter could be extremely beneficial. Even better is coming back to this chapter after reading Chapter 17.

6.1 State of the Art

The fundamental thermodynamic analysis in the preceding chapter showed us that the performance of "real" gas turbines can be related to the efficiency of a Carnot cycle operating between the two thermal reservoirs at T_3 (set equal to the turbine inlet temperature [TIT] of the gas turbine in question) and T_1 (set to the ambient; e.g., ISO conditions), respectively, via two parameters, i.e.,

1. *Cycle factor*, which is the ratio of ideal Brayton-cycle efficiency at cycle pressure PR (i.e., pressure ratio) and cycle maximum temperature T_3 to the efficiency of the said Carnot cycle.
2. *Technology factor*, which is the ratio of the actual gas turbine efficiency to the ideal Brayton-cycle efficiency.

In order to determine the technology factor, ISO base load rating data from a trade publication are utilized. For three major OEMs of heavy-duty industrial gas turbines, performance specifications for their 60–Hz product lines are summarized in Tables 6.1–6.3. The ISO base load rating is a "sticker" performance based on certain assumptions, which essentially reflect the "new and clean capability" of a particular gas turbine. Performances achieved at "real" site ambient and loading conditions and with normal wear and tear should be expected to differ significantly from those idealized numbers. Nevertheless, ISO ratings constitute the best gauge for the state of the art in gas turbine technology at a given point in time.

In Tables 6.1–6.3, the first five columns reflect the published data "as is." Specific work (SP WK), TIT, and the firing temperature, TFIRE, are estimated values.

Using the *Gas Turbine World* (GTW) 2016 rating data for advanced F-, H-, and J/HA–class turbines, the existing state of the art is summarized in Figure 6.1. TIT in the figure is an estimate from Equation 5.15 with OEM data for exhaust temperature and

Table 6.1 General Electric 60–Hz products (*Gas Turbine World* 2016 performance specifications)

	Output (Kw)	Efficiency	PR	TEXH (°F)	MEXH (lb/s)	SP WK (Btu/lb)	TIT °C	TIT °F	TFIRE (°F)
7E.03	91,000	33.9%	13.0	1,026	650	135	1,233	2,252	2,141
7F.04	198,000	38.6%	16.7	1,151	1,013	189	1,459	2,658	2,496
7F.05	241,000	39.8%	18.4	1,171	1,188	197	1,520	2,769	2,589
7F.06	270,000	41.4%	22.1	1,100	1,311	200	1,518	2,764	2,563
7HA.01	280,000	41.7%	21.6	1,159	1,266	215	1,576	2,868	2,661
7HA.02	346,000	42.2%	23.1	1,153	1,539	218	1,598	2,908	2,690

Table 6.2 Mitsubishi Hitachi Power Systems 60–Hz products (*Gas Turbine World* 2016 performance specifications)

	Output (Kw)	Efficiency	PR	TEXH (°F)	MEXH (lb/s)	SP WK (Btu/lb)	TIT °C	TIT °F	TFIRE (°F)
M501DA	113,950	34.9%	14.0	1,009	780	141	1,242	2,268	2,149
M501F3	185,400	37.0%	18.0	1,136	1,032	174	1,473	2,683	2,512
M501GAC	276,000	39.8%	21.0	1,143	1,349	198	1,545	2,813	2,614
M501JAC	310,000	41.4%	23.0	1,135	1,367	220	1,575	2,867	2,653

Table 6.3 Siemens 60-Hz products (*Gas Turbine World* 2016 performance specifications)

	Output (Kw)	Efficiency	PR	TEXH (°F)	MEXH (lb/s)	SP WK (Btu/lb)	TIT °C	TIT °F	TFIRE (°F)
SGT6-2000E	117,000	35.2%	12.0	990	811	139	1,169	2,137	2,040
SGT6-5000F	244,000	39.0%	18.9	1,114	1,270	186	1,469	2,676	2,499
SGT6-8000H	295,000	40.0%	19.5	1,166	1,410	203	1,540	2,803	2,614

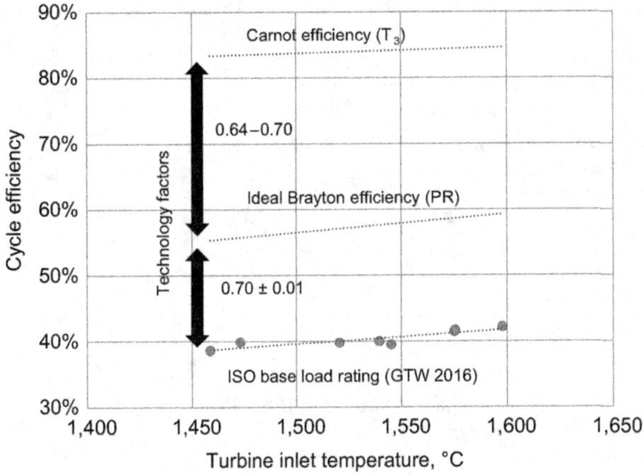

Figure 6.1 State-of-the-art gas turbine Brayton cycle and technology factors.

compressor PR. Ideal-cycle efficiency from Equation 5.2 is based on the listed PR with $k = 0.2857$ (i.e., $\gamma = 1.4$).

TIT is a technology classification parameter (e.g., see Figure 4.8 and the accompanying discussion in the chapter on gas turbine history) and, as such, it is a "public" number. Actual gas turbine *firing temperatures*, on the other hand, are closely guarded trade secrets. They are rarely (if ever) published by OEMs. One exception to that is Westinghouse (e.g., see Table 4.4). In addition, a licensed packager of industrial gas turbines by a major OEM listed the firing temperatures of their products in the 1998–1999 edition of GTW's performance specifications (Volume 19). The data in question are summarized in Table 6.4, which includes an early 1970s General Electric (GE) B-class gas turbine (1,010°C TIT), a GE Frame 5 (943°C TIT), and an old European Gas Turbines (EGT) G3142J gas turbine. Needless to say, these are pretty dated units.

Using the published values of cycle PR and exhaust temperature in Tables 4.4 and 6.4 as independent variables, the author found that the firing temperature data were reasonably well represented by a formula derived from the turbine isentropic correlation. In fact, the correlation with the published firing temperature and the associated TIT (reflecting the technology at the time) was in very good agreement with the particular OEM's technology. As such, the derived formula is deemed to be a reliable technology quantifier with a slight modification to fit the state-of-the-art gas turbines (the last term in the equation is changed from its original value of 460 to 410):

Table 6.4 Bharat Heavy Electricals gas turbines (from GTW 1998–99)

Class	PR	TFIRE °F	TFIRE °C	MEXH (lb/s)	TEXH (°F)	Output (kWe)	Efficiency
G3142J	7.1	1,730	943	115	979	10,450	25.6%
Fr. 5	10.2	1,765	963	270	909	26,300	28.5%
B	12.0	2,020	1,104	308	990	39,620	31.9%
FA	15.0	2,350	1,288	437	1,109	70,140	34.2%
E	12.3	2,055	1,124	890	1,001	123,400	33.8%

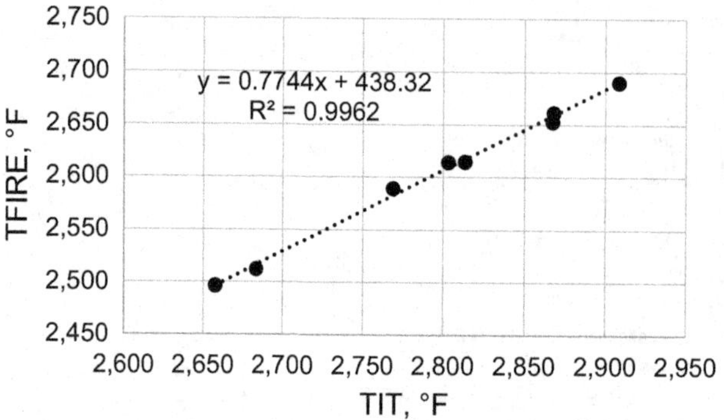

Figure 6.2 Generic correlation between turbine inlet and firing temperatures.

$$\text{TFIRE} = \frac{\text{TEXH} + 460}{0.882\left(\text{PR}^{-0.25} - 1\right) + 1} - 410. \tag{6.1}$$

Equation 6.1 can be used for a wide range of modern gas turbines (i.e., E to J class) to predict either the firing temperature from a known exhaust temperature or vice versa. As a rough check, if the TIT is known, the firing temperature should be in the following proverbial "ballpark":

- Subtract 200°F from the TIT (advanced F, H, and J class)
- Subtract up to 250°F from the TIT (early F class)
- Subtract 165°F from the TIT (for E class with PR $\leq$14)

For rough estimates of firing temperature for a given TIT, the linear transfer function in Figure 6.2 can be used. The data in the figure are taken from Tables 6.1–6.3.

Note that Equation 6.1 and the associated "rules of thumb," including the linear transfer function in Figure 6.2, are for *air-cooled* gas turbines. For H-class gas turbines with *closed-loop, steam-cooled* stator 1 nozzle vanes, the difference between TIT and TFIRE is less than 100°F.

As shown in Figure 6.1, the technology factor is a very robust number at 0.70. Since the Carnot efficiency derived from T_3 has no practical value, from now on this technology factor, which references the actual gas turbine performance to the *Carnot-equivalent*

Table 6.5 Cycle factors

	Symbol	Numerator	Denominator	Significance
Carnot Factor	CaF	Actual cycle efficiency	Carnot efficiency	Measure of "goodness" vis-à-vis an impossible theoretical ideal
Cycle Factor	CyF	Ideal Brayton-cycle efficiency	Carnot efficiency	"Possible" theoretical ideal
Technology Factor ("True Carnot Factor"	TCF	Actual cycle efficiency	Ideal Brayton-cycle efficiency	Measure of "goodness" vis-à-vis a possible theoretical ideal

performance of the ideal Brayton cycle, will be the metric representing the *true Carnot factor* (TCF). Note that, in the literature, the term "Carnot factor" is used quite frequently to quantify the "overall" deviation between a *law of physics* (i.e., the second law of thermodynamics as personified by the Carnot cycle) and a *man-made construction* (i.e., the actual gas turbine). This, however, is a fundamentally flawed approach because there is *absolutely nothing whatsoever* within the technological arsenal of humankind that can close the gap between the two cycles, which is fixed once and for all once the heat engine cycle (in this case the Brayton cycle) is chosen. Furthermore, as rigorously demonstrated in Chapter 5, the ideal Brayton cycle is in fact a *Carnot-equivalent* cycle when expressed in terms of mean effective heat addition and rejection temperatures. Thus, the TCF, as controlled by technology, must be referred to the ideal to which we *can* approach, and *not* to the ideal to which we *cannot*. A summary of the discussion in this paragraph can be found in Table 6.5 and Equations 6.2 and 6.3.

$$\mathrm{CaF} = \frac{\eta_{act}}{1 - \dfrac{T_1}{T_3}}, \mathrm{CyF} = \frac{1 - \dfrac{\mathrm{METL}}{\mathrm{METH}}}{1 - \dfrac{T_1}{T_3}}, \mathrm{TCF} = \frac{\eta_{act}}{1 - \dfrac{\mathrm{METL}}{\mathrm{METH}}}, \tag{6.2a, b, c}$$

$$\mathrm{TCF} = \frac{\mathrm{CaF}}{\mathrm{CyF}} \text{ or } \mathrm{CaF} = \mathrm{CyF} \times \mathrm{TCF}. \tag{6.3}$$

The ideal Brayton-cycle efficiency in Figure 6.1 is based on a calorically perfect gas model (i.e., the air-standard cycle), which leads to the fact that the Brayton-cycle efficiency is a function of cycle PR only. We can go one step further in "realism" and use an ideal gas model, which takes into account the temperature dependence of c_p such that the ideal Brayton-cycle efficiency is impacted by T_3 as well. In that case, the technology is found to be 0.75. With a model incorporating combustion and thus accounting for the working fluid composition change as well, the technology is calculated as 0.77. This will be looked at a bit more closely in Section 6.3.

As long as one is consistent in his or her assumptions and methods, any approach is equally valid. Since using rigorous property calculations does not add to the predictive power of technology factor methodology, due to its inherent simplicity, the ideal-cycle approach is perfectly adequate.

6.1.1 Combined Cycle

Another useful technology factor is obtained for the combined-cycle efficiency. Once again, GTW 2016 combined-cycle rating data for advanced F-, H-, and J/HA–class turbines is used for the existing state of the art, which is summarized in Figure 6.3. The ideal-cycle efficiency is based on T_1 and METH (from Equation 5.3 using listed PR, estimated TIT [T_3], and k = 0.2857). The technology factor is found to be a very solid number at 0.80. Using an ideal gas model to calculate METH does not change the value of the calculated value of the factor. Comparison of Figures 6.1 and 6.3 dramatically illustrates the power of the combined-cycle concept in bringing the "real" power plant performance rather close to the Carnot ideal. In particular, the real cycle factor is increased by a third from about 0.7 to 0.9. The key theoretical driver in this improvement is the *isothermal* heat rejection at ambient temperature. In practical terms, we can indeed get reasonably close to this ideal in the steam Rankine bottoming cycle at a condenser temperature quite close to the ambient temperature.

The data analyses summarized in Figures 6.1 and 6.3 are repeated with data from the GTW 1998–99 Handbook (Volume 19) in order to assess the advancement in technology over a period of two decades. At that time, the only 1,500°C TIT-class gas turbine was (then) Mitsubishi Heavy Industries' G–class, *steam-cooled* gas turbine. Without that engine in the mix, the technology factor for the combined cycle was 0.74, and with that engine it was 0.75 (±0.02). For the simple cycle, the technology factor was 0.65 (±0.02). Prima facie, this may not seem to be an impressive progress in technology (i.e., only an increment of 0.05 in about 20 years). Nevertheless, one should be cognizant of the absolute upper limit imposed by a law of physics (i.e., 1.0 implying a Carnot-equivalent cycle) and the significant maturity of the technology (i.e., three-quarters of a century if we use the Brown-Boveri Company (BBC) gas turbine and Jumo-004B as the starting point, or more than a century if we draw the line back to Holzwarth). As will be discussed in

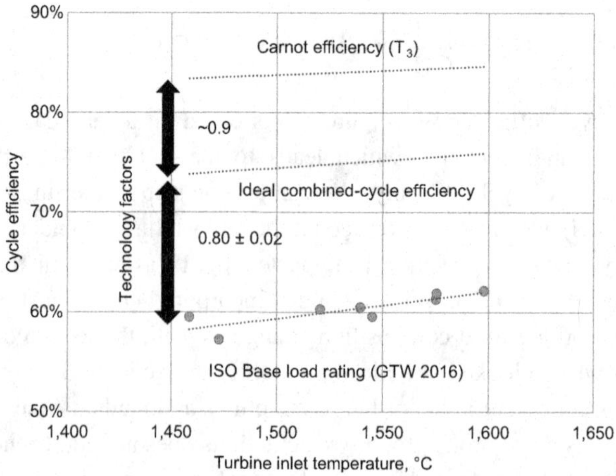

Figure 6.3 State-of-the-art gas turbine combined-cycle (CC) and technology factors.

more detail in the subsequent chapters of the book, with the exception of TIT, design engineers are rapidly running out of "knobs to turn" to bring the technology forward. This, in turn, puts the onus on materials, protective coatings, and manufacturing technologies to ensure that part lives are not adversely impacted by reduced amounts of parasitic cooling air extraction. Indeed, recent developments in ceramic matrix composite materials and advanced manufacturing (commonly known as *3D printing*), if they can be sustained and brought to commercialization, suggest that they might just do that.

6.1.2 Bottoming Cycle

What about the "real" performance of the bottoming cycle? In order to get a handle on this, we use the steam turbine power output data listed in the GTW 2016 combined-cycle ratings. Note that, due to the larger-volume flow of steam, larger steam turbines with multiple gas turbines are more efficient. Therefore, they are used as *the* reference bottoming-cycle technology. Furthermore, as will be shown later in Section 17.3, the bottoming-cycle auxiliary load (i.e., boiler feed pumps, condensate pumps, etc.) is roughly 2 percent of steam turbine generator output and must be subtracted from the listed number to arrive at the *net* bottoming-cycle output (with the exception of data for one of the OEMs, which probably already accounts for the auxiliary loads).

The exergy of the gas turbine exhaust gas, as rigorously shown in Chapter 5, is a thermodynamic property and gives the maximum possible (theoretically, that is) bottoming-cycle output. (Note that this is essentially a thermodynamic law; specifically, the second law in disguise. In other words, no specification of the bottoming cycle, working fluid, etc., is required a priori.) Exhaust gas-specific exergy for the advanced-class gas turbines in Figure 6.1 are calculated using Equation 5.60. Its product with exhaust mass flow rate (from GTW 2016 ratings in Tables 6.1–6.3) gives the total exergy in kW. Bottoming-cycle technology factors based on *net* output are plotted in Figure 6.4.

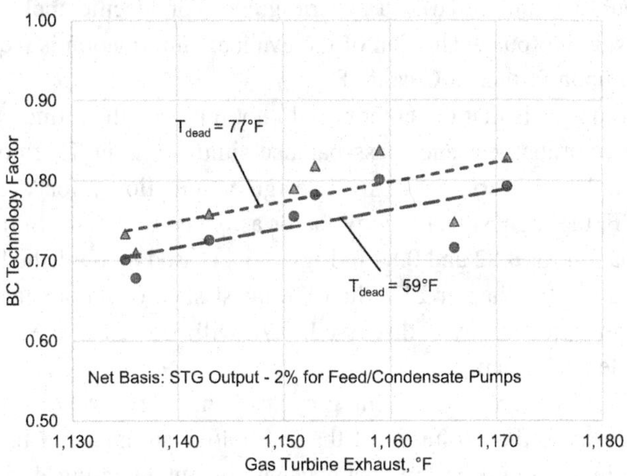

Figure 6.4 State-of-the-art steam Rankine bottoming-cycle (BC) technology factor (for different zero-enthalpy reference temperatures). STG = steam turbine generator.

Using a different zero-enthalpy reference temperature does not make a difference in exergy calculation because the formula entails enthalpy and entropy differences. Using different "dead state" definitions, however, makes a small difference (i.e., a higher dead state temperature leads to a slightly inflated exergetic efficiency). The average technology factors, representing the 2016 state of the art in steam bottoming-cycle design, are 0.74 ± 0.05 (dead state temperature of 59°F) and 0.78 ± 0.05 (dead state temperature of 77°F).

This results in a quite important revelation: on a state-of-the-art technology basis, *Brayton and Rankine cycles are equivalent to each other in terms of their respective developmental stage.*

Not only that, since a technology factor of unity implies an impossibility (i.e., an ideal Carnot-equivalent cycle), at a value of about 0.8, the unmistakable implication is that remaining opportunities for further development for either cycle are (i) very limited and (ii) very costly – *unless a step change in technology emerges.*

In regard to the bottoming-cycle technology factors in Figure 6.4, a few clarifications have to be made. First of all, gas turbine exhaust temperatures are those for simple-cycle ISO base load ratings. Those ratings are typically with zero inlet and exhaust losses (or values commensurate with simple-cycle installations). For a combined-cycle application, a three-pressure heat recovery steam generator (HRSG) imposes at least 10–12 inches of water column additional back pressure on the gas turbine. This would result in a 4–6 degrees rise in the gas turbine exhaust temperature. Rating data provided by the OEMs are not always clear on the exact assumptions used in the performance calculations (which also includes commercial margins adding to the uncertainty). Thus, splitting hairs at this point is futile.

The second thing to note in Figure 6.4 is the significant scatter of the data points. This is also a reflection of different assumptions used by the OEMs in their gas turbine and bottoming-cycle performance assumptions. Since the details provided by the OEMs are rather sparse, trying to resolve discrepancies is simply not possible. Therefore, the trend in Figure 6.4 should be considered as a generic guide. For specific applications, on a case-by-case basis, a rigorous evaluation of the available information is requisite. This will be elaborated upon further in Chapter 8.

The third point to note is that the author could not replicate the rating performance in Figure 6.4 via rigorous heat and mass balance simulations in Thermoflow Inc.'s GTPRO software. Using very aggressive design assumptions, for a gas turbine exhaust at 1,175°F, the highest technology factor achieved was 0.78. In other words, there is no explanation for 0.82 and 0.84 in Figure 6.4 (based on published rating data with 77°F dead state). In Chapter 21, when the latest state of the art in heavy-duty industrial gas turbine technology is discussed, 0.78 will be used in the evaluation of the combined-cycle performance.

How much room is there in bottoming-cycle technology growth? In order to answer that, we examine the evolution of the technology summarized in Figure 6.5. The vertical axis in the figure is the steam generator output rating divided by gas turbine exhaust exergy (using uncorrected ISO base load rating data as published in the trade journals). Thus, it reflects the *gross* bottoming-cycle technology factor. The

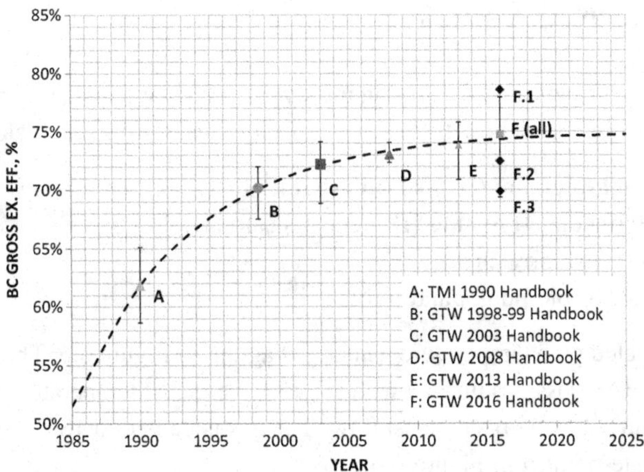

Figure 6.5 Combined-cycle bottoming-cycle (BC; as measured by steam turbine generator output) evolution (exergy dead state 59°F).

data until 2013 displayed a clearly identifiable asymptotic behavior represented by the dashed line. However, the rating data from GTW 2016 showed an inexplicable step change. In the short time period that elapsed between 2013 and 2016, there was no technology advance in steam Rankine bottoming-cycle components to explain such a radical performance improvement. In order to get a better handle on this too-good-to-be-true performance jump, 2016 is plotted for each major OEM separately (F.1, F.2, and F.3 in Figure 6.5). This clearly points to the source of the observed anomaly. One OEM essentially presented ISO base load product ratings for their combined-cycle portfolio with almost "off-the-chart" steam turbine performances (F.1 in Figure 6.5).

The reader is cautioned to consider *net* bottoming-cycle technology factors above 0.75 with extreme caution. They may be based on extremely expensive, aggressive designs (e.g., very low HRSG evaporator pinch temperature deltas) and/or aggressive component efficiencies without rigorous field validation.

One confounding factor in calculating bottoming-cycle exergetic efficiency is the presence of steam cooling of the gas turbine hot gas path and/or cooling air cooling. This applies to the following technologies:

1. GE's H-System (not commercially offered anymore)
2. MHPS's G- and J-class gas turbines
3. Sequential combustion gas turbines GT24/26 with significant cooling air cooling
4. Some Siemens gas turbines with cooling air cooling

How to account for them using the exergy concept was described at the end of Section 5.3. In the generation of Figure 6.5, the author avoided using H-System, sequential combustion gas turbines, and G/J-class gas turbines (their air-cooled GAC/JAC variants were used).

6.2 How to Use Technology Factors

Let us now look a bit closer at the bottoming cycle, which we already know to be a Rankine cycle, specifically a reheat cycle. Let us specify a state-of-the-art cycle as follows:

- 2,500 psia high-pressure (HP) steam at 1,112°F
- 400 psia intermediate-pressure (IP) reheat steam at 1,112°F
- 0.8 psia condenser pressure
- No losses, isentropic pump, and turbine

The system is depicted on a temperature–entropy diagram in Figure 6.6. The gas turbine is specified as a J/HA–class gas turbine with a 1,153°F exhaust temperature (state-point *a* in Figure 6.6) and a 1,500–lb/s exhaust flow. The gas is cooled to 180°F in an HRSG to make steam (state-point *b* in Figure 6.6).

First, we identify the three different ways to quantify the thermodynamic potential of the system in Figure 6.6 utilizing the mean effective heat addition and heat rejection concepts. For the Rankine cycle in Figure 6.6, there are *three* possible METH definitions:

1. Between gas turbine exhaust and stack temperatures (i.e., T_a and T_b, respectively).
2. Between gas turbine exhaust and ambient temperatures (i.e., T_a and T_0, respectively).
3. Between steam and water temperatures at state-points 3, 3R, and 2.

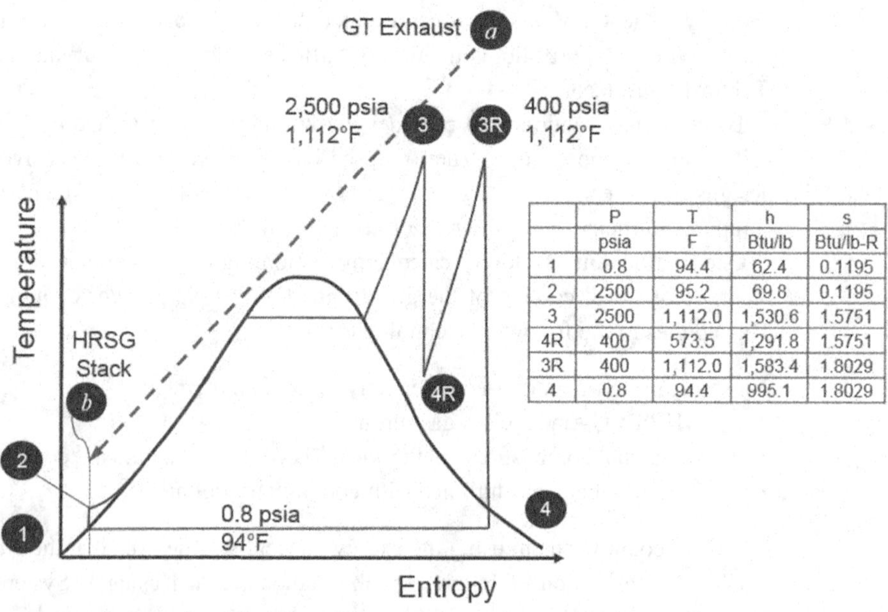

	P	T	h	s
	psia	F	Btu/lb	Btu/lb-R
1	0.8	94.4	62.4	0.1195
2	2500	95.2	69.8	0.1195
3	2500	1,112.0	1,530.6	1.5751
4R	400	573.5	1,291.8	1.5751
3R	400	1,112.0	1,583.4	1.8029
4	0.8	94.4	995.1	1.8029

Figure 6.6 Steam Rankine bottoming cycle.

Furthermore, there are *two* possible METL definitions:

1. Condenser steam temperature, $T_1 = T_4$.
2. Ambient temperature, T_0.

We calculate METH according the second definition above as follows (using the gas properties in Thermoflex):

$$METH = \frac{(h_a - h_0)}{(s_a - s_0)} = \frac{(286.7 - -4.4)}{(0.2894 - -0.0083)} - 460 = 518°F.$$

The Carnot-equivalent efficiency implied by this mean effective heat addition temperature and heat rejection at ambient temperature (i.e., the first possible METL definition above) is given by

$$\eta_1 = 1 - \frac{59 + 460}{518 + 460} = 46.9\%. \tag{6.4}$$

Note that Equation 6.4 is the translation of the gas turbine exhaust gas exergy into an efficiency term. This can be easily seen by rewriting Equation 6.4 as

$$\eta_1 = 1 - T_0 \frac{(s_a - s_0)}{(h_a - h_0)}, \tag{6.5a}$$

$$\eta_1 = \frac{(h_a - h_0) - T_0(s_a - s_0)}{(h_a - h_0)}. \tag{6.5b}$$

We immediately recognize the numerator of Equation 6.5b as the exergy of the gas turbine exhaust gas (remember, the ambient is the "dead state"). The denominator of Equation 6.5b is the maximum possible heat transfer from the exhaust gas. Thus, Equation 6.5 gives us the *theoretically possible maximum bottoming-cycle efficiency*.

Rankine-cycle METH, which is the third possible definition above, is calculated as follows using the ASME steam tables:

$$METH = \frac{(h_3 - h_2) + (h_{3R} - h_{4R})}{(s_4 - s_2)} = \frac{(1531 - 70) + (1583 - 1292)}{(1.803 - 0.12)} - 459.67 = 581°F.$$

Since METL for the Rankine cycle is the steam condensing temperature of 94°F, which is the second possible METL definition above, another definition of ideal (i.e., Carnot-equivalent) Rankine-cycle efficiency is

$$\eta_2 = 1 - \frac{94 + 460}{581 + 460} = 46.8\%. \tag{6.6}$$

According to Figure 6.6, heat taken from the gas turbine exhaust gas is

$$\dot{Q}_{exh} = \dot{m}_{exh}(h_a - h_b). \tag{6.7}$$

Similarly, heat given to the Rankine-cycle working fluid (i.e., water/steam) is

$$\dot{Q}_{stm} = \dot{m}_{stm}((h_3 - h_2) + (h_{3R} - h_{4R})). \tag{6.8}$$

The ratio of the two is the "heat recovery effectiveness," i.e.,

$$\eta_{HR} = \frac{\dot{Q}_{stm}}{\dot{Q}_{exh}}. \tag{6.9}$$

Based on the technology factor (i.e., the TCF) curve in Figure 6.4, actual bottoming-cycle work is given by

$$\dot{W}_{BC} = TCF \cdot \dot{m}_{exh} a_{exh}. \tag{6.10a}$$

From the numerator of the term on the right-hand side of Equation 6.5b, we find the exhaust gas exergy as

$$a_{exh} = (h_a - h_0) - T_0(s_a - s_0), \tag{6.11a}$$

$$a_{exh} = \eta_1 (h_a - h_0). \tag{6.11b}$$

Substituting into Equation 6.10a, we obtain

$$\dot{W}_{BC} = TCF \cdot \eta_1 \cdot \dot{m}_{exh}(h_a - h_0). \tag{6.10b}$$

Note that the term in parentheses on the right-hand side of Equation 6.10b is the *maximum possible heat transfer from the gas turbine exhaust gas per unit mass*, i.e.,

$$\dot{Q}_{exh,\ max} = \dot{m}_{exh}(h_a - h_0). \tag{6.12}$$

Based on Equations 6.6 and 6.8, the *maximum possible* bottoming cycle work is

$$\dot{W}_{stm,\ max} = \eta_2 \cdot \dot{m}_{stm}((h_3 - h_2) + (h_{3R} - h_{4R})). \tag{6.13}$$

Consequently, the *actual* bottoming-cycle work can be found via another technology factor, say, TCF′, i.e.,

$$\dot{W}_{BC} = TCF' \cdot \eta_2 \cdot \dot{m}_{stm}((h_3 - h_2) + (h_{3R} - h_{4R})), \tag{6.14a}$$

$$\dot{W}_{BC} = TCF' \cdot \eta_2 \cdot \dot{Q}_{stm}, \tag{6.14b}$$

$$\dot{W}_{BC} = TCF' \cdot \eta_2 \cdot \eta_{HR} \cdot \dot{Q}_{exh}. \tag{6.14c}$$

Combining Equations 6.10b and 6.14, the second technology factor is found as

$$TCF' = \frac{TCF \cdot \eta_1 \cdot \dot{m}_{exh}(h_a - h_0)}{\eta_2 \cdot \dot{m}_{stm}((h_3 - h_2) + (h_{3R} - h_{4R}))}, \tag{6.15a}$$

$$TCF' = TCF \frac{\eta_1}{\eta_2} \cdot \frac{1}{\eta_{HR}} \cdot \frac{(h_a - h_0)}{(h_a - h_b)}. \tag{6.15b}$$

To a very good approximation, $\eta_1 \approx \eta_2$, and typical heat recovery effectiveness is 0.995, so that TCF′ ≈ 1.11·TCF or, based on the state-of-the-art technology line in Figure 6.4, TCF′ is ~0.83 at a gas turbine exhaust temperature of ~1,150°F.

Continuing with the example, from Equation 6.7 we calculate the heat transfer from the exhaust gas as

$$\dot{Q}_{exh} = 1500 \cdot (290.7 - 30.4) = 390,400 \text{ Btu/s}.$$

Finally, using Equation 6.14c, the bottoming-cycle net (actual) output is found as

$$\dot{W}_{BC} = TCF' \cdot \eta_2 \cdot \eta_{HR} \cdot \dot{Q}_{exh}, \text{ or}$$

$$\dot{W}_{BC} = TCF \cdot \eta_1 \cdot \frac{(h_a - h_0)}{(h_a - h_b)} \cdot \dot{Q}_{exh},$$

$$\dot{W}_{BC} = 0.75 \cdot 46.9\% \cdot 1.11 \cdot 390,400 \cdot 1.05506 = 160,800 \text{ kWe},$$

where 1.05506 is the unit conversion factor.

We could have found this result fairly quickly via the property transfer function of Equation 5.60, i.e.,

$$a_{exh}[\text{Btu/lb}] = 0.1961 \cdot T_{exh}[\text{in}^\circ\text{F}] - 86.918.$$

Thus, gas turbine exhaust exergy is

$$A_{exh} = 1500 \cdot (0.1961 \cdot 1153 - 86.918) \cdot 1.05506 = 220,273 \text{ kWe}.$$

Using the TCF of ~0.75 obtained from Figure 6.4 at 1,153°F, we have

$$\dot{W}_{BC} = 0.75 \cdot 220,273 = 165,200 \text{ kWe}.$$

The difference of 4,400 kWe (about 2.6 percent) in the results is due to the property calculation using a property package versus a simple transfer function.

6.3 Is the Technology Factor a Fudge Factor?

The short answer to this question is "no." Even though the manner in which it is introduced in this chapter might evoke a feeling of incredulity, there is, of course, a rigorous approach to deriving the technology factor for any given thermodynamic cycle. Such a derivation for the Brayton-cycle technology factor can be found in the chapter written by U. Desideri, who named it *internal efficiency* (see chapter 3 of *Fundamentals of Gas Turbine Cycles* in [11] of Chapter 2).

Earlier in this chapter, the technology factor for the gas turbine Brayton cycle was defined as (Equation 6.2c)

$$TCF = \frac{\eta_{act}}{1 - \dfrac{METL}{METH}},$$

which can also be written as

$$TCF = \frac{\eta_{act}}{1 - PR^{-k}} \text{ with } k = 1 - \frac{1}{\gamma}.$$

An approximate formulation of the actual Brayton-cycle efficiency will be developed in Chapter 8 (Equation 8.3). Herein, it is presented in its final form

$$\eta = \frac{\left(1 - PR_c^{\frac{k_a}{\eta_c}}\right) + (1+f)\chi_g\tau_3\left(1 - \frac{1}{PR_t^{\eta_t \cdot k_g}}\right)}{f \cdot \ell}.$$

Ignoring the difference between compressor and turbine PRs and specific heats and specific heat ratios of air and gas, and using *isentropic* component efficiencies instead of their *polytropic* counterparts (the difference between the two will be elaborated upon in Section 7.2), we end up with

$$\eta_{act} = \frac{\left(1 - PR^k\right) + (1+f)\tau_3\left(1 - PR^{-k}\right)\eta_t\eta_c}{\eta_c f \cdot \ell}. \tag{6.16}$$

In Equation 6.16, f is the fuel–air mass ratio with

$$\ell = \frac{LHV}{c_{p,a}T_1} \text{ and}$$

$$\tau_3 = \frac{T_3}{T_1}.$$

Using the heat balance around the combustor and ignoring the contribution of f in the numerator (about 0.025 for state-of-the-art, heavy-duty industrial gas turbines), Equation 6.16 can be reformulated as

$$\eta_{act} = \frac{\left(1 - PR^k\right) + \tau_3\left(1 - PR^{-k}\right)\eta_t\eta_c}{\eta_c(\tau_3 - 1) - \left(PR^k - 1\right)}. \tag{6.17}$$

Substituting Equation 6.17 into the definition of the TCF, we obtain

$$TCF = \frac{\tau_3\eta_t\eta_c - PR^k}{\eta_c(\tau_3 - 1) - \left(PR^k - 1\right)}. \tag{6.18}$$

Equation 6.18 is the "rigorous" definition of the Brayton-cycle technology factor. For the latest F-, H-, and J-class gas turbines with

- $\eta_t = 0.84$, $\eta_c = 0.89$ (isentropic efficiencies)
- $k = 0.2857$ (i.e., $\gamma = 1.4$)

Equation 6.18 returns about 0.70, which is the average in Figure 6.1. For vintage pre-E-, E-, and F-class gas turbines, with $\eta_t = 0.83$, $\eta_c = 0.88$, on average, Equation 6.18 returns 0.62, which is in good agreement with the Westinghouse gas turbine data in Table 4.4.

Similarly with component isentropic efficiencies of 79.5 and 78.0 percent for the turbine and the compressor, respectively, Equation 6.18 returns 0.52 for Jumo-004 (see Section 4.3), with a cycle PR of 3.14 and a TIT of 1,427°F. This is essentially the same as the technology factor found from the engine data (i.e., 0.53) (see Table 4.1).

Clearly, Equation 6.18 is a very handy tool to estimate the technology factor of the Brayton cycle with the two cardinal cycle parameters (i.e., PR and TIT) and component

Table 6.6 Component isentropic efficiency progress

	1940s	1970s–1990s	1990s–2000s
Turbine	79.5%	83%	84%
Compressor	78%	88%	89%
Technology Factor	0.53	0.62 ± 0.02	0.70 ± 0.01

Table 6.7 Ideal Brayton cycles

	Perfect Gas	Ideal Gas	
Working Fluid	Generic	Atmospheric Air	
Constant c_p	Yes	No	
Combustion	No	No	Yes – 100% CH_4
TIT	2,912	2,912	2,912
PR	23	23	23
η	59.2%	55.4%	54.0%
METH	1,694	1,701	NA
METL	419	503	NA

isentropic efficiencies. The historical development trend is summarized in Table 6.6. If the values in the table seem too low for the later-vintage technology, please be aware that the isentropic efficiency for the compressor is a function of the cycle PR, whereas the turbine isentropic efficiency is a function of the cycle PR *and* the amount of chargeable cooling air. Consequently, the progress in the aerodynamic design of the components, which is reflected by the *polytropic* efficiency, is more difficult to discern in the *overall* performance of the component as reflected by the *isentropic* efficiency.

If you are still not fully convinced, please return here after reading Sections 7.2 and 10.4. For a concrete numerical example, please refer to Table 10.12, which lists the uncooled and cooled first-stage efficiencies for advanced 60-Hz gas turbines.

Note that the technology factor in this discussion (and in Figure 6.1) is predicated on the ideal, air-standard Brayton cycle, whose efficiency is based on the cycle PR only. The underlying assumption is that the working fluid (i.e., air) specific heat is constant (i.e., air is treated as a calorically perfect gas). Another ideal-cycle definition relaxes this condition, i.e.,

$$c_p = f(T).$$

Yet another ideal-cycle definition can be made with cycle heat input via combustion so that compressor and turbine working fluid compositions are different. These three types of ideal-cycle definitions are summarized in Table 6.7. (They are also covered in detail in Chapter 5.) Consequently, depending on which cycle one uses as a reference, the technology factor can vary between 0.70 and 0.77. As long as one is consistent in his or her choice of reference, either approach is equally valid. In this book, the choice of reference is the "cold" air-standard Brayton cycle (i.e., constant c_p) because of its fundamental importance and theoretical simplicity.[1]

6.4 Takeaways

An industrial gas turbine generator is defined by its four *cardinal* parameters:

1. Net generator output
2. Net generator efficiency
3. Exhaust mass flow rate
4. Exhaust temperature

Borrowing from management terminology, they are the *key performance indicators* (i.e., measurable values that demonstrate how effectively the gas turbine operates). Those four parameters are typically obtained from trade publications such as GTW or *Turbomachinery International* at an ISO base load rating point. Armed with this information, one can immediately determine the following performance information with *very high accuracy*:

1. Gas turbine exhaust exergy in kWe, which is the maximum bottoming-cycle work dictated directly by the second law of thermodynamics.
2. Actual net bottoming-cycle power output via a technology factor (see Figure 6.4).
3. Steam turbine generator output (using an approximately 2–percent auxiliary load factor; this will be elaborated upon later in the book – see Section 17.3).
4. Combined-cycle *gross* output via summation of gas turbine and steam turbine generator outputs.
5. Combined-cycle *net* output via a plant auxiliary load factor (1.6 percent for ISO base load rating purposes; higher for realistic estimates – see Section 17.3).

The consistency of the listed rating performance data (or any gas turbine performance) for air-cooled gas turbines can be checked rigorously via the following means:

1. Cycle PR and TIT via Figure 5.3 or Figure 5.5.
2. Cycle PR and specific power output via Figure 5.4.
3. Cycle PR and TIT with net simple-cycle efficiency via Figure 6.1.
4. TIT and net combined-cycle efficiency via Figure 6.3.

Gas turbines with steam cooling (i.e., MHPS G- and J–class gas turbines) and sequential combustion gas turbines (GE's GT24 and Ansaldo's GT26) require a closer look and some adjustments, which can be best assessed via rigorous but straightforward heat and mass balance analysis (see Chapter 7).

Exaggerated, "futuristic" performance claims, new "inventions," and "game-changing" technology developments that fall too far away from the established technology factors must be treated with utmost skepticism. As discussed in detail in Chapter 4,

[1] For instance, if the combustion variant is repeated with 100 percent H_2 fuel, the cycle efficiency is 55.1 percent.

technology progress over the last three-and-a-half decades is *utterly devoid* of step changes and displays a truly *evolutionary* growth in gas turbine simple- and combined-cycle efficiency (i.e., refer to Figures 4.9 and 4.10). We demonstrated this with a "deep-dive" examination of steam turbine rating data from GTW 2016 (i.e., see Figure 6.5 and the accompanying discussion).

The technology factors developed in this chapter are culminations of the second law of thermodynamics. They already indicate the high degree of maturity of the underlying technology, which, as measured by the technology factor, increased from about 0.5 to 0.7 in 75 years for the Brayton-cycle gas turbine (e.g., refer to Table 4.1 or Figure 6.1). In terms of the combined cycle, the technology factor is just short of 0.8. In short, to expect large improvements in technology is unrealistic.

There are three modifications to the basic Brayton cycle that can result in significant efficiency gain:

1. (Quasi-)constant volume combustion
2. Reheat combustion
3. Recuperation

Since the third one is a "dud" in terms of combined-cycle efficiency and since by far the largest deployment of large gas turbines for electric power generation is in the combined-cycle configuration, it is of little practical value.

As one would expect from the discussion of the thermodynamic fundamentals in Chapter 5, the first two cycle modifications attack the largest source of cycle irreversibility. Neither concept is new. In fact, reheat is a textbook cycle variant that has been available in a commercial product. *Sequential combustion* GT24 and GT26 gas turbines were introduced in the late 1990s by Asea Brown-Boveri (which became Alstom until the 2015 acquisition of the former by GE and the divestiture of GT26 to Ansaldo). Unfortunately, hasty introduction of the technology resulted in significant field issues, which, although eventually resolved, hampered wide-scale market acceptance – especially in the 60-Hz market. Furthermore, the excessive cooling air requirement of the large hot gas path (two combustors and five turbine stages) prevented the realization of the true thermodynamic potential of the concept. This made the complex and costly product even less palatable to a wide segment of users. Nevertheless, with the advent of new materials, coatings, and manufacturing technologies, a reheat gas turbine might have a chance to be the "game changer" it was supposed to be to begin with.

Constant-volume combustion was already present at the very beginning with Holzwarth's explosion gas turbines in the early twentieth century. Pulse(d) detonation, wave rotor, rotating detonation and other variants of "pressure gain combustion" have been investigated since the 1950s, primarily with an eye to applications in aircraft propulsion. The implementation of a pressure gain combustor in a stationary gas turbine for electric power generation promises significant efficiency gains (e.g., 2.0–2.5 percentage point gains in combined-cycle efficiency) without going beyond state-of-the-art H- and J-class gas turbine TITs.

7 Heat and Mass Balance

Gas turbine heat and mass balance analysis is one of the most useful tools available to practicing engineers. It is based on the application of the two conservation laws (i.e., mass and energy) to the control volume encompassing the gas turbine generator. There are many uses for it, but the ultimate goal is to ascertain the consistency of data (from performance tests, published ratings, calculations by others, etc.) based on fundamental laws of physics. As such, it deserves its own chapter in a book on gas turbines.

7.1 Control Volume

The basis of heat and mass balance analysis is the *control volume* approach. Control volume is an imaginary box drawn around the system of interest (i.e., in this case, the gas turbine). All material and energy streams crossing the control volume must be identified individually. Once this is done, then

- The first law of thermodynamics dictates that the sum of all entering and exiting energy streams in a steady-state, steady-flow (SSSF) process is exactly equal to zero.
- The conservation of mass principle dictates that the sum of all entering and exiting mass streams in a SSSF process is exactly equal to zero.

Both conservation laws hold in transient, unsteady processes with mass and/or energy storage inside the control volume as well. However, in such cases, the sums do *not* equal to zero; they are nonzero and can be negative or positive depending on the particular process. The nonzero number represents the time rate of change of energy and mass, respectively, in the control volume.

The secret to an accurate control volume heat and mass balance analysis lies in paying close attention to the following three items:

1. Ensuring that the process can indeed be approximated as SSSF.
2. Identification of *all* energy and material streams crossing the system boundary.
3. Proper evaluation of material properties via a suitable equation of state.

It must be stressed that a "true" SSSF condition does *not* exist. In order to be deemed a SSSF condition for performance testing in the field, according to ASME PTC 22-2014,

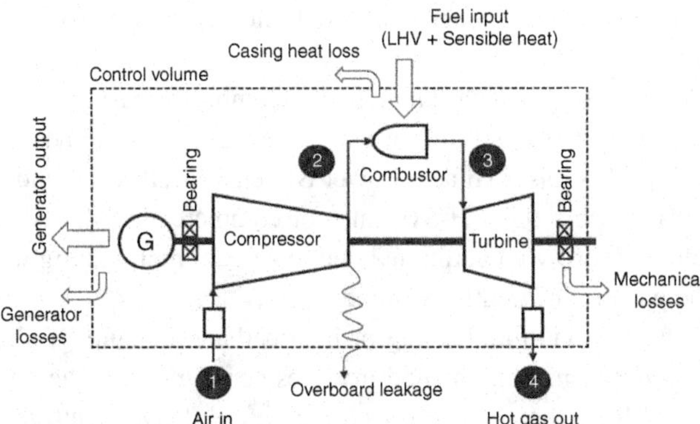

Figure 7.1 Gas turbine control volume. LHV = lower (net) heating value.

over a period of 30 minutes, average readings of power output at generator terminals and fuel flow should have a sample standard deviation of 0.65 percent each or less. For a 300−MWe H/J-class gas turbine (about 41 percent net thermal efficiency), this corresponds to 300 ± 2 MWe and 32.23 ± 0.21 lb/s natural gas flow rate.

A typical heavy-duty, industrial gas turbine control volume is shown in Figure 7.1. In addition to those shown in Figure 7.1, one should be cognizant of other possible streams crossing the gas turbine control volume, e.g.,

1. *Water or steam injection* into the combustor for NOx control. This is typically quite rare for modern gas turbines with Dry-Low-NOx (DLN) combustors. However, when present, its inclusion requires the utmost care (unless a simulation software is used) due to the differences in property packages used for water/steam (i.e., ASME steam tables) and air/gas (e.g., JANAF tables). This will be illustrated later in the chapter with a rigorously worked-out example.
2. *Exhaust frame blower* for cooling the aft bearing and the diffuser frame (typically quite small).
3. Gas turbine inlet conditioning via inlet *evaporative cooler* or *inlet fogger*. These are devices that cool the inlet air via introducing cold water into the air stream. The easiest way to handle them is to keep them *outside* the control volume (i.e., upstream of state-point 1).
4. In *wet compression*, water is injected directly into the compressor, typically downstream of stage 2. In the compression process, injected water evaporates and its latent heat of evaporation is taken from the hot compressed air with a net "intercooling" effect. Sometimes, *inlet fogging with overspray* is also referred to as wet compression. The distinction between the two processes must be understood, i.e.,
   a. Inlet fogging with overspray takes place *upstream* of state-point 1. Water sprayed into the air stream is beyond that required to fully saturate it (i.e., relative humidity of 100 percent) so that unevaporated water droplets are sucked into the compressor and evaporate during compression.

   b. In wet compression, water is injected into the compressor *downstream* of state-point 1.
5. *Steam cooling* of hot gas path components (combustor casing, transition piece, stator vanes, rotor buckets, etc.) via a *closed-loop* arrangement. This is the scheme used in Mitsubishi Hitachi Power Systems (MHPS) G- and J-class units and General Electric's (GE) H-System (not commercially offered anymore). It can be accounted for by an additional energy stream representing the heat taken out from the gas turbine control volume.
6. *Cooling air cooling* via fuel heating or intermediate-pressure (IP)/high-pressure (HP) steam generation. This method involves cooling of the chargeable cooling flows extracted from the compressor casing before delivery into the cooled hot gas path parts. Typical heat sinks are combustor fuel gas or IP or HP feed water extracted from the heat recovery steam generator (HRSG). In the latter case, IP or HP steam generated in a kettle-type evaporator is sent back into the bottoming cycle for additional power generation in the steam turbine. Similar to the steam cooling, it can be accounted for by an additional energy stream representing the heat taken out from the gas turbine control volume. Cooling air coolers (CACs) are integral parts of *sequential (reheat) combustion* gas turbines (GT24 and GT26, formerly Alstom, now owned by GE and Ansaldo, respectively) and GE's H-System. Some older MHPS frame machines have CACs with fuel gas heating. Siemens F-class gas turbines have CACs with low-pressure (LP) steam production in a kettle evaporator.

A detailed calculation methodology for gas turbine control volume heat balance calculations can be found in performance test codes ASME PTC 22 and ISO-2314. They are fairly straightforward, albeit tedious, to implement in an Excel spreadsheet. For commercial acceptance tests and other similar commercial activities, their use is usually contractually enforced.

In this section, the reason for our interest in heat and mass balance analysis is so that it can help us glean some more insight into the basic industrial gas turbine technology features. Before moving on, let us define a few important parameters using the component control volume in Figure 7.2, which depicts a stand-alone compressor driven by an electric motor. In particular,

1. The component is *adiabatic* (i.e., there is no heat transfer across the control volume).
2. *Compression work* is found by the enthalpy difference between state-points 1 and 2.
3. Mechanical losses (e.g., friction losses in the bearings) are accounted for by a mechanical efficiency term. Compression work multiplied by mechanical efficiency gives us *shaft work*.
4. Synchronous ac machine losses (electrical and mechanical) are accounted for by a motor efficiency term (generator efficiency for an expander).
   a. Compressor shaft work *divided* by motor efficiency gives *motor power* (input).

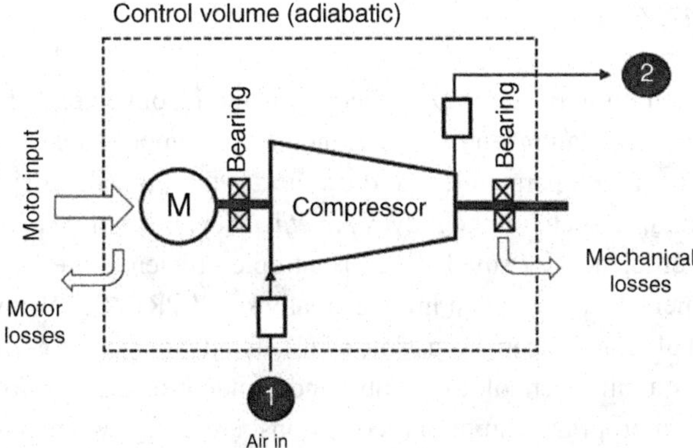

Figure 7.2 Component control volume (specifically, a compressor).

b. Expander shaft work *multiplied* by generator efficiency gives *generator power* (output).

Typical values (unless provided by the manufacturer) are

1. 99–99.5 percent for mechanical efficiency
2. 99 percent for motor or generator efficiency

Thus, on a *per unit mass flow rate* basis,

$$w_{mot} = \frac{\eta_{mech}}{\eta_{mot}}[h_2(T_2) - h_1(T_1)], \tag{7.1}$$

$$w_{gen} = \eta_{gen}\eta_{mech}[h_1(T_1) - h_2(T_2)]. \tag{7.2}$$

Note that the convention used above ensures that calculated motor (input) and generator (output) powers are *both* positive. Furthermore,

$$w_{mot} = \frac{\eta_{mech}}{\eta_{mot}}\frac{[h_{2s}(T_{2s}) - h_1(T_1)]}{\eta_{c,s}}, \tag{7.3}$$

$$w_{gen} = \eta_{gen}\eta_{mech}\eta_{t,s}[h_1(T_1) - h_{2s}(T_{2s})], \tag{7.4}$$

where $\eta_{c,s}$ and $\eta_{t,s}$ are isentropic compressor and turbine efficiencies, respectively, with reference to a (hypothetical) isentropic end state such that

$$s_{2s}(p_2, T_{2s}) = s_1(p_1, T_1), \tag{7.5a}$$

$$s_2(p_2, T_2) > s_1(p_1, T_1). \tag{7.5b}$$

In some texts, documents, etc., the isentropic efficiency is referred to as the *adiabatic* efficiency. This is not incorrect per se, but it is incomplete and misleading. An isentropic process *must* be adiabatic, whereas an adiabatic process can be irreversible and thus, as discussed in detail in Section 5.5, non-isentropic.

7.2 Polytropic or Isentropic?

Note that using the definition of *isentropic efficiency* is more convenient when using an equation of state for calculating enthalpies (function of temperature) and entropies (function of pressure and temperature). The drawback is that its value depends on the pressure ratio (PR) and, as such, unlike *polytropic efficiency*, it is not a *pure* technology number. Polytropic efficiency is "small-stage" isentropic efficiency (i.e., the isentropic efficiency of an elemental stage with an infinitesimally small PR). (See Saravanamuttoo et al. for a very detailed and clear discussion of it.) Its advantage vis-à-vis isentropic efficiency is that it is a pure technology number independent of the component PR. In fact, provided that appropriate similarity conditions are met, two compressors or turbines designed using the same basic assumptions and boundary conditions would have the same polytropic efficiency. The disadvantage is that it makes calculations quite tedious, albeit straightforward, as long as it is known a priori.

Using the polytropic efficiency, component power calculation can be done by dividing the component in question into N small stages, say, N = 10, so that each stage's isentropic calculation can be done individually. Thus, for instance, for an expander, we have

$$w_{gen} = \eta_{gen}\eta_{mech}\sum_{i=1}^{N}\Delta h_i, \tag{7.6}$$

$$PR_i = \frac{p_{i,2}}{p_{i,1}} = PR^{\frac{1}{N}}, \tag{7.7}$$

$$s_{i,2s}\left(p_{i,2}, T_{i,2s}\right) = s_{i,1}\left(p_{i,1}, T_{i,1}\right), \tag{7.8}$$

$$\Delta h_{i,s} = h_{i,1} - h_{i,2s}, \tag{7.9}$$

$$h_{i,2} = h_{i,1} - \eta_{t,p}(h_{i,1} - h_{i,2s}), \tag{7.10}$$

$$h_{i=1,1} = h_1, h_{i=N,2} = h_2, h_{i+1,1} = h_{i,2}. \tag{7.11}$$

As an example, consider a steam expander with a PR of 4 and an isentropic efficiency of 85 percent utilizing steam at 1,600 psia and 1,000°F. Using ASME steam tables, the expansion power per unit mass flow rate of steam is calculated as 151.2 Btu/lb (see Table 7.1).

Let us now divide this expander into 10 imaginary stages, each with a PR of 1.149. The calculation now becomes a series of 10 consecutive "mini-expanders" as shown in Table 7.2 with a polytropic efficiency of 83.1 percent (found by using Microsoft Excel "Goal Seek" so that the expansion power is the same as before, i.e., w = 1,486.9 – 1,335.7 = 151.2 Btu/lb).

The fundamental correlation between the isentropic and polytropic efficiencies of an expander can be derived as

$$\eta_{isen} = \frac{1 - PR^{-\eta_{poly}k}}{1 - PR^{-k}}. \tag{7.12}$$

Table 7.1 Simple steam expander

	p (psia)	T (°F)	h (Btu/lb)	s (Btu/lb-R)	w (Btu/lb)	η
1	1,600	1,000	1,486.9	1.5916		
2s	400	603	1,309.0	1.5916		
2	400	650	1,335.7			
					151.2	85.0%

Table 7.2 Stage-by-stage steam expander calculation

Stage	p (psia)	T (°F)	h (Btu/lb)	s (Btu/lb-R)
1,1	1,600	1,000	1,486.9	1.5916
1,2s	1,393	955	1,466.6	1.5916
1,2 = 2,1	1,393	961	1,470.0	1.5941
2,2s	1,213	917	1,450.2	1.5941
2,2 = 3,1	1,213	922	1,453.6	1.5965
3,2s	1,056	879	1,434.2	1.5965
3,2 = 4,1	1,056	885	1,437.5	1.5989
4,2s	919	843	1,418.7	1.5989
4,2 = 5,1	919	848	1,421.8	1.6014
5,2s	800	808	1,403.5	1.6014
5,2 = 6,1	800	813	1,406.6	1.6038
6,2s	696	773	1,388.6	1.6038
6,2 = 7,1	696	778	1,391.7	1.6063
7,2s	606	740	1,374.2	1.6063
7,2 = 8,1	606	745	1,377.2	1.6088
8,2s	528	707	1,360.1	1.6088
8,2 = 9,1	528	712	1,363.0	1.6112
9,2s	459	676	1,346.4	1.6112
9,2 = 10,1	459	681	1,349.2	1.6137
10,2s	400	645	1,333.0	1.6137
10,2	400	650	1,335.7	1.6162

Similarly, for a compressor

$$\eta_{\text{isen}} = \frac{\text{PR}^{k} - 1}{\text{PR}^{k/\eta_{\text{poly}}} - 1}. \tag{7.13}$$

As will be demonstrated by numerical examples below,

- Compressor polytropic efficiency is a few percentage points *higher* than the corresponding isentropic efficiency. This is also known as the *reheat effect.* In other words, the temperature increase associated with entropy production in stage *i – 1* is detrimental to the work consumption in stage *i.*
- Expander (turbine) polytropic efficiency is a few percentage points *lower* than the corresponding isentropic efficiency. This is also known as the *heat recovery effect.* In other words, the temperature increase associated with entropy production in stage *i – 1* is beneficial for the work production in stage *i.*

Since using one or the other efficiency definition results in the same outcome, the question becomes when to use which. As mentioned earlier, polytropic efficiency can be used as a *quantifier* of the state-of-the-art technology in component design. As such, it should be preferred over the isentropic efficiency when performing conceptual design study calculations. (The corresponding isentropic efficiency can be calculated from Equations 7.12 and 7.13.) When reverse engineering an existing design, analyzing test data, or doing stage-by-stage design calculations with rigorous treatment of stator and rotor rows, extraction flows, cooling flows, etc., the isentropic efficiency should be the preferred definition.

For the simple steam expander example here, average steam-specific heat can be found as $\Delta h/\Delta T$ = 0.4479 Btu/lb-R, using the values in Table 7.1. This results in an average γ of 1.3269 and k = 0.2464 so that, using Equation 7.12, the isentropic efficiency is calculated as 85.4 percent (i.e., 0.4 percentage points higher than the "true" value in Table 7.1).

Using the simple isentropic p–T correlation

$$w = h_1 - h_2 \approx \eta_{t,s} c_p T_1 \left(1 - PR^{-k}\right) = 0.854 \times 0.4479 \times (1{,}000 + 460) \times \left(1 - 4^{-0.2464}\right) = 161.5\,\text{Btu/lb}$$

with an error of ~7 percent, which is not too bad considering steam at 1,600 psia is a "real gas" and not an "ideal gas" (i.e., far from being calorically perfect).

A similar calculation is done for an air compressor (PR of 18.3, polytropic efficiency of 93 percent) twice: (i) in a Thermoflex model and (ii) using the polytropic compression (COMPP) code in Appendix C with JANAF properties (PROPS). The Visual Basic (VB) code is implemented in an Excel spreadsheet. The results are summarized in Table 7.3. Note that PROPS enthalpies are converted to a 77°F zero-enthalpy reference for easy comparison with Thermoflex results. From the data in Table 7.3, the compressor isentropic efficiency is calculated as

$$\eta_{c,s} = \frac{156.9 - -4.37}{174.3 - -4.37} = 90\%.$$

Lastly, we repeat the same exercise for a turbine/expander with hot combustion products. The PR is 17 and the polytropic turbine efficiency is 85.12 percent. The combustion gas composition is given as:

Table 7.3 Polytropic compression calculations

			h (Btu/lb)		
	p (psia)	T (°F)	Thermoflex	COMPP	s (Btu/lb-R)
1	14.6	59.0	–4.36	–4.37	1.6069
2s	266.3	708.5		156.9	1.6069
2	266.3	776.0	174.4	174.3	1.6214
w (Btu/lb)			178.8	178.7	

Table 7.4 Polytropic expansion calculations

		T (°F)		h (Btu/lb)		
	p (psia)	Thermoflex	TURBP	Thermoflex	TURBP	s (Btu/lb-R)
1	253.6	2,270.6	2,270.6	619.6	622.6	1.8868
2S	14.9		943.4		228.1	1.8868
2	14.9	1,085.0	1,102.2	266.3	272.6	1.9169
w (Btu/lb)				–353.3	–349.9	

$O_2 = 12.62\%$
$N_2 = 74.45\%$
$CO_2 = 3.706\%$
$H_2O = 8.327\%$
$Ar = 0.897\%$

Once again, the calculation is done twice: (i) in a Thermoflex model and (ii) using the polytropic expansion (TURBP) code in Appendix C with JANAF properties (VBA function PROPS). The results are summarized in Table 7.4. From the data in Table 7.4, the turbine/expander isentropic efficiency is calculated as

$$\eta_{t,s} = \frac{622.6 - 272.6}{622.6 - 228.1} = 88.7\%.$$

This is 0.4 percentage points lower than that calculated by Thermoflex at 89.1 percent.

Note the 17°F difference in expander exit temperature between the two calculations. The difference in exit enthalpy is 2.38 percent and the difference in specific output is 0.95 percent. From the data in Table 7.4, average c_p is 0.2981 Btu/lb-R. The average γ is 1.307, which leads to a value of 0.2347 for k. Using the isentropic p–T correlation,

$$T_2 = (2270.6 + 460) \cdot 17^{-0.8512 \cdot 0.2347} - 460 = 1,091°\text{F},$$

$$w = 0.2981 \cdot (2207.6 - 1091) = 351.8\,\text{Btu/lb}.$$

When using three different approaches with different levels of difficulty (not to mention the availability of requisite computing platforms and/or tools), quite similar results are obtained. If the ultimate goal is to estimate turbine output, using one or the other method does not have a big impact on the bottom line. Turbine exit temperature, however, is a different matter. In combined-cycle calculations, for example, a 17°F error in gas turbine exhaust temperature makes quite a significant difference in calculated steam turbine power output.

When using commercial software, these are mere details one simply does not think about. The point, however, is that, at a minimum, one should be cognizant of the property package and calculation method inside a particular software program. In cases where hand or spreadsheet calculations are necessary (or when one wishes to check the results obtained from a particular software product), whether one takes a simple route or goes the distance by undertaking rigorous property calculations is a decision to be made on a case-by-case basis. With that in mind, let us proceed to a heat and mass balance exercise.

7.3 Examples

7.3.1 F-Class Gas Turbine

For this exercise, we use published data for an older-generation (2003 vintage, to be exact) F–class gas turbine by Siemens. The gas turbine in question is V94.3A introduced in 1996 (current model designation for this particular gas turbine is SGT5-4000F). The data are specified as follows [1]:

- Fuel temperature of 392°F (210°C)
- Net generator output of 272,000 kWe with 100 percent methane
- Net efficiency of 39 percent
- Exhaust mass flow rate of 1,479 lb/s (671 kg/s)
- Exhaust temperature of 1,085°F (585°C)

Based on stated output and efficiency values, fuel mass flow rate is calculated as 30.74 lb/s. We use the simple Thermoflex model in Chapter 5 (no cooling) and run it to achieve 1,085°F and 1,479 lb/s by adjusting turbine efficiency and air inlet flow with 93 percent compressor polytropic efficiency. We tinker with component efficiencies simply because we cannot specify the "boundary conditions" in a flowsheet model like Thermoflex. We can only specify "internal parameters" like component efficiencies. There is an infinite number of compressor/turbine efficiency combinations that would give us the same end result. The reason for choosing 93 percent compressor polytropic efficiency is that it is reasonably representative of the state of the art in compressor aero. Once the compressor polytropic efficiency is specified, turbine polytropic efficiency falls out from the heat and mass balance for the requisite exhaust temperature.

The heat and mass balance from the Thermoflex model is summarized in Table 7.5. In particular,

- Thermoflex ideal gas properties are with a zero-enthalpy reference of 77°F (this is why air enthalpy at 59°F is negative).
- Pressures (and included pressure losses) do *not* impact calculated enthalpies.
- Static/total temperature differentiation is ignored.
- Published performance of 272 MWe and 39 percent efficiency is achieved with a heat loss of 0.5 percent (referenced to gas turbine heat consumption) and mechanical losses of 1.1 percent (referenced to shaft output).
- Generator efficiency is 98.9 percent.
- Heat balance error is zero – as one would expect from a physics-based model.
- Mass balance error is also zero (again, as it should be).

Let us dive a bit deeper into the numbers in Table 7.5. First note that, in the absence of other cycle inputs like water/steam injection, etc., gas turbine heat consumption can

Table 7.5 Gas turbine heat and mass balance table. CV = control volume; LHV = lower (net) heating value

					Q	
Into the CV	p (psia)	T (°F)	m (lb/s)	h (Btu/lb)	Btu/s	kWth
Air	14.7	59	1,448.3	–4.4	–6,310	
Fuel LHV at 77°F		77	30.7	21,517.6	661,451	
Fuel Sensible Heat		392		183.7	5,648	
Total In					660,788	
					Q	
Out of the CV	p (psia)	T (°F)	m (lb/s)	h (Btu/lb)	Btu/s	kWth
Exhaust Gas	14.92	1,085	1,479.0	266.3	393,928	415,617
Heat Loss					3,335	
Mechanical Loss					2,744	
Total Out					400,007	
Shaft Power (In – Out)					260,781	275,140
Heat Balance Error					0	
Generator Loss					2,976	
Generator Power					257,805	272,000

be used by itself to estimate exhaust gas energy as a function of gas turbine performance very accurately. In other words, for 272 MWe-39%,

$$\text{Heat Consumption} = 272/39\% = 697.4 \text{ MWth}$$
$$\text{Exhaust Gas Energy} = \text{HC} \times (1 - 39\%) = 425.4 \text{ MWth}$$

The "exact" number from Table 7.5 (77°F zero-enthalpy reference) is 415.6 MWth (i.e., indicating an error of only 2.5 percent). Furthermore, from the data in Table 5.6, we can convert the estimated exhaust *energy* to exhaust *exergy*, i.e.,

$$\text{Exhaust Gas Exergy} = 0.4723 \times 415.6 \text{ MWth} = 196.3 \text{ MWe}.$$

Finally, with the technology factor from Figure 6.4, we can estimate the bottoming-cycle net output, i.e.,

$$\text{BC Net Power} = 0.7 \times 196.3 \text{ MWe} = 137 \text{ MWe}.$$

Heat and mass balance analysis does not help us with the actual design of the gas turbine. It just tells us whether specified information (i.e., power output, efficiency, and exhaust conditions) is thermodynamically self-consistent or not. It does not (and cannot) tell us anything about cycle PR, component efficiencies, chargeable and nonchargeable cooling flows, etc. There is, however, one exception to this; namely, the *ISO-2314 definition of turbine inlet temperature*. With specified values of airflow and fuel flow (from power output and efficiency) and compressor efficiency (which determines combustor air inlet temperature), combustor exit temperature is essentially equal to the ISO-2314 value. It should be emphasized that the combustor in this context is *not* the "real" gas turbine

Table 7.6 Gas turbine heat and mass balance with water injection. CV = control volume; LHV = lower (net) heating value

					Q	
Into the CV	p (psia)	T (°F)	m (lb/s)	h (Btu/lb)	Btu/s	kWth
Air	14.7	59	1,448.3	–4.4	–6,310	
Water	500	350	15.4	–773.0	–11,881	
Fuel LHV at 77°F		77	30.7	21,517.6	661,451	
Fuel Sensible Heat		392		183.7	5,648	
Total In					648,908	
					Q	
Out of the CV	p (psia)	T (°F)	m (lb/s)	h (Btu/lb)	Btu/s	kWth
Exhaust Gas	14.92	1,049	1,494.4	258.2	385,901	407,149
Heat Loss					3,335	
Mechanical Loss					2,744	
Total Out					391,981	
Shaft Power (In – Out)					256,927	271,073
Heat Balance Error					0	
Generator Loss					2,976	
Generator Power					253,951	267,933

combustor, which we will cover in detail in a separate chapter later in the book. It is a simple "chemical reactor." For the heat balance in Table 7.5, with 93 percent compressor polytropic efficiency, the calculated ISO-2314 turbine inlet temperature (TIT) is 2,270.6°F or 1,244°C, which is very close to the nominal ISO-2314 TIT for this class of gas turbine (1,250°C). Repeating the same exercise for an H–class gas turbine from *Gas Turbine World* (GTW) 2016-17 Handbook (i.e., Siemens SGT6-8000H rated at 305 MWe-40%), the calculated ISO-2314 TIT is 2,485°F (1,363°C).

7.3.2 Water Injection

As stated earlier, there is really not much one can learn from the heat and mass balance analysis about the inner workings of a gas turbine. Before we conclude, however, there are a few quite important "special cases" we have to examine. The first one is water injection for NOx control. Let us assume that the gas turbine we looked at has water injected into its combustor to limit NOx emissions as dictated by applicable environmental regulations. Water at 500 psia and 350°F is supplied at a water/fuel mass flow rate ratio of 0.5. The heat and mass balance from the Thermoflex model is summarized in Table 7.6.

The number of interest in Table 7.6 is the enthalpy of water, which is listed at –773 Btu/lb (i.e., a *negative* value!). Consequently, its contribution to the control volume heat balance is negative ~12,000 Btu/s. In other words, water injected *into* the control volume takes ~12.5 MWth *out of it*! How is that possible?

First of all, from the ASME steam tables, the enthalpy of water at 500 psia and 350°F is 322.324 Btu/lb. Furthermore, the zero-enthalpy point of ASME steam tables is defined at the *triple point* of water, which is at 32°F. Thus, in order to be able to use water (or steam) enthalpies in heat balance calculations with ideal gas properties using a 77°F zero-enthalpy reference, we must make an adjustment. This is done as follows:

- Steam enthalpy at 77°F and triple-point pressure, where steam is dilute enough to be considered an ideal gas, is 1,095.3 Btu/lb.
- In order to set it to 0 Btu/lb, 1,095.3 Btu/lb must be subtracted from the ASME steam table enthalpy.
- Thus, in order to bring the enthalpy of water at 500 psia and 350°F to a 77°F zero-enthalpy reference, we must subtract 1,095.3 Btu/lb from the enthalpy of water obtained using the ASME steam tables.
- When we do that, the enthalpy that should be used in heat balance calculations is 322.3 – 1,095.3 = –773 Btu/lb.

Now that we have the numerical explanation for the *negative* enthalpy value, what is the *physical* explanation for it? In order to answer this question, we note that the water injected into the combustor evaporates and exits the gas turbine as a constituent in the exhaust gas in the form of H_2O vapor (in addition to that from the ambient air and as a result of the combustion reaction). Therefore, the amount of heat requisite to evaporate the water must be taken *out of the energy balance*, hence the negative enthalpy.

7.3.3 Steam Injection

Let us repeat the same exercise by injecting steam at 500 psia and 650°F into the combustor (corresponding enthalpy is 1,329.09 Btu/lb from the ASME steam tables). The heat and mass balance from the Thermoflex model is summarized in Table 7.7. Note the steam enthalpy in the heat balance calculation, which is given by 1,329.1 – 1,095.3 = +233.8 Btu/lb, as explained above.

7.3.4 Reheat Gas Turbine

Finally, let us look at a *sequential (reheat) combustion* gas turbine: 60–Hz GT24, which was originally designed and developed by Asea Brown Boveri (ABB) in the 1990s. ABB was a direct descendant of Brown-Boveri Company (BBC), which later became Alstom. GT24 is a high-PR gas turbine to accommodate the two combustors in series separated by a HP turbine stage. Cycle PR is 35 with a single-stage HP turbine (PR of 2) following the first combustor. After the second combustor, downstream of the HP turbine, there is a four-stage LP turbine. Due to the high cycle PR, the compressor discharge air temperature is above 1,000°F. The ideal thermodynamic design would require an intercooled compressor, which was indeed considered during the early design stages. This idea was dropped fairly quickly because the resulting machine was impractically large. Since the large hot gas path with two combustors and five turbine

Table 7.7 Gas turbine heat and mass balance with steam injection. CV = control volume; LHV = lower (net) heating value

					Q	
Into the CV	p (psia)	T (°F)	m (lb/s)	h (Btu/lb)	Btu/s	kWth
Air	14.7	59	1,448.3	–4.4	–6,310	
Steam	500	650	15.4	233.8	3,594	
Fuel LHV at 77°F		77	30.7	21,517.6	661,451	
Fuel Sensible Heat		392		183.7	5,648	
Total In					664,382	

					Q	
Out of the CV	p (psia)	T (°F)	m (lb/s)	h (Btu/lb)	Btu/s	kWth
Exhaust Gas	14.92	1,069	1,494.4	264.1	394,674	416,404
Heat Loss					3,335	
Mechanical Loss					2,744	
Total Out					400,753	
Shaft Power (In – Out)					263,629	278,144
Heat Balance Error					0	
Generator Loss					2,976	
Generator Power					260,653	275,004

stages imposes a significant cooling load on the design, a CAC is used to cool the compressor discharge air and reduce the amount of coolant flow. The heat rejected is utilized to make IP and/or HP steam to be used in the bottoming cycle.

Alstom was acquired by GE in 2015, which retained GT24, but divested GT26, the 50–Hz variant, to Ansaldo Energia. The gas turbine data are from the GTW 2014–15 Handbook:

- 235,000 kWe ISO base load
- 40 percent efficiency (8,531 Btu/kWh)
- 1,113 lb/s exhaust mass flow rate
- 1,126°F exhaust temperature
- 35.4 cycle PR

Fuel flow is calculated as 25.88 lb/s from the power output and efficiency. The heat and mass balance from the Thermoflex model is summarized in Table 7.8. Assuming 1 percent mechanical loss, 0.5 percent heat loss, and 98.9 percent generator efficiency, the heat balance is closed by rejecting heat from the control volume, ~11.5 MWth, which is debited to the CAC (about 2 percent of cycle heat consumption).

Heat rejected from the cooling air is used in a kettle-type evaporator to make IP or HP steam. Let us assume that 1,800 psia HP steam is generated in the CAC. For illustrative purposes, assume that feed water is supplied at 2,000 psia and 610°F and returned to the bottoming cycle at 1,800 psia and 1,050°F. The steam mass flow rate can be calculated as

Table 7.8 Heat and mass balance of a gas turbine with reheat combustion. CV = control volume; LHV = lower (net) heating value

					Q	
Into the CV	p (psia)	T (°F)	m (lb/s)	h (Btu/lb)	Btu/s	kWth
Air	14.7	59	1,087.1	–4.4	–4,736	
Fuel LHV at 77°F		77	25.9	21,517.6	556,875	
Total In					552,139	
					Q	
Out of the CV	p (psia)	T (°F)	m (lb/s)	h (Btu/lb)	Btu/s	kWth
Exhaust Gas	14.92	1,126	1,113.0	279.4	310,967	328,088
Cooling Air Cooler					10,875	11,473
Heat Loss					2,784	
Mechanical Loss					2,300	
Total Out					326,427	
Shaft Power (In – Out)					225,712	238,140
Heat Balance Error					0	
Generator Loss					2,477	
Generator Power					222,736	235,000

$$\dot{m}_{stm} = \frac{10{,}875\,\text{Btu/s}}{(1511-614)\,\text{Btu/lb}} = 12.1\ \text{lb/s}.$$

Using Equation 5.28 and ASME steam tables,

$$a_{stm,in} = (614-27) - 519 \times (0.8091 - 0.0536) = 195.5\ \text{Btu/lb}$$
$$a_{stm,out} = (1511-27) - 519 \times (1.5959 - 0.0536) = 684.1\ \text{Btu/lb}^{\cdot}$$

The difference between the exergy streams into the CAC and out of the CAC is the maximum power, which can be generated by a Carnot engine utilizing that steam. Using the calculated feed water and steam exergies and steam mass flow rate per Equation 5.32, we obtain

$$\dot{W}_{max} = 12.1\cdot(684.1 - 195.5)\cdot 1.05506 = 6{,}251\,\text{kW}.$$

The actual contribution of steam generated in the CAC to the bottoming-cycle steam turbine generator is a fraction of that maximum (i.e., closer to about 5.5 MWe).

Gas turbines with steam cooling can be treated in a similar manner. Note that there is no guarantee that the calculated heat transfer rates for cooling air cooling or steam cooling from commercial rating data represent "correct" values. The reason for that is the "commercial margins" used by the original equipment manufacturer (OEM) in their published ratings (see Section 20.4). Unless exact "engineering design" data from the OEM are available (highly unlikely unless you happen to be an employee of an OEM in the first place), the only way to ensure that a reasonably correct value is obtained is via rigorous stage-by-stage cooled turbine analysis.

Some very rough, order of magnitude numbers to keep in mind are provided for information:

1. For GT24 and GT26, CAC heat rejection is of the order of 4–5 percent of gas turbine heat consumption. Rejected heat is utilized to make IP and HP steam.
2. For an H-System gas turbine (not offered commercially anymore), heat rejected to the cooling steam is about 3 percent of gas turbine heat consumption. Note that, in this gas turbine, both the stationary *and* rotating parts of the first two turbine stages are steam-cooled.
3. For a J-class gas turbine, heat rejected to the cooling system is about 1.0–1.5 percent of the gas turbine heat consumption.
4. Like all other Mitsubishi F/G-class gas turbines, the J-class gas turbine also has cooled cooling air; it is about 1 percent of the gas turbine heat consumption, which is significantly smaller than that for reheat combustion gas turbines because of the lower cycle PR of J-class machines (23 vis-à-vis 35 for the reheat machines). The rejected heat is used for fuel performance heating.

7.4 Exergy Balance

Second law or exergy analysis of a gas turbine (Brayton) topping cycle is quite straightforward and its major takeaways are not too surprising. The cycle is dominated by two very large irreversibilities: heat addition and heat rejection. This was discussed in detail already in Chapter 5 (see Figure 5.1 and the associated text). Since the steam bottoming cycle takes care of the heat rejection irreversibility extremely well – as will be demonstrated below – it is the prime candidate for the second law approach. For a good understanding of gas turbine loss mechanisms via second law analysis, the reader is referred to the paper by Facchini et al. [2].

Standard chemical exergy of the gas turbine fuel is rather tedious to calculate. To a good approximation, it can be assumed to be the same as the lower (net) heating value (LHV) at the same reference temperature and pressure (e.g., 21,515 Btu/lb for 100 percent methane at 77°F). More exact values can be obtained from table C.2 in Bejan et al. [3]. Methane values are provided below (chemical exergy per methods I and II in the cited reference).

Higher (gross) heating value	Lower (net) heating value	Chemical Exergy II	Chemical Exergy I	Units
		831,650	824,348	kJ/kmol
		357,544	354,404	Btu/lbmol
23,876	21,518	22,291	22,095	Btu/lb

Depending on the method used, the chemical exergy of 100 percent methane is 2.5–3.5 percent higher than its LHV. For typical F-class gas turbines, approximately 35 percent of the fuel exergy is destroyed via three major mechanisms:

- Combustion irreversibility (~27 percent of fuel exergy)
- Turbine cooling and mixing (~4 percent)
- Non-isentropic compression and expansion (~3 percent)

As far as the combustion irreversibility is concerned, due to the nature of the combustion process, there is no possibility of avoiding it. Combustion or combustor irreversibility can be calculated using the exergy balance as follows:

$$\dot{I}_{comb} = \dot{m}_{air} \cdot a_{air} + \dot{m}_{fuel} \cdot a_{fuel} - \dot{m}_{prod} \cdot a_{prod} - \dot{Q}_{loss} \cdot \left(1 - \frac{T_0}{\bar{T}_{comb}}\right).$$

Using a simple Thermoflex model (similar to that described in Section 5.1, uncooled, with a combustor), calculations with 100 percent methane fuel are carried out for gas turbines itemized in Table 6.1 (i.e., GE's 60-Hz portfolio). The reference temperature and pressure are 77°F and 14.7 psia, respectively, with ISO conditions at the compressor inlet. Fuel temperature is also 77°F and the LHV is substituted for fuel chemical exergy. There are no heat, friction, or pressure losses in the cycle with 100 percent efficient components. The results are summarized in Table 7.9 (note that the combustor exit temperature is the same as the TIT).

What is calculated and presented in Table 7.9 is the maximum possible obtainable efficiency for a "real" gas turbine designed for TIT and PR listed in the top two rows of the table. There are two loss mechanisms, about which *nothing can be done*:

1. Combustion irreversibility or *exergy destruction* (roughly 25–30 percent of the gas turbine heat consumption)
2. *Net exergy transfer* out of the cycle with turbine exhaust (roughly 20 percent of the gas turbine heat consumption)

It is worth repeating and emphasizing that *once the cycle is specified (i.e., TIT and PR), there is nothing in a designer's arsenal that can do anything about these two loss mechanisms*. Thus, one cannot do better than the efficiency listed in the last row of Table 7.9.

In Chapter 6, a technology or Carnot factor was calculated by referencing the rating efficiency of the gas turbines in Table 6.1 to the efficiency of an ideal, air-standard

Table 7.9 Combustor irreversibility. HC = heat consumption

TIT (°F)	2,252	2,658	2,769	2,764	2,868	2,908
Cycle PR	13.0	16.7	18.4	22.1	21.6	23.1
Air Exergy In (Btu/lb)	145.8	166.8	175.0	190.0	188.6	194.5
Fuel Exergy In (Btu/lb)	500.8	628.1	661.6	642.4	684.9	694.0
Products Exergy Out (Btu/lb)	486.7	615.4	653.3	657.2	690.0	705.1
Irreversibility (Btu/lb)	159.9	179.5	183.3	175.2	183.5	183.4
Irreversibility (% of HC)	31.93%	28.58%	27.71%	27.27%	26.79%	26.43%
Compressor Inlet (% of HC)	2.23%	2.09%	2.05%	2.07%	2.03%	2.02%
Turbine Exhaust (% of HC)	22.05%	22.88%	22.74%	20.94%	21.90%	21.54%
Net Exergy Out (% of HC)	19.82%	20.79%	20.69%	18.87%	19.87%	19.52%
Maximum Possible η	48.25%	50.63%	51.60%	53.86%	53.34%	54.05%

Table 7.10 Gas turbine technology entitlement

TIT (°F)	2,252	2,658	2,769	2,764	2,868	2,908
Cycle PR	13.0	16.7	18.4	22.1	21.6	23.1
Equation 5.2	51.94%	55.26%	56.48%	58.70%	58.43%	59.22%
Actual Gas Turbine Efficiency	33.90%	38.60%	39.80%	41.40%	41.70%	42.20%
Technology Factor	0.653	0.698	0.705	0.705	0.714	0.713
Maximum Possible η	48.25%	50.63%	51.60%	53.86%	53.34%	54.05%
Maximum Possible Technology Factor	0.703	0.762	0.771	0.769	0.782	0.781

Brayton cycle, which is a function of cycle PR only and is given by Equation 5.2. This factor was a quite stable 0.7 as illustrated by the curves in Figure 6.1. Using the data in Table 7.9, we can now calculate the maximum possible value of this technology factor, which is listed in Table 7.10. Thus, using the last column of the table as an example, it is seen that, for a J–class, heavy-duty industrial gas turbine with a cycle PR of 23.1 and a TIT of 2,908°F, one cannot achieve more than 54.05 percent net cycle efficiency, which corresponds to a technology/Carnot factor of 0.781. Considering that the state of the art in gas turbine technology, at least as indicated by ISO base load rating data in a 2016 trade publication, is already at 0.713, one can easily foresee the difficulty in making significant progress toward that ideal. (That ideal, it is worth stressing again, is based on absolutely zero losses of any kind and isentropic components.)

7.4.1 Terminology Check

Combustion or combustor irreversibility is *not* the same as cycle heat addition irreversibility. The latter is a quantification of the temperature delta between cycle maximum temperature, which defines the ultimate Carnot cycle with isothermal heat addition at that temperature, and the mean effective heat addition temperature, METH, of the ideal, air-standard Brayton cycle with a constant-pressure heat addition process between T_2 and T_3. For the ideal cycle itself, heat addition irreversibility is, by definition, zero because its heat addition process is equivalent to *isothermal* heat addition at METH. As discussed in Chapter 6, the difference between the ideal Brayton cycle and the Carnot cycle is quantified by the *cycle factor*.

In the "real" gas turbine cycle, however, heat addition takes place by a highly irreversible combustion process comprising many complex and simultaneous chemical reactions and mixing of fluids. The exergy destruction of that process, as quantified by the data in Table 7.9, is equivalent to 25–30 percent of the cycle heat consumption in LHV.

Along with the net exergy transfer out of the cycle with turbine exhaust gas, this is by far the most dominant driver behind the *technology factor* defined in Chapter 6 (i.e., the ratio of "real" gas turbine-cycle and ideal Brayton-cycle efficiencies), which was a very robust 0.7 (e.g., see Figure 6.1). As demonstrated by the data in Table 7.10, it is thermodynamically impossible to do better than 0.78, and it is not realistic to hope that one will be able to do much better than 0.75 (if that).

References

1. Thien, V., Becker, B., *The Siemens VX4.3A Gas Turbine – A High Efficiency Industrial Gas Turbine for Flexible Operation and High Reliability* (Milton Keynes, UK: IDGTE Gas Turbine Conference, 2003).
2. Facchini, B., Fiaschi, D., Manfrida, G., Exergy Analysis of Combined Cycles Using Latest Generation Gas Turbines, *Journal of Engineering for Gas Turbines and Power*, **122**: 2 (2000), 233–238.
3. Bejan, A., Tsatsaronis, G., Moran, M., *Thermal Design & Optimization* (New York: John Wiley & Sons, Inc., 1996).

8 Real Cycle Analysis

While going through the basic thermodynamics to analyze the gas turbine Brayton cycle in Chapter 5, it was stated that "realism" required attention to be paid to the following *three* areas:

1. Component (i.e., compressor and turbine) efficiencies
2. Cycle losses
   a. Pressure losses
   b. Heat losses
   c. Mechanical losses
3. Hot gas path cooling flows (chargeable and nonchargeable)

The first two areas are trivial in terms of accounting for them in a straightforward manner in detailed calculations. Using polytropic component efficiencies reflecting the state of the art in aerothermodynamic design of *axial-flow* turbomachinery with a suitable property package takes care of the first one. (*Radial-flow* turbomachines are not used in electric power generation applications.)

Taking care of myriad losses, in whatever form, is a simple matter of applying relevant efficiencies (or loss factors, which simply translate to "one minus the respective efficiency or effectiveness").

The third area is the heart of real gas turbine cycle calculations. It requires stage-by-stage turbine calculations for diligent bookkeeping of chargeable and nonchargeable cooling flows. Furthermore, for G-, J-, and some H–class gas turbines, it also requires the evaluation of the closed-loop steam cooling arrangement. The underlying drivers are:

1. Cycle parameters (i.e., cycle pressure ratio [PR] and turbine inlet temperature [TIT])
2. Stage aerothermodynamics (i.e., rotor-relative stagnation temperature as a function of stage loading)
3. Available materials (i.e., nickel-based alloys)
4. Stator and rotor component ("vanes" and "buckets") manufacturing technology (i.e., directionally solidified [DS] or single-crystal [SX] casting)
5. Thermal barrier coating (TBC)
6. Cooling air cooling

The basics of compressor and turbine aerodynamics, materials, and manufacturing technologies and combustion fundamentals will be covered in requisite detail in their own chapters later in the book (Chapters 10 and 11).

In this chapter, we will make use of all of the key considerations listed above to make cycle calculations with sufficient realism (but no more) to illustrate how complicated fundamentals are translated into actual cycle performance. We will start with a simple gas turbine model without detailed treatment of hot gas path cooling.

8.1 "Real" Uncooled Turbine

First law treatment of a compressor or expander via enthalpy difference using an appropriate equation of state is the fundamental analysis method. In terms of a mathematical representation of the underlying physics, however, it is very sparse and provides little insight into the governing mechanisms. This is why we resort to simpler but richer-in-detail representations using, for example, the isentropic p–T correlation and polytropic efficiencies.

The simplest parametric representation of a gas turbine operating in a Brayton cycle can be written as follows:

$$w = \left(1 - PR_c^{\frac{k_a}{\eta_c}}\right) + (1+f)\cdot\chi_g\cdot\tau_3\cdot\left(1 - \frac{1}{PR_t^{\eta_t\cdot k_g}}\right), \quad (8.1a)$$

where the *non-dimensional* net gas turbine work is defined as

$$w = \frac{\dot{W}}{\dot{m}_a c_{p,a} T_1}. \quad (8.1b)$$

Note that the gas turbine work in the numerator of the fraction on the right-hand side of Equation 8.1b is in Btu/s so that w is non-dimensional. In SI units, the numerator is in kW (which is equal to 1.05506 Btu/s). In Equation 8.1, turbine cooling flow extraction from the compressor and its reinjection into the turbine are ignored. Furthermore,

- PR_c and $PR_t < PR_c$ are compressor and turbine PRs, respectively.
- The non-dimensional cycle maximum temperature is given by $\tau_3 = T_3/T_1$.
- k_a and k_g are isentropic exponents for air and combustion products, respectively.
- η_c and η_t are compressor and turbine polytropic efficiencies, respectively.

The first term on the right-hand side of Equation 8.1a is the compressor work (negative).

The second term on the right-hand side of Equation 8.1a is the turbine (expander) work (positive).

The ratio of the specific heats of combustion products and air is χ_g, and f is the fuel–air mass flow ratio. Heat addition to the cycle can be found from the fuel consumption, i.e.,

$$\dot{Q}_{in} = \dot{m}_f \cdot LHV, \quad (8.2)$$

where LHV is the net or lower heating value of the fuel at the reference temperature, T_{ref}. The efficiency of the gas turbine is the ratio of Equations 8.1 and 8.2, i.e.,

$$\eta = \frac{\left(\dfrac{k_a}{1 - PR_c^{\eta_c}}\right) + (1+f)\cdot\chi_g\cdot\tau_3\cdot\left(1 - \dfrac{1}{PR_t^{\eta_t\cdot k_g}}\right)}{f\cdot\ell}, \tag{8.3}$$

Where l is the non-dimensional fuel energy content, i.e.,

$$\ell = \frac{LHV}{c_{p,a}T_1}.$$

The fuel mass flow rate requisite for a specified cycle maximum temperature (i.e., T_3) as a fraction of turbine airflow is given by

$$f = \frac{\dot{m}_f}{\dot{m}_a}, \tag{8.4a}$$

$$f = \frac{\chi_g\cdot T_3 - T_2}{\ell\cdot T_1 + \chi_f\cdot T_f - \chi_g\cdot T_3}, \tag{8.4b}$$

$$f = \frac{\chi_g\cdot \tau_3 - \tau_2}{\ell + \chi_f\cdot \tau_f - \chi_g\cdot \tau_3}, \tag{8.4c}$$

where the second term in the denominator is the sensible fuel energy input and χ_f is the ratio of the specific heats of fuel and air. Equation 8.4b (or 8.4c) represents the heat and mass balance of the combustor control volume (CV).

In Equation 8.4b or 8.4c, the implicit assumption is that enthalpy is given by

$$h = c_p T,$$

with a zero-enthalpy reference temperature of 0 Rankine (–459.67°F). The generic formula for enthalpy of a calorically perfect gas is

$$h = c_p(T - T_{ref}).$$

Since the fuel LHV is typically quoted at 77°F, the logical choice for T_{ref} would be 77°F as well. The convention chosen herein (i.e., T_{ref} = −459.67°F) simplifies the equations and does not introduce an appreciable error. For a simple model, which will be calibrated using "correction factors" as shown below anyway, this is perfectly acceptable and adequate.

The fuel–air ratio obtained from Equations 8.4b and 8.4c deviates from a rigorous combustor calculation. This is quantified by running a simple combustor model in Thermoflex over a range of air and fuel temperatures and fuel flow rates. The comparison showed that Equations 8.4b and 8.4c overestimated the fuel–air ratio by about 20 percent. The range of overestimate is 20–27 percent, with an average of 23 percent. In other words, when using Equation 8.3 to calculate the gas turbine efficiency, a correction factor of about 1.25 should be applied to η (or, equivalently, a correction factor of 0.82 should be applied to f from Equation 8.4).

Note that, for clarity, mechanical and electric losses are omitted from Equations 8.1 and 8.3, which represent the "shaft" performance. Typically quantified via mechanical and generator efficiencies, η_m and η_g, respectively, these losses should be accounted for in the gas turbine generator's *net* electric output.

Equation 8.3 is an "advanced" version of the air-standard Brayton-cycle efficiency, Equation 5.2, which just tells us that the cycle efficiency is a function of the cycle PR only. Equation 8.3, however, tells us that this is only *partially* correct; cycle efficiency is *also* a function of maximum cycle temperature.

The requisite assumptions or approximations for the use of Equations 8.1–8.4 for "real" gas turbine analysis are enumerated below.

Specific heats of air and combustion products are averages across the compression and expansion processes, respectively. For air, a reasonably accurate value of $c_{p,a}$ is 0.25 Btu/lb-R. For combustion products (gas), the following linear correlation is sufficiently accurate for this type of simple calculation:

$$c_{p,g} = 0.2243 + 0.0316\frac{T_3[^\circ F]}{1000}\ \text{Btu/lb-R}. \tag{8.5}$$

Corresponding values of γ are derived from the specified specific heats and noting that

$$c_{p,g/a} = c_{v,g/a} + \frac{R_{unv}}{MW_{g/a}}.$$

The cycle maximum temperature, T_3, is the first-stage rotor inlet (i.e., "firing") temperature. This can be explained by referring to the simple gas turbine diagram in Figure 8.1, which depicts the three CVs comprising the gas turbine in question.

As shown in Figure 8.1, the combustor CV includes the nonchargeable cooling flows and stage 1 stator nozzle vanes (S1N). As such, the total temperature at state-point 3 is the rotor inlet (stagnation) temperature, which is the "firing temperature" of the particular gas turbine. Therefore, this simple "uncooled" gas turbine model excludes the "chargeable" cooling flows.

The turbine CV comprises the stage 1 rotor (S1B) and the remaining turbine stages, as well as the exhaust diffuser and the rest of the plant, which imposes an "exit loss" on the back pressure. Thus, as shown in Figure 8.1, state-points 1 and 4 have the same pressure (i.e., the ambient).

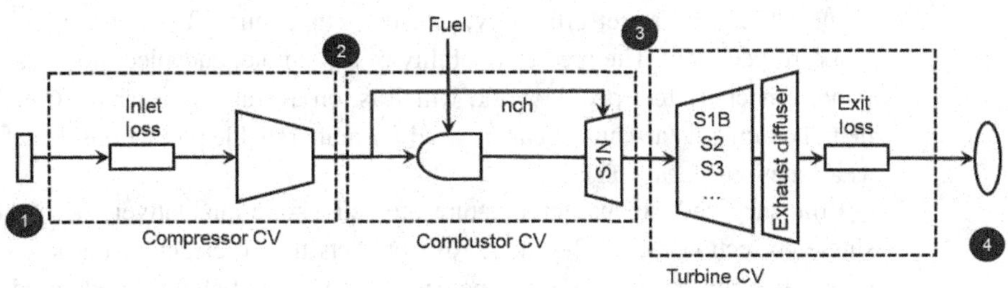

Figure 8.1 Gas turbine CVs for the simple model. nch = nonchargeable.

Similarly, the compressor CV encompasses the inlet loss and exit diffuser. The cycle PR, p_2/p_1, is thus the compressor PR, PR_c.

The turbine PR, $PR_t = p_3/p_4 = p_3/p_1$, includes the pressure loss in the combustor, which is typically around 5–6 percent, and the stagnation pressure loss across the S1N due to friction, which is around 3–5 percent. Consequently,

$$PR_t = PR_c \times (1 - \Delta p_{comb}) \times (1 - \Delta p_{S1N}).$$

With these assumptions, we can ignore the total/static dichotomy and use the pressures and temperatures that, strictly speaking, represent total/stagnation values in Equations 8.1–8.5, in a consistent manner. The polytropic compressor and turbine efficiencies in Equations 8.1 and 8.3 include *all* losses, including inlet, exit, and diffusers. They are set to 90 and 85 percent, respectively, which reasonably reflects the state of the art in compressor and turbine aero *qualitatively*.

As mentioned earlier, a "correction factor" was applied to the fuel–air ratio calculated from Equation 8.4. A correction is also needed for the exhaust temperature, T_4, calculated from

$$T_4 = \frac{T_3}{PR_t^{\eta_t \cdot k_g}}. \tag{8.6a}$$

The reason for that is the missing dilution effect of chargeable cooling flows. Thus, we have

$$f = c_f \frac{\chi_g \cdot \tau_3 - \tau_2}{\ell + \chi_f \cdot \tau_f - \chi_g \cdot \tau_3}, \tag{8.4d}$$

$$T_4 = \frac{T_3}{PR_t^{\eta_t \cdot k_g}} - \Delta T_{corr}. \tag{8.6b}$$

The correction factors, c_f and ΔT_{corr}, are determined by trying to match the ISO base load rating performance of gas turbines in Table 6.1. The resulting values are

$$c_f = 0.85,$$

$$\Delta T_{corr} = -136^{\circ}F.$$

The comparisons between the calibrated simple model predictions and data are in Figures 8.2–8.4 for net efficiency, exhaust temperature, T_4, and normalized specific work, respectively. The predictive ability of the simple, uncooled model is surprisingly good, especially for specific work, which is, on average, only about 6 percent higher than the rating data. Thus, it can be used for quite reliable predictions of net cycle work as a function of, say, TIT.

Efficiency and exhaust temperature predictions are qualitatively acceptable. With a single correction factor, the model overpredicts the efficiency at the low end (i.e., E-class gas turbines) by about two percentage points and slightly underpredicts it at the high end (i.e., H/J-class gas turbines).

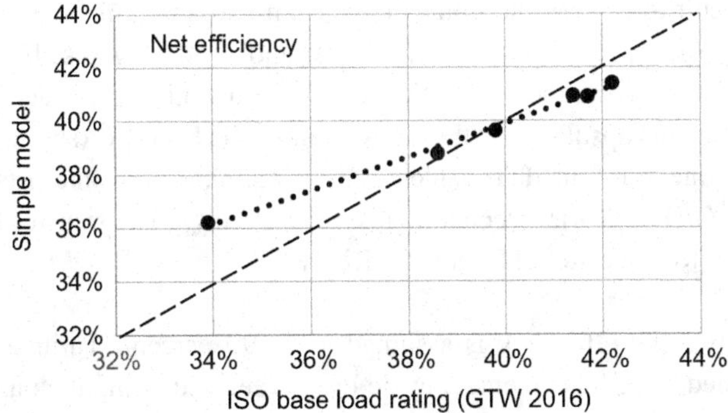

Figure 8.2 Simple model efficiency prediction. GTW = *Gas Turbine World.*

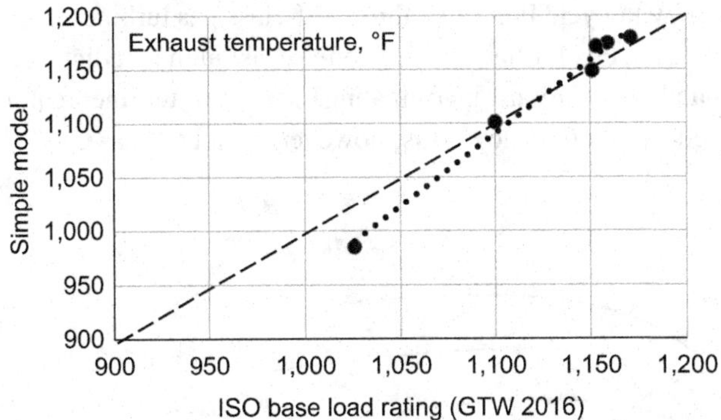

Figure 8.3 Simple model exhaust temperature prediction. GTW = *Gas Turbine World.*

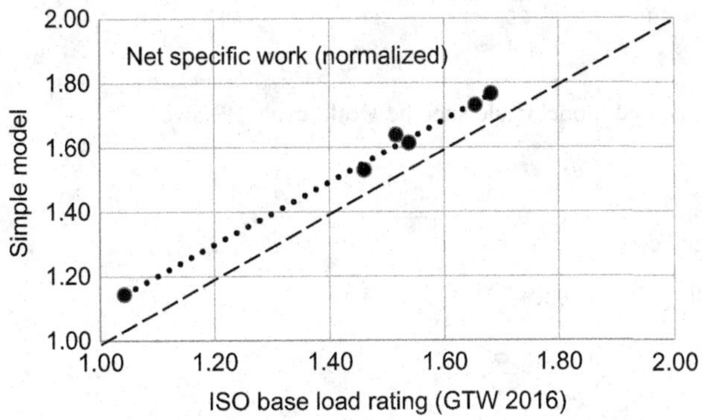

Figure 8.4 Simple model cycle-specific work prediction. GTW = *Gas Turbine World.*

For the exhaust temperature, the situation is reversed. Again, with only a single correction factor, the model underpredicts the exhaust temperature at the low end by about 40°F and slightly overpredicts it at the high end.

Obviously, much better agreement with the data can be achieved by making the correction factors, c_f and ΔT_{corr}, functions of, say, TIT. We will return to this point shortly.

While it is difficult to discern the impact of cycle maximum temperature, T_3, or, in non-dimensional terms, τ_3 on cycle-specific work and efficiency just by looking at Equations 8.1 and 8.3, a sample calculation can be illustrative. In order to do that, Equations 8.1–8.6 are encapsulated by the VBA function GTURBUC (see Appendix C). Calculations are done for four TIT values: 1,300, 1,400, 1,500, and 1,600°C (i.e., representing E, F, G/H, and J/HA technologies, respectively). The transfer function in Figure 6.2 is used to translate TIT into TFIRE, which is the T_3 to be used in the simple model.

Compressor polytropic efficiency is assumed to be 90 percent. Turbine polytropic efficiency is assumed to be 85 percent. A cycle PR "sweep" covering the range of 9–38 is done, with the results summarized in Figure 8.5.

Once again, the surprising predictive ability of this very crude model is verified by the cycle PRs corresponding to cycle-specific work maxima in Figure 8.6. The values are nearly coincident with cycle PRs of E- through J-class gas turbines as listed in trade publications. Prima facie, this further increases one's trust in the predictive ability of the model for conceptual calculations investigating, say, gas turbine requirements for 65 percent combined-cycle efficiency. This, however, is not the case.

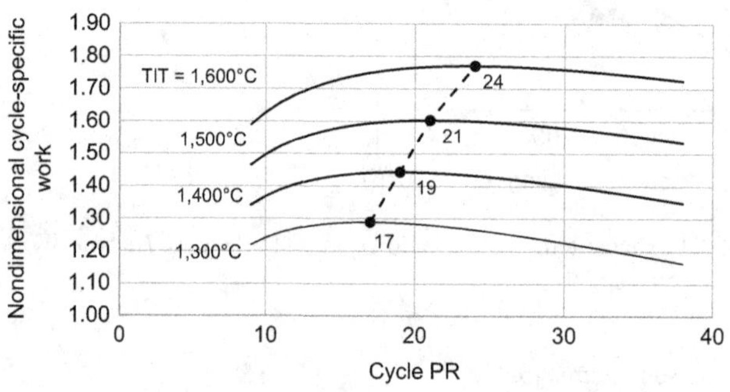

Figure 8.5 Simple, uncooled model cycle specific work, cycle PR sweep.

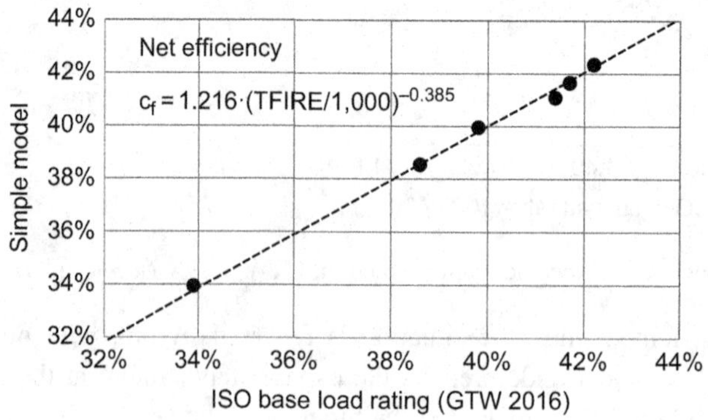

Figure 8.6 Improved simple (uncooled) model gas turbine efficiency prediction. GTW = *Gas Turbine World.*

In order to use the model to explore future combined-cycle performance improvement, we need to do some more adjustment. In particular, Figure 8.4 strongly suggests that at higher TIT–PR combinations, the simple model will severely underestimate the gas turbine efficiency. This can be remedied by making the correction factor c_f, a function of TIT, to result in a better fit to the data; in particular,

$$c_f = 1.216 \cdot (\text{TFIRE}/1{,}000)^{-0.385}. \tag{8.7}$$

The efficiency prediction of the simple model with the improved correction to the fuel–air ratio is shown in Figure 8.6.

For the combined-cycle runs, the model is calibrated to reproduce General Electric 7HA.02 simple- and combined-cycle ISO base load rating performances (42.2 and 62.3 percent, respectively) from *Gas Turbine World* (GTW) 2016 performance specifications with η_c = 0.897 and ΔT_{corr} = –156°F. Exhaust exergy is calculated using Equation 5.60. Bottoming-cycle exergetic efficiency is calculated using the following base correlation (~2008 vintage, from Gülen and Smith [1]):

$$\eta_{BC,\text{base}} = 0.2441 + 0.0746 \cdot (T_4/100) - 0.00279 \cdot (T_4/100)^2. \tag{8.8a}$$

The background of Equation 8.8a will be presented in detail in Section 17.2. A technology adder is used to represent the state of the art in bottoming-cycle technology such that

$$\eta_{BC,SOA} = \eta_{BC,\text{base}} + \Delta\eta. \tag{8.8b}$$

The adder $\Delta\eta$ is determined to be 0.02, or 2 percent, to match the ISO base load rating of the 7HA.02 combined cycle. This corresponds to a bottoming-cycle exergetic efficiency of about 75 percent. Note that, for this simple exercise with a rather simplistic model, no attempt has been made to correct the rating performance of the gas turbine for extra back pressure imposed by the heat recovery steam generator (HRSG). ISO base load performance data are used *as they are*. The error introduced by this is not important for this illustrative exercise. Steam turbine output is determined by accounting for 1.9 percent bottoming-cycle auxiliary load [2]. In non-dimensional terms, it is given by

$$w_{ST} = \frac{\dot{W}_{ST}}{\dot{m}_a c_{p,a} T_1} = \frac{(\dot{m}_a + \dot{m}_f)\eta_{BC} a_{exh}}{(1 - 1.9\%)\dot{m}_a c_{p,a} T_1} = \frac{(1 + f)}{(1 - 1.9\%)} \eta_{BC} \alpha_{exh},$$

$$\alpha_{exh} = \frac{a_{exh}}{c_{p,a} T_1}.$$

The sum total of gas and steam turbine outputs gives the combined-cycle *gross* output:

$$w_{CC,\text{gross}} = w_{GT} + w_{ST}.$$

The combined-cycle *net* output is determined by subtracting 1.6 percent for plant auxiliary load:

$$w_{CC,\text{net}} = (1 - 1.6\%) \cdot w_{CC,\text{gross}}.$$

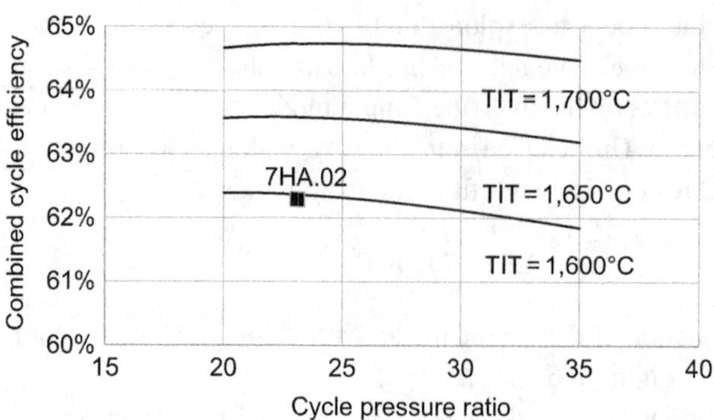

Figure 8.7 Simple (uncooled) model combined-cycle efficiency prediction.

This is representative of rating performance with a once-through, open-loop condenser and bare-bones plant auxiliaries [2].

Cycle PR "sweeps" for PR = 20–25 at TITs of 1,600, 1,650, and 1,700°C, are graphically summarized in Figure 8.7. For 65 percent combined-cycle efficiency, which is the next "goalpost" for the 2020–2025 time frame, a TIT of 1,700°C and a cycle PR of 25–26 are requisite (and bottoming cycle technology adder of 3%).

8.2 "Real" Cooled Turbine

There are very powerful modeling platforms for stage-by-stage, cooled turbine calculations. They include in-house, proprietary design tools used by the original equipment manufacturers (OEMs; e.g., "cycle decks"), as well as commercially available software programs. One example of the latter is GASCAN, which is now included in Thermoflex as a stand-alone component ("Cooled Turbine Stage"). GASCAN was originally developed by Elmasri as an MS-DOS program in BASIC language [3]. It incorporates the pioneering work done by Elmasri in the 1980s on gas turbine cooling requirements [4,5]. This program is used in this book to do sample calculations and analysis. The underlying principles will be highlighted in this chapter briefly. They will be discussed in more detail in the following chapter. They should suffice for an engineer or student to intelligently tweak all the knobs available to a turbine designer for accurate modeling of existing or conceptual products.

Unfortunately, even a user-friendly program such as Thermoflex and GASCAN within its framework requires a fairly steep learning curve. The best way to get acquainted with the intricacies of the turbine design process is to walk through the calculation process manually at least once. Since commercial or proprietary design tools hide all the "gruesome" details from the end user, this is an important step in the learning process. This is hopefully going to be a useful exercise leading to a better appreciation of the finer points of cooled turbine modeling.

In order to provide guidance for such an exercise, a simple model is developed. The model contains all the pertinent principles in the form of physics-based equations

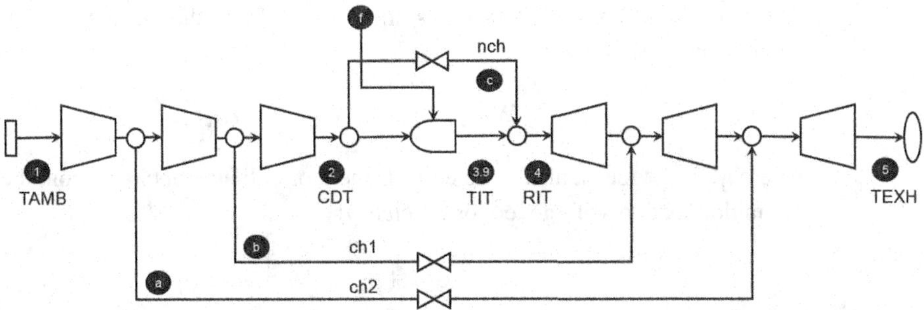

Figure 8.8 Simple cooled gas turbine model. CDT = compressor discharge temperature; ch = chargeable; nch = nonchargeable; RIT = rotor inlet temperature; TAMB = turbine ambient temperature; TEXH = turbine exhaust (exit) temperature.

amenable to spreadsheet calculations. The goal is to provide a framework to put all the building blocks together and watch them "in action." The simple model, which is shown in Figure 8.8, is an air-cooled gas turbine with the cooled turbine section utilizing air from compressor extraction ports. As shown in Figure 8.8, the compressor and the turbine sections are divided into three "hypothetical" or "pseudo" stages.

The calculations can be done utilizing basic equations with simple assumptions. For example, the PR distribution across each "pseudo" stage is equal (once you get comfortable with the model, this assumption can be relaxed as needed). The two ports between the three "pseudo" compressor stages and the compressor discharge provide the turbine chargeable cooling air. The nonchargeable cooling air is taken from the compressor discharge. As was mentioned earlier, the terms "chargeable" and "non-chargeable" refer to the location of the cooling air where it mixes with the gas stream (i.e., upstream of the first-stage rotor or downstream, respectively).

Each compressor and turbine "pseudo" stage is defined by its respective polytropic efficiency, which is the same for all stages. (Again, later on, this assumption can be relaxed as needed as well.) Consequently, for a given stage i (i = 1, 2 or 3),

$$\dot{W}_{c,i} = \left(\dot{m}_c c_p T_{in}\right)_i \left(PR_{c,i}^{k/\eta_c} - 1\right) \tag{8.9}$$

$$\dot{W}_{t,i} = \left(\dot{m}_t c_p T_{in}\right)_i \left(1 - PR_{t,i}^{-k\eta_t}\right). \tag{8.10}$$

The subscript "in" in the equations denotes the temperature at the inlet to the stage, which is the total (stagnation) temperature. For the total compressor and turbine work, stage works are added together, i.e.,

$$\dot{W}_{c,tot} = \sum_{i=1}^{3} \dot{W}_{c,i}, \tag{8.11a}$$

$$\dot{W}_{t,tot} = \sum_{i=1}^{3} \dot{W}_{t,i}. \tag{8.11b}$$

The net gas turbine shaft work is then found from the balance of the two total component work terms as

$$\dot{W}_{GT,shaft} = \eta_m(\dot{W}_{t,tot} - \dot{W}_{c,tot}), \quad (8.12)$$

where η_m is the mechanical efficiency. Generator output is obtained from the shaft work via multiplication with generator efficiency:

$$\dot{W}_{GT,gen} = \eta_{gen}\dot{W}_{GT,shaft}. \quad (8.13)$$

The turbine PR is computed from the compressor PR (a design input) by accounting for combustor, inlet, and exhaust losses:

$$PR_t = PR_c(1 - \Delta P_{comb})(1 - \Delta P_{in})(1 - \Delta P_{exh}). \quad (8.14)$$

The PR for each "pseudo" turbine stage is found as

$$PR_{t,i} = \sqrt[3]{PR_t}. \quad (8.15)$$

The PR distribution among the compressor "pseudo" stages is handled differently. They are determined from the total pressures at the exits of the first and second turbine stages. In particular,

$$PR_{c,1} = \frac{PR_{t2,ex} + 15}{P_1(1 - \Delta P_{in})}, \quad (8.16a)$$

$$PR_{c,2} = \frac{PR_{t1,ex} + 15}{P_1(1 - \Delta P_{in})PR_{c,1}}, \quad (8.16b)$$

$$PR_{c,3} = \frac{PR_c}{PR_{c,1}PR_{c,2}}. \quad (8.16c)$$

The subscript "ex" in the equations denotes the pressure at the exit of the stage, which is the total (stagnation) pressure. This ensures that the extraction ports have sufficient pressure head to supply the cooling ports in the turbine section.

The energy balance around the combustor can be written as

$$\dot{m}_2 c_{p,air} T_2 + \dot{m}_f(LHV \cdot \eta_b + c_{p,f}T_f) = \dot{m}_{3.9} c_{p,gas} T_{3.9}, \quad (8.17)$$

where η_b is the combustor efficiency (usually 0.995). Thus, for any given combustor exit temperature $T_{3.9}$ (which is essentially the TIT), we can calculate the fuel mass flow rate for a specific fuel with a requisite LHV and fuel temperature, T_f:

$$\dot{m}_f = \frac{\dot{m}_{3.9} c_{p,gas} T_{3.9} - \dot{m}_2 c_{p,air} T_2}{(LHV \cdot \eta_b + c_{p,f}T_f)}. \quad (8.18)$$

Note that Equations 8.17 and 8.18 are equivalent to Equation 8.4 in the simple, uncooled gas turbine model described in the preceding section. Thus, the gas turbine efficiency becomes

$$\eta_{GT} = \frac{\dot{W}_{GT,gen}}{HC_{GT}} \quad , \qquad (8.19)$$
$$HC_{GT} = c_f \dot{m}_f LHV$$

where c_f is a "fudge factor" that is utilized to adjust the calculated heat consumption, HC_{GT}, so that the user can calibrate the gas turbine efficiency to that of a known, "actual" gas turbine. This correction factor has the same functionality (and the same underlying reason) as the correction factor used in the simple, uncooled gas turbine model in the preceding section. For better fidelity, a TFIRE-adjusted value per Equation 8.7 can be used.

For ease of programming, portability, and usability, constant air- and gas-specific heats can be used in the calculations. Typical values are 0.25 and 0.33 Btu/lb-R, respectively. (For the specific heat of gas, an adjustment for TIT can be made via Equation 8.5.) The corresponding γ values are 1.3778 and 1.2622, respectively. The value used for fuel gas specific heat $c_{p,f}$ is 0.6 Btu/lb-R. This eliminates the need for add-ins and/or special coding for an equation of state as well as lengthy iterations. The selected values are representative of the temperatures in each component and the respective working fluid (i.e., air and hot combustion gas) for state-of-the-art industrial gas turbines. For better accuracy, one can upgrade the equations and calculation algorithm to fit a property package such as JANAF. However, it must be stated that the added computational bag and baggage will not add anything to this exercise in terms of "learning." (Forty years ago [i.e., in the 1970s], when calculations were still done with slide rules, this might have been a worthwhile exercise, but not anymore.)

Chargeable and nonchargeable cooling flow fractions are given as

$$ch1 = \frac{\dot{m}_{ch1}}{\dot{m}_1}, \quad ch2 = \frac{\dot{m}_{ch2}}{\dot{m}_1}, \quad nch = \frac{\dot{m}_{nch}}{\dot{m}_1}$$

Cooling flow fractions are determined from the "cooling effectiveness" method as follows:

$$nch = \alpha K_{nch} \frac{\dot{m}_4}{\dot{m}_1} \left(\frac{\varepsilon}{\varepsilon_{\infty,1} - \varepsilon} \right)^{\beta_1}, \qquad (8.20a)$$

$$ch1 = \alpha K_{ch1} \frac{\dot{m}_{t2,in}}{\dot{m}_1} \left(\frac{\varepsilon}{\varepsilon_{\infty,2} - \varepsilon} \right)^{\beta_2}, \qquad (8.20b)$$

$$ch2 = \alpha K_{ch2} \frac{\dot{m}_{t3,in}}{\dot{m}_1} \left(\frac{\varepsilon}{\varepsilon_{\infty,3} - \varepsilon} \right)^{\beta_3} \qquad (8.20c)$$

The parameter α in these equations is a technology factor, which is a number between 0 and 1. Thus,

- If α is 0, the turbine is uncooled (i.e., chargeable and nonchargeable flows are zero).

- If α is 1, chargeable and nonchargeable flows represent the state of the-art.
- Any value of α that is less than 1 implies an "advanced" cooling technology. For example, $\alpha = 0.8$ means that cooling flows are slashed by 20 percent.

The coefficients K_{nch}, K_{ch1}, and K_{ch2} are ideally determined from calibration to a known gas turbine with state-of-the-art cooling technology. They are assumed to be 0.05, 0.1, and 0.05, respectively, to start the calculation. Calibration adjustments are made to match a known gas turbine performance and/or boundary conditions (e.g., exhaust temperature).

The cooling effectiveness ε is defined as

$$\varepsilon = \frac{T_{gas} - T_b}{T_{gas} - T_c}, \tag{8.21}$$

where

- T_{gas} is the adiabatic wall temperature of the hot gas as represented by the incoming gas stagnation temperature for the stator vanes and the incoming gas *rotor-relative* stagnation temperature (T_{rr}) for the rotor buckets.
- T_b is the surface metal temperature of the stator vane or rotor bucket.
- T_c is the cooling air temperature.

A cooling effectiveness of unity means that the blade surface is cooled to the same temperature as the cooling air. This would be possible only for film cooling with an infinitely large cooling air flow. In the cooling flow equations, ε_∞ is the maximum cooling effectiveness, which is the asymptotic value of ε at infinitely large cooling air flow rates. Typical values of ε_∞ are 1 for film cooling and 0.85 for convective cooling.

An in-depth discussion of gas turbine hot gas path cooling in general and the cooling effectiveness method in particular will be undertaken in Chapter 9. At this point, the reader can either take Equations 8.20 and 8.21 as "given" and continue reading this section or take a brief pause and go to Chapter 9 to better understand the turbine cooling basics before coming back here to pick up the discussion where he or she left it.

The simple model described here does not divide the "pseudo" turbine stage into a stator and rotor. The rotor-relative temperature stagnation temperature is simply set equal to the rotor inlet stagnation temperature (i.e., the firing temperature). (In reality, it is several hundred degrees Fahrenheit *lower* than the corresponding inlet stagnation temperature.) For this simple exercise, all exponents, β_i, and maximum cooling effectiveness values, $\varepsilon_{\infty,i}$, are set to unity.

The calculation is basically a bookkeeping exercise to subtract and add cooling flows at their respective stations as shown in Figure 8.8 and to calculate mixed temperatures. It is encapsulated in the VB function GTURBC (in Appendix C).

8.2.1 Applying the Model

The first task is "calibration" of the model. The calibration basis is the ISO base load rating of General Electric's 60–Hz gas turbine, 7HA.02, as listed in GTW's 2016 Performance Specifications (Volume 32) – also see Table 6.1. The simple cycle rating data

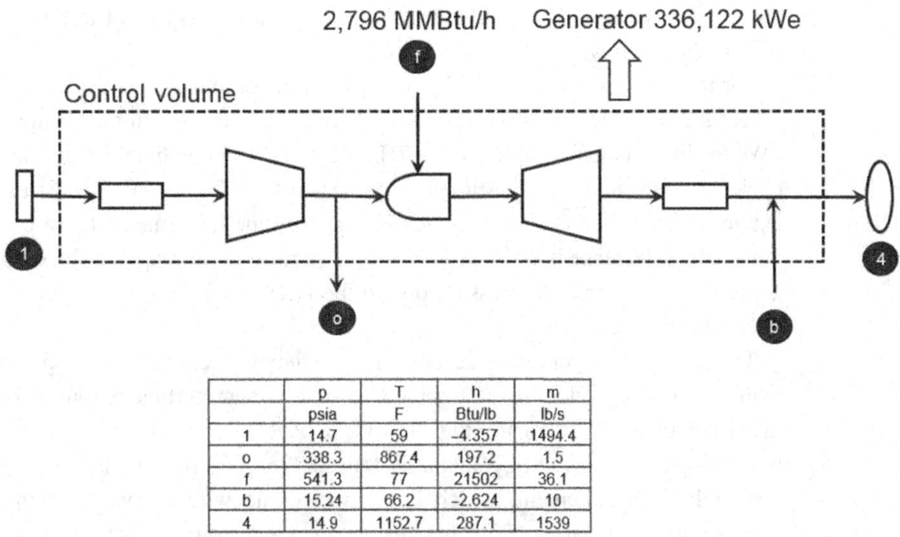

	p psia	T F	h Btu/lb	m lb/s
1	14.7	59	-4.357	1494.4
o	338.3	867.4	197.2	1.5
f	541.3	77	21502	36.1
b	15.24	66.2	-2.624	10
4	14.9	1152.7	287.1	1539

Figure 8.9 Heat balance of 7HA.02 (in Thermoflex), efficiency is 41.13 percent. o = overboard leak; b = exhaust frame blower; f = fuel.

in GTW 2016 are first put to a heat and mass balance test. For the listed exhaust flow and temperature, 1,539 lb/s and 1,153°F, respectively, and the (implicit) fuel flow rate, 36.1 lb/s, the thermodynamically consistent output and efficiency are found as 336 MWe and 41.13 percent, respectively, vis-à-vis listed values of 346 MWe and 42.2 percent (see Figure 8.9).

Potential reasons for the aggressive simple-cycle rating are myriad. First and foremost are the commercial margins hidden in the rating performances, which are practically impossible to "reverse engineer." This one is a certainty; with such scant information, everything else is simply conjecture. For example, there may be other, also commercially driven reasons as well (e.g., upgraded performance to reflect better-than-expected factory test findings, which is rushed for publication). The bottom line is that, for calibration of our model (or *any* model, for that matter), we cannot use performance data failing the heat and mass balance check. Therefore, we proceed with 336 MWe and 41.13 percent as *the* performance to be used for model calibration.

The four cardinal gas turbine parameters (i.e., output, efficiency, exhaust flow, and exhaust temperature) are matched by adjusting and/or setting the following model parameters:

- TIT 2,912°F (1,600°C)
- Compressor PR of 23.1
- Compressor polytropic efficiency of 92.8 percent
- Turbine polytropic efficiency of 88.3 percent
- Correction factor coefficient in Equation 8.7 set to 1.258 (instead of 1.216)

For the cooling effectiveness model, the blade metal surface temperatures T_{b1}, T_{b2}, and T_{b3} are set to 1,550°F, 1,300°F, and 1,200°F, respectively, to reflect state-of-the-art alloys and coatings. The coefficients K_{nch}, K_{ch1}, and K_{ch2} are set to 0.053, 0.153, and 0.076, respectively.

For the combined-cycle calculation, gas turbine output is reduced by 0.32 percent to reflect the changed boundary conditions. This brings the gas turbine output to 334,735 kWe (as listed in GTW 2016 for the 7HA.02 1 × 1 × 1 combined cycle). In a combined cycle, gas turbine fuel is assumed to be heated to 410°F (210°C) utilizing feed water extracted from the bottoming cycle. The gas turbine heat rate increase commensurate with output decrease is estimated to be 0.32 percent (assuming that the output decrease is driven by higher back pressure due to the HRSG). Corresponding exhaust temperature increase is about 1°F.

The bottoming-cycle base second law efficiency is calculated using Equation 8.8a with a technology adder of 4.81 percent to set the steam turbine output to 180,465 kWe (as listed in GTW 2016 for the 7HA.02 1 × 1 × 1 combined cycle). The overall bottoming-cycle technology factor is thus 0.795. The power plant auxiliary load is assumed to be 2 percent, which is commensurate with a once-through, open-loop, water-cooled condenser [2]. This results in a net combined-cycle output of 504.9 MWe vis-à-vis 509 MWe in GTW 2016, which corresponds to an unrealistically low 1.2 percent debit to the gross output for plant auxiliary load. The net combined-cycle efficiency is found to be 62.5 percent, which is 0.5 percentage points higher than the listed value of 62 percent. This is an indirect confirmation that the gas turbine simple-cycle performance commensurate with the listed combined-cycle performance is closer to what we found from the heat and mass balance analysis summarized in Figure 8.9. Since 62 percent seems to be what the OEM is willing to commit to at this stage, for calibration purposes, 0.5 percent is debited to the calculated combined-cycle efficiency.

A combined-cycle performance "sweep" is done with the calibrated model covering gas turbine cycle PRs of 21–35 at TIT values of 1,600, 1,650, and 1,700°C. Calculated efficiencies are shown in Figure 8.10. The chart also includes estimated combined-cycle

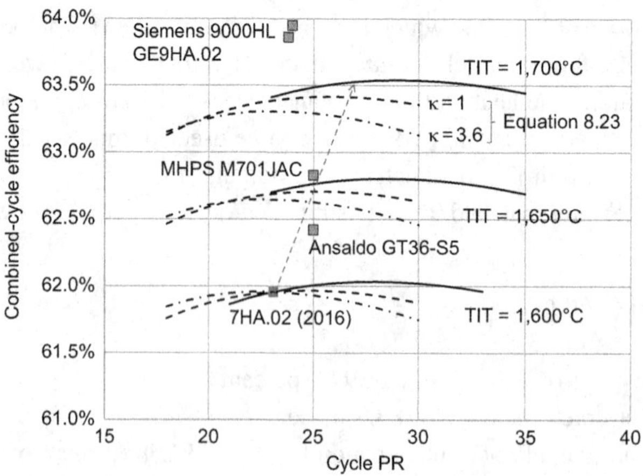

Figure 8.10 Combined-cycle efficiency sweep (TIT–PR). For the data points, see Chapter 21. GE = General Electric; MHPS = Mitsubishi Hitachi Power Systems.

Table 8.1 Combined-cycle (CC) data from the simple, cooled model (cooling flows are referenced compressor airflows). GTG = gas turbine generator; STG = steam turbine generator

Cycle PR	23.1	24.1	25.2	26.2	27.2
TIT (°F)	2,912	2,957	3,002	3,047	3,092
TFIRE (°F)	2,694	2,724	2,753	2,782	2,810
Nonchargeable Flow	8.6%	9.0%	9.4%	9.9%	10.3%
Chargeable Flow	11.7%	12.1%	12.5%	12.8%	13.2%
GTG Output (MWe)	334.7	340.4	345.8	351.0	356.0
GTG Efficiency	41.41%	41.92%	42.41%	42.85%	43.28%
Exhaust Temperature (°F)	1,153	1,154	1,154	1,155	1,156
STG Output (MWe)	180.5	180.6	180.6	181.0	181.4
CC Net Output (MWe)	504.9	510.5	515.9	521.4	526.7
CC Net Efficiency	62.0%	62.4%	62.8%	63.2%	63.5%

efficiencies for the most advanced 50-Hz gas turbines from the four major OEMs using the same assumptions enumerated above. For the actual OEM gas turbine data, please refer to Chapter 21. Qualitatively, the trends are nearly identical to those obtained with a simple uncooled model (see Figure 8.10, solid lines). Quantitatively, efficiencies predicted by the cooled model are more pessimistic at higher TITs by about one percentage point. The reason for this, of course, is the increase in cooling air consumption at higher TITs with the same technology (as represented by blade metal surface temperatures and the technology coefficient, α).

A detailed look is provided in Table 8.1. A higher TIT is accompanied by increasing cycle PR to keep the exhaust temperature constant. Increasing TIT by 100°C (180°F) increases the chargeable cooling flows from 11.7 percent (referenced to compressor airflow) to 13.2 percent. Nonchargeable flows increase from 8.6 to 10.3 percent, which increases the TIT-TFIRE drop from 218 to 282°F. In other words, a 64°F loss is incurred in the "potential" firing temperature.

The question is now "what if we recover that lost TFIRE with a better stage 1 nozzle vane alloy, coating, and cooling techniques?" This is done by increasing the blade metal temperature to 1,600°F and reducing the cooling technology factor to 0.835 (i.e., a 16.5 percent reduction in cooling air flow). The results are shown in Table 8.2.

As shown in Table 8.2, better materials and cooling techniques increase TFIRE by 64°F, leading to a 0.5–percentage point increase in combined-cycle efficiency. Note that the favorable impact of a higher TFIRE is damped by an increase in chargeable cooling flows from 13.2 to 14.1 percent. The reason for that is the higher gas temperature from the front stage (due to the higher TFIRE), which increases the cooling load of downstream stages. In order to alleviate that, better material is used in the mid-stages (represented by a 50°F increase in T_{b2}) with a 0.3–percentage point improvement in combined-cycle efficiency. Further reduction in chargeable flows to the base level increases the combined-cycle efficiency by another 0.3 percentage points to 64.6 percent. Consequently, to reach the 65 percent goal, a one-percentage point increase in turbine polytropic efficiency is needed.

Table 8.2 Combined-cycle (CC) performance improvement. GTG = gas turbine generator; STG = steam turbine generator

Cycle PR	27.2	27.2	27.2	27.2	27.2	27.2
TIT (°F)	3,092	3,092	3,092	3,092	3,092	3,092
TFIRE (°F)	2,810	2,874	2,874	2,874	2,874	2,874
T_{b1}	1,550	1,600	1,600	1,600	1,600	1,600
T_{b2}	1,300	1,300	1,350	1,350	1,350	1,350
T_{b3}	1,200	1,200	1,200	1,250	1,250	1,250
α_1	1.000	0.835	0.835	0.835	0.835	0.835
α_2	1.000	1.000	1.000	1.000	0.900	0.900
α_3	1.000	1.000	1.000	1.000	0.900	0.900
η_τ	88.3%	88.3%	88.3%	88.3%	88.3%	89.3%
Nonchargeable Flow	10.3%	7.9%	8.0%	8.0%	8.1%	8.1%
Chargeable Flow	13.2%	14.1%	12.9%	12.4%	11.5%	11.3%
GTG Output (MWe)	356.0	365.0	371.2	373.7	378.1	384.7
GTG Efficiency	43.28%	43.51%	43.63%	43.67%	43.74%	44.42%
Exhaust Temperature (°F)	1,156	1,175	1,190	1,197	1,207	1,198
STG Output (MWe)	181.4	186.8	191.1	192.9	195.9	193.3
CC Net Output (MWe)	526.7	540.8	551.0	555.3	562.5	566.4
CC Net Efficiency	63.5%	64.0%	64.3%	64.4%	64.6%	64.9%

Note that in this simple calculation, compressor and turbine polytropic efficiencies are kept constant. Polytropic compressor and turbine efficiencies are strongly dependent on the aerodynamic component size and design. For a discussion and data, including predictive equations, the reader is referred to Wilson and Korakianitis ([2] in Chapter 4) or Traupel ([6] in Chapter 2). They need to be corrected for PR and the detrimental effects of the coolant flow via mixing and momentum losses. Formulae that are qualitative in nature and are intended to provide only a sense of the real impact of the turbine cooling on turbine efficiency can be found in Gülen ([2] in Chapter 5). They can be refined, modified, or expanded via comparison or calibration with rigorous aero-thermodynamic stage design calculations. In particular,

$$\eta_c = \eta_{c,\,max} - 0.005\ln(PR_c), \tag{8.22}$$

$$\eta_t = \eta_{t,\,max} - 0.010\ln(PR_t)\left(1+\kappa\sqrt{\chi}\right), \tag{8.23}$$

where χ is the total cooling air flow as a fraction of airflow and κ is an empirical factor. Using Equations 8.22 and 8.23 with $\eta_{c,max} = 0.95$ and $\eta_{t,max} = 0.93$ ($\kappa = 1$), recalibrating the model and rerunning the combined-cycle sweeps resulted in the performance landscape shown by the dashed lines in Figure 8.10, which is not too different from that shown by the solid lines, except for a marked shift in maximum combined-cycle efficiency to lower cycle PRs.

The implication of changing polytropic efficiency with increasing PR is holding the stage count constant while increasing the stage loading. In theory, one can assume "rubber" components with the stage count continuously adjustable with changing PR. In practice, rotor dynamics and size/cost considerations make that infeasible. Strictly speaking, though, it is unavoidable that at some point between a cycle PR of 22–23 and 35, the designer will be forced to change from a four-stage turbine design to a five-stage one. (The underlying aerothermodynamic considerations will be covered in Chapter 10.) The same is true for the compressor layout. In fact, this is an even more complicated subject (e.g., see the advanced compressor for Siemens' next-generation HL–class gas turbine with *12* stages for a cycle PR of 24 in Section 21.1). Nevertheless, for a conceptual design space exploration done at a high level as in this section, such step changes are quite impractical to build into simplified relationships. Thus, these caveats should be kept in mind while interpreting technology curves similar to those in Figure 8.10.

In order to magnify the impact of cooling air on turbine polytropic efficiency, κ is set to 3.6 and $\eta_{t,max}$ is set to 0.965 (to maintain the calibration point) and the model is rerun. The results shown in Figure 8.10 as dash-dot lines display a readily discernable shift in efficiency at higher PRs (as one would expect with higher cooling air temperatures and higher cooling air flows). Nevertheless, the change is not substantial enough to dramatically shift the optimal cycle PR for maximum combined-cycle efficiency.

Comparison of constant polytropic component efficiencies and those from Equations 8.22 and 8.23 (with $\kappa = 1$) shows that the error resulting from a constant efficiency assumption is quite small and explains the closeness of the combined-cycle efficiency predictions in Figure 8.10 from an absolute value perspective. Nevertheless, the shift in optimal cycle PR (for maximum combined-cycle efficiency) is significant, so one can argue that the extra effort for better precision pays off. (Note that in actual design codes used by OEMs, stage efficiency calculations are significantly more complex than the simple corrective formulation in Equation 8.23.) This is certainly true, and it is absolutely imperative that the most rigorous, physics-based, and, most importantly of all, *data-validated* formulations are used in actual component design. From a conceptual perspective, the impact of cooling on turbine efficiency is definitely worth capturing. Nevertheless, it is also instructive to realize that even the simplest physics-based formulations can lead to quite reliable predictions with minimal computational effort. After all, considering the flatness of the constant-TIT efficiency trends with respect to cycle PR in Figure 8.10, if already armed with the knowledge of cycle PR for maximum gas turbine-specific work, one could arrive at the same conclusion regarding the optimal value for maximum combined-cycle efficiency.

8.3 Another "Real" Cooled Turbine Model

Another cooled turbine model is developed by Gülen ([2] in Chapter 5) based on the work of Khodak and Romakhova [6]. The system is shown schematically in Figure 8.11. The conventional, cooled gas turbine Brayton cycle, which is the "base" system, is envisioned as a (hypothetical) *combined* cycle comprising two Brayton cycles as follows:

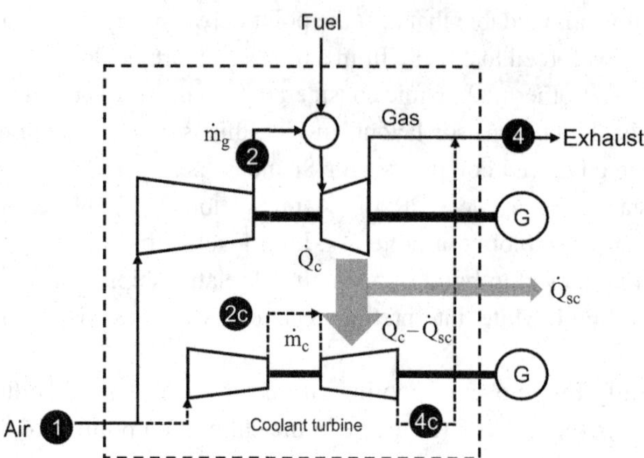

Figure 8.11 Cooled gas turbine model. (after Khodak and Romakhova [6])

1. A *topping* Brayton cycle, which is an "internal combustion" gas turbine with an expander section cooled by sensible heat transfer from the turbine casing.
2. A *bottoming* Brayton cycle, which is an "external combustion" gas turbine with a heat exchanger utilizing the heat rejected from the expander of the topping gas turbine to increase the temperature of the working fluid.

The model is pretty straightforward, using basic thermodynamic formulae developed earlier in the book. The computational "trick," in a manner of speaking, is twofold:

1. Two CVs, one around the combustor and one around the combustor and stage 1 nozzle vanes (conceptual) of the topping gas turbine, to calculate turbine inlet and firing temperatures (see Figure 3.3 in Chapter 3).
2. Two parallel turbines (expanders), one for the topping gas turbine (working fluid is gas) and one for the bottoming gas turbine (working fluid is air), with each turbine CV divided into N sections with successive adiabatic expansion and zero-work heat transfer stages.

In Figure 8.11, subscripts "g" and "c" denote "gas" and "coolant," respectively (the gas in question is the combustion product, whereas the coolant is air). The term $\dot{Q}_c$ is the total heat transfer from the topping turbine to the bottoming turbine. From an actual turbine hardware perspective, it is the total heat transfer from the hot gas path components to the cooling air. $\dot{Q}_c$ is given by

$$\dot{Q}_c = \dot{m}_c c_p (T_b - T_{2c}),$$

where $\dot{m}_c$ is the total cooling air flow rate, T_b is the bulk metal temperature (representative of the turbine technology), and T_{2c} is the temperature of the cooling air, which may be cooled by transferring heat to air, fuel (e.g., as in some Mitsubishi gas turbines), or feed water from the HRSG to make intermediate-pressure (IP) or high-pressure (HP) steam (e.g., as in GE/Alstom/Ansaldo's GT24/26 gas turbines). In steam-cooled gas turbines, some of the heat transfer from the cooled hot gas path components is dumped

Table 8.3 Gas turbines for testing GasTurbine VBA function

Class	E	F	HA
Vintage	1997	1998	2015
GT Gross Power (kW)	114,828	163,546	344,923
GT Gross LHV Efficiency	34.9%	37.4%	41.6%
Fuel Flow (lb/s)	14.5	19.3	36.5
Fuel Temperature (°F)	99.3	123.7	195.2
Heat Consumption (Btu/s)	311,979	414,523	785,072
Gas Turbine Exhaust Mass Flow (lb/s)	782.9	980.3	1,524.4
Gas Turbine Exhaust Temperature (°F)	1,036.2	1,052.9	1,171.3
Cycle PR	14.1	16.6	22.7
Nominal TIT (°F)	2,372	2,552	2,912

into an external sink (e.g., cooling steam). This heat transfer is denoted by $\dot{Q}_{sc}$ in Figure 8.11.

The model in Gülen (cited above) is encapsulated in an Excel VBA function, GasTurbine, which is listed in Appendix C. The function can be used for parametric studies of simple- and combined-cycle gas turbines similar to those discussed in the preceding section. The function can also be used as a heat and mass balance analysis tool (with the added benefit of providing information on TIT, RIT [i.e., "firing" temperature], chargeable and nonchargeable cooling flows, turbine "swallowing area," etc.). The application of the model is demonstrated by applying it to three different classes of air-cooled, heavy-duty industrial gas turbines, which are listed in Table 8.3 (from Thermoflow Inc.'s GTPRO program – engine library #85, #99, and #602 – run at ISO base load). Note that the fuel temperatures in Table 8.3 are at the exit of the fuel gas booster compressor (pipeline pressure was set to 250 psia). Nominal TITs are class-specific (please refer to the discussion in Section 4.7).

Program inputs and outputs are listed in the tables in Appendix C. Highlighted inputs are machine-specific and taken directly from the data in Table 8.3. Turbine polytropic efficiency (input #6) is adjusted to match the exhaust temperature. Nonchargeable flow fraction (input #31) is adjusted to match the TIT. Cooling and hot gas path material technologies are *qualitatively* represented by T_b (higher value means better materials and coatings) and μ_c (lower value means better cooling technology). Other inputs are program defaults are described in detail in Gülen (cited above).

8.4 Performance Derivatives

The three simple models presented herein incorporate all of the important aspects of gas turbine design for optimal simple- and combined-cycle performance. There is really not much more one can learn from thermodynamic analysis alone. With calibration to a known design, extremely robust conceptual design calculations can be performed. (Note that, in this chapter, published rating data for a particular gas turbine were used for calibration. Having access to "real" performance and design data would obviously lead to more reliable predictions and model enhancements.)

The next step in gas turbine analysis involves the following two quite challenging tasks:

1. In-depth turbine cooling analysis
2. Rigorous (ideally 3D) aerodynamic analysis

The ultimate goal is to determine whether a conceptual design emerging from a comprehensive thermodynamic-cycle study is translatable into an actual product within the limits of available materials, coatings, cooling methods, and manufacturing tools and techniques.

The third, even more challenging task involves combustion calculation and combustor design. As far as thermodynamic-cycle analysis is concerned, the design engineer can set the TIT to any value up to stoichiometric combustion with about 2,000°C. Even when material and cooling problems can be overcome (at this point in time, this is extremely unlikely), one would be stopped well before that entitlement limit due to constraints imposed by stable combustion with the extremely low emissions regulated by government agencies.

Those three areas will be covered in more detail in the following chapters.

Before moving on, however, it is illustrative to examine *performance derivatives.* Two other synonymous terms frequently encountered in the trade and academic literature are *influence factors* and *sensitivity factors* (or, simply, *sensitivities*). The four strongest knobs available to the design engineer are polytropic component efficiencies (turbine and compressor) and cooling air flows (chargeable and nonchargeable). Combined-cycle efficiency derivatives with respect to these four parameters are summarized as follows:

- A one-percentage point increase in compressor polytropic efficiency is worth 0.3 percentage points in combined-cycle efficiency.
- A one-percentage point increase in turbine polytropic efficiency is worth one percentage point in combined-cycle efficiency.
- A one-percentage point reduction in nonchargeable cooling flow is worth 0.31 percentage points in combined-cycle efficiency.
- A one–percentage point reduction in chargeable cooling flow is worth
  - 0.55 percentage points in combined-cycle efficiency ($\kappa = 3.6$).
  - 0.49 percentage points in combined-cycle efficiency ($\kappa = 1.0$).

Clearly, there are relatively limited opportunities in compressor and turbine efficiencies, which are probably quite close to their entitlement values. For example, a one-percentage point increase in compressor polytropic efficiency (a huge improvement considering that the state of the art is 93 percent) is worth only 0.3 percentage points in combined-cycle efficiency.

Reductions in chargeable and nonchargeable cooling air flows present significant challenges in materials, coatings, and cooling techniques to maintain part lives, but also offer bigger rewards. For example, slashing the nonchargeable cooling flow by a third is worth about one percentage point in combined-cycle efficiency. For modern H/HA/J/JAC-class gas turbines, this is equivalent to a relative improvement in combined-cycle efficiency or heat rate of 1/60 ≈ 1.5 percent. It is interesting to compare the chargeable cooling flow derivative cited above to a study done by an OEM (see [1] in Chapter 5). The OEM study shows a 2.5 percent (relative) drop in combined-cycle efficiency with the increase of cooling flows from 0.20 (referenced to compressor inlet airflow) to 0.24. Based on the vintage of the study (1991) and the optimal combined-cycle efficiency level of 55 percent for $\mu = 0.20$, this is equivalent to 2.5% × 55 = 1.4 percentage points. Thus, assuming that a gas turbine OEM is best placed to know the "guts" of its end product, the derivative with $\kappa = 3.6$ is most likely closer to "reality." We will return to this later in Section 10.4 and determine that $\kappa = 2.39$ using rigorous Thermoflex model runs. At the end of the day, the reader should realize that these are not bona fide hardware design formulae, but physics-based estimates to get a handle on where we are in the particular problem we are looking at.

The sensitivity of simple- or combined-cycle efficiency to hot gas path cooling flows is a very important piece of information, but it is a very tricky number to get a handle on. There is scant data available in the published literature, and only a few of them are from the "horse's mouth" (i.e., the OEMs). In most cases, the available data are very difficult to extract from the charts and tables in the papers because (being written mostly by academics) they contain many obscure fudge factors or normalized and/or bunched-together parameters, but not the proverbial "meat." Furthermore, there is usually no distinction made between chargeable and nonchargeable cooling flows, and it is never made clear what is meant by "total" cooling flows. To make things even worse, the distinction between percentage change in "relative" or "absolute" terms is rarely made, and very few bothers to state clearly whether the "cooling flow as a percentage" is referenced to the compressor inlet airflow or turbine inlet hot gas flow.

Let us start with the nonchargeable cooling flows. There are two ways to evaluate the nonchargeable cooling flow derivative of gas turbine efficiency:

1. At the same TIT (TFIRE changes)
2. At the same fuel flow rate (TFIRE is constant but TIT changes)

In the first case, the simple-cycle efficiency derivative is essentially zero. There is a slight change in combined-cycle efficiency via slightly higher/lower gas turbine exhaust temperatures, but, for all practical purposes, it can be ignored as well. In the second case, there is a slight change in the simple-cycle efficiency.

The combined-cycle efficiency derivatives cited above are compared with derivatives obtained from other sources in Table 8.4 (for a 10 percent change in cooling flow). The sources include two papers by Sir Horlock's group at Cambridge University and papers written by OEM personnel. In general,

- Gas turbine simple-cycle efficiency is impacted by 0.5–0.9 percent for each 10-percent change in chargeable cooling flow.
- Gas turbine combined-cycle efficiency is impacted by 0.5 percent for each 10-percent change in chargeable cooling flow.

Table 8.4 Efficiency derivatives with respect to hot gas path cooling flows (10–percent change in stated cooling flow). ch = chargeable; nch = nonchargeable; CC = combined cycle; GT = gas turbine

Source	ch/nch	CC Efficiency	GT Efficiency	Notes
This book	nch	0.31%		See the text
This book	ch	0.55%		κ = 3.6
[1] in Chapter 5	?	1.25%		OEM paper
Wilcock et al. [7]	?		0.49%	Table 2
Horlock et al. [8]	?		0.58%	Figure 7
Gülen ([2] in Chapter 5)	ch	0.35%	0.50%	Figures 5 and 7 (cycle PR 20)
Gülen	ch	0.45%		Unpublished
Briesch et al. [9]	?		0.86%	Figure 3

At the same TIT, as stated earlier, there is essentially no impact on simple- or combined-cycle efficiency via a change in nonchargeable cooling flows. At the same TFIRE, the impact on simple- or combined-cycle efficiency is of the order of 0.3 percent.

All derivatives are "negative," i.e.,

- Increasing cooling flow → lower cycle efficiency
- Decreasing cooling flow → higher cycle efficiency

As far as actual gas turbine hot gas path hardware is concerned, polytropic efficiency is a pretty useless parameter from a (sensitivity of) cycle thermal efficiency perspective. This is primarily due to the effect of the cooling flows on the performance of individual turbine stages (typically four, but three in some older GE machines). This will be clear after reading Chapters 9 and 10 (especially, Section 10.4). Furthermore, due to the "reheat" effect, the last stage of the turbine has a bigger impact on the cycle efficiency than the first stage. Consequently, the cycle efficiency derivative with respect to "turbine efficiency" is a murky concept. For a given product from a particular OEM, the only reliable way to evaluate it is to have access to the applicable "cycle deck."

For "back of the envelope" purposes, consider that the cooled-stage isentropic efficiency (total to total) is somewhere between 80 and 90 percent depending on the technology vintage and stage number (i.e., first, second, third, etc.). Each 1–percent "relative" change (i.e., from 85 to 85.9 percent) *in one turbine stage's efficiency* is worth anywhere from 0.15 to 0.20 percent in combined cycle heat rate. The derivative is "negative" with respect to heat rate and "positive" with respect to efficiency. Thus, for example, if the combined-cycle base efficiency is 60% (5,687 Btu/kWh in heat rate), a 1–percent improvement in *one* turbine stage's efficiency *reduces* the combined-cycle heat rate by 11–12 Btu/kWh. This is equivalent to a 0.12–*percentage point* increase in efficiency (i.e., from 60 to 60.12 percent).

Some other useful derivatives are as follows:

- Each 18°F (10°C) in TFIRE is worth about 0.18–0.23 percent in combined-cycle heat rate (–).
- Each 54°F (30°C) in fuel gas temperature is worth about 0.13 percent in combined-cycle heat rate (–).

The signs indicate whether the derivative is positive or negative. For heat rate, (–) means improvement (i.e., a reduction with the increasing value of parameter x). Similarly, (+) means deterioration (i.e., an increase with the increasing value of parameter x).

Some performance derivatives are different whether one is looking at a "rubber" design or to a "fixed" design (or off-design) case. The latter refers to the case of an actual power plant operating in the field with fixed equipment (fixed hardware). The former refers to, say, a conceptual design study where each different value of a particular parameter corresponds to a new hardware design (hence the moniker "rubber").

For example,

- Each 1 percent in compressor airflow is worth about 0.2 percent in combined-cycle heat rate for *rubber* designs (–).
- Each 1 percent in compressor airflow is worth about 0.05 percent in combined-cycle heat rate for *fixed* designs (+).

Similarly,

- Each 1 in. of H_2O (2.5 mbar) in gas turbine inlet or exhaust loss is worth about 0.13 percent in combined-cycle heat rate for *rubber* designs (+).
- Each 1 in. of H_2O in gas turbine inlet or exhaust loss is worth about 0.01 percent in combined-cycle heat rate for *fixed* designs (+).

8.5 Improving the Basic Cycle

In this section, the focus is exclusively on the basic Brayton cycle with four processes. The fundamental thermodynamic analysis in Chapter 5 revealed the following facts regarding the gas turbine Brayton cycle:

1. There are two basic mechanisms to improve the cycle efficiency:
   a. Increase mean effective cycle heat addition temperature (METH)
   b. Decrease mean effective cycle heat rejection temperature (METL)
2. To increase METH,
   a. Increase cycle PR
   b. Increase cycle T_3
   c. Combination of both
3. To decrease METL,
   a. Increase cycle PR
   b. Add a bottoming cycle
4. Cycle PR is the most important "knob" available to the design engineer because it increases METH and decreases METL *simultaneously*.
5. However, high cycle PR leads to low net cycle work *unless* accompanied by high T_3.
6. Fundamental analysis and commercial gas turbine rating data confirmed that there is an *optimum* cycle PR for a given cycle T_3 that

a. Maximizes gas turbine-specific power output and
b. Maximizes combined-cycle efficiency

These considerations, when summarized on a temperature–entropy diagram, lead to three "textbook" Brayton-cycle improvement mechanisms. They logically follow from the "turning" of the strongest knob available to the gas turbine designer; namely, the cycle PR. Here they are in their logical (thermodynamically, that is) order:

1. *Reheat* – to increase Brayton-cycle-specific work and bottoming-cycle potential at high cycle PR *without* increasing the TIT;
2. *Intercooled* compressor – to reduce compressor work input *and* the compressed air temperature (to be used in turbine hot gas path cooling); and
3. *Recuperative* heating of compressed air with hot turbine exhaust gas – to reduce fuel burn in the combustor for a specified TIT, especially with an intercooled compressor.

Redrawing the T–s diagram of Figure 5.1 for a high cycle PR (see Figure 8.12) graphically illustrates the substantial increase in cycle efficiency (via higher METH and lower METL) accompanied by an equally substantial reduction in net cycle work (the shaded area) and in combined-cycle potential (the triangular area {1–4–4C–1}). This cycle is representative of aeroderivative gas turbines, which have good efficiencies of around 40 percent or higher, but mostly (with the exception of GE's intercooled LMS100) well below 100 MWe output and low exhaust temperatures. They are primarily used in a simple-cycle configuration for peaking duties.

The solution to the problem of low combined-cycle efficiency with high-PR gas turbines presents itself as a truncated expansion followed by another heat addition process as shown in Figure 8.13. The second heat addition or "reheat" maintains the high cycle PR, increases METH over the base Brayton cycle while keeping the same cycle maximum temperature, T_3, and also preserves the high bottoming cycle potential represented by the triangular area {1–4–4C–1}.

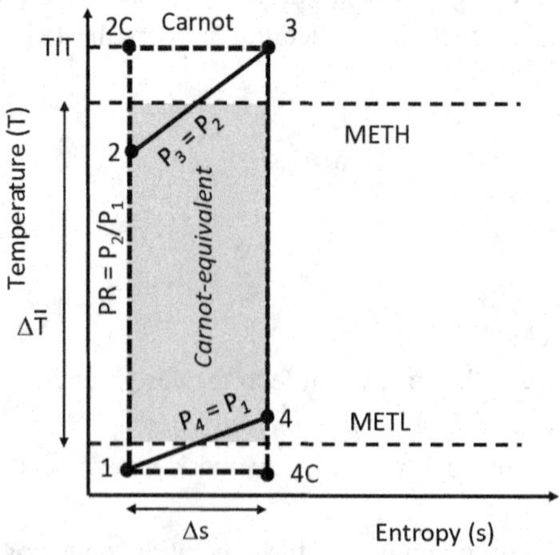

Figure 8.12 High PR Brayton-cycle T–s diagram (exaggerated for visual effect).

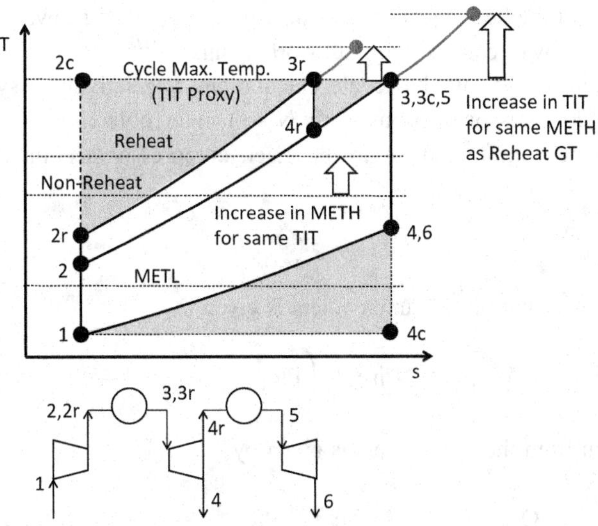

Figure 8.13 Reheat Brayton cycle T–s diagram. GT = gas turbine.

In Figure 8.13, heat addition and heat rejection irreversibilities (losses) of the ideal Brayton cycle {1–2–3–4–1} are represented by the triangular areas {2–2c–3c–2} and {1–4–4c–1}, respectively. Reheat cycle {1–2r–3r–4r–5–6–1} reduces the heat addition irreversibility as quantified by the area {2–2r–3r–4r–2}. The net effect is an increase in cycle METH and cycle efficiency without an accompanying increase in cycle maximum temperature (i.e., T_3).

Intercooling is a well-known method for reducing compressor power consumption, which takes away roughly 50 percent of turbine power output in a gas turbine (e.g., see Table 5.4). The disadvantage is a reduced compressor discharge temperature, which leads to an increase in heat consumption for a given T_3. Therefore, intercooling makes sense when it is combined with recuperation (see below) – especially when combined with reheat. Obviously, this results in an even more complex and costly system with the addition of the intercoolers to the recuperator. Thus, it is no surprise that there are no commercially available gas turbines with intercooling and recuperation. However, Rolls-Royce WR-21 (ICR) is an intercooled–recuperated engine used for marine propulsion applications (primarily used in Royal Navy ships, although development funding was provided by the US and French navies as well). It is rated at 21 MWe with an efficiency of about 38 percent.

Intercooling by itself is of interest for applications in high-PR aeroderivative gas turbines for the reduction of compression power and compressor discharge temperature, thereby eliminating the need for CAC. (Since the exhaust temperature is too low due to high cycle PR, recuperation is precluded.) GE's 100-MWe LMS100 (with a cycle PR of 40 and nearly 45 percent efficiency) is a commercially available intercooled aeroderivative gas turbine, which has been deployed successfully in peaking duty applications. For advanced heavy-duty industrial gas turbines with PRs of 20–23, intercooling is unlikely to be an economic feature. The size of the intercoolers and the added cost associated piping and the heat rejection system (e.g., a dry cooling tower), along with the latter's parasitic power consumption, all work against it.

Introducing intercooling into the cycle calculations is relatively easy. Consider that there are N compressor sections with N – 1 intercoolers in between. Each intercooler cools the compressed air to T_1. Each compressor section i has a PR, which is given by

$$PR_i = PR_{overall}^{\frac{1}{N}}. \tag{8.24}$$

Total compressor work with N – 1 intercoolers is given by

$$\dot{W}_{comp} = N \dot{m} c_p T_1 \left(PR_{overall}^{\frac{k}{N\eta_c}} - 1 \right). \tag{8.25}$$

Total heat rejection from the intercoolers is given by

$$\dot{Q}_{intc} = (N-1) \dot{m} c_p T_1 \left(PR_{overall}^{\frac{k}{N\eta_c}} - 1 \right). \tag{8.26}$$

Total compressor work without intercooling is given by

$$\dot{W}_{comp} = \dot{m} c_p T_1 \left(PR_{overall}^{\frac{k}{\eta_c}} - 1 \right). \tag{8.27}$$

Consequently, the work reduction via intercooling is represented by the fraction

$$\beta = N \frac{PR_{overall}^{\frac{k}{N\eta_c}} - 1}{PR_{overall}^{\frac{k}{\eta_c}} - 1}. \tag{8.28}$$

Implementing Equation 8.28 in Excel shows that with N = 2 (i.e., only one intercooler), β is about 0.75 (i.e., 25 percent reduction in compression work). Thereafter, the benefit diminishes rapidly with increasing numbers of intercoolers, and beyond N = 5, there is practically zero benefit. Thus, it is extremely unlikely for more than one intercooler to be economically feasible. For a quick estimate, it is sufficient to calculate β with N = 2 and use it to reduce the non-intercooled compression work.

Recuperation involves the addition of a heat exchanger (usually counter-flow) into the cycle. Hot exhaust gas from the turbine is utilized in this heat exchanger to heat the compressed air prior to its entry into the combustor. The cycle is shown in Figure 8.14. Also shown in Figure 8.14 is the power train arrangement of Solar Turbines' recuperative Mercury™ 50 gas turbine (see Section 8.5.1). Hot exhaust at state 3.1 cools down to state 4 while heating cold compressed air at state 2 to state 3.1. This "recuperative" heat exchange increases METH (now a logarithmic mean of T_3 and $T_{3.1}$ instead of T_3 and T_2) while simultaneously decreasing METL (now a logarithmic mean of T_1 and $T_{4.1}$ instead of T_1 and T_4). Thus, its beneficial impact on cycle efficiency is two-pronged and is similar to that of a higher cycle PR.

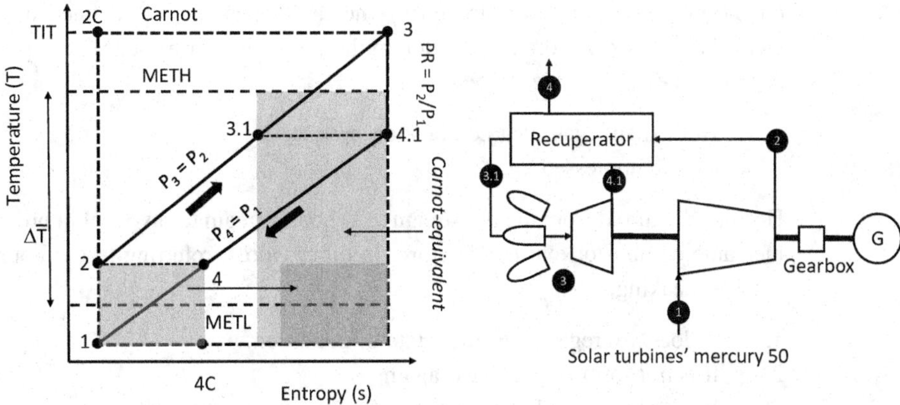

Figure 8.14 Recuperative Brayton-cycle T–s diagram.

On the debit side of the ledger, as indicated by the small triangular area {1–4–4C–1} in Figure 8.14, recuperation significantly diminishes the combined-cycle potential. In that sense, its dilemma is similar to that encountered by the aeroderivative gas turbines with high cycle PRs, but with the additional burden (in terms of cost and size) imposed on it by a heat exchanger. Solar Turbines (owned by Caterpillar) Mercury 50 is essentially the only recuperated gas turbine that is commercially available. It is a small unit rated at 4.6 MWe and 38.5 percent efficiency (see Section 8.5.1 for more on this gas turbine).

Incorporating the impact of recuperation into cycle calculations is quite easy as well. Using the concept of heat exchanger effectiveness, the temperature of the air heated prior to entry into the combustor can be calculated as (with the state-point numbering in Figure 8.14)

$$T_{3.1} = T_2 + \varepsilon(T_{4.1} - T_2).$$

An effectiveness of $\varepsilon = 1$ implies an infinitely large heat exchanger with zero temperature deltas. For a reasonably large, industrial gas turbine, using $\varepsilon > 0.85$ is probably not a judicious choice. With modern printed-circuit or plate-fin heat exchangers, $\varepsilon > 0.9$ is achievable. In any event, a cost–performance trade-off is requisite for a better answer.

The energy conservation around the recuperator can be written as

$$(1 - \text{nch} - \text{ch})c_{p,a}(T_{3.1} - T_2) = (1 + f)c_{p,g}(T_{4.1} - T_4).$$

Combined with the recuperator effectiveness, this can be rewritten as

$$\frac{1 - \text{nch} - \text{ch}}{1 + f}\frac{c_{p,a}}{c_{p,g}}\varepsilon = \frac{T_{4.1} - T_4}{T_{4.1} - T_2}.$$

This is a good place to shed some light on a minor but quite irritating terminology confusion. In some texts, recuperation is referred to as *regeneration*. Similarly, respective heat exchangers are interchangeably referred to as *recuperators* or *regenerators*. According to Merriam-Webster, "recuperate" means "to regain a former state or condition" (e.g., health or strength), whereas "regenerate" is defined as "formed or created again" or "restored to a better, higher, or more worthy state." The particular heat exchange herein involves two streams:

1. Hot exhaust gas (i.e., combustion products)
2. Cold compressed air

Heat is transferred from the hot stream to the cold stream to save fuel in order to achieve the same combustor exit temperature. In other words, referring to the hot gas stream, strictly speaking,

1. It does *not* regain a former state or condition.
2. It is *not* formed or created again.
3. It is *not* restored to a better, higher, or more worthy state.

In conclusion, linguistically, neither "regeneration" nor "recuperation" are appropriate. The process in question is "waste heat recovery" (WHR; i.e., retrieval of some [but not all] of the heat energy in a hot gas stream, which otherwise would be "wasted" via ejection through a stack to the surroundings). This brings a third term to mind – *recoup*, which, according to Merriam-Webster, means "to make good or make up for something lost." Now, this fits into the situation at hand quite seamlessly – it looks like someone, somewhere, at some long-forgotten time, decided to get rid of the "o" in "recoup" and added the "–eration" to arrive at "recuperation" (because there is no "recouperation" defined in Merriam-Webster).

In fact, according to an article that appeared in the Fall 2001 issue of *Energy Matters* published by the US Department of Energy (DOE),[1] "a *recuperator* is a gas-to-gas heat exchanger placed on the stack of the furnace," whereas "a *regenerator* is a rechargeable storage battery for heat." The latter is

> an insulated container filled with metal or ceramic shapes capable of absorbing and storing relatively large amounts of thermal energy. During part of the operating cycle, process exhaust gases flow through the regenerator, heating the storage medium. After a while, the medium becomes fully charged, so the exhaust flow is shut off and cold combustion air is admitted to the unit. As it passes through, the air extracts heat from the storage medium, increasing in temperature before it enters the burners. Eventually, the heat stored in the medium is drawn down to the point where it is necessary to recharge the regenerator.

This is a sufficiently clear description of two distinctly different devices of WHR. As such, the convention in this book is as follows:

1. A heat recovery process from the hot exhaust/stack gas of a gas turbine for heating cold combustion air is *recuperation*.
2. The heat exchanger accomplishing this duty is a *recuperator* or, alternatively, a *recuperative* heat exchanger.

[1] "Increased Efficiency through Waste Heat Recovery," by Richard L. Bennett, President, Janus Technology Group, Inc., Rockford, IL.

In conclusion, for large industrial gas turbines in utility-scale electric power generation applications, the only feasible and commercially available (albeit with a spotty record) cycle feature is reheat under the marketing name of "sequential combustion." This type of gas turbine will be covered in more detail later in the book (e.g., Chapter 21).

In terms of modern, utility-scale electric power generation, intercooled and/or recuperated gas turbines are of little interest. Having said that, it must be emphasized that, in the early days of gas turbine-based electric power generation, complex power plants with intercooling and recuperation were quite common. In fact, in 1951, only two years after the first Frame 3 gas turbine was installed at Belle Island (see Chapter 4), GE installed three gas turbine power plants in Rutland, VT, based on a two-shaft derivative of the same turbine.[2] The machinery was indoors and included twin intercoolers and recuperators (in parallel, on both sides of the power train). These gas turbines were dubbed as "kilowatt machines" by GE engineers.[3] The owner and operator of these plants was Central Vermont Public Service (CVPS), and they were operational until 1987 when, after 100,000 hours of operation, they were mothballed and later sold for scrap.[4] Overall, 10 turbines of this design were manufactured by GE, and all were installed in the early 1950s in various states.

A Thermoflex model of one of the three CVPS plants was built and run at the design site conditions of 1,000 feet altitude (about 14.2 psia) and 80°F, at which the plant could generate 5 MWe. Gross efficiency was calculated as 28.8 percent. For 1,500°F TIT and 80°F ambient, base Carnot efficiency is

$$1 - (80 + 460)/(1{,}500 + 460) = 72.45\%.$$

From Equation 5.2, the ideal Brayton-cycle efficiency is 46.72 percent (with a cycle PR of 9.1) so that the cycle factor is 46.72/72.45 = 0.64 and the technology factor is 28.8/46.7 = 0.62. The expected overall thermal efficiency for the power plant was 26.5 percent [4]. These are very respectable numbers for a 1950s vintage gas turbine with a modest TIT and cycle PR, which, of course, came at the cost of considerable size and complexity. Once advances in compressor aero-thermodynamics, turbine hot gas path materials, and coatings made compact, "jet engine-like" gas turbines with high cycle PRs and firing temperatures technically feasible, complex plants á la Rutland quickly became history.

8.5.1 Recuperated Gas Turbine

Solar Turbines' Mercury 50 recuperated turbine was developed under the auspices of US DOE's Advanced Turbine Systems (ATS) program in the 1990s. It went commercial in 2003 with about 100 of them in service around the world. Mercury 50 is a single-shaft, *industrial* gas turbine with a cycle PR of 9.9 and a shaft speed of slightly above 14,000 rpm. It is rated at 4.6 MWe-38.5% (ISO base load) with 39.2 lb/s airflow and

[2] As told by Dave Lucier in a blog on the website of PAL Turbine Services, LLC (www.pondlucier.com).
[3] Ibid. [4] Ibid.; apparently, GE was offered the units at museum prices, but unfortunately they declined.

about 370°C exhaust temperature. (When it first came out, Mercury 50 was rated at more than 40 percent efficiency.)

Mercury 50 has a recuperator for high efficiency and an "ultra-lean premix" combustor for low NOx emissions (less than 5 ppmv at 15 percent O_2). Its primary deployment has been for cogeneration applications (combined heat and power, or CHP, as Europeans prefer to call it) in university campuses and hospitals.

The two-stage turbine has a firing temperature of about 1,163°C (at the first-stage rotor inlet) [10] and an exhaust temperature of 374°C (705°F). Similar to the design philosophy adopted by GE for their "frame" gas turbines, the fully cooled first turbine stage is highly loaded. Consequently, the second stage has cooled stator vanes but uncooled rotor blades (which are shrouded). This saves on cooling air and enables the "ultra-low" lean-premix combustion in the annular combustor, which is an advanced variant of Solar's SoLoNox combustion system. It incorporates a diverter valve to let some of the airflow bypass the combustor at part load operation. (A similar system was used by Mitsubishi and Westinghouse in can-annular combustors for their frame machines.)

As an ATS gas turbine, Mercury 50 incorporated new superalloys and casting technologies. Single-crystal first-stage nozzle vanes are made from MAR-M-247, a nickel-based superalloy, which is also used for conventionally cast (equiaxed) second-stage nozzle vanes. First-stage, third-generation, single-crystal rotor blades are made from CMSX-10. Shrouded and uncooled second-stage rotor blades are also made from equiaxed MAR-M-247. A powdered metal forging of Udimet 720 was used for the second-stage wheel. The fine grain structure of the rim makes it suitable for the higher rim operating temperatures while still maintaining sufficient low-cycle fatigue strength at the hub.

The recuperator (90 percent effectiveness) comprises "air cells" constructed from 0.1−mm-thick Alloy 625 (formed into an undulating corrugated pattern, which maximizes the contact surface between hot exhaust gas and "cold" compressed air). The originally specified 347 SS was dropped due to unsatisfactory performance and durability identified in field tests. Pairs of these sheets are welded together around the circumference to form air cells. Layers of these cells are clamped together with clamping bars and the assembly is welded to the intake and discharge headers (i.e., no internal welds or joints). Stacked layers constitute multiple friction interfaces for energy absorption and prevent high-cycle fatigue. They also help with sound attenuation without a silencer and the additional pressure drop that would go with it.

Using the same assumptions as in LM2500+ (see Chapter 23), the following technology characteristics are calculated for Mercury 50:

- METH = 1,311°F (TIT is assumed to be 100°C higher than firing temperature)
- METL = 339°F
- Ideal Brayton-cycle efficiency = 54.9 percent
- Technology/Carnot factor = 38.5/54.9 = 0.70

The ever-present technology factor of 0.70 emerges again. It shows that when hunting for Brayton-cycle efficiency, there are many (proverbial) ways to "skin the cat." Aeroderivative gas turbines such as GE's LM2500+ achieve the same fraction of their thermodynamic potential with a high cycle PR, whereas Solar Turbines' Mercury 50 does so with a textbook cycle variation; namely, recuperation and a cycle PR of only ~10 (less than half of that of LM2500+). The trade-off, as always, is between performance and size, weight, materials, complexity, and cost expended to achieve it. (Interestingly, LMS100, with a much higher cycle PR, higher TIT, and intercooling, tries to do the same, but "Carnot's curse" follows – see Section 23.5.)

8.5.2 Humidified Air Turbine

The humidified air turbine (HAT) is a variant of the recuperated Brayton cycle, which got a lot of attention in the 1990s. When all is said and done, even though several very small research/pilot versions have been built and operated to collect data, the technology was too complex and expensive and consumed too much water (a scarce resource in many places in the world) to turn it into a commercially viable product. There is no prospect of this situation being different in the future and, as such, HAT would have been ignored in this book. As it turns out, though, the author started his industry career in a start-up company that developed and marketed an ingenious version of the basic HAT; namely, the "cascaded HAT" or CHAT. One might say that he has a "soft spot" in his heart for this particular cycle. In any event, the reader is referred to the section on HAT in the handbook chapter written by the author for a brief introduction to the technology and references for further reading [11].

Note that HAT is different from the *steam-injected gas turbine* (STIG),[5] in which steam generated in the HRSG is injected into the gas turbine combustor for power augmentation and NOx suppression. In the *evaporative gas turbine* (EGT), water (not steam) is injected into the compressor discharge of a recuperated gas turbine. Evaporatively cooled compressed air is then heated in the recuperator by the hot exhaust gas. For a thorough thermodynamic analysis of the EGT, the reader is referred to the paper by Horlock [12].

From a basic thermodynamic perspective, HAT is a recuperated gas turbine in which the single recuperator, which is a counterflow compact heat exchanger, is replaced by the combination of a WHR economizer and a direct-contact heat exchanger. The latter, commonly referred to as a "saturator," is a packed column.[6] In a HAT cycle, heat recovered from the hot gas turbine exhaust gas is transferred to the "cold" compressed air via *direct-contact heat and mass exchange* with hot water from the economizer. The cycle's benefits are multifold:

- Reduction of heat addition in the combustor
- Reduction of compressor load (since part of the turbine flow is provided by the water vapor absorbed in the saturator)

[5] Note that STIG is a GE trademark, but is used as a generic descriptor (akin to calling all tissue products Kleenex).

[6] Columns with random and structured packings are widely used in the chemical process industry (e.g., for distillation applications). The fuel gas saturators of the H-System gas turbines in Inland Empire Energy Center are randomly packed columns, which moisturize and heat natural gas (see Section 21.2 in Chapter 21).

Depending on the particular cycle design, the H_2O content of the hot combustion gas at the turbine inlet can be as high as 30 percent by volume. This changes the working fluid characteristics significantly. Specific heat becomes higher (about 0.37 Btu/lb-R or 1,500 J/kg-K vis-à-vis 0.33 Btu/lb-R [1,375 J/kg-K] for combustion gas in a typical F–class gas turbine with 11–12 percent H_2O by volume in the mixture), whereas γ becomes lower (about 1.06 vis-à-vis about 1.33 in typical gas turbines). Consequently, the HAT cycle requires a new turbine (expander) development, which is a very expensive and lengthy undertaking – especially if the hot gas path temperatures are high enough to necessitate vane/blade cooling.

It can be shown that, at least "on paper," with a cycle PR of 45 and a TIT of 1,600°C, a HAT turbine can achieve about 52.4 percent net efficiency (see figure 13.47 in [11] and the related discussion). While this is a very respectable performance, it does not pay off in the face of the accompanying cost and complexity. For small-scale applications, natural gas-fired, reciprocating internal combustion engines (by Wärtsila or GE Jenbacher) can easily get to about 50 percent efficiency in a much more cost-effective and flexible manner. For larger, central power generation applications, gigawatt-scale H–class combined-cycle power plants are rated at 60 percent efficiency at ISO at less than $1,000/kW.

The CHAT, however, goes beyond the basic HAT concept. In particular, it is an intercooled *and* recuperated reheat gas turbine with two combustors in series. A simplified system diagram (ignoring heat recovery from the compressor intercoolers) is shown in Figure 8.15. A more detailed version can be found in the paper by Nakhamkin et al. [13] (see figure 3 therein), which also includes a detailed description of the system and the individual pieces of equipment.

The heart of the CHAT concept is a modified F–class gas turbine, specifically Westinghouse 501F (note that this concept goes back to the early 1990s, at which time W501F was *the* state of the art). The original W501F compressor is removed and replaced by the compressor of the smaller W501D, which increases the shaft output of the new gas turbine by 140 MW (up from the original 160 MW). This requires a new generator, which is placed between the compressor and the turbine.

The overall cycle PR is about 76 (no typo, it is *seventy-six*). The compression is accomplished by a three-section compressor train with two intercoolers in between. The ingenuity of the CHAT lies in the layout of said compressor train. The IP and HP compressors are on a separate shaft from the main gas turbine, which includes the low-pressure unit (i.e., the axial W501D compressor). The IP compressor is a combined axial/centrifugal compressor, whereas the HP compressor is a barrel-type centrifugal compressor. They are driven by a HP expander on the same shaft as they are on. In essence, the IP–HP compressors and the HP expander of the CHAT cycle comprise a balanced shaft akin to the *gas generator* of an advanced aeroderivative gas turbine.

The system is heavily intercooled to save compression power. Heat recovered by the intercoolers is utilized in preheating the saturator water, which is further heated in the *economizer* section of the gas turbine WHR unit. Intercooling saves compression

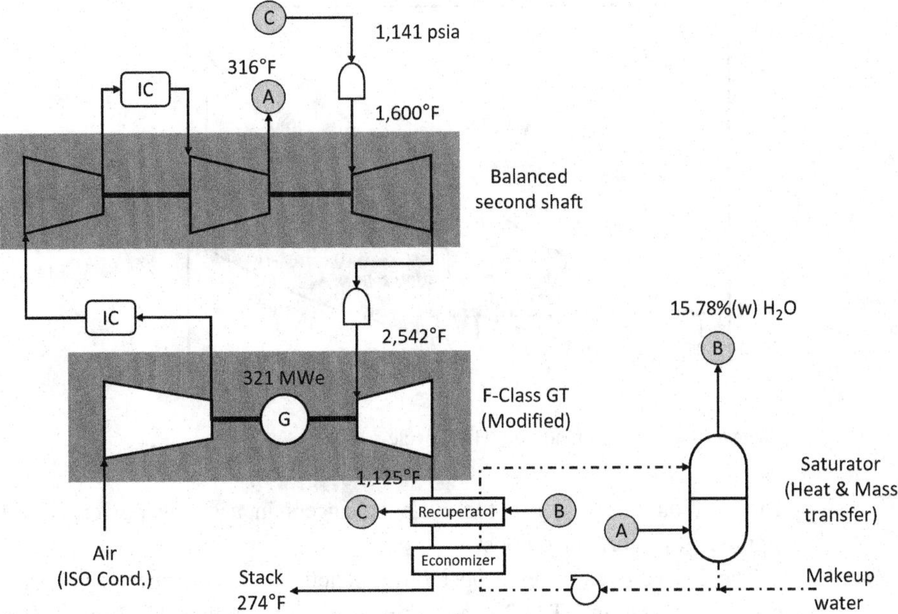

Figure 8.15 Simplified CHAT cycle. GT = gas turbine; IC = intercooler.

power, but is harmful to the cycle efficiency because of the low compressor discharge temperature (316°F in CHAT). Fuel consumption in the HP combustor is reduced by recovering heat from the W501F exhaust gas (1,125°F) in the *recuperator* section of the WHR unit.

Before the recuperator in the WHR unit, compressed air from the HP compressor discharge is heated in the saturator via direct-contact heat exchange with the hot water from the WHR economizer. The heat transfer in the saturator packing comprises two mechanisms: (i) sensible heat transfer and (ii) evaporation of water into the "dry" compressed air (low relative humidity; i.e., mass transfer). A very detailed description of the heat and mass transfer between water and air in the saturator is provided in the two-part paper by Aramayo-Prudencio and Young [14,15].

Heated and humidified air, containing water vapor by about 16%(w), which is about 23%(v), is directed to the HP combustor and expander with a TIT of 1,600°F and a turbine inlet pressure of 1,085 psia. The selected TIT ensures that the HP expander does not require cooling. It is a six-stage expander, whose design is a blend of steam and gas turbine flow path characteristics. The exhaust from the HP turbine goes to the combustor of the W501F turbine with 2,542°F TIT. Generator output is 321 MWe; net output is estimated at 317 MW with a corresponding efficiency of 54.7 percent.

The ideal, air-standard cycle for the CHAT is shown in Figure 8.16. The ideal cycle efficiency is 66.6 percent, which corresponds to a technology factor of 0.82 for the CHAT cycle (see Chapter 6 for a full discussion of the technology factors). Clearly,

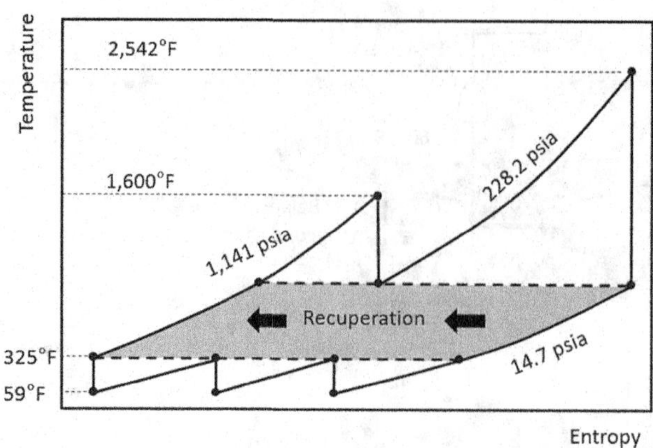

Figure 8.16 Ideal, air-standard CHAT cycle.

combination of *three* Brayton-cycle enhancers in a single power plant results in a significant performance boost.

The CHAT cycle offered operability benefits as well. Reheat combustion, as in the case of GE/Alstom GT24/26 gas turbines, ensures that the CHAT cycle would have had better part-load efficiency vis-à-vis typical combined cycles. Humidification enables much lower efficiency lapses at high ambients because a reduction in airflow is compensated for by moisture addition in the saturator. Water vapor in the combustion air acts as a heat sink and limits the NOx production.

Despite its myriad good features, CHAT failed to turn into a commercial product. In the early 2000s, variants were proposed with small industrial and aeroderivative gas turbines (e.g., Rolls-Royce Avon). With these smaller units, which are eminently unsuitable to combined-cycle applications due to their low exhaust gas temperatures, CHAT could present an opportunity for better performance at comparable capex. This is not surprising because CHAT can be thought of as a "combined cycle" that evaporates water *below* its boiling point in the saturator and uses the gas turbine as a quasi-steam turbine.

Unfortunately, even though the concept was based on existing technology, the investment required to modify the core gas turbine (Westinghouse W501F in the original version) and develop the HP turbine and combustor (Dresser-Rand was the technology partner OEM for that component and the IP–HP compressors) was simply too much. The concept eventually faded away.

8.5.3 Cheng Cycle

The Cheng Cycle is the name (and registered trademark) given by its inventor, Dr. Dah Yu Cheng, to his patented system, which is essentially a gas turbine with a single-pressure HRSG, whose steam product is fully used for injection into the gas turbine. In other words, it is a gas turbine combined cycle where the gas turbine also serves as the "steam turbine." The original invention goes back to the early 1970s. For an early paper on the Cheng Cycle, please refer to Digumarthu and Chang [16].

The inventor has a very pronounced pet peeve; namely, that his invention is *incorrectly* referred to as another version of STIG, which he strongly rejected in several trade journal articles and papers (e.g., see [17]). Alas, in the opinion of this author, this claim does not pass the "duck test" (i.e., "if it looks like a duck, swims like a duck, and quacks like a duck, then it probably is a duck"). The Cheng Cycle is essentially a STIG with a dedicated HRSG and an ingenious control philosophy (see the next section).

At the time of writing, the Cheng Cycle, primarily as a retrofit, is commercially offered by Cheng Power Systems, Inc. According to the company website, there are 135 Cheng Cycle units worldwide with Allison 501KH small industrial gas turbines. As stated on the website, the steam-injected Allison 501KH was co-developed with Detroit Diesel Allison in Indianapolis, IN. In 1984, the first commercial Cheng Cycle was installed at San Jose State University located in San Jose, CA, in a cogeneration capacity.

The system is quite straightforward, with a gas turbine exhausting to a single-pressure HRSG. A steam flow control valve, in conjunction with a control algorithm and the HRSG thermal inertia, ensures that the system responds rapidly to changing load and extra power demands so that the system always runs at peak efficiency. A performance map for 501KH San Jose State's Cheng plant can be found in the article "Combined Cycle Heat Rates at Simple Cycle $/kW Plant Costs" in GTW's March–April 2013 issue.

8.5.4 STIG™

GE's STIG systems utilize steam produced in a HRSG for power augmentation (steam injection for NOx abatement does not qualify as STIG). The HRSG can be duct-fired. Some of the steam can be sent to a process host (e.g., cogeneration application). Typically, steam can be injected with the gas turbine operating from 50 percent power to full load. The basic system description is the same as for the Cheng Cycle.

The location at which steam is injected into the gas turbine differs according to the design of the particular model. Typically, steam is injected into the HP spool at the combustor fuel nozzles (for NOx abatement) and the compressor discharge plenum (for power augmentation). STIG versions of LM1600, LM2000, and LM2500 are available. According to GE, steam injection for power augmentation is not planned for LM6000.

The impact of steam injection on gas turbine performance is shown in Table 8.5. The performances are at ISO base load with natural gas fuel (4/10 in. H_2O inlet/exit losses).

8.5.5 Air Injection

Injection of air into the gas turbine at the compressor discharge is another method to make up for the loss of compressor inlet mass flow at high ambient temperatures. In dry air injection, an industrial compressor compresses air to a pressure that, after piping and combustor losses, replicates the nominal turbine inlet pressure (i.e., at ISO base load). A gas-to-gas recuperator utilizes turbine exhaust gas to heat the compressed air to a

Table 8.5 STIG capability of GE aeroderivative gas turbines ([2] in Chapter 23)

Model	Dry MWe	Dry Efficiency	STIG MWe	STIG Efficiency
LM1600	13.3	35%	16	37%
LM2000	18	35%	23.2	39%
LM2500	22.2	35%	27.4	39%

Table 8.6 Dry and humid air injection for hot day power augmentation (GE's 7EA gas turbine)

	Base	Dry Air Injection	Humid Air Injection	
Air Moisture Content (%)	NA	NA	9	11
Net Output (MW)	74.3	88.6	103.9	106.2
Net Heat Rate (LHV) (Btu/kWh)	9,760	10,640	9,440	9,380
Incremental Output (MW)	0	14.3	29.6	31.9
Incremental Heat Rate (Btu/kWh)	0	15,210	8,640	8,495

temperature equal to the nominal compressor exit temperature (e.g., at ISO base load). All told, utilizing a separate compressor and recuperator makes up for the lost airflow and power output and some more on a hot day.

One variation of the basic air injection scheme described above is "humid air injection." In this variant, the recuperator is replaced by an economizer, which heats compressed water. Hot and compressed water is utilized in a saturator (similar to the CHAT cycle above) to heat and moisturize the compressed air. Both variants are described in detail in a GTW article [18]. The expected performance impact from the cited reference is summarized in Table 8.6.

Yet another variant of dry air injection for power augmentation is marketed under the name "Turbophase."[7] In this concept, a multi-stage intercooled centrifugal compressor is driven by a turbocharged natural gas- or diesel-fired reciprocating internal combustion engine (e.g., MTU 2-MWe engine rated at 43 percent efficiency) that compresses air to the requisite pressure depending on the gas turbine application. (In the original dry and humid air injection concepts, the compressors were driven by an electric motor.) The air then enters a recuperator that recovers heat from the engine exhaust (instead of the gas turbine exhaust as in dry air injection described above) in order to heat the clean, dry, compressed air to the requisite temperature at the compressor discharge section.

According to the article cited above, typically 5–10 percent of compressor inlet air can be added to the combustor. At 5 percent injection, for a 2 × 1 × 1 gas turbine combined cycle with GE's 7FA.04 gas turbines (about 600 MW at ISO base load), more than 40 MW additional output (36 MW from the two gas turbines and 5 MW from the

[7] See the article "Well-Timed Megawatts Offer Potential Value to Morris Cogen" in *Combined Cycle Journal*, 2Q/2014 issue, pp. 20–24.

steam turbine) is possible with this scheme. This corresponds to about an 8 percent output boost per gas turbine vis-à-vis nearly 20 percent claimed with dry air injection in Table 8.6. Then again, there is no commercial installation of dry and humid air injection as opposed to the Turbophase (the interested reader can check the company website of Powerphase).

References

1. Gülen, S. C., Smith, R. W., Second Law Efficiency of the Rankine Bottoming Cycle of a Combined Cycle Power Plant, *Journal of Engineering for Gas Turbines and Power*, **132** (2010), 011801.
2. Gülen, S. C., Importance of Auxiliary Power Consumption on Combined Cycle Performance, *Journal of Engineering for Gas Turbines and Power*, **133** (2011), 04180.
3. Elmasri, M. A. GASCAN – An Interactive Code for Thermal Analysis of Gas Turbine Systems, *Journal of Engineering for Gas Turbines and Power*, **110** (1986), 201–210.
4. Elmasri, M. A., Pourkey, F., "Prediction of Cooling Flow Requirements for Advanced Utility Gas Turbines – Part 1: Analysis and Scaling of the Effectiveness Curve," ASME 86-WA/HT-43.
5. Elmasri, M. A., "Prediction of Cooling Flow Requirements for Advanced Utility Gas Turbines – Part 2: Influence of Ceramic Thermal Barrier Coatings," ASME 86-WA/HT-44.
6. Khodak, E. A., Romakhova, G. A., Thermodynamic Analysis of Air-Cooled Gas Turbine Plants, *ASME Journal of Engineering for Gas Turbines and Power*, **123** (2001), 265–270.
7. Wilcock, R. C., Young, J. B., Horlock, J. H., The Effect of Turbine Blade Cooling on the Cycle Efficiency of Gas Turbine Power Cycles, *Journal of Engineering for Gas Turbines and Power*, **127** (2005), 109–120.
8. Horlock, J. H., Watson, D. T., Jones, T. V., Limitations on Gas Turbine Performance Imposed by Large Turbine Cooling Flows, *Journal of Engineering for Gas Turbines and Power*, **123** (2001), 487–494.
9. Briesch, M. S., Bannister, R. L., Diakunchak, I. S., Huber, D. J., A Combined Cycle Designed to Achieve Greater than 60 Percent Efficiency, *Journal of Engineering for Gas Turbines and Power*, **117** (1995), 734–741.
10. Teraji, D., "Mercury™ 50 Field Evaluation and Product Introduction," Paper 05-IAGT-1.1, 16th Symposium on Industrial Application of Gas Turbines (IAGT), October 12–14, 2005, Banff, Alberta, Canada.
11. Gülen, S. C., "Advanced Fossil Fuel Power Systems," in *Energy Conversion*, 2nd edition, Eds. D.Y. Goswami, F. Kreight (Boca Raton, FL: CRC Press, 2017).
12. Horlock, J. H., The Evaporative Gas Turbine (EGT) Cycle, *Journal of Engineering for Gas Turbines and Power*, **120** (1998), 336–343.
13. Nakhamkin, M. et al., The Cascaded Humidified Advanced Turbine (CHAT), *Journal of Engineering for Gas Turbines and Power*, **118** (1996), 565–571.
14. Aramayo-Prudencio, A., Young, J. B., "The Analysis and Design of Saturators for Power Generation Cycles: Part 1 – Thermodynamics," ASME Paper GT2003-38945, ASME Turbo Expo 2003, June 16–19, 2003, Atlanta, GA.
15. Aramayo-Prudencio, A., Young, J. B., "The Analysis and Design of Saturators for Power Generation Cycles: Part 2 – Heat and Mass Transfer," ASME Paper GT2003-38946, ASME Turbo Expo 2003, June 16–19, 2003, Atlanta, GA.

16. Digumarthi, R., Chang, C.-N., Cheng Cycle Implementation on a Small gas Turbine Engine, *Journal of Engineering for Gas Turbines and Power*, **106** (1984), 699–702.
17. Cheng, D. Y., "The Distinction between the Cheng and STIG Cycles," ASME Paper GT2006-90382, ASME Turbo Expo 2006, May 8–11, 2006, Barcelona, Spain.
18. deBiasi, V., "Air Injected Power Augmentation Validated by Fr7FA Peaker Tests," *Gas Turbine World*, March–April 2002 issue, pp. 12–15.

9 Turbine Cooling

9.1 History

Turbine or, more generically, "hot gas path" (HGP) cooling was a feature of the gas turbine as we know it today from the very beginning.[1] Eckardt traces its appearance in the gas turbine literature to the 1920 book *Die Gasturbinen* by Eyermann-Schulz (see [4] in Chapter 4). In fact, Holzwarth's fin de siècle gas turbine had water-cooled blades. When Jumo-004 development started in the 1930s, the single-stage turbine rotor blades were planned to be solid airfoils made from Krupp alloy "Tinidur" (containing Ni, Cr, and Ti). Tinidur was similar to *Nimonic 80* used by the British, which had up to 80 percent nickel content, but with only 30 percent nickel and with more than 50 percent iron. Since the Germans had a severe shortage of nickel and chromium, ideas such as increasing the nickel content to 60 percent or using cobalt were dropped. German engineers had to revert back to a "hollow" air-cooled blade design first disclosed in a Brown-Boveri patent of 1920 and utilized by Christian Lorenzen in 1929 in his experiments with car engine turbochargers. For a detailed history, the reader can consult the superb book by Eckardt.

There were three sources of cooling air in Jumo-004 as shown in Figure 4.5: (1) compressor stage 4 exit, (2) compressor discharge, and (3) between the compressor discharge and combustor cans. Overall, about 7 percent of compressor airflow was used for HGP cooling. Stage 4 bleed was used to cool the exhaust system and the cone. Compressor discharge air taken through the Hirth serrations of the last disk was channeled through the main casting for impingement cooling of the turbine disk. (In the production units with "hollow" turbine blades, this air was used to cool the blades.) Thus, blade roots were estimated not to get much above 450°C, whereas near the center temperatures could go up to about 750°C. The third cooling air extraction was used to cool the stator nozzle vanes.

Stator nozzle vanes and rotor blades of the early version of Jumo-004A were of solid material, but the production blades on 004B were hollow to save material. Initially, Tinidur was the alloy of choice for hollow blades as well. However, the Jumo-004B production engines used in most of the Me-262 aircraft in 1943–1945, when nickel scarcity was most severe, deployed hollow blades made from "Cromadur," which

[1] The hot gas path encompasses the combustion chamber, stators, rotors, disks, exhaust diffuser, and nozzle cone in aircraft engines.

contained 12 percent Cr and 18 percent Mn with trace amounts of nickel. The blades were manufactured by folding flat sheets of Cromadur and welding the trailing edge. The lower creep strength of Cromadur prohibited solid blades, but luckily it was reasonably easy to manufacture (vis-à-vis Tinidur, which had to be deep drawn from a flat circular "blank" because it was not weldable) and used fewer than 5 lb of precious Cr overall. In passing, it should be noted that the rest of the HGP components were mild steel (SAE 1010) with an aluminum coating.

Still, the average life of production engines was 25 hours. Overhauls were required after 10 hours. In comparison, early engines with solid Tinidur blades were tested for up to 150 hours in actual flight tests and for up to 500 hours on the test stand. They were, however, too heavy to be deployed in the aircraft and used up exorbitant amounts of scarce strategic metals (i.e., those imported from other countries).

The turbine inlet temperature (TIT) of Jumo-004B was 1,427°F (775°C), so if the Germans had no difficulty in getting nickel, chromium, molybdenum, and other strategic metals, there would be no need for cooling the turbine rotor blades. (Even then, a case could be made to reduce engine weight and cost by going with hollow airfoils.) In fact, in the early 1950s, this was exactly the case when TITs hovered at around 1,600°F (around 900°C). It was a different story for the combustor cans, wherein flame temperatures could go as high as 1,800°F (982°C) and cooling was a necessity.

At this point, a pause in discussion is in order to clearly delineate what is meant by "turbine cooling." There are two major components in the HGP of a gas turbine: combustor and expander (turbine). Performance-critical cooling considerations apply mostly to the latter, which has the following two major "families of components":

1. Stationary parts
    a. Turbine/stator carrier rings
    b. Stator "segments" containing more than one nozzle vane, which are circumferentially attached to the carrier rings in each stage (typically three to four stages)
    c. Stator nozzle vanes
2. Rotating parts
    a. Turbine rotor "wheel"
    b. Rotor blades (also referred to as *buckets*) circumferentially attached to the wheels in each stage
    c. Spacer disks (between the wheels)

In terms of turbine component cooling, however, there is a differentiation between *airfoil-shaped* components (i.e., rotor blades/buckets and stator nozzle vanes) and the rest. The former are cooled by main or *primary* coolant flows, whereas the latter are cooled by *secondary* coolant flows. The main focus of industrial and academic research and development is understandably on primary cooling flows, which is also the case herein. However, it must be stressed that cooling of "the rest" (e.g., turbine wheels) was present from day one (e.g., see Figure 4.5 and the "impingement cooling" of the turbine wheel by air extracted from the compressor discharge).

In modern gas turbines, with temperatures inside the combustor and at the turbine inlet pushing 3,000°F and beyond, without appropriate thermal barrier coatings (TBCs) and convective plus film cooling, the turbine parts made from the most advanced alloys would melt away within a few seconds. Thus, first two or three stages of an advanced gas turbine for utility-scale electric power generation are typically *fully* cooled (i.e., no more solid vanes or buckets).

In order to stress the importance of secondary cooling flows, consider General Electric's (GE) H-System gas turbine, whose first two stages – stator nozzle vanes *and* rotor blades – were cooled by steam in a closed loop. Even so, considerable airflow was requisite to cool the bucket platforms and wheel spaces (e.g., more than 6 percent; referenced to turbine inlet flow) in chargeable flows for the first stage. This was significantly higher than expected (i) due to very high (for the time) gas temperatures (less than 100°F temperature loss across the stator vanes) and (ii) due to the need to cool the vane trailing edge (which does not have a cooling film anymore because it is convection-cooled internally by steam).

9.2 Open-Loop Air Cooling

How TBC and cooling work together to increase the temperature capability of a super-alloy nozzle vane or rotor blade is conceptually described in Figure 9.1, where

- T_g is the "free-stream" or "main-stream" hot gas temperature (in some sources, it is designated as T_∞). It is the total/stagnation temperature of the gas and can be as high as 1,600°C (~2,900°F) for stage 1 nozzle vanes of J/HA-class gas turbines. Note that it is the "absolute" stagnation temperature of the gas for stationary components (i.e., stator nozzle vanes) and the "relative" stagnation temperature for rotating components (i.e., rotor blades).
- T_{aw} is the "adiabatic wall" temperature. It can be considered as the "driving" temperature for the heat flux across the coolant film.

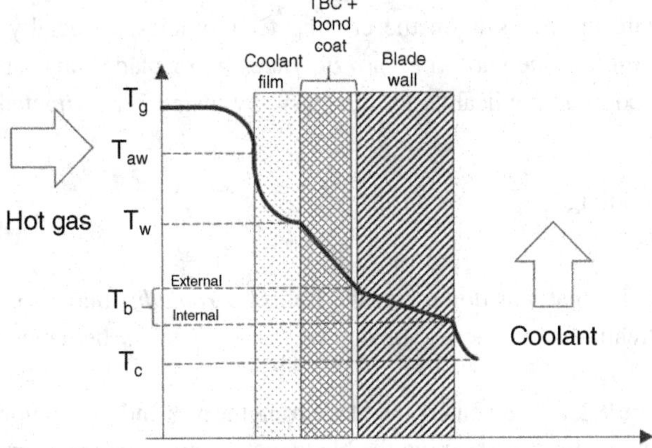

Figure 9.1 Conceptual temperature profile across the wall of a generic airfoil (not to scale).

Table 9.1 Typical values of temperatures in Figure 9.1

	With TBC°F (°C)	No TBC°F (°C)
T_g	2,600 (1,427)	2,600 (1,427)
T_{aw}	2,075 (1,135)	2,075 (1,135)
T_w	1,495 (813)	1,480 (805)
T_{bi}	1,200 (648)	1,365 (740)
T_{bx}	1,320 (716)	1,480 (805)
T_{ci}	1,100 (594)	1,100 (594)
T_{cx}	1,285 (696)	1,285 (696)

- T_w is the "wall" temperature, where the wall includes the bare metal (also referred to as the metal substrate) and the coating layers (i.e., protective thermal barrier and the bond coat).
- T_b is the bare metal temperature. It has three values: (1) external, T_{bx}, which is the same as T_w in the absence of any coating; (2) internal, T_{bi}; and (3) "bulk," which is the average of the two.
- T_c is the coolant (i.e., air bled from the compressor) total/stagnation temperature, "absolute" for stationary and "relative" for rotating components. Its value where it enters the airfoil, T_{ci}, can be different from its value at the location of extraction (due to change in flow velocities). By the time it exits, T_{cx} is higher than T_{ci} due to heat transfer from the hot airfoil (internal) service.

Typical values are provided in Table 9.1. Unlike the simple schematic in Figure 9.1, the actual heat transfer mechanism between the "source" (hot gas) and "sink" (coolant) constitutes a very complex, three-dimensional problem involving convection, conduction, and turbulent as well as laminar boundary layers (thermal and momentum). Nevertheless, simple one-dimensional treatment using definitions of "blade cooling effectiveness" and "film cooling efficiency" provides us with a reasonably complete picture of the underlying physics.

Material and coating technologies will be covered in more detail in a separate chapter (Chapter 13). Herein, the focus is on the cooling technologies, especially cooling of stationary and rotating turbine parts (i.e., nozzle vanes, rotor blades, and disks). There are two types of cooling applicable to this task by using air extracted from the compressor:

1. Convection cooling
2. Film cooling

Convection cooling is what was done with the hollow *Cromadur* blades of Jumo-004. For a simplified treatment of such a blade, we refer to the schematic diagram in Figure 9.2.

As shown in Figure 9.2, coolant enters the blade from the root and moves up through the hollow blade, picking up heat and cooling the blade metal, and then exits from the tip and mixes with the main gas flow. Inner (i.e., coolant-side) and outer (i.e., gas-side)

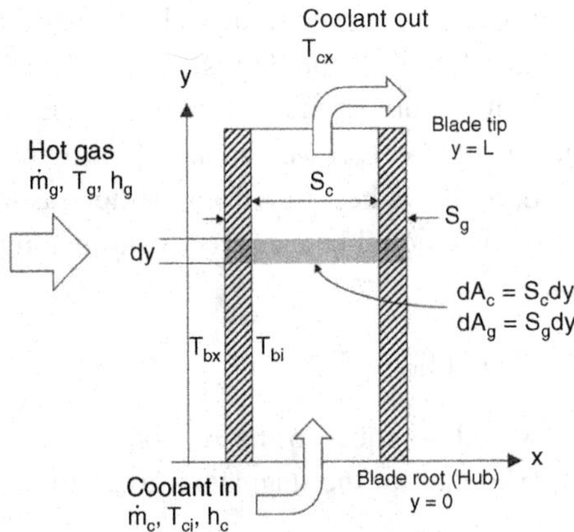

Figure 9.2 Simple convection-cooled blade and definition of parameters.

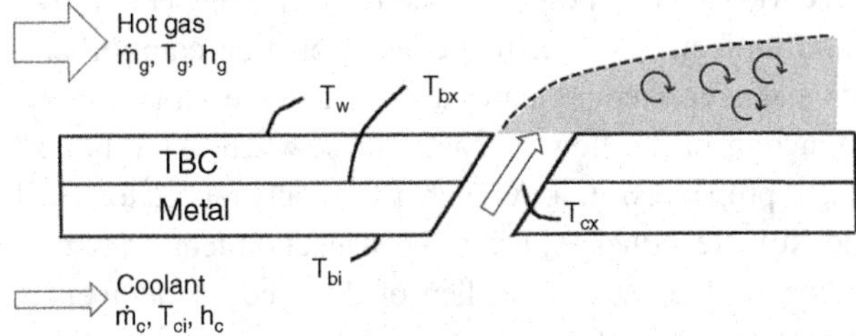

Figure 9.3 Schematic diagram of blade film cooling and definition of parameters.

circumferences of the blade are denoted by S_c and S_g, respectively. When multiplied by the differential length, dy, we obtain the differential areas, dA_c and dA_g, which, when integrated from y = 0 to y = L (blade height or length, depending on how you look at it), gives us the blade areas covered by the coolant and hot gas, A_c and A_g, respectively. Convective heat transfer coefficients on the gas and coolant sides are denoted by h_g and h_c, respectively.

The simple sketch in Figure 9.2 does not show the film cooling holes and the TBC. If we focus on a piece of the blade wall in Figure 9.2 and enlarge it, what we see will be somewhat similar to what is shown in Figure 9.3. Temperatures in Figures 9.3 are total/stagnation temperatures. For stator nozzle vanes, they are the absolute values. For rotor blades, they are rotor-relative values. As shown in Figure 9.3, the coolant jet exiting the film cooling hole forms a layer between the hot gas and the blade wall. As the distance from the hole grows, however, the film increasingly entrains the hot gas, which reduces the *film cooling effectiveness* given by

$$\eta_{fc} = \frac{T_g - T_{aw}}{T_g - T_{cx}}. \tag{9.1}$$

The *adiabatic wall temperature*, T_{aw}, in Equation 9.1 is one of the more confusing parameters in film cooling theory. Unlike the other temperatures, T_{aw} is a theoretical construction to facilitate heat flux calculations. As such, it is defined as the highest possible wall temperature, which can only be achieved if the wall is perfectly insulated. In other words, imagine that the coolant film in Figure 9.3 is a hypothetical *insulation blanket* with zero conductivity. The temperature on the surface of that blanket where it comes into contact with the hot gas would be T_{aw}. The coolant film is, of course, an *imperfect* insulation blanket. As such, it allows heat flux from the hot gas to the blade wall so that said heat flux is equal to what would be achieved with the removal of the hypothetical insulation blanket, i.e.,

$$\dot{Q} = h_g A_g (T_{aw} - T_w). \tag{9.2}$$

If there were no film cooling, T_{aw} would be equal to the *recovery temperature*, T_{gr}, which is given by an equation similar to that for the stagnation (total) temperature:

$$T_{gr} = T_s \left(1 + r \frac{\gamma - 1}{2} Ma_g^2 \right), \tag{9.3}$$

where $r \leq 1$ is the *recovery factor*. For $r = 1$, the recovery temperature is equal to the stagnation temperature. In fact, at the leading edge stagnation point of an airfoil, $r = 1$ and the recovery and stagnation temperatures are one and the same. In other locations, the value of r is a function of the flow characteristics (whether the boundary layer is laminar or turbulent, gas properties, etc.). As such, r is usually related to key characteristic numbers (i.e., Pr and Re). Nevertheless, using the stagnation temperature – absolute or rotor-relative depending on the airfoil – in lieu of the recovery temperature does not introduce a large error to this type of calculation.

Heat transferred from the hot gas to the blade wall in Equation 9.2 is eventually picked up by the coolant, i.e.,

$$\dot{Q} = \dot{m}_c (h_{cx} - h_{ci}) \cong \dot{m}_c c_{p,c} (T_{cx} - T_{ci}), \tag{9.4}$$

after being conducted first through the TBC, i.e.,

$$\dot{Q} = \frac{k_{TBC}}{t_{TBC}} A_g (T_w - T_{bx}), \tag{9.5}$$

and then through the blade metal (substrate), i.e.,

$$\dot{Q} = \frac{k_{met}}{t_{met}} A_g (T_{bx} - T_{bi}). \tag{9.6}$$

In Equations 9.5 and 9.6, k_{TBC} and k_{met} are thermal conductivities of the respective layers. Also note that, in Equation 9.4, h_{cx} and h_{ci} are coolant inlet and exit *enthalpies* (not to be confused with heat transfer coefficients). Using h and k for convective and conduction heat transfer coefficients is in line with US textbook and academic conventions. This has a slight potential for confusion with enthalpy or the isentropic exponent, but it is thought to be the more recognizable notation for most readers. In many European texts, the Greek symbols of α and λ are used for h and k, respectively.

Combining Equations 9.2 and 9.4–9.6, we obtain three important parameters. The first one is a dimensionless coolant mass flow rate:

$$m^* = \frac{\dot{m}_c c_{p,c}}{h_g A_g} = \frac{T_{aw} - T_w}{T_{cx} - T_{ci}}. \tag{9.7}$$

In Equation 9.7, m* represents the ratio of the heat absorbing capacity of the coolant to the heat transfer rate from the hot gas. The second important parameter is the Biot number for the TBC:

$$Bi_C = \frac{h_g t_{TBC}}{k_{TBC}} = \frac{T_w - T_{bx}}{T_{aw} - T_w}. \tag{9.8}$$

The Biot number is a measure of the relative magnitudes of convective and conductive heat transfer mechanisms. If we use the "electrical resistances in series" analogy to calculate the effective heat transfer coefficient across the film and the TBC, we find that

$$h_{eff} = \frac{h_g}{1 + Bi_C}. \tag{9.9}$$

According to Equation 9.9, if there is no TBC, Bi_C is zero and h_{eff} is equal to the gas-side convective heat transfer coefficient. Typical Biot numbers for the TBC for the components of a major original equipment manufacturer (OEM) range between 0.6 and 0.9 [1]. Thus, presence of the TBC reduces the heat transfer coefficient by 40–50 percent.

Finally, the third important parameter is the Biot number for the blade metal:

$$Bi_M = \frac{h_g t_{met}}{k_{met}} = \frac{T_{bx} - T_{bi}}{T_{aw} - T_w}. \tag{9.10}$$

At this point, we define another important parameter, *cooling effectiveness*, which is given by

$$\varepsilon = \frac{T_g - T_{bx}}{T_g - T_{ci}}. \tag{9.11a}$$

This is where things start becoming somewhat confusing due to different approaches used by different researchers and investigators in setting up the problem and the simplifying assumptions they use. Up to this point, we followed the approach presented by Young and Wilcock in the second part of their two-part paper, which, in the opinion of this author at least, is probably the best paper written on the subject [2]. Before continuing the discussion following [2], we will pause and take a look at a different and simpler take, which can be found, among others, in Saravanamuttoo et al., and uses the schematic diagram in Figure 9.2.

Equation 9.2 is always the starting point. One simplifying assumption, which was not made in Figure 9.3 and Equations 9.4–9.11a, is to ignore that the blade metal has finite conductivity so that its temperature is represented by a "bulk value," T_b. In other words, $T_{bx} = T_{bi}$. Thus, the definition of cooling effectiveness becomes

$$\varepsilon = \frac{T_g - T_b}{T_g - T_{ci}}. \tag{9.11b}$$

The next step is to combine Equations 9.2 and 9.4 into the following form

$$\frac{T_g - T_b}{T_g - T_{ci}} = \frac{\dot{m}_c c_{p,c}}{h_g A_g} \frac{T_{cx} - T_{ci}}{T_g - T_{ci}}. \tag{9.12}$$

Equation 9.12, in addition to m* and ε (per Equation 9.11b) defined earlier, contains a new parameter (which requires slight algebraic manipulation of the quotient on the right-hand side of the equation), *cooling efficiency*:

$$\eta_c = \frac{T_{cx} - T_{ci}}{T_b - T_{ci}}. \tag{9.13}$$

In order to differentiate the cooling efficiency (note that it is "cooling" efficiency and not "film cooling" efficiency because the blade in Figure 9.2 contains only *internal convective* cooling) in Equation 9.13 from the film cooling effectiveness, we use the subscript "c."

Now we can write the heat balance for the elementary strip of the height/length dy in Figure 9.2 as

$$\dot{m}_c c_{p,c} dT_c = h_c S_c (T_b - T_c) dy. \tag{9.14}$$

Integrating Equation 9.14 from T_{ci} to T_{cx} and 0 to L, we obtain

$$\frac{T_b - T_{cx}}{T_b - T_{ci}} = \exp\left(-\frac{h_c S_c}{\dot{m}_c c_{p,c}} L\right), \tag{9.15a}$$

$$\eta_c = 1 - \exp\left(-\frac{h_c A_c}{\dot{m}_c c_{p,c}}\right), \tag{9.15b}$$

$$\eta_c = 1 - \exp\left(-\frac{1}{m^*}\frac{h_c A_c}{h_g A_g}\right). \tag{9.15c}$$

The assumption that the bulk metal temperature does not change along the y-direction is a pretty strong one. In other words, the entire blade metal is assumed to be at T_b. Nevertheless, this simplification enables us to take a quick look at the governing physical mechanisms without undue computational baggage. Combining Equations 9.12 and 9.13 gives us

$$\eta_c m^* = \frac{\varepsilon}{1 - \varepsilon}. \tag{9.16}$$

Consequently, by substituting Equation 9.15c into Equation 9.16, the result becomes

$$\left(1 - \exp\left(-\frac{1}{m^*}\frac{h_c A_c}{h_g A_g}\right)\right) m^* = \frac{\varepsilon}{1 - \varepsilon}. \tag{9.17a}$$

The second term inside the exponential argument – let us call it α – is the ratio of the heat transfer rate on the coolant side to that on the gas side. For the simple, convection-cooled

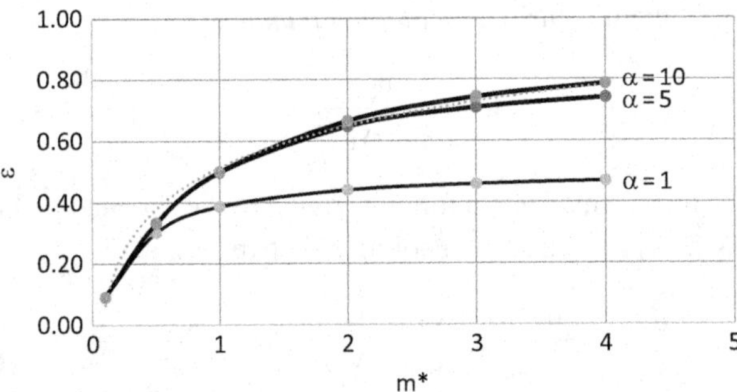

Figure 9.4 Cooling effectiveness versus dimensionless coolant mass flow rate.

blade in Figure 9.2, the area ratio is nearly unity, so that the true governing parameters are respective convection heat transfer coefficients.

Equation 9.17a is actually a quite informative correlation for determining the coolant flow rate for a known "allowable" blade metal temperature. First, we rewrite it as

$$\left(1 - \exp\left(-\frac{\alpha}{m^*}\right)\right)m^* = \frac{\varepsilon}{1-\varepsilon}, \tag{9.17b}$$

and then we use it to plot ε as a function of m* for three different values of α (i.e., 1, 5, and 10), which is shown in Figure 9.4. The best cooling technology is obviously one that can bring ε as close to unity as possible with m* as small as possible. The best way to accomplish this is to increase α (i.e., enhance the heat transfer on the coolant side by increasing the coolant–metal contact surface, A_c, as much as possible). As noted by the small difference between the cooling curves for α values of 5 and 10, however, there is definitely a point of diminishing returns, beyond which one is forced to increase m*.

Is there any practical relevance to the curves in Figure 9.4? The answer is, indeed, affirmative; the cooling curve for α >5 is quite representative of film plus convection cooling technology (e.g., see figures 17 and 18 in [3] or figure 7.30 in [4]). Thus, from Figure 9.4 and from Equation 9.17, for a very large α, we can easily confirm that

$$m^* = \frac{\varepsilon}{1-\varepsilon}. \tag{9.18}$$

Using Equation 9.7, we have

$$\dot{m}_c = \frac{h_g A_g}{c_{p,c}} \frac{\varepsilon}{1-\varepsilon}. \tag{9.19}$$

Note that gas mass flow rate around the blade is given by

$$\dot{m}_g = \rho_g V_g A', \tag{9.20}$$

where A′ is the flow area available to the gas passing through the blade row. Using Equation 9.20 and gas-specific heat, Equation 9.19 can be rewritten as

$$\dot{m}_c = \frac{h_g A_g}{c_{p,c}} \cdot \frac{c_{p,g}}{c_{p,g}} \cdot \frac{\dot{m}_g}{\rho_g V_g A'} \frac{\varepsilon}{1-\varepsilon}. \tag{9.21}$$

Using the definition of the dimensionless *Stanton number*

$$St_g = \frac{h_g}{\rho_g V_g c_{p,g}}, \tag{9.22}$$

Equation 9.21 can be transformed into a formula giving us the coolant mass flow rate (as a fraction of the gas flow) as a function cooling effectiveness, ε, i.e.,

$$\frac{\dot{m}_c}{\dot{m}_g} = St_g \frac{c_{p,g}}{c_{p,c}} \cdot \frac{A_g}{A'} \frac{\varepsilon}{1-\varepsilon}. \tag{9.23a}$$

Equation 9.23a is the fundamental equation of blade cooling theory. Consequently, the coolant mass flow rate requirement for a given metal capability, as represented by the allowable bulk metal temperature, is dependent on three dimensionless parameters:

1. The Stanton number (also known as the Margoulis number), which is a characteristic number in forced convection heat transfer problems and quantifies the intensity of energy dissipation in the gas flow.
2. The ratio of gas- and coolant-specific heats.
3. The ratio of total blade surface area (where it comes into contact with the hot gas) to the gas flow area.

These three dimensionless numbers represent the thermofluid state of hot combustion gas flowing through a turbine stage and the geometry of the particular blade row (i.e., stator or rotor). Typical values for the Stanton number are about 0.005 at the stage 1 stator inlet and about 0.002 at the stage 1 rotor inlet (assuming a heat transfer coefficient value of 3,000 W/m^2-K and an inlet absolute velocity of 150 m/s). The gas-to-coolant-specific heat ratio is about 1.25. The gas surface-to-flow area ratio is strongly dependent on the exact blade row geometry. It can be anywhere between 4–5 and 8–10. Thus, if we rewrite Equation 9.23a as

$$\frac{\dot{m}_c}{\dot{m}_g} = K \frac{\varepsilon}{1-\varepsilon}, \tag{9.23b}$$

a reasonable value for the *cooling flow factor* K is 0.05 for a stage 1 stator and 0.01–0.02 for a stage 1 rotor. Note that Equation 9.23b is derived from Equation 9.18, which is an approximation of Equation 9.17b for large values of α. For convection cooling, α is around 2, so that the approximation of Equation 9.18 is not valid anymore. Strictly speaking, one should then use Equation 9.17b as is. However, this is quite inconvenient because there is no algebraic solution for m*. A quick look at the two cases is presented in Figure 9.5. It appears that, for convection-only cooling, multiplying the cooling flow factor K by 2 is an acceptable approximation.

At this point, we can return to the analysis following [2], which we had left off at Equation 9.11a. First, two more cooling efficiencies are defined, which are similar to that defined by Equation 9.13, but they differentiate between the metal temperatures on the gas side of the blade and on its coolant side. As mentioned earlier, Equation 9.13 and the ensuing analysis assumed a "bulk" metal temperature that was the same at any

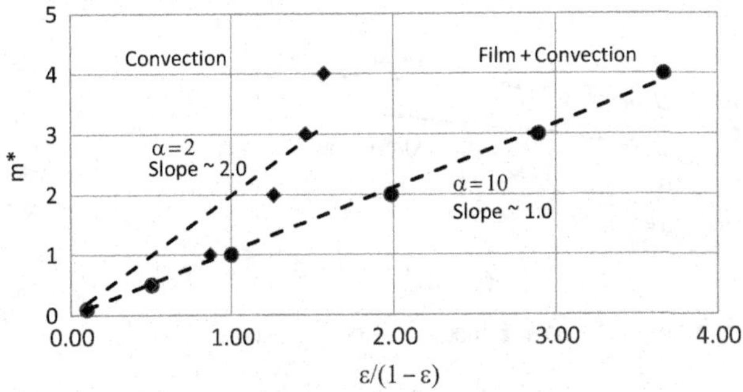

Figure 9.5 Comparison of film and convection cooling.

given location in the blade metal. Consequently, we now have external and internal cooling efficiencies, i.e.,

$$\eta_{cx} = \frac{T_{cx} - T_{ci}}{T_{bx} - T_{ci}}, \text{ and} \tag{9.24a}$$

$$\eta_{ci} = \frac{T_{cx} - T_{ci}}{T_{bi} - T_{ci}}. \tag{9.24b}$$

It can also be shown that the correlation between η_{cx} and η_{ci} is given by

$$\eta_{cx} = \frac{\eta_{ci}}{1 + \eta_{ci} m^* Bi_M}. \tag{9.25}$$

Combining Equations 9.1–9.11a with Equation 9.24 and going through the algebra, we arrive at the following correlation for m*:

$$m^* = \frac{\varepsilon}{\eta_{cx}(1-\varepsilon)} - \eta_{fc}\left(\frac{1}{\eta_{cx}(1-\varepsilon)} - 1\right), \text{ or} \tag{9.26a}$$

$$(1 + Bi_C)m^* = \frac{\varepsilon}{\eta_{cx}(1-\varepsilon)} - \eta_{fc}\left(\frac{1}{\eta_{cx}(1-\varepsilon)} - 1\right). \tag{9.26b}$$

Equation 9.26a is for an uncoated, film-cooled blade. Equation 9.26b is for a coated, film-cooled blade and it indicates that, in the presence of TBC, m* is reduced by the factor of (1 + Bi_C). For a typical heat transfer coefficient of 3,000 W/m^2-K, TBC thickness of 0.2 mm (about 8 mils – 1 mil is one-thousandth of an inch), and TBC conductivity of 2 W/m-K, Bi_C is 0.3, so that m* is reduced by 23 percent.

When there is no coating and no film cooling (i.e., η_{fc} is zero), Equation 9.26a reduces to Equation 9.16 with a bulk metal temperature, which ultimately led us to Equation 9.23. Comparison with Equation 9.26b suggests that Equation 9.23b can be expanded to include the impact of TBC in a similar fashion, i.e.,

$$(1 + Bi_C)\frac{\dot{m}_c}{\dot{m}_g} = K\frac{\varepsilon}{1-\varepsilon}. \tag{9.27}$$

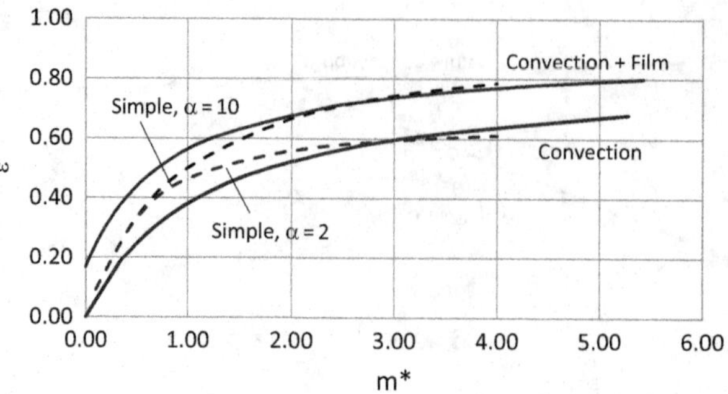

Figure 9.6 Cooling effectiveness versus dimensionless coolant mass flow rate.

Equation 9.26a is plotted for $\eta_{ci} = 0.7$, $Bi_M = 0.2$, and $\eta_{fc} = 0.4$ (η_{cx} is obtained from Equation 9.25) and compared with the predictions of the simple cooling model of Equation 9.16 in Figure 9.6. Note that for $m^* = 0$, Equation 9.26a gives a non-zero value for ε, i.e.,

$$\varepsilon = \frac{\eta_{fc}(1 - \eta_{ci})}{1 - \eta_{fc}\eta_{ci}}. \tag{9.28}$$

In other words, even if there is no internal convection cooling, non-zero film-cooling effectiveness implies that, somehow, a coolant film is present between the hot gas and the blade surface. The value of ε from Equation 9.28 is indicative of the effectiveness of "film-only" cooling, which is a practically unachievable situation (how would one introduce the coolant to the blade surface without first going through the internal blade channels and performing convective cooling in the process?). Therefore, in this chapter and elsewhere, when we talk about "film cooling," what we really mean is "convection *plus* film cooling."

Using Equation 9.7, Equation 9.26b can be translated into the same form as Equation 9.27, i.e.,

$$(1 + Bi_C)\frac{\dot{m}_c}{\dot{m}_g} = K\left\{\frac{\varepsilon}{\eta_{cx}(1 - \varepsilon)} - \eta_{fc}\left(\frac{1}{\eta_{cx}(1 - \varepsilon)} - 1\right)\right\}, \tag{9.29}$$

where the cooling flow factor K, as before, is given by

$$K = St_g \frac{c_{p,g}}{c_{p,c}} \cdot \frac{A_g}{A'}.$$

Variation in blade metal temperature from the blade root (y = 0 in Figure 9.2) to the blade tip (y = L) can be estimated by noting that, for the differential length dy in Figure 9.2, heat balance requires that

$$\dot{Q} = h_g dA_g (T_g - T_b) = h_c dA_c (T_b - T_c), \tag{9.30}$$

$$\dot{Q} = \dot{m}_c c_{p,c} dT_c = h_c dA_c (T_b - T_c), \tag{9.31a}$$

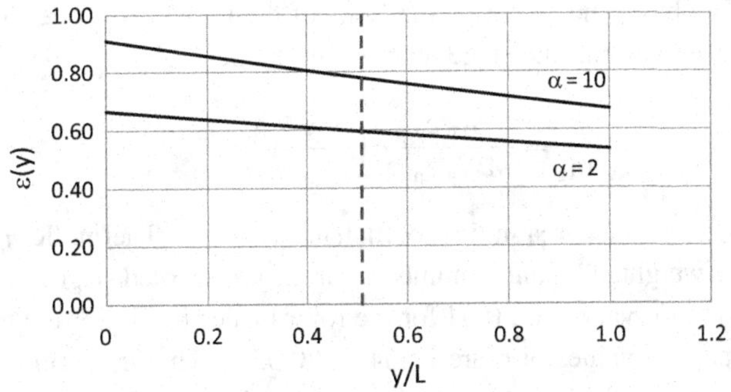

Figure 9.7 Variation of cooling effectiveness along the blade length (height).

$$\dot{Q} = \dot{m}_c c_{p,c} \frac{dT_c}{dy} = h_c S_c (T_b - T_c). \tag{9.31b}$$

Combining Equations 9.30 and 9.31 and integrating and rearranging, we obtain

$$\frac{T_b(y) - T_{ci}}{T_g - T_{ci}} = 1 - \frac{\alpha}{\alpha + 1} \exp\left(-\frac{\alpha}{\alpha + 1} \frac{1}{m^*} y\right), \tag{9.32}$$

where α is the coolant-to-gas side hS ratio. Comparing with the cooling effectiveness defined by Equation 9.11b, we recognize that Equation 9.32 is equivalent to

$$\varepsilon(y) = \frac{\alpha}{\alpha + 1} \exp\left(-\frac{\alpha}{\alpha + 1} \frac{1}{m^*} y\right). \tag{9.33}$$

Thus, Equation 9.33 describes the variation in cooling effectiveness from blade root (hub) to tip, which is shown in Figure 9.7 for the same two values of α in Figure 9.6 and $m^* = 3$. Not surprisingly, ε values in Figure 9.6 for $m^* = 3$ correspond to the average value at $y/L = 0.5$.

At this point, we recognize that the coolant flow requisite to maintain the allowable airfoil metal temperature (stator nozzle vane or rotor blade), as a fraction of gas flow, can be estimated reasonably accurately via the generic formula

$$\frac{\dot{m}_c}{\dot{m}_g} \propto f(\varepsilon). \tag{9.34}$$

The proportionality factor in Equation 9.34 is primarily a function of the stage geometry, thermodynamic properties of the gas, gas and coolant side heat transfer coefficients, and coolant and coating thickness and conductivity. The simplest form of $f(\varepsilon)$ on the right-hand side of Equation 9.34 is $\varepsilon/(1 - \varepsilon)$ and, as shown in Figure 9.6, it does as good a job of coolant flow prediction as its more rigorous counterpart. The proportionality factor can be estimated as a starting point and improved via calibration to test data or an existing stage design (as was done in Chapter 8).

Elmasri came up with a similar formulation via rigorous analysis of coated and uncoated blades with convection and film cooling in his seminal papers published in the 1980s [5–7]. His basic model forms the basis of the GASCAN code [5], which is encapsulated in the cooled turbine stage icon in Thermoflex, i.e.,

$$\frac{\dot{m}_c}{\dot{m}_g} = \frac{MW_c}{MW_g}\frac{c_{p,g}}{c_{p,c}}\alpha\left(\frac{\varepsilon}{\varepsilon_\infty - \varepsilon}\right)^\beta. \tag{9.35}$$

In Equation 9.35, α is the area ratio coefficient (ARC in Thermoflex), which is essentially an "area-weighted" Stanton number. For the film-cooled stages, α is assumed to be 0.05 for the nozzle vanes and 0.04 for the rotor blades [6,7]. For the convection-cooled stages, respective values of α are 0.064 and 0.0576. The first term on the right-hand side of Equation 9.35 is the molecular weight ratio because the original papers were done on a molar flow basis. The exponent β (PWR in Thermoflex) is 0.9 for film cooling and 1.25 for convective cooling [6]. The maximum cooling effectiveness, ε_∞, is the asymptotic value of ε at infinitely large coolant flow rates; it is 1.0 for film cooling and 0.85 for convective cooling.

For a coated blade, Equation 9.35 is modified by a correction factor (CF), i.e.,

$$\frac{\dot{m}_c}{\dot{m}_g} = CF\frac{MW_c}{MW_g}\frac{c_{p,g}}{c_{p,c}}\alpha\left(\frac{\varepsilon}{\varepsilon_\infty - \varepsilon}\right)^\beta, \tag{9.36}$$

where CF is analogous to $1/(1 + Bi_C)$ in Equations 9.27 and 9.29. Calculation of CF for film and convection cooling is described by Elmasri [7].

Predictions of the three cooling effectiveness models described herein are compared in Figure 9.8 for a film-cooled, stationary, airfoil-shaped component (i.e., stage 1 stator nozzle vane). The same comparison for a convection-cooled component is presented in Figure 9.9.

If you stopped reading Chapter 8 where chargeable and nonchargeable cooling calculations were described and started reading the turbine cooling fundamentals in this chapter, at this point you know the theory underlying, say, Equation 8.20a for the nonchargeable cooling flow, i.e.,

$$nch = \alpha K_{nch}\frac{\dot{m}_4}{\dot{m}_1}\left(\frac{\varepsilon}{\varepsilon_{\infty,1} - \varepsilon}\right)^{\beta_1},$$

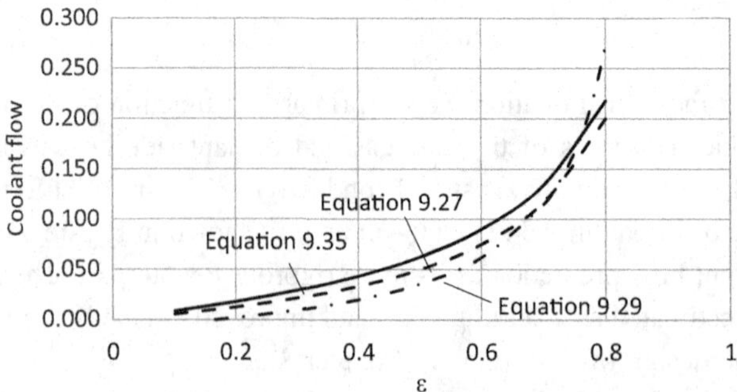

Figure 9.8 Comparison of coolant flow models for a film-cooled airfoil.

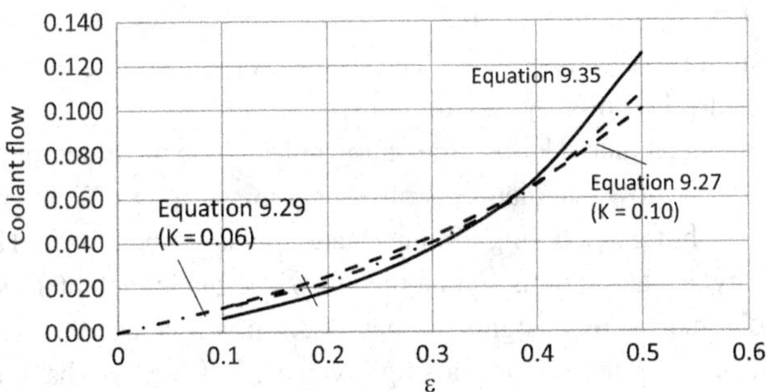

Figure 9.9 Comparison of coolant flow models for a convection-cooled airfoil.

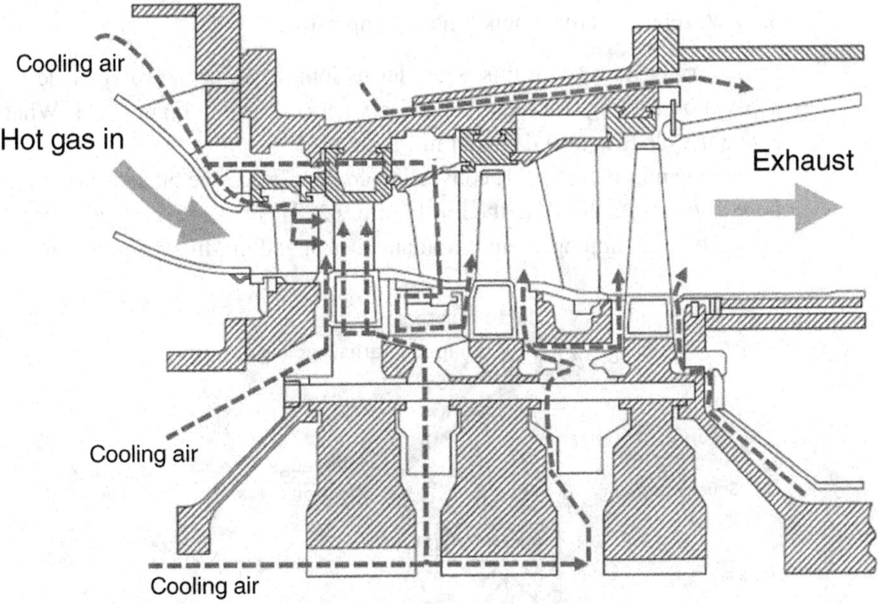

Figure 9.10 Turbine cutaway showing secondary cooling air flows.

which is equivalent to (setting the technology factor, α, to 1 for state of the art)

$$\frac{\dot{m}_{nch}}{\dot{m}_4} = K_{nch}\left(\frac{\varepsilon}{\varepsilon_{\infty,1} - \varepsilon}\right)^{\beta_1}.$$

You can now go back to Chapter 8 and keep reading on from where you left off or continue reading this chapter to its end.

Although the focus herein was on cooling flows associated with the main components (i.e., stator nozzle vanes and rotor blades), there are other "secondary" cooling flows that are much smaller in magnitude, but are absolutely essential for trouble-free operation. They are used in the cooling of other parts such as shrouds and rotor forward and aft wheel spaces between the blade rows. Figure 9.10 shows secondary flows for a typical three-stage F–class gas turbine. For each stage, such flows can amount to about 1–2 percent of incoming gas flow. For detailed coverage of gas turbine secondary flows, the reader is referred to the book by Sultanian [8].

Before moving on, let us take a look at the "allowable" blade metal temperatures. The key parameter in cooling flow calculation is obviously the cooling effectiveness, which requires knowledge of hot gas, cooling air, and blade metal temperatures. The first two are provided by the cycle calculations so that the allowable blade metal temperature, T_b, emerges as the sole hardware design parameter. At first glance, this looks like a confusing situation because there is not a single value for T_b; as we have seen in the foregoing discussion, there are several values to select from, i.e.,

1. Internal or external blade metal surface temperatures
2. At a given location along the blade length or height (i.e., $T_b[y]$)
3. A representative "bulk" metal temperature

Before trying to resolve this issue, let us look at an actual rotor blade with film and convection cooling that looks similar to what is shown in Figure 9.11. What is going on inside the blade is also depicted in Figure 9.11.

As shown in Figure 9.11, convection cooling inside the blade is not a simple channel flow as conceptually described in Figure 9.2. There are several enhancement mechanisms (i.e., wall impingement, turbulators, ribs, and pin-fins). The impingement method

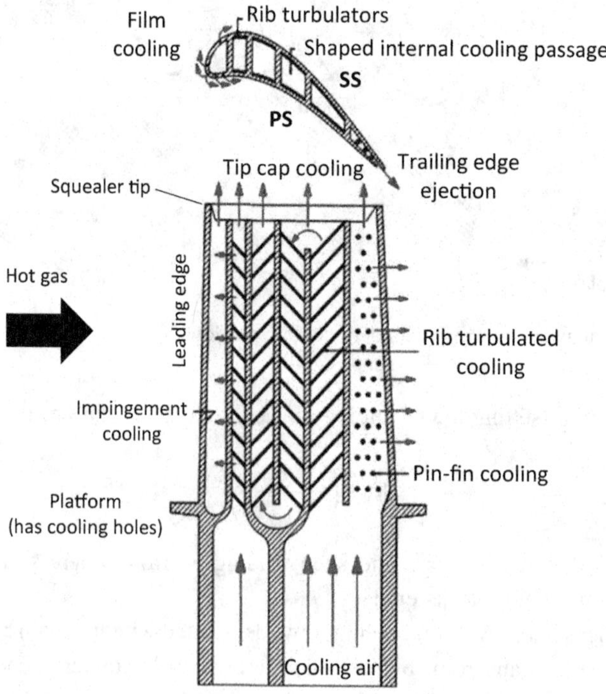

Figure 9.11 Typical rotor blade with film and convection cooling. PS = pressure side; SS = suction side.

is primarily used in stationary nozzle vanes, especially at the leading edge near the stagnation point where the highest outside temperatures are experienced. The key in impingement design is full, unobstructed impact of the coolant jet onto the metal surface. Combined impingement plus film-cooled leading edge design is commonly referred to as "showerhead film cooling."

"Serpentine" coolant flow channels and ribs (or "turbulators") enhance the heat transfer via increased heat transfer surface (i.e., higher A_c) and flow turbulence (i.e., higher h_c). Pins serve the same purpose as the ribs, but in the trailing edge ejection section of the blade. They also have a secondary benefit in that they increase the strength and solidity of the blade where the suction and pressure sides converge together.

The temperature contours for an internally cooled rotor blade with 1,210°C (2,210°F) hot gas are shown in Figure 9.12 [9]. The blade cooling passages include smooth channels, ribbed channels at the leading edge, and pin-fins in the trailing edge. As shown in Figure 9.12, the highest metal temperatures are registered at the leading edge at around 70 percent span (i.e., y/L = 0.7) and the upper portion of the trailing edge (around 1,200K or 1,700°F). In the blade root, the metal temperature is quite close to the coolant temperature, which is ~700K (~800°F).

The hot gas temperature for the internally cooled blade in Figure 9.12 is typical of an F-class gas turbine with 2,475°F TIT. The implication is that, without TBC and film cooling, the blade alloy should be capable of running safely at temperatures as high as 1,700°F (~925°C) at certain locations. Based on the temperature capability curves in

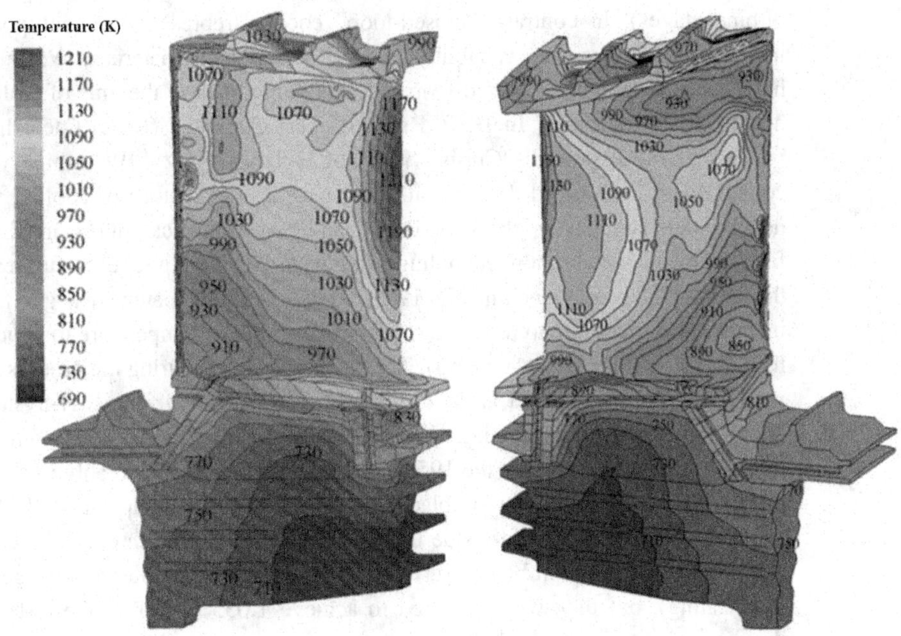

Figure 9.12 Blade temperature distribution (suction side on the left, pressure side on the right). (Reprinted from [9] with permission from ASME)

Chapter 13, this is not a tall order for fifth-generation, single-crystal superalloys, which would be exorbitantly expensive for a gas turbine of this vintage. A more likely material for this particular component is GTD-111 DS, which has a capability of less than 900°C. In other words, in an actual production unit, this blade would definitely have TBC and, probably, film cooling for a maximum allowable blade metal temperature of 1,400–1,500°F (about 800°C).

Historical trends (see Chapter 13) suggest that gas and metal temperatures go up together. In other words, as the technology advances to higher TITs, material and coating capabilities also advance to maintain service life. This trend was represented by an empirical formula in a recent paper [10], i.e.,

$$T_b[K] = \left(T_g[K] + 2190\right)/3.77. \tag{9.37}$$

Equation 9.37 can be used as a guide to specify T_b within ±50K, to be used for calculating cooling effectiveness with Equation 9.11b. While the definition itself assumes T_b to be a "bulk" temperature, one should think of it as the highest temperature a component is expected to run at during normal operation (i.e., not for a very limited time during some transient event).

9.3 Closed-Loop Steam Cooling

Convection and film cooling are also known as "open-loop" cooling techniques (i.e., once they finish their cooling job, they mix with the hot gas expanding through the turbine stages). In contrast, "closed-loop" cooling represents another possibility to increase the temperature capability of HGP component materials. Water and steam have been considered as closed-loop coolants going back to the time of Holzwarth and his "explosion turbine." In 1930, Brown-Boveri Company (BBC) fitted the first row buckets of the two-stage "Curtis wheel" of Holzwarth's 2-MW turbine with water-cooled ducts. A detailed description along with a brief history of cooling technology development (with an emphasis on BBC) can be found in Eckardt ([4] in Chapter 4). In fact, the history of component cooling using water or steam goes even further back than the BBC–Holzwarth machine. In 1903, Aegidius Elling patented a gas turbine that included water cooling to reduce the hot combustion gas temperature to about 400°C at the turbine inlet ([13] in Chapter 4). The steam generated during the process was mixed with the gas and expanded in the turbine (in essence, a *poor man's* H-System with an open-loop configuration a century before the first fire of GE's 109H combined-cycle machine in Baglan Bay). In the 1950s, Siemens invested considerably into designing a turbine rotor with water-cooled blades for a 1,000°C inlet temperature ([10] in Chapter 4). The steam generated inside the blades was routed out through the hollow rotor and piping. Myriad problems surfaced (vibration, water filter clogging, and parts overheating), but they were resolved to achieve 1,055°C TIT in the tests. However, the program eventually folded due to cost issues.

GE's 5-MWe intercooled-recuperative gas turbine with two-stage high-pressure (HP) and one-stage low-pressure (LP) turbine sections was water-cooled. About 30 gallons per minute of cooling water was used to cool the stators and rotors (but not the stator nozzle vanes or rotor blades) ([7] in Chapter 14). Turbine rings, which are the structures to which stator elements are attached, and LP turbine rotor support struts were cooled by water flowing through built-in passages. The rotors were cooled via heat radiation to stationary "cooling pads" next to the turbine wheels. The LP turbine ring was of solid construction, whereas the HP turbine ring comprised stator vane segments to allow differential expansion between the hot (austenitic steel) and cold (carbon steel) parts circumferentially.

Starting in mid-1970s, GE investigated a water-cooled stage 1 nozzle as part of the US Department of Energy's (DOE) High Temperature Turbine Technology (HTTT) program. Parts were designed and tested in cascade tests at gas temperatures as high as ~1,650°C (HTTT program's goal) at 145 psia [11]. Rig tests in an actual turbine similar to a Frame 6 gas turbine were planned, but they were never carried out. Difficulty controlling the water–steam phase change and instabilities associated with nucleate boiling in addition to limited coolant temperatures eliminated water as a turbine coolant once and for all. By the time GE joined the DOE's Advanced Turbine Systems (ATS) program, closed-loop steam cooling was the chosen path, which led to the H-System.

For all practical purposes, there is only one closed-loop steam cooled gas turbine: GE's H-System. It is true that Mitsubishi G- and J–class gas turbines also employ steam cooling for the combustor liner, transition piece, and stage 1 and 2 turbine rotor rings (J class). (Note that Mitsubishi designed and tested a fully steam-cooled "H" machine in around 2000. This machine, with a cycle pressure ratio of 25, was never offered commercially, but its compressor lives on in current G- and J–class gas turbines.) In terms of HGP "chargeable" and "nonchargeable" cooling air reduction, however, G- and J–class gas turbines are essentially air-cooled machines.

In H-System gas turbines, on the other hand, closed-loop steam cooling reduces the hot gas temperature drop across the stage 1 nozzle to less than 80°F. For the same combustor temperature and TIT, this results in an increase of 100–150°F in firing temperature vis-à-vis advanced F–class machines with air cooling (Siemens H–class gas turbines also belong in this category). The key driver of this benefit is elimination of hot gas temperature dilution via mixing with spent cooling air.

An additional benefit of steam cooling is less parasitic extraction of compressor discharge air and higher flow to the head end of the Dry-Low-NOx combustor for fuel premixing. If the firing temperature is kept at the F–class level, the benefit of steam cooling presents itself as reduced TIT and combustor temperature (i.e., reduced NOx production). This is, in fact, one of the key design challenges faced in pushing gas turbine combined-cycle performance to ever-higher levels. As fundamentally demonstrated in Chapter 5 and Chapter 8, TIT is the strongest "knob" available to the designer to improve the combined-cycle efficiency. Gas turbine cycle pressure ratio is the secondary parameter, which sets the "true" optimum for a *given* TIT. However,

1. A higher TIT requires more cooling air to be diverted from the combustor to the turbine in order to maintain allowable metal temperatures (we saw how that works out earlier in the present chapter).

2. Not doing that reduces part lives and significantly increases life cycle maintenance costs (most likely with a commensurate drop in reliability and/or availability).
3. Reduced air available for the combustor makes it difficult to control *flame temperatures* in order to limit NOx emissions, which increase exponentially with the former.

Competing air demands of the combustor and HGP components can be (relatively) easily satisfied by a closed-loop steam-cooled stage 1. Whether this will be the case or not primarily depends on the development of new materials requiring little or no cooling (e.g., ceramic matrix composites), new TBC materials and deposition techniques, and advanced manufacturing (3D printing) technologies for cost-effective production (and repair or replacement) of parts to facilitate effusion cooling.[2]

In the H-System, the first two turbine stages, including nozzles and buckets, are fully steam cooled. This reduces the amount of "chargeable" cooling air and increases gas turbine output via higher gas flow through the HGP. Heat rejected to the coolant steam is converted into additional steam turbine power output. The net benefit of full steam cooling is a two−percentage point increase in combined-cycle efficiency [12].

The starting point for steam cooling analysis is Equation 9.2, repeated below for convenience,

$$\dot{Q} = h_g A_g (T_{aw} - T_w),$$

which is rewritten as

$$\dot{Q} = h_g A_g (T_g - T_w). \tag{9.38}$$

Using the definition of the Stanton number, Equation 9.38 becomes

$$\dot{Q} = St_g (\rho V c_p)_g A_g (T_g - T_w). \tag{9.39}$$

Using the mass continuity for the gas flow with a flow cross-sectional area of A′, it follows that

$$\dot{Q} = St_g \frac{A_g}{A'} \dot{m}_g c_{p,g} (T_g - T_w). \tag{9.40}$$

Equation 9.40 is the heat transfer from the hot gas to the blade wall (including the TBC), which is picked up by the cooling steam flowing inside the blade, i.e.,

$$\dot{Q} = \dot{m}_{stm} (h_{s,out} - h_{s,in}) \approx \dot{m}_{stm} c_{p,stm} (T_{s,out} - T_{s,in}). \tag{9.41}$$

Combining Equations 9.40 and 9.41, the blade cooling steam requirement is found as

$$\dot{m}_{stm} = St_g \frac{A_g}{A'} \dot{m}_g \frac{c_{p,g}}{c_{p,stm}} \frac{T_g - T_w}{T_{s,out} - T_{s,in}}. \tag{9.42}$$

[2] "Effusion" is the term for gas flow through a hole of a diameter that is considerably smaller than the mean free path of the molecules. Effusion cooling refers to "full-coverage" film cooling through discrete, microscopically small holes in the blade wall or through a porous material blade wall.

The critical heat transfer parameter in Equation 9.42 is the "area-weighted" Stanton number

$$St'' = St_g \frac{A_g}{A'}.$$

Typical values of St" are

- Between 0.05 and 0.06 for stator nozzle vanes
- Between 0.01 and 0.02 for rotor blades

Steam-cooled gas turbines, by necessity, come only in a combined-cycle configuration. In fact, they are more aptly described as "integrated steam/gas" cycles [13]. The connection between the topping and bottoming cycles goes way beyond the exhaust gas duct between the gas turbine and the heat recovery steam generator. The network of alloy pipes and valves between the steam turbine, gas turbine, and heat recovery steam generator, in addition to the *cooling air cooler* (a *kettle reboiler*-type heat exchanger) for intermediate-pressure (IP) steam generation (not to mention the performance fuel heating), results in a veritable (and expensive) maze. The cooling air cooler is a consequence of the high Brayton-cycle pressure (23 for the H-System) requisite for an optimal design necessitated by high firing temperature (>2,600°F) and reduced hot gas dilution by coolant in the HGP (because of high compressor discharge temperatures).

A typical cooling steam source is the steam turbine exhaust or "cold reheat," which is around 500–600 psia and 650–700°F for typical combined cycles with advanced G/H-class gas turbines. The temperature rise of the cooling steam is about 300°F and the steam is typically returned to the "hot reheat" line (i.e., upstream of the IP steam turbine inlet). A critical design parameter is the steam pressure drop, which includes steam pipes between the gas turbine and the bottoming cycle and the flow channels inside the rotor and vanes/blades. It can be as high as 100–150 psi, which is a significant hurdle in H-System bottoming-cycle design. Excessive reheater pressure drop (~25 percent vis-à-vis typical 10–12 percent for modern reheat steam bottoming cycles) caused by the HGP cooling steam circuit embedded within the reheat steam piping reduces steam turbine efficiency.

It should be noted that closed-loop steam cooling does not eliminate air cooling altogether. Purge flow is still needed to prevent ingestion of hot gas into the wheel spaces. Furthermore, cooling of the trailing edges of stage 1 and 2 nozzle vanes via internal coolant flow presents a challenge. Supplementary cooling of the inner and outer side walls (platforms) and the trailing edge of the nozzle vanes with wheel space purge air is requisite to ensure adequate parts life.

9.3.1 Blade Wall Thermal Stress

Design of closed-loop, steam-cooled blades presents a significant challenge from a thermal stress perspective. The subject of thermal stress will be covered later in Chapter 13 (Section 13.6.1). However, we can introduce the basic thermal stress formula (Equation 13.1a) herein to facilitate the discussion at hand (but changing the temperature subscripts):

$$\sigma = E\frac{\alpha}{1-\nu}\left(T_{o/i} - \bar{T}_w\right)$$

In Equation 13.3a, E is Young's modulus, α is the coefficient of thermal expansion, and ν is Poisson's ratio. Subscripts "o" and "i" designate blade wall outside and inside temperatures. The second temperature term in the parentheses of the right-hand side is the average wall temperature.

The gas temperature outside the first-stage rotor blades is about 2,650°F (rotor inlet stagnation temperature). From a heat transfer perspective, the key gas temperature is the *rotor-relative* stagnation temperature, which is about 2,400°F. (See Chapter 10 on turbine aero for a detailed discussion of absolute and relative stagnation temperatures. For now, you can just take them at their *face value* for thermal stress and heat transfer calculations.)

The blade is convectively cooled by steam flowing through the channels inside the part. The inside wall temperature is of the order of 1,000°F. Let us assume that the temperature drop across the TBC is 200°C (360°F) so that the outside blade wall (metal) temperature is about 2,000°F. Consequently, we are looking at a temperature gradient of about 1,000°F across a metal "skin" of a few millimeters in thickness.

Assuming that

$$\bar{T}_w = \frac{T_o + T_i}{2},$$

we can rewrite the thermal stress formula for the (tensile) thermal stress at the outer surface of the thin blade wall as

$$\sigma = \frac{\kappa}{2}(T_o - T_i),$$

where the proportionality factor κ is a function of blade material properties. (A similar equation can be written for the [compressive] thermal stress at the inner surface wall.) The value of κ is about 0.15 ksi/F for superalloys at the temperature range of interest (e.g., see the data presented in Section 13.5.1). Thus, for a temperature delta of 1,000°F, one is looking at about 75 ksi (~500 MPa) of thermal stress. This is almost of the same magnitude as the yield strength of the superalloys in question and unacceptable.

Single-crystal superalloys have about half the longitudinal elastic modulus of polycrystalline superalloys when loaded in the [001] crystallographic direction. This is why single-crystal alloys are used in the H-System steam-cooled buckets and vanes. The casting process used in the production of single-crystal parts ensures that the single grain is very close to its [001] orientation (see Chapter 13 for more on this subject). Consequently, κ and σ can be reduced by about 50 percent via a 50 percent lower E.

In open-loop, air-cooled parts operating at high gas temperatures, film cooling is vital to maintain a low (relatively speaking) metal skin temperature (e.g., see Figure 9.1). The design of a cooled part's internal cooling channels and overall geometry is very important in order to achieve a uniform temperature distribution. Everything else being the same, ±100°F temperature variation on blade wall outer surface is equivalent to ±20 percent variation in the tensile stress. In fact, it can be shown that (see Schneider and Sommer [14])

- A ±15°C change in blade metal temperature can halve (+) or double (–) its creep life.
- A ±30°C change in blade metal temperature can halve (+) or double (–) its low cycle fatigue life.

Note that the discussion above was limited to the steady-state operation of the gas turbine. During severe transient events such as cold startup and trip or rapid load ramps (up or down), vanes and blades in the HGP are subject to high thermal stresses. This is why counts of those events are tracked diligently in order to assess the remaining pars lives (see Chapter 16).

References

1. Henze, M., Bogdanic, L., Mühlbauer, K., Schneider, M., Effect of the Biot Number on Metal Temperature of Thermal-Barrier-Coated Turbine Parts – Real Engine Measurements, *Journal of Engineering for Gas Turbines and Power*, **135** (2013), 031029.
2. Young, J. B., Wilcock, R. C., Modeling the Air-Cooled Gas Turbine: Part 2 – Coolant Flows and Losses, *Journal of Engineering for Gas Turbines and Power*, **124** (2002), 212–222.
3. Lechner, C., Seume, J. (Eds.), *Stationäre Gasturbinen*, 2nd new revised edition (Berlin, Germany: Springer Verlag, 2010).
4. Lakhsminarayana, B., *Fluid Dynamics and Heat Transfer of Turbomachinery*, (New York: John Wiley & Sons, Inc., 1996).
5. Elmasri, M. A., GASCAN – An Interactive Code for Thermal Analysis of Gas Turbine Systems, *Journal of Engineering for Gas Turbines and Power*, **110** (1988), 201–210.
6. Elmasri, M. A., Pourkey, F., "Prediction of Cooling Flow Requirements for Advanced Utility Gas Turbines – Part 1: Analysis and scaling of the Effectiveness Curve," ASME 86-WA/HT-43.
7. Elmasri, M. A., "Prediction of Cooling Flow Requirements for Advanced Utility Gas Turbines – Part 2: Influence of Ceramic Thermal Barrier Coatings," ASME 86-WA/HT-44.
8. Sultanian, B., *Gas Turbines: Internal Flow Systems Modeling* (Cambridge, UK: Cambridge University Press, 2018).
9. Alizadeh, M., Izadi, A., Fathi, A., Sensitivity Analysis on Turbine Blade Temperature Distribution Using Conjugate Heat Transfer Simulation, *Journal of Turbomachinery*, **136** (2014), 011001.
10. Jordal, K., Bolland, O., Klang, A., Aspects of Cooled Gas Turbine Modeling for the Semi-Closed O_2/CO_2 Cycle with CO_2 Capture, *Journal of Engineering for Gas Turbines and Power*, **126** (2004), 507–515.
11. Collins, M. F. et al., Development, Fabrication and Testing of a Prototype Water-Cooled Gas Turbine Nozzle, *Transactions of the ASME*, **105** (1983), 114–119.

12. Chiesa, P., Macchi, E., A Thermodynamic Analysis of Different Options to Break 60% Electric Efficiency in CC Power Plants, *Journal of Engineering for Gas Turbines and Power*, **126** (2004), 770–785.
13. Rice, I. G., The Reheat Gas Turbine with Steam-Blade Cooling – A Means of Increasing Reheat Pressure, Output, and Combined Cycle Efficiency, *Journal of Engineering for Gas Turbines and Power*, **104**: 1 (1982), 9–22.
14. Schneider, M., Sommer, T., "Turbines for Industrial Gas Turbine Systems," in: *Modern Gas Turbine Systems: High Efficiency, Low Emission, Fuel Flexible Power Generation (Woodhead Publishing Series in Energy)*, 1st edition, Ed. P. Jansohn, (Cambridge, UK: Woodhead Publishing, Ltd., 2013).

While not directly referenced in this or earlier chapters, the references listed below are highly recommended for further reading and learning.

Elderson, E. D., Scheper, G. W., Cohn, A., "Closed Circuit Steam Cooling in Gas Turbines," ASME paper 87-JGPC-GT-1, 1987 Joint Power Generation Conference, October 4–8, 1987, Miami Beach, FL.

Facchini, B., Innocenti, L., Carnevale, E., "Evaluation and Comparison of Different Blade Cooling Solutions to Improve Cooling Efficiency and Gas Turbine Performances," ASME Paper 2001-GT-0571, ASME Turbo Expo 2001, June 4–7, 2001, New Orleans, LA.

Han, J. C., Dutta, S., Ekkad, S. V., *Gas Turbine Heat Transfer and Cooling Technology*, (New York: Taylor & Francis, Inc., 2000).

Holland, M. J., Thake, T. F., Rotor Blade Cooling in High Pressure Turbines, *Journal of Aircraft*, **17**: 6 (1980), 80-4061.

Horlock, J. H., Basic Thermodynamics of Turbine Cooling, *Journal of Engineering for Gas Turbines and Power*, **123** (2001), 583–591.

Horlock, J. H., Watson, D. T., Jones, T. V., Limitations on Gas Turbine Performance Imposed by Large Turbine Cooling Flows, *Journal of Engineering for Gas Turbines and Power*, **123** (2001), 487–494.

Torbidoni, L., Massardo, A. F., Analytical Blade Row Cooling Model for Innovative Gas Turbine Cycle Evaluations Supported by Semi-Empirical Air-Cooled Blade Data, *Journal of Engineering for Gas Turbines and Power*, **126** (2004), 498–506.

Zhao, L., Wang, T., An Investigation of Treating Adiabatic Wall Temperature as the Driving Temperature in Film Cooling Studies, *Journal of Engineering for Gas Turbines and Power*, **134** (2012), 061032.

10 Turbine Aero*

Cycle thermodynamics does not tell us anything about the actual design of the gas turbine. One can come up with many combinations of cycle turbine inlet temperature (TIT) and pressure ratio (PR) that result in thermal efficiencies much higher than the current state of the art. As long as the conservation laws and the second law are not violated, the cycles in question are perfectly valid constructions. The challenge is to come up with components to translate a particular thermodynamic cycle from paper to a final product. In other words, one must ascertain that requisite compressor and turbine/expander can be designed and produced with existing materials and manufacturing techniques at a reasonable size and cost. This is where aerodynamics enters the picture. (In passing, "aero" is how the original equipment manufacturer [OEM] "experts" specializing in the design of gas turbine compressors and turbines refer to aerodynamics.)

Aerodynamics is essentially a subdiscipline of gas dynamics (which itself is a subdiscipline of fluid mechanics) dealing with the flow of air around a solid object. While the field of study is not solely limited to "air" (in Greek, *aer*) per se (i.e., consider hot combustion gas flowing around turbine stator vanes and rotor blades), from a historical (and largely practical as well) perspective, this is a rather apt name due to the two major applications that led to the discipline's coming into its own: aircraft wings and axial compressors (for gas turbine jet engines primarily). Subsequently, ever-increasing aircraft speeds into the supersonic and hypersonic realm, driven mostly by military applications, ensured rapid growth of aerodynamics, lately with the immense help of computational fluid dynamics (CFD) methods paced by exponentially increasing computing power, to its modern-day status.

What is the end goal of aerodynamic analysis and design? In essence,

- Compressor "aero" is the science and art of achieving the highest cycle PR with the lowest number of stages and the best possible efficiency.
- Turbine aero is the science and art of achieving highest turbine expansion ratio with lowest number of stages and the best possible efficiency.

* In this chapter, a distinction must be made between total/stagnation and static values of flow pressure and temperature. Therefore, "t" in the subscript denotes "total or stagnation" and "st" in the subscript denotes "static."

In terms of difficulty, turbine aerodynamics is, if not easier per se, more straightforward than compressor aerodynamics. Consider a simple pinwheel; you pick it up and blow into it (or hold it up against the wind), it will start to turn. This is essentially what a turbine rotor is and how it works. There is no counterpart of a pinwheel for creating the same effect as a compressor rotor. Designing an axial compressor to achieve a given PR in a *diffusing* flow regime without introducing excessive losses and without stalling requires extreme care. This is so because the pressure gradient acts *against* the flow direction so that it should be done in "small steps" (i.e., with a large number of stages each with a small PR) and constructed using vanes and blades of precise airfoil shapes based on aerodynamic theory and experimental "cascade" data.

In this chapter, we will take a close look at turbine ("expander") aero. In the next chapter, compressor aero will be covered. It should be mentioned at the outset that only *axial* turbines and *axial* compressors are going to be covered. Recall that this book is a *monograph* on heavy-duty industrial gas turbines for electric power generation, which do not comprise *radial* (or, using another common term, *centrifugal*) components. The latter are common to much smaller gas turbines that are used mainly in vehicle propulsion. The reader can consult many excellent textbooks and scholarly papers available in the literature pertaining to the selection of pumps, compressors, and turbines (too numerous to cite here, but as a hint, just google the phrase "Balje Chart"). The gas turbine references listed in Chapter 2 of this book (e.g., Saravanamutto et al.) also include ample details on the design of centrifugal compressors and radial turbines. Chapter 9 of Wilson and Korakianitis ([2] in Chapter 4) has an in-depth discussion of the relative merits and deficiencies of radial and axial compressor and turbine designs.

Before delving into equations and calculations, a quick review of terminology is in order. Most common turbine stage terms pertaining to a single airfoil-shaped "bucket" or "blade" of the whole rotor, which typically contains 60–70 of them circumferentially attached to the rotor "wheel," are described in Figure 10.1.[1] A cutout of a portion of the turbine hot gas path (HGP) in Figure 10.2 is used to identify most common terms used for the remaining components. Not surprisingly, the focus herein is on the rotor assembly and the individual blades because they are the "work-generating" parts and, as such, of prime importance from a performance perspective (design or off-design).

The goal herein (and in the next chapter covering the compressor aero) is *not* to give a full-blown treatise on turbine design. There are many excellent books written on that subject by world-class experts in the field. Furthermore, as stated earlier in the introductory chapters of the book, even if armed with one or more of said books (and let us assume that one has read them cover to cover and fully internalized them), the probability that one will be asked to design a turbine or compressor (intended for utility-scale electric power generation units) from a blank sheet is essentially nil. Those

[1] Strictly speaking, "bucket" is the term used by steam turbine designers, whereas "blade" is the term preferred by gas turbine engineers. The origin of the former term goes back to the airfoil shape of the impulse-type designs, which resembles the buckets of a Pelton-type hydraulic turbine (in cross-section, that is).

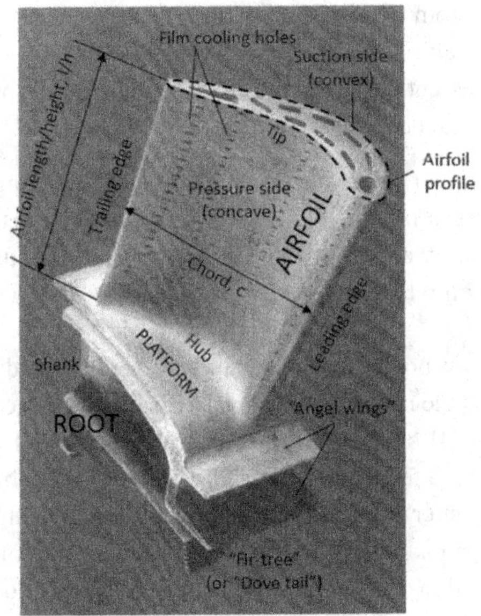

Figure 10.1 Rotor "bucket" (or "blade") terminology.

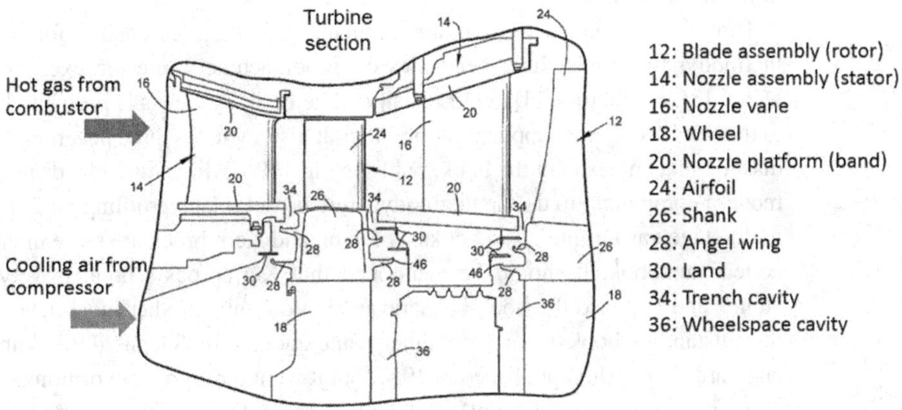

Figure 10.2 Turbine HGP terminology.

components are designed by teams of engineers, each focusing on the design of one well-defined aspect of the component in question, using in-house tools and practices deriving from decades of accumulated design and, most importantly, field experience. Even a "new" compressor or turbine represents an evolution of an existing frame (sometimes, especially in the case of compressors, borrowed from the aircraft side of the industry) with improvements reflecting the latest in research and development and field data.

The goal herein is to provide the reader with enough background in the principles of aerothermodynamics pertinent to turbomachinery design so that he or she can

1. Assess an existing design from available data (i.e., "reverse engineering") for myriad types of technical analysis, or
2. Assess a new design (i.e., essentially heat and mass balance data representing a new gas turbine cycle) for practicability.

One does not have to waste time and effort on investigating and verifying each and every claim to groundbreaking performance. All one needs to do is to determine the boundary between physical possibility and impossibility as dictated by the fundamental laws of physics and their distillation into engineering terms – in this case, aero(thermo) dynamics.

As stated earlier, there are many books that cover turbine aerothermodynamics in great detail. To begin with, there is the magnum opus by Traupel and the excellent book by Saravanamutto et al. ([6] and [3] cited in Chapter 2, respectively). The former is probably too heavy going except for a few very dedicated students of the subject (plus it is only available in German). The latter is a good starting point for young engineers and engineering students. Other good books on the subject are written by Sir John Horlock [1], Shobeiri [2], and Wilson and Korakianitis (cited above). Sir Horlock's book, although dated, is still indispensable for thoroughly understanding the subject matter. Alas, his penchant for extreme non-dimensionalization and grouping of parameters (somewhat akin to Traupel in that respect) might make reading his book somewhat difficult for beginners.

Three other references that are particularly useful, especially for self-study of thermodynamics and fluid mechanics of turbomachinery, are the excellent books by Dixon [3], Kerrebrock [4], and Lewis [5]. The book by Lewis is probably better suited to those who are more application-oriented. It also includes three programs (in a 3.5-in. diskette that comes with the book, published in 1996, which might be difficult to run on modern computers) to do the thermodynamic layout, blade profiling, and stacking for a multi-stage gas turbine.[2] The books by Dixon and Kerrebrock are more in the spirit of a college textbook. It should be mentioned that Kerrebrock's book is very similar in "tone" and depth to the book by Saravanamutto et al. Another must-have reference is the outstanding book on turbomachinery fundamentals by Shepherd [6]. Although dated and hard to find (first published in 1957, but it went through eight printings in 10 years; now, it is long out of print), the author presents a unique, unified treatment of compressors and turbines, which is quite helpful in basic guiding principles. It is highly recommended to beginners for learning the subject, who can then switch to Dixon, Kerrebrock, and/or Lewis to carry out quite elaborate front-end design and reverse-engineering calculations.

In this author's opinion, the best treatment of aerothermodynamics, from a pedagogical perspective, is the book by Mattingly [7]. This book contains a rigorous, step-by-step development of the key concepts not only with very clean and descriptive diagrams, but also with the best terminology and notation (which is extremely

[2] The programs (executables from source code in Pascal) are of no use for application to large industrial gas turbines, but they are very helpful in learning the basics in a "visual" manner.

important; otherwise most equations of aerothermodynamics become unintelligible). It contains many examples and provides key equations in a user-friendly format. The reader is strongly recommended to obtain a copy (if one is not already available in his or her library) and go through the key chapters.[3] The treatment herein follows Mattingly's terminology and diagrams, but also draws upon information drawn from the other cited references, especially [4–7].

10.1 Stage Geometry

The essence of turbine aero is the study of stage velocity "triangles," which are shown in Figure 10.3. (It is highly recommended that the reader makes a Xerox copy of Figure 10.3 and keeps it handy while reading the rest of this chapter.) The airfoils represent a circumferential cross-section of the stage at a radius r, where r is measured from the centerline of the rotor shaft. The nomenclature is as follows:

V = Velocity in a stationary coordinate system (*absolute* velocity); in some textbooks, it is denoted by c or C.
V_R = Velocity in a moving coordinate system (i.e., the rotor) – *relative* velocity; in some textbooks, it is denoted by V, w, or W.
u, v = Axial and tangential components of the absolute velocity, respectively; in some textbooks, they are denoted by subscripts (i.e., *a* or z for axial and usually θ for tangential). For example, C_a is the axial component of absolute velocity.
u_R, v_R = Axial and tangential components of the relative velocity, respectively; in some textbooks, they are denoted by the subscripts listed above. For example, w_θ is the tangential component of relative velocity.
U = Velocity of the moving coordinate system (in the tangential direction).
α = Angle between the axial coordinate and the absolute velocity.
β = Angle between the axial coordinate and the relative velocity.

The nomenclature for U, α, and β are nearly universal.

In Figure 10.3, all triangles are "vector additions"; using one of the triangles as an example,

$$\vec{V}_2 = \vec{u}_2 + \vec{v}_2,$$

$$\left|\vec{V}_2\right|^2 = \left|\vec{u}_2\right|^2 + \left|\vec{v}_2\right|^2,$$

$$\cos \alpha_2 = \frac{\left|\vec{u}_2\right|}{\left|\vec{V}_2\right|},$$

[3] As a bonus, it has a foreword written by Hans von Ohain.

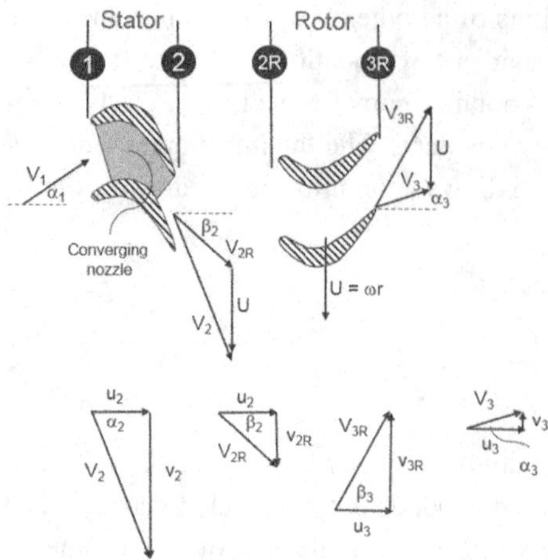

Figure 10.3 Velocity triangles of a typical turbine stage.

$$\sin \alpha_2 = \frac{\left|\vec{v}_2\right|}{\left|\vec{V}_2\right|}.$$

Since the coverage herein does not involve full-blown gas flow path calculations, vector notation is not necessary. Thus, say, V_2 or v_2 is used in lieu of $\left|\vec{V}_2\right|$ and $\left|\vec{v}_2\right|$, respectively. This makes much easier reading, but the *four building blocks* of vector algebra listed above should always be in the back of the reader's mind.

The numbering convention in Figure 10.3 is used by several authors, including Mattingly. Another numbering convention, adopted by other authors, uses 1 and 2 for rotor inlet and exit, respectively, for compressor *and* turbine stages. Thus, for a compressor stage, stator inlet and exit are labeled 2 and 3, respectively (because the rotor comes *before* the stator in the compressor stage). Similarly, for a turbine stage, stator inlet and exit are labeled 0 and 1, respectively. This usually makes using several references for comparison quite a tedious task.

Another source of confusion is the different convention used in the *orientation* of the stage diagram, which is *left to right* in the fluid flow direction and *top to bottom* in the rotational direction in Figure 10.3 (akin to looking at the "machine" from above standing on the side). Another frequently used convention is *top to bottom* in the fluid flow direction and *left to right* in the rotational direction (akin to looking at the "machine" from above standing at the downstream end). Such differences make comparative reading and interpreting of different references a real chore. In order to make it easy, keep the following two fundamental principles always in mind:

- In a *turbine* rotor, which is *downstream* of the stator, the direction of U is from the pressure/concave side of the airfoil to the suction/convex side.
- In a *compressor* rotor, which is *upstream* of the stator, the direction of U is from the suction/convex side of the airfoil to the pressure/concave side.

(Although not an ironclad rule, the standard direction of rotation for heavy-duty gas turbines is clockwise when looking into the machine from the exhaust end.) One other point to keep in the back of your mind is the sign convention. In almost all references, this point is kept somewhat obscure. In general, the magnitude of a vector in the axial and tangential directions is (i) positive if the arrow points to the right or up and (ii) negative if the arrow points to the left or down (according to the orientation in Figure 10.3). Similarly, an angle is positive if it is in the counterclockwise direction (e.g., α_3 in Figure 10.3) and negative if it is in the clockwise direction (e.g., α_2 in Figure 10.3). This can create confusion and errors in calculations unless you are 100 percent sure what convention a particular formula is based on (because, for example, tan[α] is *negative* tan [–α]). In most references, absolute values of vector magnitude angles are used and the sign convention is ignored – this will be the case herein, too. In one reference ([2] in Chapter 4), the angle sign convention is reversed (i.e., clockwise is positive [e.g., α_2 in Figure 10.3] and counterclockwise is negative [e.g., α_3 in Figure 10.3]).

Note that $U = \omega r$ is the tangential (constant) speed of the shaft with the *angular speed*

$$\omega = 2\pi N/60,$$

where N is the *rotational speed* of the shaft in rpm (i.e., 3,000 or 3,600 in most cases of interest herein). Also note that the tangential components of the absolute and relative velocities are equal to each other (i.e., $u = u_R$). There are some specific designations as well, i.e.,

- Stage exit absolute flow angle α_3 is known as the *swirl angle.*
- Tangential components of absolute velocities, v_2 and v_3, are known as *whirl* velocities.

Obviously, the rotor tangential speed is going to vary with the radius between the blade hub and tip. This will also change the gas velocity angles shown in Figure 10.3 along the rotor blades. A blade or rotor design that takes into account the variation in the velocity triangles in the radial direction is called the *vortex* design.[4] Rigorous treatment of the radial equilibrium of the three-dimensional fluid flow in the turbine annulus is too complicated for analysis via simple relationships. One simplification, known as *free vortex,* is based on the following assumptions:

- Stagnation enthalpy/temperature is constant
- u_2 = constant
- $v_2 \cdot r$ = constant

over the annulus (i.e., from blade hub to tip). Consequently, specific work done by the gas flow is constant over the annulus as well. In other words, total work can be calculated at one convenient radius and multiplied by the mass flow rate of the gas.

[4] According to Sir Frank Whittle, he first coined the term and obtained a patent for "vortex design" following an "acrimonious" debate between him and British Thompson-Houston Co. engineers who were tasked with the blade design for Sir Frank's jet engine. See pages 134–135 of his book for the full story [9]. The book is also a hidden gem with novel calculation techniques, which of course became superfluous with the advent of computers. There are still some great insights in it for the students of the subject.

In steam turbine design practice, with the exception of the very long last-stage blades, it is customary to design the *buckets* (as rotor blades are called in steam turbine parlance) at a custom *mean diameter*. This mean diameter is sometimes referred to as the *pitch-line* diameter. The losses incurred by the variation of incidence angles along the radial direction are assumed to be negligibly small for short blades. This assumption has proven to be reasonable for the gas turbines as well in carefully instrumented tests. Thus, until recently, most gas turbine OEMs resorted to using the free vortex method, at least in the preliminary design stage.

For blades with hub–tip ratios (see Figure 10.4) between 0.85 and 1.0, the free vortex approach is indeed sufficient (i.e., small radial change in velocity diagrams or radial equilibrium). For hub–tip ratios smaller than 0.75, the radial equilibrium assumption becomes untenable and two- or three-dimensional, streamline-curvature methods become requisite. For gas turbines, this becomes necessary for the last-stage blades (the longest blades in the flow path – e.g., see Figure 10.4) where hub–tip ratios become about 0.6 or less. For the largest 50-Hz H/J-class gas turbines with mass flow rates approaching 2,000 lb/s, this may be the case for the third-stage blades as well. When the free vortex assumption falls apart, one ends up with "twisted" blade profiles in the radial direction. This will be covered in detail within the context of long stage 1 rotor blades of axial compressors in Section 11.1.2 in the next chapter.

The actual turbine flow path design process starts with a free vortex or mean diameter/radius approach, in which thermodynamic cycle data are used to determine number of stages (typically three or four – very rarely five; e.g., in sequential combustion General Electric [GE]/Alstom/Ansaldo GT24/26 and GE/Alstom GT13E), hub radii, and blade lengths to accommodate the requisite mass flow rate. Along with cooling requirements and stage losses, as well as centrifugal stress limits, a first cut of an optimized flow path is determined at this stage (i.e., the dashed line, horizontal *quasi*-trapezoid in Figure 10.4). This is basically what will be covered herein, with an emphasis on important parameters and concepts.

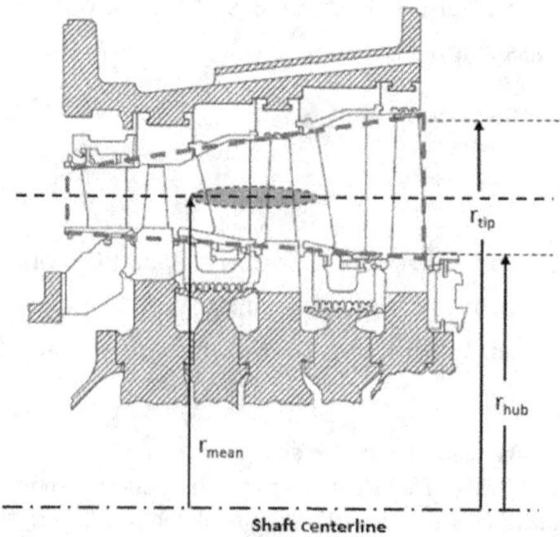

Figure 10.4 Typical three-stage gas turbine flow path arrangement.

The next step in an actual design process is most likely a two-dimensional flow analysis examining streamlines (up to 10 or so of them) between the blade hub and tip. This is also known as "throughflow" analysis. At the end of this design phase, velocity triangles are optimized and blade chord lengths and blade counts per row are determined. It should be noted that, strictly speaking, the first step described above is also a two-dimensional analysis. It uses the two coordinates (i.e., *axial* [x] and *tangential* [θ]) of a three-dimensional cylindrical coordinate system with x, θ, and r (radial) coordinates, with V (axial velocity) and U (tangential velocity, which is $r_m \cdot d\theta/dt$). The two-dimensional treatment in the second step is more correctly labeled as a "poor man's" three-dimensional treatment still with two velocity components (i.e., V and U), but with changing values and directions at different radial locations.

After the two-dimensional flow path design, airfoil profiles are determined. Using an airfoil generator, a special computer-aided design program is used to generate optimal profiles while considering aeromechanic aspects of the design in addition to cooling, stress, and manufacturability requirements. Changes to the chord lengths and blade/vane counts can be made at this stage.

This is an iterative process in which theoretical calculations are supplemented with many empirical adjustments (drawing upon decades of lessons learned by the OEM in designing previous generations of gas turbines). Nowadays, they are either replaced or validated by truly three-dimensional CFD programs, which numerically solve the fully 3D Navier–Stokes equations describing the gas flow with three velocity components (i.e., V_x, U, and V_r, in the radial direction).

Returning to the stage velocity triangles in Figure 10.3 (which, by the way, represent a partial circumferential cross-section denoted by the shaded ellipse in Figure 10.4), stage gas flow and work production can be described as follows: hot combustion gas enters the stator (stationary) at an absolute velocity V_1 and is accelerated to an absolute velocity V_2 (the passage between the two stator vanes acts as a *converging nozzle*), at which it enters the rotor, which is rotating at an angular speed of ω. Consequently, for an imaginary observer sitting on a rotor blade at radius r, the "apparent" velocity of the gas entering the passage between two blades is V_{2R}, which is found via a vector subtraction of U = ωr (tangential rotor speed at radius r) from V_2. The force imparted by the gas flow on the "pressure surface" of the rotor blade in its path with a momentum $\dot{m} v_2$ is the key mechanism of exerting torque on the blade, causing its rotation. (Note that only the tangential component, v, of the absolute velocity, V, can exert a torque on the blade.) The gas then exits the rotor (i.e., the blade passage in Figure 10.3) with absolute tangential velocity, v_3, so that the *net* torque exerted on the blade is

$$\tau = \dot{m} r (v_2 \pm v_3). \tag{10.1}$$

The summation is when the two velocities are in opposite directions; the subtraction is when they are in the same direction. The actual calculation is a "vector subtraction," i.e.,

$$\Delta \vec{v} = \vec{v}_2 - \vec{v}_3.$$

If both velocities are in the *same* direction,

$$|\Delta \vec{v}| = |\vec{v}_2| - |\vec{v}_3|.$$

If the second velocity is in the *opposite* direction,

$$|\Delta \vec{v}| = |\vec{v}_2| + |\vec{v}_3|.$$

Since for the problem herein both velocities are in the *opposite* direction (see Figure 10.3), the *plus* sign will be used exclusively. *It is impossible to overemphasize the importance of this distinction.* In many publications, Equation 10.1 is written exclusively with the minus sign without paying attention to the reference frame and the orientation of the velocity vectors. (See Mattingly [7], pp. 686, for the Euler *turbine* equation; see pp. 617 in the cited reference or Equation 10.2 for the Euler *compressor* equation, where the *minus* sign is appropriate.)

Just as the gas accelerates across the stator from V_2 to V_3, it accelerates through the rotor from V_{2R} to V_{3R}. In other words, the gas flow accelerates through *each* row of a stage, stator *and* rotor. Rate of energy transfer through the net torque gives the stage work

$$\dot{W} = \tau\omega = \dot{m}\omega r(v_2 + v_3), \tag{10.2}$$

which is also equal to

$$\dot{W} = \dot{m}(h_{2t} - h_{3t}) \approx \dot{m}c_p(T_{2t} - T_{3t}), \tag{10.3a}$$

or, per unit mass flow rate basis,

$$w = (h_{2t} - h_{3t}) \approx c_p(T_{2t} - T_{3t}). \tag{10.3b}$$

So, combining Equations 10.2 and 10.3 and noting that $U = \omega r$, we obtain the fundamental equation of turbomachinery analysis, the *Euler turbine equation*:

$$w = h_{2t} - h_{3t} = U(v_2 + v_3), \tag{10.4a}$$

$$w = c_p(T_{2t} - T_{3t}) = U(v_2 + v_3). \tag{10.4b}$$

Equation 10.4a is the "exact" Euler turbine equation, whereas Equation 10.4b is for a calorically perfect gas. Before moving on, for the sake of completeness, let us derive the physical quantity "rothalpy" by rewriting Equation 10.4a in its generic form (i.e., without making an assumption about the direction of whirl velocities) as

$$h_{2t} - U \cdot \vec{v}_2 = h_{3t} - U \cdot \vec{v}_3,$$

$$I_2 = I_3,$$

where I is the rothalpy (from "rotating enthalpy"). Rothalpy is *not* a thermodynamic quantity. It depends on the reference frame and the orientation of the velocity vectors. As shown in Figure 10.3, for instance, the magnitude of I_3 is

$$I_3 = h_{3t} + U \cdot v_3.$$

In an adiabatic (no heat transfer), steady flow (in a rotating frame of reference) with no friction, or other irreversibility (i.e., isentropic), rothalpy is constant. The reader is referred to the paper by Lyman for a rigorous discussion of rothalpy conservation in turbomachines [8].

10.2 Stage Design Parameters

The Euler turbine equation defines the most important turbine design parameter (i.e., the *stage loading parameter*):

$$\psi = \frac{h_{2t} - h_{3t}}{U^2} = \frac{v_2 + v_3}{U}, \tag{10.5a}$$

$$\psi = \frac{c_p(T_{2t} - T_{3t})}{U^2} = \frac{v_2 + v_3}{U}. \tag{10.5b}$$

Also known as the "work coefficient," the stage loading parameter is a non-dimensional measure of net torque exerted on the blade, as quantified by the ratio of "change in whirl" in the numerator of Equation 10.5 (i.e., the "cause") to the rotational speed in the denominator (i.e., the "effect"). Stages with ψ >1.5 are referred to as "highly loaded" stages. Similarly, stages with ψ <1.0 are known as "lowly loaded" stages. To put it in perspective, one should look at it this way: in a highly loaded stage with $\psi = 2$, creation of each 1 ft/s or m/s of rotational speed requires 2 ft/s or m/s of a change in whirl.

In passing, note that in some texts, ψ is defined as $\Delta h/2U^2$ (i.e., *half* of the value obtained from Equation 10.5) and in others as $\Delta h/½U^2$ (e.g., Saravanamuttoo et al.; i.e., *twice* the value obtained from Equation 10.5). Herein, the "natural" definition directly obtained from the Euler turbine equation is used. However, in reading other texts and comparing design data, one should ascertain the definition of ψ used in the calculations.

The second important parameter in turbine design is the *flow coefficient*, defined as

$$\phi = \frac{u_2}{U}. \tag{10.6}$$

The flow coefficient is the ratio of the axial component of the absolute velocity, V_2, of the gas entering the rotor to the rotor speed. The typical range of ϕ is 0.4–1.0 – the front (cooled) stages have values at the lower end of the range, whereas the last (usually uncooled) stages have values at the higher end.

The third parameter needed for a complete description of the velocity diagrams in Figure 10.3 is the *stage reaction*, R, which is defined as the *static* enthalpy change across the rotor to the *total* enthalpy change across it, which is equal to the total enthalpy change across the entire stage (since $h_{1t} = h_{2t}$). In some texts, it is referred to as the *degree of reaction* and denoted by Λ (i.e., capital lambda). In terms of the total stage work and kinetic energy change across the rotor,

$$R = 1 - \frac{\frac{1}{2}(V_2^2 - V_3^2)}{w}. \tag{10.7a}$$

In velocity terms, using Equation 10.4, we find that

$$R = 1 - \frac{(V_2^2 - V_3^2)}{2U(v_2 + v_3)}. \tag{10.7b}$$

For a stage with little or no change in the axial component of the absolute velocity (i.e., $u_2 \approx u_3$), Equation 10.7 can be shown to become (for the vector orientation in Figure 10.4 and using the absolute values of $\vec{v}_2$ and $\vec{v}_3$)

$$R = 1 - \frac{v_2 + v_3}{2U}, \tag{10.8a}$$

or in terms of velocity triangle angles

$$R = \phi \frac{\tan \beta_3 - \tan \beta_2}{2}. \tag{10.8b}$$

For a zero reaction stage, R = 0 and the change in whirl is given by

$$v_2 + v_3 = 2U$$
$$\beta_3 = \beta_2$$

In other words, the gas is accelerated across the stator only and it only changes direction across the rotor ($V_{R3} = V_{R2}$). Such a stage or turbine is called an "impulse" stage or turbine because the rotor torque fully comes from the impulse of the gas accelerating through the nozzle vanes (see Equation 10.7a). This is akin to what happens in a hydraulic turbine (i.e., the *Pelton wheel*) where the water jet from a nozzle hits the wheel "buckets." (This is also why in steam turbines, and sometimes in gas turbines as well, rotor blades are referred to as buckets.)

As an interesting sidenote, Sir Frank Whittle was opposed to the usage of "leftover" terms of *impulse* and *reaction*. He proposed to abandon them and define turbine types by the ratio of the blade speed to the change of whirl at the mean radius [9]. In other words, he proposed Equation 10.8a in a different form, i.e.,

$$R_{FW} = \frac{1}{2(1 - R)},$$

where R_{FW} designates Whittle's turbine-type parameter (for lack of a better term). Thus, 50 percent reaction was equivalent to $R_{FW} = 1.0$ and 0 percent reaction (impulse) was equivalent to $R_{FW} = 0.5$. Interestingly, 100 percent reaction (not an acceptable choice, as will be shown below) resulted in $R_{FW} \to \infty$. Maybe Sir Frank had a point after all.

There are several advantages and disadvantages to an impulse stage:

1. The static enthalpy drop of zero implies zero or little leakage across the rotor because of a small (but non-zero) pressure drop. In that sense, consider that the impulse stage (R = 0) is not truly an impulse stage because $p_2 > p_3$. (Note that the stage reaction is defined using enthalpies, *not* pressures.)
2. The high value of ψ lowers the rotor-relative stagnation temperature, which lowers the cooling flow requirement. This can be seen from equation 9.113 in

[7], which gives the rotor-relative stagnation temperature at the stage exit, T_{3R}. However, since no work is done by the gas relative to the blades, $T_{2Rt} = T_{3Rt}$ (with no blade cooling), so that, with a little algebra, we obtain

$$T_{2Rt} = T_{2st} + \frac{V_2^2}{2c_p}\left(\cos^2\alpha_2 + \left(\sin^2\alpha_2 - \frac{U}{V_2}\right)^2\right). \tag{10.9}$$

It can be shown that, depending on values of ϕ, ψ, and R, the term in the parentheses on the right-hand side of Equation 10.9 can be between 0.2 and 0.3. (For example, the absolute rotor inlet angle is 50°–60° and U/V is 0.5–0.7.)

3. However, higher values of ψ leads to poorer stage efficiency (see below).
4. Large values of α_2 can lead to a supersonic nozzle exit velocity and shock waves, which reduce the stage efficiency.
5. Similarly, a large turning angle (deflection: sum of angles β_2 and β_3 for the rotor) increases losses due to viscous boundary layer separation.

For a 100 percent reaction stage (i.e., R = 1), it is the opposite; the gas only changes direction across the stator and accelerates across the rotor. Equation 10.8 becomes

$$v_2 = v_3.$$

Since $u_2 = u_3$, with constant U, this implies $V_{R3} > V_{R2}$ and $V_2 = V_3$. As such, this case is of little practical interest because all of the static enthalpy change takes place across the rotor (unless, of course, one is designing *rotor-only* wind turbines). Physical examples used to illustrate 100 percent reaction stages are Hero's "aeolipile"[5] or common, rotating lawn sprinklers.

Note that, for an impulse stage, with a static enthalpy change of zero across the rotor, the implication is that

$$dh = Tds + vdp = 0.$$

Thus, for the isentropic case (i.e., ds = 0), dp = 0 (i.e., $p_{2st} = p_{3st}$), which, of course, is not the case. In an actual turbine stage, $p_{2st} > p_{3st}$ due to friction (albeit small).

A *negative* stage reaction (when calculated using static pressure instead of static enthalpy) implies expansion across the stator nozzle vanes and recompression across the rotor blades. This would lead to large losses and must be avoided. Note that this statement applies to the *entire blade* (i.e., from hub to tip). This is one of the pitfalls of the free vortex approach, which requires a close look at the preliminary design results to prevent negative R at the blade hub.

A typical choice for stage reaction, R, at the mean radius is 0.5 (i.e., 50 percent reaction design). Due to the change in whirl velocity with the radius as given by

[5] The ball of Aeolus, the Greek god of the air and wind, which is a steel ball with two protruding nozzles as described by the Greek mathematician Hero (or Heron) of Alexandria. The ball is filled with water, which, upon heating, becomes steam and exits through the nozzles, thereby causing the ball to turn around its axis in accordance with Newton's third law.

$$v(r) = \frac{r}{r_m} v(r_m),$$

where r_m is the mean radius. The reaction also changes from hub to tip according to

$$R(r) = 1 - (1 - R_m)\left(\frac{r}{r_m}\right)^{-2}, \tag{10.10}$$

where R_m is the reaction at r_m. According to Equation 10.10, the mean radius should be chosen such that, at the blade hub, $r_h < r_m$, R does not become negative.

In addition to ψ, ϕ, and R, the fourth important turbine aero parameter is the stage exit swirl angle, α_3. For the last turbine stage, a low swirl angle is desirable so that relative and axial gas velocities are close to each other. This reduces the amount of kinetic energy left in the gas leaving the turbine. On the other hand, at other stages, a higher exit swirl angle means larger stage work output. This can be seen from the velocity triangles, which can be used to arrive at the following correlation (after some lengthy but straightforward algebra):

$$\psi = \phi\left(\tan\alpha_2 + \frac{u_3}{u_2}\tan\alpha_3\right), \tag{10.11a}$$

which, for $u_3 = u_2$, becomes

$$\psi = \phi(\tan\alpha_2 + \tan\alpha_3). \tag{10.11b}$$

Thus, for a given flow coefficient, stage loading is proportional to α_3.

For R = 0.5, the stage loading parameter is simply

$$\psi = 2\phi\tan\alpha_2 - 1. \tag{10.12}$$

This design has lower ψ than the impulse stage, which leads to lower velocities and losses with better stage efficiency. This benefit is dampened by the increased cooling flow requirement due to a higher rotor-relative temperature (e.g., see Equation 10.9). For a zero exit swirl stage (i.e., $\alpha_3 = 0$), the stage loading parameter becomes

$$\psi = 2(1 - R), \tag{10.13}$$

so that ψ changes between 1 (50 percent reaction) and 2 (impulse).

Specification of ϕ, ψ, and R determines all the velocity triangles (i.e., angles α and β). For the general case, the resulting equations become quite tedious and lose their descriptive power. An approximation that is reasonably representative of modern gas turbine expansion flow path design can be obtained with the following simplifying assumption:

$$u_1 = u_2 = u_3,$$

which implies that

$$V_1 = V_3 \text{ and}$$
$$\alpha_1 = \alpha_3.$$

Furthermore, via mass continuity, for state-points i = 1, 2, and 3 in the turbine stage,

$$\rho_i u_i A_i = \text{constant},$$

where A_i is the flow cross-sectional area at each state-point. The simplifying assumption of constant axial absolute velocity means that

$$\rho_i A_i = \text{constant}.$$

For any value of $\{0 < R < 1\}$, velocity triangles can be determined using the following four equations:

$$\tan\beta_3 = \frac{1}{2\phi}(\psi + 2R), \tag{10.14}$$

$$\tan\beta_2 = \frac{1}{2\phi}(\psi - 2R), \tag{10.15}$$

$$\tan\alpha_1 = \tan\alpha_3 = \tan\beta_3 - \frac{1}{\phi}, \tag{10.16}$$

$$\tan\alpha_2 = \tan\beta_2 + \frac{1}{\phi}. \tag{10.17}$$

Equations 10.14–10.17 are used to generate ψ–ϕ diagrams as a function of exit swirl angle α_3. These equations (along with data from cascade tests done to predict losses and estimate stage efficiency) can be used to generate families of constant efficiency lines (contours) on a ψ–ϕ surface, which is known as a "Smith chart," after the investigator who first built those contours based on 48 uncooled turbine stages of different designs [10]. This chart will be covered in detail in Section 10.3.1. Similar charts can be found in Saravanamuttoo et al., who derived them from two sources, including Horlock's book and in Wilson and Korakianitis (both cited above). Other charts in the latter reference display optimum nozzle inlet (α_2) and exit swirl angles (α_3) as functions of ψ for different values of R.

In general, low values of ϕ and ψ imply low gas velocities and, consequently, low friction losses. The trade-off is more stages (low ψ) and a larger flow cross-sectional area (low ϕ) for a given mass flow rate, which is detrimental to aircraft engine design, but does not present an obstacle to large, stationary gas turbines.

An uncooled, isentropic stage efficiency is derived by Saravanamutto et al. by evaluating friction loss coefficients for the stator nozzle vanes and rotor blades, λ_N and λ_R, respectively. It can be used in lieu of a chart/table lookup for preliminary design or reverse-engineering estimates:

$$\eta = \frac{1}{1 + 0.5\phi\left[\dfrac{\lambda_R \sec^2\beta_3 + \tau_{32}\lambda_N \sec^2\alpha_2}{\tan\beta_3 + \tan\alpha_2 - 1/\phi}\right]}, \tag{10.18}$$

where τ_{32} is the ratio of *static* temperatures at state-points 3 and 2 given by

$$\tau_{32} = \frac{T_{3st}}{T_{2st}}, \tag{10.19}$$

$$\tau_{32} = 1 + \frac{1}{2\psi}\left(\frac{\phi}{\cos\beta_2}\right)^2 \frac{\Delta T_{stage}}{T_{2st}}\left(1 - \left(\frac{\cos\beta_2}{\cos\beta_3}\right)^2\right). \tag{10.20}$$

The ratio of stage (total) temperature loss to T_2 (static) in Equation 10.20 changes from stage to stage and has a value of 0.2–0.3. Using an average value of 0.25 is sufficient for estimate-type calculations. You can do even better (if you want to reduce computational load without losing a lot in accuracy, that is) and skip Equation 10.20 altogether and assume that $\tau_{32} \approx 1.0$. In fact, with $\tau_{32} \approx 1.0$, Equation 10.18 becomes (equation 3.41 on p. 70 of [5])

$$\eta = \left(1 + \frac{0.5}{\psi}\left(\left(\phi^2 + (\psi/2 + 1 - R)^2\right)\lambda_N + \left(\phi^2 + (\psi/2 + R)^2\right)\lambda_R\right)\right)^{-1}, \tag{10.21}$$

which is probably easier to implement in code and/or in Excel.

Loss coefficients in Equations 10.18 and 10.21 depend on stage geometry (ϕ and ψ), flow Mach, and Reynolds numbers. A detailed discussion of their experimental and analytic derivation can be found in Horlock [1]. Cascade data showed that, ignoring the compressibility effect, for a wide range of Reynolds numbers they can be expressed as a function flow deflection. A simple correlation (attributed to Hawthorne, who simplified Soderberg's original work) found in Horlock is (equations 3.22 and 3.23 on p. 88 of [1])

$$\lambda = 0.025\left[1 + \left(\frac{\varepsilon}{90}\right)^2\right]\left(1 + \frac{3.2}{AR}\right), \tag{10.22}$$

where ε is the flow deflection, i.e.,

$$\text{Stator, } \varepsilon_N = |\alpha_1| + |\alpha_2|, \tag{10.23a}$$

$$\text{Rotor, } \varepsilon_R = |\beta_2| + |\beta_3|, \tag{10.23b}$$

and AR is the blade aspect ratio (blade height divided by the axial chord). For 50 percent reaction, $\varepsilon_N = \varepsilon_R$ so that $\lambda_N \approx \lambda_R$. (In general, $\lambda_N < \lambda_R$, mainly due to the tip leakage losses in the rotor.) Note the use of *summation* to find the flow deflection when using the *absolute* values of the absolute and relative flow angles. When using the actual values, you have to do a *subtraction* with proper signs of the angles. This would give a negative value for ε_R, but the convention is to always express the deflection as a positive value.

There are *seven* cardinal stage design parameters – R, ϕ, ψ, α_2, α_3, β_2, and β_3 – with *four* equations, Equations 10.14–10.17. Thus, three parameters can be specified to fix all seven of them and estimate the isentropic stage efficiency. Possible combinations are:

- ψ, ϕ, and α_3
- R, ψ, and α_3
- R, ϕ, and α_3

Table 10.1 Turbine stage parameters

R	0.50	0.50	0.50	0.50	**R**	0.50	0.50	0.50	0.50
ψ	1.36	1.40	1.44	1.47	ψ	1.58	1.64	1.69	1.75
ϕ	0.50	0.55	0.60	0.65	ϕ	0.50	0.55	0.60	0.65
α_3	20.0	20.0	20.0	20.0	α_3	30.0	30.0	30.0	30.0
α_2	–67.1	–65.4	–63.8	–62.3	α_2	–68.8	–67.3	–66.0	–64.7
β_3	67.1	65.4	63.8	62.3	β_3	68.8	67.3	66.0	64.7
β_2	–20.0	–20.0	–20.0	–20.0	β_2	–30.0	–30.0	–30.0	–30.0
η	0.928	0.928	0.927	0.926	η	0.919	0.918	0.917	0.916
R	0.40	0.40	0.40	0.40	**R**	0.40	0.40	0.40	0.40
ψ	1.56	1.60	1.64	1.67	ψ	1.78	1.84	1.89	1.95
ϕ	0.50	0.55	0.60	0.65	ϕ	0.50	0.55	0.60	0.65
α_3	20.0	20.0	20.0	20.0	α_3	30.0	30.0	30.0	30.0
α_2	–70.1	–68.6	–67.1	–65.7	α_2	–71.4	–70.1	–68.8	–67.6
β_3	67.1	65.4	63.8	62.3	β_3	68.8	67.3	66.0	64.7
β_2	–37.4	–36.0	–34.9	–33.9	β_2	–44.3	–43.3	–42.3	–41.5
η	0.914	0.914	0.914	0.914	η	0.905	0.905	0.905	0.905
R	0.60	0.60	0.60	0.60	**R**	0.60	0.60	0.60	0.60
ψ	1.16	1.20	1.24	1.27	ψ	1.38	1.44	1.49	1.55
ϕ	0.50	0.55	0.60	0.65	ϕ	0.50	0.55	0.60	0.65
α_3	20.0	20.0	20.0	20.0	α_3	30.0	30.0	30.0	30.0
α_2	–63.8	–61.2	–59.5	–57.9	α_2	–65.3	–63.8	–62.4	–61.1
β_3	67.1	65.4	63.8	62.3	β_3	68.8	67.3	66.0	64.7
β_2	2.1	0	–1.8	–3.2	β_2	–10.1	–12.1	–13.7	–15.1
η	0.942	0.940	0.938	0.936	η	0.934	0.932	0.929	0.927

Several combinations have been calculated and tabulated in Table 10.1 for ready reference for

- Degree of reaction, R, 0.4, 0.5, and 0.6;
- Flow coefficients, ϕ, 0.5–0.65;
- Stage exit angles, α_3, 20 and 30 degrees.

Stage isentropic efficiency is estimated using Equation 10.21 (AR = 10 and $\lambda_N \approx \lambda_R$). The equation is encapsulated in the VBA code Eff_341 in Appendix C.

Simple rules of thumb for good stage design are (at least for starters):

- Stage reaction $\approx$ 0.5 (but ensure non-negative R at the hub)
- Stage/rotor exit swirl angle 20° (or lower)
- Stage loading less than 2.0
- Nozzle exit angle 60°–70°

When going from a thermal (Brayton) cycle calculation to a first estimate of the turbine layout, calculate a "turbine loading parameter" using Equation 10.5. Divide it by the estimated number of stages, say 3 or 4, to find an average stage loading parameter. As a rough rule of thumb,

- If the average stage loading is higher than ~1.7, one should consider adding another stage.
- If the stage loading is less than ~1.2, one should consider removing a stage.

This will be demonstrated with a numerical example in Section 10.3.3.

Using Equation 10.10, the first rule-of-thumb above requires that

$$\upsilon > \frac{1}{\dfrac{2}{\sqrt{1-R_m}} - 1}, \tag{10.24}$$

where $\upsilon = r_h/r_t$ is the hub–tip ratio and R_m is the stage reaction at the pitch-line radius, r_m.

Precalculated efficiency curves and/or tables are good as starting points for conceptual studies and reverse-engineering analyses. In a modern engine design, one needs more design-specific information such as blade profile, pitch–chord ratio (or solidity, σ, which is the inverse of it), leakages, and cooling flows, including wheel-space purge flows, along with extensive test and field data, for rigorous evaluation of profile losses, annulus losses, secondary flow losses, tip clearance losses, and coolant mixing losses.

10.3 Turbine Design

The focus of this book is on several key aspects of the turbine (HGP) design, which are requisite for separating the proverbial wheat from the chaff. If a particular turbine/expander energy balance with an unusual working fluid implies a turbine design with 30 stages and 5-mm-tall rotor blades, it is probably wise not to get too excited about that particular cycle. As far as real design activity is concerned, you can rest assured that nobody is going to ask you to lay out the HGP section of the next Z-class gas turbine for the OEM X.[6] Thus, the goal herein is to provide the bare minimum to grasp the physics-based principles underlying the design of the actual gas turbines so that the reader can ask the right questions at the right time (whatever they may be).

In that sense, the important questions to ask and/or answer are as follows:

- How many stages in the gas flow path and why?
- How many blades/vanes in a stage and why?
- What is the relationship with rotational speed and shaft diameter?
- What are the rotor bucket heights in each stage?
- What are the size limits?
- What are the stage efficiencies?

[6] Interestingly, in April 2018, there were reports of the catastrophic failure of a 110-MW heavy-duty industrial gas turbine during testing in the Saturn engineering plant in Rybinsk, central Russia, in December 2017. The machine was the culmination of a decision made by the Russian leadership to replace imported technology with homegrown substitutes in energy, software, aerospace, and medicine (after Western sanctions were imposed on Russia over the conflict with Ukraine). According to the anonymous sources quoted by Reuters, "the turbine fell apart." This is a dramatic illustration of the difficulties involved in designing a large gas turbine operating at high firing temperatures from scratch.

10.3.1 Smith Chart

The correlation between stage loading and turbine efficiency is given by the Smith chart, where the x axis is the stage flow parameter, ϕ, and the y axis is the stage loading parameter, ψ. The original chart in Figure 10.5 first appeared in the seminal paper by S.F. Smith in 1965 [10]. The efficiency contours (stage total-to-total isentropic efficiency) in the chart are empirical. The experimental data points (there are 70 of them) are from the four-stage "model turbine" tests, which are related to post–World War II Rolls-Royce gas turbines such as Avon, Dart, Spey, and Conway. They represent the maximum efficiency points from tests done over a range of PRs. Furthermore, they are corrected to "zero-leak" (i.e., zero tip clearance) conditions to eliminate the leakage effects. (Basically, several test points were run with different tip clearances and the data were extrapolated to zero tip clearance.) The experimental data are seen to fall within 1 percent of the empirical contours. Selected data points from the chart are listed in Table 10.2 (from [5], pp. 285–287).

Also shown in Figure 10.5 are the airfoil shapes (labeled A through D, starting from the lower left corner in clockwise direction) of the turbine rotor blades as defined by ϕ and ψ. Higher stage loading ψ is equivalent to larger flow turning via Equation 10.11 (i.e., upper versus lower airfoil shapes in Figure 10.5), which also includes the "softening" effect of ϕ on the flow turning (i.e., right versus left airfoil shapes in Figure 10.5). Corresponding relative flow and rotor flow turning angles, $\varepsilon = \beta_2 - \beta_3$, are summarized in Table 10.3 (for R = 0.5). Efficiency contours for ϕ = 0.4–0.8

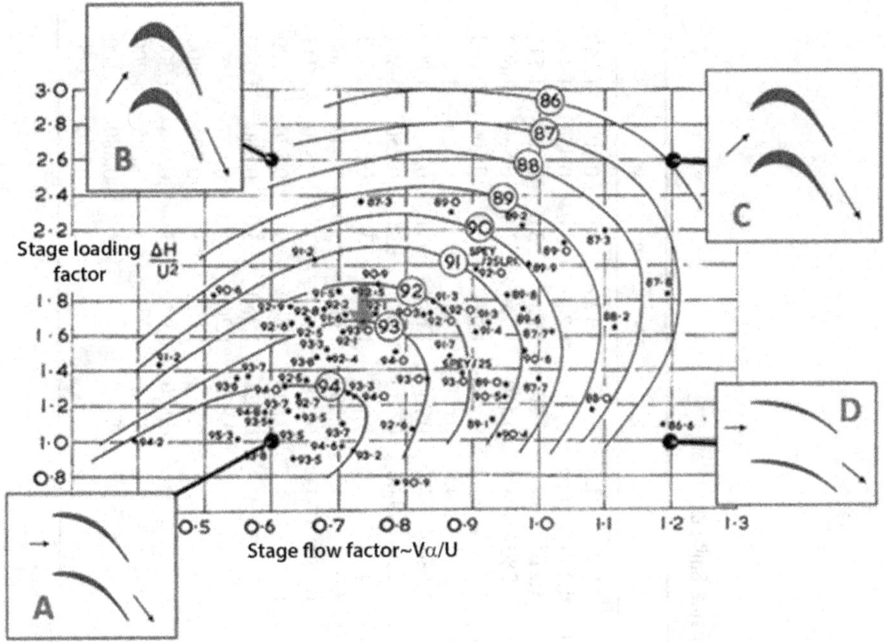

Figure 10.5 Smith chart – turbine stage efficiency (total to total) as a function of ψ and ϕ [10].

Table 10.2 Smith chart data

ϕ	ψ	η	ϕ	ψ	η	ϕ	ψ	η	ϕ	ψ	η	ϕ	ψ	η	ϕ	ψ	η
0.400	1.020	94.0	0.400	1.120	93.0	0.400	1.260	92.0	0.400	1.400	91.0	0.400	1.570	90.0	0.500	2.060	89.0
0.450	1.130	94.0	0.450	1.240	93.0	0.450	1.410	92.0	0.450	1.570	91.0	0.500	1.870	90.0	0.550	2.170	89.0
0.500	1.210	94.0	0.500	1.350	93.0	0.500	1.545	92.0	0.500	1.700	91.0	0.600	2.100	90.0	0.600	2.270	89.0
0.550	1.280	94.0	0.550	1.450	93.0	0.550	1.660	92.0	0.550	1.820	91.0	0.650	2.180	90.0	0.650	2.340	89.0
0.600	1.310	94.0	0.600	1.540	93.0	0.600	1.770	92.0	0.600	1.930	91.0	0.700	2.240	90.0	0.700	2.390	89.0
0.650	1.320	94.0	0.650	1.630	93.0	0.650	1.840	92.0	0.650	2.020	91.0	0.750	2.280	90.0	0.750	2.420	89.0
0.700	1.280	94.0	0.700	1.680	93.0	0.700	1.890	92.0	0.700	2.070	91.0	0.800	2.290	90.0	0.800	2.440	89.0
0.720	1.250	94.0	0.750	1.670	93.0	0.750	1.890	92.0	0.750	2.110	91.0	0.850	2.270	90.0	0.850	2.450	89.0
0.740	1.150	94.0	0.785	1.640	93.0	0.800	1.860	92.0	0.800	2.100	91.0	0.900	2.210	90.0	0.900	2.430	89.0
0.730	1.000	94.0	0.800	1.600	93.0	0.840	1.800	92.0	0.850	2.060	91.0	0.950	2.120	90.0	0.950	2.370	89.0
0.700	0.840	94.0	0.810	1.570	93.0	0.860	1.745	92.0	0.900	1.950	91.0	1.000	1.920	90.0	1.000	2.280	89.0
			0.820	1.500	93.0	0.870	1.700	92.0	0.950	1.780	91.0	1.018	1.800	90.0	1.030	2.200	89.0
			0.827	1.250	93.0	0.880	1.600	92.0	0.960	1.700	91.0	1.035	1.600	90.0	1.070	2.000	89.0
			0.825	1.200	93.0	0.887	1.500	92.0	0.970	1.600	91.0	1.037	1.500	90.0	1.090	1.800	89.0
			0.810	1.000	93.0	0.890	1.400	92.0	0.972	1.500	91.0	1.035	1.400	90.0	1.090	1.600	89.0
			0.800	0.920	93.0	0.888	1.300	92.0	0.970	1.400	91.0	1.027	1.300	90.0	1.080	1.400	89.0
						0.885	1.200	92.0	0.965	1.300	91.0	1.015	1.200	90.0	1.065	1.200	89.0
						0.880	1.100	92.0	0.950	1.200	91.0	0.975	1.000	90.0	1.032	1.000	89.0
						0.870	1.000	92.0	0.920	1.000	91.0						

Table 10.3 Relative flow and flow turning angles for blade profiles in Figure 10.5

	A	B	C	D
ψ	1.0	2.6	2.6	1.0
ϕ	0.60	0.60	1.20	1.20
β_3	–59.0	–71.6	–56.3	–39.8
β_2	0	53.1	33.7	0
ε	59.0	124.7	90.0	39.8

(the range of interest for the turbines covered herein) are encapsulated in the VBA code Eff_Smith in Appendix C.

Using Smith's arguments in his paper [10], Lewis [5] found the optimum stage loading, which maximizes the normalized stage work, w_s, given by

$$w_s = \frac{\Delta h_t}{V_1^2 + V_3^2} = \frac{2\psi}{4\phi^2 + \psi^2 + 1}, \tag{10.25}$$

by setting $\partial w_s/\partial \psi$ to zero as

$$\psi_{opt} = \sqrt{4\phi^2 + 1}. \tag{10.26}$$

Using Rolls-Royce data, Lewis suggested that the design rule of thumb should be

$$\psi_{opt} = 0.65\sqrt{4\phi^2 + 1}. \tag{10.27}$$

For a flow coefficient of $\phi = 0.7$, the optimum value of the stage loading coefficient (*without* the Rolls-Royce correction) is 1.72. This is indeed a good starting point for heavy-duty industrial gas turbines (for a single stage).

If Equation 10.21 is written as

$$\eta = \frac{1}{1 + L},$$

where L is the total "loss coefficient," setting $\partial L/\partial R$ to zero gives the optimum stage reaction, R_{opt}, i.e.,

$$R_{opt} = \frac{\lambda_N + 0.5\psi(\lambda_N - \lambda_R)}{\lambda_N + \lambda_R}. \tag{10.28}$$

If $\lambda_N = \lambda_R$, from Equation 10.28, the optimum R is 0.5, independent of ψ. For the values of λ_N and λ_R calculated in Saravanamutto et al. (i.e., 0.05 and 0.092, respectively), we have

$$R_{opt} = 0.35 - 0.15\psi. \tag{10.29}$$

Using the "optimum" stage loading parameter from Equation 10.26 (as a function of ϕ), the optimum stage reaction from Equation 10.29 (as a function of ψ_{opt}), and the Smith chart data in Table 10.2, stage efficiency corresponding to said optimum can be determined (see Table 10.4).

Table 10.4 Stage efficiency for optimum ψ

ϕ	ψ_{opt}	η	R_{opt}
0.40	1.28	91.85	0.54
0.50	1.41	92.67	0.56
0.60	1.56	92.90	0.58
0.70	1.72	92.81	0.61
0.80	1.89	91.89	0.63
0.90	2.06	90.58	0.66
1.00	2.24	89.12	0.68

Table 10.5 Optimum ψ from Equation 10.30

ϕ	R	ψ
0.70	0.20	1.82
0.70	0.30	1.77
0.70	0.40	1.73
0.70	0.50	1.72
0.70	0.60	1.73
0.70	0.70	1.77
0.70	0.80	1.82

We can find another optimum stage loading parameter by setting $\partial L/\partial\psi$ to zero, i.e.,

$$\psi_{opt} = 2\sqrt{\phi^2 + 0.5 + R(R-1)}, \tag{10.30}$$

which reduces to Equation 10.26 for R = 0.5. The variation in ψ_{opt} with R in the range of most interest for free vortex calculations is too small, so sticking with Equation 10.26 is perfectly adequate (see Table 10.5).

Clearly, there is some fundamental theoretical basis for $R \approx 0.5$ being the optimal choice for axial turbine design. Following this thread, using the definitions of stator (nozzle) and rotor deflection angles from Equation 10.23, i.e.,

$$\varepsilon_N = |\alpha_1| + |\alpha_2| \text{ and}$$
$$\varepsilon_R = |\beta_2| + |\beta_3|,$$

and noting that, for R = 0.5

$$|\alpha_1| = |\beta_2| \text{ and}$$
$$|\alpha_2| = |\beta_3|,$$

we can ascertain that

$$\varepsilon_N = \varepsilon_R = \varepsilon.$$

From the stage geometry, the flow deflection angle can be expressed as a function of ϕ and ψ, i.e.,

$$\varepsilon = \arctan\left(\frac{\psi\phi}{1-\frac{\psi^2-1}{4\phi^2}}\right). \tag{10.31}$$

Substituting the optimum value of ψ from Equation 10.28, it can be found that the optimum deflection angle is given by

$$\varepsilon_{opt} = \arctan(\infty) = 90°. \tag{10.32}$$

10.3.2 How Many Vanes or Blades?

In order to keep the manufacturing cost and turbine weight low, it is desirable to use as few as possible stator nozzle vanes and rotor blades in a given stage. Two items to note are:

- The number of stator nozzle vanes in a given stage must be even because vane carriers are split at the casing horizontal joint.
- There are more rotor blades than stator nozzle vanes in a given stage in order to limit the stresses acting on the blade roots.

The crucial parameter relating the number of blades/vanes to other stage geometry parameters is the *Zweifel loading coefficient*, Zw. Assuming constant axial velocity, Zw is given as

$$Zw = \frac{2}{\sigma}(\cos\beta_3)^2(\tan\beta_2 + \tan\beta_3), \tag{10.33}$$

with the solidity defined as (with *axial* blade chord, b)[7]

$$\sigma = \frac{n\cdot b}{2\pi r_m}, \tag{10.34}$$

where n is the number of rotor blades. Combining Equations 10.11b, 10.14, 10.15, and 10.33, we obtain

$$Zw = \frac{2}{\sigma}(\cos\beta_3)^2\frac{\psi}{\phi}. \tag{10.35}$$

(It can be shown that the Zweifel coefficient is proportional to the blade lift coefficient, C_L. See Section 11.1.1 in the next chapter and Equation 11.22b for a coverage of C_L in the context of compressor aerodynamics.)

Substituting for the solidity from Equation 10.34,

$$Zw = \frac{4\pi r_m}{n\cdot b}(\cos\beta_3)^2\frac{\psi}{\phi}. \tag{10.36}$$

[7] Blade chord is the length of the straight line connecting the leading and trailing edges of the airfoil (blade profile). The axial blade chord is the length of the line in the axial direction between the leading and trailing edges.

A good first guess for Zw is 0.8 (which goes back to his original 1945 publication "Die Frage der optimalen Schaufelteilung"[8] in Brown-Boveri's internal journal). Using data from the second column of Table 10.5 (typical for stage 1 of a four-stage turbine), from Equation 10.36 we find

$$n = \frac{4\pi r_m}{0.8 \cdot c}(\cos - 65.4)^2 \frac{1.7}{0.55} = 8.4135 \frac{r_m}{b}. \tag{10.37}$$

The ratio of the axial blade chord to the stage radius at the pitch-line, b/r_m, is around 0.1–0.2, decreasing from the first stage to the last. Calculated values of n for different values of b/r_m are given below.

b/r_m	n
0.10	84
0.15	56
0.20	42

A typical vane/blade count per stage is 60. Stage 1 and 2 stator nozzle vanes are probably lower (i.e., around 40). For large last-stage rotor blades, this can be as high as 90. Zweifel numbers for state-of-the-art, high-lift turbine rotor blades are closer to 1.0; for stator nozzle vanes, they are closer to 0.7.

The advantage of higher values of Zw is clear from Equation 10.36; it leads to a lower number of rotor blades (at the same loading coefficient). This is advantageous in terms of a lower number of HGP parts and a reduction in chargeable cooling air (plus, of course, lower manufacturing costs). Alternatively, it can lead to higher stage loading at the same blade count, which of course leads to reduced cooling load of the downstream stages. This was the foundation of GE's F–class turbine design philosophy for a long time, leading to three-stage turbine designs in contrast to other major OEMs.

Circumferentially arranged stator vanes attached to the casing via a "vane carrier" constitute an annular nozzle. Expansion and contraction of this annular nozzle due to changing heat loads during startup, shutdown, and load ramps create significant thermal stresses in both the individual vanes and vane carriers, which can eventually lead to low-cycle fatigue (LCF). The LCF life of the annular turbine nozzle is increased by circumferentially dividing it into segments including one, two, or three nozzle vanes. A two-vane nozzle segment is referred to as a "doublet." A three-vane nozzle segment is referred to as a "triplet." Segmenting the annular nozzle interrupts the continuity of the platforms and reduces the magnitude of thermal hoop stress. (Care must be given to the seals between the segments.)

10.3.3 How Many Stages?

Probably the most important early-stage design decision is to determine the number of stages in a given turbine. The starting point is (obviously) the cycle heat and mass

[8] "The question of optimal blade spacing."

balance. Before moving on, the reader should be aware that almost all 50- and 60–Hz heavy-duty industrial gas turbines in the market today (2017) have *four-stage* HGP designs. Notable exceptions are

- GE's vintage E- and F-class gas turbines (three stages)
- GE/Alstom GT13E2[9] (five stages)
- GE/Alstom/Ansaldo GT24 and GT26 (reheat machines, one high-pressure stage and four low-pressure stages separated by the second combustor for a total of five stages)

Consequently, if your major interest area pertains to the existing technology, what you learn here may not come in very handy. Based on the current state of the art (see Chapter 21), gone are the days of the three-stage turbine for the large power generation units. It is also unlikely that a five-stage design will be in the sights of major OEMs for a very long time to come (to keep costs down). All the latest and greatest in three-dimensional aero design tools (i.e., primarily advanced CFD with faster computers) and additive manufacturing will be deployed to design efficient, high-loaded stages to keep the turbine stage count at four while the machines become bigger and "badder." (The same goes for the compressors. See the Siemens HL class in Chapter 21 with a 12-stage compressor to achieve a cycle PR of 24.)

Since such large machines cannot be connected to the grid via a giant gearbox, one very important design degree of freedom (i.e., the rotational speed) is not available.[10] In other words, there is not much "suspense" left in laying out the HGP for a modern, heavy-duty industrial gas turbine. Nevertheless, there still are three-stage GE gas turbines on the market, and a small "vignette" is provided in the next section to put the tools learned in this chapter to work within the context of "real" machines.

With this caveat out of the way, let us start the process of selecting the stage numbers for a turbine. It starts with the definition of the loading parameter from Equation 10.5b,

$$\psi = \frac{c_p(T_{2t} - T_{3t})}{U^2} = \frac{v_2 - v_3}{U}.$$

From the cycle heat and mass balance, we have the following information for the turbine, which is uncooled (to keep things simple, but this is not a *vital* simplification):

- Inlet temperature and pressure, T_{in} and p_{in}
- Exit temperature and pressure, T_{ex} and p_{ex}
- Mass flow rate of the hot gas, $\dot{m}_{gas}$
- Turbine shaft power output, $\dot{W}_{turb}$

We assume that the working fluid properties can be represented by the perfect gas model (constant c_p and γ). (Otherwise, we have to use a full-blown equation of state and

[9] The older GT11N also had a five-stage turbine. Both turbines were first developed by Brown-Boveri Company in the 1970s. In the GT11N2 variant, under the Asea Brown-Boveri reign, turbine stages were reduced to four.

[10] The largest, geared industrial gas turbine is GT11N2 with a 3,600–rpm shaft speed for both 60- and 50-Hz (geared) applications (it is rated at about 115 MWe).

enthalpies, which makes formulae and calculation tedious but does not provide additional insight.) Furthermore, the formulae used in the narrative below are exactly as they appear in the text if you think in SI units, specifically joules (not kilojoules) for energy-related terms, meters for dimensions, and degrees Kelvin for temperatures.

If there are n stages, using Equation 10.5 as the starting point, for an *average* stage we can write that

$$\psi = \frac{c_p(T_{in} - T_{ex})}{n\left(\frac{2\pi N r_m}{60}\right)^2} = 91.189\frac{c_p(T_{in} - T_{ex})}{n(N r_m)^2}.$$

At the average stage, the flow area is given by

$$A = 4\pi r_m^2 \frac{1-\upsilon}{1+\upsilon},$$

where υ is the hub–tip ratio. Mass continuity requires that

$$\dot{m}_{gas} = \rho u_m A = \frac{p}{(R_{unv}/MW)T} u_m A,$$

where u_2 is the axial component of the absolute working fluid velocity and is defined by the flow coefficient as (Equation 10.6)

$$u_m = \phi U = \phi \frac{2\pi N}{60} r_m.$$

Substituting into the equation for mass continuity, we obtain

$$\dot{m}_{gas} = \frac{p \cdot MW}{R_{unv} T} \phi \frac{2\pi N}{60} r_m 4\pi r_m^2 \frac{1-\upsilon}{1+\upsilon}.$$

Thus, the mean or pitch-line radius is found as

$$r_m = 0.912544\sqrt[3]{\dot{m}\left[\frac{p \cdot MW}{R_{unv} T}\phi N \frac{1-\upsilon}{1+\upsilon}\right]^{-1}}. \tag{10.38}$$

For the hub–tip ratio, υ, 0.8 is a good value to use in r_m calculation.

Thus, we end up with the following formula for a first-cut estimation of the stage number as

$$n = 91.189\frac{c_p(T_{in} - T_{ex})}{\psi(N r_m)^2} \text{ in SI units and} \tag{10.39a}$$

$$n = 544.882\frac{c_p(T_{in} - T_{ex})}{\psi(N r_m)^2} \text{ in US customary units.} \tag{10.39b}$$

For cooled turbines, it is better to use the specific turbine work, i.e.,

$$n \propto \frac{\dot{W}_{turb}/\dot{m}_{gas}}{\psi(N r_m)^2}.$$

(Make sure that you convert $\dot{W}_{turb}$ to Btu/s when using US customary units.) What to use for $\dot{m}_{gas}$ presents a bit of a problem because the mass flow rate through the machine increases via addition of HGP cooling flows. It is found that using the airflow multiplied by 0.9 gives pretty reasonable results. (If you are wondering why, the full answer would require publicizing proprietary OEM data.)

Another area of uncertainty pertains to the values of p and T to be used in Equation 10.38 (to find the working fluid density in the mass continuity). Using average values of p and T (found from inlet and exit values) is not an ideal approach. Again, the wrench in the wheels is the compounding effect of the HGP cooling flows. For a given speed, N, and a representative HGP density, Equation 10.38 suggests that

$$r_m \propto \dot{m}_{gas}^{1/3}.$$

Starting from this premise, the version of Equation 10.38 recommended for cooled turbines is

$$r_m = 1.6582\sqrt[3]{\frac{\dot{m}_{gas}}{\phi N}}\,(\text{in SI units}). \tag{10.40}$$

The equations developed herein are encapsulated in the VBA function Num_Stages in Appendix C. Its use will be demonstrated in Chapter 22 when we are looking at the closed-cycle gas turbines.

10.3.4 Three or Four Stages? (A Case Study)

With the exception of the steam-cooled H-System, which had a cycle PR of 23, GE's heavy-duty industrial gas turbines (the so-called frame machines) had three-stage turbines. The design philosophy is explained in detail in the paper by Brandt [11]. In a nutshell, the highly loaded first stage resulted in a large temperature drop and reduced the cooling load of the subsequent two stages. The trade-off was the lower turbine efficiency vis-à-vis the lightly loaded four-stage turbines preferred by the competition. Another advantage of the three-stage design manifested itself in combined-cycle performance via higher exhaust gas temperature and higher exhaust gas *exergy* (see Section 5.3). The gain in bottoming-cycle output compensated for the lower "topping-cycle" efficiency and resulted in competitive combined-cycle products. This dynamic between gas turbine exhaust exergy and combined-cycle efficiency will become clearer to the reader when he or she finishes reading Chapter 17.

In 2015, GE announced their new air-cooled H-class lineup (designated as "HA" class) with a four-stage turbine. The increase in the number of stages from three to four can be best assessed by making use of the stage loading parameter, which is defined by Equation 10.5.

For a 3,600–rpm (60–Hz) unit, rotor speed is U = 377·r, where r is the radius (as measured from the turbine rotor centerline) between the blade hub and tip. (We have seen that, for a free vortex design, U is evaluated at the pitch-line, which is the midpoint of the blade.) A typical value of r for a HGP similar to that for 7HA.01 is between 3 and

Table 10.6 Average stage loading for GE's Frame 7 gas turbines

	7FA.05	7HA.01
TIT (°F)	2,769	2,908
TEXH (°F)	1,171	1,153
Average Pitch-Line Radius (ft)	3.65	3.65
Turbine Work (Btu/s)	448,182	519,381
Average Gas Flow (pps)	1,045	1,113
Stage Number	3	4
Average ψ	1.89	1.54

4 ft (i.e., U is between 1,100 and 1,500 ft/s). (For 50–Hz [i.e., larger] Frame 9 machines with 3,000 rpm, U = 314·r and r is between 4 and 5 ft. Thus, U is between 1,200 and 1,600 ft/s.) Indeed, considering that the speed of sound within the HGP is typically >2,000 fps and sonic or supersonic tangential velocities leading to shock waves are not desirable, a good HGP design should not result in rotor speeds above those ranges.

The correlation between the average pitch-line radius (in ft) and average stage loading is represented by the equation below (N is the shaft rpm and n is the number of turbine stages), which is derived from Equation 10.39 developed in the preceding section:

$$r_m = \frac{30}{\pi N}\sqrt{\frac{g_c J}{n\psi}\frac{\dot{W}_{turb}}{\dot{m}_{gas}}}. \tag{10.41}$$

Using data from Table 6.1 (N is 3,600 rpm), for 7FA.05 and 7HA.01, estimates of average values of ψ are shown in Table 10.6. A pitch-line radius of 3.65 ft (1.113 m) is used for both turbines. Turbine work is approximated as the sum of rated output (ignoring losses) and the compressor work. The latter is estimated by using a polytropic efficiency of 92.5 percent. Average turbine gas flow is estimated by multiplying the compressor airflow by 0.9.

Based on the data in Table 10.6, assuming an average flow coefficient of 0.75, per the Smith chart in Figure 10.5, the reduction in stage loading coefficient can be worth as much as one percentage point in stage efficiency (η_t) (i.e., see the arrow in Figure 10.5).

10.3.5 Blade Stresses

There are three types of mechanical stresses acting on a turbine blade:

- Centrifugal (steady-state, tensile)
- Gas bending (fluctuates with *nozzle passing frequency* [NPF])
- Centrifugal bending

In addition to the mechanical stresses, there is also the thermal stress acting on the blade due to temperature variation across the blade wall. This had been discussed earlier in Section 9.3.1.

Detailed derivation and discussion of mechanical stresses can be found in Traupel, Saravanamutto et al., and Kerrebrock [4]. Herein, the focus will be on simplified expressions to get an idea of their magnitude and design impact. The maximum value of the centrifugal stress acting on the blade is given by

$$\frac{\sigma_c}{\rho} = K \cdot \frac{\pi^2 N^2 r_t^2}{3{,}600}\left(1 - \left(\frac{r_h}{r_t}\right)^2\right), \tag{10.42}$$

where K is the taper factor (1 if the blade is untapered; i.e., of uniform cross-section) and ρ is density of the blade metal. For a linear taper,

$$K = 1 + \frac{A_{x,t}}{A_{x,h}}. \tag{10.43}$$

Subscripts "x," "t," and "h" denote cross-section, tip, and hub, respectively. Noting that the flow annulus area is given by

$$A_{ann} = \pi r_t^2\left(1 - \left(\frac{r_h}{r_t}\right)^2\right),$$

the centrifugal stress formula becomes

$$\frac{\sigma_c}{\rho} = \frac{\pi}{3{,}600} A_{ann} N^2 \left(1 + \frac{A_{x,t}}{A_{x,h}}\right). \tag{10.44}$$

The middle term on the right-hand side of Equation 10.44 is commonly used in the industry as a measure of the centrifugal stress and referred to as "A-N-squared" with the units of 10^{-9} in^{2}rpm^2. In using Equation 10.44

- with SI units, divide it by 1E6 and the result will be in MPa;
- with US customary units,
  - use slugs/ft^3 for density (1 kg/m^3 = 0.00194032 slug/ft^3)
  - use in^2 for the area A (1 m^2 = 1,550 in^2)
  - divide it by 2.0736E7 and the result will be in ksi.

Gas bending stress (tensile in leading and trailing edges and compressive in the back of the blade) arises from the force imparted by the gas flow with changing angular momentum (which, of course, creates the torque acting on the blades and the useful shaft work). It has two components in tangential and axial directions. Actual formulae are quite tedious and difficult to use for quick evaluations. Saravanamutto et al. and Kerrebrock provide approximations for the more dominant axial component of the gas bending moment, which are useful for preliminary design purposes. Although the formulae in both references are quite similar, the one given by Saravanamuttoo et al. is more suitable to our discussion. The maximum value of the gas bending stress can be approximated as

$$\frac{\sigma_{gb}}{\dot{m}_g} \cong U_m \psi r_t \frac{(1 - \upsilon)}{2n} \frac{1}{zc^3}, \tag{10.45a}$$

where z is the smallest value of the hub section modulus of a blade of unit chord and n is the number of blades in the rotor. (Typically, the maximum value of σ_{gb} occurs at the leading or trailing edge of the blade hub/root.) Equation 10.45a can be rewritten as

$$\frac{\sigma_{gb}}{\dot{m}_g} \cong \frac{\pi}{60} N\psi r_m r_t \frac{(1-\upsilon)}{nzc^3} \quad \text{or} \tag{10.45b}$$

$$\frac{\sigma_{gb}}{\dot{m}_g} \cong \frac{\pi}{60} N\psi r_m \frac{AR}{nzc^2}, \tag{10.45c}$$

where AR is the blade *aspect ratio*

$$AR = \frac{h}{c} = \frac{r_t - r_h}{c} = r_t \frac{1-\upsilon}{c}.$$

Same unit considerations as in Equation 10.44 apply to Equation 10.45 as well. A VBA function to calculate z as a function of blade thickness–chord ratio (t/c) and blade *camber* angle is provided in Appendix C. Note that the camber angle is practically equal to the turning angle, i.e.,

$$\varepsilon_R = |\beta_2| + |\beta_3| \, .$$

Using typical stage 1 values from Table 10.13 (below in section 10.5), i.e.,

- N = 3,600 rpm
- $\psi = 1.8$
- $r_m = 1.07$, $r_t = 1.14$, $\upsilon = 0.89$
- turning angle of 111.6°
- gas flow of ~550 kg/s

and assuming t/c = 0.2, n = 90, and AR = 1.5 (typical for a short stage 1 bucket), from Equation 10.45 we calculate that σ_{gb} is ~80 MPa (11.5 ksi), which is not negligible but is reasonably low. From Equation 10.44, we find that σ_c is ~135 MPa (20 ksi), which is also reasonably low. A sample centrifugal stress calculation for the long last-stage bucket (LSB) can be found in Section 10.6.

Equation 10.45 demonstrates the strong dependence of the gas bending stress on the aspect ratio. If one visualizes a blade as a cantilever beam fixed at its hub (root), it is easy to see how longer and thinner blades are more susceptible to bending stresses. In the context of axial compressors (see Chapter 11), same considerations apply. Low-aspect ratio blades have lower susceptibility to flutter. Since the blade height, h, is limited by air mass flow rate, mean radius, r_m, and axial flow speed (i.e., Mach number), a low aspect ratio implies a longer chord, c. A larger blade chord allows large spacing (pitch) for a given solidity, σ, which results in fewer blades per stage and lower manufacturing costs.

Note that whereas the centrifugal stress is steady during normal operation, the gas bending stress fluctuates with a frequency dependent on the number of stator vanes and rotor blades. One could add the shaft rotational speed to the list for the variable-speed gas generator spools of the aeroderivative gas turbines. (Shaft speed is not a variable for single-spool, heavy-duty industrial gas turbines synchronized to the grid.) Stress fluctuation is tied to the excitation caused by the aerodynamic force fluctuations seen by the rotating blade as it passes each stationary stator nozzle vane. This excitation occurs at the NPF, which is given by

$$\mathrm{NPF} = \mathrm{h} \times \mathrm{f} \times \mathrm{n}_{\mathrm{SV}},$$

where f is the shaft rotational frequency (i.e., 50 or 60), n_{SV} is the number of stator vanes, and h = 1, 2, 3 ... designates the multiples (harmonics) of the NPF.

Total stress acting on a blade is a combination of the three mechanical stresses and the thermal stress. A rigorous evaluation and analysis requires specialty finite-element analysis software to evaluate the temperature and stress distribution in the entire blade. Some qualitative observations are still possible. Based on the calculations presented in this section and in Section 9.3.1, for the blades in turbine stage 1, thermal stress is the determining component. Due to the low value of the centrifugal stress, creep is not a dominant failure mechanism. Since the rotational speed is constant during normal operation, high cycle fatigue is easier to avoid by fine-tuning the blade's natural frequencies (to prevent resonance at NPF harmonics). Oxidation and thermal fatigue emerge as the dominant failure mechanisms.

For the longer blades of the last stage, the situation is different. Due to high centrifugal stresses (see Section 10.6), creep is the dominating failure mechanism. Thermal stress is not a big factor for the uncooled LSBs operating at the lowest HGP temperatures (except the latest H/J-class gas turbines – see Chapter 21).

10.4 Impact of Cooling Flow

The detrimental impact of cooling flow on stage efficiency can be significant. A comparison of stage efficiencies for different stator nozzle vane configurations showed that the convection-cooled solid (uncooled) blade turbine with the coolant ejected through the trailing edge shows a minor efficiency hit [12]. This is indicative of a small wake with both coolant and main gas flows in the same direction. In contrast, film-cooled blades with discrete holes along the blade show much greater efficiency loss (about a one-percentage point debit for each 1 percent in cooling flow as referenced to the inlet gas flow) [12]. Kerrebrock estimated "3.2 percent loss in turbine efficiency per percent of cooling flow" [4]. He also referred to a paper by Japanese researchers (ASME paper 75-GT-116), from which he concluded that "2.5 percent rotor cooling flow reduced the efficiency [by] about 6.5 percent."

Losses associated with stage cooling flows can be classified as:

- Coolant flow bypassing the rotor blades
- "Pumping" work done on the coolant flow by the rotor blades
- Flow disturbance caused by coolant exiting the stator vanes
- Rotor blade profile loss increase due to coolant film (due to increasingly thick wake built across the pressure and suction surfaces of the blade)

Table 10.7 Model turbine stage design parameters

Stage	1	2	3
R	0.45	0.50	0.47
ψ	1.9	1.6	1.2
ϕ	0.69	0.71	0.75
α_3	30.0	23.0	5.0
α_2	65.2	61.5	56.5
β_3	63.7	61.5	54.9
β_2	35.8	23.0	10.2

Table 10.8 Impact of cooling air on stage efficiency

Case	Base	Higher Cooling Air
Stage 1 Cooling Flow (lb/s)	109.1	116.5
Stage 2 Cooling Flow (lb/s)	64.3	68.0
Stage 3 Cooling Flow (lb/s)	9.4	9.0
Total/Compressor Airflow	0.195	0.207
Net Gas Turbine Output (kW)	162,908	159,502
Net Gas Turbine Efficiency (%)	36.73	36.48
Stage 1 Cooled Efficiency (%)	87.89	87.46
Stage 1 Cooled Efficiency (%)	90.14	90.06
Stage 1 Cooled Efficiency (%)	92.34	92.35

For stage-by-stage turbine calculations, isentropic efficiency is the appropriate definition to use, along with the detailed enthalpy (temperature)–entropy diagram of the stage to account for all the losses enumerated above rigorously. The polytropic efficiency is still useful for the simple-cycle calculations described in Chapter 8. Consequently, accounting for the impact of coolant flow increase or decrease in simplified models requires appropriate debits to the polytropic efficiency (albeit quite modest vis-à-vis the debit to the isentropic efficiency). In order to illustrate this, we use a vintage F–class machine with a three-stage turbine as an example. The model of the turbine is built in Thermoflex and run for two cases of coolant flow (i.e., base and 1 percent higher).

The stage design used in the model is summarized in Table 10.7. Uncooled stage efficiencies are 92.5, 93, and 93 percent, respectively. Calculated cooled-stage efficiencies and other results are shown in Table 10.8.

Note that the cooling loss debit to the cooled isentropic efficiency is highest for stage 1, at about 0.43 percentage points, which sees the highest gas temperatures and consequently has the highest cooling air consumption (about 8 and 5 percent of the incoming gas flow for the stator and the rotor, respectively). This is a more modest loss than the prediction in [12] (which dates from the early 1970s), but is more in line with modern design practice. In comparison, the numbers cited in Kerrebrock [4] seem to be quite overblown. The stage flow diagram is shown in Figure 10.6. Calculation of the coolant loss coefficients is described in Elmasri's original paper [13] and briefly touched

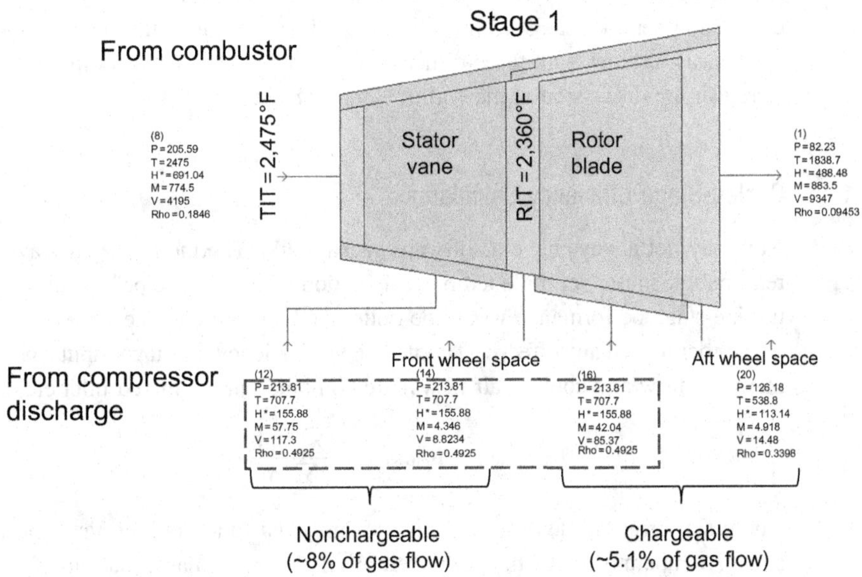

Figure 10.6 Stage 1 flow diagram for a vintage F–class gas turbine.

upon in Chapter 9. The model described in the paper (GASCAN) is embedded in the cooled turbine module of Thermoflex, which is used throughout this book for illustrative calculations.

The higher cooling flows for stages 1 and 2 in Table 10.8 are achieved by increasing cooling flows for the stage 1 rotor and stage 2 stator and rotor. Note the favorable impact of the increased gas temperature dilution on the stage 3 cooling flow requirement. Lower gas temperature reduces the cooling air flow necessary to maintain the design metal temperatures.

In order to translate this finding into a corresponding loss estimate for a simple turbine model with simple stages defined by polytropic efficiency and cooling air flows mixed with the main gas flow before and after each stage, the stage-by-stage model is modified and rerun. A polytropic efficiency of 88.3 percent was required (same for each stage) to match the original performance with the same temperatures, pressures, and flow rates. With the increased cooling air fraction (0.195–0.207 as referenced to compressor airflow), reduced performance was matched by a polytropic efficiency of 88.2 percent. The reason for the much smaller debit is due to the fact that, in simple models, the impact of the *entire* stage cooling flow bypassing the work-generating expansion takes care of a larger portion of performance loss.

Running similar scenarios to the one described above resulted in the formula presented earlier in Chapter 8 (i.e., Equation 8.23),

$$\eta_t = \eta_{t,\,max} - 0.010\ln(PR_t)\left(1 + \kappa\sqrt{\chi}\right),$$

which quantifies the cooling air impact on turbine polytropic efficiency for simple turbine calculations. The factor κ should ideally be determined from available data (which are quite difficult to obtain unless you work for an OEM). For the sample calculation here, κ is 2.39 with $\eta_{t,max} = 0.93$. It can be lower for the more modern H- or J-class gas turbines with three-dimensional aero design. (At this point, you might want to revisit the discussion at the end of Chapter 8, Section 8.4.)

10.4.1 Cooled-Stage Efficiency Calculation

You may not always have a software package like Thermoflex readily available to do reasonably simple but quite tedious calculations. Thus, a "recipe" for digging deep into turbine-stage performance would be quite handy, so here is one.

In order to find an "effective" cooled-stage efficiency, all stage input streams (i.e., the main gas flow and cooling air flows) are combined into a mixed inlet enthalpy, i.e.,

$$h_{mixed,in} = \frac{\sum_i \dot{m}_i h_i}{\sum_i \dot{m}_i},$$

where subscript "i" denotes the inlet stream i (gas and cooling air). The mixed inlet stream temperature can be calculated from the particular equation of state or h–T transfer function used in the model, i.e.,

$$T_{mixed,in} = f(h_{mixed,in}).$$

If an equation of state is used in the model, isentropic stage exit temperature can be found as

$$T_{exit,s} = f(s_{mixed,in}, p_{exit}).$$

(Mixed inlet entropy is determined in a manner similar to that for mixed inlet enthalpy.) Exit pressure is, of course, found from the division of the stage inlet pressure by the stage PR. For simplified calculations such as in the VBA function TURB_STG in Appendix C, we can use the isentropic p–T correlation

$$T_{exit,s} = T_{mixed,in}\left(\frac{1}{PR_{stage}}\right)^{\frac{\gamma-1}{\gamma}}.$$

The specific heat ratio γ can be estimated from a simple transfer function, e.g.,

$$\gamma = 2.5127\,T[K]^{-0.089},$$

which is used in TURB_STG. The temperature to be used in the formula can be chosen as an average of static stage inlet and exit temperatures (as was done in TURB_STG). The cooled-stage efficiency is then calculated as

$$\eta_{stg,cool} = \frac{h_{mixed,in} - h_{exit}}{h_{mixed,in} - h_{exit,s}},$$

when using an equation of state in the model (as is done in a software package like Thermoflex), or simply as

$$\eta_{stg,cool} = \frac{T_{mixed,in} - T_{exit}}{T_{mixed,in} - T_{exit,s}}.$$

Effective cooled-stage polytropic efficiency is estimated from

$$\eta_{poly,eff} = \frac{\gamma}{\gamma - 1} \frac{\ln\left(\frac{T_{exit}}{T_{mixed,in}}\right)}{\ln\left(PR_{stage}^{-1}\right)}.$$

Effective polytropic efficiencies are even lower than cooled-stage efficiencies by several percentage points (typically in the low 80s). Comparing with compressor polytropic efficiencies (in the low 90s), it should become obvious that HGP cooling has a very strong detrimental impact on turbine-stage aerothermodynamics via irreversible mixing losses (increasing entropy), momentum losses, and working fluid temperature drops.

Note that the "effective-stage polytropic efficiency" calculated for a "real" turbine stage as described is *not* the same as the polytropic efficiency used in the simple models in Chapter 8. Earlier in this section, it was shown that the efficiencies used in such models were in the high 80s. The reason for that is the simplistic treatment of the HGP cooling flows in those models. (Cooling flows *bypass* the pseudo-stage doing the expansion work and mix with the stage exit stream before entering the next pseudo-stage.)

10.5 Stage Enthalpy–Entropy Diagram

The enthalpy–entropy (h–s) diagram of an uncooled stage is trivial. It can be found in elementary textbooks and does not convey a wealth of information pertaining to the aerodynamic design of the stage in question. The h–s diagram of a real cooled stage comprising stator and rotor as well as wheel spaces with secondary cooling flows, on the other hand, is extremely crowded with many state-points, static and total values, and dynamic contributions. The reader can refer to the many books cited in the text (e.g., Saravanamutto et al.) to appreciate the complexity of a "real" h–s diagram. Detailed knowledge of static and total values of enthalpy during expansion stages, including mixing with cooling flows, and changes in absolute and relative gas velocities is requisite for component design, and referring to the most finely detailed h–s diagram describing the expansion process is thus unavoidable.

However, a much simpler approach is perfectly adequate to describe the underlying physics and accomplish reasonably straightforward calculations in order to estimate the stage performance. Such an h–s diagram, which is adopted from Elmasri [13], is shown in Figure 10.7.

As is shown in Figure 10.7, expanding gas experiences two "losses" during its passage across the stator nozzle vanes and mixing with cooling air coming out of the cooled vanes and the forward wheel space: (i) reduction in enthalpy (temperature), state-points 1 to 2′; and (ii) reduction in total pressure (via friction and mixing), state-points 2′ to 2.

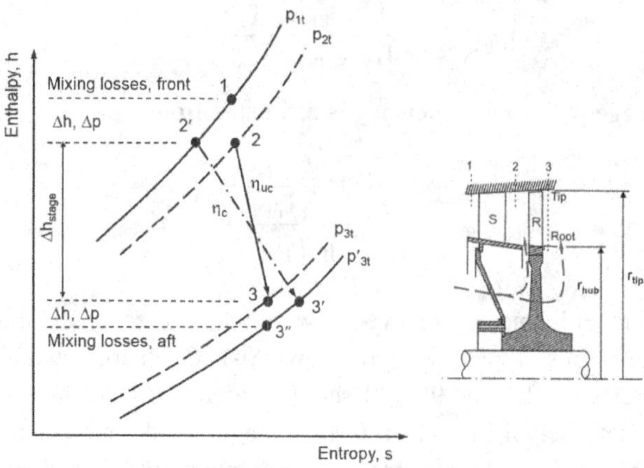

Figure 10.7 Turbine stage enthalpy–entropy diagram.

After expanding through the rotor blades and generating work, the gas experiences a second set of two "losses" (i.e., mixing with cooling air coming out of the cooled blades and the aft wheel space): (i) reduction in enthalpy (temperature), state-points 3′ to 3″; and (ii) increase in total pressure (via friction and mixing), state-points 3″ to 3.

There are two approaches to calculating the stage work per the model shown in Figure 10.7:

1. *Uncooled expansion*: the effective total PR across the rotor is reduced by the two pressure losses described above. The stage total-to-total efficiency, η_{uc}, is assumed to be unaltered as it operates at the reduced PR.
2. *Cooled expansion*: the total PR is assumed to be unaffected by cooling, whose effects are manifested as decrements to the uncooled stage efficiency and resulting in the cooled-stage efficiency, η_c.

The first approach was adopted by Elmasri in his GASCAN code [13]. Total enthalpy loss calculation is straightforward via coolant and mainstream gas mixing energy and mass balances. Forward and aft mixing pressure losses are calculated by specifying a momentum loss parameter, which is derived from the fundamental gas dynamic case of mixing a small amount of quiescent fluid into a stream flowing at a given Mach number. Details can be found in Elmasri's original paper and the references listed therein. This calculation forms the backbone of the cooled turbine stage icon (model) in Thermoflex, which is used throughout this text.

As discussed earlier, the stage is completely specified by ϕ, ψ, or R and α_3. Note that, if ϕ and ψ are given, stage reaction R can be found as (note $\alpha_3 = \alpha_1$)

$$R = 1 + \phi \cdot \tan \alpha_3 - \psi/2.$$

With the known rotational speed, N (rpm), all velocity triangles are established at the mean or *pitch-line* radius, r_m, which is determined from the mass (or volume) flow rate and definition of ψ, i.e.,

$$\dot{V}_{gas} = \frac{\dot{m}_{gas}}{\rho_{gas}} = \frac{\dot{m}_2}{p_{2,st}} \frac{R_{unv}}{MW_{gas}} T_{2,st}, \tag{10.46a}$$

$$\dot{V}_{gas} = \left(4\pi r_m^2 \frac{1-\upsilon}{1+\upsilon}\right) V_2 \cos\alpha_2, \text{and} \tag{10.46b}$$

$$\psi = \frac{\Delta h_{stage}}{(\omega r_m)^2},$$

along with υ, which is the hub–tip ratio, $\upsilon = r_h/r_t$. (When using SI units, do *not* forget to multiply $p_{2,st}$ [bar] by 100,000 to convert it to N/m^2. When using British units with $p_{2,st}$ [psi], to prevent unit conversion headaches, use R_{unv} = 1,545 ft-lbf/lbmol-R to find $R_{gas} = R_{unv}/MW_{gas}$. Otherwise, the very simple Equation 10.46 can lead to huge errors, and debugging the calculation can be quite tedious.)

Absolute and relative velocities, Mach numbers, and flow angles at the inlet and exit of each rotor can be calculated at hub, pitch, and tip radii. (Axial velocity is usually assumed to be constant across the stage; i.e., $u_1 = u_2 = u_3$.) A graphical representation of a stage enthalpy calculation is shown in Figure 10.8. The missing piece of information is the energy and mass balances, which can be quite tedious (albeit straightforward) to obtain if one chooses to use actual air and gas compositions. Using empirical $c_p(T)$ polynomials and integration thereof to determine air/gas enthalpies, calculation of stage performance and geometry from a simplified model like this is quite trivial.

Such a simple model is encapsulated in the VBA function TURB_STG (see Appendix C), which is essentially a "light" version of the GASCAN code that incorporates all the salient aspects of turbine aerothermodynamics covered up to this point.

The code is demonstrated using the 60-Hz GE products in Table 6.1, which is reproduced herein with additional data in Table 10.9. The nomenclature in the table is as follows:

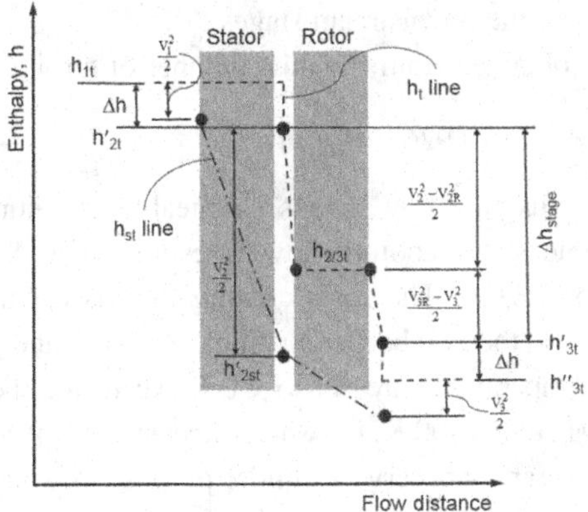

Figure 10.8 Stage enthalpy (static and total) and kinetic energy sequence.

Table 10.9 GE 60-Hz product line (*Gas Turbine World 2016 Handbook*)

		7E.03	7F.04	7F.05	7F.06	7HA.01	7HA.02
Output (kW)		91,000	198,000	241,000	270,000	280,000	346,000
EFF		33.90%	38.60%	39.80%	41.40%	41.70%	42.20%
MFUEL (lb/s)		11.8	22.6	26.7	28.7	29.6	36.1
MAIR (lb/s)		638.2	990.4	1,161.3	1,282.3	1,236.4	1,502.9
PR_{comp}		13.0	16.7	18.4	22.1	21.6	23.1
CDT (°F)		686	778	816	890	881	909
CDT (K)		637	688	709	750	745	760
WCOMP (Btu/s)		100,051	178,096	219,759	266,460	253,993	319,286
WTURB (Btu/s)		186,302	365,763	448,182	522,370	519,381	647,229
# Turb Stages		3	3	3	4	4	4
TEXH (°F)		1,026	1,151	1,171	1,100	1,159	1,153
MEXH (lb/s)		650	1,013	1,188	1,311	1,266	1,539
SP WK (Btu/lb)		135	189	197	200	215	218
TIT	°C	1,233	1,459	1,520	1,518	1,576	1,598
	°F	2,252	2,658	2,769	2,764	2,868	2,908
TFIRE (°F)		2,141	2,496	2,589	2,563	2,661	2,690

- EFF = efficiency
- MFUEL = fuel mass flow rate (100 percent CH_4 fuel with 21,515 Btu/lb)
- MAIR = compressor inlet airflow
- CDT = compressor discharge temperature (92.5 percent polytropic efficiency)
- WCOMP = compressor power consumption
- WTURB = turbine power generation
- SP WK = gas turbine-specific power output

Program input data are listed in Table 10.10. A flow coefficient of 0.6 and a swirl angle of 25° is assumed for all cases. The stage inlet absolute flow angle is 5°. The loading parameter is calculated from Equation 10.41 using pitch-line radii estimates. (A value of 3.65 ft is assumed for 7FA.05; the others are scaled using specific work in Table 10.9.) While somewhat high at 1.8, this is not unreasonable for the first turbine stage (to reduce the cooling load of the downstream stages).

The stage pressure estimate is obtained from (n is the number of turbine stages)

$$PR_{stg} = \left(0.92625\,P_{comp}\right)^{\frac{1}{n}}.$$

The stage inlet flow is airflow plus fuel flow minus chargeable and nonchargeable cooling flows. The nonchargeable stator cooling flow is estimated as 8 percent of airflow (except for 6 percent for 7E.03). The rotor cooling flow is calculated using Equation 9.36 with CF = 1.0 (i.e., no thermal barrier coating), $\alpha = 0.05$, and $\beta = 0.9$ (see Table 10.11). Cooling airflow temperatures are set to CDT except for the aft wheel space (AWS), which is set to CDT minus 100 K. Forward wheel space (FWS) and AWS purge flow and momentum loss parameters serve a similar purpose to what they do in Elmasri's GASCAN [13]. Unlike in GASCAN, in this simple code, mass flows instead of mole flows are used as a basis. Uncooled efficiencies are calculated using the VBA

Table 10.10 TURB_STG inputs

		7E.03	7F.04	7F.05	7F.06	7HA.01	7HA.02
1	N (rpm)	3,600	3,600	3,600	3,600	3,600	3,600
2	ψ	1.80	1.80	1.80	1.80	1.80	1.80
3	ϕ	0.60	0.60	0.60	0.60	0.60	0.60
4	α_3	25	25	25	25	25	25
5	Inlet flow (kg/s)	258.1	389.3	455.5	492.7	480.2	579.5
6	T_{1t} (K) (TIT)	1,506	1,732	1,793	1,791	1,849	1,871
7	P_{1t} (bar)	12.4	15.9	17.5	21.0	20.5	21.9
8	Stage PR	2.29	2.49	2.57	2.13	2.11	2.15
14	Stator Cooling Flow (kg/s)	17.4	35.9	42.1	46.5	44.9	54.5
15	Stator Coolant Temp. (K)	637	688	709	750	745	760
16	Rotor Cooling Flow (kg/s)	14.1	34.3	41.2	55.5	49.2	64.0
17	Rotor Coolant Temp. (K)	637	688	709	750	745	760
18	FWS Coolant Temp. (K)	637	688	709	750	745	760
19	AWS Coolant Temp. (K)	537	588	609	650	645	660
21	α_1	10	5	5	5	5	5

Table 10.11 Rotor cooling flows (Equation 9.36)

	7E.03	7F.04	7F.05	7F.06	7HA.01	7HA.02
Allowable Blade Temp., T_b (°F)	1,350	1,475	1,500	1,500	1,550	1,550
T_b (K)	1,005	1,075	1,089	1,089	1,116	1,116
ε	0.45	0.50	0.52	0.57	0.56	0.58
χ (Specific Heat Ratio)	1.18	1.23	1.25	1.25	1.26	1.26
m_c/m_g	0.050	0.062	0.068	0.082	0.080	0.086
Gas Flow, m_g (kg/s)	282.2	434.9	506.9	552.2	534.4	646.0
Coolant Flow, m_c (kg/s)	14.1	27.0	34.7	45.5	42.8	55.6

function Eff_341. Please refer to the code listing in Appendix C for more details. Results are listed in Table 10.12. Note that the calculation sequence in TURB_STG is slightly modified, i.e.,

$$\alpha_1 \neq \alpha_3 \text{ so that}$$
$$u_1 \neq u_2 = u_3 \text{ and}$$
$$V_1 \approx V_3.$$

Stage reaction is dictated by α_1, i.e.,

$$R = 1 + \phi \cdot \tan \alpha_1 - \psi/2.$$

The reason for this slight modification is that it gives more *realistic* stage angles (i.e., closer to those in "real" machines).

Stage 1 geometry and performance data for the six 60-Hz gas turbines calculated from public rating data using TURB_STG and presented in Table 10.12 should be considered as very rough estimates. The goal is to illustrate the basic principles and how they lead to

Table 10.12 TURB_STG outputs

		7E.03	7F.04	7F.05	7F.06	7HA.01	7HA.02
1	α_2	70.5	71.1	71.1	71.1	71.1	71.1
2	α_3	25.0	25.0	25.0	25.0	25.0	25.0
3	β_2	49.2	51.2	51.2	51.2	51.2	51.2
4	β_3	61.5	60.3	60.3	60.3	60.3	60.3
5	V_2 (m/s)	695.3	803.7	832.4	747.4	756.7	769.3
6	V_3 (m/s)	256.1	288.0	298.3	267.8	271.1	275.7
7	V_{2R} (m/s)	355.0	416.9	431.8	387.7	392.5	399.1
8	T_2 (Total – Relative) (K)	1,306	1,456	1,501	1,540	1,591	1,607
9	Firing Temp. (K)	1,457	1,651	1,709	1,706	1,759	1,781
10	Rotor Inlet Flow (kg/s)	282.2	434.9	506.9	552.2	534.4	646.0
11	Rotor Exit Flow (kg/s)	298.0	464.4	544.6	601.0	580.4	705.5
12	Stage Total Δh (J/kg)	269,426	340,646	365,416	294,521	301,585	311,865
13	T_2 (Static) (K)	1,252	1,384	1,425	1,479	1,529	1,544
14	T_3 (Static) (K)	1,200	1,340	1,378	1,444	1,494	1,506
15	T_3 (Static – Mixed) (K)	1,176	1,297	1,329	1,384	1,431	1,440
16	T_3 (Total) (K)	1,229	1,375	1,415	1,473	1,523	1,537
17	Stage PR (With Losses)	2.15	2.33	2.41	1.99	1.98	2.02
18	Pitch-Line Radius (m)	1.03	1.15	1.20	1.07	1.09	1.10
19	Hub–Tip Ratio	0.90	0.90	0.90	0.88	0.89	0.88
20	Degree of Reaction (R)	0.21	0.15	0.15	0.15	0.15	0.15
21	Uncooled Efficiency	90.0%	89.8%	89.8%	89.8%	89.8%	89.8%
22	Cooled Efficiency	79.1%	83.3%	84.1%	83.1%	84.0%	84.1%
23	Effective Polytropic Efficiency	77.3%	81.7%	82.6%	81.8%	82.8%	82.8%
24	U/a (at pitch-line)	0.55	0.59	0.61	0.54	0.53	0.54
25	V_1/V_3	0.92	0.91	0.91	0.91	0.91	0.91

actual product design. (Unfortunately, comparison with stage geometry from actual products is not possible because such information is highly proprietary.) Further refinements can certainly be made. The first step is to ensure the heat and mass balance integrity of the rating data. The next step is to build a stage-by-stage model. This can be done by repeating the TURB_STG calculation for stages 2, 3, and, if necessary, 4. An easier way is to use Thermoflex with built-in GASCAN code or a specialized flow path design software such as AxStream. Before moving on, important stage 1 rotor design parameters are summarized in Table 10.13. The general trend of increase in pitch-line radius and blade height with increasing turbine size is as expected. (The absolute values, vis-à-vis actual designs, are probably on the low side by about 20 percent.) Solidities are set so that the rotor blade count is around 90 (a good rule of thumb).

10.6 Exit Annulus Area

At any given point in the expansion path, annulus area can be calculated as a function of the flow properties at the mean radius and the gas mass flow rate. There are two important considerations in laying out the expanding flow channel:

Table 10.13 Stage 1 rotor geometry from TURB_STG

	7E.03	7F.04	7F.05	7F.06	7HA.01	7HA.02
Blade Height (mm)	107	126	132	132	131	146
Hub Radius (m)	0.97	1.09	1.13	1.01	1.02	1.03
Tip Radius (m)	1.08	1.22	1.26	1.14	1.15	1.18
Pitch-Line Radius (m)	1.03	1.15	1.20	1.07	1.09	1.10
Zweifel Number	0.80	0.80	0.80	0.80	0.80	0.80
Solidity	0.14	0.14	0.14	0.14	0.14	0.14
Flow Deflection Angle (°)	110.7	111.6	111.6	111.6	111.6	111.6
Number of Rotor Blades	77	83	83	83	83	83
$R(r_m)$	0.21	0.15	0.15	0.15	0.15	0.15
$R(r_h)$	0.12	0.05	0.05	0.04	0.04	0.03
$R(r_t)$	0.28	0.24	0.24	0.25	0.25	0.25

1. The angle of divergence of the annulus walls, which should be less than 25° to prevent flow separation;
2. The exit annulus area, A_{ex}, which determines the LSB size.

The *exit annulus area* can be calculated from

$$A_{ex} = \frac{\dot{m}\sqrt{T_{3t}}}{p_{3t}\cos\alpha_3 \mathrm{MFP}}, \tag{10.47}$$

where the *mass flow parameter* (MFP) is a function of the Mach number and given by

$$\mathrm{MFP} = \frac{Ma_3\sqrt{\gamma/R}}{\left\{1+\frac{\gamma-1}{2}Ma_3^2\right\}^{\frac{\gamma+1}{2(\gamma-1)}}}. \tag{10.48}$$

The Mach number is based on the absolute velocity of the gas and the speed of sound (which is based on the *static* temperature) as follows:

$$Ma_3 = \frac{V_3}{\sqrt{\gamma R T_{3st}}}, \tag{10.49a}$$

$$\frac{Ma_3}{U_3} = \frac{\phi_3}{\cos\alpha_3\sqrt{\gamma R T_{3st}}}. \tag{10.49b}$$

The size of the LSB is a critical parameter due to the fact that it is the longest blade with the highest tip speed. The large centrifugal force acting on that component sets the critical mechanical design and manufacturing limits. Larger LSB size as quantified by the large A_{ex} determines the selection of the bucket material and has a large impact on turbine cost. (The centrifugal stress acting on the LSB, σ_c, can be calculated using Equation 10.44.)

In general, the following limitations should be kept in mind:

1. Although a large A_{ex} is beneficial for performance due to the lower exit kinetic energy, for $Ma \approx 0.35$ or less, the size/cost disadvantage overcomes the performance advantage.

2. Mechanical limitations may force even higher Mach numbers and smaller A_{ex}, especially for uncooled LSBs with high rotor-relative gas temperatures.
3. The LSB size and AN^2 ("A-N-squared," product of exhaust annulus area and the square of the shaft speed in rpm, typically expressed in 10^{-9} in^2rpm^2 – proportional to the blade centrifugal stress, see below) is determined by the worst of the two: cold-day performance (i.e., highest mass flow rate) and expected flow rate growth capability.
4. In any event, the exit Mach number should not exceed 0.8.
5. A feasible range for the exit Mach number is 0.5–0.7. Conceptual designs or reverse-engineered cases with turbine last-stage exit Mach numbers much lower or higher should be reconsidered.
6. Vintage E- and F-class gas turbine AN^2 values are around 70–80. H-class gas turbines or gas turbines for integrated gasification gas turbine applications have higher values (due to a higher fuel gas flow rate and diluent injection).

A numerical example using actual gas turbine data should illustrate the salient aspects of the foregoing discussion. The gas turbine in question is a vintage F-class gas turbine used as an example earlier in the book (e.g., see Table 7.5). The data are repeated below again for convenience:

- Fuel temperature of 392°F (210°C)
- Net generator output of 272,000 kWe with 100 percent methane
- Net efficiency of 39%
- Exhaust mass flow rate of 1,479 lb/s
- Exhaust temperature of 1,085°F

This is a 50-Hz, nominal 1,400°C TIT (~2,550°F) gas turbine (Siemens V94.3A back in 1996, now SGT5-4000F). Based on the inlet exhaust temperatures, the stage loading coefficient can be calculated as shown below. Note that nothing is known (to us, that is) about the actual turbine geometry, so estimation of $U = \omega r$ via a known pitch-line radius is not possible. However, considering that the speed of sound within the expansion path is typically >2,000 ft/s and sonic or supersonic tangential velocities leading to shock waves are not desirable, assuming U = 1,500 ft/s is a quite reasonable choice for this exercise. This implies a pitch radius of about 4–5 ft at 3,000 rpm and is typical of 50-Hz F-class gas turbines. Thus, assuming 2,600°F at the turbine inlet, the stage loading parameter is found from Equation 10.5 as

$$\psi = 0.3 \times 32.174 \times 778.17 \times (2{,}550 - 1{,}085)/1{,}500^2 = 5.1.$$

(Do not forget the unit conversion factors g_c and J!). This value is clearly too high to be accommodated by a single-stage turbine design. Determining the actual number of stages is a process of optimization. It is known that the V94.3A is a four-stage turbine design, so an average value of $\psi = 5.1/4 \approx 1.275$ is implied. This is indeed a very reasonable average value, providing a good balance of aerodynamic efficiency and cooling flow distribution.

Assuming a reasonable value of 0.6 for Ma_3 and a specific heat ratio of $\gamma = 1.3$, MFP is

$$\text{MFP} = \frac{0.6\sqrt{1.3 \cdot 32.174/0.0685/778.17}}{\left\{1 + \frac{1.3-1}{2} 0.6^2\right\}^{\frac{1.3+1}{2(1.3-1)}}} = 0.4345 \frac{\text{lb}\sqrt{\text{R}}}{\text{s} \cdot \text{lbf}}.$$

Therefore, assuming a reasonable 5° of exit swirl angle for the last stage and 15 psia pressure (total), the exit-stage annulus area can be calculated as

$$A_{ex} = \frac{1479\sqrt{1085 + 460}}{(15 \cdot 144)\cos\left(\frac{5\pi}{180}\right)0.4345} = 62\,\text{sqft},$$

which translates into an AN^2 value of 81 for the rotational speed of 3,000 rpm. The stage exit axial velocity can be found from the mass flow rate definition as

$$u_3 = 1{,}479/(0.025 \times 62) = 945\ \text{ft/s}$$

(assuming 0.025 lbm/ft^3 for the gas density). This corresponds to a flow coefficient of 945/1,500 = 0.63 at the pitch-line radius.

Using Equation 10.44 and assuming 1.8 for the term including blade taper on the right-hand side, we can find the centrifugal stress as

$$\sigma_c = \rho \frac{\pi}{3{,}600} A_{ann} N^2 \left(1 + \frac{A_{x,t}}{A_{x,h}}\right),$$

$$\sigma_c = (7{,}850 \times 0.00194032)\left(62 \times 144 \times 3{,}000^2\right)\frac{\pi}{3{,}600}\frac{1.8}{2.0736\text{E}7} = 92.7\ \text{ksi}.$$

Blade metal density is assumed to be 7,850 kg/m^3 (see Chapter 13). The calculated stress of 640 MPa in SI units is rather high. Bear in mind, however, that this is a very high-level estimate. Using the approximate formula given by Saravanamuttoo et al, i.e.,

$$\sigma_c = \frac{4}{3}\pi\rho A_{ann}\left(\frac{N}{60}\right)^2,$$

we calculate the maximum centrifugal stress as ~475 MPa (69 ksi), which is still rather high. The higher estimate is at about 50 percent the tensile strength of conventionally cast René 80 at room temperature (also see Figure 13.7 for the rupture strength of Inconel 738 to get an idea). One would probably require a material similar to GTD-444DS (see Table 13.2). This calculation illustrates the material limits one can run into when designing large industrial gas turbines with high airflows and high power outputs.

10.7 Exhaust Diffuser

In Section 3.3.1, it was mentioned that the gas turbine exit (i.e., state-point 4) was defined at the end of the exhaust diffuser (i.e., *not* at the exit of the last turbine stage).

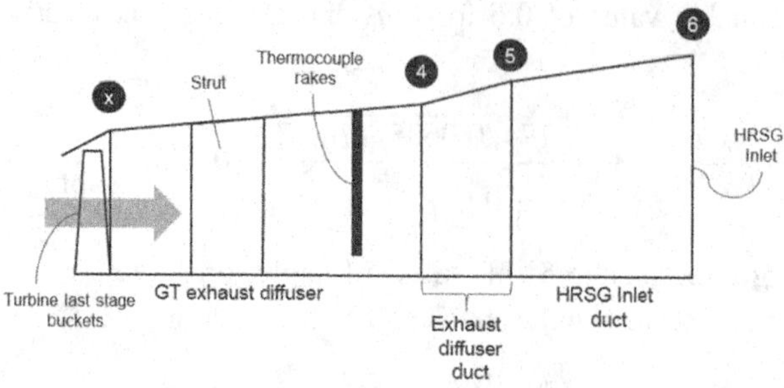

Figure 10.9 Gas turbine (GT) exhaust diffuser configuration.

The distinction was further elaborated upon via a closer look at total (stagnation) and static values of temperature and pressure within the context of thermocouples and temperature measurement. It is now time to take a "deep dive" into this subject.

At the exit of the last turbine stage (typically the fourth stage for most modern, heavy-duty industrial gas turbines), the velocity of gas is still high (i.e., of the order of 300 m/s [~1,000 ft/s], which corresponds to a Mach number of 0.5–0.6). The exhaust diffuser, which is, in effect, simply a diverging nozzle, converts the kinetic energy of the gas coming out of the last turbine stage to a rise in static pressure and drops the Mach number (i.e., the gas velocity) to a level more suitable for delivery to the exhaust stack (simple cycle) or to the heat recovery steam generator (HRSG; combined cycle). A schematic diagram of the arrangement for the latter is shown in Figure 10.9.

The performance of the exhaust diffuser is quantified by the *recovery factor*, RF, which is analogous to the recovery factor covered within the context of a thermocouple in Chapter 3 (e.g., see Equation 3.7b). The definition of diffuser RF, using the state-point numbering in Figure 10.9, is given by

$$\mathrm{RF} = \frac{p_4 - p_x}{p_{x,0} - p_x}. \tag{10.50}$$

The denominator of Equation 10.50 is the difference between static and total/stagnation pressures at the exit of the turbine and is referred to as *dynamic head*. Therefore, using Equation 3.11, i.e.,

$$p_0 = p \times \left(1 + \frac{\gamma - 1}{2} Ma^2\right)^{\frac{\gamma}{\gamma-1}},$$

and noting that

$$\frac{\gamma - 1}{2} Ma^2 \ll 1,$$

which, using the Taylor series expansion, leads to the following approximation

$$\left(1 + \frac{\gamma - 1}{2} Ma^2\right)^{\frac{\gamma}{\gamma-1}} \approx 1 + \frac{\gamma}{2} Ma^2,$$

and, via substitution into Equation 3.11, to

$$p_0 - p = \frac{1}{2}\gamma p Ma^2,$$

so the definition of the diffusion RF becomes

$$RF = \frac{p_4 - p_x}{\frac{1}{2}\gamma p_x Ma_x^2}. \quad (10.51)$$

One can show that, for an *ideal* diffuser with incompressible flow and zero losses (i.e., constant total pressure), the diffuser RF would simply be

$$RF = 1 - \left(\frac{A_x}{A_4}\right)^2,$$

where A_x and A_4 are the flow areas of the diffuser at the inlet and at the exit, respectively. Thus, if $A_4 \gg A_x$, the ideal diffuser RF would be nearly unity. The key to a good diffuser design is to achieve the area increase and, consequently, static pressure rise over an optimal length to ensure smooth flow without boundary layer separation and flow stall. The difference between a suboptimal and optimal exhaust diffuser design is conceptually depicted in Figure 10.10.

A well-designed diffuser with high RF (i.e., about 0.9) improves the gas turbine performance significantly. This can be understood by examining the expansion of gas in a simple turbine on a temperature–entropy diagram (see Figure 10.11). As shown in Figure 10.11, the specific work done by the gas in the turbine is given by

$$w = c_p(T_{0,3} - T_{0,4}).$$

The flow process from the turbine inlet (state-point 3) to the turbine exit (state-point 4) has two parts:

- Expansion in the turbine stage from pressure, temperature, and speed p_3, T_3, and V_3, respectively, to p_x, T_x, and V_x;
- Recovery of the dynamic head in the diffuser by
  - Slowing the flow from V_x to $V_4 \approx 0$,
  - Increasing the static pressure from p_x to p_4, with
  - No change in total temperature (i.e., no work done), $T_{0,x} = T_{0,4}$.

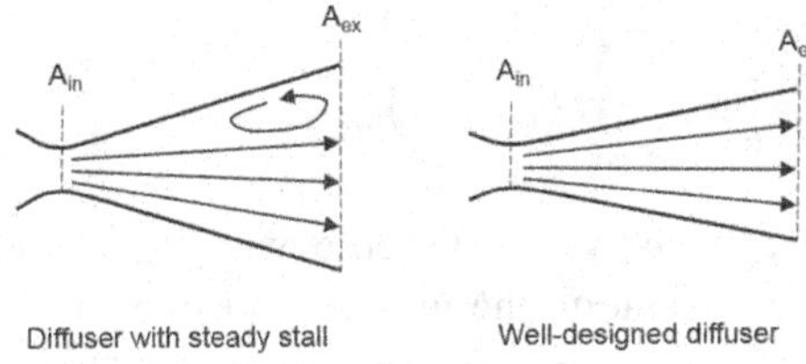

Figure 10.10 Diffusers with suboptimal and optimal flow regimes.

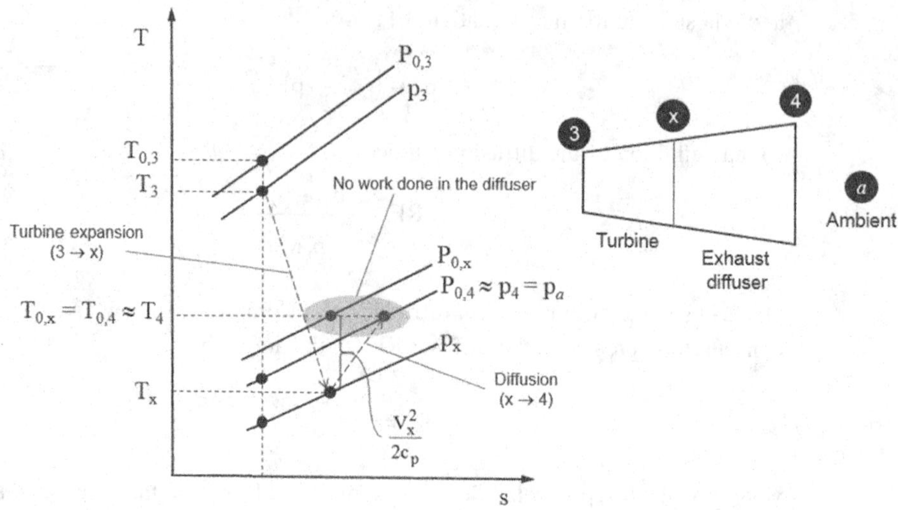

Figure 10.11 Simple turbine expansion.

From a practical perspective and for most heat and mass balance analysis, there is no benefit to be gained from an in-depth consideration of the exhaust diffuser because

$$w = c_p(T_{0,3} - T_{0,4}) = c_p(T_{0,3} - T_{0,x}).$$

For turbine aerothermodynamic design, of course, it is a different matter, leading to a plethora of turbine-stage efficiency definitions, i.e.,

- Total to static (from state point $p_{0,3}$ to p_x)
- Total to total (from $p_{0,3}$ to $p_{0,x}$)
- Static to static (from p_3 to p_x)

In a very simple analysis, where the turbine is examined as a single unit (e.g., a chemical process gas expander) with inlet and exit state-points 3 and 4 (i.e., the ambient) and with a specified turbine efficiency, the implicit assumption is that the turbine PR and the efficiency are in "total-to-static" terms, i.e.,

$$PR = \frac{p_{0,3}}{p_4} = \frac{p_{0,3}}{p_a},$$

$$\eta_t = \frac{T_{0,3} - T_{0,4}}{T_{0,3}\left(1 - \left(\frac{p_a}{p_{0,3}}\right)^{\frac{\gamma-1}{\gamma}}\right)}.$$

The obvious question, of course, is: Why not dispense with the exhaust diffuser altogether? After all, it does not contribute to the turbine work (i.e., $T_{0,4} = T_{0,x}$). Or does it? A close examination of the turbine expansion on the T–s diagram in Figure 10.11 reveals that, in fact, it *does* by lowering the (static) pressure at the exit

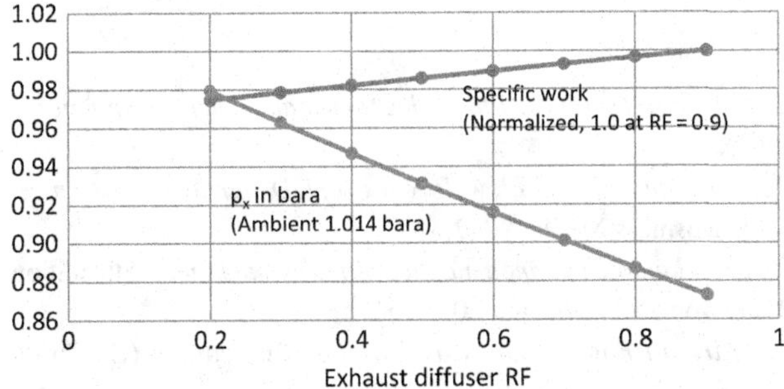

Figure 10.12 Impact of diffuser RF on turbine performance.

of the last turbine stage from p_a to p_x. In order to quantify this, consider a simple turbine where

- The inlet state is 1,550°C and 20 bara in total/stagnation terms with a gas velocity of 800 m/s;
- Turbine static-to-static efficiency is 95 percent;
- At the end of the expansion the gas velocity is 300 m/s; and
- The exhaust diffuser recovers the kinetic energy of the gas by slowing it down to ~0 m/s.

Assume a perfect gas with $\gamma = 1.333$ and $c_p = 1{,}150$ J/kg-K. Calculation results are shown in Figure 10.12. In essence, by imposing a subatmospheric pressure at the turbine exit, the exhaust diffuser increases the turbine output (by about 3 percent for this simple numerical example).

Design of a good diffuser is quite difficult and requires advanced three-dimensional CFD modeling and scale experiments (fortified by data from equipment operating in the field over a long time period at varying site ambient and loading conditions). The reason for this can be easily appreciated by recognizing that the gas velocity at the last stage exit is *not* axial (e.g., see Figure 10.3). A good design practice is to keep the "swirl" angle α_3 as low as possible for all stages, but definitely for the last stage (i.e., about 5°, definitely less than 10°). Nevertheless, some swirl is always present. From Figure 10.10, it is intuitively clear that a good diffuser design has to strike a balance between length and inlet/exit area (i.e., radius or diameter). In other words, since the diffuser is essentially a cone, the wall angle should not be excessive. (In practice, it is somewhere between 5° and 10°.) The reader is referred to the paper by Kline for the fundamentals governing the optimum design of conical diffusers [14] or chapter 4 of [3]. Clearly, OEMs have their proprietary in-house design practices for the design of the actual hardware, whose performance (over a wide range of operating conditions) depends on the interaction with other upstream and downstream equipment and/or components. The reader is referred to a paper by authors from the industry (Ansaldo in Italy) and academia on salient aspects of exhaust diffuser operation at design and part-load conditions [15].

References

1. Horlock, J. H., *Axial Flow Turbines: Fluid Mechanics and Thermodynamics* (London, UK: Butterworths, 1966).
2. Schobeiri, M. T., *Turbomachinery Flow Physics and Dynamic Performance* (New York: Springer Science & Business Media, 2012).
3. Dixon, S. L., *Fluid Mechanics, Thermodynamics of Turbomachinery*, 5th edition (Burlington, MA: Elsevier Butterworth–Heinemann, 2005).
4. Kerrebrock, J. L., *Aircraft Engines and Gas Turbines*, 2nd edition (Cambridge, MA: MIT Press, 1992).
5. Lewis, R. I., *Turbomachinery Performance Analysis* (London, UK: Arnold, 1996).
6. Shepherd, D. G., *Principles of Turbomachinery* (New York: The Macmillan Company, 1968).
7. Mattingly, J. D., *Elements of Gas Turbine Propulsion* (New Delhi, India: McGraw-Hill Education (India) Pvt Ltd., 2005).
8. Lyman, F. A., On the Conservation of Rothalpy in Turbomachines, *ASME Journal of Turbomachinery*, **115** (1993), 520–525.
9. Whittle, F., *Gas Turbine Aero-Thermodynamics* (Oxford, UK: Pergamon Press Ltd., 1981).
10. Smith, S. F., A Simple Correlation of Turbine Efficiency, *Journal of Aeronautical Science*, **69** (1965), 467.
11. Brandt, D. E., The Design and Development of an Advanced Heavy-Duty Gas Turbine, *Journal of Engineering for Gas Turbines and Power*, **110**: 2 (1988), 243–250.
12. Glasman, A. J., Moffitt, T. P., *New Technology in Turbine Aerodynamics, Proceedings of the First Turbomachinery Symposium* (College Station, TX: Texas A&M University Press, 1972).
13. El-Masri, M. A., GASCAN – An Interactive Code for Thermal Analysis of Gas Turbine Systems, *Journal of Engineering for Gas Turbines and Power*, **110** (1988), 201–209.
14. Kline, S. J., Abbott, D. E., Fox, R. W., Optimum Design of Straight-Walled Diffusers, *Journal of Basic Engineering*, **81** (1959), 321–331.
15. Dossena, V., et al., "Investigation of the Flow Field in a Gas Turbine Exhaust Diffuser at Design and Part Load Conditions," GPPF-2017-32, Proceedings of the 1st Global Power and Propulsion Forum, January 16–18, 2017, Zürich, Switzerland.

11 Compressor Aero*

Development of efficient compressors has been critical in the coming of age of gas turbines as prime movers in aviation and as reliable and efficient electric power producers. The compressor design itself has evolved hand in hand with the science of aerodynamics during the first quarter of the twentieth century. Design of axial, multi-stage compressors with aerodynamically optimized airfoils is crucial to gas turbine performance and cost (i.e., size and weight). This is dramatically illustrated by the simple comparison of a modern gas turbine compressor (14–15 stages to achieve a pressure ratio [PR] of 20 or more) with that of an early jet engine (e.g., Jumo-004B). The latter required eight stages for a modest PR of only three. For an original equipment manufacturer (OEM)-specific perspective (General Electric [GE]) on the evolution of axial compressor design, please refer to the review paper by L. Smith, Jr., with an emphasis on aircraft engine compressors [1]. This is a quite appropriate place to start because the exact same aerothermodynamic principles govern the design of aircraft and heavy-duty industrial gas turbine compressors.

The aerothermodynamic design analysis of axial compressors is very similar to that for axial turbines covered in Chapter 10 (see Figure 11.1). The nomenclature and definition of key parameters are analogous (sometimes even identical) to those for the turbines. Just like a turbine stage, a compressor stage comprises a rotor and a stator (in that order; i.e., *not* stator and rotor as in the turbine stage). In terms of absolute velocities, air is accelerated by the rotor and decelerated in the stator, where the kinetic energy imparted on the gas by the rotor is converted to static pressure. In terms of relative velocities, however, the flow decelerates in the rotor as well.

In order not to clutter the diagram in Figure 11.1, one angle, which is encountered quite frequently in the literature, is not shown (i.e., the *incidence* angle, i). This is the angle between the extension of the *camber line* beyond the leading edge of the airfoil and the relative velocity vector of air flowing into the rotor. The camber line is an imaginary line that is drawn from the leading edge of the airfoil to its trailing edge and at each point it is halfway between the top and the bottom of the airfoil. It is not to be confused with the *angle of attack*, which is the angle between the extension of the *straight* line connecting the leading and trailing edges of the airfoil and the relative

* In this chapter, a distinction must be made between total/stagnation and static values of flow pressure and temperature. Therefore, "t" in the subscript denotes "total or stagnation" and "st" in the subscript denotes "static."

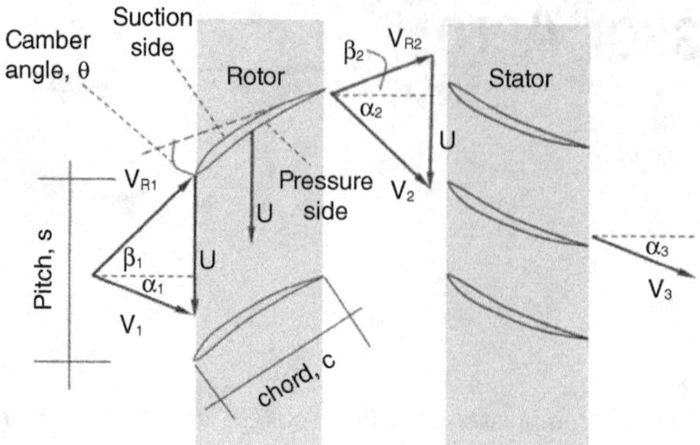

Figure 11.1 Compressor-stage velocity triangles.

velocity vector of air. Angle of attack is commonly used in aircraft wing theory. Within the context of the coverage herein, air inlet relative flow angle, β_1, serves the same purpose adequately.

The velocity triangles of a compressor stage are similar to those for a turbine stage (see Figure 10.3). As shown in Figure 11.1, with the exact same nomenclature and definitions used in Chapter 10 (and caveats regarding them), the only difference is in the order of stator and rotor blade rows. (It is highly recommended that the reader make a Xerox copy of Figure 11.1 and keep it handy while reading the rest of this chapter.) As mentioned earlier, the absolute velocity of the air increases in the rotor while its velocity relative to the rotor decreases. This diffusing flow can only be stable for a small pressure rise in order to prevent the stalling or flow-reversing effect of the growing boundary layer (BL) on the annulus walls. Unlike the turbines, where the pressure gradient and the BL growth are favorable (i.e., in the flow direction), compressors require many more stages (e.g., 14 or 15 for modern heavy-duty gas turbines; up to 18 for vintage E- and F-class units), as opposed to three or four turbine stages.[1] What happens to key aero-thermodynamic flow parameters in a compressor stage is summarized in Table 11.1.

Consequently, it is much more difficult to design an efficient axial compressor than to design an efficient axial turbine. Interestingly, however, modern gas turbine compressors with highly loaded *transonic* front stages and *controlled diffusion airfoils* (CDAs) have higher efficiencies than their expander counterparts (typically, around 90 percent isentropic vis-à-vis around 88 percent isentropic, respectively). There are three reasons for this seemingly anomalous situation:

- The debit to the turbine efficiency due to cooling (see Section 10.4);
- Lower stage loading of individual compressor stages vis-à-vis individual turbine stages;

[1] The need for a "moderate rate of change" in a diffusing flow was already discussed in Section 10.7 within the context of the exhaust diffuser of a gas turbine. Analogous to a "long" exhaust diffuser is the "long" axial compressor with a large number of stages and gradually decreasing annulus area.

Table 11.1 Compressor-stage aerothermodynamic process

		Rotor	Stator
Static Pressure	p_{st}	↑	↑
Total Pressure	p_t	↑	↓
Static Temperature	T_{st}	↑	↑
Total Temperature	T_t	↑	const.
Relative Velocity	V_R	↓	NA
Absolute Velocity	V	↑	↓
Total Enthalpy	h_t	↑	const.
Density	ρ	↑	↑

- Significantly more research and development effort poured into compressor design to overcome the considerable thermofluid problem of flow in adverse pressure gradient.

Compressor design algorithms involve empirical knowledge obtained from a large body of rig tests and 3D numerical computations using advanced computational fluid dynamics (CFD) software. Fortunately, for the purposes of performance calculations in design and off-design conditions, one does not need to go into the aerodynamic design details of the compressor stages. The pertinent design parameters are the overall PR, number of stages, stage PR, and polytropic efficiency. In cases such as a conceptual design exercise, when the compressor stages are not known a priori, estimates can be based on certain principles. More advanced 2D throughflow models and 3D CFD validation usually constitute the second part of the full detailed design process.

A significant part of axial compressor design is efficient and stable off-design performance within the operability envelope (i.e., without surge or stall). Achievement of this goal typically requires,

- *Variable inlet guide vanes* (VIGVs) to adjust airflow with changing site ambient and loading conditions;
- Three or four stages of *variable stator guide vanes* (VSGVs) to maintain optimal velocity triangles (as much as possible) in off-design conditions;
- Interstage *blow-off ports* to facilitate surge-free startup.

This is a quite onerous task with many iterations and extensive testing for validation and *performance map* building. The latter is extremely important to facilitate gas turbine control and it is the primary “lookup table” inside the controller to activate VIGVs, VSGVs, blow-off valves, and inlet bleed heat recirculation at the precise point of operation. A recent example of a successful test campaign undertaken by a major OEM to validate and finalize the design of their latest, most advanced axial compressor is described in [2]. A full-size prototype compressor (14 stages with three VSGVs) was run in a factory test bed across a wide operating envelope (actually well beyond expected field operation). The new tested compressor was driven by a full-size frame gas turbine assisted by a 50-MWe electric motor.

Note that, in the case of VIGVs, the qualifier "variable" is superfluous because the IGVs are always variable, otherwise they would not be needed. Consequently, from here on, the acronym IGV will be used. (In the trade and academic literature, both acronyms are used.)

In the case of VSGVs, the qualifier "variable" is indeed necessary, but the qualifier "guide" is superfluous because guiding the airflow into the next stage is one of the primary functionalities of the stator vanes. Therefore, from here on, the acronym VSV will be adopted. (Once again, in the trade and academic literature, both variants are used frequently.)

11.1 Design Considerations

The starting point of compressor design is the thermodynamic cycle data; specifically, compressor efficiency. The polytropic efficiency has the useful property of being constant for individual stages and the overall compressor (see Section 7.2 for an in-depth discussion of this). For a possible range of compressor polytropic efficiencies, the reader is referred to the chart in figure 19 in the review paper by L. Smith, Jr. [1]. The majority of test data displayed in the chart covers 88–91 percent (for actual product compressors up to 1982). One striking outlier is the rightmost point with nearly 93 percent, which is the 14-stage research compressor designed in 1947! In all fairness, "...its small clearances and other features made it impractical for direct engine application" [1].

The entitlement value for compressor polytropic efficiency is unlikely to be much higher than 94 percent. It is not recommended to use values higher than this for conceptual studies or reverse-engineering analyses. Scaling with number of stages and/or overall PR is a common way of adjusting the polytropic efficiency for design studies, e.g.,

$$\eta_p = \eta_{p,\,max} - \alpha \ln PR. \tag{11.1}$$

In Wilson and Korakianitis ([2] cited in Chapter 4), α is 0.0053 and $\eta_{p,max}$ is 93 percent (or 0.93) – this is what we used in Chapter 8. For $\eta_{p,max}$, 94 percent is recommended herein for the most advanced transonic compressors. The value of α can be adjusted by the reader who has access to actual hardware information.

Using the velocity triangles in Figure 11.1 (and the same nomenclature as in Figure 10.3 for axial and tangential components of fluid velocity, V; i.e., u and v, respectively) and following the derivation of the turbine Euler equation in Chapter 10, the total temperature rise across the compressor stage can be written as follows:

$$\Delta h_{stage} = U(v_2 - v_1), \tag{11.2a}$$

$$\Delta T_{stage} = \frac{U}{c_p}(v_2 - v_1), \tag{11.2b}$$

$$\Delta T_{stage} = \frac{UC_a}{c_p}(\tan\beta_1 - \tan\beta_2). \tag{11.3}$$

Noting that

$$C_a = u_1 \approx u_2 \text{(i.e., constant axial velocity)},$$

the stage loading parameter in terms of relative flow angles is given by

$$\psi = \frac{C_a(\tan\beta_1 - \tan\beta_2)}{U}. \tag{11.4a}$$

Note that Equation 11.4a can also be written in terms of the absolute flow angles; using the stage geometry in Figure 11.1, we obtain

$$\psi = \frac{C_a(\tan\alpha_2 - \tan\alpha_1)}{U}. \tag{11.4b}$$

Combining Equations 11.3 and 11.4, the result is

$$\Delta T_{stage} = \frac{\psi U^2}{c_p},$$

where $U = \omega r_m$ is the tangential speed at the mean radius. The theoretical value thus obtained is "tempered" by the *work-done factor*, λ, which quantifies the reduction in the theoretical work capacity due to variation in axial velocity from hub to tip, BL effects at the annulus walls, and tip-gap leakage:

$$\Delta T_{stage} = \lambda \frac{\psi U^2}{c_p}. \tag{11.5}$$

The variation of λ with number of compressor stages is given by

$$\lambda = 0.85 + 0.15 \exp\left(-\frac{n-1}{2.73}\right), \tag{11.6}$$

where n is the number of stages. For modern gas turbines with 14 or more stages, using $\lambda = 0.85$ is sufficiently accurate for most practical purposes.

The stage loading parameter, ψ, for a compressor is analogous to that for the turbine (see Equation 10.5). Results of the cascade tests have shown that, for the most efficient operation, the design point value of ψ is limited to the range of 0.3–0.4 and is almost always less than 0.45 for a subsonic stage in order not to overload it and to limit stall susceptibility.

"Smith charts" similar to those for the axial turbines (see Section 10.2.1) can be constructed for axial compressors as well. A detailed coverage of this can be found in the book by Lewis ([5] in Chapter 10), which determines the "optimum" and "maximum" values of ψ as (equation 4.20 in the cited reference)

$$\psi_{opt} = 0.185\sqrt{4\phi^2 + 1} \text{ and}$$

$$\psi_{max} = 0.32 + 0.2\phi, \text{ respectively.}$$

A "sensible" design range for axial compressors is stated to lie between the two correlations.

The caveat made in Chapter 10 is repeated here again because it is a major source of confusion. In some texts, ψ is defined as $\Delta h/2U^2$ (i.e., *half* of the value obtained from Equation 11.4) and in others as $\Delta h/½U^2$ (e.g., Saravanamuttoo et al.; i.e., *twice* the value obtained from Equation 11.4). Thus, in some key references (e.g., [12] cited in Chapter 2), the range of ψ from E class to H class (e.g., figure 7-31 in the cited reference) will be listed as 0.5–0.6, which is 0.25–0.30 in this chapter's definition.

Similar to the axial turbine stage, the stage loading parameter and flow coefficient, ϕ, of an axial compressor stage are linked together via the stage reaction, R, as well, i.e.,

$$\phi^2 = \frac{Ma_{R1}^2\left[\left(\frac{a_0}{U}\right)^2 - \frac{\gamma-1}{2}\left(1 - R - \frac{\psi}{2}\right)^2\right] - \left(R + \frac{\psi}{2}\right)^2}{1 + \frac{\gamma-1}{2}Ma_{R1}^2}, \tag{11.7}$$

where the Mach number is the ratio of the inlet relative velocity to the *stagnation* speed of sound, a_0 (i.e., an *artificial* parameter because the "true" speed of sound is defined by *static* properties). Typical values for Ma_{R1} and (a_0/U) are 0.9 and 0.9–1.0, respectively. For quick estimates, they can be set to unity without a big loss in accuracy. Note that Equation 11.7 is taken from the paper by Benini [3] (equation 2 on p. 11 of the cited work), which provides a very useful, concise perspective on the design of modern axial compressors.

The key to efficient design of multi-stage axial compressors is equal distribution of enthalpy/temperature rise between the rotor and the stator. For an axial compressor, stage degree of reaction, R, is defined as the ratio of static enthalpy rise across the rotor to the static enthalpy rise across the stage. Under the assumption of nearly constant axial velocity and "repeating stages" (i.e., $\alpha_1 = \alpha_3$), we can write that (see p. 130 in [3] cited in Chapter 10)

$$V_1 = V_3,$$

$$R = \frac{1}{1 + \frac{\Delta V_{Sta}}{\Delta V_{R,Rot}}} = \frac{1}{1 + d},$$

$$d = \frac{1}{R} - 1,$$

where ΔV_{Sta} is the absolute value of diffusion in the stator and $\Delta V_{R,Rot}$ is the relative velocity diffusion in the rotor. Thus, the degree of reaction is a measure of diffusion in the stator relative to the diffusion in the rotor. Consequently, an R value of around 0.5 is a good starting point because it means that the diffusion process in the compressor stage is equally shared between the rotor and the stator. This, in turn, minimizes the risk of BL separation on the rotor blade surface with large losses in stagnation pressure and efficiency. With R set to 0.5, it is found that, per Equation 11.7, the flow coefficient, ϕ, does not deviate much from 0.6.

Calculation of U requires the mean or *pitch-line* radius. With known values of U, ψ, and R, the flow coefficient, ϕ, can be determined from Equation 11.7. Knowledge of these parameters leads to the calculation of the *elementary* compressor PR, PR_c, which can be developed by a single compressor stage. The value of PR_c is derived from the Euler equation for an axial stage at the root mean square radius as

$$PR_c = \left(1 + (\gamma - 1)Ma_{R1}^2 \cos^2\beta_1 \frac{\psi}{\phi^2}\right)^{\frac{\eta_c\gamma}{\gamma-1}}, \tag{11.8}$$

where β_1 is the inlet flow angle (see Figure 11.1). (Equation 11.8 is equation 1 on p. 11 of [3].) Maximum theoretical stage efficiency is shown to be achieved by $\beta_1 \approx 45°$ and maximum flow turning, $\beta_1 - \beta_2 \approx 20°$ [3,4]. Note that the two key angles determining the flow turning are given by

$$\tan\beta_1 = \frac{1}{2\phi}\left(2R + \frac{\psi}{\lambda}\right) \text{ and} \tag{11.9a}$$

$$\tan\beta_2 = \frac{1}{2\phi}\left(2R - \frac{\psi}{\lambda}\right). \tag{11.9b}$$

Combining the two equations, we obtain another representation of the stage reaction, i.e.,

$$R = \frac{\phi}{2}(\tan\beta_1 + \tan\beta_2), \tag{11.10}$$

which can be translated into another, more informative version using the velocity triangles as

$$R = 0.5 + \frac{\phi}{2}(\tan\beta_2 - \tan\alpha_1). \tag{11.11}$$

This equation tells us that, if R = 0.5, the velocity triangles at the rotor inlet and exit in Figure 11.1 are symmetrical ($\alpha_1 = \beta_2$). The stage enthalpy rise is equally distributed between the rotor and stator rows. In the same vein,

- If $R > 0.5$, then $\beta_2 > \alpha_1$ and the static enthalpy *and* pressure rise in the rotor exceeds that in the stator.
- If $R < 0.5$, then $\beta_2 < \alpha_1$ and the static enthalpy *and* pressure rise in the stator exceeds that in the rotor.

Noting that

$$\phi = \frac{1}{\tan\alpha_1 + \tan\beta_1}, \tag{11.12}$$

Equation 11.9a can also be written as

$$\tan\alpha_1 = \frac{1}{2\phi}\left(2(1 - R) - \frac{\psi}{\lambda}\right). \tag{11.13}$$

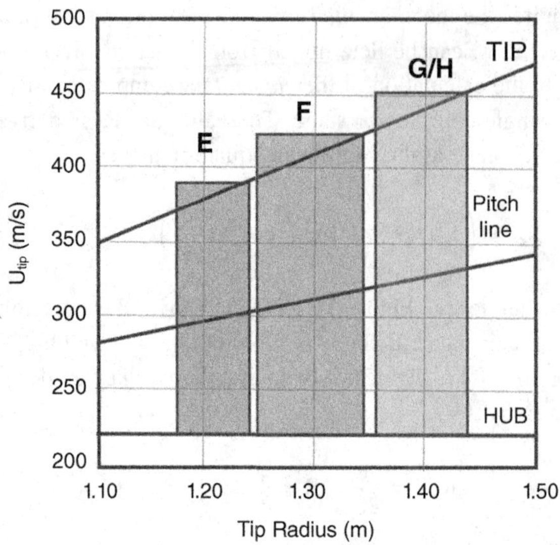

Figure 11.2 Typical 50–Hz "frame" gas turbine radii and tip speeds.

Compressor first-stage rotor blade radii (as a function of their size and airflow) for 50-Hz E-, F-, G-, and H–class heavy-duty industrial gas turbines are provided in Figure 11.2. A typical hub radius for these machines is around 0.6–0.7 m (about 2.0–2.3 ft) so that, with known airflow and inlet Mach number, all requisite preliminary design information can be readily found. The specified range of hub radii ensures a shaft diameter with a size and strength commensurate with the requisite torque transmission from the gas turbine to the ac generator. The compressor shaft can be of solid construction or composed of independent wheels and tie-bolts, depending on the construction method adopted by the particular OEM. (Note that almost all modern heavy-duty industrial gas turbines are "cold end-driven"; i.e., they are coupled to the generator on the compressor end.)

For the aforementioned preliminary design information, the requisite equations are

$$\dot{m}_{air} = \rho u_1 A, \tag{11.14a}$$

but since $C_a = u_1 \approx u_2$ (i.e., constant axial velocity)

$$\dot{m}_{air} = \rho C_a A, \tag{11.14b}$$

$$\dot{m}_{air} = \frac{Ma_V\sqrt{\gamma/R_a}\,A}{\left\{1+\frac{\gamma-1}{2}Ma_V^2\right\}^{\frac{\gamma+1}{2(\gamma-1)}}}\frac{p_{1t}}{\sqrt{T_{1t}}}, \tag{11.15}$$

$$C_a = Ma_V\sqrt{\gamma R T_{1st}}, \tag{11.16}$$

$$A = \pi\left(r_{tip}^2 - r_{hub}^2\right) \tag{11.17}$$

$$r_m = \frac{1}{2}\left(r_{hub} + r_{tip}\right). \quad (11.18)$$

For ISO conditions, the specific heat ratio, γ, is 1.4 and the gas constant for air, R_a, can be assumed to be 287 J/kg-K (0.06855 Btu/lb-R). For the Mach number, subscripts "V" and "U" denote axial and tangential flow directions, respectively. Implicit in Equations 11.12–11.16 is the ideal gas equation of state

$$p = \rho R_a T,$$

with static values of p and T. (See the cautionary remarks in Chapter 10 regarding the units when using the ideal gas equation of state.) The calculation sequence is as follows:

1. Absolute inlet velocity is determined from Equation 11.16.
2. Flow cross-sectional area, A, is determined from Equations 11.14 or 11.15 (which is Equation 11.14 with substitution of total/stagnation properties from Equations 3.7 and 3.11 in Chapter 3).
3. Tip radius is found from Equation 11.17 and pitch-line radius is found from Equation 11.18.

Using the V94.3A gas turbine used in Chapter 10 (50-Hz machine, 3,000 rpm) as an example again, the results are summarized in Table 11.2.

Using the information in Table 11.2 (i.e., r_m = 1.0 m [3.3 ft]), a series of design calculations are carried out by changing ψ from 0.3 to 0.45 (R = 0.5). The compressor temperature rise for a PR of 17 is estimated as follows: assuming a polytropic efficiency of 92.5 percent, first the compressor discharge temperature (CDT) is found as

$$\text{CDT} = T_{in} PR^{\frac{\gamma-1}{\eta_c \gamma}},$$

$$\text{CDT} = (59 + 460) \cdot 17^{\frac{1.4-1}{0.925 \cdot 1.4}} - 460 = 790°\text{F}.$$

Table 11.2 V94.3A (2003) compressor first-stage dimensions

Parameter	Units	Value	Notes
m_{air}	kg/s	657	Given
T_{0in}	K	288	Given
p_{0in}	bara	1.01325	Given
Ma_V		0.50	Assumed
r_h	m	0.70	Assumed
T_{in}	K	274	Equation 3.9
p_{in}	bara	0.85419	Equation 3.13
ρ	kg/m^3	1.09	From equation of state
a	m/s	332	Equation 11.16
V	m/s	166	Equation 11.16
A	m^2	3.65	Equation 11.14
r_t	m	1.28	Equation 11.17
r_m	ft	3.3	
	m	1.0	Equation 11.18

The temperature rise across the entire compressor is thus 731°F. An inlet Mach number of 0.5 and a stage reaction, R, of 0.50 are assumed. The axial component of the absolute velocity is assumed to be unchanged between stage inlet and exit. For given ψ and R, ϕ is calculated from Equation 11.7. Those three parameters fix the velocity triangles and everything else can be calculated from that point on by using those triangles. Number of stages is estimated as (rounded to the nearest integer)

$$n_{stages} = \Delta T_{tot} / \Delta T_{stage}.$$

Inlet flow angle, β_1, and exit flow angle, β_2, are calculated from Equations 11.9a and 11.5b. Stage PR is found from

$$PR_{stage} = PR_{tot}^{1/n_{stages}}.$$

Comparison of β_1, β_2, and PR_{stage} with their maximum theoretical/elementary counterparts (e.g., PR_c per Equation 11.8) should be one indication of reasonably good preliminary design. Typically, β_1 values beyond 55° should require a reevaluation. The results are summarized in Table 11.3.

Table 11.3 V94.3A (2003) compressor preliminary design ψ sweep

Parameter	Units	1	2	3	4
T_{0in}	K	288	288	288	288
Ma_V		0.50	0.50	0.50	0.50
T_{in}	K	274.3	274.3	274.3	274.3
PR		17	17	17	17
ΔT_{tot}	°F	731	731	731	731
N	rpm	3,000	3,000	3,000	3,000
ω	rad/s	314	314	314	314
r_m	ft	3.3	3.3	3.3	3.3
	m	1.0	1.0	1.0	1.0
$U=\omega r$	ft/s	1,023	1,023	1,023	1,023
	m/s	312	312	312	312
a	m/s	332	332	332	332
Ma_U		0.94	0.94	0.94	0.94
R		0.50	0.50	0.50	0.50
ψ		0.30	0.35	0.40	0.45
ϕ		0.67	0.65	0.63	0.61
ΔT_{stg}	°F	43	50	57	64
# stages		17	15	13	11
PR_{stg}		1.18	1.21	1.25	1.28
β_1		45	47	50	52
β_2		26	24	23	21
DF		0.30	0.37	0.45	0.52
PR_c		1.39	1.46	1.52	1.59
de Haller #		0.78	0.74	0.71	0.67

There are two important design-check parameters at the bottom of Table 11.3:

1. Diffusion factor (DF)
2. de Haller number

The *de Haller number* is defined as the ratio of absolute rotor exit and inlet velocities, V_2/V_1, which is equivalent to $\cos\beta_1/\cos\beta_2$ for $v_2 = v_1$. It is a measure of allowable diffusion and, in the past, was widely used to prevent excessive losses with a minimum value set to 0.72 (which was later relaxed to 0.69).

Modern design practice uses the DF, also known as Lieblein's DF, which, for $v_2 = v_1$, is given by

$$DF \approx 1 - \cos\beta_1\left(\frac{1}{\cos\beta_2} - \frac{\psi}{2\sigma\lambda\phi}\right). \tag{11.19}$$

In Equation 11.19, σ is the blade *solidity* defined as the ratio of the blade *chord*, c, to the blade *pitch*, s (see Figure 11.1). Solidity, common in US practice, is the inverse of the *pitch–chord ratio* (also referred to as the *space–chord ratio*) preferred by Europeans. Originally developed by NACA (the precursor of today's NASA), DF gives a measure of friction loss. Data obtained from a large number of cascade tests for the rotor hub and stator indicate that, up to DF ≈ 0.6, friction losses are reasonably low and constant. Near the rotor tip, however, friction losses increase rapidly beyond DF ≈ 0.4.

The values of de Haller number and DF (evaluated at r_m with a solidity of 1.11) in Table 11.3 suggest that 15 or 17 stage designs with ψ values of 0.3 and 0.35, respectively, are acceptable. Both are reasonable with good values of β_1 and β_2 and with stage PR reasonably close (and below) PR_c. Size and cost considerations obviously support the lower stage count, which is indeed the case for the actual compressor in the V94.3A (2003) that was used as an example. The compressor in question was, at that time, a new subsonic design (in collaboration with Pratt & Whitney to derive upon their aircraft compressor technology) with CDA airfoils.

Another critical design parameter is the maximum centrifugal tensile stress, which occurs at the blade root and is given by (in Pascals when using SI units)

$$\sigma_{max} = \frac{\rho_b}{2}(2\pi N r_t)^2\left(1 - \left(\frac{r_h}{r_t}\right)^2\right), \tag{11.20}$$

with the simplifying assumption that the blade cross-sectional area is constant from hub to tip. (Note the analogy to Equation 10.46.) For the blade metal density, a good assumption is 7,800 kg/m^3. From Equation 11.20, we note that the centrifugal stress is proportional to

- The square of the blade tip tangential speed;
- The "hub–tip radius ratio."

In particular, a reduction in hub–tip ratio (i.e., longer blades) increases the blade centrifugal stress. Consequently, the longest stage 1 rotor blades experience the highest stress. Typically, with tip speeds around 350 m/s, centrifugal stress is not likely to be a limiting design factor. For the example herein, with the hub–tip ratio of 0.54 and tip radius of 1.28 m from Table 11.2, U_{tip} and σ_{max} are calculated as 404 m/s and 447 MPa, respectively. In reality, the cross-sectional area of the blade decreases with increasing radius so that Equation 11.20 overestimates the stress by about 40–50 percent. Still, this corresponds to a maximum centrifugal stress of 240 MPa at the blade, which is equivalent to about 50 percent of the ultimate tensile strength of stainless steel (see Chapter 13 for material properties). Thus, in state-of-the-art large gas turbine compressors (e.g., those of GE's HA-class products) with airflows pushing 1,000 kg/s in 50-Hz units, titanium first-stage rotor blades are used. For a multiplicative correction to Equation 11.20 for linearly decreasing blade cross-sectional area, A_x, from hub to tip, use the following correlation

$$C = 1 - \left(\frac{1}{2} - \frac{x\,(2 - \upsilon - \upsilon^2)}{3(1 - \upsilon^2)}\right), \tag{11.21}$$

where υ is the hub–tip ratio and x is the ratio of the cross-sectional area at the blade tip to that at the blade root. Typical values of the correction factor range from 0.55 to 0.65 for tapered blades.

Evolution of the compressor technology from E-class to modern F/H/J-class gas turbines can be described by the principles and representative parameters discussed above. Using the GE class designation and cycle PRs, as illustrated in Table 11.4, a rising trend of CDT and temperature rise across the compressor over the period 1980–2015 is evident (ISO inlet with 92 percent polytropic efficiency). Increasing stage loading and a reduction in stage numbers with increasing cycle PR is strong evidence for significant advances made in 3D CFD software, hardware, and understanding of complex flow phenomena.

The trend of ever-larger gas turbines (i.e., higher airflows) and the "stress" imposed on the hardware is summarized in Figure 11.3. Centrifugal stress imposed on the first-stage rotor (commonly known as R1) blade of a modern HA-class gas turbine compressor is nearly 150 percent higher than that of an upgraded E-class gas turbine. This brings maximum tensile stress values closer to 400 MPa and introduces titanium R1 blades. The larger size of those blades is marked by the decreasing hub–tip ratio.

Design data associated with Figure 11.3 are presented in Table 11.5. Assumptions are a 60-Hz (3,600-rpm) gas turbine with ISO inlet, 92 percent polytropic efficiency, inlet

Table 11.4 Compressor design data of "frame" gas turbines

		E	F.04	F.05	F.06	HA.02
Cycle PR		13.0	16.7	18.4	22.1	23.1
# Stages		17	18	14	14	14
Stage PR		1.16	1.17	1.23	1.25	1.25
CDT (Total)	°F	691	784	822	897	916
	°C	366	418	439	481	491
ΔT (Total)	°F	632	725	763	838	857
	°C	351	403	424	466	476

Table 11.5 Compressor design parameters of 60-Hz "frame" gas turbines

Parameter	Unit	E.03	F.04	F.05	F.06	HA.02
PR		13	16.7	18.4	22.1	23.1
ΔT_{tot}	°F	632	725	763	838	857
N	rpm	3,600	3,600	3,600	3,600	3,600
ω	rad/s	377	377	377	377	377
r_m	ft	2.5	2.7	2.8	2.9	3.0
	m	0.8	0.8	0.9	0.9	0.9
$U = \omega r$	ft/s	948	1,035	1,074	1,100	1,145
	m/s	289	316	327	335	349
a	m/s	332	332	332	332	332
Ma_U		0.87	0.95	0.99	1.01	1.05
R		0.50	0.50	0.35	0.30	0.30
ψ		0.30	0.28	0.35	0.36	0.34
ϕ		0.76	0.66	0.70	0.69	0.66
ΔT_{stg}	°F	37	40	55	60	61
# stages		17	18	14	14	14
PR_{stg}		1.16	1.17	1.23	1.25	1.25
Ma_{VR}		0.9	0.9	0.9	0.9	0.9
a_0/u		1.18	1.08	1.04	1.01	0.98
β_1		41.7	44.9	38.5	36.6	37.4
β_2		23	27	12	7	8
DF		0.35	0.36	0.41	0.41	0.42
de Haller #		0.81	0.79	0.80	0.81	0.80

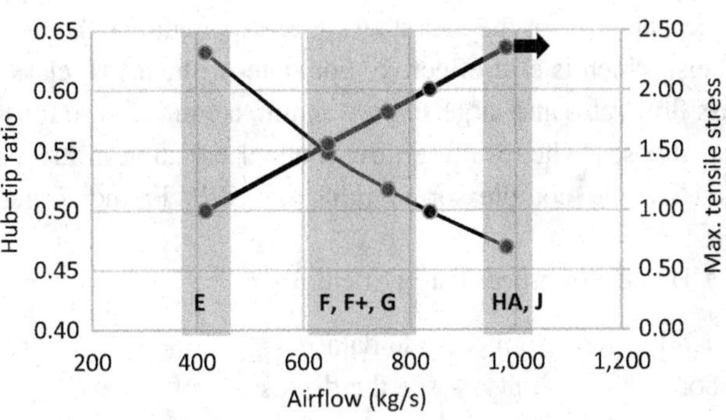

Figure 11.3 Hub–tip ratio and maximum tensile stress for 50-Hz "frame" gas turbines.

Mach number of 0.5, and hub radius of 0.6 m. For a given R, which is first assumed to be 0.5 and then reduced to result in reasonable DF and de Haller numbers with a solidity of 1.11, ψ is adjusted to result in the desired (known) number of stages.

In actual production compressors, stage reaction and PR are higher for the front stages. Typically, in large, advanced designs with high mass flow rates and cycle PRs, R is 0.8–0.9 and stage PR is about 1.4 for the front-end *transonic* stages (see Section 11.1.4). Rear, subsonic stages have R = 0.5 and stage PRs of around 1.2.

One-dimensional (1D) axial compressor design is finished by "stage stacking." As discussed earlier, for the large, stationary gas turbines in question herein, rotational speed is fixed by the grid (3,000 or 3,600 rpm). Based on the number of stages selected, stage-by-stage design is accomplished by establishing the velocity triangles and calculating the flow angles for each stage at the mean or pitch-line radius. This process establishes the compressor flow channel or annulus. There are three ways to go about it:

1. Constant inner diameter (tapering casing, cylindrical shaft)
2. Constant outer diameter (cylindrical casing, tapering shaft)
3. Constant mean diameter (tapering casing, tapering shaft)

For a given area contraction ratio between compressor inlet and outlet,

- Variant 2 (constant D_o) gives the shortest last-stage blade height, largest mean radius/diameter (i.e., highest mean blade speed), and smallest relative tip clearance.
- Variant 1 (constant D_i) gives the longest last-stage blade height, smallest mean radius/diameter (i.e., lowest mean blade speed), and largest relative tip clearance.

Since the allowable hub–tip ratio has an upper limit (0.92 is a typical value) to keep BL and tip clearance losses to acceptable values, variant 2 must have fewer stages than variant 1. Thus, higher PRs can be achieved with variant 1. Variant 3 (constant D_m) is, of course, a compromise between the other two.

All three variants have been used in the design of compressors for actual gas turbines in different applications at one time or another. In early gas turbines, variant 2 was adopted (e.g., see the Jumo-004 compressor in Figure 4.5), which limited the growth capability towards higher PRs. Constant inner diameter is the most appropriate for industrial gas turbines, at least for the rear stages, because it allows the use of constant-diameter rotor wheels, which is cost-effective. For state-of-the-art H–class gas turbines with very high mass flow rates and large stage 1 annuli, a compromise must be made to limit the centrifugal stresses. The resulting flow channel variation in the axial direction is nicely illustrated by the compressor annulus of GE's Frame 7 gas turbine in Figure 14.6.

Next steps in the 1D design process are to determine

1. The blade airfoil shapes from cascade data;
2. The pitch–chord ratio (solidity) – see the discussion of DF above;
3. The number of blades per stage (from the pitch and stage geometry).

For details, the reader can refer to Saravanamuttoo et al., which also provides a "poor man's" 2D (throughflow) design analysis via investigation of the flow angles in the radial direction using three methods, i.e.,

1. Free vortex
2. Constant reaction
3. Exponential

For subsonic stages (i.e., airflow $Ma < Ma_{crit}$), blade airfoil shapes are selected from existing profiles such as NACA-65. For transonic stages, special "thin airfoil" designs are developed to minimize shock losses and prevent blockage[2] (in conjunction with high solidity). Another more recent approach (i.e., CDAs) is aimed at eliminating shock waves altogether. These aspects of compressor design will be covered in Section 11.1.4.

It is hoped that this brief coverage has given the reader an appreciation of the difficulty of designing an efficient gas turbine compressor, which can compress large amount of air (pushing 1,000 kg/s or more than 2,000 lb/s for the most advanced, 50-Hz units – see Chapter 21) to very high pressures (cycle PRs as high as 25 in state-of-the-art H/J–class gas turbines, and more than 30 for sequential combustion GT24/26).

Designing a new compressor from a proverbial "blank sheet" is a lengthy and expensive undertaking. The technology flow from aircraft engine compressor design certainly helps. Even so, due to totally different mission definitions for the two application classes, significant numerical analysis and CFD work must be followed by extensive full-scale testing. Since turbine design and upgrade via increased firing temperatures happen at a steady rate, paced by advances in metallurgy and casting technologies, advanced coatings, and film-cooling techniques, the compressor design cycle is not fast enough to keep up with commensurate cycle PR and airflows.

Intermediate solutions include adding a "zero stage" (R0) to the front end and/or "flaring" the front stages. A good example of the former is GE's upgrade of a 17-stage E–class compressor to an 18-stage FA–class compressor in 1990. In order not to change the existing stage numbering for R1 to R17, the new stage added to the front of the compressor was labeled R0. In order to accommodate the subsequent uprates with higher airflows (the so-called FA+e class), the outer diameters of the first five stages (R0 to R4) of the compressor were "flared" to a slightly larger value (i.e., adding 0.35 inches to R0 radius, essentially making the R0 blade 0.35 inches longer) and tapering to 0 inches at R5. In order to accommodate this change, compressor casing and affected stator blades were modified as well. Furthermore, the inlet casing was recontoured and IGV aerodynamics were optimized for the new setting. This "flared" compressor, introduced in 2000, was known as the "Snowflake" compressor. (A similar flaring was applied to the HA.02–class compressor to handle the higher airflow vis-à-vis the HA.01 class in 2014.)

11.1.1 Lift and Drag Coefficients

Those who are unfamiliar with sailing and the principles of aerodynamics are under the impression that a sailboat moves when the wind coming from behind pushes at the sail. This, of course, is not the case. (In fact, the described situation is quite undesirable from a perfect sailing perspective.) What happens is that the sail of a sailboat acts exactly as the wing of an airplane does. Air flowing across the sail creates a pressure difference between the "suction" and "pressure" sides of the sail and the resulting "lift" pushes the sail and, consequently, the sailboat forward. This is why a sailboat actually moves *against* the wind.

[2] Blockage refers to the reduction in flow area (e.g., in the passage between the blades caused by boundary layer displacement as a result of the shock wave boundary layer interaction).

In an axial compressor, air flows *into* the airfoil-shaped blade rather than the other way. The flow of air is induced by the rotation of the blades, which accelerates it from the ambient (air at rest) through a convergent duct (commonly known as the "bell mouth" duct) to the inlet of the compressor and from there on "squeezes" it at quasi-constant axial velocity through successive stages to the exit diffuser and combustor inlet.

In this process, the drag force or its non-dimensional expression, the drag coefficient, translates into the blade profile loss and efficiency. The lift force or its non-dimensional expression, the lift coefficient, translates into the work done on the fluid. Thus, from the basic aerodynamics covered earlier, the lift coefficient is a measure of the flow turning across the rotor blade and total pressure increase. In fact, it can be shown that (see the excellent book by Dixon, p. 153, [3] in Chapter 10)

$$\psi = \frac{\phi}{2}\sigma \sec \beta_m (C_L + C_D \tan \beta_m), \tag{11.22a}$$

where σ is the blade solidity and β_m is the *mean* relative flow angle defined as

$$\tan \beta_m = \frac{1}{2}(\tan \beta_1 + \tan \beta_2). \tag{11.23}$$

The lift and drag coefficients, C_L and C_D, respectively, are representative of the forces exerted by the blade on the fluid, which are exactly equal and opposite to the forces exerted by the fluid on the blade. Since, $C_D \ll C_L$ (i.e., C_D/C_L is 0.04–0.05), we can simplify as

$$\psi \approx C_L \frac{\phi}{2}\sigma \sec \beta_m. \tag{11.22b}$$

As a small sidenote, cascade studies have shown that for maximum efficiency, β_m is 45° (in Dixon, cited above), so that the optimum stage loading parameter becomes

$$\psi_{opt} \approx C_L \frac{\phi}{\sqrt{2}}\sigma. \tag{11.24}$$

From equation 3.19 in Dixon (cited above), the lift coefficient is given by

$$C_L = \frac{1}{\sigma}\cos \alpha_m \left(C_f - \zeta \frac{\sin 2\alpha_m}{2} \right), \tag{11.25}$$

where

$$\tan \alpha_m = \frac{1}{2}(\tan \alpha_1 + \tan \alpha_2). \tag{11.26}$$

The tangential force component C_f is given by

$$C_f = 2(\tan \alpha_1 - \tan \alpha_2). \tag{11.27}$$

The total pressure loss coefficient ζ is defined as

$$\zeta = \frac{\Delta p_t}{\frac{1}{2}\rho C_a^2}. \tag{11.28}$$

A typical range for ζ is 0.02–0.05.

For the sample data generated earlier and displayed in Table 11.5, lift coefficients and optimum stage loading parameters are calculated using Equations 11.25–11.28. Solidity, σ, is assumed to be 1.11 and ζ is taken as 0.05. The results are displayed in Table 11.6.

Next, assuming $\alpha_m = 45°$, $C_f = 0.90$, and $C_L = 0.56$ (from Equation 11.25), optimum stage loading parameters are calculated using Equation 11.24. The results are listed in Table 11.7.

The optimum values of ψ in Table 11.7 are quite close to the actual values of ψ used in the sample stage design, even though the actual β_m is quite different from 45°. Note that, in generating the data in Table 11.5, with given R (which was first assumed to be 0.5 and then reduced to result in reasonable DF and de Haller numbers with a solidity of 1.11), ψ was adjusted to result in the known number of stages for the respective gas turbines.

In the case of an airplane wing or sailboat sail, lift and drag forces are easy to envision in terms of their functionalities. Airflow created by the forward motion of the airplane via a prime mover exerts an upward force (lift) on the wing with a given airfoil profile (e.g., NACA-65). This force first overcomes the weight of the airplane (i.e., gaining

Table 11.6 Lift coefficient C_L

	E.03	F.04	F.05	F.06	HA.02
R	0.50	0.50	0.35	0.30	0.30
ψ	0.30	0.28	0.35	0.36	0.34
ϕ	0.76	0.66	0.70	0.69	0.66
β_1	41.7	44.9	38.5	36.6	37.4
β_2	22.8	26.9	11.8	7.0	8.4
β_m	33.3	37.0	26.7	23.4	24.3
α_1	23.0	27.4	32.3	35.2	36.9
α_2	39.3	43.3	48.6	50.9	51.7
α_m	31.9	36.2	41.4	44.1	45.2
C_f	0.79	0.85	1.00	1.04	1.03
C_L	0.59	0.60	0.66	0.66	0.64

Table 11.7 Optimum ψ

	E.03	F.04	F.05	F.06	HA.02
ϕ	0.76	0.66	0.70	0.69	0.66
ψ_{opt}	0.33	0.29	0.31	0.30	0.29

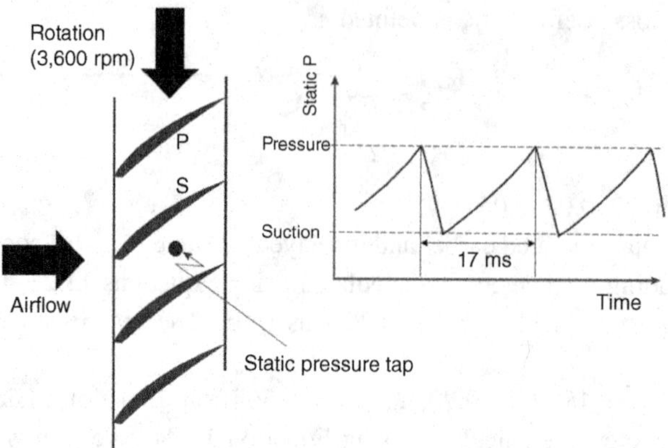

Figure 11.4 Static pressure variation in axial compressor rotor. (after figure 1.12 in Dixon [3] in Chapter 10)

altitude) and then balances it (i.e., level flight). In the case of the sailboat, the wind blows the air around the sail, which is essentially a very thin airfoil, and the force (lift) pushes the boat forward. In either case, the airfoil (wing or sail) moves in a direction opposite to that of the airflow. In either case, the lift force is caused by the pressure difference between the suction and pressure surfaces of the airfoil.

In the case of a compressor rotor blade attached to a rotating disk-shaft frame, the motion of the airfoil and the airflow are perpendicular to each other. The motion of the airfoil is due to the motion of the shaft, which is turned by the turbine or an external driver (e.g., an electric motor). The lift force exerted by the air on the airfoil is opposed by the force of the same magnitude and opposite direction exerted by the airfoil on the air. What is happening inside the rotor is actually quite interesting and counterintuitive, as explained with the help of Figure 11.4.

As shown in Figure 11.4, there is a static pressure field in each blade passage, which changes as a function of time in an oscillating (periodic) manner with a frequency of 60 Hz for a rotational speed of 3,600 rpm and with 60 blades circumferentially arranged on the rotor disk. The static pressure obtained from a tap in the compressor casing would record a signal similar to that shown in Figure 11.4. There would be a maximum every 0.017 ms when the pressure side of a blade/airfoil passes across the tap. Similarly, there would be a minimum every 0.017 ms when the suction side of a blade/airfoil passes across the tap. Thus, even though the axial airflow *across* the rotor is *steady* (i.e., time derivatives are equal to zero), the pressure variation *inside* the rotor is *unsteady*. In fact, without this unsteadiness, the compressor (and the turbine) would not work. In order to illustrate this puzzling fact, consider the momentum equation in frictionless and irrotational (i.e., reversible) flow in three dimensions (from equations 4.114 and 4.115 in White [5] – ignoring body forces):

$$\rho\left(\frac{\partial \vec{V}}{\partial t} + \left(\vec{V}\cdot\nabla\right)\vec{V}\right) = -\nabla p. \tag{11.29a}$$

This scary-looking equation is actually quite mundane in only one spatial dimension, say x, i.e.,

$$\rho\left(\frac{\partial V}{\partial t}+V\frac{\partial V}{\partial x}\right)=-\frac{\partial p}{\partial x}. \quad (11.29b)$$

Since the flow is reversible and adiabatic (i.e., no heat transfer), it is isentropic. Therefore, from the first Tds equation, we have

$$\frac{\partial h}{\partial x}=\frac{1}{\rho}\frac{\partial p}{\partial x} \text{ and} \quad (11.30a)$$

$$\frac{\partial h}{\partial t}=\frac{1}{\rho}\frac{\partial p}{\partial t}. \quad (11.30b)$$

Using the definition of the stagnation/total enthalpy in Equation 3.9 (denoted by the subscript "t" below)

$$h_t=h+\frac{V^2}{2},$$

we can write its total derivative (again, in one spatial dimension only) as

$$\frac{Dh_t}{Dt}=\frac{\partial h}{\partial t}+V\frac{\partial h}{\partial x}+V\frac{\partial V}{\partial t}+V^2\frac{\partial V}{\partial x}.$$

Substitutions from Equations 11.29 and 11.30 result in

$$\frac{Dh_t}{Dt}=\frac{1}{\rho}\frac{\partial p}{\partial t}+\frac{V}{\rho}\frac{\partial p}{\partial x}-\frac{V}{\rho}\frac{\partial p}{\partial x},$$

$$\frac{Dh_t}{Dt}=\frac{1}{\rho}\frac{\partial p}{\partial t}. \quad (11.31)$$

Equation 11.31 tells us that total enthalpy can only change (i.e., increase as in the case of a compressor rotor) *when there is a change in static pressure with time.* Said change in static pressure is caused by the motion of rotating rotor blades.

11.1.2 Blade Twist

The discussion in the preceding section was implicitly limited to a particular circumferential plane in the annular flow channel of the compressor. To be specific, it was limited to a circumferential plane at a mean or pitch-line radius, r_m, where the tangential velocity of the rotor blade was

$$U=\omega\cdot r_m.$$

Since U changes from the hub, at r_h, to the tip of the blade, r_t, optimal flow angles determined at r_m will not be optimal at those planes in the blade passage. Due to the finicky nature of the compressor airflow *against* an adverse pressure gradient, slight

Table 11.8 Change in angles in the radial direction

r (m)	U (m/s)	R	ϕ	ψ	β_1	α_1	α_2
0.70	219.9	0.82	0.96	0.60	29.5	25.9	48.0
1.00	314.2	0.50	0.67	0.30	45.2	25.9	42.9
1.28	402.1	0.36	0.52	0.18	55.1	25.9	39.5

disturbances to the velocity triangles can result in significant losses, poor performance, and flow "stall," which can ultimately degenerate to a complete breakdown of the axial flow in violent, back-and-forth oscillations (i.e., "surge"). This is especially true for the long-bladed front-end stages of modern gas turbines.

In order to illustrate the basic guiding principle, let us return to our example with V94.3. As displayed in Tables 11.2 and 11.3, the following data were estimated for compressor stage 1:

- Hub and tip radii of 0.7 and 1.28 m, respectively (mean radius 1 m);
- Stage design parameters R = 0.5, ψ = 0.3, and ϕ = 0.67 at the mean radius;
- Stage temperature rise of 43°F (17 stages).

Using Equations 11.4b, 11.5, 11.10, and 11.11, for the hub and tip radii, we obtain the values listed in Table 11.8 (including the values at the mean radius). The absolute inlet angle α_1 does not change (it would be 0 without IGVs). The change in the relative angle β_1 indicates the "twist" in the blade. The calculations are made with the assumption of constant C_a in the radial direction (210.5 m/s). Due to the viscous nature of the real flow, C_a would be actually lower at the hub and the tip (due to contact with the end wall). For this simple, illustrative example, the assumption of a flat velocity profile is adequate.

In order to get a visual feel for the numbers in Table 11.8, several β_1 values can be computed from r_h to r_t and stacked together. In essence, relative flow angles are used as proxies for blade incidence angles. A visual presentation of a similar blade construction is provided in Figure 11.5, which shows a stage 1 blade with a substantial amount of twist from the blade root/hub to the blade tip. The tip of the blade includes a recessed channel to minimize the amount of metal at the tip, which is referred to as a "squealer tip." The goal of squealer tip design is to minimize the gap between the blade tip and the end wall (i.e., the "clearance") for increased stage efficiency. In the event of a rub, reduced metal content prevents the blade from friction-caused overheating.

During operation, two distinct forces operate on the rotor blade: the centrifugal force imposed by the shaft rotation and the lift force exerted by the airflow. The former pulls the blade and forces it to straighten (i.e., "untwist"). The latter bends the blade like a cantilever anchored at the platform. Combined pull, untwist, and bend loadings create a concentrated, steady stress near the root leading edge. Unsteady stress due to vibration can also occur and compound the total stress loading on the blade. Since the leading edge of the stage 1 blade is most susceptible to erosion by microscopic particles and/or droplets (e.g., when the inlet fogging system is on), there is an increased likelihood of blade failure with catastrophic end results.

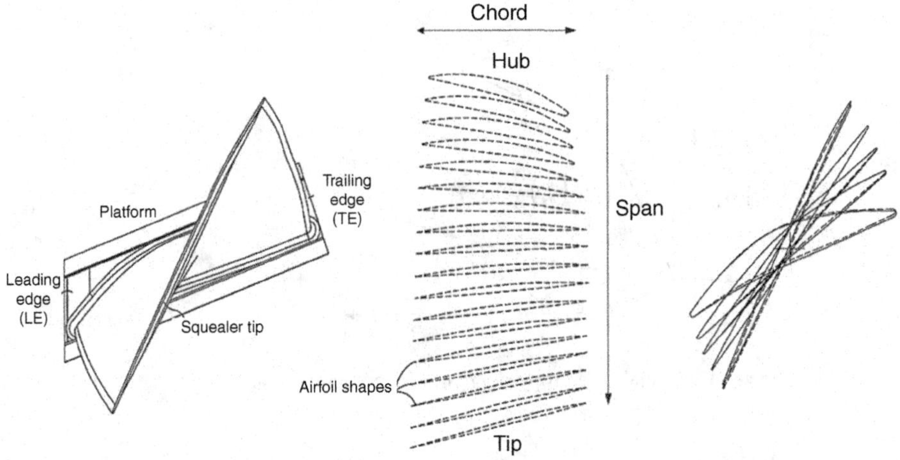

Figure 11.5 Stage 1 rotor blade design.

The twin goals of (1) stiffening the blade near the root (i.e., less subject to bending) and (2) moving the blade's natural frequency away from the resonant frequency result in the airfoil shapes shown in Figure 11.5. Another option is to attach a second platform to the tip of the blade, which is known as a "shroud." Shrouds of neighboring blades interlock with each other and make up a rigid rotor. This design is commonly used in much longer third or fourth (i.e., the last) turbine stage rotor blades, which have the same qualitative geometry with twist and varying airfoil shapes across the blade span.

Leading edge stress concentration near the root is a serious problem. For example, "Snowflake" compressors mentioned earlier in the chapter (with a flared R0 stage) experienced R0 blade "liberation" due to crack initiation by corrosion, erosion, and/or foreign object damage and propagation to a full break via high cycle fatigue. Several remedies implemented by the OEM to fix the problem included design enhancements such as "P cut," which aimed to reduce the leading edge root stress by redistributing it, and alloy changes. These measures were ultimately not fully successful and so were phased out. A complete redesign was requisite, along with measures (identified by the Electric Power Research Institute [EPRI]) such as an optimized IGV schedule to minimize rotating stall during startup/shutdown, laser shock peening to offset mean stress and improve durability, and retuning of the natural frequencies to minimize the potential for resonance.[3]

11.1.3 Compressor Efficiency

Strictly speaking, for the purposes of this book in general and this chapter in particular, we do not need to go beyond Equation 11.1, repeated below for convenience, i.e.

[3] "Compressor Dependability: General Electric FA Inlet Blade," EPRI Report 1022350, December 7, 2010.

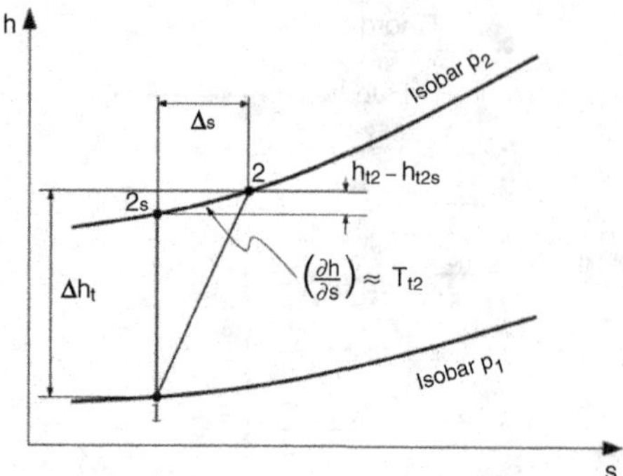

Figure 11.6 Enthalpy–entropy diagram of a compressor stage.

$$\eta_p = \eta_{p,\,max} - \alpha \ln PR.$$

For the entitlement value of efficiency, $\eta_{p,max}$, 94 percent was recommended earlier. Nevertheless, a closer look into achievable compressor stage efficiency is useful in order to arm the reader with basic knowledge to weed out "outlandish" performance claims. Let us start with the basic h–s diagram of a compressor stage as shown in Figure 11.6.

The isentropic stage efficiency from Figure 11.6 (about the same as the polytropic efficiency since the stage PR is small enough) is given by

$$\eta = \frac{\Delta h_{t,\,s}}{\Delta h_t}, \tag{11.32a}$$

where subscript "t" denotes total or stagnation and subscript "s" denotes the isentropic process. From the geometry in Figure 11.6 and Maxwell's relations (see Section 5.7), the difference between the numerator and the denominator of the term on the right-hand side of Equation 11.32a can be found as

$$\frac{i}{\Delta s} = \left(\frac{\partial h}{\partial s}\right)_p \approx T_{t,2},$$

$$i \approx T_{t,2}\Delta s,$$

which is the entropy generation, also known as *irreversibility* or the "lost work," in the compression process from state 1 to state 2 (see Section 5.3). Thus, Equation 11.32a becomes

$$\eta \approx 1 - \frac{T_{t,2}\Delta s}{\Delta h_t}. \tag{11.32b}$$

Consequently, computation of compressor-stage efficiency is a process of enumeration of all mechanisms of entropy generation (i.e., *loss* mechanisms) and computation of each of them one by one. Summation of them gives us the stage efficiency, i.e.,

$$\eta \approx 1 - \frac{T_{t,2} \sum_i \Delta s_i}{\Delta h_t}. \tag{11.32c}$$

A comprehensive treatment of individual loss mechanisms can be found in the landmark paper by Denton [6]. A similar but less comprehensive approach was adopted by Hall et al. to estimate the performance limit of axial compressor stages [7], which considered the following loss mechanisms:

1. Viscous dissipation in the BL (*profile* losses);
2. Wake mixing (downstream of blade rows);
3. Leakage flow mixing with main flow (*tip clearance* losses).

Their analysis with an incompressible flow assumption found that viscous dissipation in the BL between the fluid flow and rotor blade and stator vane surfaces was the dominant mechanism (worth about two percentage points). Including rotor and stator end walls, viscous dissipation amounted to 3.7 percentage points in efficiency loss. Total losses from the three mechanisms added up to 4.6 percentage points. Thus, the maximum stage efficiency was 95.4 percent. This maximum occurred in a flow coefficient (ϕ) range of 0.4–0.6 and a stage loading coefficient (ψ) range of 0.2–0.3, which corresponded to a de Haller number of around 0.75 (50 percent reaction stage).

For detailed design work, the basic profile loss factor for incompressible flow is modified for Reynolds and Mach number (i.e., compressibility) effects. Shock losses with supersonic inlet flows are accounted for by an additional term. Off-design deviations can be included by an additional loss term based on the incidence angle deviation. A high-level discussion of these aspects of stage efficiency calculation can be found in [3]. For more information, references cited in that paper can be consulted.

A bottom-up approach to calculating design or off-design compressor efficiency via first principles (i.e., by stacking loss coefficients from detailed design data) is not practical. For the type of analysis covered in this book, all we need is a representative number so that we can calculate the compressor power consumption and have an idea about the temperature of the hot gas path cooling flow, which is typically extracted from the compressor discharge for turbine stage 1 components.

In addition to Equation 11.1, another useful correlation for estimating axial compressor polytropic efficiency as a function of stage loading is given by Walsh and Fletcher ([8] cited in Chapter 2). For large, advanced technology engines, the curve from chart 5.1 in the cited reference is represented by the following formula:

$$\eta_p = 93.4605 + 2.71 \cdot \left(1 - e^{2.147673\psi}\right). \tag{11.33}$$

For the gas turbines in Table 11.5, compressor efficiencies estimated with Equations 11.1 and 11.33 are compared in Table 11.9. Note that cycle PR is a "concrete" piece of information. Stage loading is estimated using basic stage geometry correlations. Nevertheless, it looks like Equation 11.33 is somewhat "obsolete" for most advanced F- and H-class compressors (i.e., it underestimates their efficiency, which should be somewhere in between 92 and 93 percent – maybe a bit higher).

Table 11.9 Compressor polytropic efficiencies

		E.03	F.04	F.05	F.06	HA.02
PR		13.0	16.7	18.4	22.1	23.1
ψ		0.30	0.28	0.35	0.36	0.34
Equation 11.1 ($\eta_{p,max} = 93\%$)	%	91.6	91.5	91.5	91.4	91.3
Equation 11.1 ($\eta_{p,max} = 94\%$)	%	92.6	92.5	92.5	92.4	92.3
Equation 11.33	%	91.0	91.2	90.4	90.3	90.5

11.1.4 Transonic Compressors

As we have seen in the preceding chapters and sections of this chapter, heavy-duty industrial gas turbines have been growing steadily in size and efficiency. This puts the onus on the axial compressors to improve in efficiency with increasing airflow and PR.

Early-generation F–class gas turbines had 18-stage compressors with PRs of about 16–17. Today's H–class gas turbines achieve PRs of 20 or above with only 14 stages. Without the advanced design, more than 20 stages would be requisite to achieve the same overall PR or higher stage loading, which is detrimental to efficiency. Considering that the axial compressor constitutes a large part of the flange-to-flange gas turbine, in terms of length and weight, compact size (i.e., lower number of stages) without sacrificing efficiency is of prime importance. Indeed, a cycle PR of 24 with only 12 stages is already in the works (see Section 21.1).

Axial compressor design improvements used to flow from aircraft engines, which are subject to frontal area and weight constraints that do not constitute a concern for the stationary gas turbines used in electric power generation. Ever-increasing fuel costs force aircraft engine designers to strive for high efficiencies within those constraints. This led to the development of highly loaded, *transonic* compressors. These advanced compressor designs eventually found their way into the compressors of large, heavy-duty industrial gas turbines.

The mass continuity equation

$$\dot{m} = \rho V A$$

clearly illustrates the need for increasing compressor inlet velocities (V) with increasing mass flow rates. Due to the constraints placed on the inlet area by centrifugal forces acting on stage 1 rotor blades and the shaft diameter (defined by the torque transmitted to the generator and the journal bearing size), the design space for accommodation of the mass flow rate increase via increased A only is limited.

Rewriting the mass continuity equation as

$$\dot{m} = p\pi\sqrt{\frac{\gamma}{RT}} Ma \cdot r_t^2 \left(1 - \upsilon^2\right), \tag{11.34}$$

where υ is the hub–tip ratio, we can combine/compare it with the maximum stress equation, Equation 11.20, which is rewritten below for convenience

Table 11.10 Compressor inlet Mach numbers

		E.03	F.04	F.05	F.06	HA.02
PR		13	16.7	18.4	22.1	23.1
MAIR	lb/s	638.2	990.4	1,161.3	1,282.3	1,502.9
	kg/s	289.5	449.2	526.8	581.6	681.7
Ma		0.3	0.5	0.6	0.6	0.7

$$\sigma_{max} = C\frac{\rho_b}{2}(2\pi N r_t)^2(1-\upsilon^2),$$

where C is the correction factor from Equation 11.21. Thus, Equation 11.34 becomes

$$\dot{m} = \frac{p}{2\pi}\sqrt{\frac{\gamma}{RT}}Ma\cdot\frac{\sigma_{max}}{C\rho_b N^2}. \qquad (11.35)$$

(Note that p and T in Equation 11.35 are *static* values at the stage 1 inlet.) Thus, for fixed speed (i.e., N is 3,000 or 3,600 rpm) and inlet conditions, material properties (via σ_{max}) dictate that compressor airflow is proportional to the inlet Mach number. In other words,

$$\dot{m} = K\cdot Ma, \qquad (11.36)$$

where the proportionality factor K is about 950 kg/s for ISO ambient conditions and 3,600 rpm with maximum allowable stress of ~250 MPa (assuming $C \approx 0.5$). Consequently, using Equation 11.36, compressor inlet axial Mach numbers are estimated as shown in Table 11.10, which indicates that, for most modern axial compressors, the inlet Mach number is about 0.6–0.7.

For 59°F (15°C) ambient, we can calculate the static temperature at the compressor inlet with $Ma = 0.6$ using Equation 3.7a, i.e.,

$$T_0 = T\left(1+\frac{\gamma-1}{2}Ma^2\right),$$

$$T = \frac{288}{1+0.2\cdot 0.6^2} - 273 \approx -4.4°C(24.2°F).$$

(In passing, this is why compressor inlet heating is very important when ambient temperatures drop to about 40–45°F. It becomes *very* cold in there.) Sound speed at the calculated static temperature is

$$a = \sqrt{\gamma RT} = \sqrt{1.4\cdot 286.7\cdot 268.8} = 328.4\,\text{m/s}\ (1{,}085\,\text{ft/s}).$$

Comparing with the estimated values in Figure 11.2, it is clear that stage 1 rotor tip speeds easily become supersonic.

Using the velocity diagrams in Figure 11.1, the stage 1 rotor inlet relative velocity can be calculated as

$$V_{R,1} = \sqrt{U^2 + V_1^2\left(1 - \frac{2U}{V_1}\sin\alpha\right)} \text{ or} \tag{11.37a}$$

$$Ma_{R,1} = \sqrt{Ma_U^2 + Ma_1^2\left(1 - \frac{2Ma_U}{Ma_1}\sin\alpha\right)}. \tag{11.37b}$$

Typical values for $Ma_{R,1}$ are calculated for different values of α (which is controlled by the IGV position) using Equation 11.37b and are listed in Table 11.11.

From Equation 11.13, rewritten below for convenience, selecting α is equivalent to setting the degree of reaction, R.

$$\tan\alpha_1 = \frac{1}{2\phi}\left(2(1-R) - \frac{\psi}{\lambda}\right)$$

For the ψ–ϕ values in Table 11.5 with R = 0.5, absolute inlet angle values calculated from Equation 11.13 are listed in Table 11.12. For the advanced F- and H-class gas turbine axial compressors with $Ma_U \approx 1.1$ and $Ma_1 \approx 0.7$, judging from the numbers in Table 11.11, the stage 1 rotor inlet relative Mach numbers will definitely be high enough to result in supersonic flow at some point on the suction side of the rotor blade.

Indeed, in most advanced axial compressors, the first 1–3 stages are "transonic," with the remaining 10–15 being subsonic stages. The term "transonic" is a contraction of "trans-sonic" and refers to the fact that, on the suction side of the rotor in a transonic stage, the fluid flow Mach number accelerates to supersonic and, (ideally) avoiding a

Table 11.11 Stage 1 rotor inlet relative Mach numbers

Ma_1		0.50	0.60	0.50	0.60
Ma_U			0.90		1.10
α	0	1.03	1.08	1.21	1.25
	10	0.95	0.99	1.13	1.16
	20	0.87	0.89	1.04	1.06
	30	0.78	0.79	0.95	0.95

Table 11.12 Compressor inlet angles

R = 0.5					
ψ	0.30	0.28	0.35	0.36	0.34
ϕ	0.76	0.66	0.70	0.69	0.66
R	0.5	0.5	0.5	0.5	0.5
α_1	23.1	26.9	22.8	22.7	24.4
R = 0.8					
ψ	0.30	0.28	0.35	0.36	0.34
ϕ	0.76	0.66	0.70	0.69	0.66
R	0.8	0.8	0.8	0.8	0.8
α_1	1.8	3.1	–0.5	–1.0	0

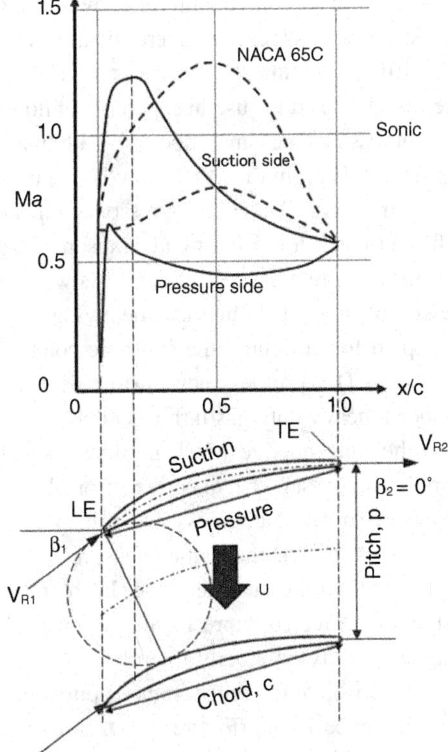

Figure 11.7 Comparison of CDA and NACA profiles. LE = leading edge; TE = trailing edge.

shock wave, decelerates (i.e., diffuses) in a subsonic regime. In other words, the flow Mach number passes through M*a* = 1.0 (sonic).

Controlling the deceleration on the suction side of the rotor blade is of prime importance to prevent BL separation and minimize profile losses. This is not possible with conventional NACA-65 airfoil profiles (originally developed for aircraft wings), which are used in subsonic stages, and led to the development of CDAs. In order to understand the difference, please refer to Figure 11.7. As shown by the Mach number distribution in the upper part of the figure,

1. There is a continuous acceleration from the leading edge of the blade from a very low value to a peak Mach number (relative to the rotating blade) near the point where the BL transitions from laminar to turbulent.
2. Said peak is limited to 1.3 to avoid BL separation, which can be induced by shock wave–BL interaction (at off-design conditions when a shock wave develops).
3. After reaching the peak Mach number, the flow decelerates continuously, passing through M*a* = 1 without a shock wave, to the trailing edge, while maintaining a turbulent BL (no separation and low skin friction).
4. The Mach number distribution on the pressure surface is nearly constant and subsonic.

The introduction of transonic stages in axial compressors of heavy-duty gas turbines goes back to the 1980s [8,9]. The existence of shockless supercritical flows was demonstrated by Whitcomb et al. in 1965 [10]. At the time, the goal was to develop supersonic aircraft wing sections. CDAs were first designed for use in supercritical flow cascades in the 1970s. A comprehensive review of loss mechanisms associated with transonic flow in compressors was provided by Kerrebrock [11]. In the 1980s, development of CDAs for shockless transonic flow regimes in multistage, axial compressor applications was in progress [12]. CDA design was first utilized in the front-end, transonic stages of axial compressors in the F–class gas turbines. Starting in the 1990s, CDAs were also used for the design of the entire compressor blading [13]. In most heavy-duty gas turbines, NACA-65 or CDA profiles are adopted for designing the subsonic compressor stages. Recently, in addition to conventional CDA profiles, new airfoil designs have been developed for applications in advanced, heavy-duty gas turbines [14].

In the absence of CDAs, when the inlet relative Mach number exceeds a critical value, Ma_{crit}, a shock wave occurs in the passage between the rotor blades. Using the shock adiabat correlations [15], it can be shown that, across a stationary shock front in a supersonic flow (upstream of the shock front), the static pressure and temperature increase, whereas the fluid velocity and total pressure decrease. In the reference frame of the rotor, however, this gas dynamic effect is more than compensated for by the diffusion of the fluid flow (via higher density). Velocity triangles of a transonic stage show that the turning of the flow in the absolute reference frame is purely a result of the flow slowing down in the relative reference frame (Figure 11.8). In comparison, in a subsonic rotor, work input to the flow is achieved by turning the flow in both the relative *and* the absolute frames of reference.

Transonic (at the inlet, $Ma > Ma_{crit}$) or supersonic (at the inlet, $Ma > 1$) stages are made from very thin rotor blades in order to reduce the flow blockage. Typically, the thickness-to-chord ratio of the blades is only a few percent. Furthermore, in order to reduce the peak Mach number on the blade surface, the blades have very low camber with only a few degrees of turning (i.e., $\beta_2 \approx \beta_1$). Thus, blade cross-sections for transonic stages are special designs that are not based on standard airfoils such as NACA-65. Blades for high subsonic stages are constructed using *double circular arc* or *multi-circular arc* (MCA) profiles. For compressor stages with transonic and supersonic inlets (based on V_R), so-called S profiles with very sharp leading edges are used. They are generated using MCAs with very small curvatures. As shown in Figure 11.8, the blade profile is slightly concave in order to reduce the flow area and slow down the supersonic flow Mach number to about 1.2 (to minimize shock losses) prior to the normal shock at the entry to the blade passage.[4] The reader is referred to chapters 10 and 16 of the excellent book by Schobeiri ([3] cited in Chapter 10) for rigorous details of this subject. In effect, transonic rotor blades resemble flat plates near the tip section. Another feature of such blades is the low aspect ratio (i.e., relatively long chord length compared to the blade height). This ensures flutter-free operation and helps with the stall delay at low speeds. (This was touched upon briefly in Section 10.3.5 on blade stresses in the context of axial turbines.)

[4] In compressible flow regimes, subsonic flows decelerate (accelerate) in a diverging (converging) channel, whereas supersonic flows decelerate (accelerate) in a converging (diverging) channel [13].

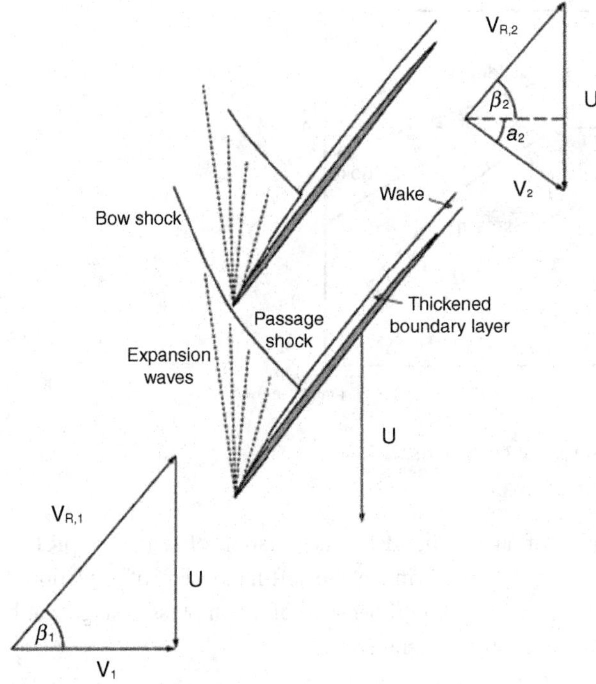

Figure 11.8 Transonic stage velocity diagrams.

11.2 Operability Considerations

The compressor-stage loading parameter, ψ, also known as the temperature rise coefficient, which is given by Equation 11.2, can be related to the flow coefficient, ϕ, for an ideal "repeating" stage with the following assumptions (refer to Figure 11.1):

$$\alpha_3 \approx \alpha_1$$
$$V_3 \approx V_1\text{,}$$

as

$$\psi = 1 - \phi(\tan\alpha_1 + \tan\beta_2), \tag{11.38a}$$

$$\psi = 1 - \phi\kappa. \tag{11.38b}$$

Using the isentropic p–T correlation for the compressor stage and expanding it by means of the binomial theorem (since $\Delta T/T_{in} \ll 1$), one obtains

$$\psi_p = \frac{\Delta p_t}{\rho U^2} = \eta_s\psi, \tag{11.39a}$$

$$\psi_p = \frac{\Delta p_t}{\rho U^2} = \eta_s(1 - \phi\kappa), \tag{11.39b}$$

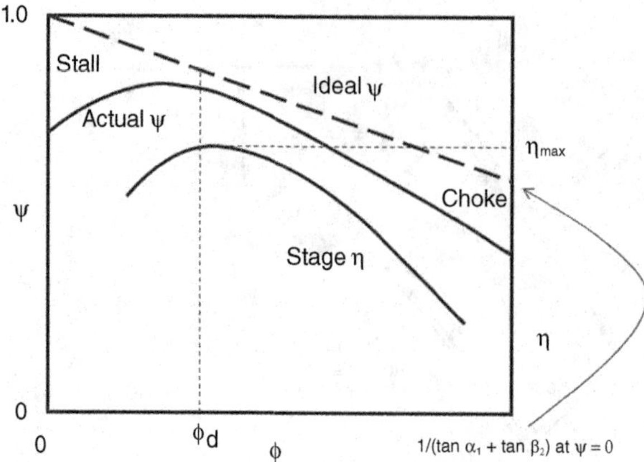

Figure 11.9 Compressor stage characteristic.

where ρ is the density of air via the ideal gas equation of state, $p = \rho RT$, and η_s is the stage isentropic efficiency. The term on the left-hand side of Equation 11.39 is the pressure coefficient, ψ_p. Thus, the performance of a compressor stage can be expressed in terms of the following three parameters:

1. Stage loading or temperature rise coefficient, ψ;
2. Flow coefficient, ϕ;
3. Pressure coefficient, ψ_p, or stage isentropic efficiency, η_s.

When the flow coefficient is close to zero (i.e., the axial velocity is close to zero, which means the flow is essentially stalled), stage loading is close to unity. From the stage geometry in Figure 11.1 (and in analogy to the turbine stage geometry in Figure 10.3 of Chapter 10) and the derivation of Equation 11.2 above, $\psi = 1$ means that the change in the tangential gas velocity is exactly equal to the peripheral speed of the rotor at the pitch diameter. This is equivalent to excessive diffusion in the stage (because DF increases with increasing ψ – see Equation 11.19), which is detrimental to stage efficiency and stability. The correlation between ψ and ϕ is shown conceptually in Figure 11.9. The pressure coefficient is not depicted separately, but since it is a product of the stage loading coefficient and stage efficiency, it is qualitatively similar to the "actual ψ" curve. For numerical values of ψ_p and η_s as a function of ϕ from test data, see figure 223 in [16].

The stage characteristic in Figure 11.9 is useful in the qualitative evaluation of the behavior of a multi-stage compressor in flow speed conditions significantly different from design. One way to do this is to imagine several *identical* stages in series, all with the design point corresponding to ϕ_d in Figure 11.9. This is a technique known as "stage stacking." Now consider the following sequence of events:

1. Compressor inlet flow decreases.
2. Thus, ϕ_1 (stage 1) decreases (i.e., $\phi_1 < \phi_d$).
3. According to Equation 11.39b, stage pressure rise increases.

4. Consequently, density at entry to stage 2 increases.
5. Per continuity equation, stage 2 inlet velocity decreases.
6. Therefore, $\phi_2 < \phi_1$.
7. This sequence repeats itself so that, at some stage n, ϕ_n will be so small that the flow *stalls*.

On the flip side:

1. Compressor inlet flow increases.
2. Thus, ϕ_1 (stage 1) increases (i.e., $\phi_1 > \phi_d$).
3. According to Equation 11.39b, stage pressure rise decreases.
4. Consequently, density at entry to stage 2 decreases.
5. Per continuity equation, stage 2 inlet velocity increases.
6. Therefore $\phi_2 > \phi_1$.
7. This sequence repeats itself so that, at some stage n, ϕ_n will be so large that the flow *chokes*.

For the axial compressor of an industrial gas turbine, the full envelope of operating characteristics of the entire multi-stage component is distilled into a "performance map," which is a collection of non-dimensional pressure rise flow curves for different values of rotational speed (also non-dimensional). This will be discussed in detail below.

11.2.1 Compressor Performance Map

At constant rotational speed, the gas turbine is a constant volumetric flow machine. In particular, using fundamental aerothermodynamics of an axial compressor, it can be shown that the velocity triangles for a given stage as depicted in Figure 11.1 are similar if the *non-dimensional* rotational speed parameter, v, as defined by Equation 11.40, is constant.

$$v = \left(\frac{2\pi N}{60}\right)\frac{D}{\sqrt{\frac{R_{unv}}{MW}T_1}}, \qquad (11.40)$$

where

- N is the rotational speed in rpm (revolutions per minute);
- R_{unv} is the universal gas constant (8,314 J/kmol-K or 1,545 ft-lbf/lbmol-R);
- MW is the molecular weight of air (kg/kmol or lb/lbmol);
- D is the diameter of the stage (stator $\approx$ rotor; m or ft);
- T_1 is the stage inlet stagnation temperature (K or R).

(The reader can easily verify that v is indeed non-dimensional – it is easier to do so in SI units.) It can be shown (e.g., see Saravanamutto et al.) that v is equivalent to (*not* equal to) the *Mach number* at the tip of the stage rotor.

In practice, the dimensional version of v is used to characterize the entire gas turbine compressor and is referred to as the *reduced* or *corrected* rotational speed, i.e.,

$$\nu = \frac{N}{\sqrt{\frac{R_{unv}}{MW} T_1}}, \tag{11.41}$$

with T_1 denoting the compressor inlet (stagnation) temperature.

The second parameter of interest is the non-dimensional mass flow, which is given by

$$\mu = \dot{m}_1 \frac{\sqrt{\frac{R_{unv}}{MW} T_1}}{D^2 p_1}, \tag{11.42}$$

where

- $\dot{m}_1$ is the stage mass flow rate (kg/s or lb/s);
- p_1 is the stage inlet stagnation pressure (bar or psi).

(Once again, it is left to the reader to verify that μ is indeed non-dimensional.) It can be shown that μ is equivalent to the flow Mach number at the stage inlet (e.g., see Saravanamuttoo et al.). In practice, the dimensional version of μ is used to characterize the entire gas turbine compressor and is referred to as the *reduced* or *corrected* mass flow, i.e.,

$$\mu = \dot{m}_1 \frac{\sqrt{\frac{R_{unv}}{MW} T_1}}{p_1}, \tag{11.43}$$

with all parameters evaluated at the compressor inlet.

For the same fluid (i.e., air in this case) and constant reduced/corrected speed, it can be shown that the reduced/corrected mass flow is also constant. Consequently, a non-dimensional, *characteristic* compressor performance map can be constructed as shown in Figure 11.10. This is a typical compressor performance map based on figure 8 of [2] cited in Chapter 2. Non-dimensional reduced parameters are denoted by the superscript "*" and found via division by their respective values at a *reference point.* The y-axis of the map is the parameter π^*, which is found by dividing the compressor PR, which itself is dimensionless by definition, by its value at the reference point. The operating lines correspond to constant (normalized) turbine inlet temperature, τ_3, which is the ratio of T_3 to T_1. Thus, at the reference point

$$\nu^* = \mu^* = \pi^* = 1.0,$$

whereas the value of τ_3 varies from about 5.0 to 6.5 (based on ISO inlet conditions; i.e., T_1 = 59°F) for older E–class through to state-of-the-art H/J–class gas turbines.

It should be emphasized that the aforementioned reference point is *not* the same as the compressor *design point.* The reference point is typically chosen as the full load performance point at ISO conditions. The compressor design point (i.e., where the stage-by-stage aerodynamic design of the compressor [i.e., the velocity triangles] is laid out) is typically at $\nu^* < 1$ (e.g., 0.95) and $\mu^* < 1$.

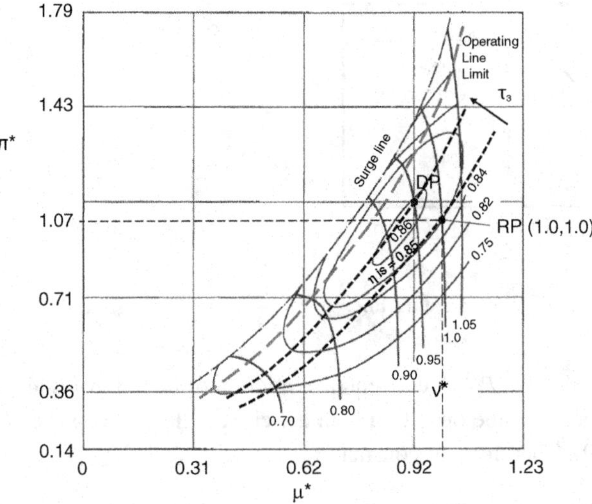

Figure 11.10 Compressor performance map. DP = design point; RP = reference point.

For the explanation of the constant τ_3 lines in Figure 11.10, one has to consider the following qualitative facts governing the operation of compressors and expanders (turbines):

- Conceptually, if one considers the compressor as a turbomachine inside a long pipe, it operates against a *downstream pressure resistance.*
- Said resistance is created by a given fluid flow trying to "squeeze" through a restriction (e.g., orifice) in the pipe.
- The expander can be considered as a *choked nozzle* inside the aforementioned pipe downstream of the compressor.
- Thus, the inlet of the expander (choked nozzle) can be considered as the restriction (orifice) creating the pressure resistance against which the compressor operates.

From the basic gas dynamics, then, at the inlet of the turbine (station 3), one obtains:

$$\frac{\dot{m}_3\sqrt{\frac{R_{unv}}{MW}T_3}}{p_3A} = f\left(\frac{p_3}{p_4}, \gamma\right), \tag{11.44}$$

where A is the stage 1 nozzle *throat area* (note that, for our purposes, we are not interested in the exact form of the right-hand side of the equation). For a given back pressure p_4 (ultimately determined by the exhaust diffuser design and the ambient pressure for multi-stage turbines), beyond a certain value of inlet gas flow rate, p_3 will stay constant at a value denoted by "*" such that

$$\frac{p_3^*}{p_4}$$

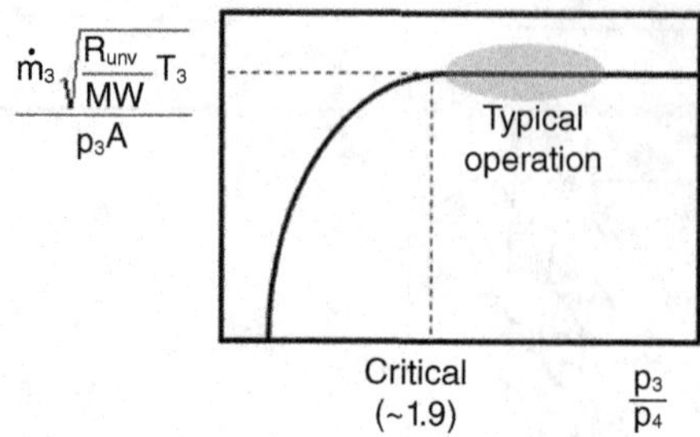

Figure 11.11 Conceptual turbine operating line.

is referred to as the *critical PR*. A conceptual depiction of Equation 11.44 is provided in Figure 11.11. Typical turbine operation is in the choked flow region (i.e., turbine stage 1 nozzle PR > 1.9). Therefore, at constant T_3,

$$\dot{m}_3 \propto p_3,$$

so that, since the turbine nozzle throat area is fixed by design, Equation 11.44 becomes

$$\frac{\dot{m}_3\sqrt{\frac{R_{unv}}{MW}T_3}}{p_3} = \text{const.} \tag{11.45}$$

In essence, Equation 11.45 is a quantification of the turbine's "swallowing capacity," and the term on the left-hand side is referred to as the *turbine flow parameter*. As such, it is the backbone of off-design performance calculations. For quick estimates, it can be simplified as

$$\frac{\dot{m}_3\sqrt{T_3}}{p_3} = \text{const.}$$

From the perspective of the compressor, since $p_2 \approx p_3$ (except for 5–6 percent combustor pressure drop) and $\dot{m}_1 \propto \dot{m}_3$ (accounting for fuel flow and chargeable and nonchargeable cooling air flows), for a given value of τ_3,

$$\dot{m}_1 \propto p_2.$$

This suggests that compressor PR is proportional to compressor inlet airflow. This linear correlation defines the "operating line" on the compressor map.

For a typical heavy-duty industrial gas turbine (50 or 60 Hz), physical speed N is fixed (3,000 or 3,600 rpm). Hence, variation in v* is caused by compressor inlet temperature, which is about the same as the ambient temperature in the absence of inlet conditioning (e.g., evaporative cooling, mechanical chilling, or inlet heating via compressor discharge bleed air). Inlet airflow will increase or decrease with decreasing or increasing ambient temperature, respectively, i.e.,

$$\dot{m}_1 \propto \frac{1}{T_1}.$$

Furthermore, from Figure 11.10, at constant τ_3, compressor PR will go up or down with inlet airflow. Since turbine PR and gas flow also vary in lockstep with corresponding compressor parameters, it follows that the net gas turbine output is proportional to compressor inlet airflow.

The reduced speed range in the map in Figure 11.10 is 0.70–1.05. Since the physical speed is fixed, the change in reduced/corrected speed (except during startup and shutdown) is caused by (i) change in grid frequency (which is an "upset" situation) and (ii) change in ambient air temperature. Grid code requirements (see Section 19.5 for more on this) stipulate that gas turbines must be able to stay online (i.e., synchronized to the grid and generating power), at least for a limited time, between 94 and 103 percent of grid frequency (i.e., between 47 and 51.5 Hz for a 50-Hz grid). Gas turbines are also designed to operate in ambient temperatures ranging from –40°C (–40°F) to as high as 50°C (122°F). Consequently, using Equation 11.41,

- At 50°C ambient and 94 percent grid frequency, reduced speed is ~0.89.
- At –40°C ambient and 103 percent grid frequency, reduced speed is 1.15.

Grid overfrequency events imply more generation than demand, which is extremely unlikely to happen at such a low ambient temperature. Therefore, the *maximum* reduced speed range of a heavy-duty industrial gas turbine covering all possible operating scenarios is 0.89–1.08. The reduced speed range during normal operation is probably closer to 0.95–1.05.

As shown in Figure 11.10, to ensure safe operation, there is an *operating limit line* (OLL) about 15–20 percent to the right of and below the surge line. As expected, each reduced speed line is qualitatively similar to the stage characteristic shown in Figure 11.9. This is so because π^* is related to ψ_p and μ^* is related to ϕ. However, there are significant *quantitative* differences. For instance, at constant reduced speed, the range of reduced mass flow is very narrow, becoming almost vertical at the higher reduced speeds.

For a given compressor, developing a comprehensive performance map requires extensive tests to obtain data points at each combination of flow, speed, and PR. In order to be able to do this in the factory before a unit is shipped to, installed, and operated on site, a variable-speed driver is needed. Examples are variable-speed electric motors or diesel engines. For the compressors of modern gas turbines rated at 300 MWe, this is practically impossible without using a gas turbine of comparable rating. One OEM uses a full-size gas turbine, which is connected to a water break. As another example, consider the full-speed, full-load validation test facility in GE's gas turbine factory in Greenville, SC, USA, which is described in detail in [2]. The drive train configuration of the test facility consists of a starter motor, a 58-MWe electric motor (to supplement the driver gas turbine), an electric drive torque converter, a gearbox, a load compressor (test article), and the driver gas turbine. The second compressor is either the test article or the "load" if the gas turbine itself is the test article. The system is thus capable of fully mapping the compressor, especially to the surge line and beyond, at full-scale, steady-state operating conditions. The inlet throttle provides flow measurement accuracy, especially at low flows with an ASME flow nozzle and 36 in. valve.

The exit throttle provides the ability to back-pressure the compressor to achieve any desired operating point.

The compressor map in Figure 11.10 is for the entire unit comprising multiple stages (i.e., 14 in modern H–class gas turbines, 17–18 in older E/F–class units). At low speeds, due to reduced PR and density, increased flow velocity in the rear stages (via mass continuity and decreasing annulus area) and choking determine the mass flow rate. At higher speeds, higher density allows the rear stages to pass all the flow coming from the front stages. Therefore, choking at the inlet of the compressor limits the mass flow rate. This is the key mechanism leading to nearly vertical speed lines. For thorough discussions of axial compressor operability considerations, the reader is referred to chapter 7 of [12] cited in Chapter 2, chapter 5 of Saravanamuttoo et al. ([3] cited in Chapter 2), section 5.2 in Walsh and Fletcher ([8] cited in Chapter 2), and chapter 8 of Wilson and Korakianitis ([2] cited in Chapter 2).

For off-design performance calculations, a "lookup table" can be created by reading points off of the performance map in Figure 11.10. One technique of discretizing map data for easy lookup is described in section 5.2.5 in Walsh and Fletcher. For normal operation of heavy-duty industrial gas turbines, as indicated earlier, only a limited number of reduced speed line data is sufficient. The problem is in accounting for the effects of IGV and variable VSV scheduling. In order to ensure stable operation of modern high-PR and high-flow compressors with highly loaded stages in a relatively thin strip of map bounded by choke and surge (or, rather, OLL) lines, the flow angles of the three to four front stages are modulated by the gas turbine controller based on site ambient and loading conditions. This is so because at lower reduced speeds (i.e., high ambients), surging is primarily caused by stalling of the front stages of the compressor.

A map similar to that shown in Figure 11.10 is only valid for a specific setting of the IGVs (e.g., fully open). Ideally, one needs similar maps for each setting of the IGVs. Apart from increasing the magnitude of already tedious data entry and management tasks, this is practically impossible unless one is actually involved in compressor design for one of the major OEMs and can readily access requisite information. The prognosis for an independent researcher or analyst is not encouraging. Detailed maps for state-of-the-art, heavy-duty or aeroderivative gas turbine compressors in the open literature are essentially non-existent. On top of that, the availability of realistic conceptual maps, which can be used as *proxies* for conceptual studies, is also scant and limited to a single IGV position.

Another problem facing the engineer who wants to model an axial gas turbine compressor for conceptual studies or reverse engineering is interstage bleeds to supply turbine hot gas path cooling air. This precludes modeling the compressor as a single unit. The common approach to this problem is dividing the compressor into three (or four) sections and treating each section as a separate compressor with a performance map of its own. Obviously, this adds significantly to the data research, modeling, and computing burden. Commercial heat and mass balance software such as Thermoflex reduce the modeling burden by enabling the user to model the compressor in several sections in series and to specify the compressor map information for each section separately.

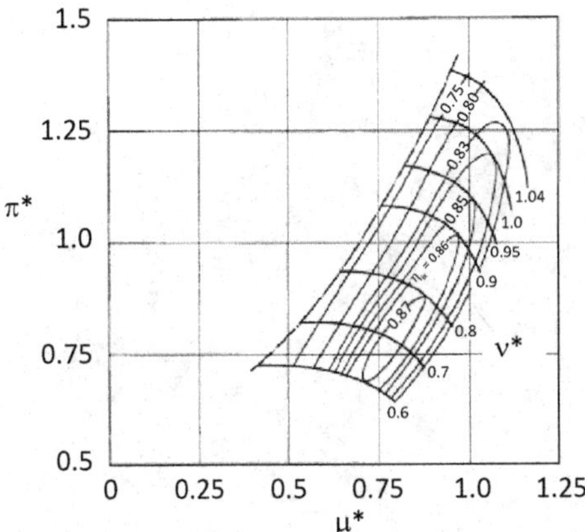

Figure 11.12 Performance map for low-PR sections.

The default axial compressor map in Thermoflex is similar to that in figure 5.41 of Saravanamuttoo et al. and should be a good proxy for sections with a PR of 3–4. Another alternative is the performance map in figure 7 (p. 55) of the book by Münzberg and Kurzke. The performance map in Figure 11.10 is a reasonably good proxy for sections with a PR of 7–8. For low-PR sections (i.e., PRs of about 2), the performance map in figure 10 (p. 58) of the book by Münzberg and Kurzke is recommended. Since the cited reference is a book that is long out of print (and in German as well, which limits the potential readership), a non-dimensional version of it is reproduced in Figure 11.12.

When modeling a gas turbine axial compressor in three or more sections, attention must be paid to the definition of reduced/corrected and normalized parameters for the second and the third sections. This is so because their reference point flows, inlet temperatures, and inlet pressures are different from those for the whole compressor (gas turbine) and the first section. When using a commercial software like Thermoflex, the program takes care of that automatically, which is another reason for using such software (when available and/or affordable, of course) in lieu of "homemade" codes.

Incorporation of IGV modulation into the model, even with a software like Thermoflex, is a chore in and of itself. Taking care of the 100 percent IGV position by using generic maps as described above is one thing. What to do about 90 or 70 percent open IGV cases? All modern gas turbines, without exception, are equipped at least with variable IGVs upstream of stage 1, which can reduce the compressor mass flow rate down to about 70 percent of its full load value (fully closed or 0 percent open position). Furthermore, state-of-the-art, advanced-class gas turbines with very large flows and high PRs contain, in addition to the IGVs, three or four stages of VSVs, which can reduce the mass flow rate down to 50 percent. This is extremely advantageous for part load operation because it ensures that combustor fuel–air ratios stay nearly constant to maintain low NOx emissions even at low loads. The second significant benefit is the high exhaust temperature (due to lower PR), which is

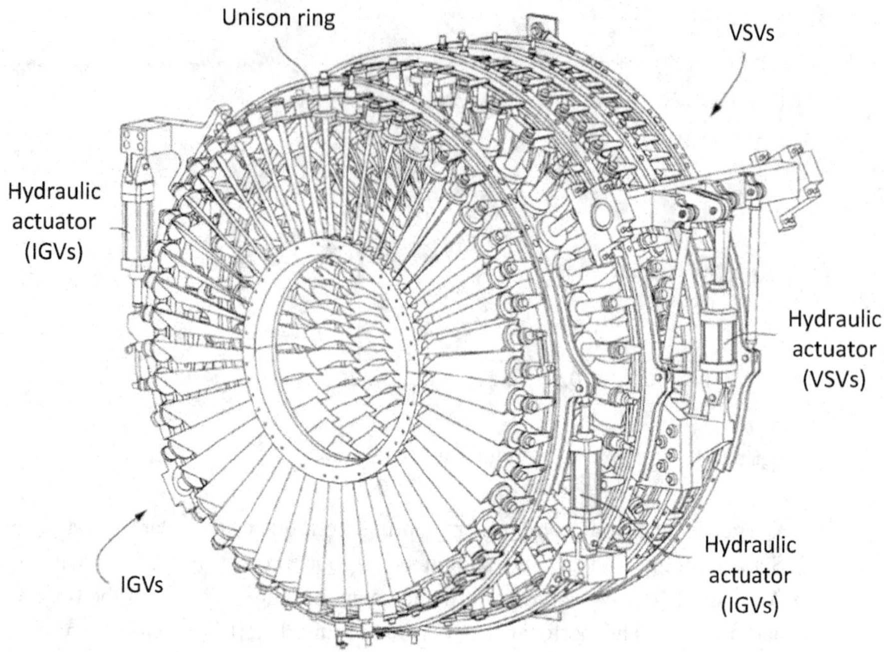

Figure 11.13 Hydraulically activated axial compressor IGVs and VSVs (three stages).

beneficial to the bottoming-cycle performance in the combined-cycle configuration. Variable vanes in a given row are operated by hydraulic or electric actuators in unison, either each row independently or all rows in unison (see Figure 11.13 for an example[5]).

Since finding multiple performance maps for multiple IGV positions is next to impossible, the best approach to modeling is using an approximation. This can be illustrated by the compressor map in Figure 11.14 (drawn after figure 7-40 in chapter 7 of [12] in Chapter 2).

As shown in Figure 11.14, as a first approximation, when IGVs are fully closed (i.e., 0 percent open), the normalized flow parameter at a given speed is estimated by the formula

$$\frac{\mu_C^*}{\mu_O^*} = (1.55 - 0.85v^*), \tag{11.46}$$

where subscript "O" designates the normalized flow parameter read from the 100 percent open IGV map (see Figure 11.14). At IGV openings between 0 and 100 percent, flow scaling of Equation of 11.46 is prorated as follows:

5 From US Patent Publication 9,068,470 B2, "Independently-controlled gas turbine inlet guide vanes and variable stator vanes," by Mills and Ruddy (June 30, 2015).

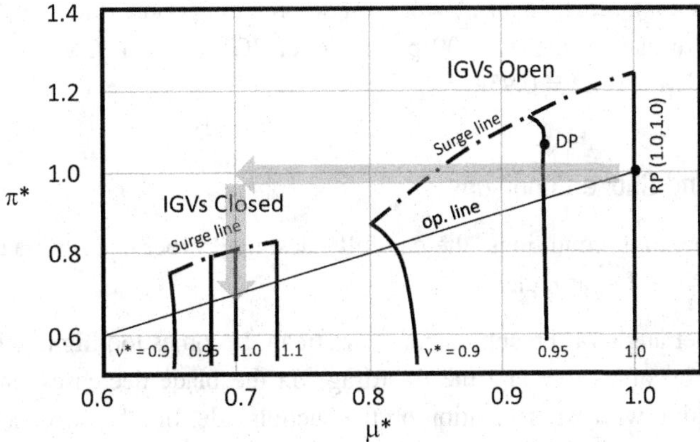

Figure 11.14 Compressor map with IGVs fully open and fully closed. RP = reference point; op. line = operating line.

$$\frac{\mu_C^*}{\mu_O^*} = C_{IGV}, \tag{11.47a}$$

$$C_{IGV} = 1 + (K - 1)\left(1 - \frac{Y_{IGV}}{100}\right). \tag{11.47b}$$

In Equation 11.47b, Y_{IGV} is the percentage open setting of the IGVs (e.g., 70 for 70 percent) [17]. The constant K gives the fraction of the 100 percent open IGV flow at a given v*; in other words, K is the value obtained from the right-hand side of Equation 11.46. For example, for v* = 0.9 and IGV 60 percent open,

- From Equation 11.46, fully closed flow scaler is 0.79;
- From Equation 11.47b, with K = 0.79, C_{IGV} is 0.91;
- Thus, from Equation 11.47a, flow scaler at 60 percent IGV is 0.91.

Continuing with the scaling exercise, when IGVs are fully closed (i.e., 0 percent open), normalized PR at a given speed is estimated by the following formula (same subscripts as in Equation 11.46):

$$\frac{\pi_C^*}{\pi_O^*} = 0.62 + \frac{9E+08}{2^{35.31v^*}}. \tag{11.48}$$

For prorating the normalized PR between 0 and 100 percent open IGVs, use Equation 11.47b where K is the value obtained from the right-hand side of Equation 11.48. For example, for v* = 0.9 and IGV 60 percent open,

- From Equation 11.48, fully closed normalized PR scaler is 0.86;
- From Equation 11.47b, with K = 0.86, C_{IGV} is 0.95;
- Thus, normalized PR scaler is 0.95.

Concluding our example, then, at $v^* = 0.9$, for an operating point with $\mu^* = M$ and $\pi^* = P$ from the performance map for 100 percent open IGVs, when IGVs are 60 percent open, $\mu^* = 0.91M$ and $\pi^* = 0.95P$.

11.2.2 Stall, Surge, and Choke Conditions

At different operating conditions, the carefully designed process described in Section 11.1.1 is perturbed. In particular,

- If the inlet angle of the approaching air stream becomes too high, axial absolute velocity becomes low and the lift (drag) on the blade decreases (increases) in combination with BL separation on the suction side. In other words, the airflow "hits" the blade on the pressure side. This phenomenon is referred to as "stall" and is schematically illustrated in Figure 11.15.
- If the inlet angle varies too far in the other direction, axial absolute velocity becomes high and the throat at the inlet of the blade-to-blade passage is "choked." Put another way, the airflow "hits" the blade on the suction side. The lift (drag) on the blade decreases (increases) in this case as well.
- If the velocity of the airflow exceeds a critical value (i.e., $Ma > Ma_{crit}$), the airflow accelerating to pass around the airfoil will become supersonic and a shock wave will result, with turbulent downstream flow and again an increase in drag. This was already discussed in Section 11.1.4.

If the stall and flow blockages via increased (separated) BL thickness are limited to one blade passage, this is referred to as "single-cell rotating stall." The term "cell" refers to the region of flow blockage with vortices downstream of the BL separation point on the suction surface of the stalled blade (see Figure 11.15). The term "rotating" refers to the fact that the separation moves from passage to passage in the direction of the rotation. The reason for that is the distribution of the blocked airflow to the neighboring blade

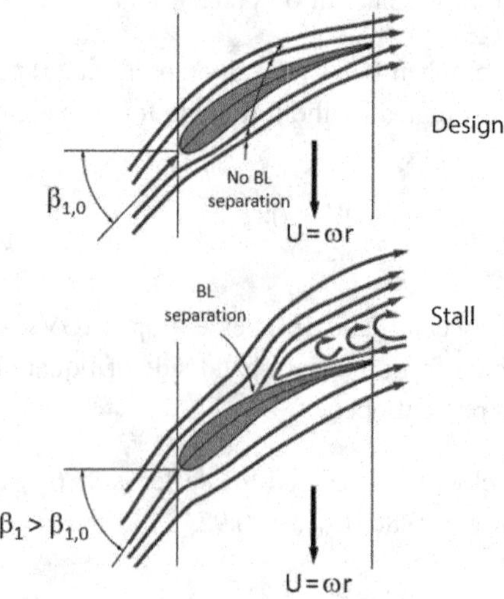

Figure 11.15 Stall phenomenon.

passages as dictated by mass continuity. In conjunction with higher flow, neighboring passages experience higher inlet angles as well. If this increase in the inlet flow angle is high enough, the passage on the side of the rotation direction will stall. The stall cell travels in this direction circumferentially at about 50–70 percent of the rotational speed. Depending on the degree of BL separation, the cell can be limited to the tip or hub region or cover the entire passage. If the stall happens in more than one blade passage, it is referred to as "multiple-cell rotating stall." In essence, stall is a flow disturbance in the *tangential* direction (i.e., as far as the gas turbine airflow is concerned, it is a "localized" disturbance affecting one stage [although it can affect several stages, too] with steady airflow through the machine).

When *all* of the blade passages in the rotor of a compressor stage are stalled, the result is complete flow blockage. In other words, the flow disturbance is now in the *axial* direction. This can quickly lead to a flow reversal due to excessive pressure buildup, which, if the original cause of the rotating stall is still present, can degenerate into a repeating pattern with pulsating flow and violent vibrations, which can destroy the compressor. This phenomenon is known as compressor "surge."

In stationary gas turbines, occurrence of rotating stall phenomena is mostly a startup issue. In fact, an EPRI study of FA-class compressor failures associated with R0 leading edge cracks cited in Section 11.1.2 found that a condition of non-synchronous vibration associated with aerodynamic stall did occur during startup and shutdown. Evidence obtained in EPRI tests indicated that the stall condition could become severe on an apparently random basis, exacerbated by the excitation of the axial vibration mode.

Surge, however, is *not* limited to startup/shutdown events and can happen without being triggered by rotating stall. For example, at low ambient temperatures (i.e., at high corrected speeds) with increased airflow, back pressure imposed by the turbine inlet can lead to a high PR, which may overload the rear stages of the compressor if they are fouled. Similarly, at high ambient temperatures (i.e., at low corrected speeds) with reduced airflow, fouled front stages can be overloaded. In either case, flow reversal and compressor surge with high dynamic forces acting on the blades can happen. Unless the control system detects the surge and counteracts by opening the bleed valves, reducing the load, or tripping the unit, catastrophic damage to the machine is unavoidable.

The problem with the last statement above is the mismatch in timescales of guide vane and bleed valve actuators and rotating stall or surge development. The former is of the order of 200 ms. In comparison, for a rotating stall to develop and morph into a surge takes a few revolutions of the turbine shaft. At 3,000/3,600 rpm, one revolution is completed in about 20 ms; 10 revolutions in 200 ms. Thus, it is very difficult for the turbine controller to detect a precursor of an impending stall/surge event from existing instrumentation and then generate a control signal to operate the IGVs, VSVs, or bleed valves to intervene.

In light of that, the preferred approach to surge control in heavy-duty, industrial gas turbines is to "stay away" from the compressor surge line by a margin of safety. This margin defines the OLL, which sets the upper bound for stable operation of the compressor. The OLL is determined from the factory tests run to characterize the operating of the compressor, which identify the operating conditions for the onset of surge, and a stacking of various instability factors such as machine-to-machine variation, compressor blade tip clearance variations, etc.

As discussed earlier, rotating stall cells can morph into full-blown surges. If a distinct limit line can be determined from the factory test data points corresponding to compressor operation with rotating stall, it can be used as the OLL because it is a precursor of a full-blown surge. The surge or stall margin (as a percentage) is typically calculated as

$$\mathrm{SM} = \left(\frac{\mathrm{PR_{lim}}}{\mathrm{PR_{oll}}} - 1\right) \times 100,$$

where the subscript "lim" designates the limiting PR corresponding to the particular instability (i.e., stall or surge), whereas the subscript "oll" designates the OLL. This correlation is used to determine the surge/stall margin (SM) on a corrected speed line (see Section 11.2.1). A typical value for SM is 15 percent or 0.15. Thus, the controller *prevents* the gas turbine from operating at a condition where, on the particular corrected speed line, corrected flow corresponds to a PR value, which makes the calculated SM less than 0.15.

During normal operation, IGV/VSV schedules ensure that proper SM is maintained. During startup, bleed valves are opened to reduce the compressor loading until the low corrected speed region is passed through. The air blown off through the bleed valves is typically discharged into the gas turbine exhaust. In comparison, in centrifugal process compressors, active control of interstage bleed valves, which recirculate the bled air into the compressor inlet, is the primary means of surge prevention during normal operation at varying load and/or suction conditions.

Surge or stall events during shutdown (verified with vibration readings) and during startup (less likely) can lead to "clashing." Clashing happens when the tip of a (stationary) stage 1 stator vane moves (downward for an observer looking at the compressor from the side) to a point such that it comes into contact with stage 1 (moving) rotor blades near their platform. This typically happens when the clearance between the S1 tip and R1 base disappears due to casing deformation caused by differential cooling and contraction during shutdown. However, severe vibrations caused by a surge event can cause the same clearance disappearance, too. Clashing can ultimately lead to cracks and blade/vane liberation and catastrophic failure.

References

1. Smith, Jr., L. H., Axial Compressor Aerodesign Evolution at General Electric, *Journal of Turbomachinery*, **124** (2002), 321–330.
2. Martin, R., Forry, D., Maier, S., Hansen, C., *GE's Next 7FA Gas Turbine "Test and Validation"* (Schenectady, NY: GE Power & Water, 2011).
3. Benini, E., "Advances in Aerodynamic Design of Gas Turbines Compressors," in *Gas Turbines*, Ed. G. Injeti (London, UK: InTech, 2010).
4. Howell, A. R., *The Present Basis of Axial Flow Compressor Design: Part 1 Cascade Theory and Performance* (Richmond, UK: HM Stationery Office, 1942).
5. White, F. M., *Fluid Mechanics* (New York: McGraw-Hill Book Co., 1979).

6. Denton, J. D., Loss Mechanisms in Turbomachines, *Journal of Turbomachinery*, **115** (1993), 621–656.
7. Hall, D. K., Greitzer, E. M., Tan, C. S., "Performance Limits of Axial Compressors," ASME Paper GT2012-69709, ASME Turbo Expo 2012, June 11–15, 2012, Copenhagen, Denmark.
8. Becker, B., Kwasniewski, M., von Schwerdtner, O., Transonic Compressor Development for Large Industrial Gas Turbines, *ASME Journal of Engineering for Power*, **105** (1983), 417–421.
9. Farkas, F., "The Development of a Multi-stage Heavy-duty Transonic Compressor for Industrial Gas Turbines," ASME Paper 86-GT-91, International Gas Turbine Conference and Exhibit, June 8–12, 1986, Düsseldorf, West Germany.
10. Whitcomb, R. T., Clark, L. R., "An Airfoil Shape for Efficient Flight at Supercritical Mach Numbers," NASA TMX-1109, May 1965.
11. Kerrebrock, J. L., Flow in Transonic Compressors, *AIAA Journal*, **19** (1980), 4–19.
12. Hobbs, D., Weingold, H., Development of Controlled Diffusion Airfoils for Multistage Compressor Application, *ASME Journal of Engineering for Gas Turbines and Power*, **106** (1984), 271–278.
13. Becker, B., Schulenberg, T., Termühlen, H, "The "3A-Series" Gas Turbines With HBR® Combustors," ASME Paper 95-GT-458, ASME 1995 International Gas Turbine and Aero-engine Congress and Exposition, June 5–8, 1995, Houston, TX.
14. Köller, U., Mönig, R., Küsters, B., Schreiber, H.-A, Development of Advanced Compressor Airfoils for Heavy-Duty Gas Turbines – Part I: Design and Optimization, *ASME Journal of Turbomachinery*, **122** (1999), 397–405.
15. Thompson, P. A., *Compressible Fluid Dynamics* (New York: McGraw-Hill, 1988).
16. Robbins, W. H., Dugan, Jr., J. F., "Prediction of Off-Design Performance of Multistage Compressors," in *Aerodynamic Design of Axial-Flow Compressors* (Washington, DC: National Aeronautics and Space Administration, 1965).
17. Nakhamkin, M., Gülen, S. C., "Transient Analysis of the Cascaded Humidified Advanced Gas Turbine (CHAT)," ASME Paper 95-CTP-28, ASME 1995 Cogen-Turbo Power Conference, August 23–25, 1995, Vienna, Austria.

While not directly referenced in the chapter, the references listed below are highly recommended for further reading and learning.

Camp, T. R., Horlock, J. H., An Analytical Model of Axial Compressor Off-Design Performance, *Journal of Turbomachinery*, **116** (1994), 425–434.

Cumpsty, N. A., Some Lessons Learned, *Journal of Turbomachinery*, **132** (2010), 041018.

Dickens, T., Day, I., The Design of Highly Loaded Compressors, *Journal of Turbomachinery*, **133** (2011), 031007.

Hall, D. K., Greitzer, E. M., Tan, C. S., "Performance Limits of Axial Compressor Stages," ASME Paper GT2012-69709, ASME Turbo Expo 2012, June 11–15, 2012, Copenhagen, Denmark.

Kappis, W., "Compressors in Gas Turbine Systems," chapter 4 in [11] in Chapter 2.

NASA Report SP-36, *Aerodynamic Design of Axial Flow Compressors*, (Washington, DC: Scientific and Technical Information Division, NASA, 1965).

In addition to the references listed above, [1,3–5] cited in Chapter 10 also include in-depth coverage of compressor aerothermodynamics.

12 Combustion

Combustion of natural gas with air as the oxidizer is the basic gas turbine heat addition process. Consequently, having a good handle on basic combustion concepts is essential. For a given fuel and fuel–air ratio, calculation of flame temperature from equilibrium chemistry is relatively straightforward. Unfortunately, due to the strongly non-equilibrium nature of actual chemical reactions taking place in the flame zone, combustor design and emissions calculations are based on mostly empirical correlations and experimental data. For comprehensive coverage, the reader should consult specialized treatises on the subject such as [1] and [2]. Relatively brief but practically helpful coverage of the basic equations can be found in the first chapter of [3]. For in-depth coverage of emissions from the gas turbine combustion, the reader is directed to the book edited by Lieuwen and Yang [4]. Similarly, for the subject of combustion instabilities, the book edited by the same researchers is an authoritative source [5].

12.1 Basics

For simple estimates, natural gas can be assumed to be 100 percent methane (CH_4) and air can be approximated by $O_2 + 3.76N_2$ (based on the assumption of 21 percent O_2 and 79 percent N_2 by volume). Thus, *stoichiometric* combustion (that is, all oxygen in the air is used up) of 1 mole of methane (16 lb/lbmol, lower (net) heating value (LHV) of 21,515 Btu/lb) is given by

$$CH_4 + 2\cdot(O_2 + 3.76\cdot N_2) \rightarrow CO_2 + 2\cdot H_2O + 7.52\cdot N_2. \tag{12.1}$$

Based on this simple chemical reaction equation,

1. Stoichiometric combustion of 1 mole of methane requires about 9.5 moles of air (about 275 lb).
2. Stoichiometric air-to-fuel ratio, on a mass basis, is about 17 (i.e., 275/16); consequently, the stoichiometric fuel-to-air ratio is about 0.06.
3. Combustion products include 1 mole (44 lb/lbmol) of CO_2 per mole of CH_4 or 44/16 = 2.75 lb of CO_2 per lb of CH_4.

The *equivalence ratio* is the actual fuel–air ratio to the stoichiometric fuel–air ratio

$$\phi = (f/a)/(f/a)_{stoich}. \tag{12.2a}$$

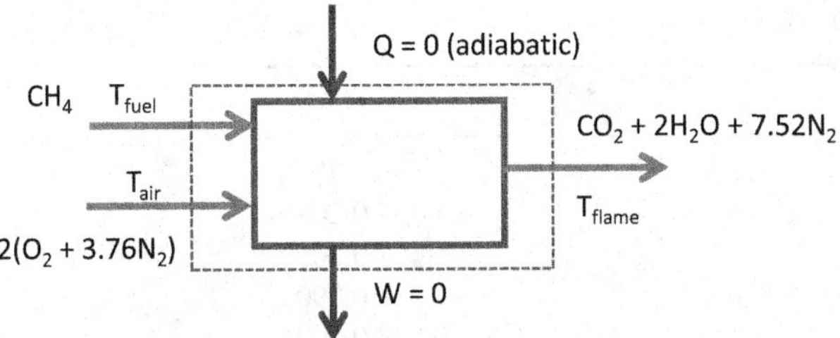

Figure 12.1 Stirred reactor control volume for combustion of methane.

The inverse of the equivalence ratio is the *excess air ratio*, i.e..

$$\lambda = 1/\phi \tag{12.2b}$$

Thus,

$\phi < 1$ $(\lambda > 1) \rightarrow$ *Fuel-Lean* combustion
$\phi > 1$ $(\lambda < 1) \rightarrow$ *Fuel-Rich* combustion

The *flame temperature* is calculated by solving the mass and energy balance problem for the adiabatic combustion reaction shown in Figure 12.1. This is commonly known as a *stirred reactor* model. In other words, the control volume representing the combustor is an idealization where combustion products are back-mixed with reactants so quickly that the reaction zone is distributed uniformly in space. The stirred reactor model is essentially the only practical approach to combustion calculations.

Calculation of the temperature of combustion products via enthalpy balance of reactants (i.e., fuel and air) and products is a straightforward but tedious process. How it can be done in a simple spreadsheet can be found in many references (e.g., see [6]). In any event, today there are many commercially available software tools to perform such mundane calculations on a computer in a fraction of second (e.g., Thermoflex). Nevertheless, a VBA function encapsulating the combustion calculations as described in [6], COMBUST, is provided in Appendix C.

In general, for combustion reaction calculations, the following information is requisite:

1. Air composition
2. Fuel gas composition

Typical dry air composition is given in Table 12.1 [6].

In order to find the "moist" air composition, one has to know the *relative humidity* (RH) of air as a fraction or percentage. Dry air has RH = 0; saturated air has RH = 1 or RH = 100 percent. ISO conditions specify that RH = 0.6 (i.e., 60 percent). RH is the ratio of partial pressure of water vapor in the air to the *vapor pressure* at the air temperature, which is also known as the "dry bulb" temperature, T_{db}. Vapor pressure, $P_v(T_{db})$, also known as the *saturation* pressure, can be obtained from the ASME Steam Tables. Selected values are provided in Table 12.2, which can be used to make a

Table 12.1 Dry air composition

	Dry Air
N_2	0.7808
O_2	0.2095
H_2O	0.0000
CO_2	0.0003
Ar	0.0094

Table 12.2 Vapor pressure of H_2O

Temperature		Vapor Pressure	
°F	°C	psi	bar
40	4.4	0.122	0.008
50	10.0	0.178	0.012
60	15.6	0.256	0.018
70	21.1	0.363	0.025
80	26.7	0.507	0.035
90	32.2	0.698	0.048
100	37.8	0.949	0.065
110	43.3	1.275	0.088
120	48.9	1.693	0.117

curve fit (e.g., using Excel's *Trend* feature) or as a lookup table with the function INTP1 (see Appendix C).

Once the vapor pressure is obtained, the weight fraction of dry air is found as

$$x_{da} = 1 - \frac{RH \cdot P_v(T_{db})}{P_{air}}. \tag{12.3}$$

The mole fractions of air constituents other than H_2O are determined as

$$y_{i,ma} = y_{i,da} \cdot x_{da}, \tag{12.4}$$

so that mole fraction of H_2O in moist air is given by

$$y_{H2O,ma} = 1 - \sum_i y_{i,ma}. \tag{12.5}$$

In Equation 12.5, the summation is over all constituents i except H_2O. The *humidity ratio* (also known as *specific humidity*) is calculated as

$$\omega = \left(\frac{1}{x_{da}} - 1\right) \cdot \frac{MW_{H2O}}{MW_{da}}, \tag{12.6a}$$

Table 12.3 Typical compounds found in natural gas

Compound	MW (lb/lbmol, kg/kmol)	LHV (Btu/lb)	LHV (kJ/kg)
CH_4	16.04276	21,503	50,016
C_2H_6	30.06964	20,432	47,525
C_2H_4	28.05376	20,278	47,167
C_3H_8	44.09652	19,923	46,341
C_3H_6	42.08064	19,678	45,771
C_4H_{10}-n	58.12340	19,587	45,559
C_4H_{10}-iso	58.12340	19,659	45,727
C_4H_8	56.10752	19,450	45,241
C_5H_{12}-iso	72.15028	19,456	45,255
C_5H_{12}-n	72.15028	19,498	45,352
C_5H_{10}	70.13440	19,328	44,957
C_6H_{14}	86.17716	19,353	45,015
N_2	28.01348	0	0
CO	28.01040	4,342	10,099
CO_2	44.09800	0	0
H_2O	18.01528	0	0
H_2	2.01588	51,566	119,943
H_2S	34.08188	6,534	15,198
He	4.00260	0	0
O_2	31.99880	0	0
Ar	39.94800	0	0

where the molecular weight of the dry air is given by

$$MW_{da} = \sum_i y_{i,da} MW_i. \tag{12.7}$$

Molecular weights and lower heating values (in Btu/lb) of the substances most commonly encountered in combustion calculations are given in Table 12.3. For the dry air composition in Table 12.1, MW_{da} = 28.9652, so that Equation 12.6a becomes

$$\omega = \left(\frac{1}{x_{da}} - 1\right) \cdot 0.62198. \tag{12.6b}$$

For saturated air (RH = 1), the humidity ratio is

$$\omega_{sat} = \frac{0.62198}{\dfrac{P_{air}}{1.0039 \cdot P_v(T_{wb})} - 1}, \tag{12.8}$$

where T_{wb} is the wet bulb temperature (also known as the *adiabatic saturation* temperature). Thus, for known ω (i.e., moist air with given RH and T_{db}), the wet bulb temperature can be found from the solution of the "adiabatic saturation" process (see pp. 586–587 in Moran and Shapiro), which requires the solution of the following nonlinear equation:

$$\omega = \frac{(1093 - 0.556T_{wb})\omega_{sat} - 0.24(T_{db} - T_{wb})}{1093 + 0.444T_{db} - T_{wb}}. \tag{12.9}$$

In Equation 12.9, ω is from Equation 12.6 and ω_{sat} is from Equation 12.8. When there is an *evaporative cooler* (EC) in front of the gas turbine compressor inlet, T_{wb} from Equation 12.9 can be used as the compressor inlet temperature for 100 percent effective EC (typically, EC effectiveness is 85–90 percent). EC effectiveness is defined as

$$\varepsilon_{EC} = \frac{T_{db,in} - T_{db,out}}{T_{db,in} - T_{wb,out}}.$$

Moist air composition is calculated from Equations 12.3–12.5 with RH = 1 and x_{da} in Equation 12.3 from Equation 12.6a with ω from Equation 12.8. For reference, calculated wet bulb temperature values in US customary and SI units are shown in Tables 12.4 and 12.5 (for atmospheric pressure). The data in the tables can be used to find other values via interpolation or extrapolation. Selected moist air compositions are listed in Table 12.6.

The third important temperature definition is the *dew point temperature*, T_{dp}, which is the saturation temperature corresponding to the partial pressure of the water vapor in

Table 12.4 Wet bulb temperatures (US customary units)

T_{DB}	RH = 40%	RH = 60%	RH = 80%
59	47.2	51.4	55.3
80	63.4	69.6	75.0
100	78.9	87.0	93.9

Table 12.5 Wet bulb temperatures (SI units)

T_{DB}	RH = 40%	RH = 60%	RH = 80%
15.0	8.5	10.8	13.0
26.7	17.5	20.9	23.9
37.8	26.0	30.6	34.4

Table 12.6 Moist air composition

	Dry Air	59°F – 60%	80°F – 60%	100°F – 60%
N_2	0.7808	0.7730	0.7647	0.7506
O_2	0.2095	0.2074	0.2051	0.2014
H_2O	0.0000	0.0101	0.0207	0.0388
CO_2	0.0003	0.0003	0.0003	0.0003
Ar	0.0094	0.0093	0.0092	0.0090
SO_2	0.0000	0.0000	0.0000	0.0000
Mol. Wt.	28.97	28.85	28.74	28.54

Table 12.7 Dew point temperatures (US customary units)

T_{DB}	RH = 40%	RH = 60%	RH = 80%
59	34.7	45.1	52.8
80	53.5	64.9	73.3
100	71.3	83.6	92.7

moist air. For reference, calculated dew point temperature values in US customary units are shown in Table 12.7. (Note that dew point temperature is *always* lower than the wet bulb temperature; the difference is higher at lower RHs.) The dew point temperature is important in gas turbine inlet chiller applications. Cooling inlet air below T_{dp} will result in water being "knocked out" from the cold air stream. Unless the water is effectively drained prior to entry into the compressor, impingement of water droplets on stator and rotor blades may lead to erosion.

For a given fuel gas composition, LHV can be calculated as

$$LHV_f = \frac{\sum_i y_i MW_i LHV_i}{\sum_i MW_i}. \tag{12.10a}$$

In Equation 12.10a, summations in the numerator and the denominator on the right-hand side of the equation are overall fuel constituents, whose molecular weight and LHV can be found in Table 12.3. Note that the denominator gives the molecular weight of the fuel gas, MW_f. Thus, Equation 12.10a can be written as

$$LHV_f = \sum_i x_i LHV_i \tag{12.10b}$$

where the mass fraction of each constituent is given by

$$x_i = \sum_i y_i \frac{MW_i}{MW_f}. \tag{12.11}$$

The COMBUST code makes use of table 5.2.2.3 in [6] to solve for the chemical reaction in Equation 12.1 for a gaseous fuel, whose composition comprises the compounds listed in Table 12.3. (The table itself is adapted from table 5.2.4.3 in [6].) The moist air composition used in COMBUST is calculated as described above (also see section 5-2.1 of [6]). The function COMBUST returns the temperature and composition of the combustion products.

By definition, stoichiometric combustion is with 0 percent excess air (i.e., $\phi = \lambda = 1$). Modern gas turbine combustors operate with about 100 percent excess air (i.e., $\phi \approx 0.5$, $\lambda \approx 2$). Stoichiometric combustion of methane with 77°F air results in ~3,700°F products (~2,300 K). This is also known as the *flame temperature*. It can be shown that,

1. Combustion with hotter air results in higher flame temperatures.
2. For fuel-lean combustion, the flame temperature is lower.
3. For fuel-rich combustion, the flame temperature is higher.

When using the VBA function COMBUST, the fuel–air ratio can be adjusted so that the product's O_2 content is 0 percent and the product's temperature returned by the function will be the *flame temperature*. Note that combustion calculations in COMBUST and in almost all heat balance simulation software such as Thermoflex are equilibrium calculations with *no dissociation*, which assume *complete combustion*. The reality, except in rare cases, is *incomplete combustion*, which results in lower temperatures for the products. At very high temperatures, some of the combustion products may *dissociate endothermically*, which is another factor in reducing the temperature of combustion products from that for complete combustion.

Before moving on, it should be emphasized that stoichiometric combustion of methane with compressed air at around 900°F air (cycle pressure ratio [PR] of about 21) will result in ~4,200°F products (~2,600 K), which is higher than the actual temperature reduced by dissociation, which is ~2,400 K. This should be considered as the *ultimate* limit for the turbine inlet temperature (TIT) in a hypothetical gas turbine burning natural gas. The exact value can be different by a few degrees depending on the exact composition and chemical property package employed. Nevertheless, this value should be memorized by everyone seriously engaged in gas turbine technology for electric power generation. Modern J-class gas turbines are nominally rated at 1,600°C TIT (i.e., 1,873 K or 2,912°F). It is quite conceivable that some of them are running at values somewhat higher. At the time of writing, one original equipment manufacturer (OEM; Mitsubishi Hitachi Power Systems [MHPS] to be specific) has already declared that 1,650°C (i.e., 1,923 K or 3,002°F) was achieved in factory tests and was planned to be commercialized by 2020. The same OEM's next-generation target, which is also supported by a national research and development program, is 1,700°C (i.e., 1,973 K or 3,092°F). The difficulty in achieving this target within stringent environmental regulations and rapid disappearance of available dilution air (to be discussed in detail later in the chapter) is obvious. As such, claims of future performance requiring TITs to get closer to the ultimate stoichiometric limit must be viewed with a very skeptical eye.

12.1.1 CO Emissions

In addition to lower-than-expected hot gas temperature, another undesirable impact of dissociation is the formation of pollutants (i.e., *carbon monoxide* [CO] and *nitric oxide* [NO]). At very high combustion temperatures (i.e., close to the adiabatic flame temperature), there is a tendency for CO_2 to dissociate:

$$CO_2 \rightarrow CO + \tfrac{1}{2}O_2. \tag{12.12}$$

Consequently, the right-hand side of the chemical reaction given by Equation 12.1, which is repeated below for convenience,

$$CH_4 + 2\cdot(O_2 + 3.76\cdot N_2) \rightarrow CO_2 + 2\cdot H_2O + 7.52\cdot N_2,$$

becomes

$$CO_2 + 2 \cdot H_2O + 7.52 \cdot N_2 \leftrightarrow (1 - z) \cdot CO_2 + z \cdot CO + (z/2) \cdot O_2 + 2 \cdot H_2O + 7.52 \cdot N_2, \quad (12.13)$$

where z is the amount of CO formed via dissociation of CO_2. In order to solve Equation 12.13 for the temperature of combustion products, we need a *second* equation, which comes from the definition of the *equilibrium constant*, K, for reacting ideal gas mixtures:

$$K = \frac{\prod_i n_i^{\upsilon_i}}{\prod_j n_j^{\upsilon_j}} \left(\frac{(p/p_{ref})}{\left(\sum_i n_i + \sum_j n_j\right)} \right)^{\sum_i \upsilon_i - \sum_j \upsilon_j}. \quad (12.14)$$

In Equation 12.14, K is the equilibrium constant for a chemical reaction similar to that given by Equation 12.12 and is based on the assumption that the tendency of CO and O_2 to form CO_2 is just balanced by the tendency of CO_2 to form CO and O_2. (In other words, the reaction direction → in Equation 12.12 is actually ↔.) Thus,

- n_i is the amount of component i in the gas mixture that appears on the right-hand side of Equation 12.12 or similar.
- n_j is the amount of component j in the gas mixture that appears on the left-hand side of Equation 12.12 or similar.
- υ_i is the stoichiometric coefficient of n_i.
- υ_j is the stoichiometric coefficient of n_j.
- p is the mixture pressure.
- p_{ref} is a reference pressure (usually 1 atm).

Using Equations 12.12 and 12.13 as examples, Equation 12.14 becomes

$$K = \frac{z \cdot \sqrt{0.5z}}{(1 - z)} \left(\frac{(p/p_{ref})}{(10.52 + 0.5z)} \right)^{0.5}. \quad (12.15)$$

Note that the values for n and υ are taken from Equation 12.13. As far as the dissociation of CO_2 is concerned, H_2O and N_2 are "inerts." As such, they inflate the total mixture moles and thus reduce the extent to which the reaction proceeds toward completion. The question now becomes: What is the value of K? From the basic theory of chemical equilibrium (e.g., see chapter 14 in Moran and Shapiro), K is a numerical proxy for the change in the *Gibbs function* for a reaction similar to that given by Equation 12.12, which takes place at temperature T and is in equilibrium (i.e., ↔ instead of →):

$$K = e^{-\frac{\Delta G^\circ}{RT}}.$$

Consequently, for a given reaction, K values are calculated and tabulated. The common tabulation method is $\log_{10}K$ versus temperature (e.g., table A-22 in Moran and Shapiro). For example, for the reaction in Equation 12.12, at 2,200 K, K is found to be 0.006.

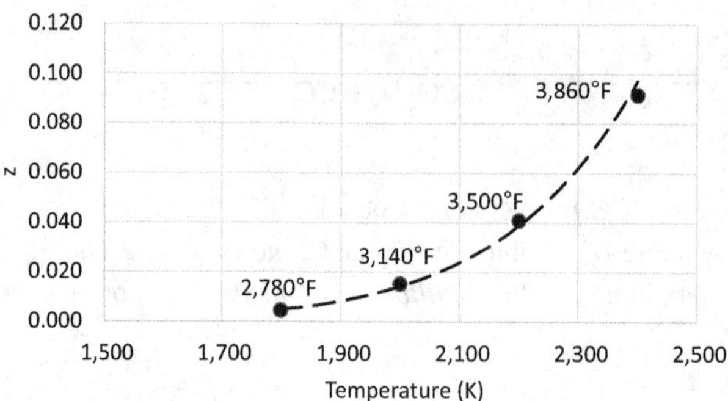

Figure 12.2 Equilibrium conversion of CO_2 to CO (z moles per mole of CO_2).

If the reaction takes place at 10 atm (roughly the old E–class gas turbine combustor pressure), from Equation 12.15, z is found as 0.041. Thus, Equation 12.13 becomes

$$CO_2 + 2 \cdot H_2O + 7.52 \cdot N_2 \leftrightarrow 0.959 \cdot CO_2 + 0.041 \cdot CO + 0.021 \cdot O_2 + 2 \cdot H_2O + 7.52 \cdot N_2.$$

Values of z as a function temperature are shown in Figure 12.2. Clearly, above 1,800 K (2,780°F), one should expect significant CO production. Considering that the adiabatic flame temperature for combustion of methane is well above that, simple complete combustion calculations should be considered as reasonably close estimates of the "real" non-equilibrium *and* incomplete process.

Even equilibrium dissociation equations become quite tedious when more than one species are considered. Certainly, one can program the CO_2 dissociation described above as well as others in a VBA function, but this would be a wasted effort. The reason for that is the strongly *non-equilibrium* nature of many chemical reactions taking place in an actual combustor flame zone. This can only be done by specialized software such as CHEMKIN, which is a widely used tool for solving complex chemical kinetics problems. Originally developed at Sandia National Laboratories (now a commercial product[1]), CHEMKIN can solve thousands of reaction combinations. This is quite impractical for gas turbine calculations for design or off-design performance analysis, optimization studies, etc. Typically, this difficulty is circumvented by assuming a "heat loss" to account for flame temperature reduction via a *combustion efficiency* factor, which is typically 0.995–0.997 (i.e., 0.3–0.5 percent loss), and applied to the combustor heat input, which is found by the product of fuel mass flow rate and LHV.

12.1.2 NOx Emissions

There are *three* mechanisms for NOx production in the combustor of a gas turbine: thermal, nitrous oxide, and prompt. Each mechanism is described by different chemical reaction paths. Of these three, when flame temperatures are above 2,780°F (1,800 K), the dominant mechanism is the *thermal NOx* or the *extended Zel'dovich* mechanism [1].

[1] CHEMKIN is a product of Reaction Design, San Diego, CA, USA (www.reactiondesign.com).

Below this temperature, the thermal reactions are relatively slow. Beyond about 3,100°F (1,700°C), thermal NOx production grows exponentially. This can be considered as an upper limit for Dry-Low-NOx (DLN) combustion, which will be covered in detail below. The extended Zel'dovich mechanism is given by

$$O_2 \leftrightarrow 2O \text{ (Dissociation of } O_2\text{)}$$
$$O + N_2 \leftrightarrow NO + N \text{ (Oxidation of } N_2\text{)}$$
$$N + O_2 \leftrightarrow NO + O \text{ (Oxidation of N)}$$

According to this mechanism, free oxygen from the equilibrium dissociation of unburned oxygen is the trigger of the chain. Note that equilibrium dissociation of nitrogen

$$N_2 \leftrightarrow 2N \text{ (Dissociation of } N_2\text{)}$$

is *not* achieved at the prevailing combustor temperatures. (From table A-22 in Moran and Shapiro, $\log_{10}K$ for oxygen dissociation is –5.142 at 2,200 K, whereas it is –15.81 for nitrogen dissociation. In other words, the latter is 10 orders of magnitude smaller.) Consequently, the only source of N atoms is the second reaction in the chain. Calculations show that the equilibrium NO level rises

- Either as ϕ is reduced at constant temperature, or
- Temperature is increased at constant ϕ.

In reality, both T and ϕ are connected to each other in a combustor, and as a result of both fuel and nitrogen competing for oxygen, peak NO formation is found on the *lean side* of the stoichiometric (whereas peak flame temperature occurs on the *rich side*). The well-known qualitative dependence of the rate of NOx formation and the flame temperature on ϕ is shown in Figure 12.3.

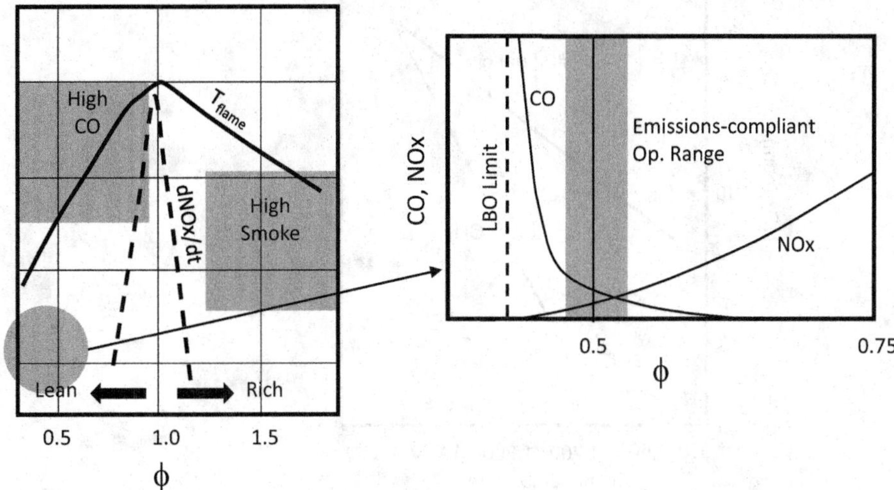

Figure 12.3 NOx production in the flame zone. LBO = lean blowout.

If ϕ is too low (i.e., below 0.8), the drop in the flame temperature dominates the effect of more free O_2 in the combustion products and the NOx formation rate goes down accordingly. On the rich side (i.e., $\phi > 1.0$), due to the higher rate of the exothermic fuel–oxygen reaction, consumption of O_2 for NOx formation is dominated by the former and the rate goes down again.

Simple estimation of NOx and CO in the flue gas is not possible. Even highly detailed chemical kinetics models require significant adjustment using empirical constants. In general, NOx formation is a function of three parameters: residence time in the combustion zone, chemical reaction rate, and mixing rate [1]. For the gas turbine combustors, the three reaction parameters can be related to the turbine operating conditions and combustor size. From that, it is found that the NOx production rate for a given system can be expressed in an empirical formula that can be written as [1]

$$NO_x \propto p^n \cdot e^{\alpha T_f}, \tag{12.16}$$

where p is combustion pressure, T_f is the stoichiometric flame temperature, and n and α are empirical constants. Experimental data for number 2 fuel oil and methane are summarized by the linear curves in Figure 12.4 [7]. Note that NOx production via combustion of either fuel becomes comparable at high flame temperatures, whereas liquid fuel combustion generates much more NOx at lower temperatures. (In addition to its higher cost, this is one more reason why it is a rarely used backup fuel.) The gray area next to the methane curve in Figure 12.4 signifies the variation to expect from different combustor designs and conditions. Results of calculations done using CHEMKIN are shown in Figure 12.5 [8], which highlights the narrow zone of flame temperature (i.e., equivalence ratio, ϕ) for low CO *and* NOx emissions.

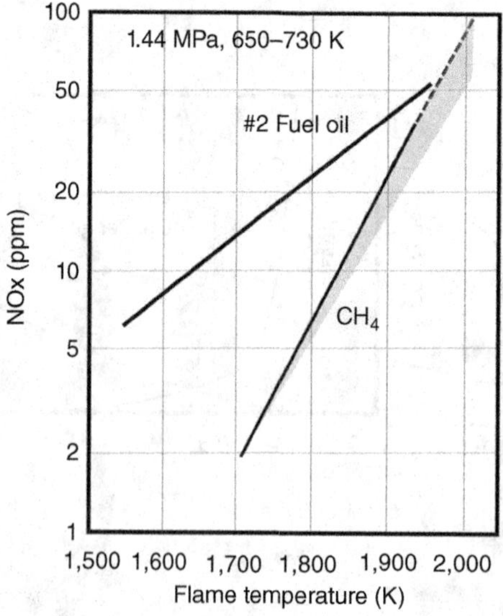

Figure 12.4 NOx generation as a function of flame temperature [7].

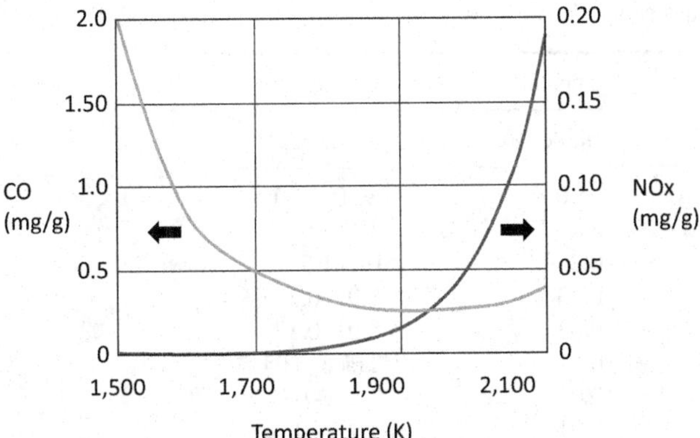

Figure 12.5 CO and NOx emissions from CHEMKIN (combustor inlet 600 K, 15 bar combustor pressure) [8].

The NOx generation characteristic in Figure 12.4 is typical of modern gas turbine DLN combustors. This curve can be used to estimate NOx emissions of a given gas turbine combustor with known (or guessed or estimated) performance (denoted by the subscript "0" in Equation 12.17). Gas turbine inlet or firing temperature (e.g., TIT) can be used as a *proxy* for the flame temperature, i.e.,

$$\text{NOX} = \left(\frac{\text{TIT}}{\text{TIT}_0}\right)^{x}\left(\frac{\text{PCOMB}}{\text{PCOMB}_0}\right)^{1/3}\text{NOX}_0. \tag{12.17}$$

The exponent x in Equation 12.17 is 24.585 for the methane curve in Figure 12.4. For the lower end of the range marked by the gray triangle, x is 20.724. Thus, a value of 20–25 for x should cover most DLN combustors' NOx generation characteristics. The lower end of the range is suitable for older-generation combustors, whereas the higher end is representative of modern ones. For a given piece of hardware, all one needs is a "base" data point (i.e., TIT_0, PCOMB_0, and NOX_0). Nevertheless, it must be kept in mind that, for combustor NOx emissions, one should always try and obtain OEM data in the first place.

Equation 12.17 considers only two of many variables affecting the production of NOx. It is adequate for comparing NOx emissions at ISO base load with a specific fuel (e.g., 100 percent methane). Among other variables, ambient humidity has the largest single effect on NOx production. A change in specific humidity from, say, 0.0063 lb/lb of dry air (i.e., ISO conditions) to 0.02 lb/lb on a humid day will result in a decrease of 30–40 percent in NOx production (in ppmv). (Think of it as a form of "water injection.")

A change in ambient air temperature, via the change in compressor discharge (i.e., combustor air inlet) temperature, will also have an impact on NOx production. In general, a change of ambient temperature from 0°F to 100°F will result in an increase of 15–20 percent in NOx production (ppmv).

The other two NOx formation mechanisms are "prompt NO" and "fuel NO." The former, also known as the *Fenimore mechanism*, is important in low-temperature, fuel-

Table 12.8 Natural gas composition

Component	Typical (mole %)	Range (mole %)
Methane	95.0	87.0–97.0
Ethane	3.2	1.5–7.0
Propane	0.2	0.1–1.5
i-Butane	0.03	0.01–0.3
n-Butane	0.03	0.01–0.3
i-Pentane	0.01	Trace–0.04
n-Pentane	0.01	Trace–0.04
Hexane	0.01	Trace–0.06
Nitrogen	1.0	0.2–5.5
Carbon Dioxide	0.5	0.1–1.0
Oxygen	0.02	0.01–0.1
Hydrogen	Trace	Trace–0.02
LHV (Btu/lb)	20,780	19,690–21,980
LHV (kJ/kg)	48,335	45,800–51,125

rich flames. There is no well-established theory leading to compact predictive equations like Equation 12.17. Its primary reaction involves the *methylidyne radical*, CH, resulting from the breakup of the fuel molecule, combining with N_2 to form hydrogen cyanide (HCN) and N, which subsequently oxidizes to NO. It can be as high as 30 ppmv under fuel-rich conditions [1]. The moniker "prompt" refers to the very short-lived presence of fuel fragments such as CH in the flame front.

Fuel NO, as the name suggests, refers to the formation of NO from nitrogen-bearing compounds in the fuel. For typical natural gas, as can be judged from the data in Table 12.8, this should not be a big concern. In general, conversion of fuel-bound nitrogen to NO is complete for fuel-lean flames, but decreases with increasing concentration of it and/or increasing ϕ. The conversion also increases with increasing flame temperature.

The other oxide of nitrogen in NOx is nitrogen dioxide (NO_2). Since oxidation of NO to NO_2 requires low temperatures, at full load, the fraction of NO_2 in total NOx is very small, but it can be as high as 50 percent at very low loads (e.g., full-speed, no-load [FSNL]). In fact, it is after the ejection through the power plant's smokestack, in the presence of air, that NO is rapidly oxidized to NO_2. NO_2 is a reddish-brown pollutant with high reactivity. It is also a strong oxidizing agent that can form the corrosive nitric acid, as well other toxic organic nitrates. Furthermore, NO_2 is also a pulmonary irritant affecting the human respiratory system, which may exacerbate asthma and other respiratory conditions. Due to its ability to absorb light, NO_2 is a main component of smog, and it is the source of the *yellow plume* emanating from a power plant's smokestack.

"Smoke" is the colloquial term for "soot" or "soot particles" in combustion products (i.e., the exhaust or flue gas). It consists mostly of carbon (in excess of 90 percent) and a mixture of hydrogen, oxygen, and other species. It is mainly a problem in the combustion of liquid fuels, which are, to a great extent, rarely used backup fuels for modern, utility-scale gas turbine power plants.

Similar to the CO, *unburned hydrocarbons* are primarily caused by incomplete combustion in local fuel-rich zones.

In conclusion, reducing the combustion (flame) temperature is the key to NOx control. There are two types of flames: *diffusion* and *premixed*. These two types of flames characterize the two main types of gas turbine combustor as well (i.e., diffusion and DLN combustors), which will be discussed in detail in the next section.

12.1.3 Carbon Dioxide Emissions

Simple combustion calculations help to illustrate the dramatic advantage of natural gas-fired gas turbine combined-cycle power plants vis-à-vis coal-fired steam power plant in terms of CO_2 emissions. For example, given fossil-fired power plant net output and efficiency (i.e., heat rate), CO_2 in the flue gas can be found as follows:

$$e_{CO2} = 1{,}000 \cdot \frac{HR}{LHV} \cdot \frac{MW_{CO2}}{MW_f} \quad \frac{lb}{MWh} \text{ (gas fuel)},^{2} \tag{12.18a}$$

$$e_{CO2} = 1{,}000 \cdot \frac{HR}{LHV} \cdot c \cdot \frac{MW_{CO2}}{MW_C} \quad \frac{lb}{MWh} \text{ (coal)}, \tag{12.18b}$$

where HR is the plant heat rate in Btu/kWh, LHV is the lower heating value of the fuel in Btu/lb, c is the fraction of carbon in coal per ultimate analysis, and MW refers to the molecular weights of fuel gas (~16 lb/lbmol), CO_2 (44 lb/lbmol), and carbon (12 lb/lbmol). Calculations show that the difference between coal and natural gas is striking. Even with high-quality bituminous coal, CO_2 in the flue gas of the most advanced coal-fired power plants is almost *twice* that of the least advanced gas-fired gas turbine combined-cycle (GTCC) power plants.

12.2 Combustor Hardware

Gas turbine combustor design requires a complex balance between several important (and frequently conflicting) requirements:

1. Materials and coatings commensurate with prevailing pressures and temperatures across the entire range operation (i.e., gas turbine loads).
2. Low emissions (NOx, CO, and others) across the widest possible range of gas turbine loads.

[2] This formula is similar to that in Federal Emissions Regulation Code for Continuous Emissions Monitoring (CEM), CFR 75.10(3)(ii): W_{CO2} (tons/hour) = ($F_c \times H \times U_f \times MW_{CO2}$)/2,000, where the F_c factor is 1,040 scf/MMBtu for natural gas, H is the heat input in MMBtu/h, and U_f is 1/385 scf CO_2/lbmol at 14.7 psia and 68°F. For typical natural gas (19,900 Btu/lb, 17.75 lb/lbmol), Equation 12.18a gives 5 percent higher values.

3. Low combustion noise (also known as "humming") caused by pressure oscillations.[3]
4. Safety concerns (i.e., prevention of flash-back and autoignition).
5. Cost-effective service life (i.e., durability with mechanical integrity) within a defined operability envelope (number of start–stop cycles, equivalent operating hours).

Achieving this balance requires careful study, analysis, resolution, and/or accommodation of the following physical mechanisms and operational considerations:

1. Highly exothermic chemical reaction with a large heat release (i.e., 200 $MWth/m^3$ [about 20 $MMBtu/h\text{-}ft^3$] or higher).
2. Complex, non-equilibrium chemical kinetics involving a very large number of simultaneous reactions (touched upon briefly above).
3. Thermo-acoustic vibrations (pressure pulsations) due to variations in heat release locations in the flame front.
4. Fuel variability (e.g., natural gas composition) and backup fuel (e.g., number 2 fuel oil for most industrial gas turbines).
5. Gas turbine load range with commensurate cycle PR, compressor airflow, and firing temperature.

From a physical construction perspective, there are three types of gas turbine combustors:

1. Silo (i.e., large single- or double-cylindrical chamber)
2. Annular (or ring)
3. Can-annular (or *cannular* or tubo-annular)

Similar to the rotor, which will be covered in Section 14.1, each OEM has its unique combustor design philosophy and design practices. For example, it is quite rare for one OEM to change from one combustor design (say, can-annular) to another (say, annular) for a new product line. (Silo-type combustors, with the exception of some legacy designs such as Siemens' E–class gas turbine, are largely no longer used – see below.) Recently, however, especially with ever-larger frames, Siemens and Alstom (Ansaldo) switched from annular to can-annular combustors in their flagship products (see Chapter 21). In general, work on the new combustor design is carried out concurrently with other design activities. Based on the airflow, cycle PR, and firing temperature requirements for the new gas turbine, the combustion team embarks on its development activities starting from one of the already available designs closest to the new specifications. The development process is heavily dependent on testing in order to achieve requisite hot gas temperature without exceeding emissions and combustion dynamics limits.

From a flame perspective, there are two categories:

1. Diffusion flame
2. Premixed flame

[3] Referred to as "hot tones" at high loads (lean premix) and "cold tones" in very lean conditions.

What is really meant by the term "flame" is the "flame front," which is a very narrow zone of a few micrometers wherein the chemical reaction between the fuel and the oxidizer (i.e., the combustion) takes place. The most common examples of diffusion and premixed flames are the wax candle and the Bunsen burner, respectively.

In a *diffusion*-type flame, fuel and oxidizer are separated by the flame front, wherein the mixing of the two takes place via diffusion of both into the reaction zone encompassed by it. The combustion products formed by the reaction, on the other hand, force themselves into either reactant's zone. Eventually, on the oxidant side, the concentration of products increases until the fuel is completely depleted. In general, a diffusion flame can be laminar or turbulent; in gas turbine combustors, it is always of the latter type. From a practical application perspective, the advantage of the diffusion flame or diffusion combustion is its stability due to the near-stoichiometric reaction in the flame front ($\phi \approx 1$) over a broad range of operability. Consequently, the flame temperature in the reaction zone can approach its theoretically maximum value, independent of excess air mixed downstream of the reaction zone. The only way to control NOx via diffusion flame temperature reduction is by injecting a *diluent* (i.e., water or steam) into the reaction zone. There are several problems associated with this method, including reduced performance, water consumption, operability issues, etc. The only alternative is post-combustion exhaust gas cleanup using *selective catalytic reduction* (SCR).

Premixed flame is the enabler of DLN technology. The moniker "dry" obviously refers to the absence of water/steam as diluent because in DLN combustors this role is played by atmospheric nitrogen in the combustion air. In DLN combustors, fuel is mixed with air upstream of the reaction zone in fuel-lean conditions ($\phi \approx 0.5$) to prevent high flame temperatures that create thermal NOx. This is the Achilles' heel of the premixed flame, whose stable operating range is typically extended by a diffusion-type *pilot flame*. Otherwise, it would be very difficult to maintain a stable flame at low loads due to "flame-out" at $\phi \approx 0.4$.

There are three types of premix combustion: (1) lean–lean or (2) rich–lean staged and (3) lean premixed combustion. Staging refers to the introduction of fuel into the combustor in the axial direction. Rich–lean staged combustion is advantageous for fuels with a high fuel-bound nitrogen content, which is not the case for most heavy-duty industrial gas turbine applications where natural gas is the primary fuel. Lean–lean staged combustion is implemented in General Electric (GE)-Alstom and Ansaldo's sequential (reheat) combustion GT24 and GT26 gas turbines, respectively, and in GE's new HA–class gas turbines (axial fuel staging, which is described in detail below). Lean premixed combustion refers to parallel or radial fuel staging, which is the hallmark of GE's DLN technology. A fourth variant is "air-staged" combustion, which is not common due to the complexity of controlling the large volume of bypassed air.[4]

Silo-type combustors are diffusion combustors of very large volume with burners located at the top. The name refers to a grain silo, which the large cylindrical combustor

[4] Limited air bypass is utilized by some OEMs for improved combustion efficiency and reduced CO emissions at low loads (see Section 21.3).

resembles, especially in a vertical orientation. Burner numbers increase with gas turbines size (e.g., 6–24 for Siemens' old V series gas turbines). There are two main types adopted by two OEMs: two silos on either side of the gas turbine (e.g., Siemens SGT5-2000E and SGT6-2000E), either vertically or horizontally oriented, or one silo on top of the gas turbine (e.g., GE-Alstom's GT 8, 9, and older 11 and 13 types).

A large silo-type combustor frees the designer from space restrictions imposed by the gas turbine frame so that its length and volume can be optimized for complete and uniform combustion. They also shield the turbine from radiation heat transfer from the hot flame zones in the combustor. Their disadvantage lies in the large, cylindrical connector (which also requires cooling) from the combustor to the gas turbine casing and annular turbine inlet with a uniform temperature profile. Although originally intended for diffusion burners, they can be used with premixed burners as well (e.g., the "hybrid burner" by Siemens, which can accommodate both diffusion *and* premixed flames). In the latter variant, however, the combustor is much shorter in order to accommodate the reduced cooling air flow (which goes to the premix burners).

The increasing TITs of modern, high-efficiency gas turbines while still complying with ever more stringent emissions regulations rendered silo-type combustors less and less suitable to the achievement of uniform temperature profiles. In order to further reduce residence times (to minimize NOx emissions) and cooling air consumption, along with the desire to achieve a more uniform temperature profile at the turbine inlet, major European OEMs (US OEMs never adopted the silo-type design configuration in the first place[5]) switched over to the *annular* (or ring) combustor design. As the name suggests, in this configuration, the burners are distributed circumferentially at the inlet of a ring-shaped combustor shell. The dimensions of the ring (i.e., its length and height) are determined from a balance of lowest possible residence time, complete combustion, and minimal cooling air consumption. The number and spacing of the burners are determined after a careful analysis of their interaction with each other with respect to crossfire (i.e., "jumping" of the flame from burner to burner) and mixing of flame products. Typical burners are Siemens' hybrid burner and Asea Brown-Boveri (ABB)/Alstom's (now GE) conical venturi-shaped EV burner.

There are two types of cooling schemes applied to the annular combustors: open (adopted by Siemens until the H class) and closed (ABB/Alstom, now GE). In the open-cooling scheme, cooling air flows over the ring wall, which is covered with ceramic (Al_2O_3–SiO_2) heat shields, whose surface temperatures can go as high as 1,500°C. Clearances between the shields are cooled by small cooling air streams. In the closed-cooling scheme, compressor air is directed into the channels built into the combustor ring wall, which it cools via impingement and convection cooling.

The oldest combustor type is the single can, with a number of them (12–14 in modern heavy-duty industrial gas turbines; as many as 18 in the older 50-Hz variants) circumferentially arranged in front of the turbine and over the compressor exit

[5] The exception is GE's MS1002, which is now GE10 (formerly GT 10 or PGT 10), manufactured by Nuovo Pignone. This machine has a single-can combustor in vertical or horizontal (on one side of the gas turbine) orientation.

("reverse flow" arrangement). The arrangement is commonly known as *can-annular*. This is the configuration primarily adopted by US OEMs (GE and now-defunct Westinghouse) and MHPS (formerly Mitsubishi Heavy Industries [MHI]). Siemens' 60-Hz units (essentially upgraded Westinghouse 501F designs), H class, and Ansaldo's new GT36 also use this configuration. The major advantage of the can-type combustor is that a single can may be tested in the laboratory under actual operating conditions for rigorous thermal and mechanical design of the component. The whole system is built by copying the new can design. The system also enables rapid replacement of one (or a few) defective can for reduced outage time or the entire set with updated design cans.

For a good overview of commercially available combustor systems for heavy-duty industrial and aeroderivative gas turbines, the reader is referred to [8]. While the paper is somewhat dated (being from 2003), its coverage is still relevant to the combustors used in modern engines (which are essentially *advanced* versions of the systems described therein) with the exception of *axial fuel staging* in GE's new HA-class (see Section 21.2) and Ansaldo's GT36 gas turbines (see Section 21.4).

12.3 DLN Combustors

The main problem in designing a DLN combustor is to keep the combustion fuel lean at *all* operating conditions. In order to fully appreciate this difficulty, assume that there is only *one* nozzle designed for $\phi = 0.5$ at the design point (full load) operation. To reduce the gas turbine load by reducing the fuel flow and keeping the same fuel-lean condition with this single degree of freedom is not possible. The obvious solution is to increase the degrees of freedom (i.e., the number of nozzles and stage the fuel flow). This is the premise behind the modern DLN combustor design. This is sometimes referred to as radial or parallel fuel staging. Radial staging includes the use of pilot flames and reducing/eliminating fuel from some injectors completely [1].

As an example, consider GE's DLN 2.6+ combustor, which has four fuel passage manifolds: one diffusion passage (D5) and three premix passages (PM1, PM2, and PM3). Each passage has an individual *gas control valve* for controlling gas fuel delivery. Gas control valves control the desired gas turbine fuel flow in response to the "fuel stroke reference" command from the turbine controller. Each fuel passage requires a certain percentage of the total fuel, which is a function of the *combustion reference temperature* (TTRF), which is essentially the gas turbine firing temperature, and the DLN operating mode. Each combustion can (the total number depends on the gas turbine frame, usually between 12 and 18 for large, heavy-duty industrial machines) has six fuel nozzles arranged in a circular configuration, with one in the middle (see Figure 12.6).[6]

[6] They are referred to as "swozzles" by the OEM, which is a combination of the terms "swirler" and "nozzle." Swirlers are essentially vanes built into the fuel passage to introduce a "swirl" or "vortex" into the flow for better mixing of air and fuel.

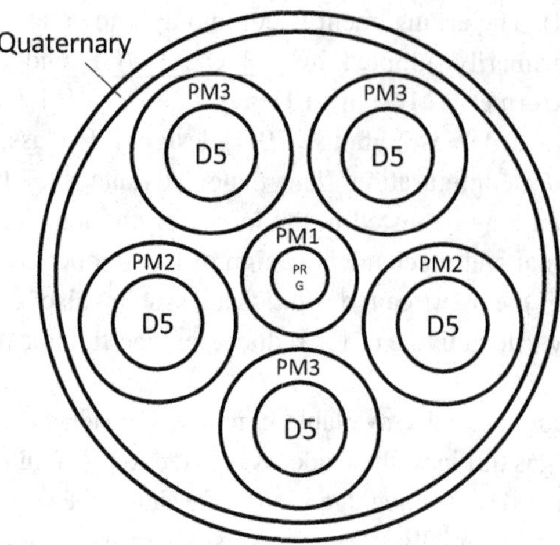

Figure 12.6 DLN combustor fuel nozzle arrangement. PRG = purge.

For each combustor can,

- The D5 gas fuel delivery system consists of five diffusion-type fuel nozzles.
- The PM2 and PM3 gas fuel delivery system consists of two and three premix-type fuel nozzles, respectively.
- The PM1 gas fuel delivery system consists of one premix-type fuel nozzle.
- A *quaternary* fuel manifold is located on the circumference of the combustion casing, injecting a small amount of fuel into the airstream upstream of the fuel nozzles through pegs located radially around the casing (also known as the "quat").

The "quat" manifold is used to mitigate combustor instability, to provide better mixing of fuel and air, and to improve the flame-holding margin of downstream fuel nozzles by accommodating up to 30 percent of total combustor fuel.

In addition to diffusion and premix circuits for gas-fueled operation, there is a liquid fuel circuit and a water circuit (for NOx control in liquid fuel operation). Number 2 fuel oil and water from these circuits are injected through the outer five nozzles.

There are several steady-state DLN modes of operation. There is also a transient load-rejection model where all the fuel is diverted to the premix nozzles PM1 and PM2. At certain modes, some passages have no fuel flow scheduled and must be purged to prevent condensate from accumulating and to limit the potential for autoignition. Compressor discharge air (cooled prior to entry into the gas fuel system) is used for this purpose.

Activating and deactivating PM and D passages according to gas turbine load and/or speed is called "staging." In general, staging goes as follows [9] (also see Figure 12.7):

1. The gas turbine is started in diffusion mode (D5 only, all others purged).
2. After flame is established (at about 15 percent load) and the turbine warmup cycle is complete, the unit accelerates to FSNL.

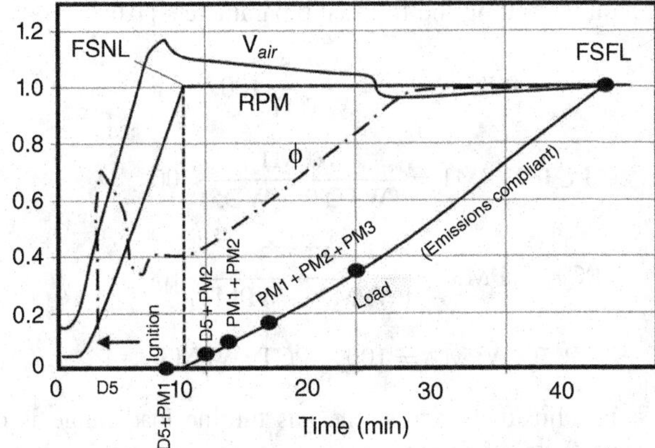

Figure 12.7 Starting a gas turbine and loading it to FSFL – generic combustor-mode transfer points.

3. At 95 percent speed, the unit transfers to the next mode (PM1 and D5). This is usually referred to as the "lean–lean" mode.
4. At rated speed (i.e., 3,000 or 3,600 rpm), the unit is synchronized and the breaker is closed. The unit starts to load to FSFL.
5. At the first TTRF switch (the exact value depends on the specific unit), the unit transfers to the next mode (D5 and PM2). This is known as the "premix transfer" mode.
6. At the next TTRF point, the diffusion mode is deactivated (i.e., no D5); the unit transfers to the first premixed mode, PM1 plus PM2.
7. At *minimum emissions compliance load* (MECL), the unit transfers to the second, fully premixed mode, PM1 plus PM2 plus PM3.
8. Above a certain load, the "quat" passage is activated.

The generic description above is not intended to be "exact." The first reason for this is obvious; specifying exact TTRF transfer points would be publicizing proprietary OEM data. In addition, there will be differences between different generations of the DLN technology as well (i.e., DLN 1, DLN 2, DLN 2.6, and DLN 2.6+). The sequence above corresponds *qualitatively* to DLN 2.6+. In some versions, there is a "piloted premix" mode where all PM passages as well as the diffusion passage are fueled. This is typically an intermediate step between premix transfer and fully premixed modes [9]. In DLN 2.6+, this is the mode during load rejection.

Furthermore, for a given gas turbine operating in the field, exact transfer points and fuel distributions between the individual nozzles (commonly referred to as the "fuel split") are determined after a careful "tuning" exercise. Assuming that fuel flows to each circuit are WD5, WPM1, WPM2, and WPM3, total fuel flow is

$$\text{WTOT} = \text{WD5} + \text{WPM1} + \text{WPM2} + \text{WPM3}.$$

Thus, the fuel split requires setting the following parameters [10]:

$$PCT\_WD5 = \frac{WD5}{WTOT} 100,$$

$$PCT\_WPM1 = \frac{WPM1}{(WTOT - WD5)} 100,$$

$$PCT\_WPM3 = \frac{WPM3}{(WPM2 + WPM3)} 100,$$

$$PCT\_WPM2 = 100 - PCT\_WPM3.$$

The effect of fuel split adjustment across the gas turbine load range is qualitatively illustrated in Figure 12.8. The first thing to note is the fuel–air ratio for the entire combustor control volume based on the air and fuel flows. For most of the load range, it is well below the lean blowout (LBO) limit. The fuel–air ratio of interest is that in the flame zone, which controls the NOx emissions. Fuel split (i.e., mode) changes at the transfer points ensure that the premixed fuel–air mixture is always just above the LBO limit. When looking at the sequence described in Figure 12.8, it is easy to imagine that the same effect could also be accomplished by simply adjusting the airflow (instead of adjusting the fuel flow via manipulating different manifolds or circuits). Air staging is indeed common in coal-fired boilers. However, its application in gas turbine combustors is rather difficult due to the complexity of controlling and bypassing the requisite amount of air.

Tuning is imperative for emissions compliance and stable combustion with minimal dynamic pressure pulsations ("humming"). This is typically done by a qualified OEM technician who is trained for and experienced in the particular system. Strictly speaking, specific fuel split for a given combustor and gas turbine at base load operation (e.g., 100 percent load at ISO conditions) includes margins with respect to NOx emission, dynamics, and component durability limits. This is illustrated by the boundaries of those limits observed during a field test in Figure 12.9 [10]. Target splits

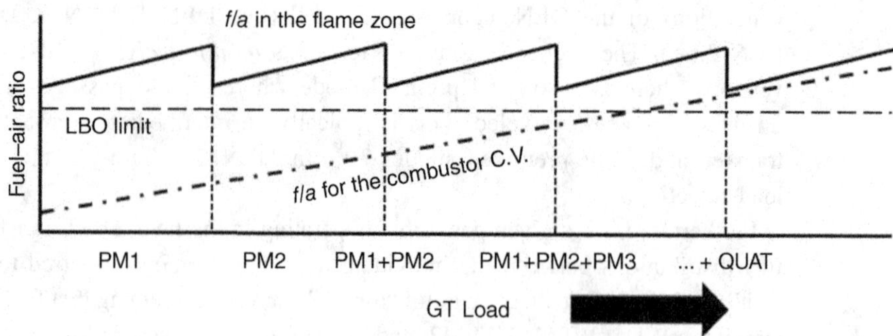

Figure 12.8 Typical DLN combustor fuel staging sequence. C.V. = control volume; GT = gas turbine; QUAT = quaternary.

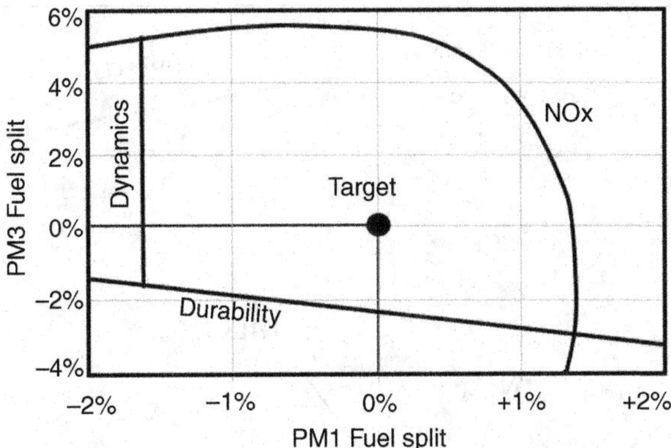

Figure 12.9 Fuel split operability window at base load [10].

at other load conditions are tuned as a trade-off between emissions, dynamics, durability, and LBO margin.

How frequently to tune a DLN combustor cannot be prescribed a priori; it depends on the site ambient and loading specifics. Nowadays, OEMs and third-party (after-market) suppliers offer "auto-tune" systems, which automatically tune gas turbine operating parameters to maintain emissions and combustion dynamics within specified limits under varying ambient conditions, engine deteriorations, and fuel gas compositions. In essence, by using dynamics and emissions signals from transducers, the auto-tune software performs the requisite calculations and changes the fuel split parameters in the gas turbine controller (GE's Mark series of controllers, Mark V, Mark VI, etc.) to maintain dynamics and emission levels within allowable limits. Over time, the auto-tune software can "remember" the best fuel splits for different combinations of gas turbine inlet conditions and loads so as to reset the fuel split as needed without needing a tuning step.

As the foregoing discussion makes clear, even with multiple nozzles and fuel staging, there are only a limited number of degrees of freedom. (After all, there is only so much space at the combustor end cap to squeeze a large number of fuel passages onto.) This limitation presents itself in MECL, which is illustrated in Figure 12.10 [11]. Gas turbine and GTCC load is controlled via gas turbine firing temperature (i.e., fuel flow) and airflow (via *inlet guide vanes* [IGVs]). For maximum GTCC efficiency at part load, load is first reduced by closing the IGVs and reducing gas turbine airflow until the exhaust temperature reaches its maximum value (typically 1,200–1,250°F). Thereafter, load is controlled via combined control of fuel flow and IGVs while keeping the exhaust temperature at its maximum. Once the IGVs reach their minimum position, further reduction in load is only possible by reducing the fuel flow. Major pollutant emissions display large discontinuities as the combustion mode changes (e.g., one premix passage is closed) at a sufficiently low airflow and fuel flow combination as dictated by requested gas turbine load and exhaust temperature (or IGV opening) limit. The first such "jump" takes place at MECL, below which the combustor emissions cannot be maintained at their environmentally regulated levels.

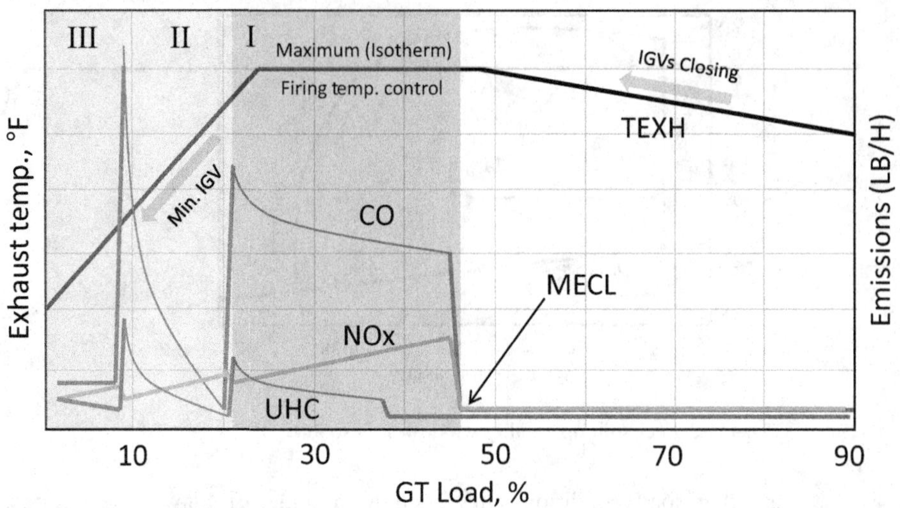

Figure 12.10 Typical gas turbine DLN combustor emissions (CO, NOx, and unburned hydrocarbons [UHC]) and exhaust temperature (TEXH) as a function of gas turbine (GT) load. (adapted from [11])

For older-generation F–class gas turbines, MECL used to be at 50 percent or even higher load. In more modern variants, it is more likely to be between 40 and 50 percent load. In modern H–class gas turbines, emissions compliance is achievable at loads as low as 30–40 percent (typical OEM "guarantee" value; even lower loads with very low [i.e., "single-digit"] emissions have been achieved in factory tests). Low MECL is essential for emissions-compliant "low-load parking" of a power plant at times of low demand or overnight instead of going through frequent start–stop cycles, which is detrimental to component life expenditure. In the light of increasingly stringent government regulations and cyclic operation duties imposed on gas turbine power plants as a result of the grid fluctuations caused by renewable generation (i.e., wind and/or solar) interruptions, this is a very significant design consideration.

From a *full load* (i.e., 100 percent load at given site ambient conditions with IGVs at their fully open position) operation performance perspective, the limitations of DLN combustors with fuel staging and a single reaction zone can be seen in Figure 12.11. For high combustor exit (i.e., turbine inlet) temperatures, the reaction zone temperature increases to the point that the increase in thermal NOx becomes exponential. This is a significant impediment to further increases of gas turbine TIT for even higher thermal performance. Today's state of the art is 1,600°C, with 1,700°C TIT under active development by one OEM (i.e., MHPS) [12]. According to MHPS (formerly MHI), the development focuses on a 1,700°C-class DLN combustor with steam-cooled casing and an *exhaust gas recirculation* (EGR) system [13]. With nearly 30 percent EGR (i.e., about 30 percent of the heat recovery steam generator [HRSG] stack gas is recirculated to the gas turbine inlet) and 17 percent (by volume) combustor inlet air O_2 (cf. about 21 percent O_2 in ambient air), NOx

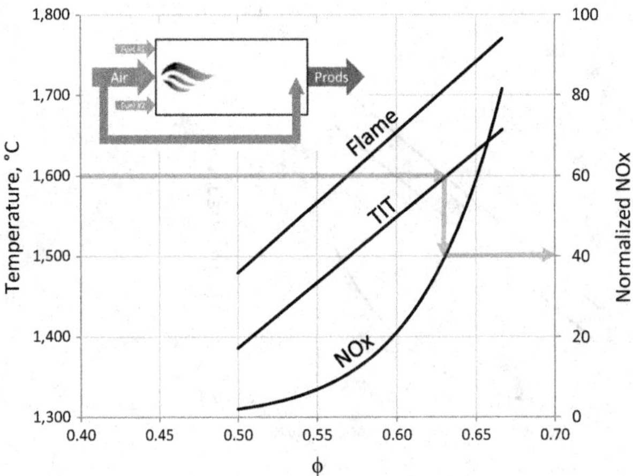

Figure 12.11 Flame temperature and normalized NOx as a function of equivalence ratio (DLN combustor with radial/parallel fuel staging and single reaction zone). Note that ϕ is based on the airflow through the flame zone (*not* the total combustor airflow). Calculations are done with Thermoflex.

emissions are reduced by 40 percent. The significance of this can be appreciated by using Equation 12.17 with x = 20.724 or referring to Figure 12.4. Assuming 25 ppmvd[7] NOx (15 percent O_2) at 1,600°C TIT, going to 1,700°C (ignoring the change in compressor PR) would result in nearly 100 ppmvd NOx, which corresponds to 60 ppmvd with a 40 percent reduction via EGR. Indeed, the OEM-targeted NOx concentration is 50 ppm (15 percent O_2) [13].

Axial staging injects fuel at two places along the combustor flow path (see Figure 12.12). Products from the first combustion zone are mixed with fuel and air in a second, subsequent combustion zone, providing an advantage for fuel-lean operation of the second zone. Combustion with axial fuel staging can be approximated by two stirred reactors in series. Assuming that 25 percent of total fuel flow is sent to the second reactor, the results in Figure 12.12 clearly illustrate the advantage of axial fuel staging (e.g., compare the curves in Figure 12.12 with those in Figure 12.11).

One promising method for achieving high TIT in DLN combustors without excessive NOx generation is *fuel moisturization*. The method was introduced to utilize the lowest-grade energy in the HRSG (i.e., flue gas downstream of the low-pressure [LP] section) to heat the fuel in a direct-contact heat exchanger (called a *moisturizer* or *saturator*), which is a randomly packed column [14]. Similar to steam or water injection in diffusion combustors, water vapor in the fuel gas acts as a heat sink, reducing the flame temperature and lowering thermal NOx production. Performance improvement, subject to cycle optimization, is about a 0.2-percentage point increase in net efficiency and about a 1.25-percentage point increase in net output. The system was deployed

[7] Parts per million by volume on a dry basis.

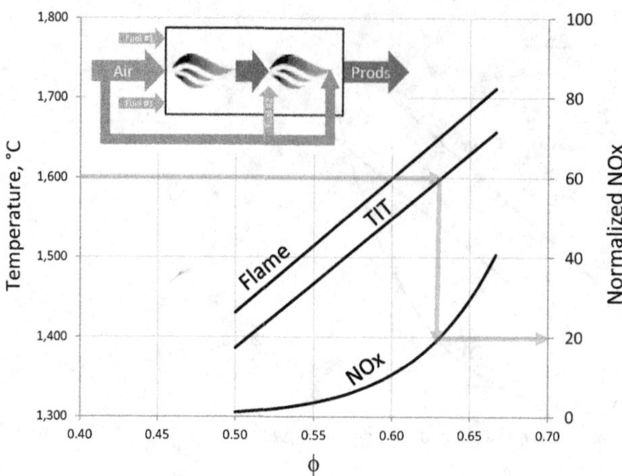

Figure 12.12 Flame temperature and normalized NOx as a function of equivalence ratio (DLN combustor with axial fuel staging and two reaction zones). Note that normalized NOx at 1,600°C TIT is ~20 vis-à-vis ~40 in Figure 12.11 for a reduction of about 50 percent.

successfully in a GTCC power plant in California,[8] but complexity and cost prevented widespread commercial acceptance.

12.4 Combustor Operability

Gas turbine combustors used to be very simple devices; just a metal cylinder with a fuel nozzle at one end (e.g., see Figure 4.3, Brown-Boveri Company's Neuchâtel gas turbine). In some designs, several of them were arranged circumferentially between the compressor exit and turbine inlet and that was that (e.g., see Figure 4.5, the cutout of Jumo-004). There were, of course, some design features for practicality, which are described in Figure 12.13. In any event, the combustion itself (i.e., the "diffusion flame") required enough air for an equivalence ratio of about unity (i.e., $\phi \approx 1$); the remainder was routed around the liner, first performing the cooling duty for the liner and protecting the can from high temperatures, and then being reintroduced into the main stream to dilute the hot gas and achieve the desired TIT. Originally, since emissions were main concerns (smoke was the major "undesirable"), the equivalence ratio was rarely below 0.8 and, as described above in more detail, the diffusion flame did not present a major stability concern across the entire operating envelope. Consequently, since there was ample air left for dilution and liner cooling, component durability was not a big concern either.

[8] The plant, Inland Empire Energy Center in southern California, comprises two 107H single-shaft combined-cycle blocks. For the last several years, it was in the top 20 list of GTCC power plants with the least NOx emissions published in *Electric Power & Light* magazine every year in the November/December issue (available online).

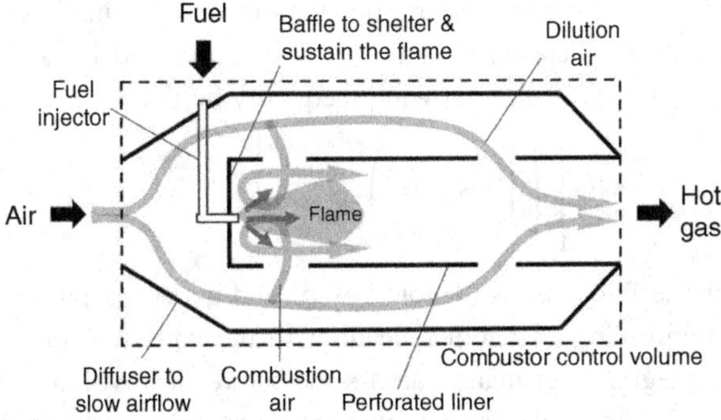

Figure 12.13 Simple can-type combustor.

Due to very stringent emissions regulations, modern combustors do not have the luxury of simplicity afforded by the diffusion flame. Premixed flames require very lean operation with air–fuel mixtures in the primary zone only slightly above the LBO limit (i.e., ϕ in the range of 0.45–0.55; e.g., see Figure 12.3 for a qualitative depiction).

As a result, more than 90 percent of the air coming into the combustor is reserved for the flame, with less than 10 percent available for liner cooling. Let us translate this into some numbers. As an example, consider GE's 7FA.05, an advanced 60-Hz gas turbine rated at 241 MWe and 39.8 percent efficiency at ISO base load (see Table 6.1). The exhaust mass flow rate is 1,188 lb/s. For this class of gas turbine, total chargeable and nonchargeable cooling flows are 23 percent of the compressor airflow. Thus, doing the algebra, with 26.7 lb/s of 100 percent methane fuel (obtained from rated output and efficiency with an LHV of 21,515 Btu/lb), only 894.2 lb/s air reaches the combustor inlet. If one looks at the entire combustor control volume, this translates into a fuel–air ratio of about 0.03. Since the stoichiometric fuel–air ratio for combustion of methane is about 0.06, this corresponds to an equivalence ratio of about 0.5. This, however, is roughly the requirement for lean premixed combustion, which explains why more than 90 percent of the air coming into the combustor (i.e., 894.2 lb/s in this example) is actually used in the combustion. For the sake of argument, let us assume 93 percent, which results in 831.6 lb/s combustion air (corresponding ϕ is 0.55; i.e., on the high side). The compressor discharge air temperature is 810°F for a cycle PR of 18.4 and assumed compressor polytropic efficiency of 93 percent ($\gamma = 1.4$). Substituting these numbers into COMBUST, the flame temperature is found as 2,889°F. Thus, 894.2 – 831.6 = 62.6 lb/s air at 810°F is available for cooling the combustor liner and diluting the hot gas to the TIT of 2,769°F (see Table 6.1).

Consequently, in modern DLN combustors with exit gas temperatures in the range of 1,500–1,600°C (about 2,700–3,000°F) and very limited cooling air budgets, design compromise between performance, emissions, and durability leads to compact designs with very high heat release (in MWth per cubic meter or MMBtu per cubic foot). As a result of this, chemical reaction-caused spatial and temporal fluctuations (also referred to as *oscillations*) in the rate of heat release (which are largely unavoidable) lead to amplified pressure or acoustic fluctuations, which require some attention.

The correlation between heat and pressure fluctuations was originally developed by Lord Rayleigh. The criterion bearing his name is frequently used for assessing combustor instabilities. In essence, if the following inequality holds:

$$\int_T \left(\iiint_V p'q'dV \right) dt > 0,$$

pressure and heat release fluctuations (denoted by p' and q') are in phase (i.e., *resonance*) and thus are reinforcing each other, leading to the amplification of the initial instability [5]. (The integral is evaluated across the entire flow volume, V, and one period of oscillation, T.) In order for the inequality to hold, the phase difference between the two fluctuations or oscillations should be less than 90° ($\pi/2$ in radians). If the phase difference is greater than 90° (i.e., when the oscillations are out of phase), heat release fluctuations dampen the pressure fluctuations. The feedback mechanism at work is described as follows:

- Oscillation of premixed fuel–air flow (i.e., ϕ or λ) leads to
- Oscillation of the flame front, which leads to
- Oscillation of the rate of heat release, which leads to
- Pressure oscillations, which lead to oscillations in ϕ/λ, and so on and so forth.

Combustion instability is known by different names (e.g., pulsations, dynamics, humming, or screech; i.e., low- and high-frequency noise for humming and screech, respectively). Rate of heat release variations are caused by the natural turbulence of the flame and variations in flame speed due to variations in the equivalence ratio, ϕ, fuel and/or air temperature, and/or fuel composition. Those variations produce hot and cold spots, which are then carried with the bulk gas flow. Their coupling with particle acceleration in the bulk gas flow leads to pressure perturbations. Under conditions of resonance (i.e., the heat and pressure oscillations are in phase), oscillations grow, leading to discrete tones at resonant frequencies associated with the acoustic characteristics of the combustor. Such undamped thermoacoustic instabilities manifest themselves as noise. Ultimately, vibrations induced by the pressure oscillations lead to fatigue cracks in the liner, significantly reducing the component life.

The mechanism underlying the process described above can be illustrated via simple mathematics. In turbulence modeling, parameters fluctuating about a mean value are described as

$$p(x,t) = \bar{p} + p',$$

where the second term on the right-hand side denote fluctuations superimposed on the mean. Using this convention, we can expand the equation defining the equivalence ratio as follows:

$$\bar{\phi} + \phi' = \frac{\dfrac{\dot{m}'_f}{\bar{\dot{m}}_f} - \dfrac{\dot{m}'_a}{\bar{\dot{m}}_a}}{1 + \dfrac{\dot{m}'_a}{\bar{\dot{m}}_a}},$$

where subscripts "f" and "*a*" denote fuel and air mass flow rates, respectively. For fixed fuel flow, this can be approximated as

$$\frac{\phi'}{\bar{\phi}} \approx -\frac{\dot{m}'_a}{\bar{\dot{m}}_a}.$$

Since the fuel flow rate is too small in comparison to the air flow rate and mass flow rate is proportional to flow velocity, we can rewrite this correlation as

$$\frac{\phi'}{\bar{\phi}} \approx -\frac{V'}{\bar{V}},$$

where V is the mixture velocity. Noting that

$$\rho V' \propto -p' \text{ and}$$

$$\bar{V} = Ma \cdot a = Ma\sqrt{\gamma R\bar{T}},$$

we obtain

$$\frac{\phi'}{\bar{\phi}} \approx \frac{1}{\gamma Ma}\frac{p'}{\bar{p}}.$$

Thus, with $\gamma \approx 1.33$ and $Ma \approx 0.05$, a 1 percent fluctuation in static pressure inside the combustion chamber can lead to a 15 percent fluctuation in the equivalence ratio. Such an amplification of small acoustic perturbances means that, even though diffusive and turbulent mixing processes will tend to homogenize the mixture as it flows from the fuel injector into the flame zone, fluctuations in ϕ will persist at the flame and modulate the heat release inside the combustion chamber. If the resulting q' is in phase with the original acoustic perturbance, p', this will lead to resonance and growing of the initial disturbance.

Pressure oscillations become critical when their root mean square amplitudes are of the order of 1–2 percent of the static pressure in the combustion chamber. (Since typical flow Mach number in the chamber is about 0.05, using Equation 3.11, it can be verified that static and total pressures in the chamber are roughly equal.) For an advanced gas turbine with a cycle PR of 20, amplitudes of 3–6 psi indicate undesirable combustion dynamics. The acceptable value is about a tenth of that (i.e., 0.3–0.6 psi). These are "ballpark" numbers; typically, cold tone limits are lower than those for hot tones (see below) because they are more harmful to the combustor and downstream hot gas path hardware. The control system of a gas turbine is equipped to monitor the pressure of amplitudes of the first few acoustic mode and take corrective action (i.e., the "auto-tune" systems).[9] In other units, manual tuning by the OEM technician is necessary when the

[9] Typically piezo-electric dynamic pressure transducers installed as close as possible to the combustor chamber.

dynamics approach the alarm limits. If there is no OEM-installed combustor dynamics monitoring system, it is prudent to install one, which may not be cheap, but can prevent much more costly equipment failure.

Combustor dynamics (i.e., pressure oscillations) can be grouped into the following three categories:

- Low frequency (up to 50 Hz), also known as "cold tones" because their amplitude increases with decreasing flame temperature (lean conditions, almost LBO).
- Medium frequency (typically about 100–150 Hz, but can be up to 1,000 Hz), also known as "hot tones" because their amplitude increases with firing temperature and gas turbine load.
- High frequency (>1,000 Hz, but can be lower – most likely indicating resonance due to excitation of a natural frequency by the oscillations), also known as "screech," which is rather uncommon in heavy-duty industrial gas turbines (but is rapidly destructive when it happens because it takes a very short time to reach 1 million cycles at very high frequencies).

Prevention of combustion instability is a prime concern of combustor designers, who resort to passive and active measures in this endeavor. The former require design changes to impact flame response (e.g., the "quat" circuit in the DLN 2.6+ combustor) whereas the latter require changes to the control logic (e.g., auto-tune algorithms). Tight control and/or monitoring of fuel composition and temperature are imperative for the prevention of thermoacoustic instabilities.

In this last regard, an important parameter is the modified *Wobbe index* (MWI), which is defined as

$$\mathrm{MWI} = \frac{\mathrm{LHV}}{\sqrt{\mathrm{SG} \cdot \mathrm{T}_{\mathrm{fuel}}}} = \frac{\mathrm{WI}}{\sqrt{\mathrm{T}_{\mathrm{fuel}}}},$$

where WI is the Wobbe index, SG is the *specific gravity* of the fuel relative to air, LHV is in Btu/scf (i.e., it is the volumetric heating value), and fuel temperature is in degrees Rankine. (In some references, MWI is referred to as the gas index [GI].) Thus, MWI is a dimensional parameter and, as such, using US customary or SI units will result in different values of MWI for exactly the same conditions. For an ideal gas, the formula becomes

$$\mathrm{MWI} = \frac{\mathrm{LHV}}{\sqrt{\frac{\mathrm{MW}_{\mathrm{fuel}}}{28.96} \cdot \mathrm{T}_{\mathrm{fuel}}}}.$$

The Wobbe index is a relative measure of the energy injected into the combustor at a fixed fuel nozzle PR. The original definition of the Wobbe index excluded the fuel temperature. In modern gas turbines in combined-cycle applications with performance fuel gas heating, which can take the fuel to temperatures well above its pipeline value, including the temperature results in a more accurate measure of the fuel energy content.

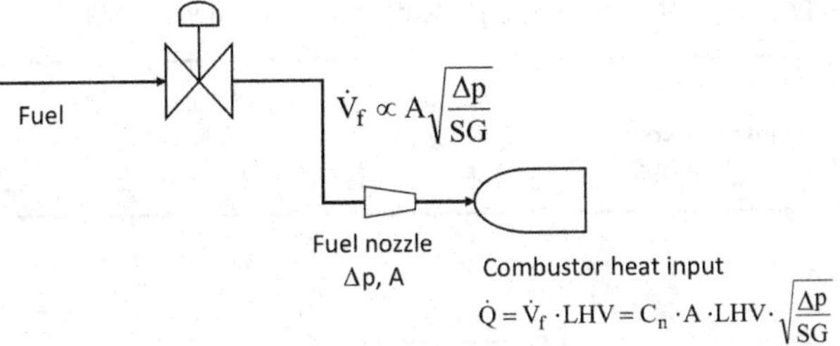

Figure 12.14 Simplified gas turbine fuel delivery system.

The importance of MWI can be understood by using a very simple (and partial) schematic of the gas turbine fuel delivery system as shown in Figure 12.14.

For a given fuel nozzle with cross-sectional area A and pressure drop Δp, the volumetric flow rate of the gas turbine fuel (i.e., scf/s or "scuff" per second) is given by

$$\dot{V}_f \propto A\sqrt{\frac{\Delta p}{SG}}.$$

Consequently, heat input to the combustor (in Btu/s) is given by

$$\dot{Q} = \dot{V}_f \cdot LHV = C_n A \cdot LHV\sqrt{\frac{\Delta p}{SG}},$$

where C_n is a nozzle flow factor. In other words, heat input is proportional to the product of the Wobbe index and the square root of the nozzle pressure drop, i.e.,

$$\dot{Q} \propto MWI\sqrt{\Delta p}.$$

Alternatively, for constant heat input, the nozzle pressure drop is inversely proportional to the square of MWI, i.e.,

$$\Delta p \propto \frac{1}{MWI^2}.$$

The effect of MWI variation on heat input and the nozzle pressure drop is illustrated by the data in Table 12.9. Typically, gas turbine OEMs specify an allowable MWI range to ensure that required fuel nozzle PRs are maintained during all combustion/turbine modes of operation. Going beyond the allowable MWI range in either direction will result in combustion instability.

Note that MWI also describes the interaction between air and fuel as a function of fuel nozzle geometry, i.e.,

$$\frac{d}{D} \propto \frac{1}{MWI}\frac{\dot{Q}}{A},$$

where d is the penetration depth of the fuel jet and D is the fuel nozzle diameter. Since $A \propto D^2$,

Table 12.9 Combustor heat input and fuel nozzle pressure drop variation with MWI

MWI Variation	±5%	±10%	±15%
Heat Input Variation (Constant Nozzle Δp)	±5%	±10%	±15%
Δp Variation (Constant Heat Input)	–9% to +11%	–17% to +23%	–24% to +38%

$$d \propto \frac{1}{MWI}\frac{\dot{Q}}{D} \text{ or}$$

$$d \propto \frac{\sqrt{T_{fuel}}}{WI}\frac{\dot{Q}}{D},$$

so that, for a given heat input, if D × MWI is constant, fuel jet penetration depth is constant. Therefore,

- A fuel with a lower-than-design Wobbe index can be compensated by a lower T_{fuel} (to the extent possible during operation) or a larger fuel nozzle (hardware change).
- A fuel with a higher-than-design Wobbe index can be compensated by a higher T_{fuel} or a smaller fuel nozzle.

As discussed earlier, each combustor has a characteristic frequency set primarily by its geometry and other design factors. When the pressure inside the combustor rises, the fuel flow through the fuel nozzle is reduced due to the lower fuel nozzle pressure drop. Similarly, a decrease in the combustor pressure leads to an increase in the fuel flow. As a result, combustor fuel flow fluctuates in phase with the pressure fluctuations inside the combustor. How the combustor might transition from a stable to an unstable regime via an increase in MWI is illustrated in Figure 12.15.

Allowable MWI variation used to be ±5% in the early F–class gas turbines. Modern H- and J–class gas turbines have an extended operability range commensurate with the fuel flexibility requirements (blending natural gas with ethane or propane,[10] limited natural gas reserves in some countries, etc.) up to ±15% or even higher. Calculated MWI values for typical hydrocarbon fuels are listed in Table 12.10.

The other three key combustor operability considerations are blowout (or, more specifically, lean blowout), flashback, and autoignition.

Blowout, as the term suggests, describes a situation where the flame is detached from its anchoring location and physically blown out of the combustor. Blowout occurs when the chemical reaction timescale, τ_C, is larger than the characteristic residence time, τ_R. The former is a measure of the reactivity of a propagating flame, i.e.,

$$\tau_C = \frac{\lambda}{S_L^2},$$

[10] Ethane and propane, for example, are becoming available in increasing quantities due to shale gas production in the USA.

Table 12.10 MWI of typical hydrocarbon fuels (fuel temperature 77°F)

	Units	Natural Gas	Methane	Ethane	Propane
Fuel Density	lb/cuft	0.0445	0.0424	0.0794	0.1165
Fuel LHV	Btu/scf	925	911	1,622	2,320
	MJ/m^3	34.5	33.9	60.5	86.4
SG		0.58	0.56	1.04	1.53
MWI	Btu/scf	52.2	52.8	68.7	81.1
	MJ/m^3	1.95	1.97	2.56	3.02

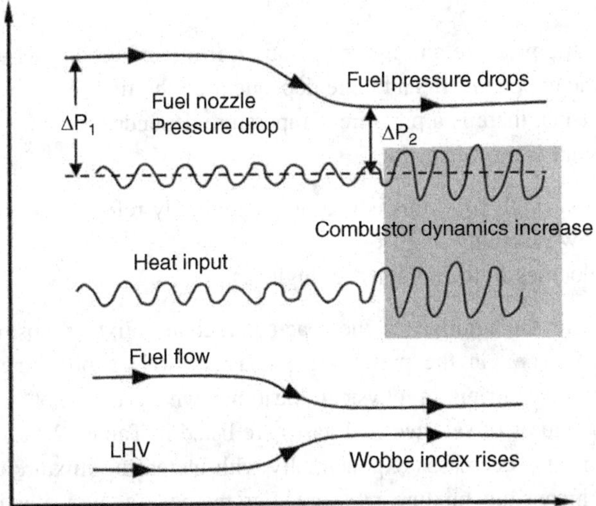

Figure 12.15 Schematic description of combustor dynamics with MWI change.

where λ is thermal conductivity (m^2/s) and S_L is the laminar flame speed of the fuel–air mixture (m/s). In other words, τ_C is a measure of the time needed to complete the combustion reaction via heat conduction from the burned gas to the unburned fuel–air mixture.

One definition of characteristic residence time is the ratio of recirculation zone length to the bulk velocity in the combustor, i.e.,

$$\tau_R = \frac{L}{V}.$$

The ratio of the two timescales is the *Damköhler* number,

$$\text{Da} = \frac{\tau_R}{\tau_C} \text{ or}$$

$$\text{Da} = \frac{LS_L^2}{V\lambda}.$$

The blowout limit is defined by a limiting Damköhler number (i.e., Da_{bo}). In other words, blowout occurs when

$$Da < Da_{bo} \text{or}$$

$$V > \frac{S_L^2}{\lambda} \frac{L}{Da_{bo}}.$$

The parameter L/Da_{bo} is characteristic of a combustor design. Furthermore, $\lambda = f(T)$ and the laminar flame speed is

$$S_L \propto \frac{T^m}{p^n},$$

where p and T are the pressure and the temperature of the unburned fuel–air mixture, respectively, and the exponents m and n are dependent on the fuel (e.g., m = 2 and n = 0.5 for methane). Thus, there is a pressure–temperature dependence, too.

In general, blowout is a concern for

- Lean operation (high τ_C) – this is why it is commonly referred to as LBO;
- Fuels with low reactivity (high τ_C);
- High gas velocities in the combustor (high τ_R).

Autoignition refers to the ignition of the reactive fuel–air mixture upstream of the combustion chamber (i.e., in the premixing passages). It is a phenomenon of the combustion reaction occurring in physical locations where it is not supposed to. Autoignition temperatures of selected fuel gases are listed in Table 12.11.

Autoignition time, τ_{ai}, decreases exponentially with increasing mixture temperature and linearly with increasing mixture pressure. For methane at typically F-class gas turbine compressor discharge conditions, τ_{ai} is of the order of 200 ms. Natural gases containing higher hydrocarbons (C2+) have lower values. Note that, in Table 12.11, higher hydrocarbons have lower autoignition temperatures, too.

Typical residence times in heavy-duty industrial gas turbine combustors between the premixer and the flame zone are less than 5 ms. Thus, autoignition is not a big concern (*except* in the reheat or LP combustor of sequential combustion GT24–26).

Flashback is a problem characteristic of premixed combustion, where the flame propagates upstream into the premixing passages, which are not designed for high

Table 12.11 Autoignition temperatures for stoichiometric fuel–air mixtures at 1 bar

	°C	°F
CO	605	1,121
H_2	560	1,040
CH_4	610	1,130
C_2H_6	525	977
C_3H_8	470	878

temperatures. The flame is anchored where the turbulent flame speed and local gas velocity balance each other. Turbulent flame speed, S_T, can be related to the laminar flame speed as

$$S_T = S_L f(u', \ell, y_i),$$

where u' is the turbulent velocity fluctuation, l is the turbulent length scale, and y_i is the composition of the fuel–air mixture. Turbulent flame speeds in gas turbine combustors burning natural gas are about 10–20 m/s. (Since the turbulent flame speed is much more difficult to estimate than the laminar one, the latter is frequently used in the assessment of flashback probability.) Downstream of the compressor outlet diffuser, air velocity is typically slowed down to about 30–35 m/s. Much lower velocities increase the likelihood of flashback, whereas much higher velocities increase the likelihood of blowout. As will be discussed later, the much higher flame speed of H_2 is a significant concern in DLN combustor design for 100 percent H_2 fuel.

12.5 Emission Regulations

According to the Clean Air Act (CAA), the US Environmental Protection Agency (EPA) is required to set National Ambient Air Quality Standards (NAAQS; 40 CFR Part 50) for pollutants considered harmful to public health and the environment. According to the CAA definition, there are two types of national ambient air quality standards:

1. *Primary* standards provide public health protection, including protecting the health of "sensitive" populations such as asthmatics, children, and the elderly.
2. *Secondary* standards provide public welfare protection, including protection against decreased visibility and damage to animals, crops, vegetation, and buildings.

The EPA has set NAAQS for six principal pollutants, which are called "criteria" air pollutants. The latest standards, current at the time of writing this book, are listed in Table 12.12. Units of measure for the standards are parts per million (ppm) by volume, parts per billion (ppb) by volume, and micrograms per cubic meter of air (μg/m^3). The reader should bear in mind that different standards may be applicable in different countries. Furthermore, these standard are reviewed periodically and are subject to change.

Clearly, in order to maintain the "air quality" commensurate with the values in Table 12.12, the EPA has to regulate the emissions from the "sources of pollutants" as well. Gas turbines and duct burners in HRSGs in combined-cycle power plants are subject to Subpart KKKK of the Standards of Performance for New Stationary Sources, 40 CFR 60.4300 through 60.4420 (Standards of Performance for Stationary Combustion Turbines). In particular, table 1 of Subpart KKKK of Part 60 specifies nitrogen oxide emission limits for new stationary gas turbines and is reproduced in Table 12.13.

Table 12.12 National Ambient Air Quality Standards for six principal pollutants, which are called "criteria" air pollutants. P = primary; S = secondary; PM = particulate matter

Pollutant		Type	Averaging Time	Level	Form
CO		P	8 hours	9 ppm	Not to be exceeded more than once per year
			1 hour	35 ppm	
Lead (Pb)		P & S	Rolling 3-month average	0.15 μg/m^3	Not to be exceeded
NO_2		P	1 hour	100 ppb	98th percentile of 1-hour daily maximum concentrations, averaged over 3 years
		P & S	1 year	53 ppb	Annual mean
Ozone (O_3)		P & S	8 hours	0.07 ppm	Annual fourth-highest daily maximum 8-hour concentration, averaged over 3 years
PM	PM2.5	P	1 year	12.0 μg/m^3	Annual mean, averaged over 3 years
		S	1 year	15.0 μg/m^3	Annual mean, averaged over 3 years
		P & S	24 hours	35 μg/m^3	98th percentile, averaged over 3 years
	PM10	P & S	24 hours	150 μg/m^3	Not to be exceeded more than once per year on average over 3 years
SO_2		P	1 hour	75 ppb	99th percentile of 1-hour daily maximum concentrations, averaged over 3 years
		S	3 hours	0.5 ppm	Not to be exceeded more than once per year

A new advanced class gas turbine rated at, say, 300 MWe and 41% burns 300,000 / 0.41 / 1.05506 × 0.0036 = 2,500 MMBtu/h. (You can also refer to Table 3.1.) Thus, according to 40 CFR Part 60 and table 1 therein (reproduced in Table 12.13), the applicable standard is 15 ppm of NOx at 15 percent O_2 (see below for an explanation of the units used in NOx emission quotes). In many cases, state and/or local authorities may impose even more stringent requirements. Furthermore, developers of power plant projects frequently propose "single-digit" emission limits above and beyond those required for compliance (e.g., 2 ppm at 15 percent O_2) in order to make the permitting process as painless as possible.

In addition to NOx, 40 CFR 60.4330 sets sulfur dioxide (SO_2) emission limits to be met by gas turbines. In the continental USA, except Alaska, for compliance, emission of exhaust gas containing SO_2 in excess of 110 ng/J (0.90 lb/MWh) is prohibited. Furthermore, any fuel that contains sulfur content that could result in sulfur emissions in excess of 26 ng/J SO_2 (0.060 lb/MMBtu) is also prohibited.

Pipeline-quality natural gas typically has a sulfur content of 0.4 grains per 100 scf (standard cubic feet or "scuff"). Potential SO_2 emissions for this level of sulfur content are 0.011 lb/MMBtu. Typically, owners/operators of natural gas-fired power plants enter into a contract with the fuel supplier requiring the sulfur content not to exceed a specified level. Usually, in accordance with this stipulation, fuel suppliers provide a certification indicating the sulfur content and conduct periodic fuel sampling to demonstrate compliance.

Table 12.13 Nitrogen oxide emission limits for new stationary combustion turbines (from table 1 of Subpart KKKK of Part 40 CFR Part 60); ng/J is nanograms per Joule

Combustion Turbine Type	Combustion Turbine Heat Input at Peak Load (HHV)	NOx Emission Standard
New turbine firing natural gas, electric generating	≤50 MMBtu/h	42 ppm at 15% O_2 or 290 ng/J of useful output (2.3 lb/MWh)
New turbine firing natural gas, mechanical drive	≤50 MMBtu/h	100 ppm at 15% O_2 or 690 ng/J of useful output (5.5 lb/MWh)
New turbine firing natural gas	>50 MMBtu/h and ≤850 MMBtu/h	25 ppm at 15% O_2 or 150 ng/J of useful output (1.2 lb/MWh)
New, modified, or reconstructed turbine firing natural gas	>850 MMBtu/h	15 ppm at 15% O_2 or 54 ng/J of useful output (0.43 lb/MWh)
New turbine firing fuels other than natural gas, electric generating	≤50 MMBtu/h	96 ppm at 15% O_2 or 700 ng/J of useful output (5.5 lb/MWh)
New turbine firing fuels other than natural gas, mechanical drive	≤50 MMBtu/h	150 ppm at 15% O_2 or 1,100 ng/J of useful output (8.7 lb/MWh)
New turbine firing fuels other than natural gas	>50 MMBtu/h and ≤850 MMBtu/h	74 ppm at 15% O_2 or 460 ng/J of useful output (3.6 lb/MWh)
New, modified, or reconstructed turbine firing fuels other than natural gas	>850 MMBtu/h	42 ppm at 15% O_2 or 160 ng/J of useful output (1.3 lb/MWh)
Modified or reconstructed turbine	≤50 MMBtu/h	150 ppm at 15% O_2 or 1,100 ng/J of useful output (8.7 lb/MWh)
Modified or reconstructed turbine firing natural gas	>50 MMBtu/h and ≤850 MMBtu/h	42 ppm at 15% O_2 or 250 ng/J of useful output (2.0 lb/MWh)
Modified or reconstructed turbine firing fuels other than natural gas	>50 MMBtu/h and ≤850 MMBtu/h	96 ppm at 15% O_2 or 590 ng/J of useful output (4.7 lb/MWh)
Turbines located north of the Arctic Circle (latitude 66.5° north), turbines operating at less than 75% of peak load, modified and reconstructed offshore turbines, and turbines operating at temperatures less than 0°F	≤30 MW output	150 ppm at 15% O_2 or 1,100 ng/J of useful output (8.7 lb/MWh)
Turbines located north of the Arctic Circle (latitude 66.5° north), turbines operating at less than 75% of peak load, modified and reconstructed offshore turbines, and turbines operating at temperatures less than 0°F	>30 MW output	96 ppm at 15% O_2 or 590 ng/J of useful output (4.7 lb/MWh)
Heat recovery units operating independent of the combustion turbine	All sizes	54 ppm at 15% O_2 or 110 ng/J of useful output (0.86 lb/MWh)

In almost all cases, a state-of-the-art power plant is equipped with a *continuous emissions monitoring system* (CEMS) to comply with 40 CFR Part 60 requirements and demonstrate compliance with NOx and other emission limit requirements. Note that there is other equipment in a natural gas-fired power plant (i.e., in addition to the gas turbine[s] and the HRSG duct burner[s]; e.g., the auxiliary steam boiler and the "black start" unit – typically, a gas- or fuel oil-fired reciprocating engine). They are also required to comply with applicable 40 CFR Part 60 standards.

OEMs of most advanced H- or J-class gas turbines with DLN combustors can offer NOx emissions as low as 9 ppm at 15 percent O_2 so that achieving an ultra-low 2 ppm level requires an SCR system. In combined-cycle applications, the SCR is placed between the HRSG tube bundles (typically downstream of the high-pressure [HP] evaporator section), where the gas temperature is suitable for SCR operation. The optimum operating temperature is a function of the SCR catalysts; typically, it is above 350°C (570°F). In simple-cycle (i.e., gas turbine-only) applications, either a high-temperature catalyst is employed or the gas turbine exhaust gas is cooled via air injection. The exact configuration is subject to a cost–performance trade-off involving catalyst efficiency (i.e., desired NOx reduction; e.g., 80 percent, 90 percent, etc.), cost, and catalyst life. Ammonia (NH_3) is injected upstream of the SCR catalyst, where it combines with oxides of nitrogen in the exhaust gas to form nitrogen and water vapor. Often, there is a separate catalyst layer designed to oxidize CO to CO_2 upstream of the *ammonia injection grid*. The reason for this is the strong dependence of the CO conversion rate on gas temperature, which means that it is particularly high at low gas turbine loads.

A "multi-function" SCR catalyst can achieve NOx and CO/VOC[11] reduction in one layer. Elimination of the separate catalyst section reduces the exhaust gas pressure drop with beneficial impact on gas turbine performance. Note that an SCR system can add 2–3 inches of water column to the exhaust gas pressure loss in the HRSG. (Each extra inch of water column in gas turbine exhaust pressure is worth about 0.4 percent in lost output and 0.1 percent in higher heat rate.) An extra inch of water column pressure loss can be attributed to the CO catalyst.

Emissions of NOx are typically quoted at a reference condition on a volumetric basis (e.g., 25 ppmvd [parts per million by volume dry] at 15 percent O_2 [for gas turbines] or 5 percent O_2 [for some gas engines]). For the fraction of NOx in the gas turbine exhaust gas, conversion from a mass basis to a volume basis (dry) can be done as follows:

$$v = \frac{\dfrac{\dot{m}_{NO}}{\dot{m}_{exh}} \cdot \dfrac{MW_{exh}}{MW_{NO}}}{1 - y_{H2O}}, \tag{12.19}$$

where the terms in the numerator refer to the mass flow rates of NO (proxy for NOx in the exhaust gas), gas turbine exhaust gas, and the molecular weight thereof. The term in the denominator is the fraction of water vapor in the exhaust gas on a volume basis. Correction to X percent O_2 results in (assuming air is 21 percent oxygen)

$$v_c = 10^6 \frac{\left(0.21 - \dfrac{X}{100}\right) \cdot v}{0.21 - (1 - y_{H2O}) \cdot y_{O2}}. \tag{12.20}$$

For a typical heavy-duty industrial gas turbine, exhaust gas molecular weight is about 29 lb/lbmol; exhaust gas oxygen and water vapor volume fractions are 12 and 9 percent, respectively (for 100 percent methane fuel). Thus, for a machine rated at 15 ppmvd

[11] Volatile organic compound.

NOx (15 percent O_2; i.e., in compliance with 40 CFR Part 60), from Equation 12.20, v is found as 0.0000252. If the exhaust gas mass flow rate is 1,175 lb/s, from Equation 12.19, the NOx mass flow rate is about 100 lb/h. For a 230–MW unit, this corresponds to about 0.44 lb/MWh NOx emissions (as NO). Thus, as a rule of thumb, about 0.4 lb/MWh is a good NOx estimator for modern units with DLN combustors. (The same calculations can be done for CO emissions by using the molecular weight of CO, 28, instead of that of NO, 30, in Equation 12.19.) Note that, if we had used NO_2 as a proxy for NOx in Equation 12.19, the result would have been twice as high (i.e., about 1 lb/MWh).

Another method, provided by the EPA, is based on an *emissions index* (EINOx) in units of lb NOx per 1,000 lb fuel, which is proportional to the exhaust NOx emission levels in ppmv by a constant, K[12]:

$$\frac{v_c}{\text{EINOx}} = K. \tag{12.21}$$

Equation 12.21 and K values (for v_c at 15 percent O_2) were provided by OEMs for different gas turbines and fuels (e.g., K = 11.6 for methane, K = 12.1 for pipeline-quality natural gas, and K = 13.2 for number 2 fuel oil). The calculation above with NO_2 as a proxy for NOx suggests a K value of 8.9.

The EPA's Method 19 allows for the use of the following estimating factors:

- 1 ppmv NOx (at 15 percent O_2) = 0.0036 lb/MMBtu (natural gas fuel)
- 1 ppmv NOx (at 15 percent O_2) = 0.0040 lb/MMBtu (distillate)

Note that Equation 12.21 with the given K values returns a value of 0.004 in lb/MMBtu of NOx emissions for each ppmv of NOx. Using a weighted average molecular weight of 31.6 lb/lbmol for NOx in Equation 12.19 (i.e., a mixture of mostly NO and some NO_2), one obtains a K value of 12.9 and 0.0036 lb/MMBtu for each 1 ppmv of NOx from the sample calculation above.

12.5.1 Emissions Guarantees

The foregoing discussion emphasized the prime importance given to reduction of emissions by regulatory agencies and customers, which became a key factor in obtaining a permit to build a gas turbine power plant (or any fossil fuel-fired power plant for that matter). Consequently, the engineering, procurement, and construction contractor (in close coordination and cooperation with the gas turbine OEM) guarantees the stack emissions and proves them during the performance test at the end of the commissioning process. In the case of supplementary or duct-fired units, the HRSG OEM and their guarantees (with or without an SCR) enter the picture as well.

[12] From "Alternative Control Techniques Document – NOx Emissions from Stationary Gas Turbines," US Environmental Protection Agency, EPA-453/R-93–007, January 1993, Office of Air and Radiation, Office of Air Quality Planning and Standards, Research Triangle Park, NC.

Guarantees are typically made for unfired and duct-fired (if applicable) operation over a load and ambient range. In the unfired case, emissions guarantees cover the gas turbine load range from MECL to 100 percent load within the ambient temperature range between a minimum and maximum value (e.g., 5°F and 100°F, respectively). Guarantees cover NOx, NH_3 slip (if there is an SCR), CO, VOC, and particulate matter (PM) and are based on testing at the exhaust stack (the HRSG stack for a combined-cycle power plant) in accordance with EPA test methods.

For example, demonstration of an NOx guarantee can be based on the average of three test runs at minimum and maximum gas turbine loads per EPA Test Method 7E. For NH_3 slip, the applicable method is EPA Test Method 027. In addition, note the following requirements:

- For CO guarantees, EPA Test Method 10
- For VOC guarantees, EPA Test Methods 25A and 18
- For PM guarantees, EPA Test Methods 5 and 202
- O_2 measurements for ppmvd corrections in accordance with EPA Method 3A

12.6 Fuel Flexibility

In the 1950s and 1960s, gas turbines were primarily used for peaking power and burned liquid fuels such as residual distillates and blends. A gas turbine book written back then would have had to devote a good part of its coverage to the subject of fuel treatment and related issues. Modern gas turbines for electric power generation rarely use liquid fuels and, if they do, it is only as a secondary (backup) fuel for limited duration. (Due to the existing fuel supply infrastructure and the ability to easily store it on site, number 2 fuel oil [i.e., essentially number 2 diesel fuel] is by far the most common backup fuel.) Therefore, the focus herein is on gaseous fuels.

Once a gas turbine (and combustor) is specified and delivered for a contractually specified fuel per OEM requirements, variations in the following parameters during field operation should be carefully monitored:

- Fuel composition
- Fuel pressure at the combustor inlet
- Fuel temperature at the combustor inlet

The key parameter for gauging the combined impact of the variations in the parameters listed above on combustor stability is the MWI, which was covered in detail in Section 12.4. The typical operability range for MWI is ±5 percent for stability. Modern combustor designs claim to extend this range to ±10 percent or even higher. However, one should also be cognizant of the impact of MWI variation on NOx, CO, and CO_2 emissions, which may render a stated capability moot in practice.

One aspect of fuel flexibility is the ability of the DLN combustor to continue operating within its stability limits even when the composition and heating value of the fuel gas is subject to significant variation. Two major gaseous fuels for modern gas turbines are liquefied natural gas (LNG) and pipeline natural gas. While there is no single international standard pertaining to either product across the globe, in the USA, natural gas must have more than 70 percent methane [15], while the EU standard requires at least 80 percent methane [16]. Goldmeer et al. [17] conducted a study to evaluate the variation in US shale gas and LNG compositions and heating values. They found that the LNG, with a mean value of 993 Btu/scf and a standard deviation of ~34 Btu/scf, was more uniform vis-à-vis the shale gas with corresponding values of 951 and ~78 Btu/scf, respectively.

According to the same source, based on a reference fuel, which was defined as an average fuel composition for LNG and shale gas and a fixed fuel temperature of 400°F, all but one of the LNG samples were within ±3 percent of the average. For the US shale gas, the largest MWI deviation from the average was ~14.5 percent. However, many of the shale gas compositions were within ±5 percent, and all but four of the wells in the data set were within ±10 percent. These variations are well within the MWI range of modern DLN combustors.

Natural gas from the production well often contains inert gases such as nitrogen and carbon dioxide, or heavier hydrocarbon species such as ethane, propane, and butane. There are strict specifications on the composition of natural gas transported through the pipeline network so that the raw product gas is treated in order to provide the "pipeline-quality" gas. In the process, natural gas liquids (NGLs) are created as a by-product at the gas processing plant. Propane, butane, and their mixtures are a subset of NGLs commonly referred to as LPG. LPG specifications and compositions vary greatly around the world.

Properties of selected gaseous fuels are listed in Table 12.14.

- 100 percent CH_4 is the most commonly used gas turbine rating fuel.
- LNG in the table has 87 percent (v) methane with 12.3 percent higher hydrocarbons (C2+). This composition is typical of pipeline natural gas, too.

Table 12.14 Properties of selected gaseous fuels

		Units	CH_4	LNG	LPG	H_2	CO
Density	ρ	kg/m^3	0.718	0.822	2.536	0.090	1.251
		lb/cuft	0.045	0.051	0.158	0.006	0.078
Lower Heating Value	LHV	MJ/kg	50.1	48.8	46.0	120.0	10.1
		MJ/m^3	35.9	40.1	116.6	10.8	12.6
		Btu/lb	21,522	20,993	19,772	51,578	4,342
		Btu/cuft	965	1,077	3,130	290	339
Wobbe Index	WI	MJ/m^3	48.2	50.3	80.3	40.9	12.9
Air–Fuel (ϕ = 1)			17.4	16.9	15.7	34.5	2.5
Flame Temperature	T_F	°C	2,206	2,214	2,253	2,376	2,370
		°F	4,003	4,017	4,087	4,309	4,298
Laminar Flame Speed	S_L	cm/s	54.0	62.0	89.0	770.0	2.7
(ϕ = 0.5)			12.0	14.0	20.0	43.0	0.8
Chemical Reaction Time	τ_c	s	2.0E–05	1.0E–05	5.0E–06	2.0E–07	6.0E–03
(ϕ = 0.5)			4.0E–04	3.0E–04	1.0E–04	4.0E–05	8.0E–02
Autoignition Time at 1,000°C	τ_{ai}	s	1.0E–03	3.0E–04	3.0E–04	2.0E–05	4.0E–02

- LPG in the table is a 50–50 propane–butane mixture (Grade B LPG in Europe; HD5 and HD10 grades are 90 percent propane [18]).
- Hydrogen and CO are the major constituents of the "synthetic gas" (syngas) product in coal gasification. Hydrogen is also a widely researched gas turbine fuel due to there being zero CO_2 in the combustion products.

Propane's LHV is 2,244 Btu/scf; butane's LHV is 2,919 Btu/scf (at 77°F, 14.7 psia). In contrast, methane's LHV is less than 900 Btu/scf; natural gas, including LNG, has an LHV range of 800–1,200 Btu/scf (methane is the major component). High-Btu (in volumetric terms; i.e., Btu/scf or kJ/scm) hydrocarbon fuels can and have been used in gas turbines. When LPG is used as a gaseous fuel, it must be heated to high temperatures to prevent condensation of the constituent heavy hydrocarbons. Special injectors are required to ensure proper fuel metering. In liquid form, the design of the fuel system requires special care due to the very low viscosity of the LPG (e.g., special-design pumps). Other possible problems to watch out for are waxing (if the fuel temperature becomes too low) and vapor lock (due to premature vaporization). Propane and LPG in gaseous form can be useful as "bridging" fuels until natural gas becomes available. Due to their lower costs and favorable emissions characteristics, they are more suitable in that capacity vis-à-vis diesel and fuel oils.

Examples of low-Btu gaseous fuels used in gas turbines for electric power generation are:

- Syngas from gasification (in integrated gasification gas turbine [IGCC] power plants) with H_2 and CO as the major components contributing to the heating value.
  - Air-blown gasifier syngas, 100–150 Btu/scf.
  - O_2-blown gasifier syngas, 200–400 Btu/scf
- Steel production gases, i.e.,
  - Blast furnace gas (BFG), ~100 Btu/scf.
  - Coke oven gas (COG), ~500 Btu/scf.

A gas turbine can burn any of these gaseous fuels (and their blends; e.g., BFG + COG in steel mill applications) with or without modifications. The "magic" fuel for gas turbine utilization is, of course, 100 percent hydrogen, which, however, creates difficulties, especially for the DLN combustors.

For almost all low-Btu fuels, the combustor is of the diffusion flame type. The key enabler in DLN combustor technology is lean premixed flame. The presence of a high hydrogen content in most low-Btu syngas fuels and the concomitant high flame speeds preclude the utilization of DLN combustors in such applications (this might change – see Section 12.6.3). Furthermore, a second conventional fuel (e.g., natural gas or fuel oil) is typically required for startup and shutdown. The combustion and control systems are designed to operate over the entire load range on either fuel. Depending upon the quantity of syngas available (e.g., during the construction phase or when the gasifier unit is down), the gas turbine can be operated in a variety of fuel conditions ranging from co-firing (i.e., conventional fuel and syngas) to full syngas or conventional fuel (natural gas but not liquid fuel) firing at rated load conditions [19].

In very low-Btu and low-hydrogen fuels such as steel mill gases, LBO is the major design problem. This is a direct result of very high volumetric fuel flow, which leads to low residence times (high velocity) and is exacerbated by reduced chemical reactivity (low hydrogen). Reduced residence times and reaction rates combined with CO coming from the fuel is problematic from a CO emissions perspective. However, due to the same reasons, NOx emissions are less of a concern. In fact, fuel-bound NH_3 can be the leading source of NOx emissions rather than the thermal mechanism. This is why a diffusion flame combustor is the preferred solution in steel mill gas turbines, where typically a blend of BFG and COG is the main fuel. The reader is referred to the papers from two major OEMs for more details on steel mill applications [20,21].

For a perspective on fuel flexibility from two major OEMs, the reader is referred to Welch and Igoe [22] and Jones et al. [23]. From a small industrial gas turbine perspective, the paper by Meier et al. is a good source of information [24]. For an excellent overview of fuel flexibility in gas turbines, please consult the chapter written by Huth and Heilos [25].

12.6.1 Burning Heavy Fuels

There are many good references available in the literature focusing on heavy fuel treatment and operating concerns (e.g., [26–28]). The user is encouraged to consult them for more information on hot corrosion problems in liquid fuel-burning gas turbines and how to prevent them.

All liquid fuels are produced in an oil refinery via distillation of crude oil. Classification of distillation products, from the top of the distillation column (low boiling point and low viscosity) to its bottom (high boiling point and high viscosity), goes as follows: butane and lighter products, gasoline blending components, naphtha, kerosene or jet fuel, distillates (diesel, heating oil), heavy gas oil, and residual fuel oil.

From a gas turbine application perspective, they can be divided into two groups: relatively "clean" distillates and truly "nasty" residual fractions. Examples of the former are number 2 fuel oil (chemically similar to the diesel fuel used in cars and trucks), which is the most common backup fuel and naphtha.

Heavy residuals and crude oil are "last resort" fuels. Therefore, their use is limited to certain regions in the world (i.e., near the oil fields in Saudi Arabia). There is only one residual oil-fired gas turbine combined-cycle power plant in the USA, which is located in Hawaii.[13] Kalaeloa cogen plant comprises two ABB/Alstom E–class gas turbines (GT11NM) rated at 86.4 MWe each, which burn low-sulfur heavy fuel oil that is supplied by the refinery, which is the steam host of the cogen plant. (Propane is used for ignition; number 2 fuel oil is used during startup and shutdown.) Net output and efficiency at guarantee conditions are 193 MW and 42.5 percent. Emission limits are 130 ppmv NOx, 98 ppmv SO_2, and 14 ppm CO.

[13] See the article "Kalaeloa Cogeneration Plant: Continual Improvement Defines Black-Oil-Fired Combined Cycle," in *Combined Cycle Journal* 2Q, 2009 issue. Available online at www.ccj-online.com/2q-2009/kalaeloa-cogeneration-plant-continual-improvement-defines-black-oil-fired-combined-cycle/.

Table 12.15 Comparison of ASL with number 2 fuel oil (diesel fuel)

Parameter	Number 2 Fuel Oil	ASL (Typical)
Flashpoint (°C)[a]	60	–25
Vapor Pressure at 40°C (kPa)	7	52.7
Kinematic Viscosity at 40°C (mm^2/s)	3	1.25
LHV (kJ/kg)	43,000	42,494
Density at 15°C (kg/m^3)	840	780.8

[a] The temperature at which a liquid fuel will produce vapors sufficient to support combustion once ignited.

Heavy residuals and crudes contain high amounts of alkali metals (Na, K) and heavy metals (V, Ni, etc.), which result in "ash" deposits and high-temperature corrosion in gas turbine hot gas path components. Treatment includes "fuel washing" of the raw fuel to dilute and extract water-soluble salts (if the Na + K content is more than 2.1 ppm) and addition of magnesium-based additives to inhibit the vanadic oxidation processes (if Va content is more than 0.5 ppm).

In principle, only older E–class gas turbines with diffusion combustors are suitable for burning such fuels because the TIT must be limited to about 2,000°F to minimize the formation of hard ash deposits on turbine blades. Nevertheless, GE managed to verify the applicability of at least one special class of crude, Arabian Super Light (ASL), to an F–class gas turbine with a DLN combustor [29]. Twelve 7FA.05 gas turbines capable of burning ASL are in operation in Saudi Electricity Company's (SEC) Power Plant (PP) 12 in Riyadh, Saudi Arabia (the first four units went commercial in 2014).

As illustrated by the data in Table 12.15, apart from the flashback risk due to its lower flashpoint, ASL is a feasible alternative to diesel fuel. Siemens also verified that the DLN combustor in their SGT6-5000F gas turbine (Siemens refers to it as ULN – ultra-low NOx) can accommodate ASL by varying the water–fuel ratio. Their rig testing at full pressure and full firing established that combustion dynamics, flashback, and emissions with ASL are similar to those with diesel fuel.[14]

12.6.2 Dual Fuel Blues

As mentioned earlier, the most common backup fuel is number 2 fuel oil. It is typically used in emergencies when the natural gas supply is unexpectedly interrupted but the unit must run to meet the demand. In some of the older F–class units, the liquid fuel system required constant attention (if it was not used regularly) to prevent coking in the nozzles and to ensure that the air atomizer worked fine and the fuel oil in the storage tank was usable. In some cases, the idle period can extend to several years and may

[14] From a presentation made by Adam Foust, 5,000F Package Frame Manager, at the 3rd SEC–Siemens Technology Forum in March 2014.

require a recommissioning of the system, which would entail cleaning and repair of the oil delivery system and the storage tank, getting rid of the unusable fuel in the tank, etc. The less costly alternative is, of course, to remove the fuel oil system (if permitted by the Regional Transmission Organization [RTO] – see below) and revert to gas-only mode. Removal of the fuel oil system components can be done at no additional cost during a major outage. Additional work at some cost includes replacement with gas-only fuel nozzles and turbine control system changes.

There are several reasons for not making much use of the dual fuel capability. For one, number 2 fuel oil is much more expensive than natural gas (i.e., about $12 versus about $3 for natural gas per million Btu in HHV in the USA in 2017). In many sites, existing emissions regulations put a limit on the number of hours that a gas turbine can run on liquid fuels. Finally, dual fuel capability can be a significant capex adder.

Recently, the polar vortex event of the winter of 2013–2014 resulted in 22 percent of PJM's generation capacity going on forced outage. Although equipment issues were the biggest factor, natural gas delivery interruptions played a significant role as well (25 percent of all outages). This, of course, led to exorbitant natural gas and electricity wholesale prices. As a result, PJM enforces the requirement of firming up fuel supply on the operators of power plants who clear its annual capacity auctions. Compliance requires establishing a dual fuel capability, installing on-site fuel storage, or connecting to a second nearby pipeline.

12.6.3 Burning Hydrogen

Hydrogen is the "Holy Grail" of gas turbine fuels. Why that is so is not hard to see. Its complete combustion products are water vapor and nitrogen (which comes from combustion air). There is no carbon dioxide, no carbon monoxide, and no SOx (but a lot of NOx; remember, N_2 comes in with air). Why not use hydrogen instead of natural gas regularly, then? On paper (i.e., going through the tedious but straightforward stoichiometric calculations manually or using a heat balance simulation software such as Thermoflex), there is no apparent problem.[15]

The reality, of course, is quite different. First and foremost, 100 percent hydrogen is not available naturally (i.e., it must be produced using quite complicated and energy-intensive thermochemical processes [or electrolysis]). Put another way, hydrogen is *not* an energy "resource." It is an energy "carrier." For a high-level overview of the processes generating hydrogen for gas turbine combustion, the reader is referred to the review chapter by Gülen ([10] in Chapter 8). In addition, even assuming that H_2 is readily available, burning it in a modern gas turbine with a DLN combustor is not easy.

The key enabler of current DLN technology (i.e., premixing) becomes very difficult with 100 percent H_2 due to the latter's much higher flammability limits and lower

[15] Hans von Ohain chose a hydrogen combustor for his demonstration engine He S-1, which was successfully tested in 1937. He chose it over a liquid fuel for design simplicity and stable combustion on the quickest path to a successful demonstration.

ignition temperatures (vis-à-vis natural gas). Even with syngas fuels with large H_2 content (say, 20–40 percent by volume), diffusion combustors are used. As this section was being reviewed (January 2018), MHPS announced that they were successful in firing a 30 percent hydrogen fuel mix in a premix combustor at J–class pressure and temperature conditions. Stable combustion was demonstrated while satisfying NOx emission limits and with 10 percent less CO_2 production vis-à-vis natural gas operation. No information is available about commercial deployment of the technology in production gas turbines.

MHPS's high-H_2 DLN combustor development is part of a project of Japan's New Energy and Industrial Technology Development Organization (NEDO) aimed at developing technologies for realizing a "hydrogen society." In the USA, GE carried out similar work under the sponsorship of the Department of Energy's (DOE) Advanced IGCC/Hydrogen Gas Turbine Program. Their report covering the work done in the 2005–2015 period (phases I and II) is available on the DOE National Energy Technology Laboratory (NETL) website.[16] The key component developed by GE for high-hydrogen DLN combustion is the multi-tube micro-mixer. Scale-up to "full-can" (multi-nozzle) combustor tests followed with several hundred hours of high-H_2 operation and single-digit NOx measured at F–class conditions. No word is available about actual commercial deployment. However, especially with the additive manufacturing technology enabling precise machining of the micro-mixer fuel injection part, it should not be surprising to see it in the next-generation, advanced-class GE machines. (Siemens also participated in the aforementioned DOE program. Their development activities are described in a paper by Bradley and Fadok [30].) Power Systems Mfg (PSM) (part of Ansaldo Energia) reported a demonstrated capability of DLN combustion with up to 65 percent (v) H_2 content at F–class conditions in a test rig with their FlameSheet™ technology [31].

It should be noted that reheat or sequential combustor ABB/Alstom/GE/Ansaldo GT24/26 gas turbines are amenable to lean-premixed (i.e., DLN) combustion of high-H_2 fuels due to their unique architecture. In particular, when burning highly reactive fuels such as H_2 or a 70/30 H_2/N_2 mixture, the first combustor (or the HP combustor, which is referred to as the "EV" combustor by the OEM) can operate at a low flame temperature. This reduces the inlet temperature to the second (LP or "SEV") combustor to counteract the reduced auto-ignition delay time of the high-H_2 fuel to allow for proper premixing. Nevertheless, modifications to the existing burner to optimize residence time and pressure drop are necessary. The details on this can be found in the paper by Poyyapakkam et al. [32].

As discussed above, NOx control in diffusion combustors requires water or steam injection. In syngas applications, nitrogen from the air separation unit (ASU) can be used as a diluent as well. GE's experience with burning H_2-containing fuels (as high as 37 percent by volume) in diffusion combustors with different diluents can be found in

[16] Advanced IGCC/Hydrogen Gas Turbine Development, Final Technical Report, Reporting Period Beginning October 1, 2005, Ending April 30, 2015, July 30, 2015, DE-FC26-05NT42643, GE Power and Water, 1 River Road, Schenectady, NY 12345.

Shilling and Jones [33]. The combustor in question is the Multi-Nozzle Quiet Combustor (MNQC), the design details of which can be found in Miller [34]. For Siemens' experience and research and development efforts, please refer to the paper by Wu et al. [35]. How a gas turbine is modified to burn the H_2-containing syngas and integrated with the ASU to utilize N_2 as the diluent is described in Wimer et al. [36].

For a very good discussion on using H_2 instead of methane as a gas turbine fuel and the difference between using diffusion and DLN combustors, the reader is referred to two excellent papers by Gazzani et al. [37] and Chiesa et al. [38]. Key aspects to consider with diffusion combustors are as follows:

- Accommodation of higher volume flow rate at the turbine inlet (additional fuel and diluent) to maintain compressor operating limit line via closing IGVs and/or reducing TFIRE;
- Blade cooling effect (higher heat transfer rate, larger enthalpy drop);
- Hardware modification (e.g., opening up the S1N inlet – see Figure 21.4 in Chapter 21) to accommodate higher PR;
- Compressor uprate (e.g., adding a "zero" stage).

Difficulties associated with H_2-fueled DLN combustion are

- Higher flammability and the lower ignition temperature (already mentioned);
- Very high flame speed (flashback, especially at part load);
- High gas velocity (excessive combustor pressure loss).

Assuming a DLN combustor is developed for hydrogen, the thermodynamic effects of the increase of the water vapor content in the product gases on the turbine hot gas path will be similar to those encountered in the diffusion combustion. Detailed performance comparison of either variant can be found in [37].

Several cases were run in Thermoflex for a vintage F–class gas turbine (14.5 PR, 2,475°F TIT) to illustrate the salient aspects of using H_2 as a gas turbine fuel. The cases are

- 100 percent methane (base case)
- 30 percent H_2 (remainder CH_4)
- 100 percent H_2 (no dilution)
- 100 percent H_2 (with steam injection)

The results are summarized in Table 12.16.

As can be seen in Table 12.16, if the original combustion can handle the hydrogen fuel without running into operability and NOx emissions issues, the same hardware for natural gas can be used with improved performance and reduced CO_2 emissions. With a 30 percent H_2 blend, a 10 percent reduction is CO_2 emissions is possible. Significant performance improvement in the steam injection case is misleading because it does not include the detrimental effect of steam extraction on the bottoming-cycle performance.

Table 12.16 F-class gas turbine with H_2 fuel

	100% CH_4	30% H_2	100% H_2	100% H_2
Turbine Inlet Pressure (psia)	205.33	205.35	205.69	212.13
TIT (°F)	2,475	2,475	2,475	2,475
Turbine Inlet Flow (lb/s)	780	779	765	766
Turbine Inlet Flow (Volume) (cuft/s)	4,232	4,243	4,332	4,417
Turbine Inlet Enthalpy (Btu/lb)	Base	+2.7	+24.1	+80.9
S1B Inlet Temperature (°F)	2,358	2,358	2,357	2,362
S1B Exit Temperature (°F)	1,842	1,842	1,838	1,861
S1 Efficiency (%)	87.99	87.98	87.88	87.71
S1 Output (kW)	127,860	128,162	130,596	136,937
Exhaust Flow (lb/s)	954.8	953.4	943.5	953.8
Exhaust CO_2 Flow (lb/s)	54.6	48.5	0.4	0.4
Exhaust Temperature (°F)	1,092.6	1,091.5	1,082.7	1,103.1
Fuel Flow (lb/s)	19.75	18.47	8.393	9.193
Steam Flow (lb/s)	–	0	0	59.75
Blow-Off (lb/s)	–	0	0	50
Chargeable + Nonchargeable (lb/s)	174.3	174.9	178.8	188.0
Nonchargeable (lb/s)	63.4	63.6	65.0	68.4
Output (kWe)	165,725	166,411	172,742	193,061
Efficiency (%)	36.96	37.04	37.79	38.56

12.7 Combustor Calculations

Unfortunately, very few parameters related to the combustion of a given fuel (mostly natural gas in electric power generation applications) in a gas turbine combustor can be calculated reasonably accurately from the *first principles* (energy and mass balance, etc.). They are:

- Moist air composition for given dry air composition and relative humidity;
- Fuel gas LHV, HHV, and molecular weight for a given composition;
- Hot product gas temperature and composition (including CO_2);
- Wobbe index.

Accurate prediction of CO and NOx emissions with even the most sophisticated chemical kinetic software is not possible. For a given gas turbine model and combustor, an OEM's proprietary formulae based on extensive test data are the only reliable sources. Usually, OEMs provide only the guarantee numbers for air quality permits driven by applicable regulatory requirements, which vary from site to site. Thus, only the data collected by the CEMS can be used to come up with a formulation, which will only apply to a particular application and, somewhat less reliably, to other applications with the same hardware and similar site ambient and loading conditions. (Of course, plant operators would never share their CEMS data with outsiders.)

The only viable option is using Equation 12.17 (or a similar formula) with a known data point for calibration, which then can be used to estimate or predict emissions from the *same or very similar hardware* (i.e., gas turbine combustor) at other operating points. The reader is encouraged to consult the book edited by Lieuwen and Yang [4] and the references therein for more details.

The system of equations governing the combustor dynamics includes three-dimensional, non-linear Navier–Stokes equations combined with transport and chemical reaction equations and they require sophisticated numerical solution methods. The obvious method of attack (i.e., *direct numerical solution* of the governing equations [there are 2N + 9 of them for N species] with a resolution encompassing the smallest flow scales [length, time, and velocity] as well as flame scales [i.e., reaction zones]) is not practical for realistic situations of practical relevance.[17] Models with simplifying assumptions based on the comparison of turbulent flow and chemical reaction time-scales have been developed to reduce the astronomical computation time to manageable levels. One such method, which emerged in the 1990s, is *large-eddy simulation* (LES). LES of turbulent combustion has been applied to a broad range of combustion problems, including emissions predictions, flashback, and blowout in stationary gas turbines and combustion instabilities. Nevertheless, while quite valuable to a fundamental understanding of complex physical phenomena, LES (or other numerical simulation methods, such as Reynolds-averaged Navier–Stokes simulations) is not used in the design of combustors, which is still based on extensive testing and in-house empirical formulae developed by the OEMs over decades.

Another possibility is analytical treatment of gas turbine combustion instabilities based on linear (i.e., acoustic) one- or three-dimensional treatment of the governing equations. Detailed discussion of such methods can be found in different chapters of the volume edited by Lieuwen and Yang [5]. The interested reader is encouraged to consult it and the references therein. Unfortunately, even for test problems with simple geometry, the requisite mathematical formulae and solution thereof are within the capability of only a select group of highly trained specialists. The author is certainly not within that group and is not aware of their use by OEMs in the design cycle of their hardware or in online monitoring software (or in similar analytical tasks). As such, no well-defined engineering calculation methods can be provided herein to predict combustor stability under given boundary conditions.

Still, a lot can be accomplished with the analytical tools available to mere mortals. Let us start with the liquid fuel. While gas turbines were originally developed for burning liquid fuels (which, of course, they still do in aircraft propulsion applications), in modern times, mainly driven by emissions and price[18] considerations, natural gas is the dominant fuel in stationary gas turbines for electric power generation. Liquid fuel, in particular number 2 fuel oil (also known as *diesel* or *heating oil*), is relegated to a backup role. Data for combustion of number 2 fuel oil (18,297 Btu/lb LHV, 19,489 Btu/

[17] The requisite number of operations increases roughly in proportion to the third power of the Reynolds number. Thus, DNS with detailed chemistry for a practical problem can take decades.

[18] According to the US Energy Information Administration, natural gas prices for electricity generation varied between $2.65 and $3.36 per MMBtu in 2016. For distillate fuel oil, the price range was %9.00–$12.14 per MMBtu.

Table 12.17 Combustion of liquid fuel in a DLN combustor. f/a = fuel–air ratio; w/f = water–fuel ratio; w/a: water–air ratio; CDT = compressor discharge (combustor inlet) temperature; RH: relative humidity of air; gas composition is in percentage by volume at the combustor exit

f/a	0.0374	0.0379	0.0337	0.0378	0.0366	0.0377	0.0368	0.0362	0.0360	0.0351	0.0342
w/f	1.4565	1.3814	1.1348	1.4002	1.1524	1.4309	1.1543	1.1218	1.1216	1.0618	1.1654
w/a	0.0545	0.0523	0.0383	0.0530	0.0422	0.0539	0.0425	0.0406	0.0403	0.0373	0.0398
RH	0.60	0.90	0.90	0.85	0.85	0.80	0.80	0.60	0.45	0.50	0.14
PR	22.0	22.6	14.5	22.6	14.0	22.6	13.8	13.4	13.4	13.1	12.5
CDT (°F)	863.4	708.4	563.0	758.1	593.9	805.9	632.1	682.6	716.1	737.7	753.1
TIT (°F)	2,590	2,525	2,316	2,552	2,449	2,572	2,484	2,497	2,509	2,495	2,457
N_2	68.91	69.75	71.43	69.61	70.78	69.34	70.54	70.37	70.19	69.79	70.54
O_2	8.51	8.54	9.89	8.53	9.01	8.52	8.9	9.01	9.04	9.15	9.59
CO_2	6.96	7.1	6.48	7.08	6.96	7.04	7.0	6.89	6.84	6.68	6.52
H_2O	14.77	13.76	11.33	13.92	12.38	14.26	12.7	12.87	13.08	13.52	12.49
Ar	0.83	0.84	0.86	0.84	0.85	0.84	0.85	0.85	0.85	0.84	0.85

lb HHV) in an advanced gas turbine DLN combustor with water injection across a range of operating conditions are summarized in Table 12.17. Fuel is heated to 50°F, whereas injected water is at 500 psia and 80°F. Fuel composition is 86.6 percent carbon and 12.7 percent hydrogen by mass, with trace amounts of oxygen, nitrogen, and sulfur (specific gravity 0.8654). If needed, the specific heat of the liquid fuel can be estimated as

$$c_p = 0.0485 \cdot (\text{T in °F}/100) + 0.4127 \text{ Btu/lb-R.}$$

The data in Table 12.17 are a good guide for most combustion calculations involving liquid fuel combustion in advanced DLN combustors with water injection. The fuel–air ratio can be represented by the following transfer function:

$$\text{FA} = 0.011 \cdot \text{X} + 0.0144$$
$$\text{X} = 0.001 \cdot (\text{TIT} - \text{CDT})/\text{WF}^{0.5},$$

with the water–fuel ratio given by

$$\text{WF} = 0.1686 \cdot \text{Y} + 1.0658$$
$$\text{Y} = (\text{PR}/20)^{3.5} \cdot (\text{TIT}/2,500)^{5.0}/\text{RH}^{0.75}.$$

(Note that the airflow in the fuel–air definition above is the airflow *into the combustor*, which is less than the airflow *into the compressor* by the sum of chargeable and nonchargeable turbine cooling air.) Another rough estimation for water injection for NOx control is given by

$$\text{RNOX} = 0.09 + 0.9 \cdot \exp(-0.6 \cdot \text{WA}),$$

where RNOX is the reduction in NOx emissions as a fraction (with respect to the no-injection case) and WA is the water–air ratio for the combustor.

The exact schedule of combustor water injection for NOx control when firing the backup liquid fuel in a DLN combustor can only be determined by the OEM after careful consideration of combustion dynamics and flame-out limits.

One interesting takeaway from Table 12.17 is the high water vapor content of the combustion products. This increases the heat load on hot gas path components with a detrimental impact on part lives. Since materials and coatings are specified based on mostly natural gas-fired operation, this requires a reduction in firing temperature in liquid fuel-fired operation. Another driver of firing temperature reduction is the increase in turbine airflow, which forces the cycle pressure up due to constant inlet area (i.e., the "choked nozzle" effect). A generic comparison of the gas- and liquid fuel-fired operation of an advanced, heavy-duty industrial gas turbine is provided in Table 12.18.

Natural gas combustion calculations with COMBUST can also provide valuable information about the gas turbine design. In order to illustrate this, consider the gas turbine data in Table 6.1, which is repeated below for convenience as Table 12.19. Four cardinal parameters (i.e., output, efficiency and exhaust conditions [flow and temperature], and cycle PR) are from *Gas Turbine World* (GTW) 2016. Inlet airflow can be found by subtracting fuel flow (determined from output and efficiency with fuel LHV) from the exhaust flow. Firing temperature is estimated from cycle PR and exhaust temperature using Equation 6.1. TIT is also estimated from cycle PR and exhaust temperature using Equation 5.12.

Table 12.18 Gas turbine natural gas and liquid (number 2 fuel oil) operation (gas composition is in percentage by volume at the gas turbine exhaust)

	Natural Gas	Number 2 Fuel Oil
Airflow	1.00	1.00
Exhaust Flow	1.00	1.05
Exhaust Temperature	Base	~–100°F
TIT	Base	~–300°F
Fuel Heat Input	1.00	1.08
Generator Output	1.00	0.99
Heat Rate	1.00	1.09
N_2	74	70
O_2	11.5	10
CO_2	4.3	5.8
H_2O	9.3	13.3

Table 12.19 GE 60-Hz products (GTW 2016 Performance Specifications)

	Output (kW)	Efficiency	PR	TEXH (°F)	MEXH (lb/s)	TIT °C	TIT °F	TFIRE (°F)
7E.03	91,000	33.9%	13.0	1,026	650	1,233	2,252	2,141
7F.04	198,000	38.6%	16.7	1,151	1,013	1,459	2,658	2,496
7F.05	241,000	39.8%	18.4	1,171	1,188	1,520	2,769	2,589
7F.06	270,000	41.4%	22.1	1,100	1,311	1,518	2,764	2,563
7HA.01	280,000	41.7%	21.6	1,159	1,266	1,576	2,868	2,661
7HA.02	346,000	42.2%	23.1	1,153	1,539	1,598	2,908	2,690

Table 12.20 Combustion calculations to find total turbine cooling flow. f = gas turbine overall fuel–air ratio; AF = combustor air–fuel ratio; FA = combustor fuel–air ratio

		7E.03	7F.04	7F.05	7F.06	7HA.01	7HA.02
Total Cooling Flow		0.177	0.195	0.230	0.219	0.222	0.228
Combustion Airflow	lb/s	525.4	797.7	894.2	1001.0	961.6	1160.8
	kg/s	238.3	361.8	405.6	454.0	436.2	526.6
Fuel Flow	lb/s	11.83	22.61	26.69	28.75	29.60	36.14
	kg/s	5.37	10.26	12.11	13.04	13.43	16.39
f		0.0185	0.0228	0.0230	0.0224	0.0239	0.0240
AF		44.41	35.28	33.50	34.82	32.49	32.12
FA		0.0225	0.0283	0.0298	0.0287	0.0308	0.0311
ϕ		0.3855	0.4853	0.5110	0.4917	0.5270	0.5330
λ		2.59	2.06	1.96	2.03	1.90	1.88
Excess Air		159%	106%	96%	103%	90%	88%
CDT	°F	681.2	772.5	809.8	883.3	873.9	901.7
	°C	360.7	411.4	432.1	472.9	467.7	483.2
TIT	°F	2,252	2,658	2,769	2,764	2,868	2,908
N_2		0.74225	0.73485	0.73297	0.73438	0.73180	0.73136
O_2		0.12130	0.10015	0.09476	0.09881	0.09144	0.09018
H_2O		0.08833	0.10739	0.11225	0.10860	0.11524	0.11637
CO_2		0.03922	0.04879	0.05124	0.04940	0.05274	0.05331

Table 12.21 Combustion calculations to find chargeable turbine cooling flow

		7E.03	7F.04	7F.05	7F.06	7HA.01	7HA.02
Chargeable Cooling Flow		0.103	0.103	0.136	0.107	0.113	0.113
Nonchargeable Cooling Flow		0.073	0.091	0.094	0.112	0.109	0.115
Combustion Airflow	lb/s	572.2	887.9	1,003.8	1,145.0	1,096.9	1,333.0
	kg/s	259.6	402.8	455.3	519.4	497.6	604.7
Fuel Flow	lb/s	11.83	22.61	26.69	28.75	29.60	36.14
	kg/s	5.37	10.26	12.11	13.04	13.43	16.39
TFIRE	°F	2,141	2,496	2,589	2,563	2,661	2,690
N_2		0.74462	0.73849	0.73707	0.73895	0.73657	0.73641
O_2		0.12807	0.11055	0.10648	0.11186	0.10506	0.10459
H_2O		0.08223	0.09801	0.10169	0.09684	0.10296	0.10338
CO_2		0.03615	0.04408	0.04593	0.04349	0.04657	0.04678

With known fuel flow rate and TIT, combustor inlet airflow can be found using COMBUST. The combustor air inlet temperature is equal to the compressor discharge temperature (CDT), which can be estimated from the cycle PR using a compressor polytropic efficiency of 93 percent and $\gamma = 1.4$. With known gas turbine inlet airflow, calculated combustor airflow gives one a very good estimate of total, chargeable, and nonchargeable turbine cooling flows. The results are summarized in Table 12.20. (Cooling flows are expressed as a fraction of gas turbine inlet airflows.)

Similarly, one can estimate the chargeable turbine cooling flow by using the firing temperature (TFIRE). This approach is based on the assumption that stage 1 nozzle vanes and the combustor itself constitute a single control volume. Thereafter, by using the total cooling values in Table 12.20, the nonchargeable turbine cooling flow can be found as well. The results of this calculation are summarized in Table 12.21. (Cooling flows are expressed as a fraction of gas turbine inlet airflows.)

References

1. Lefebvre, A. H., *Gas Turbine Combustion* (Philadelphia, PA: Hemisphere Publishing Corporation, 1983).
2. Strehlow, R., *Combustion Fundamentals* (New York: McGraw-Hill, 1984).
3. Khartchenko, N. V., *Advanced Energy Systems* (Washington, DC: Taylor & Francis, 1998).
4. Lieuwen, T. C., Yang, V. (Editors), *Gas Turbine Emissions* (New York: Cambridge University Press, 2013).
5. Lieuwen, T. C., Yang, V. (Editors), *Combustion Instabilities in Gas Turbine Engines* (Reston, VA: AIAA, Inc., 2005).
6. *Gas Turbine Heat Recovery Steam Generators*, ASME Performance Test Code PTC 4.4-2008, The American Society of Mechanical Engineers, Three Park Avenue, New York.
7. Lefebvre, A. H., The Role of Fuel Preparation in Low-Emission Combustion, *Journal of Engineering for Gas Turbines and Power*, **117** (1995), 617–654.
8. Røkke, P. E., Hustad, J. E., Røkke, N. A., Svendsgaard, O. B., "Technology Update on Gas Turbine Dual Fuel, Dry Low Emission Combustion Systems," ASME Paper GT2003-38112, proceedings of ASME Turbo Expo, June 16–19, 2003, Atlanta, GA.
9. Davis, L. B., Black, S. H., "Dry Low NOx Combustion Systems for GE Heavy-Duty Gas Turbines," GER-3568G, 2000, available online at http://citeseerx.ist.psu.edu/viewdoc/download?doi=10.1.1.468.8577&rep=rep1&type=pdf.
10. Venkataraman, K. et al., "F-Class DLN Technology Advancements: DLN 2.6+," ASME Paper GT2011-45373, ASME Turbo Expo 2011, June 6–10, 2011, Vancouver, BC, Canada.
11. Smith, G. R., Kulkarni, P., Hoskin, R. F., "Transient Operation of Combined Cycle Power Plants and the Associated Air Emissions," Powergen International 2012, Orlando, FL.
12. Ito, E. et al., Development of Key Technologies for an Ultra-High-Temperature Gas Turbine, *Mitsubishi Heavy Industries Technical Review*, **48**: 3 (2011).
13. Tanaka, Y. et al., Development of Low NOx Combustion System with EGR for 1700°C-Class Gas Turbine, *Mitsubishi Heavy Industries Technical Review*, **50**: 1 (2013).
14. Smith, R. W. et al., "Fuel Moisturization for Natural Gas Fired Combined Cycles," GT2005-69012, ASME Turbo Expo 2005, June 6–9, 2005, Reno, NV.
15. Standards for Performance of Combustion Turbines; Final Rule, Federal Register, 40 CFR 60, Part III Environmental Protection Agency, July 6, 2006, p. 38505.
16. European Union (large combustion plants) regulations (2012), Statutory Instruments, S.I. No. 566 of 2012.
17. Goldmeer, J., Vandervort, C., Sternberh, J., "New Capabilities and Developments in GE's DLN 2.6 Combustion Systems," POWERGEN International, December 5–7, 2017, Las Vegas, NV.
18. Welch, M., "Can Ethane and Propane Displace Diesel and Fuel Oils as Fuels for Power Generation?" POWERGEN International, December 5–7, 2017, Las Vegas, NV.

19. Brdar, R. D., Jones, R. M., "GE IGCC Technology and Experience with Advanced Gas Turbines," GER-4207, 2000, available online at www.ge.com/content/dam/gepower-pgdp/global/en_US/documents/technical/ger/ger-4207-ge-igcc-technology-experience-advanced-gas-turbines.pdf.
20. Hall, J. M. et al., "Development and Field Validation of a Large-Frame Gas Turbine Power Train for Steel Mill Gases," ASME Paper GT2011-45923, ASME Turbo Expo 2011, June 6–10, 2011, Vancouver, BC, Canada.
21. Takano, H., Kitauchi, Y., Hiura, H., Design for the 145-MW Blast Furnace Gas Firing Gas Turbine Combined Cycle Plant, *Journal of Engineering for Gas Turbines and Power*, **111** (1989), 218–224.
22. Welch, M., Igoe, B., "An Introduction to Combustion, Fuels, Emissions, Fuel Contamination and Storage for Industrial Gas Turbines," ASME Paper GT2015-42010, ASME Turbo Expo 2015, June 15–19, 2015, Montreal, QC, Canada.
23. Jones, R., Goldmeer, J., Monetti, B., "Addressing Gas Turbine Fuel Flexibility," GER-4601, 2011, available online at www.ge.com/content/dam/gepower-pgdp/global/en_US/documents/technical/ger/ger-4601b-addressing-gas-turbine-fuel-flexibility-version-b.pdf.
24. Meier, J. G., Hung, W. S. Y., Sood, V. M., Development and Application of Industrial Gas Turbines for Medium-Btu Gaseous Fuels, *Journal of Engineering for Gas Turbines and Power*, **108** (1986), 182–190.
25. Huth, M., Heilos, A., "Fuel Flexibility in Gas Turbine Systems: Impact on Burner Design and Performance," 2013, chapter 14 in [11] in Chapter 2.
26. Tomlinson, L. O., Alff, R. K., "Economics of Heavy Fuels in Gas Turbines and Combined Cycles," ASME Paper 81-GT-45, Gas Turbine & Products Show, March 9–12, 1981, Houston, TX.
27. Bunz, W. J. et al., Crude Oil Burning Experience in MS5001P Gas Turbines, *Journal of Engineering for Gas Turbines and Power*, **106** (1984), 812–818.
28. Bromley, A. F., Attention to Detail Critical for Maintaining High Availability When Firing Liquid Fuels, *Combined Cycle Journal*, **Third Quarter** (2006), 33–39.
29. Goldmeer, J. et al., "Evaluation of Arabian Super Light Crude Oil for Use in a F-Class DLN Combustion System," ASME Turbo Expo 2014, June 16–20, 2014, Düsseldorf, Germany.
30. Bradley, T., Fadok, J., "Advanced Hydrogen Turbine Development Update," ASME Paper GT2009-59105, ASME Turbo Expo 2009, June 8–12, 2009, Orlando, FL.
31. Stuttaford, P. et al., "FlameSheet™ Combustor Engine and Rig Validation for Operational and Fuel Flexibility with Low Emissions," GT2016-56696, ASME Turbo Expo 2016, June 13–17, 2006, Seoul, South Korea.
32. Poyyapakkam, M. et al., "Hydrogen Combustion within a Gas Turbine Reheat Combustor," GT2012-69165, ASME Turbo Expo 2012, June 11–15, 2012, Copenhagen, Denmark.
33. Shilling, N. Z., Jones, R. M., "The Impact of Fuel-Flexible Gas Turbine Control Systems on Integrated Gasification Combined Cycle Performance," ASME Paper GT2003-38791, ASME Turbo Expo 2003, June 16–19, 2003, Atlanta, GA.
34. Miller, H. E., "Development of the GE Quiet Combustor and Other Design Changes to Benefit Quality," GER-3551, 1994, available online at www.ge.com/content/dam/gepower-pgdp/global/en_US/documents/technical/ger/ger-3551-development-ge-quiet-combustor.pdf.
35. Wu, J. et al., "Advanced Gas Turbine Combustion System Development for High Hydrogen Fuels," ASME Paper GT2007-28337, ASME Turbo Expo 2007, May 14–17, 2007, Montreal, QC, Canada.

36. Wimer, J. G., Keairns, D., Parsons, E. L., Ruether, J. A., Integration of Gas Turbines Adapted for Syngas Fuel with Cryogenic and Membrane-Based Air Separation Units: Issues to Consider for System Studies, *Journal of Engineering for Gas Turbines and Power*, **128** (2006), 271–280.
37. Gazzani, M. et al., Using Hydrogen as Gas Turbine Fuel: Premixed versus Diffusive Flame Combustors, *Journal of Engineering for Gas Turbines and Power*, **136** (2014), 051504.
38. Chiesa, P., Lozza, G., Mazzocchi, L., Using Hydrogen as Gas Turbine Fuel, *Journal of Engineering for Gas Turbines and Power*, **127** (2005), 73–80.

13 Materials

In Chapter 5, it was shown that fundamental thermodynamics dictate the highest possible cycle heat addition temperature (METH) for maximum thermal efficiency.[1] In order to achieve the highest possible METH, the highest possible cycle working fluid pressure and temperature are requisite. If design engineers had access to a fictional material "unobtanium" that would preserve its strength and integrity under arbitrarily high pressures and temperatures as well as in any corrosive environment, performances envisioned for advanced gas turbines in the not-so-near future would have been achieved decades ago.

In Chapter 9, within the context of hot gas path cooling, we touched upon the trade-offs made by German engineers in World War II because they lacked the metallurgical resources to match the nickel content of British Nimonic (80 percent). In fact, the Soviets had started their own gas turbine jet engine development in the 1930s, but did not achieve significant progress due to their lack of knowledge in metallurgy. After the war, among other things, they quickly realized that Tinidur used in German jet engines did not allow for acceptable engine life (only tens of hours). Consequently, they turned to the British for engine technology and purchased 55 Rolls-Royce engines (Nene and Derwent), which were state of the art in the late 1940s. Since the British did not share the composition of Nimonic with the Soviets, they resorted to shady tactics to obtain samples such as stealing parts and metal shavings from the floor next to the lathes for chemical analysis back at home.[2] Having access to strategic materials and knowledge of advanced metallurgy were the keys to successful gas turbine design and manufacturing from day one.

Unobtanium, of course, does not exist. What engineers have to work with are *metals* (i.e., mainly steels and alloys). Metals used in the construction of gas turbine equipment and structures lose their strength and resistance to corrosion as the temperature of the environment they are in increases and they tend to undergo structural changes. Degradation of metals leads to a shortening of component design life and premature failure. Failure mechanisms in power plant components can be physical (i.e., stress rupture) or chemical (i.e., corrosion). The end result of failure is deformation and fracture. The

[1] Equally important is the lowest possible cycle heat rejection temperature (METL); however, from a materials availability perspective, the higher temperature is the determining factor.

[2] See the book by Kotelnikov and Buttler ([7] in Chapter 4). One imaginative Soviet visitor wore soft-soled shoes and "carefully trampled over some metal [chips]" to collect alloy samples.

resistance of metals to deformation and fracture decreases with increasing temperature (*creep*) and pressure (*stress*) of the environment they exist in. Creep is defined as time-dependent, thermally assisted deformation of components under load (stress). The critical parameter is the *time to rupture*, which is a function of the *yield strength* of the metal in question (i.e., its inherent strength) and its environment (temperature and stress). An increase in either or both of the latter two will shorten the time to rupture. Obviously, short-term phenomena such as extreme overheating due to an accidental loss of coolant flow can accelerate the failure process. From an advanced-cycle design perspective, the main concern is long-term, high-temperature creep. Thus, the material selected should comply with the design life with a sufficient margin of safety under the requisite pressure and temperature. This is the basic premise of advanced materials for advanced power plants.

The other key feature from an advanced materials perspective is resistance to corrosion. *Corrosion* and oxidation are chemical reactions and, as such, their rates increase with temperature. They either occur uniformly over the metal surface or are limited to small areas (pitting). The rate of corrosion manifests itself in the loss of material, expressed as loss in weight per unit area and time (e.g., milligrams per square centimeter per day) or penetration in length per unit time (e.g., inches per year). Pitting corrosion is, in general, a much more serious problem than general corrosion – its rate can be as high as 1,000 times the uniform corrosion rate.

A quick summary of the materials used in gas turbine construction is provided in Figure 13.1. For those readers who are reasonably well-versed in metallurgy, the soup of letters and numbers designating various steels and alloys in Figure 13.1 is probably not an unusual sight. For others, it presents a daunting challenge to form a basic understanding of the materials used in state-of-the-art gas turbines for electric power generation. Therefore, a brief introduction to steels and supperalloys is provided to form a foundation that is firm enough to follow the rest of the discussion in the present chapter.

The progress in metallurgy and thermal barrier coating (TBC) technology over the 50-year period covering the second half of the twentieth century and the first 15 years of the twenty-first century is summarized in Figure 13.2. With no TBC and convection cooling of the airfoils, gas temperatures up to about 1,100°C (2,000°F) were possible. This was about 200°C or 360°F above the material capability defined as the temperature with a rupture time (or creep life) of 100,000 hours at 140 MPa (20 ksi). Note that, when quoting material "capability," one must pay attention to the underlying assumptions, especially the rupture time. There are different life definitions one might come across in the literature (e.g., 1,000-hour creep life at 20 ksi). The difference in metal temperature for 1,000 and 100,000 hours of creep life is nearly 150°C.

A life expectancy of 100,000 hours is equivalent to about 11 years (continuous) operation or, more realistically, 15–20 years in a more typical duty cycle. To prove that a selected material for the new advanced steam cycle conditions will indeed survive that long is a difficult proposition. This is one of the biggest hurdles in the ASME code of approval for new materials considered for use in plant equipment construction and, ultimately, widespread commercial acceptance. Usually, tests are conducted in the

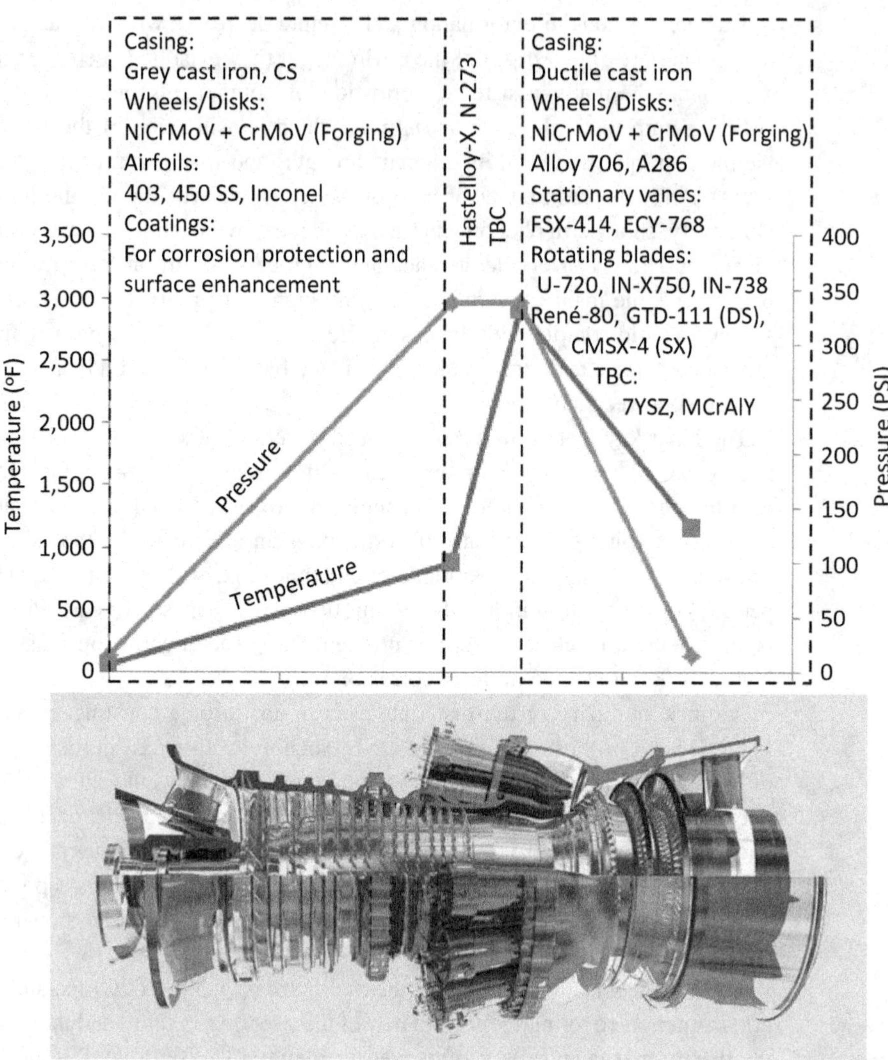

Figure 13.1 Gas turbine materials and coatings [1].

laboratory for shorter durations and expected creep is extrapolated using accepted theoretical approaches (e.g., *Larson–Miller* [L-M] curves for creep rupture, *Manson–Coffin* S–N curves for low cycle fatigue, and *Goodman* diagrams for high cycle fatigue [3]).

L-M curves for two superalloys are plotted in Figure 13.3. The L-M parameter, P, is calculated from temperature (T in degrees Rankine) and time to rupture (t in hours) – see the formula on the x-axis label of Figure 13.3. For a specified stress level, combinations of T–t can be read from the curve. Thus, for a specified allowable stress level (usually with a hefty design margin), higher alloy temperatures can be tolerated only for a reduced time period at constant P. To maintain the same part lives at higher temperatures, one has to reduce the allowable stress or the design margin.

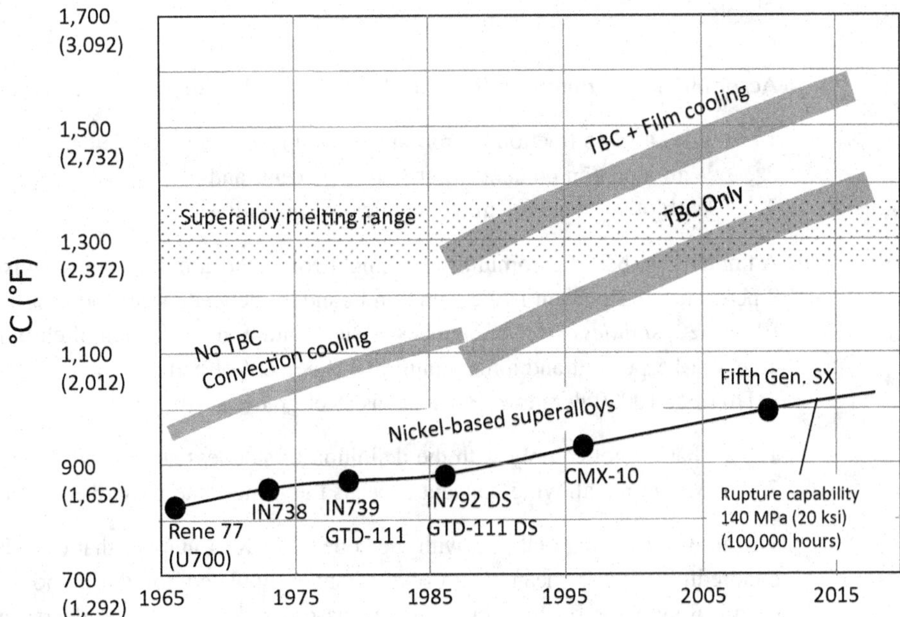

Figure 13.2 Temperature capability of nickel-based superalloys (adapted from a chart in Clark et al. [2]). SX = single-crystal.

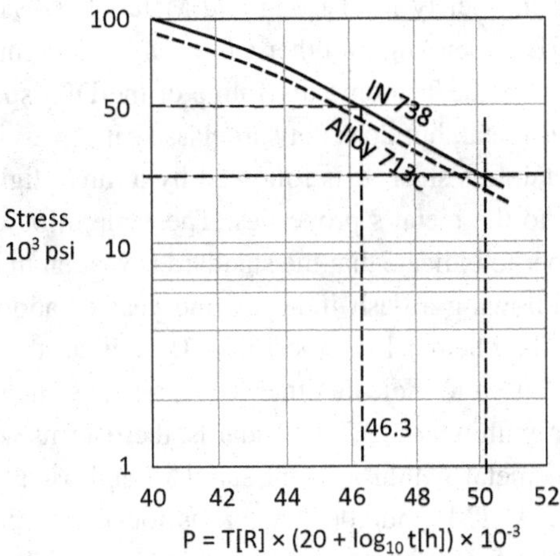

Figure 13.3 Larson–Miller curves [4].

For example, in Figure 13.3, for IN 738, a creep life of, say, 20,000 hours at 50 ksi stress is only possible at operating temperature of 986°F. This is how that value is found:

- At 50 ksi, P is read as 46.3
- 46.3 = (T + 460) × (20 + log 20,000)/1,000
- 46,300 = (T + 460) × 24.3
- T = 1,445, R = 986°F

13.1 Steels

According to the European Standard DIN EN 10020 (July 2000), *steel* is a material

1. With a mass fraction of iron higher than of every other element;
2. With a carbon content lower than 2 percent; and
3. With other elements in its chemical composition.

A limited number of chromium steels might contain a carbon content that is higher than 2 percent, but 2 percent is the common boundary between steel and cast iron. Per DIN EN 10020, *stainless steels* are grades of steel with a mass fraction of chromium (Cr) of at least 10.5 percent and a maximum of 1.2 percent of carbon.

DIN EN 10020 defines *alloy steels* as steel grades

1. That do not comply with the definition of stainless steels; and
2. Where one alloying element exceeds the limit value prescribed in the standard.

The alloyed steels are labeled with the code letter X, a number that complies with the hundredfold of the mean value of the range stipulated for the carbon content, the chemical symbols of the alloying elements ordered according to decreasing contents of the elements, and numbers that, in sequence of the designating alloying elements, refer to their content. For example, *X10CrNi18-10* has 10/100 = 0.1 percent C, 18 percent Cr, and 10 percent Ni in its chemical composition. As such, it is equivalent to *SA-213* per ASME or American Society for Testing and Materials (ASTM) specification (grade TP321H or TP347H depending on other elements in its composition).

ASTM specifications do not have the Teutonic discipline of the DIN specifications. The letter "A" describes a ferrous metal, but does not sub-classify it as cast iron, carbon steel, alloy steel, tool steel, or stainless steel. It is followed by a three-digit sequential number without any correlation to the metal's properties. There might be a letter "M" appended to the sequential number to indicate that the standard is written in rationalized SI units. A two-digit number following a dash indicates the year of adoption or last revision. Grade is used to describe chemical composition. Type is used to define the deoxidation practice; and class is used to indicate other characteristics such as strength level or surface finish. (However, within the ASTM standards, these terms were adopted and used to identify a particular metal within a metal standard and used without any strict definition.) Another use of ASTM grade designators is found in pipe, tube, and forging products, where the first letter P refers to pipe, T refers to tube, TP may refer to tube or pipe, and F refers to forging. For example, ASTM *A335/A335-03, Grade P22* refers to seamless ferritic alloy-steel pipe for high-temperature service.

The most common alloying elements are:

- Manganese (Mn) – manganese is present in most commercially made steels (it acts as a deoxidizer in the manufacturing of steel). It increases the strength and hardness of the steel.
- Chromium (Cr) – chromium increases the toughness and wear resistance of steel as well as its corrosion resistance.

- Silicon (Si) – silicon is also used as a deoxidizer in the manufacture of steel. It slightly increases the strength of steel, and when used in conjunction with other alloys, it can help increase the toughness and hardness of steel.
- Nickel (Ni) – nickel increases the strength steel. It is used in low-alloy steels to increase toughness and hardenability. Nickel forms the base of high-temperature superalloys because of its ability to develop an adherent oxide and precipitation hardening phases based on Ni_3Al.
- Molybdenum (Mo) – molybdenum increases the hardness and high-temperature tensile strength of steel (due to its high melting point). It improves resistance to pitting corrosion in chloride environments and to crevices in both Fe–Cr alloys and Fe–Cr–Ni alloys.
- Vanadium (V) – vanadium increase the toughness and strength of steel.

Obviously, the element with the largest share in steel's chemical composition is iron (Fe). At room temperature, iron is composed of atoms arranged in a *body-centered cubic* (BCC) lattice; irons of this type of crystal structure are referred to as ferrite or *alpha* (α) iron. The crystal structure changes to *face-centered cubic* (FCC) at around 1,670°F; at that temperature, alpha iron is transformed to *austenite* or *gamma* (γ) iron and is nonmagnetic. Alloys of BCC and FCC crystal structure are thus referred to as *ferritic* and *austenitic*, respectively.

13.2 Superalloys

The Superalloy Committee (SAC) of the Specialty Steel Industry of North America is a voluntary trade association representing producers of high-performance alloys in North America. According to the Committee, *specialty metals* are defined as follows:

- Steel
  - With a maximum alloy content exceeding one or more of the following limits: manganese, 1.65 percent; silicon, 0.60 percent; or copper, 0.60 percent; or
  - Containing more than 0.25 percent of any of the following elements: aluminum, chromium, cobalt, molybdenum, nickel, niobium (columbium), titanium, tungsten, or vanadium.
- Metal alloys consisting of
  - Nickel or iron–nickel alloys that contain a total of alloying metals other than nickel and iron in excess of 10 percent; or
  - Cobalt alloys that contain a total of alloying metals other than cobalt and iron in excess of 10 percent.
- Titanium and titanium alloys; or
- Zirconium and zirconium alloys.

Metal alloys have diverse chemical compositions, properties, and applications. According to the SAC, they are defined by

1. The ability to perform under exceedingly rigorous service conditions; or
2. Properties (e.g., electrical, thermal, magnetic) that are finely tuned to suit particular requirements.

Thus, the SAC identifies four metal alloy groups:

1. Corrosion-resistant alloys
2. Heat-resistant alloys
3. Superalloys
4. Alloys with special physical properties

Most of the alloys in these four groups are high-nickel alloys, with nickel content ranging from 25 to nearly 100 percent. However, some of the alloys are cobalt-based (or high in cobalt) and a few are iron-based. In virtually all of the alloys, other elements – including chromium, niobium, molybdenum, aluminum, and titanium – significantly influence strength or corrosion resistance. *Superalloys* have good creep and oxidation resistance. They are ideally suited to functioning under high temperatures and mechanical stress. There are three groups of superalloys:

- Cobalt-based
- Nickel-based
- Iron-based

They can be used at temperatures as high as 85 percent of their melting point [5]. The melting temperature range of most superalloys is 1,260–1,370°C (2,300–2,500°F) [5]. Thus, all three types of superalloys can be used at temperatures well above 540°C (1,000°F).

Due to their high-temperature strength and hot-corrosion resistance, nickel-based "superalloys" are the materials of choice for manufacturing turbine hot gas path components. Combined with corrosion-resistant coating and TBC and with manufacturing techniques such as directional and single-crystal solidification, these materials enable high turbine inlet temperatures (TITs) up to 1,600°C (even beyond) with long maintenance intervals and high reliability while limiting cooling air consumption to acceptable levels.

The origin of nickel-based superalloys can be traced to Nimonic 80, which was developed in the UK in the early years of World War II as a suitable material for jet engine applications. Nimonic 80 had 20 percent Cr content for hot-corrosion resistance. In later years, the reduction in Cr content in favor of refractory metals necessitated development of aluminide and platinum aluminide diffusion coatings.

Superalloys are defined by their patented chemical compositions.[3] For example, one of the most widely used superalloys in aircraft and power generation gas turbines, Alloy 718, was patented by Herbert Eiselstein and assigned to The International Nickel Company, Inc., in 1962 (US patent 3,046,108 "Age-Hardenable Nickel Alloy"). The chemical compositions of two commonly used superalloys are listed in Table 13.1.

[3] There is no definitive information on the origins of the moniker "super." The term really stuck in the technology jargon around 1960.

Table 13.1 Chemical compositions of selected superalloys

Elements	Inconel 617	Haynes 263
Nickel, Ni	48.85–62.00	52
Chromium, Cr	20.0–24.0	20
Cobalt, Co	10–15	20
Molybdenum, Mo	8–10	6
Manganese, Mn	≤1	0.6
Silicon, Si	≤1	0.4
Carbon, C	≤0.15	0.06

- *Inconel 617* is a solid-solution strengthened alloy and has good oxidation resistance and a wide variety of corrosive resistance. Its tensile strength is >70 ksi and its yield strength is >40 ksi. Its melting point is 1,363°C (2,485°F).
- *Haynes 263* is a nickel–cobalt–chromium alloy that also contains molybdenum to provide it with hot-temperature strength. It is amenable to precipitation hardening (a heat treatment technique to increase the yield strength of the material) and is comparatively easy to form with a good high-temperature performance. This alloy is primarily used for applications up to 900°C (1,650°F). Its tensile strength is >150 ksi and its yield strength is >90 ksi. Its melting point is around 1,350°C (2,450°F).

For an alloy or superalloy, rigorous testing is necessary to demonstrate that the material has a stable microstructure at elevated temperatures and that it has adequate stress–rupture properties after long-term, elevated temperature exposure to accommodate the design stresses. Creep testing usually requires at least 30,000 hours and data are then extrapolated to 100,000 hours. (For example, a base load plant operating 8,000 hours per year for 20 years is expected to have a design life of 160,000 hours. Obviously, not all parts are expected to survive such a long time without replacement, and allowance must be made for periodic replacement. However, some parts, such as turbine casings and rotors, must be able to reach the end line in one piece.)

Cast, nickel-based superalloys are primarily used to manufacture gas turbine hot gas path components (e.g., nozzle vanes and rotor blades [stationary and rotating airfoils, respectively]). There are three types of casting techniques used in hot gas path component manufacturing:

1. Conventional (equiaxed grain) investment casting
2. Vacuum casting with *directional solidification* (DS)
3. *Single-crystal* (SC or SX) [5].

Conventional and DS grain microstructures are shown in Figure 13.4. Even to an eye untrained in metallurgy, the advantage of columnar grains aligned parallel to the longitudinal (i.e., the primary stress) axis in terms of creep strength and ductility (and thermal fatigue resistance to boot) is intuitively obvious. Scientifically rigorous reasoning for this can be found in [5]. Production of a casting containing only a single

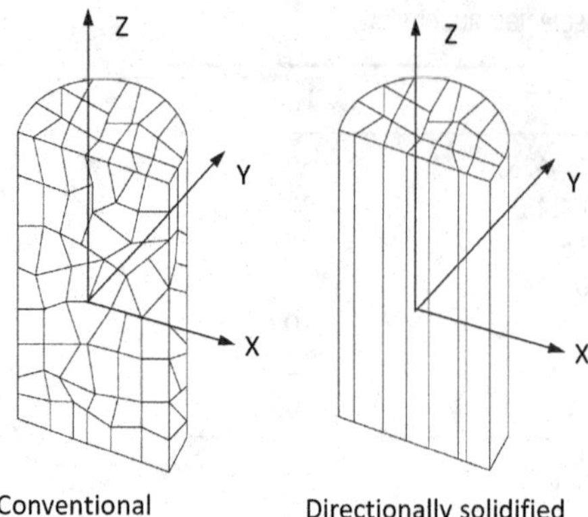

Figure 13.4 Conventional and DS grain microstructures.

columnar grain was a natural follow-up to the DS casting. Both DS and SX casting technologies were pioneered by Pratt & Whitney to be used in aircraft jet engines. The former was first introduced in 1969 for military aircraft engines and in 1974 for commercial ones. SC work was also started in parallel to DS development, but was discontinued until the mid-1970s because the early attempts did not improve significantly over DS castings. The PWA 1480 SX superalloy used in F-100-PW-220 (first and second turbine stage vanes and buckets) was introduced in the early 1980s and was instrumental in the success of the engine. The first introduction of SX turbine vanes and blades in heavy-duty industrial gas turbines took place in Alstom's GT24/26 reheat gas turbines (CMSX-4) and shortly thereafter in General Electric's (GE) steam-cooled H-System (René N5).

Vacuum casting with DS or SC is described in many treatises (e.g., [6]). After the furnace is evacuated, liquid superalloy melt is poured into the basin. Solidification begins from the water-cooled copper plate at the bottom on which the ceramic "mold" is sitting. The mold is withdrawn at a specific rate into the "cold zone" of the furnace. Consequently, the solidification "front" moves upward in the mold. If the solidification is forced to go through a thin, spiral grain selector (also known as a "pigtail"), the result is a single grain with a preferred crystallographic orientation to enter the component portion of the mold. For productivity, typically 10–30 components are cast simultaneously in groups called "clusters," oriented in a circular pattern for thermal uniformity [6].

Vacuum casting with DS or SX provides higher resistance to thermal fatigue at elevated temperatures. As such, they are used for manufacturing vanes and blades in the first two stages of the turbine, which experience the highest gas temperatures. (Wrought superalloys are used for turbine discs, on which the airfoils are mounted.)

There are several trademarks, which appear frequently when it comes to nickel-based superalloys in advanced industrial (and aircraft, of course) gas turbines:

- Inconel (e.g., In-617, In-738)
- Udimet (e.g., U-500, U-720)
- Hastalloy/Haynes

- Waspalloy
- René
- CMSX

Monel, Incoloy, Inconel, and Udimet are trademarks of Special Metals, owned by Precision Castparts Corporation (in the 1960s, it was a subsidiary of Allegheny Ludlum). It incorporates trademarks of the former Inco, which traces its origins to International Nickel Company, Ltd., which was created in New York in 1902 (a charter member of the Dow-Jones Industrial Average when it was formed in 1928). Hastalloy and Haynes are trademarks of Haynes International in Indiana, USA.[4] Waspalloy is a trademark of United Technologies Corporation in Connecticut, USA, which used to own Pratt & Whitney (now owned by Mitsubishi).[5] René superalloys are developed and trademarked by GE. The SC superalloy CMSX is the registered trademark of Cannon Muskegon Corporation in Michigan, USA (founded in 1952).

Note that alloys are typically identified by their number (e.g., Alloy 718). However, manufacturers sometimes change the chemistry of an existing nickel-based alloy and give it their own trade name without changing the alloy's number. The confusion this creates is significant; for example [7]:

- *Inconel 700* with Ni 44%, Mo 3%, Cr 14%, Co 29.5%, C 0.1%, Ti 2.5%, and Al 3.0%
- *Udimet 700* with Ni 53%, Mo 5%, Cr 15%, Co 18.5%, C 0.12%,Ti 3.5%, Al 4.25, and B 0.08%

The most ubiquitous trademark in the superalloy realm is arguably *Inconel.* Most of the currently manufactured products of the Monel, Incoloy, and Inconel families of nickel alloys were developed by the researchers in Huntington Alloys works in Huntington, West Virginia. The company began in 1922 as a processing mill for nickel alloys melted at the Orford Works of the International Nickel Company in New Jersey. Eventually, melting operations were added to the Huntington Works, and it became a fully integrated nickel alloy processing mill. Research and development became a key component of the Huntington operation. The alloys developed by Huntington included Alloy 718 (the aforementioned patent by Eiselstein) and its derivatives.

Today, the trademark is applied to alloys such as 600 and 718 in a manner very similar to calling all facial tissues in the marketplace "Kleenex" (a unique brand). The original Inconel alloy and Alloy 600 were quite close in chemistry, i.e.,

- *Inconel* with Ni 78%, Cr 14.5%, Fe 7%, and C 0.05%
- *Alloy 600* with Ni 72%, Cr 15.5%, Fe 8%, and C 0.075%
- *Alloy 718* with Ni 50–54%, Co 0–1%, Cr 17–21%, and Mo ~3%

[4] The founder, Elwood Haynes, patented nickel- and cobalt-based alloys as early as 1907. Founded as Haynes Stellite Co., at one point it was owned by Union Carbide for almost 50 years.

[5] Waspalloy was named after Pratt & Whitney's popular aircraft engine with radially arranged pistons, the Wasp.

In addition, there are superalloys developed by the original equipment manufacturers (OEMs), e.g.,

- FSX,[6] GTD (Gas Turbine Division) by GE (also aforementioned René)
- MAR-M by Martin Marietta Corp.
- MGA2400 (stator vanes) and MGA1400 (rotor blades) by Mitsubishi Heavy Industries (MHI) [8]

Nickel-based superalloys can be divided into three basic groups:

1. Wrought alloy (Nimonic 80, 105, etc.)
2. Mechanical alloy (Inconel MA 760, etc.)
3. Cast alloy
   a. Conventional (equiaxed)
   b. DS
   c. SC or SX

Conventional-cast superalloys include GTD-111, GTD-222, In-713, In-738, MAR-M-247, U-500, and U-700. DS superalloys include GTD-111, René 80H, René 142, etc. SC alloys are classified into five generations, i.e.,

- First-generation SX superalloys include CMSX-2, -3, and -6 and René N4;
- Second-generation SX superalloys include CMSX-4 and René N5;
- Third-generation SX superalloys include CMSX-10 and René N6;
- Fourth- and fifth-generation SX superalloys are developed by National Institute of Material Science (NIMS) in Tsukuba, Japan, and are designated by TMS (e.g., TMS-196, a fifth-generation SX superalloy).

Cobalt-based superalloys are conventionally cast and include FSX-414, X-40 and -45 (invented by Rudolf Thiellemann), and MAR-M-509 and -302.

A historical summary of superalloy development going back to the 1940s is provided in [5], specifically figure 7 on p. 8 (which can be viewed on the Internet, too). In almost all such curves, superalloys are quantified by their "temperature capabilities." Note that the capability in the cited reference is defined as the temperature in degrees Celsius for 100 hours at 140 MPa. For 1,000 hours at the same stress level, an approximately 5 percent reduction (in Celsius scale) in temperature is requisite; for 100,000 hours, the requisite reduction in temperature is about 20 percent. One should pay attention to capability definitions when comparing data from different sources.

Performance and operability requirements of the gas turbine lead the designer in materials selection for major gas turbine components. The most critical component, due to its operating pressure, temperature, and environment (potentially corrosive hot gas), is the turbine "hot gas path." Material selection for the most cost-effective design to strike a balance between performance and part lives is based on the following considerations:

[6] F for A.D. Foster, S for C.T. Simms, and X for R. Thiellemann.

- Type of component (i.e., stator vane or rotor blade);
- Hot gas path location (i.e., stage 1, 2, 3, or 4);
- Output and efficiency (i.e., airflow, which sets the parts size, and TIT);
- Rotational speed (which determines the centrifugal forces);
- Cooling scheme (if applicable; i.e., film, convective, etc.);
- Fuel type (corrosive components in combustion products);
- Maintenance inspection intervals (initial cost versus maintenance cost).

The considerations enumerated above translate into material properties, i.e.,

- Physical properties
  - Density, ρ (typically about 0.3 lb/in^3 or 8,300 kg/m^3)
  - Melting range (typically 2,250–2,400°F)
  - Stability
  - Specific heat, c_p
  - Thermal conductivity, k
  - Thermal expansion coefficient, α
  - Modulus of elasticity, E
- Chemical properties
  - Oxidation
  - Sulfidation
- Mechanical properties
  - Tensile properties
  - Impact properties
  - Axial (push–pull) fatigue properties
  - Thermal fatigue properties
- Fabrication
  - Machining and grinding
  - Castability
  - Welding

The superalloys frequently used in turbine hot gas path components are listed in Table 13.2. There are two groups. The first group ("Legacy") comprises E–class and older, first-generation F–class gas turbines from different OEMs. The second group ("Advanced") comprises newer-generation F-, H-, and J–class gas turbines. Table 13.2 is not intended to be comprehensive. OEM-specific alloys (specifically for MHI/Mitsubishi Hitachi Power Systems [MHPS] and some for Siemens) are not included. In any event, in chemical composition and properties, they would be quite similar to those listed in Table 13.2 anyway.

13.3 Future Materials

Combined with DS or SX casting and advanced film cooling, TBC has been a strong enabler of state-of-the-art TITs exceeding 1,600°C. Nevertheless, going to even higher

Table 13.2 Turbine hot gas path materials used in older- and newer-generation gas turbines. LC = low carbon; NV = nozzle vane; RB = rotor blade

Stage	Component	Legacy	Advanced
1	NV	In-939, In-738, FSX-414, GTD-111	MAR-M-509, CMSX-4, MAR-M-247, René 108, MGA2400
	RB	In-738LC, GTD-111DS	PWA 1483SC, CMSX-4, MGA1400DS
2	NV	In-939, FSX-414, GTD-111	René 80, MAR-M-247, René 108, MGA2400
	RB	In-738LC, U-520, GTD-111DS	In-6203DS, CM-247LC, René 108DS, MGA1400DS
3	NV	In-939, X-45, GTD-222	René 80, René 108, MGA2400
	RB	In-738LC, GTD-111DS	René 80, MAR-M-247, GTD-444DS, MGA1400DS
4	NV	In-939, X-45	René 80, GTD 262, MGA2400
	RB	U-520, In-792	René 80, GTD-444DS, MGA1400CC

temperatures (e.g., 1,700°C TIT) is pushing the limits of current metallurgy. Refractory metal alloys have been looked at as possible substitutes for nickel- and cobalt-based superalloys. However, they have several deficiencies, such as a severe lack of oxidation resistance, which preclude them from being viable alternatives at this point.[7]

Although there is no "unobtanium," of course, there is a reasonably close substitute: *ceramic matrix composites* (CMCs). CMCs are produced from ceramic fibers embedded in a ceramic matrix. For gas turbine applications, the most promising candidate is SiC/SiC, a CMC made of silicon carbide fibers and a silicon carbide matrix. There is also good potential for oxide–oxide CMCs with *friable graded insulation* (FGI), which consists of hollow ceramic spheres and filler bonded by aluminum phosphate. For a comprehensive overview of CMC materials, the reader is referred to the handbook edited by Krenkel [9].

Just like their monolithic forebears (Si_3N_4 and SiC), CMCs have excellent creep resistance and show high stiffness at a third the weight of nickel-based alloys. Furthermore, their high fracture toughness solves the main shortcoming of the monolithic ceramics, namely susceptibility to brittle fracture and low thermal shock resistance. SX nickel-based alloys with TBC have a temperature capability of around 1,150°C, whereas CMCs can go up to 1,300°C with *environmental barrier coatings* (EBCs) for protection from oxidation without air cooling (or very little of it).[8]

CMC materials have been in development since the 1990s. Static (non-rotating) parts such as combustor liners and stage 1 shrouds made from SiC/SiC with EBCs by different OEMs have survived several thousands of hours in field installations [10]. One OEM has accumulated about 25,000 hours on a single hybrid oxide CMC (aluminosilicate/alumina) with FGI on a combustor liner [10]. Another OEM tested

[7] Other important considerations are the dwindling supplies and increasing prices of refractory metals such as rhenium, which are predominantly imported materials.

[8] About 20 percent of compressor airflow is utilized for component cooling. The resulting dilution of combustion exit gas temperature (by about 100°C) across the stage 1 nozzle and loss of shaft output via airflow bypassing the turbine stages severely hamper the gas turbine efficiency.

low-pressure turbine blades made from CMCs in an aircraft test engine in early 2015. Thus, it appears that a breakthrough has been made. How long it will take for full aircraft engine implementation and technology transfer into heavy-duty industrial gas turbines remains to be seen. Cost and durability under fire, specifically component life demonstration (25,000–30,000 hours) under the combined centrifugal and thermal stresses plus high aerodynamic loading (rotating parts), are the key hurdles to commercialization.

Advanced gas turbine combined-cycle power plants require efficient bottoming cycles. Along with the increase in gas turbine firing and exhaust temperatures, optimal thermodynamic design dictates commensurately high steam temperatures. For the most advanced J–class machines with exhaust temperatures pushing 1,200°F (650°C), steam temperatures of 600°C are requisite for good performance. Combined with high steam pressures of around 2,400–2,500 psi (160–170 bars), T91 and P91 are requisite as heat recovery steam generator (HRSG) tube, header, and steam pipe materials. Higher steam pressures (even supercritical) are possible with once-through evaporator designs. With supplementary firing, steam conditions similar to those in advanced, ultra-supercritical, coal-fired power plants can be achieved. Whether such a design with requisite materials including superalloys would be cost-effective is difficult to predict.

13.4 Thermal Barrier Coatings

Corrosion resistance of the superalloys due to limited amounts of Cr (to make room for refractory metals in the composition) has been supplemented by special coatings. Protective coatings are primarily of two types: *diffusion* (e.g., platinum-aluminide) and *overlay* (e.g., MCrAlY, where M is Co, Ni, or Co/Ni). There are several different processes to applying these coatings on the component. For the overlay coatings, they are electron beam physical vapor deposition (EB-PVD), low-pressure plasma spraying (LPPS), vacuum plasma spraying (VPS), or air plasma spraying (APS) and high-velocity oxyfuel (HVOF) spraying. The choices for diffusion coating are pack cementation and chemical vapor deposition (CVD).

In addition to coatings used for corrosion protection, starting in the 1980s, TBCs have been used to increase the temperature capability of superalloys by up to 200°C (360°F). TBCs are essentially ceramic "insulation blankets" protecting the bare metal surface (or metal surface with corrosion-protective coating) from the hot gas. They were first deployed in aircraft engines in the 1960s and eventually found their way into land-based industrial gas turbines in the 1980s. TBC is a multilayer coating comprising an insulating ceramic *top coat* and a metallic *bond coat*. The most common top coat material is 7 percent (wt.) yttria partially stabilized zirconia (7YSZ) with MCrAlY as the bond coat. Gas turbine OEMs and third-party (aftermarket) suppliers actively develop their own TBC and application technologies. One example of the latter is dense vertical cracking, wherein the thermal spray process produces vertical cracks in the microstructure of the ceramic coating layer to provide strain compliance as well as resistance to material loss via chipping, erosion, etc. This coating was first deployed by

GE in the steam-cooled H-System, whose hot gas path materials experienced the highest temperatures at the time (early 2000s).

Research and development in the TBC area includes sensor coatings (enabling remote detection of coating temperature), thermal memory coatings, gadolinium zirconate coatings, addition of rare earth elements to the coating composition to lower thermal conductivity, and application technologies such as suspension plasma spray.

13.5 Typical Material Properties

The data in Table 13.3 are handy sources for typical steels used in power plant equipment and construction. For nickel- or cobalt-based alloys, thermal expansion coefficient values are closer to the high end of the listed range. Elastic modulus values are at room temperature (see Table 13.4 for the temperature effect). For later-generation SX nickel-based alloys, the density can be as high as 9,000 kg/m^3. For more detailed tables and charts, the reader is referred to chapter 25 of [11] in Chapter 2. In fact, this chapter, written by Berger and Grünling (from the Technische Universität Darmstadt in Germany), is, in this author's opinion, the best compact compilation of high-

Table 13.3 Typical steel properties for quick reference

Properties	Carbon Steels	Alloy Steels	Stainless Steels	Tool Steels
Density (1,000 kg/m^3)	7.85	7.85	7.75–8.10	7.72–8.00
Modulus of Elasticity (GPa)	190–210	190–210	190–210	190–210
Poisson's Ratio	0.27–0.30	0.27–0.30	0.27–0.30	0.27–0.30
Thermal Expansion Coefficient (10^{-6}/K)	11.0–16.6	9.0–15.0	9.0–20.7	9.4–15.1
Melting Point (°C)	1,370–1,450			
Thermal Conductivity (W/m-K)	24–65	13–28	11–37	20–48
Specific Heat (J/kg-K)	450–2,100	450–1,500	420–500	–
Tensile Strength (MPa)	276–1,882	750–1,400	515–827	640–2,000
Yield Strength (MPa)	190–750	350–1,200	200–550	380–440

Table 13.4 Young's modulus (E). GPa = gigapascals; ksi = 1,000 psi

Temperature		E	
°F	°C	GPa	ksi
300	149	199.9	29,000
500	260	193.1	28,000
700	371	186.2	27,000
900	482	175.8	25,500
1,100	593	162.0	23,500

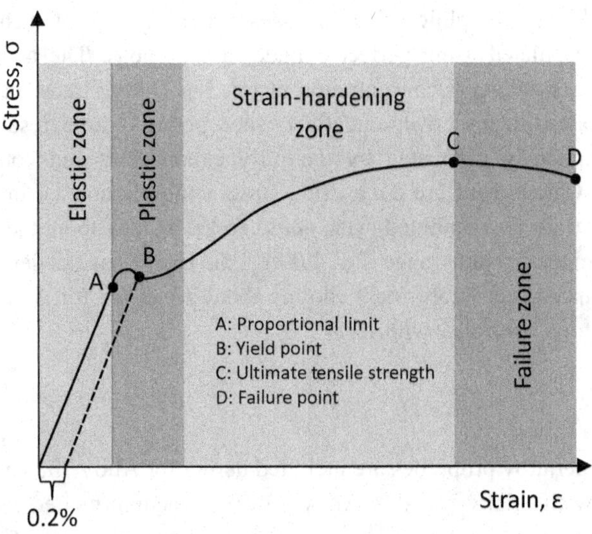

Figure 13.5 Conceptual stress–strain diagram.

temperature materials used in gas turbine hot gas path components with an extensive bibliography.

The slope of the stress–strain diagram within the elastic zone is called the "modulus of elasticity" or "Young's modulus," E, which depends on the material type and temperature. A typical stress–strain diagram is shown in Figure 13.5. For most materials, there is no sharp transition from the elastic region to the plastic region. Therefore, a small strain, such as 0.1 or 0.2 percent, is chosen to determine the yield stress. Such a stress is termed as 0.1 or 0.2 percent offset yield strength. The peak in the stress–strain curve is the ultimate tensile stress or the tensile strength of the material. The stress at which the material fractures is the fracture strength. Tensile yield strengths (defined as 0.2 percent offset yield strength) of selected alloy steels can be found in [1].

Young's modulus as a function of temperature is given in Table 13.4, which is for CrMo (pronounced "chromoly") steels. It is represented by the second-order polynomial below (in GPa):

$$E = -3 \cdot 10^{-5} T^2 - 0.0034 \cdot T + 203.36.$$

Values of E for most other steels show little variation for most practical purposes (not for detailed design, of course).

In a gas turbine combined-cycle power plant, in terms of thermal stress management, the most critical component is the steam turbine rotor. For large combined-cycle steam turbines, rotors can weigh about 200,000 lb or more. (By far the heaviest one is the low-pressure turbine rotor, which accounts for about 75 percent of the total weight.) Their diameters can be more than 20 inches – up to 24 inches or even more – for the high-pressure rotor and 30 inches or more for the low-pressure rotor.

A steam turbine rotor is not a perfect cylinder. Variations in the outer surface geometry of the rotor such as a step change in the diameter (e.g., due to the presence of a turbine wheel) cause amplification of stresses in the locations of such discontinuities from those calculated using perfect cylinder assumptions. The amplification is quantified by the thermal stress concentration factor, K_T. During transient events with significant temperature changes over a relatively short period of time, resulting thermal stresses are continuously calculated by the turbine controller and compared with allowable values. Calculations are done using stress concentration factors, which are typically around 2.0 for the combined-cycle units. To get an idea about the magnitudes, in the typical steam temperature range (i.e., 1,000 ± 50°F), the tensile yield strength for a CrMoV (pronounced "chromoly-vee") alloy is about 70–80 ksi (cf. 57 ksi allowable stress in turbine stress controller with $K_T = 2.0$).

13.5.1 Inconel 738

Representative superalloy properties are included herein for Alloy 738 (Inconel or In-738) of Inco (now Specialty Metals). Alloy 738 is a vacuum-melted, vacuum-cast, precipitation-hardenable nickel-based alloy specially designed for gas turbine applications. It has a good creep strength up to 1,800°F and can withstand long-term exposure to hot and corrosive environments. Alloy 738 has high and low carbon variants. Its nominal composition contains 16 percent Cr, 1.75 percent Mo, 8.5 percent Co, 2.6 percent W, 1.75 percent Ta, and 3.8 percent each of Al and Ti. Low carbon (0.11 versus 0.17 percent) is necessary for improved castability for larger parts and does not impact tensile and stress–rupture properties appreciably. The best combination of mechanical properties is achieved after heat treatment, which contains the following two steps:

1. 2,050°F for 2 hours then air-cooling
2. 1,550°F for 24 hours then air-cooling

Machining, grinding, and castability of In-738 is similar to other high-temperature, high-strength nickel-based alloys. Similar to other superalloys containing aluminum and titanium, which form the major strengthening γ' phase in the alloy, In-738 is not ordinarily considered "weldable." (The reason for this is its susceptibility to intergranular cracking in the heat-affected zone.) A weldability chart of superalloys is shown in Figure 13.6. Progress made in crack-free welding of In-738 and similar difficult-to-weld nickel-based alloys using techniques such as laser arc hybrid welding, friction welding along with pre-weld thermal treatment, and using filler alloys with less Al + Ti + Nb + Ta concentrations can be found in the literature [11].

For most of the nickel-based superalloys, physical property data presented for In-738 (α, ρ, k) are representative. The range of values for different superalloys is pretty narrow. Note, however, the modulus of elasticity of a DS alloy is about 35–40 percent smaller than that for its conventionally cast counterpart (same composition). This, of course, is the driver behind the much longer thermal fatigue life of DS and SX superalloys vis-à-vis conventionally cast ones (i.e., by a factor of 3–5 for DS and about 10 for SX under comparable boundary conditions). From [3], specific heat, thermal conductivity, dynamic modulus of elasticity (E for tension and G for torsion, for

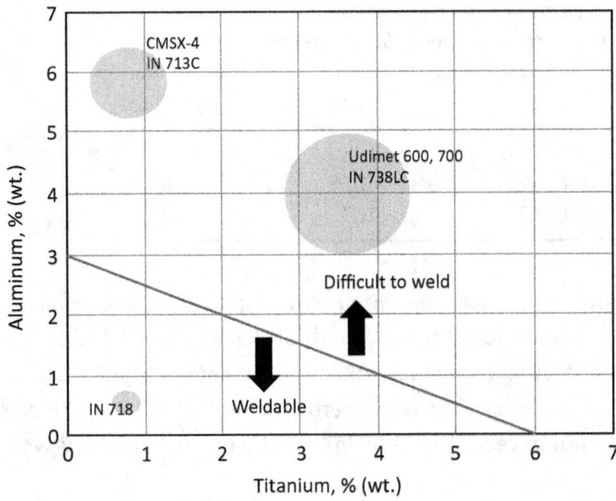

Figure 13.6 Weldability of superalloys. (from figure 10 on p. 512 in [12])

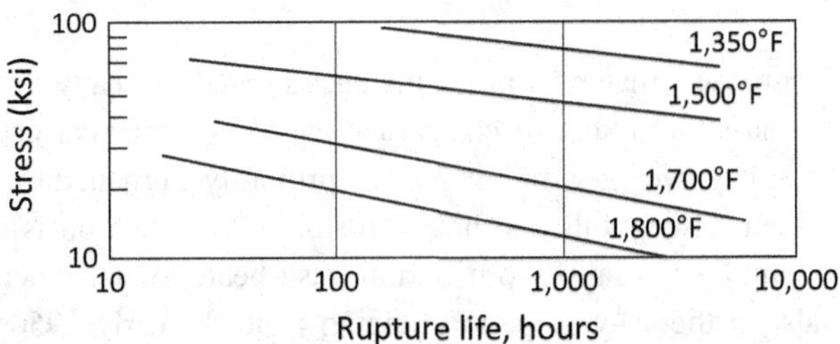

Figure 13.7 Stress–rupture properties of heat-treated In-738 [4].

Poisson's ratio $\nu \approx 0.3$), and average coefficient of linear expansion as a function of temperature are represented by the simple linear relationships listed below. Stress–rupture curves are presented in Figure 13.7. The thermal fatigue resistance characteristic is summarized in Table 13.5.

$$c_p[\mathrm{Btu/lb-F}] = 0.0355\left(\frac{\mathrm{T[F]}}{1000}\right) + 0.1015,$$

$$\mathrm{k}\left[\mathrm{Btu-in/ft^2-hr-F}\right] = 0.0671\,\mathrm{T[F]} + 55.167,$$

$$\mathrm{E[ksi]} = 29620 - 4.771\,\mathrm{T[F]},$$

$$\mathrm{G[ksi]} = 11582 - 1.9305\,\mathrm{T[F]},$$

$$\alpha\left[\mathrm{x10^{-6}/F}\right] = 6 + 0.0025\,\mathrm{T[F]}.$$

In terms of "stability," the biggest concern is the formation of "undesirable phases" during service. These phases are called topologically close-packed phases, with the most commonly found in nickel-based alloys being "sigma" and "mu" (i.e., σ and μ).

Table 13.5 Thermal fatigue resistance of selected alloys [4], as defined by thermal fatigue cycles to first crack in three different peak temperatures (°F)

Alloy	1,472	1,652	1,832
In-738	1,790	150	13
In-100	>1,372	107	29
In-713C	>1,462	107	27

Sigma is a good source for crack initiation and propagation, and it has a detrimental effect on the high-temperature rupture strength of the superalloy [12].

For prediction of sigma formation tendency in superalloys, "electron vacancy" calculations are used. In order to ensure microstructural stability, the electron vacancy number, Nv, should not exceed 2.36. For In-738LC, for example, Nv was found to be 2.31 [4].

13.6 When Materials Fail

For most practical purposes, materials problems encountered in heavy-duty industrial gas turbines can be classified into two categories: thermo-mechanical and corrosion. Gas turbines in electric power generation service primarily burn natural gas, which drastically reduces their susceptibility to hot corrosion.[9] The situation is different for liquid fuel-burning units, especially those burning ash-bearing fuels such as crudes, blends, and residuals. In the 40-year period ranging from the early 1950s to the late 1980s, burning such fuels in myriad applications (i.e., transportation prime mover, pipeline, chemical process plant, etc., as well as in electric power generation) constituted the bulk of gas turbine deployment.

While the ash-bearing fuels are the primary sources of contaminants such as sodium, potassium, and vanadium, which play key roles in hot corrosion, they are not the only ones. Contaminants in ambient air and system water can find their way into the machine through the air inlet system, via evaporative cooler carryover, compressor wash solutions, and water injection for NOx control (e.g., when burning the backup fuel oil in Dry-Low-NOx [DLN] combustors). Thus, inlet air filter selection and maintenance as well as proper water treatment are extremely important in order to prevent hot corrosion. Even small amounts of sodium in the air can be an issue (sodium sulfate is a key deposit on hot gas path components).

In the absence of performance augmentation technologies such as inlet fogging and wet compression, the compressor operates in a dry environment. In an extended lay-down, water condensation on the cold surfaces of stator vanes and rotor blades can be problematic. Hygroscopic[10] substances such as chlorides and sulfur trioxide (a key

[9] Hot corrosion is a form of oxidation accelerated by a high-temperature environment. It results from the chemical reaction between the turbine component and molten salts deposited on its surface.

[10] An adjective indicating a tendency to absorb moisture from the air.

pollutant and the primary agent in acid rain) can absorb the condensing water, forming an aqueous solution that can lead to corrosion of compressor components (pitting, stress corrosion cracking, etc.) [13]. If chlorides are present in the ambient air with high relative humidity, static temperature drop at the compressor inlet can result in the formation of a corrosive liquid during operation as well. In addition to using a proper inlet air filtration system with high-efficiency filters and a self-cleaning pulse-jet system, use of corrosion-resistant base metals and protective coatings in manufacturing the compressor components is imperative.

In the thermo-mechanical category, there are three primary damage mechanisms (i.e., creep, fatigue, and brittle failure). The creep mechanism has been discussed earlier. Creep rupture is a significant issue for the parts operating at very high temperatures and stresses for a long time (i.e., turbine stage 1 rotor blades). This is the key factor in selecting superalloys and protective coatings for those components.

Fatigue results from the application of repetitive or fluctuating stresses at levels generally lower than the tensile strength of the material in question. There are two types of fatigue: high-cycle fatigue (HCF) and low-cycle fatigue (LCF). LCF is the result of stresses applied once per engine operating cycle. The most significant LCF cause is steep temperature gradients and resulting thermal stresses in components with thick metal walls during startup. Such stresses can go up to 60 or 80 percent of the tensile strength of the material. HCF occurs at much higher frequencies associated with resonant vibrations. Furthermore, the magnitude of fluctuating stresses is much lower than that encountered in LCF conditions. For plastic deformation, crack initiation, and crack propagation mechanisms in creep, LCF, and HCF, the reader is referred to the excellent book by Viswanathan [3]. The subject of fatigue classification will be touched upon again in Section 16.2 when covering failure mechanisms.

From a practical perspective, a gas turbine engineer must be thoroughly familiar with the following two mechanical concepts:

- Thermal stress (LCF)
- Vibrations (HCF)

This is so because during the operation of the gas turbine, the controller continuously measures, calculates, and monitors these two parameters and compares them with prescribed alarm and trip levels.

13.6.1 Thermal Stress

Thermal stresses in a solid body made of a particular material result from changes in temperature in the body and can lead to fracture or plastic deformation. There are two primary causes:

1. Restrained thermal expansion or contraction
2. Temperature gradients

Consider a cylindrical, solid body made of steel (e.g., similar to a turbine rotor).

1. When the cylinder is heated or cooled and there are no constraints at either end, it can expand (i.e., become longer) or contract (i.e., become shorter) freely.
2. When the cylinder is *heated* and there are constraints preventing it from getting longer (expansion), a *compressive* thermal stress is produced.
3. When the cylinder is *cooled* and there are constraints preventing it from getting shorter (contraction), a *tensile* thermal stress is produced.

The magnitude of the stress σ resulting from a temperature change from T_i (initial) to T_f (final) is given by

$$\sigma = E\frac{\alpha}{1-\nu}(T_i - T_f), \tag{13.1a}$$

where E is the elastic or Young's modulus, α is the linear coefficient of thermal expansion, and ν is Poisson's ratio, which converts the *linear* thermal expansion coefficient to a *volumetric* thermal expansion coefficient (for isotropic substances).

Let us do a quick calculation. Assume that the low-alloy steel rotor of a steam turbine is at 700°F and subjected to a steam flow at 900°F. Ignoring the construction details of the rotor and assuming it to be a solid cylinder, if the rotor were constrained from both sides, using data from Section 13.5, the resulting stress would be

$$\sigma = 200\,\text{GPa} \times 12{\cdot}10^{-6}\text{K}^{-1} \times (700 - 900)/1.8\,\text{K} = -0.027\,\text{GPa} = -270\,\text{MPa}.$$

Since $T_f > T_i$ (i.e., heating), σ is a negative value, indicating a compressive stress. If T_f were lower than T_i (i.e., cooling), σ would be a positive value, indicating a tensile stress. In the 700–900°F range, the average tensile strength is about 550 MPa (note that 1 ksi ≈ 7 MPa). Thus, the magnitude of the calculated compressive stress is about 50 percent of the tensile strength of the material.

This is why turbogenerators (gas or steam) have only one thrust bearing to allow the rotor to expand or contract freely in the axial direction. Likewise, turbine casings are also supported such that only one end is fixed in the axial direction (the other end can slide along the built-in guides). Similarly, in order to accommodate the thermal expansion or contraction of a large steel pipe, thermal expansion joints (bellows) and loops are utilized.

The second cause of thermal stress is temperature gradients across a body that is heated or cooled. For illustrative purposes, consider a cylinder with a thick wall. Initially, the cylinder is at a uniform temperature and free of any stresses. Upon rapid heating, however, the exterior of the cylindrical wall becomes hotter than its interior (i.e., a temperature *gradient*) and, as such, expands more. This induces compressive stresses on the exterior surface of the cylinder, which are balanced by tensile stresses on the interior surface. Upon rapid cooling, on the other hand, the situation is reversed. This process induces tensile stresses on the exterior surface of the cylinder, which are balanced by compressive stresses on the interior surface. The operative word in this explanation is *rapid*. *Gradual* changes in temperature over a long time period in which the ambient and the solid body are in thermal equilibrium do *not* generate thermal stresses.

For this case, the stress formula becomes

$$\sigma = E\frac{\alpha}{1-\nu}(T_{s,1} - T_{s,2}), \tag{13.1b}$$

where $T_{s,1}$ and $T_{s,2}$ refer to the metal temperatures at the interior and the exterior/surface of the solid body in question, respectively, at a given time, t.

This is why the rate of temperature changes during severe plant operation transients such as startups and load ramps are of extreme importance, especially for *thick-walled* components such as the steam drum (essentially a cylindrical tank) of the high-pressure evaporator in the HRSG or the casing of the high-pressure section of the steam turbine. The measure of "thickness" is typically a "characteristic" length (e.g., the wall thickness of cylindrical components). For bodies of irregular shape, which makes identifying a characteristic length difficult, it is customary to define it as $L_c \equiv V/A_s$ (i.e., volume divided by surface area). In relatively *thin-walled* components with sufficiently small L_c (e.g., the gas turbine casing vis-à-vis the steam turbine casing), the absolute value of the temperature difference, $|T_{s,1} - T_{s,2}|$, is small so that σ is small.

In summary, solid bodies experience thermal stresses as follows:

1. Upon cooling ($T_{surface} < T_{interior}$),
   a. Solids contract.
   b. The interior constrains the surface from contracting.
   c. Tensile stress occurs on the surface.
   d. Compressive stress occurs in the interior.
2. Upon heating ($T_{surface} > T_{interior}$),
   a. Solids expand.
   b. The interior constrains the surface from expanding.
   c. Compressive stress occurs on the surface.
   d. Tensile stress occurs in the interior.

Overall, due to the requirement of mechanical equilibrium, tensile and compressive stresses inside the body always balance each other.

Equation 13.1 is a very simple yet powerful tool. It can be used to gain an excellent *qualitative* understanding of many complex problems. In the hands of an experienced and knowledgeable engineer, it can even be helpful in obtaining approximate but useful *quantitative* information. All one needs is basic information on the materials and geometry of the component under investigation (turbine rotor, casing, etc.) and the time history of the component temperature and the temperature of its surrounding (steam, hot gas, etc.).

How Does It Work?

The bulk metal temperature of a turbine component, T_m, and, more precisely, its variation in a metal structure across a characteristic dimension, L_c, is the key determinant of thermal stress via the Equation 13.1.

In practical problems, the critical component from a thermal stress perspective is the steam turbine high-pressure rotor (in a gas turbine combined cycle). For the rotor, the temperature difference in Equation 13.1 is the difference between rotor surface or bore and mean body (bulk) temperatures for surface and bore stresses, respectively. The characteristic dimension is the diameter of the rotor, which is about 20–25 in. for the high-pressure rotor of modern gas turbine combined-cycle units.

For a given steam temperature, T_{stm}, bulk rotor body T_m varies according to the exponential decay law

$$\frac{T_m - T_{m,0}}{T_{stm} - T_{m,0}} = 1 - e^{-\frac{t}{\tau}}, \qquad (13.2)$$

with a characteristic time constant, τ, which is a function of rotor material (e.g., 1 percent CrMoV) and size cum geometry represented by L_c, i.e.,

$$\tau = \frac{\rho \cdot L_c \cdot c}{h}, \qquad (13.3)$$

where h is the convective *heat transfer coefficient* (HTC) between steam and metal. Equations 13.1–13.3 tell the entire steam turbine thermal stress management story in the concise language of mathematics:

1. Thermal stress is determined by the temperature gradient in the rotor (essentially a cylinder) via Equation 13.1.
2. The latter is determined by the initial steam–metal ΔT (denominator of the left-hand side of Equation 13.2) with a time lag.
3. The time lag itself is dictated by the HTC in Equation 13.3.
4. Everything hinges on the initial value of T_m, $T_{m,0}$, which is a function of the cooling period (see Section 19.3.2 later in the book).

In physical terms, this translates into a mechanism to control steam flow, pressure, and temperature (FPT) into the steam turbine at initial values sufficient

1. To roll the unit from turning gear speed to full-speed, no load;
2. To warm the steam turbine (ST) rotor until steam–metal ΔT decreases to an acceptable level; and
3. To ramp them up at acceptable rates to their rated levels while ensuring that thermal stresses do not exceed prescribed limits.

Steam flow enters the picture via the HTC in Equation 13.3, which controls the rate of heat transfer between steam and the rotor surface as described by the *heat flux* balance at the steam–metal boundary (x = 0)

$$\dot{q} = h \cdot (T_{stm} - T_m) = k \cdot \left. \frac{dT_m}{dx} \right|_{x=0}. \qquad (13.4)$$

This equation introduces the dimensionless *Biot* number, $Bi = h \cdot L_c/k$, which is a relative measure of the uniformity of temperature gradients inside a heated or cooled body. The heat transferred from steam to the rotor at the surface increases the rotor's bulk temperature according to Fourier's law,

Table 13.6 Representative values of major parameters characterizing the transient heat transfer during steam turbine warmup for typical steam flow, pressure, and temperatures (1 Btu/h-ft^2-°F = 5.678263 W/m^2-K, 1 ft^2/h = 2.58064E-5 m^2/s)

m/m_o	P (psia)	T (°F)	h (Btu/h-ft^2-°F)	Bi	δ (ft^2/h)	τ (min)
1.0	120	700	116	7	0.26	37
1.0	120	1,050	100	6	0.21	54
1.0	1,200	700	958	56	–	5
1.0	1,200	1,050	701	41	–	8
0.2	120	700	32	2	–	135
0.2	120	1,050	28	2	–	196
0.2	1,200	700	264	15	–	16
0.2	1,200	1,050	193	11	–	28

$$\frac{dT_m}{dt} = \frac{k}{\rho \cdot c} \cdot \frac{d^2T_m}{dx^2}. \tag{13.5}$$

Equation 13.5 introduces the *thermal diffusivity*, δ = k/ρc, which quantifies the *speed* with which the temperature of a heated or cooled body changes. Typical values for the key parameters governing ST rotor thermal transients are given in Table 13.6.

For ferritic steels used in modern combined-cycle units, k and ρ do not show significant variation. Thus, δ is primarily a function of temperature and changes by about 25 percent between 700 and 1,050°F (i.e., the rate of change of metal temperature is 25 percent faster at the *lower* temperature). The data in Table 13.6 can be summarized as follows: *higher steam flow and/or pressure result in higher rates of heat transfer between steam and metal, which is quantified by higher Biot numbers and shorter time constants (i.e., faster heating or cooling).*

In conjunction with the data in Table 13.6, Equations 13.4 and 13.5 identify the two distinct phases in ST start with thermal stress control:

1. Low flow and high steam–metal ΔT with low HTC until temperature gradients settle down (*non-stationary* phase or Phase I).
2. Increasing steam FPT to load the unit with high HTC and nearly constant, low steam–metal ΔT (*quasi-stationary* phase or Phase II).

Equation 13.6 describes Phase I via its simplified solution for a cylindrical geometry given by [14]:

$$\sigma_{max} = E' \cdot \alpha \cdot \left[\frac{Bi}{2.8 + Bi + \sqrt{Bi}}\right] \cdot K_T \cdot \Delta T, \tag{13.6}$$

which gives the maximum thermal stress implied by a given step rise in T_{stm} at time t = 0 (with a time lag characterized by the Biot number). Note that in Equation 13.6, K_T is the *stress concentration factor* and

$$E' = \frac{E}{1 - \nu}.$$

Similarly, Equation 13.5 describes Phase II via its simplified form given by

$$\frac{dT_{stm}}{dt} = \frac{\delta}{\varphi_F \cdot E' \cdot \alpha \cdot L_c^2} \cdot \sigma_{max}, \tag{13.7}$$

where ϕ_F is the *form factor* (0.125 for a cylinder [14]).

Equation 13.7 gives the allowable T_{stm} ramp rate for a given maximum *allowable* stress, σ_{max}, which is dependent on rotor material and typically lies in a range of 50–80 ksi. For the cited range, with the data in Table 13.6, Equation 13.6 suggests that

- For low HTC (~100 Btu/h-ft^2-°F [~570 W/m^2-K] or less), steam–metal ΔT can range from 200–300°F (high K_T) to 500°F and higher (low K_T).
- For high HTC (~650 Btu/h-ft^2-°F [~3,700 W/m^2-K]), steam–metal ΔT can range from 100–200°F (high K_T) to about 400°F (low K_T).

Similarly, using Equation 13.7 with Table 13.6, it can be seen that allowable values for dT_{stm}/dt range from 3–6 to 8–10°F/min.

The allowable stress is *not* a precisely defined material property. (For ferritic steels used in steam turbine rotor construction, 0.2 percent tensile yield strength lies between 70 and 90 ksi for temperatures of 600 and 1,000°F.) It is derived from the *S–N curves* relating total strain to cycles to failure, which give the fatigue life of the material in question (for the LCF life of CrMoV alloy, see Figure 13.8, which is from equation 4.11 in [3]). Based on the correlation between stress and strain, ε, via the modulus of elasticity,

$$\sigma = \frac{E'}{K_T} \cdot \varepsilon, \tag{13.8}$$

this curve is used to determine σ_{max} for a defined fatigue life. For example, for a cyclic life requirement of 5,000 cycles, the total strain range is 0.0047, which is roughly 2ε. Thus, from Equation 13.8, the maximum allowable stress for K_T = 1.5 is found as 54 ksi.

In practice, the correlation between σ and ΔT allows the translation of the S–N curve into *cyclic life expenditure* (CLE) curves, which determine the allowable T_{stm} ramp rates (Figure 13.9). Depending on the rotor material, size and geometry, and its

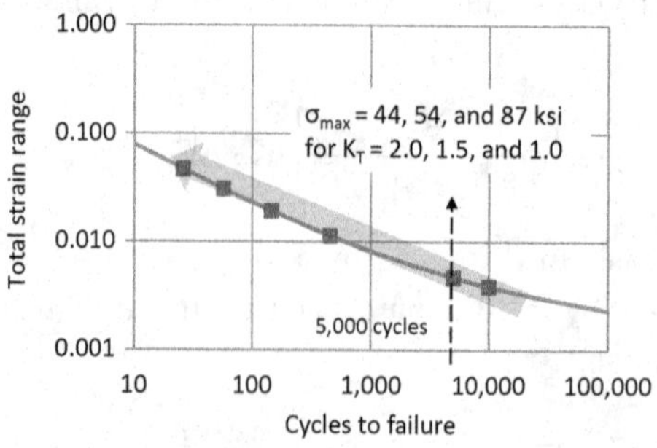

Figure 13.8 Typical S–N curve for steam turbine rotor LCF (CrMoV).

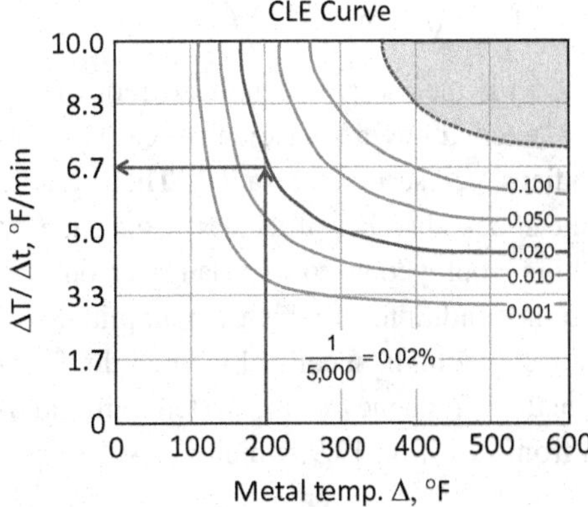

Figure 13.9 Typical CLE curve for steam turbine rotor LCF (CrMoV) [3]. Metal ΔT in the chart represents the *total* temperature change between initial and final states (beyond 600°F, curves are flat).

temperature at start initiation, the range is limited to about 5–10°F/min, except for very hot "restarts" after a few hours of downtime. In Figure 13.9, for example, for a total metal temperature change of 200°F and 5,000 cycles, the allowable ramp rate is 6.7°F/min.

13.6.2 Vibrations

At least a third of gas turbine failures are attributed to compressor or turbine blade problems (see Section 16.2). Such failures almost always include "liberation" of one or more blades at very high speed and subsequent "corn-cobbing" of the downstream stages. Repair and downtime costs of such events can be astronomic. Just to give an idea of them, consider that a single row of superalloy blades with TBC in a modern, large industrial gas turbine can cost several millions of dollars. The single most significant cause of compressor and turbine blade failures is vibration. Consequently, an adequate understanding of vibration fundamentals within a turbomachine context is extremely important.[11] (See Section 16.5 for vibration *monitoring*.)

In vibration analysis of rotating components such as turbine blades, one frequently comes across the term "mode," usually in association with "natural" frequencies (especially when discussing the Campbell diagrams). The mode and natural frequency of an object are functions of the object's geometry. For complex three-dimensional objects (e.g., the long last-stage bucket [LSB] of a modern steam turbine), it is very difficult to visualize and fully grasp them. But a basic understanding can be gained from a simple geometry. The simplest one is the simple mass and spring, which has only one mode (oscillating up and down) and has a natural frequency of

[11] Please refer to section 1 of API Recommended Practice 684 (API Standard Paragraphs Rotordynamic Tutorial: Lateral Critical Speeds, Unbalance Response, Stability, Train Torsionals, and Rotor Balancing) for a very thorough overview.

$$f_n = \sqrt{k/m},$$

where k is the spring constant and m is the mass. The unit of frequency is s^{-1} or Hz. Now, consider that this system is "excited" by a sinusoidal force, $F = F_0 \cdot \sin(\omega t)$ with $\omega = 2\pi f$ (*angular* frequency, radians per second or rad/s). This system also has a "viscous damper" analogous to the shock absorber of a car's suspension system. From everyday experience (recall pushing your young son or daughter on a swing in the park), we know that there are certain conditions in which the amplitude of the oscillations (vibrations) becomes uncontrollably high, specifically when the frequency of the exciting force, ω, is equal to the natural frequency of the system with no damping, ω_n. This can be mathematically seen from Equation 13.9, which gives the *amplitude* of the oscillations, A, i.e.,

$$A = \frac{F_0}{\sqrt{k^2\left(1 - \frac{\omega^2}{\omega_n^2}\right)^2 + c^2\omega^2}}. \tag{13.9}$$

From Equation 13.9, when there is no damping (i.e., c is 0) and $\omega = \omega_n$, the denominator of the fraction on the right-hand side of the amplitude formula becomes 0 and $A \rightarrow \infty$. However, this is not the case when there is damping of the system (i.e., $c \neq 0$). In that case, the denominator is never zero.

This very basic theory can be translated into the much more complex vibration problems such as "flutter" in the large LSBs of a gas or steam turbine. As mentioned earlier, while the simple mass–spring–damper system has only one mode of oscillation, the complex-shaped LSB has many modes of oscillation (i.e., vibration). The key point is that each vibration mode, i, has a natural frequency, f_i, and if the periodic force acting on the blade has a frequency of $f = f_i$, the amplitude of vibrations becomes very high (i.e., resonance[12] occurs). *Resonance* leads to component failure via HCF.

How can there be more than one mode of vibration? In order to answer this question, consider a beam pinned on both sides. The first three modes of vibration for this system correspond to the beam taking the shape of a sinusoid covering 0.5, 1, and 1.5 periods, respectively. Furthermore, the vibration frequency of each mode *n* can be calculated algebraically, i.e.,

$$f_i = \frac{\pi}{2} \cdot \left(\frac{n}{L}\right)^2 \sqrt{\frac{E \cdot I}{m/L}}, \tag{13.10}$$

where m is the beam's mass, L is its length, E is the modulus of elasticity, and I is the area moment of inertia. Thus, for example, if the natural frequency of the first mode is 50 Hz, then the natural frequency of the second mode is 200 Hz and the natural frequency of the third mode is 450 Hz. For turbine blade and disk vibration modes, please refer to the paper by Meher-Homji [15].

[12] Per API 684, resonance is defined as "the manner in which a rotor vibrates when the frequency of a harmonic (periodic) forcing function coincides with a natural frequency of the rotational system."

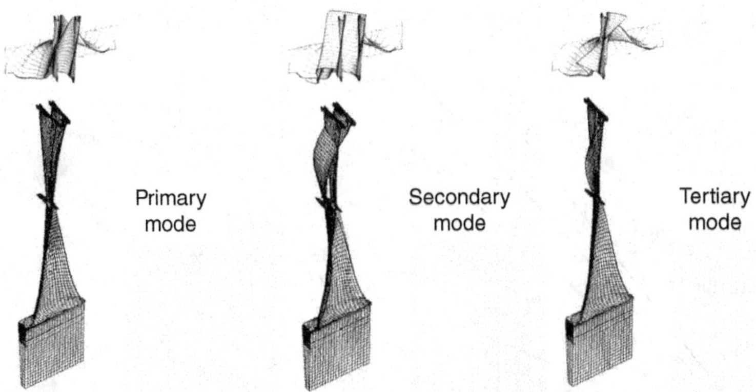

Figure 13.10 Vibration modes from the three-dimensional FEA for a single blade [16].

The only difference between the simple beam pinned on both ends and the LSB is that the modes in the latter case are much more complicated and the natural frequencies, f_i, cannot be determined from closed-form formulae similar to Equation 13.10. They can only be determined using cyclic-symmetry finite-element analysis (FEA) and nonlinear forced response analysis, which are then verified by rotating vibration tests (i.e., in the vacuum chamber or in the spin pit). (A note for the mathematically inclined: natural frequencies and mode shapes are the *eigenvalues* and *eigenvectors* of the second-order differential equation describing the harmonic motion of a vibrating object. For complex geometries, the equation can only be solved numerically; i.e., via FEA.) Other than that, however, the principles are the same.

How complicated are the vibration modes of a large gas or steam turbine blade? Examples of the vibration analysis results for Mitsubishi's new 3,600-rpm (60-Hz) ultra-long LSBs (50 inch ≈ 1.25 m) are shown in Figure 13.10 [16]. Note, however, that the vibration of a fully bladed disk is much more complicated than is suggested by the characteristics of a single cantilever blade. There is a multiplicity of modes of vibration in the turbine working frequency range (they are referred to as *nodal diameters*).

Large blades such as the LSBs have natural frequencies that are lower than those of the shorter blades (higher mass, m, with $f \propto \frac{1}{\sqrt{m}}$). Typically, the first mode is around 100 Hz, the second mode around 200 Hz, and so on. Thus, they are susceptible to excitation by the lower harmonics[13] (below the eighth harmonic or *engine order*). Vibration characteristics are verified via tests, and the results are plotted on a *Campbell diagram* (Figure 13.11). In the Campbell diagram, where natural frequency lines and harmonics cross, there is a prospect of resonance in service. It is usual to confine attention to ±6 percent of synchronous speed (e.g., 2,820–3,180 rpm for 3,000-rpm [50-Hz] units). A common specification is that in this range there should be no resonance up to the

[13] A harmonic of a wave is a component frequency of the signal that is an integer multiple of the fundamental frequency (i.e., if the fundamental frequency is f, the harmonics have frequencies 2f, 3f, 4f, etc.). In turbomachinery applications, it is also known as the *engine order*.

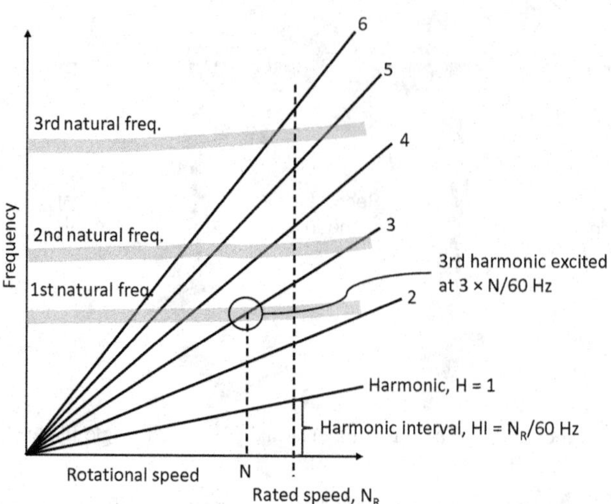

Figure 13.11 Generic Campbell diagram.

eighth order. (In general, the higher the harmonic, the lower the amplitude of vibration.) Problem modes can be "detuned" by adjusting the blade mass near the tip or by adding or removing mass to or from the shrouding.

Note the increase in the natural frequencies in Figure 13.11 as the rotational speed increases. This is the "stiffening" effect of the centrifugal force on the long blades, which increases the value of stiffness, k (and note that $f \propto \sqrt{k}$). Due to the stiffening effect, testing is done at running speed in an evacuated wheel chamber. This is so because (i) the presence of air would require a large amount of power to drive the disk and (ii) windage near the blade tips would cause overheating and make the results difficult to interpret. Typically, the test (also known as a *wheel box* test) is conducted in a vacuum chamber by first increasing the rotational speed up to about 110 percent and measuring the static (centrifugal) stress. Air or oil jet excitation is applied while decreasing the speed and the stress, and displacement amplitudes generated at resonance are measured using dynamic strain gauges and blade vibration monitoring.

What exactly is "excitation" of a mode? A mode will only be excited if (i) the excitation frequency coincides with the resonant frequency of the mode and (ii) the loading has the necessary component of spatial distribution. The best-known example is the collapse of the Tacoma Narrows Bridge in 1940 due to *aeroelastic flutter*. (The reader is strongly encouraged to watch a YouTube video of the event for a visual appreciation of the phenomena discussed herein.) When the wind blew (the forcing event) at a certain speed, air buffeting (vortex shedding) that apparently had the same frequency as the bridge's natural frequency at a certain mode shape caused the bridge to oscillate. This led to resonance (i.e., the vibration became large and uncontrolled) until the bridge broke up and collapsed into the Puget Sound.

Sources of vibration excitation in gas or steam turbines are

1. Non-uniform flow (e.g., caused by fluid entering over only a portion of the circumference), complex axial to radial flow behavior (recently addressed via 3D advanced-aero design codes), or flow distortion;
2. Periodic effects due to manufacturing constraints (e.g., inexact matching at fir-tree roots, eccentricity of diaphragms, ellipticity of stationary parts, non-uniformity of manufacturing thicknesses);
3. Nozzle wake excitation as a rotating blade passes a stationary blade.

Sources 1 and 2 cause excitation at the low harmonics of the rotational speed (50 or 60 Hz). Source 3 causes excitation at a much higher frequency (referred to as the *nozzle passing frequency* – see Section 13.6.4). For a detailed discussion of turbine and compressor component failure modes, their causes, and troubleshooting, please refer to the paper by Meher-Homji [17].

There are also sources that can cause *self-excitation* at frequencies that are unrelated to rotational frequency, such as acoustic resonances (in inlet passages, extraction lines, or other cavities), unsteady flow separation from stationary blades, unsteady shock waves in blade passages (e.g., condensation shocks in steam turbines when the expansion rapidly crosses the *Wilson line*), surface pressure fluctuations from impingement of turbulent flow, and vortex shedding from bluff bodies. Note that the aforementioned Tacoma Narrows Bridge collapse also falls into this category (vortex shedding).

In long LSBs, damping is achieved by tip shrouds and/or snubbers (or stubs), whose geometries and positions are decided based on analysis and testing, and by structural damping via geometry and size optimization. The design goal for the blade is to clear resonance in a band around the design rotational speed (50 or 60 Hz) and low-cycle operational range. Having said that, it should be understood that passage through resonance zones with increased vibration amplitudes is unavoidable during startup and shutdown (e.g., see the excitation of the first mode by the third harmonic of the rotational speed at 2,000 rpm in the Campbell diagram in figure 13 of [16]). Typically, harmonics below the eighth order are cleared at normal operating speed.

13.6.3 Fatigue Fracture

Fatigue is the phenomenon leading to fracture under repeated or fluctuating stresses having a maximum value less than the tensile strength of the material. Fatigue fractures are progressive, beginning as minute cracks that grow under the action of fluctuating stresses.

There are two types of fatigue: HCF and LCF. HCF occurs at relatively large numbers of cycles or stress applications. The numbers of cycles may be in the hundreds of thousands, millions, or even billions. As an example, consider the mode 1 natural frequency of a 50-Hz (3,000–rpm) steam turbine LSB, which is between 100 and 150 Hz (i.e., 100–150 cycles per second, depending on speed). At 2,000 rpm during startup, it is found that this mode is excited by the third harmonic of the system.[14] If the system (i.e., the LSB) spends 10 minutes at 2,000 rpm when this mode is excited, it will go through $10 \times 60 \times 100 = 60{,}000$ cycles. The critical question is this: How severe is this resonance from a fatigue failure (fracture) perspective?

The failure of the material due to fatigue fracture is a function of

- *Fatigue limit*, which is the maximum stress below which a material can presumably endure an infinite number of stress cycles; or
- *Fatigue strength*, which is the maximum stress that can be sustained for a specified number of cycles without failure.

An example of a low cycle is the start–stop cycle of a power plant. Depending on the nature of the operating regime of the plant (e.g., base load or cycled), this can be anywhere between 10 to a few hundred per year. Each start–stop cycle subjects the equipment to thermal stresses, which can ultimately lead to LCF.

There is no sharp dividing line between LCF and HCF. For practical purposes, however, HCF is not accompanied by *plastic* (permanent) deformation; in other words, it is characterized by *brittle fracture* (as opposed to *ductile fracture*). LCF may be accompanied by some plastic deformation, which remains after removal of the load or force that caused the deformation (shape of change). The example most familiar to the layman is the permanent deformation and eventual breakup of a paper clip or wire coat hanger upon repeated bending with high stress relatively quickly (i.e., maybe 10 or 20 bends or "cycles").

A fracture has three distinct phases: initiation or origination, propagation, and final rupture. Initiation is the most complex and difficult-to-understand stage. Final rupture, however, is the easiest to understand; it is essentially the proverbial *last straw* that breaks the camel's back. Propagation is the most readily identifiable area of the fatigue fracture with distinct microscopic and macroscopic marks and features.

Fatigue strength and limit are typically quantified by the S–N curve (S for stress, N for number of cycles), which was shown for the CrMoV material earlier in Figure 13.8. The y-axis is the alternating stress amplitude (in MPa or psi) and the x-axis is the number of cycles before failure occurs. Note that the S–N curve does not distinguish between crack initiation and crack propagation. At low stresses (to the right), N_f denotes primarily initiation, whereas at high stresses (to the left), N_f corresponds primarily to crack propagation. Corrosion, erosion, surface marks (via *foreign object damage*), and stress concentration location (corners, discontinuities, etc.) severely reduce the fatigue life.

Definitions of alternating and mean stress amplitudes can be found in Figure 13.12. One specific case of interest of the general situation depicted in Figure 13.12 is when the mean stress (σ_m) is zero (known as the *fully reversed* case). This is the case for typical S–N testing leading to the development of S–N curves. For a combination of mean and alternating stresses, determination of cycles to failure (i.e., the *fatigue life* of the

[14] Note that 2,000 rpm is 33.3 Hz, so that its third harmonic is $3 \times 33.3 = 100$ Hz and thus "excites" the mode 1 natural frequency (also 100 Hz).

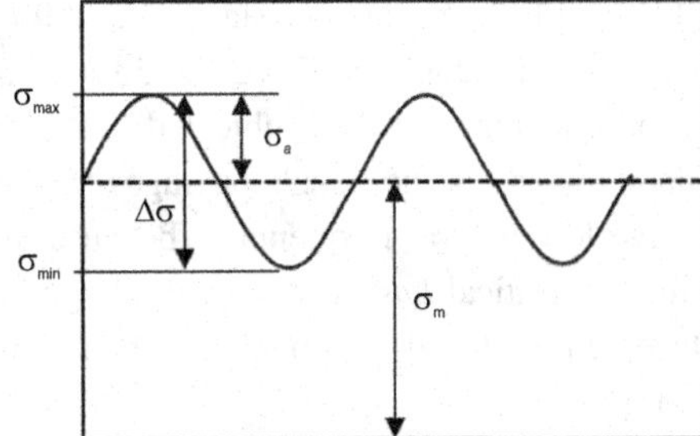

Figure 13.12 Alternating stress, σ_a, due to vibration – general case.

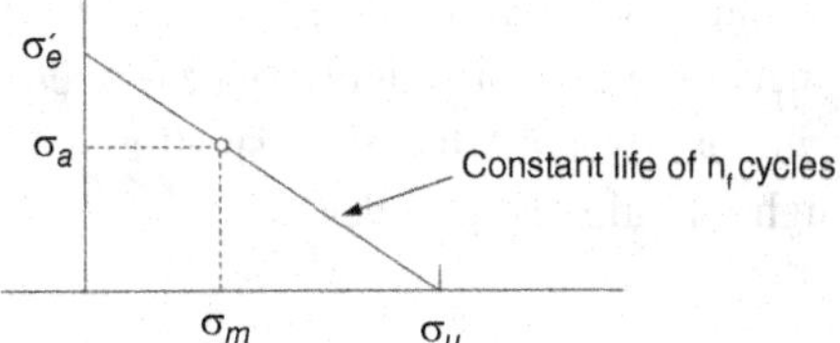

Figure 13.13 Goodman line.

material) requires the combined use of the S–N curve and the Goodman line, which is the most commonly used empirical curve to estimate the effect of mean stress on fatigue life (see Figure 13.13).[15]

In the Goodman line, σ'_e is the effective, purely alternating stress that will cause failure at N_f cycles and σ_u is the ultimate tensile strength of the material. Using σ'_e, one can determine the lifetime for a given combination of alternating and mean stresses from the S–N diagram for the given material, similar to that in Figure 13.8. If the combination of mean and alternating stresses from Figure 13.12 indicates a point on the diagram in Figure 13.13 that lies under the Goodman line, it means that the part will survive. If the point is above the Goodman line, then the part will fail.

In terms of the total stresses on a turbine blade, the important quantity is the *combined* static (centrifugal) and vibratory (steam or gas bending) stresses (i.e., σ_m and σ_a, respectively). In particular,

- Selection of blade material, size, and aerodynamic design (length, mass, airfoil shape) determines the blade natural frequency.
- Selection of the number of blades in a stage (arranged circumferentially on the rotor) determines the *nozzle passing frequency* (NPF), which in turn determines the amplitude of the oscillations (i.e., A in Equation 13.9; how the fluid force on the blade, F_0, is *amplified* by the blade vibrations).

[15] The Goodman line is similar to the Soderberg line, which terminates at the yield stress on the horizontal axis and thus is very conservative (seldom used). The parabolic version is the Gerber curve, which is less conservative than the Goodman line. (Test data fall between the Gerber curve and Goodman line.) The name Haigh line is also used in the literature.

For example, for a typical large LSB with 53 blades per stage at 3,000 rpm (50 Hz), assuming the same number of nozzles for the stage, NPF is 50 × 53 = 2,650 *cycles per second.* In other words, in one second, a single LSB is subjected to 2,650 steam flow "distortions" resulting from the blade traversing the nozzle passages from which steam is discharged. Since large LSBs have low natural frequencies (because of large mass, m), NPF is not a concern for them. The critical *harmonics* for the large blades are the first few caused by relative displacement of nozzles to buckets, diaphragm joints, and structural supports in the flow path.

During the design phase, large blades such as the LSBs in a steam turbine are *detuned* to avoid resonance at lower harmonics at normal operating speed. However, aeroelastic disturbances such as unstalled flutter can cause excitation at frequencies corresponding to those harmonics (e.g., up to several hundred Hertz). Since freestanding blades, unlike coupled blades, have no damping mechanisms of any significance of their own, during the passage through resonance excitation, dynamic forces as small as 1 percent of the static forces are enough to cause high vibration amplitudes.

13.6.4 Flutter

Flutter is an aeroelastic instability that occurs when energy is exchanged between the fluid and the structure (e.g., steam/gas and the rotating blade) in a manner that creates *self-excitation.* The key term here is "self-excitation," which separates flutter from the synchronous (harmonic) excitation depicted in a Campbell diagram at the intersection of a natural frequency with the harmonics of rotational speed (i.e., 50 or 60 Hz) at a given speed, which leads to resonance (see Figure 13.11). Flutter is a special case of non-synchronous vibration at low to moderate "reduced" frequencies. Reduced frequency, k, is a measure of the unsteadiness of the problem defined as

$$k = \frac{2\pi f\, L_c}{V_{flow}},$$

where f is the vibration frequency. For steady flows, k = 0; k = 0.2 designates a highly unsteady case (with the characteristic length, L_c, taken as the airfoil semi-chord).

There are two types of flutter: *stalled* and *unstalled.* (Note that in many cases the qualifier "unstalled" is dropped when discussing this phenomenon and the term "flutter" is dropped when discussing the "stall" phenomenon.) Stalled or stall flutter (also known as *aerodynamic buffeting*) is similar to the well-known stall of an airplane wing. In gas or steam turbines, it happens when the LSBs experience a negative angle of attack at low loads (i.e., low fluid flows) and/or high exhaust/back pressures (i.e., condenser pressure in steam turbines). The region of stall that forms at the trailing edge of the blade results in a form of unstable vibration called stall flutter. Stall flutter can result in accumulation of a considerable number of cycles in a very short period of time, which leads to blade failure via HCF (see the Section 13.6.2).

This phenomenon is typically limited to the upper portion of the blade due to the action of the centrifugal forces. The fluid in the remaining region is "churned" by the blades and recirculates. In steam turbines, stall flutter is limited to low loads (well below 50 percent) and high back pressures.

Unstalled flutter (also known as *negative aerodynamic damping*), as the name suggests, occurs without stall conditions at high flow rates. There is no flow separation and the flow is attached to the blade at all times. It is more likely to happen in freestanding blades that lack frictional damping. The mechanism is typically vortex shedding downstream of the blade's trailing edge. When the shedding frequency excites a natural frequency of the blade, resonance happens and can lead to HCF. A common yardstick for the susceptibility of the component in question to flutter is the *Strouhal* number, which is defined as

$$St = f_s L_c / V,$$

where f_s is the shedding frequency, L_c is a characteristic length (e.g., the airfoil thickness), and V is the fluid velocity. (Note the analogy between the Strouhal number and the reduced frequency.) In general, if St above 0.35 for the first bending mode, the long LSB is deemed safe from a flutter perspective [18]. For the first torsional model, St has to be above unity [18].

Flutter becomes a problem because of the possible phase differences between the blades when they are vibrating. If the blades are identical, they vibrate at the same frequency, but with a constant *phase angle* between their neighbors. Each vibration mode (known as the *traveling wave mode*) has a different *inter-blade phase angle*. The inter-blade phase angle affects the phase between the local unsteady flow and local blade motion, which in turn affects the unsteady aerodynamic work done on the blades. Adverse phase angles can lead to positive work being performed on the blades (i.e., *negative aerodynamic damping*, which is unstable).

References

1. Schilke, P. W., *Advanced Gas Turbine Materials and Coatings, GER-3569G*, (Atlanta, GA: GE Energy, 2004).
2. Clark, D. R., Oechsner, M., Padture, N. P., Thermal Barrier Coatings for More Efficient Gas Turbine Engines, *MRS Bulletin*, **37** (2012), 891–897.
3. Viswanathan, R., *Damage Mechanisms and Life Assessment of High-Temperature Components* (Metals Park, OH: ASM International, 1989).
4. "Alloy IN-738, Technical Data," Inco, The International Nickel Company, Inc., New York, NY.
5. Davis, J. R. (Editor), *Heat-Resistant Materials, ASM Specialty Handbook* (Materials Park, OH: ASM International, 1999).
6. Thomas, B. G., Goettsch, D. D., "Modeling the Directional Solidification Process," Proceedings of the Fifth International Conference on Modeling of Casting and Welding Processes, September 16–21, 1990, Davos, Switzerland.
7. Kracke, A., "Superalloys, the Most Successful Alloy System of Modern Times – Past, Present and Future," 7th International Symposium on Superalloy 718 and Derivatives, The Minerals, Metals & Materials Society, 2010.

8. Okada, I., Torigoe, T., Takahashi, K., Izutsu, D., "Development of Ni Base Superalloy for Gas Turbine," Proceedings of Superalloys 2004 (The Minerals & Materials Society), 2004.
9. Krenkel, W. (Editor), *Ceramic Matrix Composites*, (Weinheim, Germany: Wiley-VCH Verlag GmbH, 2008).
10. Van Roode, M., Ceramic Gas Turbine Development: Need for a 10 Year Plan, *Journal of Engineering for Gas Turbines and Power*, **132** (2010), 011301.
11. Ola, O. T., Ojo, O. A., Wanjara, P., Chaturvedi, M. C., Crack-Free Welding of IN 738 by Linear Friction Welding, *Advanced Materials Research*, **238** (2011), 446–453.
12. Simms, C. T., Stoloff, N. S., Hagel, W. C., *Superalloys II: High Temperature Materials for Aerospace and Industrial Power*, (New York: John Wiley & Sons, Inc., 1987).
13. Kolkman, H. J., Mom, A. J. A, "Corrosion and Corrosion Control in Gas Turbines, Part I: The Compressor Section," ASME Paper 84-GT-255, ASME 1984 International Gas Turbine Conference and Exhibit, June 4–7, 1984, Amsterdam, The Netherlands.
14. *VGB PowerTech Guideline, "Thermal Behaviour of Steam Turbines, Revised 2nd Ed.," VGB-R105e* (Essen, Germany: VGB PowerTech Service GmbH, 1990).
15. Meher-Homji, C. B., "Blading Vibration and Failures in Gas Turbines: Part A – Blading Dynamics and the Operating Environment," ASME Paper 95-GT-418, ASME 1995 International Gas Turbine and Aeroengine Congress and Exposition, June 5–8, 1995, Houston, TX.
16. Fukuda, H. et al., Development of 3,600-rpm 50-inch/3,000-rpm 60-inch Ultra-long Exhaust End Blades, *MHI Technical Review*, **46**: 2 (2009).
17. Meher-Homji, C. B., "Blading Vibration and Failures in Gas Turbines: Part B – Compressor and Turbine Airfoil Distress," ASME Paper 95-GT-419, ASME 1995 International Gas Turbine and Aeroengine Congress and Exposition, June 5–8, 1995, Houston, TX.
18. Schneider, M., Sommer, T., "Turbines for Industrial Gas Turbine Systems," chapter 6 in [11] in Chapter 2.

14 The Hardware

Let us pause for a moment and see where we are at this point in the book. (Assuming, of course, that you have read the chapters of this book in the order that they appear in.) Thermodynamic cycle calculations are finished (Chapter 5). They are translated into compressor and turbine hardware design parameters (Chapters 10 and 11). Combustor design, development, testing, and manufacturing is a separate endeavor in and of itself (see Chapter 12 for more on this). Material selection is done (i.e., superalloys for turbine stator nozzle vanes and rotor blades – see Chapter 13), and drawings are completed and sent to the manufacturing facility. Vanes and blades are typically precision cast (directional solidification or single crystal as specified) and then machined to the final dimensions before applying coatings (where specified).

Before moving on, it should be emphasized that the design of the three major components of the gas turbine, individually or as a whole system, is rarely started from the proverbial "blank sheet." It is typically an evolutionary process wherein one or more components are "tweaked" within an existing "architecture" to improve performance and operability. Over the last three quarters of a century (roughly 1940–2015), there have been, of course, instances when original equipment manufacturers (OEMs) came up with quasi-revolutionary design changes and/or "new product" introductions. Some prominent examples are the F–class gas turbine [1], the Dry-Low-NOx (DLN) combustor [2], the reheat or sequential combustion gas turbine [3], and steam-cooled turbine stages [4]. In many cases (e.g., the DLN combustor and the steam-cooled hot gas path), as already discussed in Chapter 4 (in particular, see Section 4.8), government-supported national and international projects have been instrumental in such "more-than-evolutionary" technology advances. Nevertheless, more often than not, major OEMs' approaches to the heavy-duty industrial gas turbine design process are quite conservative, with many incremental design changes and extensive testing.

The synchronous alternating current (ac) generator is such an important part of the whole package (after all, the raison d'être of the gas turbine in question herein is *electric* power generation) that its coverage is postponed to a full chapter of its own (Chapter 15). Even a whole chapter on the synchronous ac generator can barely scratch the surface. In terms of size, weight, cost, design complexity, and other features, it is a separate "machine" in its own right. In fact, the "gas turbine generator" is a bona fide two-machine system that requires two tomes for proper coverage. This book is mainly dedicated to the flange-to-flange gas turbine. There are many excellent references on synchronous ac machines (some of them are listed in Chapter 15), which will have to be consulted by the reader interested in learning more about them.

14.1 The Rotor

This is a good place to say a few words about the key mechanical parts of the gas turbine: the rotor and the bearings (casings will be discussed in Section 14.3). Neither component enters the steady-state cycle performance and aerodynamic calculations. Lost power via friction between the rotor and the bearings is accounted for by the mechanical efficiency term (see Chapter 7). In unsteady-state analysis (e.g., for estimating the power and time to roll the unit from the turning gear speed [about 5 rpm] to full-speed, no-load [FSNL; 3,000 or 3,600 rpm]), rotor inertia is an input to the torque–speed formula (see Chapter 19). This is pretty much all there is to mechanical parameters used in performance calculations.

This lack of attention does not alter the simple fact that mechanical design is as important (if not more so) as thermodynamic and aerodynamic design. In terms of the modern gas turbine design process, however, it is rare that, say, a bearing is designed as a unique component for the particular gas turbine. In other words, it is pretty much an "off-the-shelf" product procured from external suppliers. Why this is so will be clarified in the following paragraphs.

To begin with, pretty much all OEMs of large industrial gas turbines for electric power generation adopted a two-bearing design:

- Combined thrust–journal bearing (for compactness) or separate thrust and journal bearings on the "cold end" (i.e., near the compressor inlet);
- Journal bearing on the "hot end" (i.e., near the turbine exit).

(There are some old E-class machines [e.g., General Electric's (GE) 7EA] with three bearings [i.e., one "hot bearing" between the compressor and turbine].) The hydrodynamic, radial journal bearings as well as the axial thrust bearing are typically of the "tilting pad" type. Radial bearings carry the static load of the rotor and provide damping for vibrations. The bidirectional, axial (thrust) bearing balances the net hydrodynamic thrust (transmitted to it through the "thrust collar" – see Figure 14.1) created by the working fluid inside the gas turbine and thus maintains the relative position of the stator and rotor within allowable limits. Once the key mechanical design parameters are known (total rotor weight, rotational speed, etc.), requisite sizing specifications commensurate with applicable part lives and maintenance requirements are prepared and the part is ordered from a suitable bearing OEM. The reason is obvious; just like the gas turbine itself, the bearing is a special product to be ideally procured from an OEM with decades-worth of design and test experience under its belt.

In a similar fashion, rarely is a gas turbine rotor designed from a blank sheet. Unlike the case with bearings, however, there is no independent "rotor OEM." While a bearing only "sees" a short part of the rotor (i.e., the "journal"), separated by a thin oil film

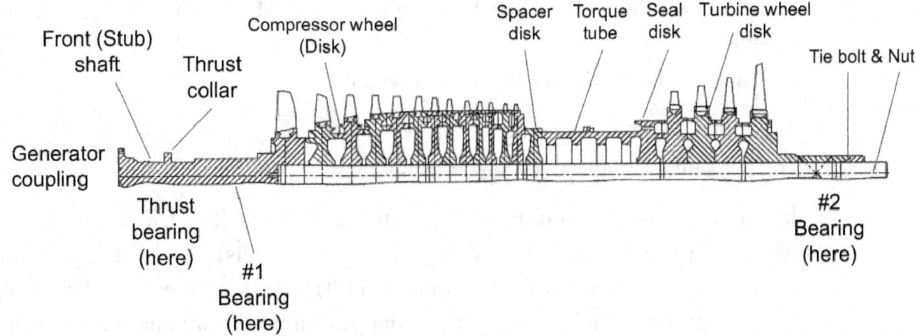

Figure 14.1 Rotor cross-section (Siemens SGT6-5000F, formerly Westinghouse 501FD2).

rotating between its pads, and does not care one bit about the rest of the gas turbine, the rotor is intimately connected with the other parts. In particular, the rotor of a gas turbine

- Carries the compressor and turbine blades;
- Transmits the torque from the turbine to the compressor and the generator (coupled to the gas turbine on the "cold end");
- Provides channels for the cooling air bled from the compressor to reach the turbine parts;
- Provides the platform for the labyrinth seals.

Based on the available thermodynamic and aerodynamic design information translated into mechanical parameters, the rotordynamic team develops the design and sizing criteria for the gas turbine rotor. Critical design inputs are the following:

- Tensile stress from centrifugal forces acting on the rotor;
- Thermal stress due to changes in temperature during transient events (startup and shutdown, load changes, etc.);
- Bending stress caused by the weight of the rotor.

Based on this information, the design team specifies the size and materials for the compressor and turbine rotor. A particular rotor construction is one of the few basic types adopted by the OEM over the course of many years spent in the evolutionary design of their product line and experience gained in field applications.

There are two major types of gas turbine compressor and turbine rotor construction:

1. Bolted disk (or "wheel")
2. Welded disk

The former is adopted by most major OEMs (i.e., GE, Siemens, Mitsubishi [MHI/MHPS], formerly Westinghouse [it continues to "live" in certain Siemens and MHPS designs], and Ansaldo [mainly building upon Siemens and Alstom technologies]). There is a third type (i.e., mono-block shaft with shrunk-on disks), which is not seen in industrial gas turbines (although it is common in steam turbines). A variation of this type of design is a drum-type (i.e., hollow) rotor with shrunk-on disks. This design approach was adopted in old Westinghouse and Mitsubishi 501D gas turbines for the compressor rotor.

The bolted disk design has two variants:

1. Single (central) tie-bolt
2. Multiple, radially arranged tie-bolts

In this construction, torque is transmitted either via face friction between individual disks (i.e., GE design with radially arranged tie-bolts) or wheels via Hirth serrations (i.e., Siemens design with a central tie-bolt). (As discussed in Chapter 4, the latter technology goes back to the birth of the gas turbine as a bona fide aircraft propulsion engine in Jumo-004.) For enhanced torque transmission via face friction, flange surfaces are grit-blasted (GE practice). Westinghouse (now Siemens SGT6-5000F – see Figure 14.1) design practice uses Gleason curvic coupling with the teeth radially arranged between the outer rim and the center of the wheels.

Typically, turbine blades are attached to the wheels using axial, fir-tree dovetails (e.g., see Figure 10.1) that fit into matching slots on the wheel rim. Compressor blades are attached to the wheels using broached axial slots or circumferential grooves around the wheel periphery.

The most common materials used in rotor construction are the following:

- Inconel 706 (turbine wheels, spacers, aft shaft) and 718 (tie-bolts);
- NiCrMoV and CrMoV for the compressor rotor.

The welded disk rotor design is a hallmark of Brown-Boveri Company (BBC)/Asea Brown-Boveri (ABB) gas turbines going back to the glory days of BBC in the 1950s and 1960s. Initially, only the turbine rotor was constructed with wheels welded at the rim (like in the 1939 Neuchâtel unit). Compressor blades were attached to a drum-type rotor via insertion into machined grooves. In the 1950s, when the machines grew in capacity with commensurately higher stresses, welded disk designs were adopted throughout the entire unit. Later inherited by Alstom and finally by GE in 2015, this rotor construction technology continues to live in legacy ABB/Alstom gas turbines. A detailed development history and manufacturing techniques for welded disk rotor design can be found in the book *Gas Turbine Powerhouse* by Eckardt (see the references section in Chapter 4).

Once the rotor is designed, the next task is a rotordynamic analysis to determine critical modes of vibration (i.e., lateral, torsional, and axial) and ensure that they are not excited at the operating speed. This would lead to resonance and very large vibrations well beyond the damping ability of the bearings. Nevertheless, some resonance is usually unavoidable during turbine transients (e.g., coast-down to turning gear speed upon a turbine trip or shut-off). The design goal in that case is to ensure that those critical speeds are passed through as quickly as possible.

After the welding (followed by heat treatment) or stacking of the disks is completed, the new rotor is ready for balancing in the "spin pit." This procedure ensures that any unbalance in the rotor assembly is eliminated or reduced to minimize unit vibration during operation. (If necessary, after installation on site, a "trim balancing" can be done as well.) Good information on rotor design for heavy-duty industrial gas turbines can be found in the papers by Florjancic et al. [5] and Endres [6].

14.2 This Is How It Was Back Then ...

Today (meaning in the second decade of the twenty-first century), a highly experienced mechanical engineer in the electric power generation industry is probably in his or her late 40s or 50s. In other words, he or she was born in the 1960s and graduated from college with a BS degree in the late 1970s or 1980s. It is thus extremely inconceivable to this engineer that a gas turbine could look anything other than what is shown in Figure 14.6 (in Section 14.3; i.e., a jet engine “on steroids”). (There may be those unique individuals, of course, who are interested in finding out about the now-obscure past of their profession.) Alas, in the 1950s, most gas turbine power plants for electric power generation (mainly used for peaking duties) looked nothing like that.

In Chapter 8, as an example of an intercooled-recuperated gas turbine Brayton cycle, the three power plants designed by GE and supplied to the Central Vermont Public Service (CVPS) were mentioned. These gas turbine power plants (dubbed as “kilowatt machines” by GE engineers) were installed in Rutland, VT, and they were operational until 1987. (Overall, 10 turbines of this design were manufactured by GE.) According to the quarterly of the Rutland Historical Society (Volume X, Number 3, summer of 1980), the first of these three power plants (under the name *Weybridge Project*) went online in October 1951. The total estimated cost of the three 5-MWe units was $3.6 million (resulting in rate increases for the customers). Between 1951 and 2016, the US dollar experienced an average annual inflation rate of about 3.5 percent (i.e., $1 in the year 1951 was worth $9.25 in 2016). Doing the math, it is seen that Rutland gas turbines came at an installed cost of $2,200 per kilowatt in 2016 dollars.

At the time of writing, a 5-MWe gas turbine is around $800–$900/kW installed (at an efficiency of around 30 percent; i.e., not much better than the Rutland gas turbines). In comparison, a 300-MWe advanced F/H/J–class gas turbine with 40 percent efficiency costs around $250/kW installed. A 600-MWe combined-cycle power plant based on those advanced gas turbines is around $700/kW. (These are *Gas Turbine World* “budget prices” – actual price at the time of commercial operation can be significantly higher due to site-specific owner’s costs and other commercial adders.) It is no wonder that gas turbines à la Rutland became extinct.

Why this is so can be easily gauged from the conceptual drawing of the gas turbine system in Figure 14.2 (after figure 9-19 in Skrotzki and Vopat [7]). The combined high-pressure (HP)–low-pressure (LP) turbine itself is dwarfed by the recuperators and the intercoolers.

The specifications of the Rutland gas turbine power plant machinery are as follows [7]:

- Nine-stage, axial-flow LP compressor, 7,200 rpm (driven by LP turbine, polytropic efficiency estimated as 85 percent)

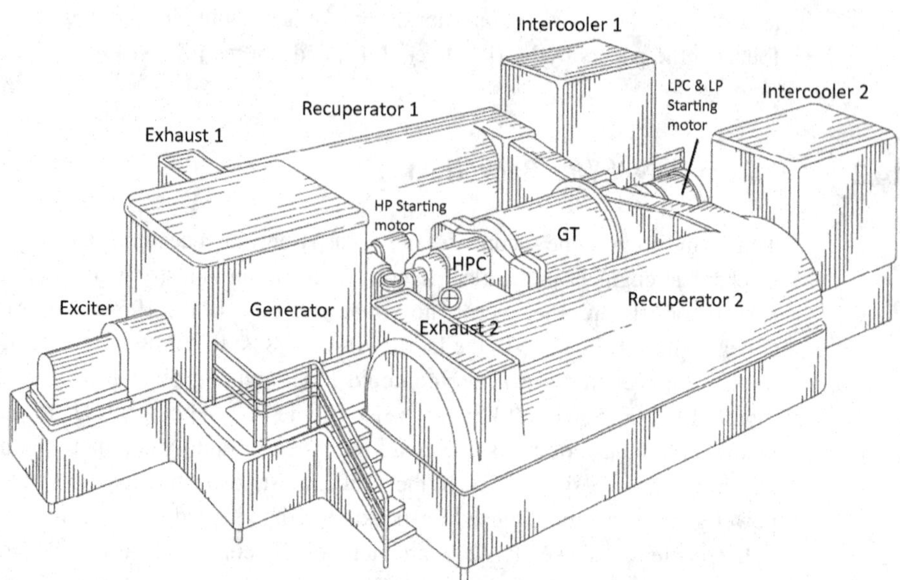

Figure 14.2 Artist's conception of GE's 5-MWe gas turbine plant with intercoolers and recuperators. GT = gas turbine; HPC = high-pressure compressor; LPC = low-pressure compressor.

- Eleven-stage, axial-flow HP compressor, 8,694 rpm (driven by HP turbine), with inlet guide vanes (IGVs); polytropic efficiency estimated as 83.5 percent
- Six combustors (liquid fuel; the plant would start on number 2 fuel oil and later transfer to the "bunker C" fuel[1])
- One-stage LP turbine (7,200 rpm when at 100 percent rated speed, polytropic efficiency estimated as 89.5 percent)
- Two-stage HP turbine (8,694 rpm when at 100 percent rated speed, 1,500°F turbine inlet temperature [TIT] with no cooling, polytropic efficiency estimated as 90 percent)
- Reduction gear (8,694–3,600 rpm) driven by HP turbine
- 5,000-kW generator, 6,250 KVA, 0.8 power factor, three-phase, driven by HP turbine via the gearbox
- Two recuperators (connected in parallel)
- Two intercoolers (connected in parallel)
- Two starting motors for both for LP and HP compressors

Kilowatt machines were indeed well ahead of their time in terms of performance, which, unfortunately, came at the expense of considerable complexity and size as shown by the Rutland plant layout in Figure 14.3. Sixty years later, the same generator output is delivered by a Solar Mercury 50 (also recuperated) packaged gen-set, which is about the same size as the recuperator in Figure 14.2 (4.6 MWe, 38.5 percent with 690°F exhaust).

[1] See Footnote 2 in Chapter 8 for the source.

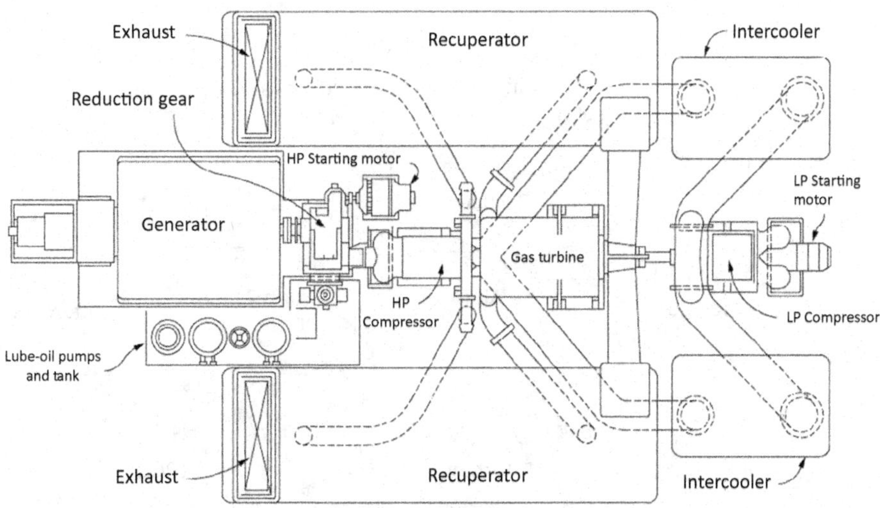

Figure 14.3 CVPS intercooled-recuperated gas turbine power plant layout [7].

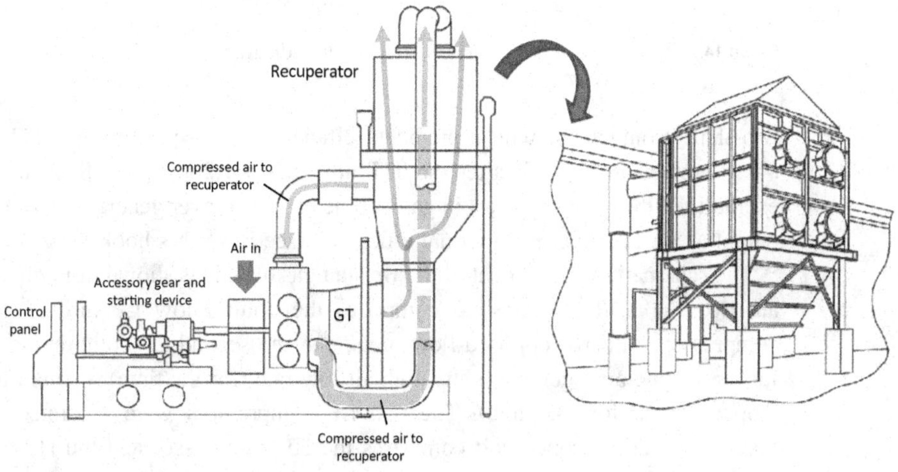

Figure 14.4 Schematic diagram of recuperative GE MS3002. GT = gas turbine.

However, it must also be pointed out that GE's MS3002 (i.e., two-shaft Frame 3) recuperative (but no intercooling) gas turbine packages of the same vintage as the complex unit in Figure 14.2 were much more compact and similar to Solar's Mercury 50 in size and arrangement (see Figure 14.4). Developed in the 1950s for gas pipeline and petrochemical applications, quite a large number of these machines are still used by US pipeline transmission facilities, chemical process industry plants, and oil refineries.

Around the same time as GE designed and deployed the 5-MWe intercooled-recuperative gas turbines, BBC in Switzerland had already designed and built the world's largest gas turbine power plant to date: the 27-MWe Beznau unit II (Beznau unit I was rated at 13 MWe). This machine had intercooling, recuperation, and reheat in a

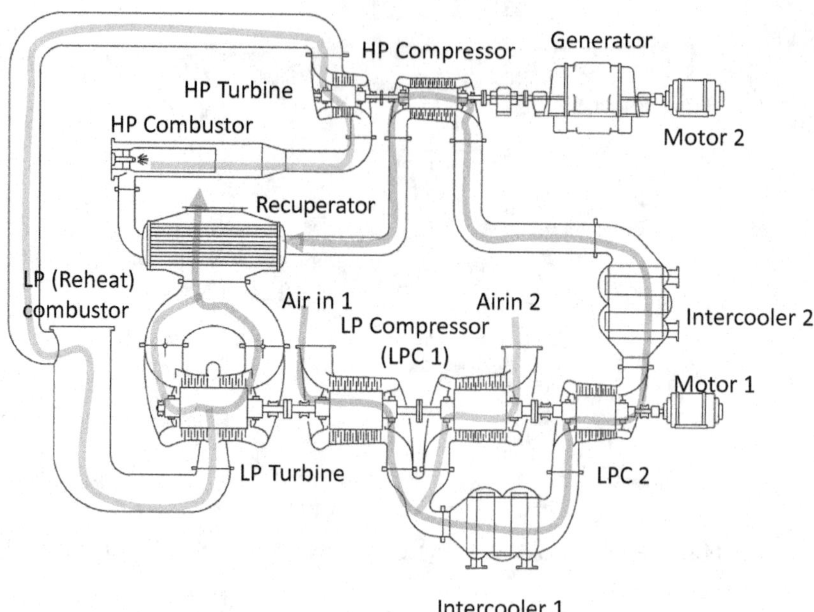

Figure 14.5 BBC Beznau gas turbine schematic flow diagram [7].

two-shaft arrangement with a predicted efficiency of 34 percent at a TIT of 1,100°F (41°F ambient). In 1953, after the TIT upgrade to 1,200°F, the Beznau power plant exemplified the state of the art in gas turbine electric power generation with a 40-MWe output at about 30 percent thermal efficiency (see Eckardt's book).

The schematic arrangement of the Beznau gas turbine is shown in Figure 14.5. There are two shafts. The first shaft comprises the double-flow LP turbine, the first LP compressor with two opposed-flow units, and the second LP compressor. The reason for the unique arrangement is the high airflow rate (for the time) required for the large output (again, for the time). The two LP compressors were separated by the first intercooler. The second shaft comprises the HP compressor and the HP turbine. This is also the output shaft to which the generator is connected. The recuperator takes the double-flow hot gas from the LP turbine and heats the air discharged from the HP compressor prior to entry into the HP turbine.

The actual plant arrangement is in two parallel power trains. The second (HP) shaft is on the left with the starting motor, HP compressor, HP turbine (in that order), and the two combustors. The first (LP) shaft is on the right. The intercoolers are below the turbine hall.

Note that the Beznau gas turbine was only the largest and most efficient iteration of BBC's intercooled-recuperated gas turbines with reheat. Between the Neuchâtel unit and 1945, several of these gas turbines, with arrangements very similar to that in Beznau, were designed and actually ordered by customers from across the world, including Rumania (no recuperator) and Peru (see Eckardt's book). Due to post-war

Table 14.1 Beznau gas turbine power plant technical data

	Beznau I	Beznau II
Generator Output (MW)	13	27
Airflow (kps/pps)	90/198.4	180/396.8
Air Inlet Temp. (°C/°F)	5/41	5/41
Cooling Water Temp. (°C/°F)	5/41	5/41
HP TIT (°C/°F)	650/1,200	650/1,200
LP TIT (°C/°F)	600/1,112	600/1,112
LP Pressure ratio	3.8	3.9
Overall Pressure ratio	8.8	8.1
Exhaust Temp. (°C/°F)	180/356	180/356
Recuperator Effectiveness (%)	80	80
HP rpm	4,750/3,000	3,000
LP rpm	1,600–3,000	1,600–3,000

difficulties, however, Beznau was the first unit to go commercial. The total plant cost, including units I and II, was 17.4 million Swiss Francs (CHF) or 435 CHF per kilowatt. The average annual inflation rate in Switzerland between the early 1950s and 2016 was about 2.5 percent. Thus, Beznau's installed cost of 435 CHF/kW corresponds to about $2,200/kW in 2016 dollars. This is exactly the same as what was estimated earlier for GE's intercooled-recuperated 5-MWe gas turbine.

Technical data for the Beznau units I and II are presented in Table 14.1. In their time, they represented the state of the art. As mentioned earlier, at the time of its commissioning, Beznau II was the world's largest and most efficient gas turbine. At a cycle pressure ratio of about 8 and a TIT of 1,200°F, its closest match today is the aeroderivative SGT-A30 RB by Siemens with an RB211 gas generator (based on the Rolls-Royce RB211 aircraft engine) and an RT62 power turbine rated at about 27 MWe and 36.4 percent efficiency (ISO base load) with a budgetary price of less than $500/kW. The cycle pressure ratio is 20.6 (more than twice that of Beznau II), airflow is 91 kg/s (half of Beznau II), and exhaust temperature is 501°C (934°F). With a Dry-Low-Emissions combustor, NOx emissions are less than 25 ppmvd at 15 percent O_2.

The significant advantage of cycle pressure ratio (i.e., >20 for SGT-A30 with an RB211 gas generator vis-à-vis 8 for Beznau II) and materials/coatings allowing much higher TITs (1,100–1,200°C for RB211 vis-à-vis 650/600°C for Beznau II) is obvious. Gas turbine designers at BBC had to go to great lengths to squeeze every bit of efficiency from the Brayton cycle with limited cycle pressure ratios and cycle maximum temperatures. The end result was a whole power plant instead of a compact engine in a skid.

However, there is an interesting question that comes to mind: What if the BBC designers had access to compressor aerodynamics knowledge, hot gas path materials, and coatings available to the gas turbine designers in the modern era? For example, what if they could design Beznau II with an overall cycle pressure ratio of 23 and TITs of 900°C/870°C for the HP/LP turbines (to obviate the need for turbine hot gas path cooling)? Without changing component polytropic efficiencies (probably around 85 percent) and the recuperator effectiveness, the answer is 67 MWe (same airflow) at ~43 percent efficiency. At F–class airflow, one is looking at a 200 MWe-43 percent simple-cycle gas turbine power plant with a much lower than E–class TIT and very low emissions. An impressive performance, undoubtedly, but it would still be economically infeasible. However, the implication is unmistakable: there might be a scenario for going "back to the future" pretty soon when a plateau is reached in firing temperature with acceptable emissions and the attempt to squeeze all the performance into a single frame becomes quite untenable.

In the 1950s, BBC supplied "tailor-made" gas turbines similar to the ones in Beznau all over the world (see Eckardt's *Gas Turbine Powerhouse* for a complete list). They were typically small units (10 MWe or less) either for peaking duty power stations or mechanical drives in refineries and oil fields. All types of fuels were utilized, including natural gas and blast furnace gas. The oil-fired Beznau power plant operated as a peaker until 1988. It was replaced by an ABB GT8C, which, rated at 54 MWe, occupied only *a quarter* of the floor space of the old turbine hall. In the early 1960s, however, it had become clear that they were no match for the compact "jet engines on steroids" one tends to associate with the term "gas turbine" these days. The last "world-record" plant by the BBC was the 100–MWe Port Mann (Vancouver, BC, Canada) facility comprising four 25–MWe two-shaft, intercooled-reheat gas turbines (no recuperator). The power plant, owned by the British Columbia Electric Co., went commercial in 1959.[2] The powerhouse containing the four units and the control room had a length of 140 meters with a width of 25 meters and a height of 18 meters. The four 25–MWe units lined up in a row along the turbine hall. In comparison, the entire 7FA.05 installation (see Figure 14.7 in the Section 14.3) is less than 40 meters long with a rating of almost 250 MWe (the gas turbine itself, without the generator, is about 10 meters long).

14.3 ... But This Is Now

Once all of the parts are manufactured and assembled in the factory, the end product looks similar to the machine shown in Figure 14.6. Compressor and turbine sections are enclosed by horizontally split casings. What is not seen in the picture in Figure 14.6 includes compressor and turbine "vane carriers." These are essentially rings on which stator vanes are attached. In the compressor, there are typically two vane carriers, whereas in the turbine, each stage has its own vane carrier. Vane carriers are mounted on their respective casings and are adjustable during assembly.

[2] See the article "Intercooling Key to the Commercial Success of Gas Turbines," by Septimus van der Linden, in 1Q–2012 issue of *Combined Cycle Journal* (can be accessed online at www.ccj-online.com/1q-20012/gas-turbine-historical-society/, last accessed September 19, 2018).

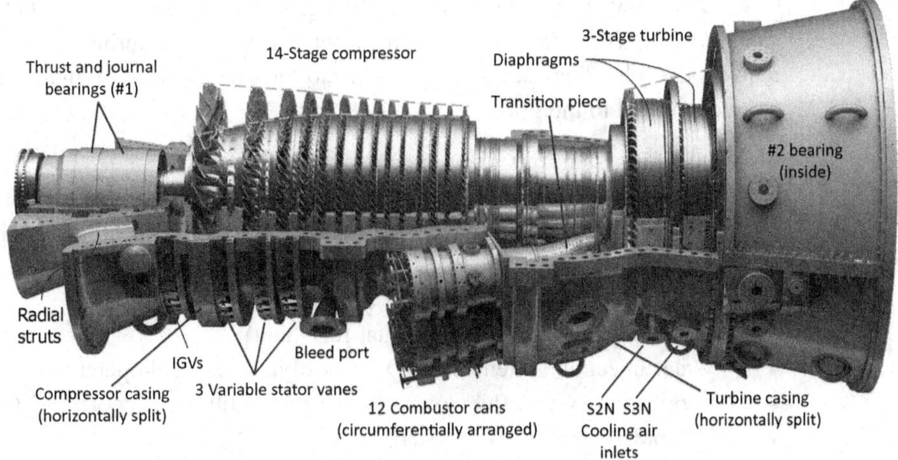

Figure 14.6 Heavy-duty industrial gas turbine – GE's Frame 7. (courtesy: GE)

There is one row of IGVs and typically three rows of variable stator vanes (VSVs), each with its own actuator to adjust their angle as needed during operation. The actuators are either hydraulically or electrically activated with a push-rod that connects to an adjustment ring. The push-rod rotates the adjustment ring in the circumferential direction and, in turn, rotates each vane to its desired angle (see Figure 11.13).

Also not shown in Figure 14.6 is the compressor outlet diffuser, which reduces the airflow velocity so that the kinetic energy of the compressed air is transformed into static pressure. Excessive turbulence at the compressor exit is prevented by the outlet guide vanes at the entry of the diffuser section, which "straighten" the airflow.

The function of the gas turbine casing is to form a pressure-retaining outer shell. There are different casing configurations adopted by each OEM. Typically, there are four or five casing sections:

1. Inlet casing
2. Compressor casing
3. Turbine casing
4. Exhaust casing

The first three or four casings are horizontally split, whereas the exhaust casing is a single piece (see Figure 14.6). The compressor casing can be in two pieces or, as shown in Figure 14.6, comprise only one piece. In two-piece compressor casing designs, the second one also houses the combustion system (i.e., together with the turbine casing, the back section of the compressor casing forms the shell around the annular combustor).

Unlike the one shown in Figure 14.6, in advanced machines with very high firing temperatures and fast-start ability, the turbine shell consists of a double-wall design for improved maintainability and reduced turbine tip clearances (due to better thermal matching between the rotor and the stator). Stator nozzle vanes are attached to the inner shell.

What is shown in Figure 14.6 is the "flange-to-flange" gas turbine. On the cold-end side, it is coupled to the generator. This assembly is the "gas turbine generator," which requires the following "accessory" systems to form an operational, final package ready to be connected to the grid:

- Inlet air system (filter and silencer)
- Exhaust diffuser, which connects the gas turbine generator to it stack (simple cycle) or to the heat recovery steam generator (HRSG) inlet duct (combined cycle)
- Fuel gas system
- Fuel gas performance heater (in a combined-cycle unit)
- Liquid fuel system (if it is a "dual-fuel" unit)
- Water injection system (for NOx control in liquid fuel operation)
- Lubrication system (lube oil reservoir, pump, filter and cooler, mist eliminator)
- Compressor water wash system
- Instrumentation and control system
- Hydraulic oil system (for the actuators)
- Fire protection system
- Starting system (LCI[3] and turning gear)
- Pipe rack
- Enclosure

In addition to these systems common to almost all heavy-duty industrial gas turbines, there is also OEM-specific equipment, e.g.,

- Exhaust frame blower (GE gas turbines)
- Rotor air cooler (some Siemens and Mitsubishi units)
- Cooling air cooler (GE's H-System, ABB/Alstom/Ansaldo GT24, and GT26 sequential combustion gas turbines)
- Cooling steam system (GE's H-System and Mitsubishi's G- and J-class gas turbines)
- Clearance control system (e.g., HCO by Siemens – see Section 21.1)

Finally, the synchronous ac generator has its own accessories, e.g.,

- Excitation system
- Current and voltage transformers
- Generator gas cooler (for hydrogen-cooled generators)
- Generator lube oil system
- Grounding equipment

The list is actually much longer with myriad control hardware and software, protective systems, small motors, etc. The user manual and system description documents that come with the purchase of a gas turbine generator typically run to several thousand pages. Nevertheless, the list of items above covers the most important auxiliaries.

[3] Load commutated inverter (also known as the static starter); it runs the ac generator as an ac motor to roll the gas turbine to its self-sustaining speed during startup.

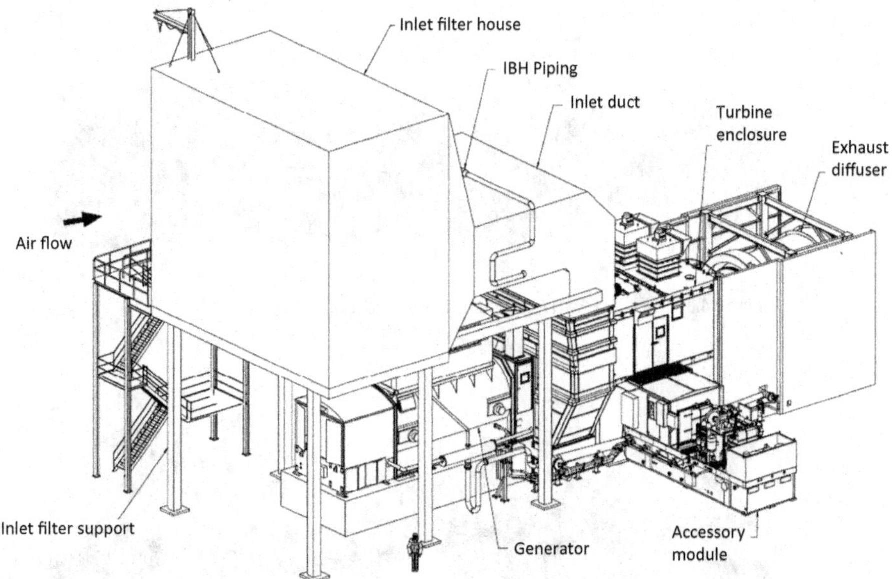

Figure 14.7 GE's 7FA.05 gas turbine generator outdoor installation. (courtesy: GE) IBH = inlet bleed heat.

Packaging of the main gas turbine generator assembly along with its accessories (typically in prepackaged modules) for ease of shipping and installation is of prime importance. When the installation is completed, the system looks quite different from the bare "flange-to-flange" engine shown in Figure 14.6. An outdoor installation of a gas fuel-only GE Frame 7 gas turbine is shown in Figure 14.7. The small human figure in the bottom gives an idea about the size of the entire structure. The accessory module contains lube oil and hydraulic auxiliaries. In addition to what is depicted in Figure 14.7, there are several compartments and "skids" (not shown; i.e., water wash skid, fuel gas performance heater, control compartment, isolation and excitation transformers, fuel gas processing skids, etc.).

For the sake of completeness, it should be pointed out that by far the most common architecture for heavy-duty industrial gas turbines from all major OEMs comprises a four-stage turbine (hot gas path) section. The example shown in Figure 14.6 reflects the early GE design philosophy (three-stage turbine with a highly loaded first stage). A state-of-the-art, heavy-duty industrial machine with a four-stage turbine is shown in Figure 14.8. A notable exception is GE's GT13E2 (formerly Alstom, formerly ABB) with a *five-stage* turbine, which is shown in Figure 14.9. This is an E–class gas turbine with a high cycle pressure ratio for an E class (18.2 for the 203–MWe 2012 version with 38 percent efficiency). For its development history, the reader is referred to Eckardt's book ([4] in Chapter 4).

In terms of gas turbine architecture, ABB/Alstom GT24/26 reheat (sequential combustion) gas turbines are unique. (After 2015, the 50-Hz version, GT26, was inherited by Ansaldo Energia; the 60-Hz version, GT24, came over to GE with the Alstom acquisition.) A cross-section of the GT26 gas turbine is shown in Figure 14.10.

Figure 14.8 GE's 9HA.01/.02 gas turbine (50 Hz). (courtesy: GE)

Figure 14.9 GE's GT13E2 gas turbine (50 Hz). (courtesy: GE)

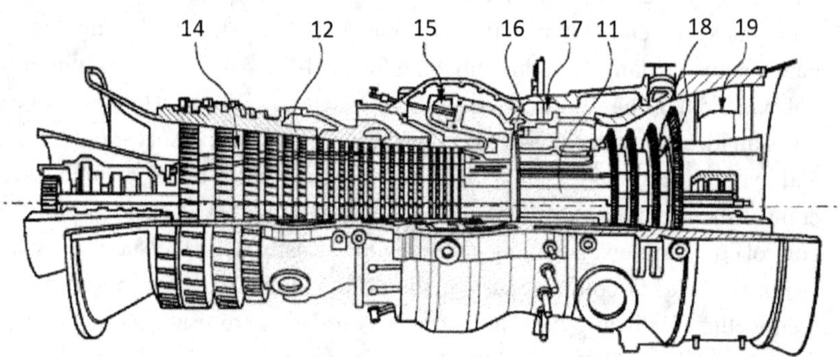

Figure 14.10 Ansaldo's (formerly ABB and Alstom) GT26 reheat (sequential combustion) gas turbine.[4]

The single-stage HP turbine is an impulse design. This is a good occasion to put the learning in Chapter 10 to use. Why an impulse turbine and why one stage? The answer is buried in two of the equations from Chapter 10 (i.e., Equations 10.4 and 10.8):

[4] From the European Patent EP 3 023 697 B1 "Fuel Lance Cooling for a Gas Turbine with Sequential Combustion" by Gao and Benz.

$$w = U(v_2 + v_3) \text{and}$$

$$R = 1 - \frac{v_2 + v_3}{2U}.$$

Consequently, for an impulse design (i.e., R = 0), stage work, w, is twice that for a 50 percent reaction design (i.e., R = 0.5). In other words,

$$w_{imp} = 2U^2 > w_{rxn} = U^2.$$

In order to squeeze out the maximum useful work from the hot gas products of the first (EV in ABB/Alstom parlance – for "environmental") combustor, an impulse stage is thus the logical choice. As an interesting aside, with a *velocity-compounded* (Curtis) stage, w would be $8U^2$ (e.g., see pp. 135–137 in Sir Frank Whittle's book – [9] in Chapter 10). In fact, the first three ABB design versions of the reheat gas turbine had a two-stage HP turbine (plus an intercooler!) – see Eckardt's book for the full story. In any event, the two-stage design was eventually dropped due to size and cost considerations.

14.4 Erection

Erection means the site construction and installation of the electrical and mechanical equipment and includes unloading, storage, erection and installation on site, performance of construction tests, and other work associated with a system, equipment, or component, including the supervision and direction to carry this out.

The flange-to-flange gas turbine is typically shipped from the factory with all combustor cans installed. It looks similar to what is shown in Figure 14.6. With the ever-increasing size of the latest, advanced heavy-duty industrial gas turbines (e.g., almost 500 MWe for the largest 50-Hz unit), ensuring that they all fit within standard commercial shipping limitations is becoming a challenge.[5] (The exhaust diffuser is shipped separately as a package.) Usually, the base is supplied by the OEM with the gas turbine. The machine is installed onto the foundation pedestals designed and supplied by the engineering, procurement, and construction (EPC) contractor. Typically, the gas turbine and the generator are shipped by rail; the rest of the equipment can be shipped by truck.

[5] This was highlighted in 2018 when the first GE 9HA.02 (544 MW–430 tons vis-à-vis 446 MW–390 tons for 9HA.01) was scheduled to be transported from the factory in Belfort, France, to Greenville, SC, for the factory test. In 2013, GE and the Territoire de Belfort had already invested €800,000 to improve the Belfort–Strasbourg route. For the 9HA.02, a new improvement effort was required (widening of roundabouts, leveling of embankments, etc.). After a six-day journey from Belfort to Strasbourg, the machine was transported by river barge to the port of Antwerp to be shipped to its final destination. Due to the limitations of the railroads in the USA, however, the first 9HA.02 was shipped to Charleston, SC, in two parts (the rotor separately).

Table 14.2 Typical gas turbine dimensions

	Old F	New F	H/J OEM A	H/J OEM B	H/J OEM C
Output (MWe)	185	241	289	305	370
Length (ft)	28.5	28.5	30.2	36.3	~50
Weight (st)	172	222	274	309	349
Diffuser Length (ft)	13.5	13.5	13.5	27.0	~20
Diffuser Weight (lb)	20,000	17,500	17,500	20,000	NA
Height (ft)	13	13	16	16	18

The generator may be shipped from the factory, with or without the collector compartment assembled – this depends on the distance from the factory and location of the project site. For overseas destinations, the collector compartment is usually shipped separately from the generator.

Typical dimensions for advanced, heavy-duty industrial gas turbines are listed in Table 14.2. Clearly, as the machine becomes more "powerful" (i.e., in terms of airflow and generator output), it also becomes larger and heavier. In fact, output–weight data are very well represented by a straight line with the slope of 0.96 st/MWe (st stands for "short ton" or 2,000 lb).

Apart from the list presented in the Section 14.3 (typical for air-cooled gas turbines), additional special equipment is required for advanced gas turbines with steam-cooled parts. Some gas turbines required cooling air cooling, which requires additional piping and heat exchangers. In combined-cycle applications, the fuel gas performance heater is almost always provided by the OEM.

A typical fuel gas system includes

- Temperature control valve
- Flow control valves
- Pressure control valves
- Shutoff valve
- Vent valve
- Flow meter
- Strainer
- Booster compressor (if required)
- Regulation equipment (if required)
- Filter/separator
- Condensate knockout drum (if required)
- Preheater (usually electric heater)
- Metering station

A typical fuel oil system (or fuel oil skid) includes

- Control valve
- Flow meter
- Main pump and suction strainer

- Drain pump and tank
- Forwarding pump
- Unloading pump
- Metering station
- Treatment equipment
- Oil storage tank
- Accumulators

It is interesting to note that, although it is used very rarely, the fuel oil system adds quite significantly to the overall gas turbine installation in terms of equipment (including water injection and, at least for some OEMs in the past, *atomizing air* systems), installation labor, and materials and schedule. Combined with the extra piping, hoses, valves, and combustor fuel nozzles on the gas turbine itself, plus acceptance test, commissioning, and three-day fuel inventory, this can add more than $20 million to the total installed cost of the gas turbine. Unless required specifically by the system operator (e.g., PJM in the USA), this is why owners in the USA or Europe would much prefer to drop the dual-fuel capability from their shopping list. In other locations, on the other hand, plant owners do indeed demand dual-fuel capability, and not only as a backup to be used only for a few days when natural gas supply is interrupted. Drivers for this demand include natural gas supply infrastructure readiness and/or reliability, price fluctuations, and other site-specific economic criteria.

In addition to the gas turbine equipment, there are other materials requisite for the erection of a gas turbine on site. These materials, depending on the contract, are either supplied by the OEM or by the EPC contractor. They include

- Layout and loading data of equipment and structure for the civil engineering and layout designing
- Foundation design criteria
- Pipe supports and bolts
- Anchor bolts for equipment and steel work
- Embedded materials for equipment and steel work
- Enclosure for equipment
- Special tools for the erection and maintenance of plant equipment
- Supporting structures, access platforms, handrails, stairways, and ladders
- Lagging and cladding for equipment, pipe work, and ducting
- Driving motors installed on the common base plates of auxiliaries
- Motorized valves for remote operation
- Local instrumentation and control, local panel, etc.
- Bolt, nuts, washers, and gaskets for terminal flanges of piping and ducts

Site equipment and materials requisite for erection are

- Special and ordinary tools
- Rotor lifting beam
- Test and commissioning equipment
- Temporary instruments for performance test

- Scaffolding
- Consumables (e.g., oil for initial fills, flushing oil, lubricants, grease, water treatment chemicals)

The EPC contractor responsible for the erection of the gas turbine equipment provides communications, electrical supplies, raw and demineralized water supplies (as necessary), potable water supplies, waste disposal, compressed air for instruments and tools, firefighting facilities, medical (first aid) stations, warehouses, and site security.

Prior to installation of the gas turbine equipment, site preparation is done by the EPC contractor. This includes clearing, rough grading, excavation and fill, water and earth retaining structures, and site security fencing. Site finishing includes finish grading, yard and site drainage, dewatering and piling, landscaping, excavation, and backfill for yard piping. The site is completed by the construction of roads and parking.

According to the layout provided by the OEM, the EPC contractor then erects the equipment and, if required, turbine building foundations. Subsequently, the building superstructure (steel and concrete) and equipment supporting steel are erected.

In modern designs, most auxiliary equipment comes in factory-assembled *modules* to speed up the installation schedule. To a great extent, erection of a gas turbine system in the field turns into putting "Lego bricks" together and connecting the pipes and cables in between. Each unique system (e.g., fuel gas system or fuel oil system) becomes a "Lego brick" (usually referred to as a "skid") with factory packaging and fixed locations. Similarly, several smaller items are combined into a single box (Lego brick; e.g., the electrical package, which contains uninterruptible power supply batteries, the control system, motor control centers, generator protection, and fire protection). This significantly reduces labor man-hours, which in the past were spent on erecting scaffolding, installing piping supports, putting loosely shipped internal pipes and valves together, and routing instrument air tubing, electrical wiring, conduits, commissioning valves and instrumentation.

Heavy equipment, such as the generator, must be lifted or handled in accordance with the OEM's instructions (e.g., pertaining to temporary support points and other temporary devices utilized during the lifting and/or handling of the equipment).

Setting the equipment involves cleaning and inspecting soleplates prior to rigging and lifting it from the truck transporter and placing it down onto the foundation (designed and built by the EPC contractor). Checking for flatness and finish is crucial.

In the case of the generator and the gas turbine, the latter is installed first. Then, the generator equipment is lifted and placed down over the anchor bolts on shim packs located on the sub-soleplates in the foundation. After rough alignment of the generator base to the foundation center lines and turbine load coupling, in accordance with the OEM's alignment instructions, all anchor bolts are installed. Alignment is done by shimming under the generator feet.

In a combined-cycle power plant with a single-shaft (S-S) arrangement, the alignment process becomes even more complex and vital. In that case, three major pieces of rotating equipment – the steam turbine, gas turbine, and generator – must be connected to each other with very tight tolerances.

14.4.1 Shaft Alignment

Probably the most important part of gas turbine installation (or of any rotating machine installation, for that matter) is to ensure that the gas turbine and its generator are aligned precisely in accordance with OEM's requirements.

There are *three* types of misalignment encountered in coupling two rotating machines:

- Parallel offset misalignment (shaft centers unaligned in vertical direction)
- Angular misalignment (shafts are not horizontal)
- A combination of the two

In some gas turbines, there may be a "spacer shaft" between the turbine and the generator. In most cases, there is a "coupling" between them. In aeroderivative gas turbines and some small gas turbines, there may be a gearbox between the generator and the turbine. Obviously, in S-S, combined-cycle power trains, the alignment of three rotating machines (four, actually, if one counts the LP turbine of the steam turbine as a separate rotating machine, which, of course, it is) is a significant endeavor.

Misalignment is the second-biggest cause of vibration (after unbalance) in rotating equipment. It significantly decreases bearing and shaft life and also impacts performance. Acceptable misalignment tolerances for gas turbines with 3,000/3,600 rpm are

- 0.7/0.5 mils per inch of offset (e.g., spacer shaft) at 3,000/3,600 rpm (parallel)
- 0.7/0.5 mrad (1/1,000 radians) per inch of offset at 3,000/3,600 rpm (angular)

There are several alignment methods, e.g.,

- Rim and face
- Reverse dial indicator (RDI) or laser

The latter is probably the best method. However, both shafts (i.e., the gas turbine and the generator in this case) must be rotated together. Both rim–face and RDI methods utilize dial gauges (dial indicators) mounted on the shaft or coupling flanges (hubs) and the turning of one or both shafts to four separate 90° positions. For best accuracy, separate sets of readings are made, which requires turning the massive shafts of the generator and/or the turbine. This, of course, is easier said than done (i.e., see the weights of gas turbines in Table 14.2; corresponding generators weigh about the same). The lift oil system must be on (so that the shaft "floats" on the bearing oil film) and the turning gear must be able to operate in a start–stop sequence to be precisely stopped at 90° intervals of rotation. In lieu of this, strap and hook wrenches for manual shaft turning are utilized. Accomplishing this requires a skilled operator team and lots of time.

The alternative is a laser shaft alignment system, which can make continuous measurements while the system is turned on the turning gear. Start–stop rotation in 90° increments is not needed. The system is much more expensive, but compensates for this with its higher accuracy and reduced downtime and labor costs.

As stated earlier, an S-S combined-cycle power plant has a more complex alignment of all shafts than does a *multi-shaft* (M-S) one (or a simple-cycle power plant). Increased quantities and complexity also involve more construction time. An S-S plant's construction schedule may not be significantly longer than that of an M-S plant; however, the complexity of the S-S plant's civil/structural aspects and its single equipment foundation require an earlier field mobilization date than does an M-S plant. This affects time-related construction costs associated with field nonmanual staff, construction equipment and facility rentals, and escalation of craft wages. This in turn increases project costs and schedule slippage potential.

14.5 Commissioning

The next step after completing the installation of the gas turbine is to commission it. *Commissioning* is the term for pre-operational testing and setting to work of all systems and equipment such that acceptance tests upon completion can be performed as set out in the contract. There are several distinct steps to the commissioning process, which can be summarized under two headings (i.e., "cold" and "hot") [8]. The "cold" phase refers to the commissioning activities undertaken prior to the "first fire" of the gas turbine. All commissioning tasks are performed by plant operators and EPC startup engineers under the guidance of the OEM's technical advisors and engineers (they are known as "site TAs").

The "cold" phase activities include:

- Ensuring that all components and enclosures are clean
- Ensuring all component and enclosure openings are secure
- System checks and calibration
    - Exciters and LCI
    - Protective devices
    - Motors
    - Piping
    - Lube oil system
    - Cooling water system
    - Inlet and exhaust systems
    - Instrumentation (transducers, switches, etc.)
- Energizing equipment
- Energizing control panels
- Control system (hardware and software) checks
- Mock runs with simulation software

Once all control and monitoring systems have been thoroughly checked, calibrated, and preset, the gas turbine is started up and rolled to the FSNL condition. This is the "first fire" and the start of the "hot" phase of the commissioning. During this operation, OEM engineers look for signatures of any mechanical disturbances (noise, vibration, etc.), tune the combustion system, record key temperatures (turbine exhaust, turbine inlet, lube oil, etc.), and verify the operation of the control system and instrumentation. They ensure that, once the gas turbine is "heat soaked," expansion of the casing and rotor does not lead to mechanical rubs and leakages. This run usually takes 1 or 2 hours.

Once the startup team is satisfied with the operation of the gas turbine at FSNL, the machine is ready for restart and synchronization to the grid followed by loading to its full load condition. During this period of the "hot" phase, final combustion tuning is done. The DLN combustor is taken through all of its transfer points and necessary tuning is done to ensure that NOx and CO emissions are compliant. The machine is started several times to check operability items such as full load rejection, several starts in quick succession, and hot and cold starts. Key temperatures, pressures, flows, clearances, vibration trends, etc., are recorded and analyzed to verify that all systems operate properly. Other tests undertaken during this period include water or steam injection (if applicable), inlet conditioning (i.e., evaporative cooler, inlet fogger, etc., if applicable), and grid support tests. If the gas turbine is a dual-fuel unit, switchover to the backup fuel (usually number 2 fuel oil) is also tested. Compressor IGV and VSV operation is tested and optimized. Inlet bleed heat (anti-icing) system operation is checked and its satisfactory operation is verified (if one is present).

In the "hot" phase, the machine is online and generating power. However, the control is still with the EPC contractor, who is putting the gas turbine through its paces according to the protocols drawn up by the OEM and monitored by the site TAs. The owner's personnel, although participating in the process, do not "run the show." In commercial terms, the gas turbine (and the entire plant) has not been turned over to the owner.

14.5.1 Acceptance Tests

The final commissioning activity before turning over the proverbial "keys" to the owner is the *acceptance test*. Another term for the acceptance test is the *guarantee performance test*. Acceptance tests are strictly commercial activities dictated by minutely detailed contracts. There are typically two types of acceptance tests:

- Guarantee output, efficiency, and emissions at reference site conditions specified in the contract.
- Operability tests to prove that the gas turbine conforms to the specifications.

Conductance of the acceptance test is usually according to ASME Performance Test Codes (PTCs) or similar codes adopted by other countries and/or supranational organizations as specified in the contract. Those codes prescribe all of the necessary details of test conduct and data recording, analysis, correction to reference conditions, and reporting. Typically, an offline water wash is done prior to the acceptance test.

Since a typical gas turbine-based electric power generation plant comprises several major pieces of equipment (i.e., gas turbine[s], steam turbine[s], HRSG[s], and heat rejection system components), a distinction is necessary between "plant" and "equipment" performances. Consequently,

- Plant performance tests are performed to demonstrate that the plant meets the guaranteed performance offered by the EPC contractor to the plant owner.

- Equipment performance tests are performed to demonstrate that the equipment vendors (OEMs) have met their guaranteed performance parameters to the EPC contractor.

Performance test requirements are highly dependent on the guaranteed performance and commercial terms of the contract. Consequently, each test program is carefully designed to balance the testing costs, project risk (i.e., "liquidated damages"), and technical requirements of the test. The process can be summarized as follows:

- Development
  - Review contract to determine basic guarantee requirements
  - Define/develop performance calculation model/algorithm
  - Determine requisite measurements
  - Determine measurement instruments
  - Finalize the test procedure
- Planning
  - Form the test team
  - Integrate test requirements with the project schedule
  - Install temporary test equipment (transducers, transmitters, data acquisition and storage)
  - Do a "dry run" to verify the proper operation of the test equipment and "tune" the system
  - Prepare the plant for the test
- Execution
  - Carry out the test
  - Prepare the test report

In preparing the test report, probably the most important task is to determine the uncertainty in measured performance parameters. There are two related concepts for this, which must not be confused in this endeavor:

- "Test uncertainty" is a statistical concept that provides an *estimate* of the accuracy of a test. It is usually expressed within a 95 percent confidence interval (CI). Underlying principles and calculation methodologies are provided in detail in ASME Test Code PTC 19.1.
- "Test tolerance" is a commercial concept. It is *not* the same as "test uncertainty." Although one would rationally except that the test tolerance should somehow be linked to the test uncertainty, this may or may not be the case.

As an example, consider that the guaranteed output, per contract, is 100 MW. Furthermore, the contract stipulates that the test tolerance is the same as the test uncertainty as calculated per ASME PTC 19.1. Let us assume that the test uncertainty for the measured output is ±1 percent (95 percent CI). Therefore, as long as the measured output *plus* 1 percent is equal to or greater than 100 MW, no liquidated damages are paid. In other words, the performance test should result in a plant output of 99.01 MW or higher. In order to earn a "bonus," measured output *minus* 1 percent is equal to or greater than 100 MW. In other words, the performance test should result in a plant output higher than 101.01 MW.

Most important measurements and specifications to be checked in gas turbine (simple or combined cycle) performance tests are:

- Generator power output at low-voltage terminals
- Fuel flow, composition, and heating value
- Ambient temperature and relative humidity
- Barometric pressure
- Generator power factor
- Evaporative cooler or inlet chiller status (on/off, if any)
- HRSG duct burner status (on/off, if any)
- HRSG blowdown (typically set to 0 percent)
- Demineralized water system (typically off, if any)
- Cooling tower blowdown and recovery system (typically on)
- Shaft speed (i.e., system frequency)
- Auxiliary loads (OEM and EPC scopes)

Generator power output is measured with temporary high-accuracy power meters using station instrument transformers, which are installed on the low-voltage terminal of the generator. High-accuracy, calibrated flowmeters are installed to ensure proper natural gas flow measurement. (Care must be taken to ensure that upstream and downstream pipe length requirements are met during the plant design stage.)

If the plant distributed control system (DCS) is used as a data acquisition system, all of the following items must be verified for each data point:

- Deadbands
- Modbus[6] list of data tags
- Plant data historian (e.g., OSIsoft's PI System™) list of data tags
- Significant figures
- Data type (average, snapshot, etc.)
- DCS–Excel (or another similar data analysis tool) interface

Test preparation activities may include plant troubleshooting and optimization, performance checks, data validation, temporary instrumentation installation, water wash, cycle isolation (for combined-cycle power plants), etc. Cycle isolation, or identification and fixing (or accounting for) water and steam losses, has a major impact on steam turbine performance. It is important to identify and repair any leaks before a performance test.

Typically, guaranteed performance data are based on "new and clean" conditions. What the term "new and clean" exactly means is another contractual stipulation. When one buys a new car from the dealership, in many cases, there can be up to 100 miles or so on the odometer due to test drives, transfer from another dealership, etc. (It is

[6] Modbus is a communication protocol developed by Modicon systems. It is a method used for transmitting information over serial lines between electronic devices.

practically impossible to have a brand new car delivered to you at exactly 0 miles on the odometer anyway.) Similarly, a gas turbine accumulates operating hours during startup and commissioning phases prior to the performance test. Therefore, a typical contract stipulates that the performance test must take place *prior to* accumulating X hours. The "hours" in question are *not* straightforward clock hours, though. They are *equivalent degradation hours* calculated via an OEM-supplied formula, which weights part load, fuel oil-fired (if any), and peak-fired operating hours higher than base load operating hours with natural gas. If the performance test takes place *after* X hours, an OEM-specified degradation correction must be applied to the measured output and heat rate.

In addition to output and heat rate, the EPC contractor must prove that guaranteed emissions and sound levels are met as well. Emissions guarantees are covered in Section 12.5.1. Acoustical guarantees are two-fold: *near-field* and *far-field*. Near-field refers to the area that surrounds the noise source (e.g., the gas turbine), which is typically defined by a *source envelope contour*. Far-field refers to a free environment (e.g., at the plant property boundaries) where the sound field is spreading spherically. A typical near-field sound level guarantee is 85 dB(A) (i.e., 85 "A-weighted" decibels). Typical far-field guarantees are around 50–60 dB(A). What does that mean? The threshold of hearing is generally given as 20 micropascals at 1 kHz, at which point the logarithmic decibel scale is set to zero and, strictly speaking, is referred to as "dB SPL."[7] For instance, an unsilenced turbine exhaust is 130 dB SPL, whereas the sound level at a rock concert can easily exceed 100 dB SPL. (Note that the threshold of pain is 140 dB SPL.) Since hearing sensitivity varies at different frequencies, A-weighting simulates the frequency response of the human ear.

Note that there are many noise sources in a typical power plant (e.g., a gas turbine combined-cycle power plant). Therefore, even at a point that can be defined as near-field for equipment A (say, the gas turbine), contribution of another equipment, B, can result in a sound level higher than the guarantee sound level for equipment A. Consequently, the gas turbine (and its enclosure) must be designed to a level several decibels below the guarantee value (e.g., 80 dB[A] for an 85 dB[A] guarantee) in order for sound levels within the turbine building to meet the project requirement at all locations.

Applicable US and international standards are the following:

- ANSI/ASME PTC 36-1985, "Measurement of Industrial Sound"
- ANSI B133.8-1977, "Gas Turbine Installation Sound Emissions"
- ANSI S1.1-1994, "American National Standard Acoustical Terminology"
- ISO 3746, "Acoustics – determination of sound power levels of noise sources using sound pressure survey method using an enveloping measurement surface over a reflecting plane"
- ISO 6190, "Acoustics – measurement of sound pressure levels of gas turbine installations for evaluating environmental noise – survey method"
- ISO 10494-1993, "Gas turbine and gas turbine sets – measurement of emitted airborne noise – engineering/survey method"

[7] Sound pressure level.

For more information, the reader is referred to the following GE reference documents (available at www.gepower.com):

- GER-4221, "Power Generation Equipment and Other Factors Concerning the Protection of Power Plant Employees against Noise in European Union Countries"
- GER-4239, "Power Plant Near Field Noise Considerations"
- GER-4248, "Acoustic Terms, Definitions and General Information"

Operability tests are also specified in minute detail in the contract, but there are no specific test codes governing them. This is so due to the site and plant duty-dependent, variable nature of such tests. They may include:

- Several days-long operation according to the customer's program, with all modes of operation, including up–down load ramps, single-unit heat rate (for multi-gas turbine combined cycles), part load runs, etc. This exercise is referred to as the "reliability test" or "demonstration run."
- Demonstration of "fast start" ability.
- Full load rejection by manually opening of the generator breakers and letting the unit run at FSNL while observing the performance of speed governor and automatic voltage regulator.
- Overspeed test (run at no load until trip).

For combined-cycle power plants, the list can be longer by inclusion of balance of plant equipment (cooling tower drift test, etc.).

Once all of the tests have been completed, the EPC contractor (in some cases, it can be the OEM) provides the owner with a written final test report. Receipt of the final test report and concurrence of the owner with its conclusion (that all guarantees have been met) via a written notice is one of the requirements of "substantial completion." Inability to meet guaranteed performance leads to "liquidated damages" as prescribed in the contract. This is usually a dollar amount per shortcoming (i.e., say, $1,000 per each kWe output shortfall or $100,000 per each Btu/kWh above the guaranteed heat rate).

References

1. Brandt, D. E., "The Design and Development of an Advanced Heavy-Duty Gas Turbine, *Journal of Engineering for Gas Turbines and Power*, **110**: 2 (1988), 243–250.
2. Davis, L. B., Washam, R. M., "Development of a Dry Low NOx Combustor," ASME paper 89-GT-255, ASME Gas Turbine and Aeroengine Congress and Exposition, June 4–8, 1989, Toronto, ON, Canada.
3. Joos, F., Brunner, P., Schulte-Werning, B., Syed, K., Eroglu, A., "Development of the Sequential Combustion System for the ABB GT24/GT26 Gas Turbine Family," ASME paper 96-GT-315, ASME Gas Turbine and Aeroengine Congress and Exposition, June 10–13, 1996, Birmingham, UK.

4. Paul, T. C., Schonewald, R. W., Marolda, P. J., "Power Systems for the 21st Century – 'H' Gas Turbine Combined Cycles," GER-3935A, 39th GE Turbine State-of-the-Art Technology Seminar, 1994.
5. Florjancic, S. S., Pross, J., Eschbach, U., "Rotor Design in Industrial Gas Turbines," ASME Paper 97-GT-75, ASME Gas Turbine and Aeroengine Congress and Exposition, June 2–5, 1997, Orlando, FL.
6. Endres, W., "Rotor Design for Large Industrial Gas Turbines," ASME Gas Turbine and Aeroengine Congress and Exposition, June 1–4, 1992, Köln, Germany.
7. Skrotzki, B. G. A., Vopat, W. A., *Steam and Gas Turbines* (New York: McGraw-Hill Book Company, Inc., 1950).
8. Oegerli, R., Gas Turbine Commissioning Procedure, *ASME Journal of Engineering for Power*, 101 (1979), 125–129.

Part III

Extras

15 The Alternating Current Generator

The subject matter of this book is not the "gas turbine"; it is the "gas turbine *generator*." The generator in question is a *synchronous alternating current (ac) machine*. Note the usage of the term "machine"; this is not a "slip of the keyboard." The same ac machine can run as a generator as well as a *motor*. In fact, this is exactly how it runs during gas turbine startup. The load commutated inverter (LCI) runs the generator as a motor to crank the engine to its ignition speed and then assist it until it becomes self-sufficient. Not only that, the gas turbine generator can also be run as a *synchronous condenser*, in which case, separated from the gas turbine by a synchronous self-shifting (SSS) *clutch*, the machine either supplies *reactive power* to or absorbs it from the grid.

Practically, in all textbooks on gas turbines, the generator is a mere afterthought (if it is mentioned at all). The only attention it attracts is limited to its efficiency (about 99 percent for modern machines). In other words, of each 100-MW worth of shaft power generated by the gas turbine engine, 1 MW is lost during the conversion of mechanical power to electric power and only 99 MWe is supplied to the low-voltage terminals of the generator transformer.

Low-voltage terminals? Transformer? LCI? There is obviously a lot to cover here. Unfortunately, within the space of a single chapter, there is only so much that can be handled. The reader is strongly encouraged to consult the excellent references on the electric machinery by Fitzgerald et al. [1], Stevenson [2], Kundur [3], and Grigsby [4]. In order to grasp the key attributes of a gas turbine's synchronous ac generator and how it affects the performance of the entire gas turbine power plant, a crash course on ac machine fundamentals is presented below. The goal is to arm the reader with sufficient background to understand the origins and importance of terms like excitation, lagging or leading power factor (PF), etc. Nevertheless, the reader should be cautioned that the coverage below barely scratches the surface of the subject matter. The references cited above or their most recent editions (among others) should be consulted for comprehensive coverage.

15.1 Synchronous Machines

Synchronous machines belong to a special class of rotating machines, whose speed is proportional to the frequency of the ac in its *armature* (also known as the *stator*). The magnetic field created by the armature currents rotates at the same speed as that created by the field current on the rotor (commonly known as the *excitation current*), which is rotating at the synchronous speed, and a steady torque results.

Synchronous machines are widely used as electric power *generators* connected to constant-speed prime movers such as steam or gas turbines. Since the *reactive power* generated by a synchronous machine can be adjusted by controlling the magnitude of the excitation current, unloaded synchronous machines are also often installed in power systems solely for PF correction or for control of reactive power flow (commonly known as *reactive volt-amperes* or VAR). Such machines, known as synchronous *condensers*, may be more economical in the large sizes than static capacitors.

The same synchronous ac machine can be used as a generator or condenser. For instance, consider that a generator is brought up to the speed necessary to synchronize it with the local electricity grid. At that point, the generator is disconnected from its driver (e.g., a gas turbine or an electric motor). Thereafter, the generator acts as a motor driven by the electrical grid – supplying VAR to the grid while drawing *leading* current from it when *overexcited* and absorbing VAR from the grid while drawing *lagging* current from it when *underexcited.*[1]

15.1.1 Synchronous Alternating Current Generator

Mechanical shaft power generated by the gas turbine (or any prime mover) is transformed into electric power in an ac generator. The fundamental physical principle governing this transformation is *Faraday's law of induction*, which states that "the *induced electromotive force* (emf) or *voltage* in any closed circuit is equal to the negative of the time rate of change of the *magnetic flux* through the circuit." In mathematical terms, for a coil of wire with N identical turns, each having the same magnetic flux ϕ, emf is given by:

$$\varepsilon = -N\frac{d\phi}{dt}. \tag{15.1}$$

The magnetic flux in Equation 15.1 is the product of

$$\phi = B \cdot A, \tag{15.2}$$

where B is the *magnetic flux density* and A is the cross-sectional area of the magnetic flux. The interested reader can consult any introductory physics textbook for more on the subject. In fact, a good way to grasp the underlying principles is to watch a YouTube video – many such videos are available online – demonstrating the ac generator operating principles in a cartoon format. Most of them include a simple setup, where the coil of wire (referred to as the *armature*) rotates within a magnetic field. In other words, the armature is the *rotor* and the magnet is the *stator*.

The rotor/armature/coil of wire is rotated by a mechanical source of energy, which, in a real power plant, is a prime mover such as a gas or steam turbine. A voltage (emf) is

[1] Strictly speaking, a synchronous condenser is a "borderline" motor. Its power consumption, to cover its full load losses, is about 1.5%–3% of its MVA rating.

created in the armature coil and it is proportional to the rate of change of the area, A, facing the magnetic field (constant B), i.e.,

$$\frac{d\phi}{dt} = B \cdot \frac{dA}{dt}. \tag{15.3}$$

Due to the rotational movement of the armature, the voltage generated, when plotted as a function of time, is of a *sinusoidal* form.

When the frequency of mechanical rotation is equal to the frequency of the generated voltage/current in the ac generator, the machine is referred to as a *synchronous ac generator.*

In an *actual* synchronous ac generator, the situation is different from the simple setup discussed above. In particular,

1. The armature winding is on the stator.
2. It is the magnet that rotates.

(In a direct current [dc] machine, however, it is indeed the armature that rotates.)

A simplified description is given in Figure 15.1. Note the field *winding* on the rotor (i.e., the magnet), which is excited by the dc conducted to it by means of carbon *brushes* on *slip rings* or *collector rings*. It is the field winding that creates the magnetic flux (i.e., the rotor is *not* a permanent magnet, although they may be used in small machines).

As shown in the very simple diagram in Figure 15.1, the rotor has two *salient* (projecting) poles with concentrated windings. A salient pole construction is characteristic of hydroelectric generators, which rotate at relatively low speeds (in rpm) so that a large number of poles, P, is required to produce the desired frequency, f (e.g., 60 Hz in the USA). Note that

$$f = \frac{P}{2} \cdot \frac{rpm}{60}. \tag{15.4}$$

Thus, for a large hydroelectric turbo-generator with 360 rpm speed, P = 20 poles are needed for 60-Hz frequency. For steam or gas turbines, high speeds are requisite for better efficiency (e.g., 3,600 rpm), so that their generators are two-pole *cylindrical rotor* machines, which are also known as *uniform air gap* machines. (Exceptions are very big steam turbines in large fossil or nuclear power plants with 1,800 rpm low-pressure [LP] turbines, which are connected to four-pole generators.)

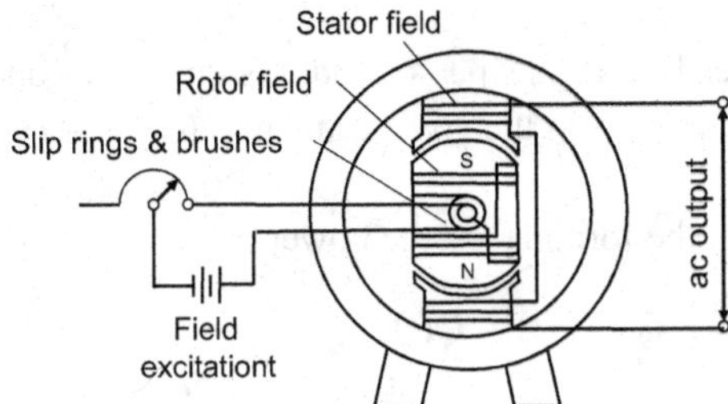

Figure 15.1 Alternating current generator cross-section (one phase).

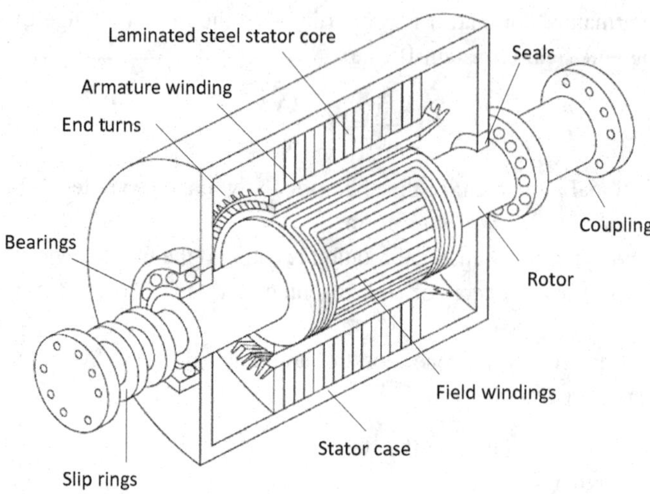

Figure 15.2 Cutaway view of a synchronous ac generator.

A simplified drawing of an actual synchronous ac generator is shown in Figure 15.2. The rotor is typically forged from a single steel ingot for a two-pole machine. Axial slots are cut into the rotor for the field windings. They can be as large as several hundred tons for large-utility machines. The number, shape, size, and spacing of the winding slots are chosen to obtain the maximum magnetic flux density, B.

Synchronous generators are *three-phase* machines with production of a set of three voltages separated by 120 electrical degrees in time. This *poly-phase* arrangement is especially advantageous for power transmission and heavy-duty motor design (e.g., the horsepower rating of three-phase motors and the kVA rating of three-phase transformers are about 1.5 times greater than their single-phase counterparts).

Electric power is the product of voltage and current. In an ac generator, each is a *sinusoidal* function of time and, consequently, so is the generated power, P(t). For the *instantaneous* values of single-phase voltage, current, and power, one can write

$$v(t) = V_{max} \cos(\omega t), \tag{15.5}$$

$$i(t) = I_{max} \cos(\omega t - \theta), \tag{15.6}$$

$$p(t) = v(t) \cdot i(t) = V_{max} I_{max} \cos(\omega t) \cos(\omega t - \theta), \tag{15.7}$$

where $\omega = 2\pi f$ is the angular speed in radians per second, t is time in seconds, and θ is the *phase angle* (positive when current lags the voltage and negative when current leads the voltage).

Using trigonometric identities, the formula for the power can be written as

$$p(t) = P(t) + Q(t), \tag{15.8}$$

$$P(t) = \frac{1}{2} V_{max} I_{max} \cos(\theta)(1 + \cos(2\omega t)), \tag{15.9}$$

$$Q(t) = \frac{1}{2} V_{max} I_{max} \sin(\theta) \sin(2\omega t). \tag{15.10}$$

P(t) is always positive and has an average value of

$$P_{avg} = \frac{1}{2} V_{max} I_{max} \cos(\theta). \tag{15.11}$$

(For the exact derivation of Equations 15.12–15.16 and an understanding of their physical meaning, the author's recommendation is chapter 2 in [2].) Defining the *root mean square* (rms) values of voltage and current as

$$V_{rms} = \frac{V_{max}}{\sqrt{2}}, \; I_{rms} = \frac{I_{max}}{\sqrt{2}}, \tag{15.12}$$

the result is

$$P_{avg} = V_{rms} I_{rms} \cos(\theta). \tag{15.13}$$

The average power thus calculated is the *real power* and has the units *watts* (W) – usually used in larger units of kilowatts (kW) or megawatts (MW). (Note that 1 watt is equal to 1 VAR.)

- The cosine of the phase angle between the voltage and the current is called the PF.
- If v(t) and i(t) are in phase, as they are in a purely *resistive* load, P(t) is never negative and P_{avg} is simply V·I (cosine of 0° is 1).
- If they are out of phase by 90°, as in a purely *inductive* or purely *capacitive* ideal circuit element, p(t) will have equal positive and negative half-cycles and P_{avg} will be zero (cosine of 90° is 0).

When plotted as a function of time, Q(t) fluctuates between positive and negative values and has an average value of zero. Its *maximum* value is given by

$$Q_{max} = V_{rms} I_{rms} \sin(\theta) \tag{15.14}$$

and is known as *the reactive power*, which has the units *reactive VAR* or *vars* (strictly speaking, *vars* is the same as *watts*). It represents the flow of energy alternately toward the load and away from the load.

Equations 15.12–15.14 are for a *single phase*. After a bit of lengthy algebra involving trigonometric identities, the three-phase real power is then found as

$$P(t) = P_A(t) + P_B(t) + P_C(t), \tag{15.15}$$

$$P(t) = 3 V_{rms} I_{rms} \cos(\theta) = \text{const}. \tag{15.16}$$

The square root of the sum of the squares of P_{avg} and Q_{max} is equal to the product of V_{rms} and I_{rms} (recall the theorem of Pythagoras). This leads to the well-known *power triangle* (see Figure 15.3).

As shown in Figure 15.3, the angle θ is positive – the current is lagging the voltage. Therefore, the PF, which is the cosine of θ, is referred to as the *lagging PF*.

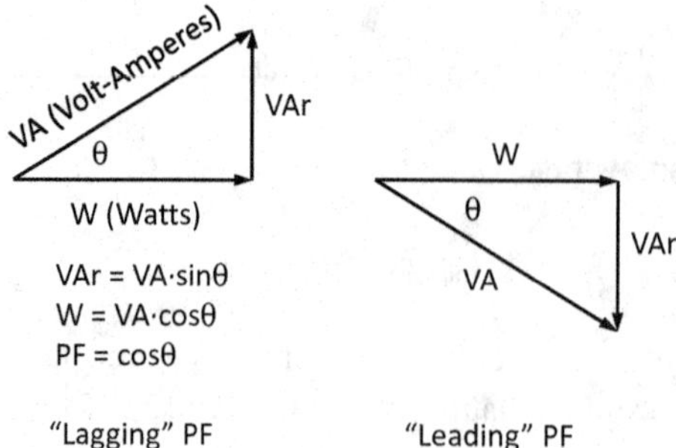

Figure 15.3 Leading and lagging PFs.

If the angle θ is negative, then the current is leading the voltage. Therefore, the PF is referred to as the *leading PF.*

- An inductive circuit has a lagging PF. Reactive power, Q_{max}, is positive (θ is positive). In other words, an *inductor* represents
  - A load drawing lagging current or
  - A generator of negative reactive power supplying leading current.
- A capacitive circuit has a leading PF. Reactive power, Q_{max}, is negative (θ is negative). In other words, a *capacitor* represents
  - A load drawing leading current or
  - A generator of positive reactive power supplying lagging current.
- A purely resistive circuit has unity PF (reactive power and θ are zero).

Sidenote

In order to understand the leading/lagging and inductive/capacitive dichotomies, consider that a capacitor is similar to a water tank. A stream of water has to fill the tank in order to create the water level. (The water level will create a stream in the bottom of the tank when the bottom plug is taken off.) Analogously, in order to create a voltage over a capacitor, an electric current has to charge a plate of that capacitor (i.e., the current has to be earlier than the voltage). In other words, the current *leads* the voltage.

An inductor (essentially a coil of wound conductive wire) can be compared to an electric car motor. To give motion to the car (i.e., a *mass*), a force has to be applied to the mass before speed results. The speed always *lags* behind the driving force. Analogously, the current through a pure inductor *lags* behind the driving voltage.

The power triangle also explains the phasor notation used in electromechanical theory. The complex power of the generator (the *hypotenuse* of the power triangle) is called S, which is also known as the *apparent power* or *generator MVA*. Thus, mathematically

$$S = \sqrt{P^2 + Q^2} \text{ and} \tag{15.17}$$

$$\theta = \tan^{-1}\left(\frac{Q}{P}\right). \tag{15.18}$$

In phasor notation, this can be written as

$$P + j \cdot Q = |S| \cdot e^{j\theta}, \tag{15.19}$$

where j is the *imaginary number*, $j = \sqrt{-1}$, |S| is the *complex modulus*, and the phase angle is the complex argument. The angle θ is the counterclockwise angle from the positive real axis (i.e., the x-axis). Thus,

- When θ is 90° (π/2 in radians), P = 0 and Q is positive.
- When θ is –90° (–π/2 in radians), P = 0 and Q is negative.
- When θ = 0°, Q = 0.

The next question is obvious: What determines the PF and whether it is lagging or leading? In order to answer this question, one must consider the generator load. In particular,

- The generator PF is lagging when the load is *inductive*.
- The generator PF is leading when the load is *capacitive*.

Inductive loads such as transformers and motors (i.e., any type of wound coil) consume reactive power with current waveform lagging the voltage.

Capacitive loads such as capacitor banks or buried cables generate reactive power with the current phase leading the voltage.

At either condition, the phase angle and its cosine (i.e., the PF) can be changed by changing the amount of *excitation* current (see Section 15.1.4).

15.1.2 Synchronous Alternating Current Condenser

In this section, a brief overview of the lesser-known application of the synchronous ac machine technology (i.e., operation as a synchronous condenser) is provided.

Rated PF is the PF specified for operation at rated megawatt load and rated VAR (lagging). Specifying a lower PF results in an increased capability to supply VAR to the system, but requires a generator with a larger MVA rating. For an air-cooled generator, the PF at output rating is typically set at 0.90, 0.85, or 0.80 lagging (overexcited). For a hydrogen-cooled generator, the PF at output rating is typically set at 0.90 or 0.85 lagging (overexcited). For both air-cooled and hydrogen-cooled generators, Institute of Electrical and Electronics Engineers (IEEE) Standard C50.13 recommends that the generator should be capable of providing 0.95 leading (underexcited) PF at rated MW. About 35 percent more reactive power can be produced in the condenser mode vis-à-vis in generator mode (e.g., see Figure 15.6 later in the chapter).

There are several methods to inject VAR into the system. Two widely employed "VAR generators" are shunt capacitor banks and synchronous condensers. Newer technologies are static VAR compensators (SVCs), which comprise *thyristor*-switched reactors and capacitors to provide rapid and variable reactive power, and self-commutated VAR compensators.

A synchronous condenser, as described earlier, is essentially an unloaded and overexcited synchronous motor. Synchronous condensers can be used at both distribution and transmission voltage levels to improve stability and to maintain voltages within desired limits under varying load conditions and contingency situations. They are superior to capacitor banks in terms of harmonics (no resonance), robustness, and smooth response (no voltage spikes) and they can be cost-effective at large sizes. The advantage of synchronous condensers vis-à-vis static compensators lies in their ability to handle high temporary overloads and to provide short-circuit support.[2] However, they cannot be switched on/off as fast as SVCs and they cost much more (not to mention the size and weight requiring reinforced concrete foundations).

Regional transmission organizations (RTO)[3] in the USA such as PJM continuously monitor and manage reactive power. In particular, RTOs

- Gather real-time information about voltage levels and the need for reactive power at various locations on the grid;
- Limit the amount of energy that can move from point to point if there is insufficient generation locally to produce the needed reactive power;
- Adjust the output of generating stations under their control to increase the supply of reactive power when it is needed;
- Pay generation owners to compensate them for lost energy revenue when they must increase their output of reactive power at the expense of megawatts (see the numerical example above);
- Require new generators connecting to the grid to agree to specific reactive power obligations, with financial penalties for noncompliance.

Thus, there is clearly a financial incentive for generators to provide reactive power generation ability without adversely affecting their real power generation capability, which, after all, is their main revenue source.

The following two factors must be considered in a cost–performance trade-off to justify the acquisition of reactive power generation capability:

1. Initial investment versus frequency of use (revenue stream).
2. Balance of real-reactive power generation.

[2] An important benefit of a synchronous condenser is that it contributes to the overall short circuit capacity in the network node where it is installed. This, in turn, improves the chances that equipment connected to the network will be able to "ride through" network fault conditions.

[3] A regional transmission organization (RTO) in the United States is an organization that is responsible for moving electricity over large interstate areas. An RTO coordinates, controls and monitors an electricity transmission grid that is larger than the typical power company's distribution grid with much higher voltages.

A recently implemented solution is conversion of an idle synchronous ac generator of a decommissioned turbine generator plant into a synchronous condenser.[4] In fact, one original equipment manufacturer (OEM) offers a packaged engineering solution for conversion of an existing synchronous generator to a synchronous condenser.[5] In 2003, General Electric (GE) Aeroderivative and Package Services (APS) in Houston retrofitted an LM6000 aeroderivative gas turbine at ATCO Power's Valleyview Generating Station in Alberta, Canada, with an SSS or "Triple-S" (synchronous self-shifting) clutch, which enables the unit to be used in power generation *or* synchronous condenser modes depending on the grid demand. (The SSS clutch is a product of SSS Clutch Company, Inc., in Delaware, USA. It is a freewheel-type, overrunning clutch that transmits torque through concentric, surface-hardened gear teeth.) The SSS clutch allows the gas turbine to be shut down while the generator remains synchronized to the grid, supplying or absorbing VAR. When real power (MW) is needed, the plant distributed control system restarts the gas turbine and engages the generator via the SSS clutch.

15.1.3 Reactive Power

Reactive power is probably the most difficult-to-understand concept in engineering (with the possible exception of the second law of thermodynamics). The difficulty mostly stems from the absence of a robust analogy to simple mechanical systems (e.g., fluid flow in a pipe), which captures all salient features of the ac systems, of which reactive power is a vital component, in generation as well as transmission. The best analogies involve simple hydraulic systems (e.g., flow of water in a channel feeding a waterwheel). The flow rate of water in, say, gallons per minute is analogous to electric current. The voltage is analogous to the hydraulic head between the water reservoir and the waterwheel. Only a portion of the total flow in the channel pushes the wheel and does useful work (analogous to the *real* or *active power* in ac systems). The remainder flows around the wheel without doing any work at all (analogous to the reactive power). If, however, the channel was designed such that only the amount of water that did work was allowed to flow through it, the stream would not be deep enough for the wheel to turn.

Another analogy is particularly helpful due to its similarity to the electric *power triangle* in terms of vector representation. As shown in Figure 15.4, a boat in a canal is pulled by a horse on the bank. Since the rope is pulling at the flank of the horse and not straight behind it, the horse's capacity to deliver work is limited. At the same time, the turned rudder (otherwise the boat would be pulled towards the bank) leads to extra losses.

Reactive power establishes and sustains the electric and magnetic fields of ac equipment. The ac system consumes reactive power to keep electricity flowing. Unfortunately, reactive power does not do the work of electricity (e.g., keeping the lights on and the TV running). As the amount of electricity flowing in a transmission line increases, so does the amount of reactive power needed to move the additional

[4] R. Peltier, "Teaching Old Generators New Tricks," POWER, November/December 2003, pp. 33–38.

[5] J.M. Fogarty, R.M. LeClair, "Converting Existing Synchronous Generators into Synchronous Condensers," Power Engineering, October 2011.

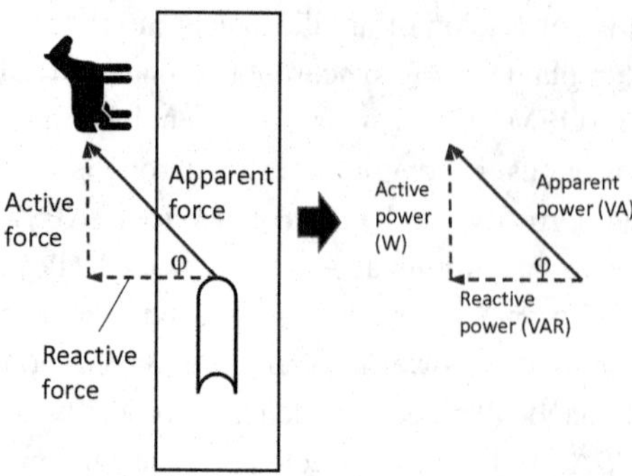

Figure 15.4 The horse-and-boat analogy of real/reactive power.

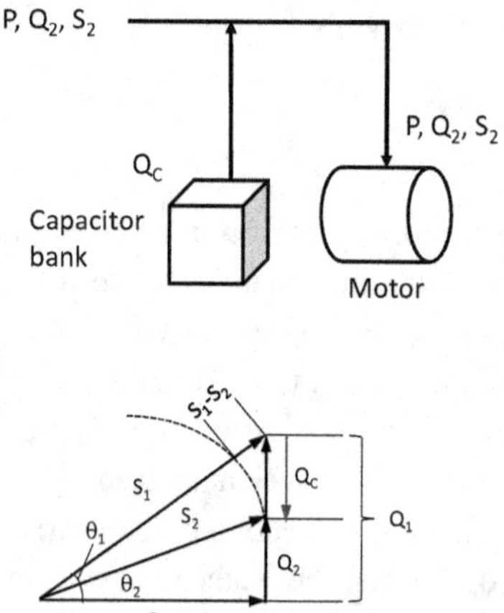

Figure 15.5 Synchronous ac motor with capacitor bank.

electricity and maintain the proper voltage. The longer the distance (i.e., the transmission line) between the power source (i.e., the generating station) and the load (i.e., factories, houses, schools, etc.), the more reactive power is consumed.

When reactive power is insufficient, voltage drops. If it continues to drop, protective equipment will shut down affected power plants and lines to protect them from damage. Eventually, the system comes to a screeching halt. In fact, according to a report from Cornell University's Engineering and Economics of Electricity Research Group, reactive power shortages played a key role in the Northeastern Blackout on August 14, 2003. Similarly, a Power Systems Engineering Research Center report comes to a similar conclusion with regard to the 1996 Western Electricity Coordinating Council (WECC) blackouts.

The need (benefit) of reactive power control can be best appreciated by considering an inductive load such as an ac motor as shown in Figure 15.5. Assume that the motor is running in a plant connected to a 400−V, three-phase bus and consumes 300 kW at a PF of 0.8.

Using the power triangle in Figure 15.5, θ_1 is ~37°, P = 300 kW, S_1 = 375 kVA, and Q_1 = 225 kVAR. The current drawn by the motor is

$$I_1 = 375 \times 1{,}000/\left(400 \times \sqrt{3}\right) = 540\,\text{A}.$$

Assume that a capacitor bank is connected to the terminals of the motor to inject Q_C = 100 kVAR reactive power into the circuit. Now the reactive power is Q_2 = 225 – 100 = 125 kVAR, so that the new phase angle is

$$\theta_2 = \tan^{-1}(125/300) = 23^{\circ},$$

and the new PF is

$$\text{PF} = \cos 23^{\circ} = 0.92,$$

so that the apparent power and current drawn by the motor drop to

$$S_2 = 300/0.92 = 325\,\text{kVA}$$

$$I_2 = 325 \times 1{,}000/\left(400 \times \sqrt{3}\right) = 469\,\text{A}.$$

In other words, PF correction via reactive power compensation leads to

- Better utilization of electrical machines
- Better utilization of electrical lines
- Reduction of losses
- Reduction of voltage drops

Reactive power shortages are caused by a variety of factors:

- Plant retirements
- Plant trips
- Transmission line failures
- Peak electricity demand

The ability of a synchronous ac generator to absorb power is described by a *reactive capability curve* (see Figure 15.6). In this curve, the VAR produced or absorbed is on the y-axis (positive going up). The x-axis shows real power in kW (positive to the right). VAR and kW are shown as *per-unit* quantities based on the rating of the generator (not necessarily the generator set, including the prime mover driving the generator, which may have a lower rating). Operation when the generator supplies VAR to the power system is *overexcited* operation; operation when the generator absorbs VAR from the power system is *underexcited* operation.

15.1.4 Excitation

The field winding on the generator rotor, which is *excited* by a dc, supplies the magnetic flux leading to the generation of the emf in the stator (armature) winding. The direct current is supplied from a dc generator called an *exciter* (whose power consumption is equivalent to about 0.2–0.4 percent of the megawatt rating of the synchronous machine).

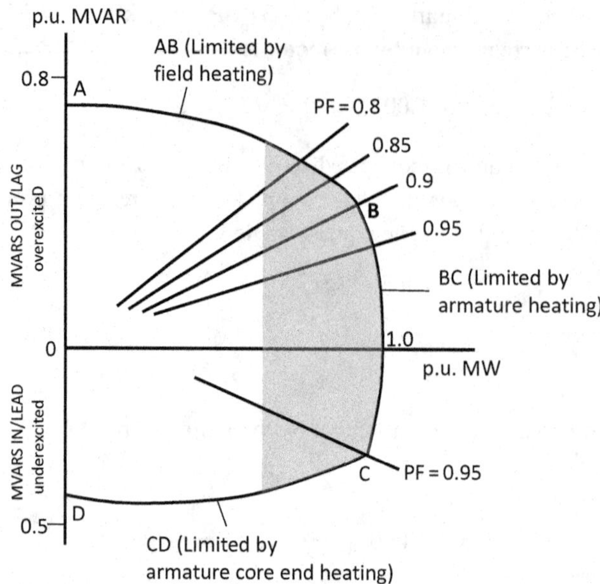

Figure 15.6 Typical synchronous ac generator capability curve.[6] Synchronous condenser capability is indicated on the y-axis by labels A and D. Reactive power capability of a synchronous generator is described by the shaded area.

As shown in Figure 15.7, the exciter current is fed to the field winding of the rotor via brushes and slip rings (in effect, a *mechanical rectifier*). Typically, the exciter voltage is 125–600 V. Excitation currents range from a few hundred amperes (e.g., 400 A at 200 V for a 30–MW machine) to several thousand (e.g., 5,700 A at 640 V for a 1,000–MW machine). The excitation is varied automatically to respond to load changes or to control the amount of reactive power delivered to the grid.

Generators are usually designed to deliver any load from zero to rated output over a voltage range of ±5 percent, at any PF between rated (usually 0.8–0.9 lagging) and 0.95 to 0.9 leading. The *automatic voltage regulator* setting controls must provide the corresponding range of excitation, and also provide about 85 percent of rated voltage on no load. Accuracy of control is usually within ±2.5 or ±1.0 percent of the set value over the load range.

Modern generators use either *brushless* or *static* excitation systems. Brushless systems comprise a three-phase stationary field generator and a bridge rectifier to convert its ac output into dc current fed into the field winding (see Figure 15.7). The dc control current from the pilot exciter, I_c, regulates the main exciter output current, I_x. The frequency of the main exciter is typically two to three times that of the generator.

A static exciter is a digital excitation system with no rotating parts, which comes equipped with a full-wave thyristor bridge, which supplies excitation power to the

[6] From IEEE Standard C50.13–2014 (Figure 3 on page 22)

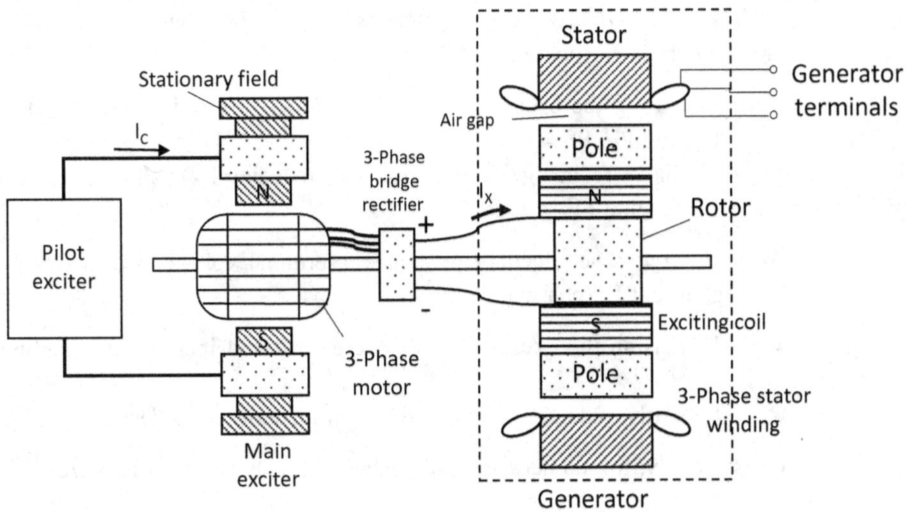

Figure 15.7 Brushless exciter.

rotating field winding of the main ac generator through slip rings and brushes. The source is stepped-down voltage from the station auxiliary bus (typically 4 kV). All control and protective functions are implemented through the system software. A static exciter is the preferred type for gas turbines with an LCI to run the generator as a motor during startup. Requisite current is supplied through the slip rings and brushes of the static starter. In general, slip ring/brush systems are more maintenance intensive, but not to a degree to make one exciter type superior to the other. For example, in order to use an LCI with a brushless exciter, a slip ring/brush add-on, which is engaged by a motor during startup and disengaged thereafter, is necessary.

Note that, according to the IEEE standard C50.13 (2014), if the generator excitation system power is supplied from the generator terminals, the net power delivered to the system will be lower than the generator rated output by the amount of excitation power being supplied.

The rms value of internal generated voltage E_A induced in the stator of the generator is

$$E_A = \sqrt{2}\pi N_C \phi f, \tag{15.20}$$

where N_C is the number of windings in each coil and f is the frequency in Hertz and f is the frequency.

Since flux in the machine ϕ depends on I_F, E_A is also a function of the field current I_F (controlled by the exciter's dc current). Note that E_A in a single phase equals V_ϕ (the output voltage of the phase) only when there is no armature current in the machine. Otherwise, it differs from it due to (i) distortion of the air-gap magnetic field caused by the current flowing in the stator I_A (the armature current - goes with V_{fi}), (ii) self-inductance, and (iii) resistance of the armature coils. In general, $V_\phi > E_A$ by the amount of armature reaction voltage.

An *overexcited* ac generator generates a terminal voltage, V_ϕ, in one phase with a lagging current I_A (phase angle θ).

- Increasing the excitation (i.e., increasing E_A) will increase θ and decrease the lagging PF, PF = cos θ.
- Similarly, reducing the excitation (i.e., reducing E_A) will decrease θ and increase the lagging PF.

An *underexcited* ac generator generates a terminal voltage, V_ϕ, in one phase with a leading current I_A (phase angle θ).

- Increasing the excitation will decrease θ and increase the leading PF, PF = cos θ.
- Similarly, reducing the excitation will increase θ and decrease the lagging PF.

Consequently, from a network/grid perspective, if a generator is *overexcited*:

- It supplies lagging current and reactive power.
- Phase angle θ and reactive power Q are positive.

If a generator is *underexcited*:

- It supplies leading current and absorbs reactive power.
- Phase angle θ and reactive power Q are negative.

15.1.5 Generator or Motor?

A synchronous ac generator and a synchronous ac motor are the opposite sides of the same coin. In other words:

- The ac machine is a *generator* when the input is mechanical (i.e., shaft work) and the output is electrical (i.e., ac voltage and current).
- The ac machine is a *motor* when the input is electrical (i.e., alternating current) and the output is mechanical (i.e., shaft work).

Thus, the ac generator concepts discussed above are equally valid for an ac motor. Consequently, applying a "linguistic mirror" to the terminology, from a network/grid perspective, one arrives at the following conditions: if a motor is *overexcited*:

- It absorbs leading current and supplies reactive power.
- Phase angle θ and reactive power Q are negative.

If a motor is *underexcited*:

- It absorbs lagging current and reactive power.
- Phase angle θ and reactive power Q are positive.

Synchronous ac motors are often run at no active load as synchronous condensers for the purpose of PF correction.

15.1.6 Power and Torque

To complete the discussion, the connection between "mechanical" and "electrical" phenomena in synchronous ac machines needs to be established in terms of *power* and *torque*. The *Sankey* diagram in Figure 15.8 illustrates this connection concisely.

The mechanical power supplied by the prime mover (e.g., a gas turbine) is given by

$$P_{in} = \tau_{app}\omega_m, \tag{15.21}$$

where the angular speed of the prime mover shaft, ω_m, is $2\pi\cdot 50$ or $2\pi\cdot 60$ (depending on the grid frequency) and τ_{app} is the "applied" torque, which is partially converted to the "induced" torque, i.e.,

$$P_{conv} = \tau_{ind}\omega_m, \tag{15.22}$$

$$P_{conv} = 3E_A I_A \cos\gamma, \tag{15.23}$$

where γ is the angle between E_A and I_A.

Using considerable algebra and geometry, the *real* power output of the three-phase synchronous ac generator can be written as

$$P_{out} = 3\frac{X_{ad}}{X_S}V_\phi I_F \sin\delta, \tag{15.24}$$

$$P_{out} = 3V_\phi I_A \cos\theta, \tag{15.25}$$

$$P_{out} = \sqrt{3}V_T I_L \cos\theta, \tag{15.26}$$

where θ is the angle between V_ϕ and I_A and δ is the *torque angle*. Typically, generators have a full load torque angle (also known as *load angle*) of between 15° and 20°. The angle γ is the sum total of θ and δ.

Similarly, the *reactive* power output of the three-phase synchronous ac generator is

$$Q_{out} = 3\frac{X_{ad}}{X_S}V_\phi I_F \cos\delta - \frac{V_\phi^2}{X_S}, \tag{15.27}$$

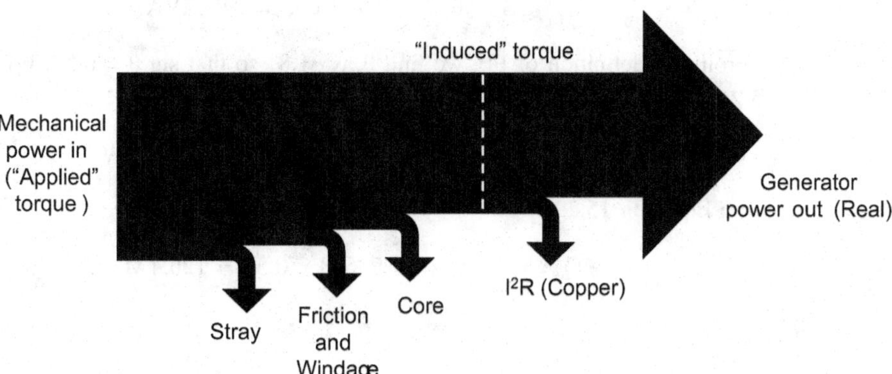

Figure 15.8 Sankey diagram of ac generator electromechanical power conversion.

$$Q_{out} = 3V_{\phi}I_A \sin\theta, \quad (15.28)$$

$$Q_{out} = \sqrt{3}V_TI_L \sin\theta, \quad (15.29)$$

where V_T is the *terminal voltage* (also known as the *line-to-line* voltage, V_L) and I_L is the *line current* (also known as the *armature current*, I_A). The apparent power, S, can be found from the power triangle as discussed earlier (see Equation 15.17) or

$$S = \sqrt{3}V_LI_L. \quad (15.30)$$

Note that, from a generator electromechanical design perspective, S is proportional to the physical dimensions and the electromagnetic characteristics via the *Esson formula*, i.e.,

$$S \propto D^2L{\cdot}B{\cdot}A_S{\cdot}\omega, \quad (15.31)$$

where D^2L is proportional to the volume of the rotor winding and A_s is the *linear current density* (in amperes per unit length, also known as the *electrical loading*). Magnetic flux density B is 0.75–1.05 T for cylindrical-rotor synchronous generators.[7] Linear current density is 250–300 kA/m for directly hydrogen-cooled stator and rotor designs. For directly air-cooled designs, linear current density is 160–200 kA/m.

For a typical F-class gas turbine generator, the following data are available:

- Two-pole, 3,600 rpm, 240,000 kVA, 0.85 PF
- Rated armature voltage (V_T or V_L) 18,000 V (18 kV)
- Rated armature current (I_A or I_L) 7,698 A
- Field current at rated load (I_F) is 1,694 A
- Excitation voltage 355 V
- Hydrogen-cooled (40°C cold gas at 30 ft altitude)
- Coolant flow 1,820 gpm (50 percent propylene glycol)

From Equation 15.30,

$$S = \sqrt{3} \times 18 \times 7,698 = 240\,\text{MVA}.$$

From the definition of PF, we find θ as 31.8° so that sin θ = 0.53. From Equation 15.26,

$$P_{out} = \sqrt{3} \times 18 \times 7,698 \times 0.85 = 204\,\text{MW}.$$

From Equation 15.29,

$$Q_{out} = \sqrt{3} \times 18 \times 7,698 \times 0.53 = 126.4\,\text{MVAr}.$$

[7] Tesla (T) is the unit of magnetic flux density in SI units; one Tesla is equal to one Weber per m^2, where Weber (Wb) is the SI unit of magnetic flux. 1 Wb = 1 V·s = 1 J/A (see Faraday's Law, Equation 15.1). Also, 1 Wb = 1 H·A, where H (Henry) is the SI unit for inductance.

Combining Equations 15.28 and 15.29 (or Equations 15.25 and 15.26),

$$V_\phi = 18/\sqrt{3} = 10.4\,\text{kV}.$$

Combining Equations 15.24 and 15.27, we have

$$Q_{out} = P_{out}\cot\delta - \frac{V_\varphi^2}{X_S}. \quad (15.32)$$

Let us assume that δ is 20°. Ignoring the resistance loss (i.e., assuming $P_{conv} \approx P_{out}$), it can be found that

$$\gamma = \delta + \theta = 20 + 31.8 = 51.8\,\text{degrees},$$

$$\cos\gamma = 0.62, \text{ and}$$

$$E_A = 204{,}000/(3 \times 7{,}698 \times 0.62) = 14.28\,\text{kV}.$$

From Equation 15.32, the synchronous reactance X_S is calculated as

$$X_S = (10{,}400)^2/(204{,}000 \times 2.75 - 126{,}400) = 0.249\,\Omega.$$

Finally, using Equation 15.22, the shaft torque can be found as

$$\tau_{ind} = 204{,}000/(2\pi 60) \times 1{,}000 = 541{,}127\,\text{Nm}.$$

When plotted on a P–Q diagram, the expression for the generator MVA, S, developed earlier in Equation 15.19,

$$P + j \cdot Q = |S| \cdot e^{j\theta},$$

determines the *armature heating limit*. The expressions for P and Q above determine the *field current heating limit* (when plotted as a circle with its center on the Q-axis at $-V_\phi^2/X_S$ and the radius of $(X_{ad}/X_S)V_\varphi I_F$).

The localized heating in the end region of the armature imposes a third limit on the operation of the synchronous ac machine. This is known as the *end-region heating limit* and affects the capability of the machine in the underexcited condition. The combination of the three effects above determines the synchronous ac generator's *reactive capability curve* (see Figure 15.6).

A three-phase synchronous ac generator can be connected in two configurations: Y (wye or star) and Δ (delta). The naming stems from the arrangement of the phase leads; the same leads for each phase (plus the neutral/ground line) connected together to form a Y (star) or the leads connected end to end to form a triangle (delta). For a wye connection:

$$V_T = \sqrt{3}V_\phi, \quad (15.33)$$

$$I_L = I_A. \quad (15.34)$$

For a delta connection:

$$V_T = V_\phi, \tag{15.35}$$

$$I_L = \sqrt{3}I_A. \tag{15.36}$$

Thus, the equations for P and Q are the same for either type of connection.

Practically all large ac generators, including gas turbine ac generators, have wye connections and delta–wye transformer connections (see Figure 15.11 below). On the 18-kV generator side, the windings are connected in a delta configuration. The reason for this is to avoid ground fault currents that could damage the generator stator core. Loads such as large motors can be of either type (or a combination thereof).

During startup, when the LCI runs the synchronous machine as an ac motor, the neutral line from the wye is disconnected from the ground. Once the startup is complete and the synchronous machine is back in the ac generator mode, neutral-to-ground connection is reestablished.

15.2 The Hardware

Gas turbine generators can be classified into three groups based on the cooling medium used: air-, hydrogen-, or water-cooled. How well the armature winding of a generator is cooled has a significant influence on the overall size of the generator. The cooling of the armature winding is dependent on the following factors:

- Cooling medium (air, hydrogen, or water);
- Insulation thickness;
- Overall electrical losses (see Figure 15.8).

The relative properties and capabilities of cooling medium types are summarized in Table 15.1.

Air-cooled generators come in two basic configurations:

1. Open ventilated (OV)
2. Totally enclosed water-to-air cooled (TEWAC)

In the OV design, outside air is drawn directly from outside the unit through filters, passes through the generator, and is discharged outside the generator. In the TEWAC design, air is circulated within the generator, passing through frame-mounted air to

Table 15.1 Relative comparison of generator coolants

Fluid	Specific Heat	Density	Volumetric Flow	Heat Transfer Rate
Air	1.00	1.00	1.00	1.00
H_2 at 30 psig	16.36	0.21	1.7	5.00
H_2 at 45 psig	16.36	0.26	1.7	7.50
Water	4.16	1,000	0.03–0.17	140–700

water heat exchangers. As such, TEWAC designs are more effective at keeping the sensitive internal hardware of the generator clean. Air-cooled generators (both stator and rotor) are typically used for ratings up to 350 MVA.

In the 300–550–MVA range, hydrogen-cooled machines are the economic choice because they are typically

- 20 percent smaller than air-cooled machines at the same rating)
- More efficient (e.g., 99 percent vis-à-vis 98.8 percent for air-cooled machines)[8]

Beyond that (rare for a gas turbine except for the most recent, very large 50-Hz H- or J-class products), water-cooled stator designs (with hydrogen-cooled rotors) are available.

As can be deduced from the data in Table 15.1, hydrogen's advantage as a coolant comes from its lower density vis-à-vis air (i.e., lower windage losses) and higher heat transfer capability with comparable volumetric flow rate (due to higher thermal conductivity and convective heat transfer coefficients). Its disadvantage is the explosive potential upon mixing with air (between 4 and 75 percent H_2 in the mixture), which requires special measures to be taken to prevent leakages during operation, as well as gassing and degassing (purging).[9] As a side benefit, advanced seals requisite to contain hydrogen also help to keep the generator interior clean. Furthermore, due to the absence of oxygen inside the sealed and pressurized casing, armature insulation deterioration caused by *coronas* is less in hydrogen than in air.[10]

The internally recirculating, once-through hydrogen stream dissipates generator heat through a gas-to-water heat exchanger. Ventilation fans are mounted at each end of the rotor. The fans facilitate the circulating gas flow for cooling the generator. Cooling of the stator core is accomplished by forcing the coolant through the radial ducts formed by space blocks. The axial length of the core is made up of many individual segments separated by the radial ventilation ducts. The rotor winding, which is typically a directly cooled radial or axial flow design, is largely self-pumping. The rotor is cooled externally by the coolant flowing along the gap over the rotor surface and internally by the coolant that flows through sub-slots under the field coils within the rotor body and passes directly through radial cooling ducts in the copper coils and wedges.

Water is an even better heat transfer medium than hydrogen (plus it is not explosive). In theory, it can be used in any size class (i.e., 300 MVA or smaller). However, due to the complex cooling configuration (i.e., increased manufacturing cost) and additional equipment (auxiliary water cooling and deionization skid, pipes, valves, etc.), water cooling is not cost-effective except for at very large ratings. However, in applications to large industrial gas turbines, water *is* used as the stator coolant, whereas hydrogen is used to cool the rotor. Cooling water must be of a very high purity and very low conductivity (about 5 μS/cm). This requires a dedicated water treatment system with

[8] Plus, hydrogen-cooled generators maintain their efficiency at part load better than air-cooled ones.

[9] For example, an inert gas, carbon dioxide, is used as an intermediate gas so that air and hydrogen do not mix inside the generator. The purging procedure for the generator has carbon dioxide introduced first to displace air, then hydrogen introduced to displace the carbon dioxide.

[10] Corona refers to an electrical discharge caused by the ionization of the cooling medium (air or hydrogen) surrounding the electrically charged conductor.

redundant filters, circulation pumps, coolers, etc., and an ion exchanger that continuously deionizes a stream of circulating cooling water to ensure low conductivity.

In terms of the "method of cooling," generators can be classified into the following two groups:

1. Direct cooling
2. Indirect cooling

In *direct* cooling, the coolant, flowing through the cooling passages in the copper conductors, comes into direct contact with them, picks up the heat generated by them, and disposes of it in the external heat exchangers.

In *indirect* cooling, the coolant cools the stator core or the rotor shaft, which is in contact with the exterior of the conductor insulation. In other words, the heat generated by the conductors, before being picked up by the coolant, travels through the insulation layer first.

Representative heavy-duty industrial gas turbine generator design (electric and physical) and performance data are summarized in Table 15.2. Note that, according to the IEEE standard C50-13 (2014), a generator's rating is its apparent output (MVA or kVA), which is available continuously at the terminals at rated frequency, voltage, PF, and primary coolant temperature. For hydrogen-cooled generators, the rating also includes hydrogen pressure and purity. For air-cooled generators, the rating includes altitude (it should not degrade up to 1,000 meters above sea level).

Table 15.2 Typical gas turbine generator design and performance data

	Units	Type A	Type B	Type C	Type D
Rotor Coolant		Air (Radial)	Air (Radial)	H_2 (Radial)	H_2 (Axial)
Stator Coolant		Air (Indirect)	Air (Indirect)	H_2 (Indirect)	H_2O (Direct)
Hydrogen Pressure	psig			45	45
Rating	MVA	140	300	450	550
Power Output	MW	112	240	360	440
Voltage	kV	10.5	15.75	22	21
Exciter Voltage	V	293	412	426	487
Exciter Current	A	1,194	1,359	3,503	4,153
Temperature Class		F	F	F	F
Application Limit		B	B	B	B
Subtransient Reactance		0.18	0.16	0.19	0.19
Transient Reactance		0.26	0.24	0.28	0.26
Short-Circuit Ratio		0.48	0.47	0.54	0.48
Heat Loss, Rotor	kW	316	490	1,410	1,940
Heat Loss, Stator	kW	335	418	680	1,035
Total Losses	kW	1,528	3,470	4,275	5,465
Efficiency	%	98.65	98.58	98.83	98.77
Rotor Weight	mt	34	57	54	52
Stator Weight	mt	150	260	303	293

Table 15.3 NEMA temperature classifications. SF = service factor

	Maximum Allowable Operating Temperature	Allowable Temperature Rise	
Class	**°C/°F**	**°C (SF 1.0)**	**°C (SF 1.15)**
B	130/266	80	90
F	155/311	105	115
H	180/356	125	–

Electrical insulation of motors and generators is rated by standard American National Standards Institute (ANSI)/National Electrical Manufacturers Association (NEMA) classifications (per ANSI/NEMA Standard MG 1) according to maximum allowable operating temperature (see Table 15.3). Allowable temperature rises are based upon a reference ambient temperature of 40°C. Maximum allowable operation temperature is the sum total of reference temperature, allowable temperature rise, and allowance for "hot spot" winding. Thus, for an F–class machine, maximum allowable operating temperature is 40 + 105 + 10 = 155°C. For a more detailed table and other requirements, the reader should consult table 5 (air-cooled generators) and table 6 (hydrogen-cooled generators) in IEEE C50.13 (2014). For example, for hydrogen-cooled generators, no values are prescribed by the IEEE standard; those values are to be reached by agreement.

As shown in Table 15.2, the armature and field windings of the generators are designed with ANSI/NEMA class F insulation materials. (Stator winding and gas temperatures are monitored and recorded via transducers installed inside the machine.) However, as a safety margin, throughout the allowable operating range, temperature rises are limited to class B specifications.

In some countries, especially those with 50-Hz grids, the practice is to specify generator voltages in proportion to rated power output; in other words, for generators rated at

- Less than 200 MVA, 10.5 kVA
- Between 200 MVA and ~350 MVA, 15.75 kVA
- Between 350 MVA and ~900 MVA, 21 kVA
- More than 900 MVA, 27 kV

ANSI standards used to prescribe 13.8 kVA for generators rated at less than 150 MVA. Up to about 300 MVA, 16.5 and 18 kVA were popular with the US OEMs (i.e., 60-Hz machines). In IEEE C50.13 (2014), there is no voltage recommendation.

There are three electrical design parameters in Table 15.2: transient and subtransient reactances (strictly speaking, *direct-axis* reactances) and the *short-circuit ratio* (SCR). As the monikers "transient" and "short-circuit" imply, they define the stability characteristics of the particular synchronous ac machine. Even a cursory treatment of their physical origins and mathematical derivation is well beyond the scope of this chapter (not to mention the author's limited knowledge of electric machinery fundamentals).

The reader is referred to [1] and [3] for comprehensive coverage (caveat emptor: this is extremely hard going for those who do not have at least a BS degree in electrical engineering).

For our purposes, suffice it to say that a short-circuit is a *fault* condition. The generator must survive that fault condition without significant damage to its hardware. When the fault happens, the ac component of the short-circuit current jumps to a very high value.[11] The *subtransient period* identifies the very early stage of the fault condition (the first few cycles), during which the short-circuit current drops very rapidly. The transient period identifies the next several seconds, during which the short-circuit current drop rate is more moderate. Finally, the short-circuit current reaches a *steady-state* value.

Transient and subtransient reactances determine the currents in their respective period. Synchronous reactance, X_S, determines the steady-state ac fault current, i.e.,

$$I_{SS} = \frac{E_A}{X_S},$$

where, as discussed in Section 15.1.4, E_A is the internally generated voltage. The SCR is the inverse of the synchronous reactance. More specifically, SCR is the ratio of the field current required for the rated voltage at open circuit to the field current required for the rated armature current at short-circuit. A higher SCR will improve the steady-state stability characteristics of the generator, but it will result in a larger machine. A lower SCR will result in a smaller machine, but at the expense of stability. In most turbogenerators, the SCR value is set at 0.5–0.6.

A generator's capability is its highest acceptable continuous output of apparent power (MVA or kVA) at prevailing site ambient and loading conditions. The OEM is required to supply curves of generator capabilities under site conditions over the specified range of generator primary coolant temperature (see Figure 15.9). For a machine with OV air cooling, for example, this is the same as that of the ambient air surrounding the generator. For a machine with TEWAC air cooling or hydrogen cooling, this is the water inlet temperature of the external cooler.

Gas turbines typically have both a base load capability and a peaking capability (with a higher firing temperature for a limited duration). Thus, the OEM is required to specify the same capabilities for the generator as well. The generator should be designed to match the capabilities of the gas turbine over the ambient temperature range. This is especially important for performance at low ambient temperatures when the gas turbine output substantially increases due to much higher density of the surrounding air. For OV-type air-cooled generators, the coolant temperature rises in lockstep with the site ambient temperature, which is detrimental to the generator capability. Fortunately, gas turbine output drops with increasing site ambient temperature as well, such that this does not create a big handicap.

[11] Before the fault, only ac voltages and currents are present, but immediately after the fault, both ac and dc currents are present.

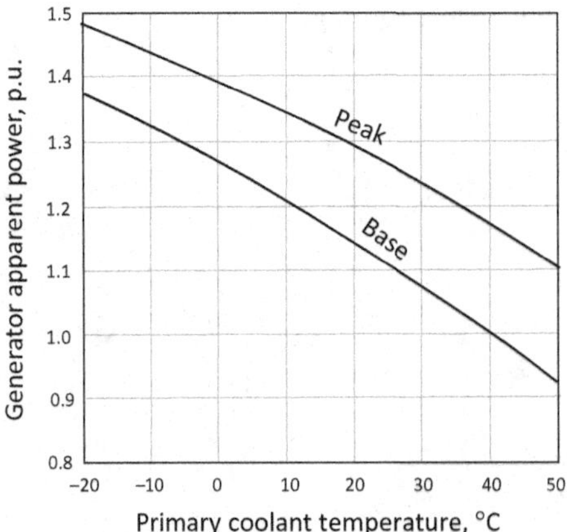

Figure 15.9 Typical generator base and peak capability curves.

Figure 15.10 Hydrogen-cooled, two-pole synchronous ac generator. (courtesy of Siemens Energy, Inc. © 2018. All rights reserved)

The salient features of a modern hydrogen-cooled, two-pole generator for heavy-duty industrial gas turbines can be seen in Figure 15.10.

The parts and features numbered in Figure 15.10 are as follows:

1. Exciter end (opposite the gas turbine coupling)
2. Outer (pressurized oil) seal
3. Bearing oil catcher

4. Rotor shaft (single piece, alloy steel forging)
5. Carbon seal bracket
6. Carbon seal
7. Bearing (tilting pad)
8. Ventilation fan hub (one on each end)
9. Fan shroud
10. Gas cooler (one on each end)
11. Fan blades
12. Stator windings
13. Finger plates[12]
14. Winding brace
15. Core spring support system

OEMs design their generators in accordance with established standards (e.g., IEEE C50.13 – 2014 [which replaced ANSI C50] in the USA or IEC 60034[13]). IEEE C50.13 governs operational requirements, rating and performance characteristics, insulation systems, temperature limits, and test procedures. Familiarity with this standard or IEC 60034 is highly recommended.

In addition to rated performance and electromechanical design data, generator OEMs typically supply the following four types of curve:

1. Reactive capability curve (see Figure 15.6)
2. Output vs. coolant temperature curve (see Figure 15.9)
3. "Vee" curve
4. Impedance (short-circuit) curves

The "Vee" curve (or the V curve) shows the relation between armature current (y-axis) and field current (x-axis) at constant terminal voltage and constant active power output for different values of the PF. Consequently, the y-axis also gives the apparent power, S, in "per-unit" (pu or p.u.) terms.

The impedance curve shows the stator (armature) current as a function of the field current with the stator winding terminals short-circuited and the generator operating at the rated speed. It also includes the open circuit, which shows the open circuit stator terminal voltage as a function of the field current at the rated generator speed.

The interested reader should consult Kundur's book for detailed coverage [3].

15.2.1 Load Commutated Inverter

By operating the ac generator as a synchronous ac motor, the LCI (which is the key component of a *static starter*) accelerates the gas turbine according to a preset schedule that provides optimum starting conditions for the gas turbine. The LCI eliminates the

[12] Finger plates are heavy-duty end plates, which are used to help compress the stator core, holding it together during the final generator assembly.

[13] International Electrotechnical Commission

need for separate starting hardware, such as an electric motor or diesel engine, torque converters, and associated auxiliary equipment. For a very detailed description of the LCI system architecture, the reader is referred to [5].

In the most basic arrangement, one LCI starts one prime mover generator (i.e., one gas turbine). A single LCI can be used to start multiple generators in sequence (i.e., one gas turbine at a time, not simultaneously) to save space and cost. Since there is no redundancy, if the LCI goes down, the generator(s) cannot be started. Different configurations with inherent redundancy are discussed in [5]. More coverage of LCI and its operation can be found in Section 19.3.

15.3 Synchronization

During the gas turbine startup, when the unit reaches full-speed, no load, generator breakers are closed and the unit is *synchronized* to the grid. Synchronization is the process of matching the speed and frequency of the ac generator to those of the grid. In order to synchronize a generator to the grid, the following four conditions must be met:

1. The phase sequence (or phase rotation) of the three phases of the generator must be the same as the phase sequence of the three phases of the electrical system (grid).
2. The magnitude of the sinusoidal voltage produced by the generator must be equal to the magnitude of the sinusoidal voltage of the grid.
3. The frequency of the sinusoidal voltage produced by the generator must be equal to the frequency of the sinusoidal voltage produced by the grid.
4. The phase angle between the voltage produced by the generator and the voltage produced by the grid must be zero.

IEEE Standard C50.13 specifies the following limits for synchronization:

- Breaker closing angle $\pm 10°$
- Generator side voltage difference relative to system 0 to +5 percent
- Frequency difference ± 0.067 Hz

Outside these limits, *faulty synchronizing* occurs. Faulty synchronizing can cause intense, short-duration currents and torques, which may cause considerable damage to the generator.

If the generator voltage is higher than the grid voltage at the synchronization, when the breakers are closed, the generator will be *overexcited* and it will supply *reactive* power to the grid. Conversely, if the generator voltage is less than the grid voltage at the synchronization, when the breakers are closed, the generator will be *underexcited* and it will absorb *reactive* power from the grid.

If the generator breaker is closed when the generator is slower than the grid (i.e., its frequency is lower), the generator would be out of step with the external electrical system. It would behave like a motor and the grid would try to bring it up to speed. In doing so, the rotor and stator would be slipping poles and damage (possibly destroy) the generator.

The same problem would occur if the generator was faster than the grid (i.e., its frequency was higher). The grid would try to slow it down and lead to slipping of poles.

If the generator and the grid have matching frequencies but are separated by a phase angle at the time of breaker closure, the grid would pull the generator into lockstep, but the large current inrush to the generator and high stresses on the rotor/stator would do damage to the generator. If the generator was leading the grid, it would try to immediately push power into the grid with the same destructive force. Therefore, the generator must be brought to a point where the grid voltage waveform exactly matches what it is producing (i.e., the two waveforms would fit like the teeth of two interlocking gears).

15.4 Electrical Configuration

A three-phase power system such as a gas turbine generator connected to the bulk electrical system or a sub-transmission system is represented by a single-line or one-line diagram. A typical single-line diagram for a gas turbine generator is shown in Figure 15.11 [6]. The output voltage of the generator (typically 18 kV for large industrial gas turbines) is increased by the generator step-up transformer to the transmission system voltage (typically 230 kV). A medium-voltage (4−kV) auxiliary power bus is fed by the unit auxiliary transformer (UAT). A low-voltage (480−V) auxiliary power bus is fed from the 4−kV bus through the station service transformer (SST). Obviously, UAT and SST are "step-down" transformers.

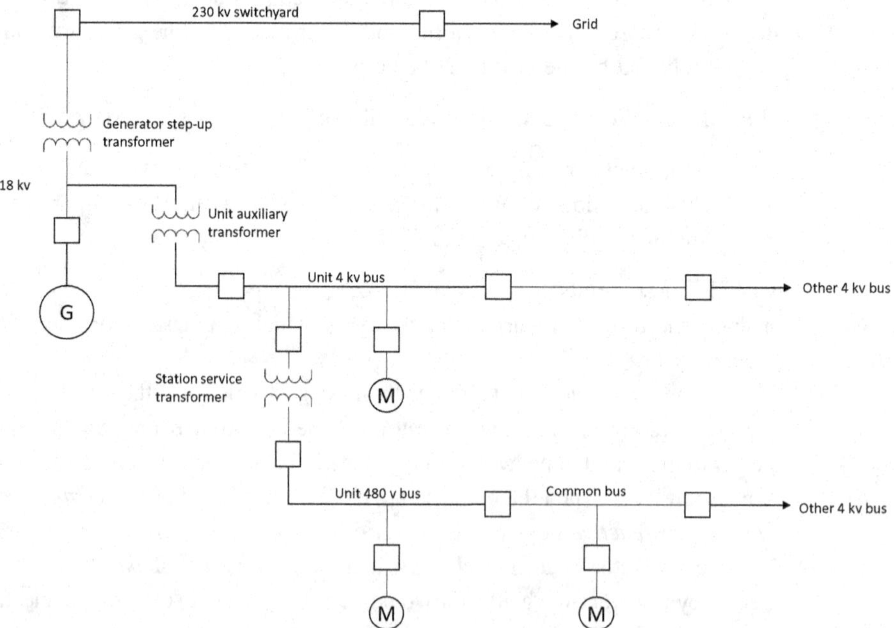

Figure 15.11 Simple single-line diagram of a gas turbine power plant.

In general, for switchgear-fed motors, the following generic rules apply:

- Those rated at less than 300 hp are fed by a 480-V (low-voltage) bus.
- Those rated up to 5,000 hp are fed by a 4-kV (medium-voltage) bus.
- Those rated at more than 5,000 hp are fed by a 6.9 or 13.8-kV bus.

Motors rated at more than 5,000 hp (~3,750 kW) are quite rare in gas turbine power plants (except, maybe, for very large, combined-cycle power plants).

A gas turbine power plant is a small part of a much larger system (i.e., the *bulk electric system* [BES]). The North American Energy Reliability Council (NERC) Rules of Procedure and the NERC Glossary of Terms Used in NERC Reliability Standards define the BES as follows:

"Bulk Electric System" means, as defined by the Regional Entity, the electrical generation resources, transmission lines, interconnections with neighboring systems, and associated equipment, generally operated at voltages of 100 kV or higher. Radial transmission facilities serving only load with one transmission source are generally not included in this definition.

(The actual BES definition document is five pages long and contains several inclusions and exclusions.) The NERC is the electric reliability organization certified by the Federal Energy Regulatory Commission (FERC) to establish and enforce reliability standards for the North American bulk power system.

Note that the BES is not to be confused with the *bulk power system* (BPS), which is defined by the NERC Rules of Procedure as follows:

"Bulk Power System" means, depending on the context: (i) Facilities and control systems necessary for operating an interconnected electric energy supply and transmission network (or any portion thereof), and electric energy from generating facilities needed to maintain transmission system reliability. The term does not include facilities used in the local distribution of electric energy.

According to NERC, "BPS" is the term to use when generally speaking about the interconnected network or power grid. BES, however, is "the portion of the bulk power

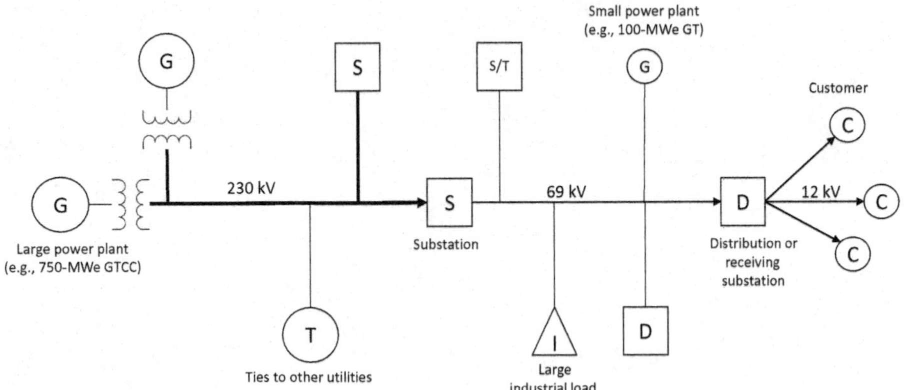

Figure 15.12 Typical utility system. GT = gas turbine; GTCC = gas turbine combined cycle.

system to which the standards apply and should be used when that specific meaning is intended."

A BES can encompass multiple utilities and states. A typical configuration is shown in Figure 15.12 [6]. In this example, the BES is the 230-kV system denoted by the thick line to which a gas turbine combined cycle is connected (among other large electric power generation plants). A smaller gas turbine power plant (e.g., a peaker plant comprising one or more aeroderivative gas turbines) can also be connected to the 69-kV sub-transmission system.

References

1. Fitzgerald, A.E., Kingsley, Jr., C., Kusko, A., *Electric Machinery*, 3rd edition (New York: McGraw Hill Book Co., 1971).
2. Stevenson, Jr., W.D., *Elements of Power System Analysis*, 3rd edition (New York: McGraw Hill Book Co., 1975).
3. Kundur, P., *Power System Stability and Control. The EPRI Power System Engineering Series* (New York: McGraw Hill Book Co., 1994).
4. Grigsby, L.L., *The Electric Power Engineering Handbook* (Boca Raton, FL: CRC Press, 2001).
5. General Electric Company, "LS2100 Static Starter Product Description," GE Industrial Systems Document GEI-100539, 2002.
6. Baker, T.E., *Electrical Calculations and Guidelines for Generating Stations and Industrial Plants* (Boca Raton, FL: CRC Press, 2012).

16 Reliability, Availability, and Maintainability

You can have the best gas turbine in the world (in terms of performance; i.e., output and heat rate), but if it can barely run a few hours at a stretch before tripping and requires a lot of maintenance labor and parts to restart, it has essentially zero value. In economic terms, cost of electricity becomes exorbitantly high (e.g., see Equation 20.1; as running hours, $H \rightarrow 0$, cost of electricity [COE] $\rightarrow \infty$). In technical terms,

- It is not *reliable* (trips a lot while running).
- It is barely *available* (spends a lot of time in the proverbial "shop").
- It is not *maintainable* (requires constant attention and immense labor and materials to upkeep).

In World War II, the Germans had arguably the best medium tank (Panther, PzKw-V) and the most feared heavy tank, the awesome Tiger (PzKw-VI) [1]. A single Tiger could annihilate an entire armored column by itself (google Michael Wittmann). The Allies figured that four or five Shermans lost in order to knock out a Tiger was a fair trade-off. Even the rumor of a German heavy tank battalion with Tigers and Panthers in the neighborhood was enough to shatter the morale of the Allied soldiers.

Alas, Tiger in particular was prone to mechanical mishaps, had a very low speed with a monstrous appetite for fuel, and was quite tricky to operate and maintain. When they broke down, due to their immense weight, they were very difficult to tow and/or transport to a repair location. Many Tigers were lost simply because their crews had to abandon them after a relatively minor breakdown or shell damage.

When the campaign record of Tiger was examined after the war on a rational basis, despite its superlative performance on a unit or crew basis, it could not be deemed to be a truly successful weapons system. This was in direct contrast to its erstwhile rival at the Eastern Front, the Soviet T-34. Compared to Tiger, and most other Allied tanks for that matter, T-34 was in fact a rather primitive vehicle. (Its early variants did not even have wireless sets in them.) However, it was its simple design and ruggedness with an emphasis placed on the systems that really counted (i.e., the high-velocity 76.2-mm gun, sloped armor, and "slack-track" tread system), ensuring that T-34 was cheap and simple to produce (and deploy) in large numbers and able to "get the job done" in the most efficient manner.

A similar "high-tech" versus "low-tech" dichotomy would emerge two decades later in the Vietnam War between the newly introduced American M-15 assault rifle and the Soviet-designed AK-47 Kalashnikov. The former was prone to frequent jams in the humid climate and muddy terrain, whereas the latter, with its ridiculously loose tolerances, was simply unbreakable. American soldiers were known to throw away their M-15s and pick up AK-47s from dead enemy soldiers [2]. Obviously, over time, the teething problems were solved and M-15 turned out to be a fine infantry weapon. Nevertheless, the lesson was unmistakable: the number one priority in any system, in the battlefield or in an industrial plant, is maximum availability (i.e., being ready to fire), reliability (i.e., firing bullets whenever the trigger is pulled without jamming), and maintainability (i.e., sturdy and quick to disassemble, clean, and reassemble in field conditions). As such, it is not an exaggeration that RAM (reliability, availability, and maintainability) is the most important acronym in gas turbine technology. Consequently, the subject deserves a whole chapter in a book on gas turbines.

16.1 RAM Metrics

Definitions of key RAM metrics can be found in the following two industry standards:

1. IEEE 762 "IEEE Standard Definitions for Use in Reporting Electric Generating Unit Reliability, Availability, and Productivity"
2. ISO 3977-9:1999 "Gas turbines – Procurement – Part 9: Reliability, Availability, Maintainability and Safety"

Several key RAM metrics defined in those standards are covered below.

The reliability factor (RF) is the probability that a unit, major piece of equipment, or component will not be in a forced outage condition at a point in time. It is given by the following formula:

$$\mathrm{RF} = 1 - \frac{\mathrm{FOH}}{\mathrm{PH}},$$

where FOH is forced outage hours and PH is period hours. The fraction on the right-hand side of the equation is the forced outage factor (FOF). In other words, the RF is the complement of the FOF.

FOH is the time, in hours, during which the unit or a major item of equipment was unavailable due to forced (unplanned) outages. In IEEE 762, there is a bit more granularity to the definition of unplanned outage hours (UOH), which, of course, is the complement of planned outage hours (POH), i.e.,

$$\mathrm{UOH} = 1 - \mathrm{POH}.$$

In particular,

$$\mathrm{UOH} = \mathrm{FOH} + \mathrm{MOH},$$

where MOH is the maintenance outage hours. This is based on the IEEE 762 classification of the basic states of unit operation, i.e.,

1. In service
2. Reserve shutdown
3. Planned outage (basic)

4. Planned outage (extended)
5. Unplanned outage Class 0 (starting failure)
6. Unplanned outage Class 1 (immediate)
7. Unplanned outage Class 2 (delayed)
8. Unplanned outage Class 3 (postponed)
9. Maintenance outage (basic)
10. Maintenance outage (extended)

In accordance with this classification, per IEEE 762, FOH represents the number of hours a unit was in a Class 0, Class 1, Class 2, or Class 3 unplanned outage state. A planned outage is scheduled well in advance and is of a predetermined duration; it can last for several weeks (i.e., extended outage) and occurs only once or twice a year. A maintenance outage, on the other hand, can occur at any time during the year; it has a flexible start date, may or may not have a predetermined duration, and is usually much shorter than a planned outage.

The availability factor (AF) is the probability that a unit, major piece of equipment, or component will be usable at a point in time, based on the past experience with that specific gas turbine. It is given by the following formula (ISO 3997):

$$\mathrm{AF} = 1 - \frac{\mathrm{FOH} + \mathrm{POH}}{\mathrm{PH}} = \frac{\mathrm{AH}}{\mathrm{PH}}.$$

The service hours (SH) is the accumulated period of time from main flame ignition through to flame extinction (ISO 3997). Put another way, SH is the number of hours a unit was in the in-service state, where a unit is electrically connected to the system and performing generation function (IEEE 762). It is also known as SH per start.

The service factor (SF) is the ratio of SH to period hours in a period under consideration. It is given by the following formula:

$$\mathrm{SF} = \frac{\mathrm{SH}}{\mathrm{PH}}.$$

Net actual generation (NAG) is the actual amount of energy (in megawatt hours) supplied by the unit during the period being considered, minus any energy supplied by the unit for that unit's own station services or utilities.

Gross maximum capacity (GMC) is the maximum capacity a unit can sustain over a specified period of time when not restricted by seasonal or other deratings.

Net maximum capacity (NMC) is the GMC minus the unit capacity utilized for that unit's station services or auxiliaries.

Consequently, the net output factor (NOF) is given by the formula

$$\mathrm{NOF} = \frac{\mathrm{NAG}}{\mathrm{SH} \times \mathrm{NMC}},$$

whereas the net capacity factor (NCF) is given by

$$\mathrm{NCF} = \frac{\mathrm{NAG}}{\mathrm{PH} \times \mathrm{NMC}}.$$

(One can drop the moniker "net" for most practical purposes, because it is the net plant performance that is of most interest in many cases.)

SF and SH determine the "duty cycle" or the "mission profile" of the power plant in question (e.g., gas turbine simple or combined cycle).

Capacity and output factors quantify the electric output contribution of the operating assets.

The AF represents the fraction of time that a generating asset is available for service, either actually operating (i.e., SH) or in a state of ready reserve. The complement of availability (i.e., 1 – AF) is *unavailability* (i.e., the fraction of time the unit is out of service).

These metrics are used by the North American Electric Reliability Corporation (NERC), which maintains the Generating Availability Data System (GADS), and by the ORAP® (Operational Reliability Analysis Program) system offered by Strategic Power Systems, Inc. (SPS).

GADS was initiated by the electric utility industry in 1982; it expands and extends the data collection procedures begun by the industry in 1963. GADS maintains complete operating histories on more than 7,700 generating units, representing over 90 percent of the installed generating capacity of the USA and Canada. As of January 1, 2013, GADS became a mandatory industry program for conventional generating units of 20 MW and larger.

ORAP is a RAM-reporting system with a specific focus on gas turbines in simple- and combined-cycle arrangements, various applications and duty cycles, and across the various manufacturers. ORAP was initiated by General Electric (GE) in 1976 in response to major design issues encountered in Frame 7 gas turbines. Field data were required to characterize issues experienced by gas turbine operators and measure the effectiveness of the engineering fixes developed to solve those issues. At the time, of course, gas turbines were mostly relegated to peaking service with SFs of about 10 percent. In 1978, ORAP was extended to cover simple- and combined-cycle power plants on an equipment basis, which required addition of many more measurements to the system. In 1980, ORAP transitioned from being a customer support system to a bona fide engineering design support system. At present, ORAP is a commercial product supported by all major original equipment manufacturers (OEMs).

Representative NERC GADS data for simple- and combined-cycle gas turbines are summarized in Tables 16.1–16.3 [3]. ORAP data for simple-cycle E- and F-class gas

Table 16.1 Simple-cycle E class (50–125 MWe); CF $\approx$ 1–5 percent (NERC GADS)

	AF	FOF	RF
1984–1988	0.911	0.036	0.964
1989–1993	0.902	0.027	0.973
1994–1998	0.917	0.018	0.982
1999–2003	0.917	0.022	0.978
2004–2008	0.932	0.025	0.975

Table 16.2 Simple-cycle F class (126–300 MWe); CF ≈ 1–7 percent (NERC GADS)

	AF	FOF	RF
1984–1988	0.857	0.026	0.974
1989–1993	0.886	0.026	0.974
1994–1998	0.801	0.062	0.938
1999–2003	0.892	0.026	0.974
2004–2008	0.940	0.019	0.981

Table 16.3 Combined-cycle plant (200–299 MWe); CF ≈ 30–40 percent (NERC GADS)

	AF	FOF	RF
1984–1988	0.889	0.013	0.987
1989–1993	0.863	0.011	0.989
1994–1998	0.890	0.019	0.981
1999–2003	0.883	0.037	0.963
2004–2008	0.902	0.027	0.973
	0.872[a]	0.055[a]	0.945[a]

[a] Combined-cycle gas turbine.

Table 16.4 Simple-cycle E class (ORAP, 2004–2008)

	AF	SF	FOF	RF
Peaking	0.967	0.050	0.007	0.993
Cycling	0.947	0.354	0.007	0.993
Base Load	0.933	0.771	0.019	0.981

Table 16.5 Simple-cycle F class (ORAP, 2004–2008)

	AF	SF	FOF	RF
Peaking	0.970	0.058	0.005	0.995
Cycling	0.943	0.474	0.013	0.987
Base Load	0.932	0.664	0.018	0.982

turbines are listed in Tables 16.4 and 16.5 [3]. Similar data are also presented by Steele et al. based on ORAP data [4].

From the data in the tables, some general observations can be made as follows:

- There is a favorable impact of "technology maturity" (i.e., compare E- and F-class availabilities for the 1980s and 1990s in Tables 16.1 and 16.2).

- "Stuff happens" (i.e., the E- and F-class forced outage factors for the 1980s and 1990s in Tables 16.1 and 16.2).
- There have been "teething problems" (i.e., the jump in FOF for F class in the 1994–1998 period – this is when significant rotor issues were discovered in the field for new 9FA-class gas turbines).
- There is an impact of "wear and tear" (i.e., the more one runs a machine, the more issues pop up – from Tables 16.4 and 16.5).

Market expectations based on the performance of mature technologies such as the E class and expectations set by the Electric Power Research Institute (EPRI) and the US Department of Energy (DOE; in the Advanced Technology Systems [ATS] program of the 1990s) are 93 and 98 percent for availability and reliability, respectively [4]. Clearly, after a rocky start beset by numerous field issues, F–class technology met those expectations.

Analogies to the observations made above can be made for passenger automobiles. Everything else being the same, on average, a 2000s model of any brand is more available and more reliable than its, say, 1970s predecessor. Still, you are liable to "discover" more mechanical issues in your 2015 model of your favorite wheels when you drive it 30,000 miles a year vis-à-vis when you drive it only 10,000 miles a year. Sometimes, a new model is typically more mechanical and/or electrical issue-prone (e.g., more "factory recalls") in its year of introduction vis-à-vis subsequent years. Finally, no matter how good a car is or how well it is maintained, a small amount of "forced outages" are bound to happen (but, as a fraction, much less than the FOFs observed in gas turbines) – although more in some brands compared to the others, of course.

At the time of writing (2017), major OEMs claim reliabilities of 99 percent or higher for their advanced-class gas turbines. Reported numbers are claimed to be based on the ORAP database, but typically with specified (or unspecified) modifications (e.g., "... modified by [OEM] to rolling 12-month data between dates ..."). For example,

- OEM A claims 98.8 and 92.7 percent for reliability and availability, respectively, of its F–class fleet vis-à-vis 97 and 87.9 percent for other OEMs' F/G/H/J-class fleet (composition of the fleet unspecified).
- OEM B claims 99.5 percent reliability for its J–class gas turbine (in 11,500 hours of commercial operation – duty type unspecified).

In any event, whatever the OEM-advertised RAM numbers might be, when the time comes to sign a contract, the owner requests "guaranteed" numbers from the OEM. This is where commercial considerations take over and standard definitions become somewhat irrelevant. For instance, for the owner/operator of a gas turbine power plant, highest revenue generation takes place in the hottest and coldest parts of the year (i.e., the 4 or 5-month period in mid-summer and mid-winter). As such, regardless of what the average fleet availability of the gas turbine in question might be, the owner demands 100 percent availability in those times of the year. In statistical terms, however, this is equivalent to a request of effectively 100 percent *reliability*.

This, of course, is an impossibility. Note that RF is a statistic; it indicates a *probability*, not a certainty. A value of, say, 99 percent means that, *on average*, there is a chance of losing 1 hour out of every 100 period hours to an unexpected malfunction based on X units observed over Y hours of field operation. This does *not* guarantee that unit X + 1 will not be down 30 hours out of every 100 period hours (due to whatever unique reason). Conversely, unit X + 1 could go hundreds and hundreds of hours without losing a single hour to forced outage as well.

This is why OEMs use sophisticated statistical models to determine availability guarantees. Such models quantify the probability distribution for a given model of gas turbine (e.g., 60-Hz F–class gas turbine of OEM X) with mean, mode, median, and confidence intervals. Assume that there is a 10–unit sample culled from all such units operating in the field. Also assume that the single-year, single-unit distribution of AFs for this sample from the ORAP database is as follows:

Unit	AF (%)
1	92
2	93
3	88
4	93
5	94
6	93
7	93
8	92
9	89
10	91

The statistics for this particular AF distribution are as follows:

Mean	91.8
Mode	93.0
Median	92.5
Std. Dev.	1.93

Based on these numbers, the 95 percent confidence interval is 90.4–93.2 – in other words, one can be 95 percent confident that the *population* mean of *all* similar units operating in the field falls between those two values. Consequently, based solely on the raw data,

- An extremely *risk-averse* OEM would guarantee 90.4 percent AF (only 5 percent chance of not meeting it);
- A *smart* owner would ask for 92 percent AF (only 3 out of 10 units did fall below this);
- A *greedy* owner would ask for 93 percent AF (the mode);
- The number we would read (maybe) in the trade publications would be 91.8 percent.

The search for "truth" (if one can use that term) requires a case-by-case investigation of such factors as site differences, operator practices, unusual events, etc. For example, the poor-performing units 3 and 9 (88 and 89 percent reliability, respectively) might be installed in sites with extremely harsh environments (salt water, desert sand, etc.). Such data are only available to the OEMs and, to some extent, subscribers to the ORAP database.

The table below is an *actual* OEM guarantee for availability for a "long-term service agreement" (LTSA) that starts at "substantial completion" and terminates at the end of 16 years or earlier if the gas turbine in question reaches a specified number of FFH or EOH:

Operation Year	AF (%)	Operation Year	AF (%)
1	90.1	9	94.9
2	95.5	10	94.8
3	94.9	11	95.5
4	94.8	12	92.8
5	95.5	13	92.7
6	92.8	14	95.5
7	92.7	15	94.9
8	95.5	16	94.8

The average AF over 16 years is 94.2 percent. Guarantee numbers are tied to bonuses (if exceeded) or "liquidated damages" (if not met) with specified caps and myriad contractual caveats such as excessive factored starts, etc.

Large industrial gas turbines for electric power generation are mostly deployed in combined-cycle configuration. This brings the RAM of the other two major pieces of equipment (i.e., the heat recovery steam generator [HRSG] and the steam turbine generator) into the picture. Average F-class gas turbine combined-cycle outage factors in Table 16.6 provide a good measure of technology reliability (they are from the ORAP database). As shown in Table 16.6, the gas turbine-only unexpected outage rate is

Table 16.6 Average outage factors for F–class gas turbines in combined-cycle configuration and other major pieces of equipment (based on units with a minimum of 6,500 annual operating hours) [5]

	1995–1999	2000–2004	2005–2009
Gas Turbine Subsystem			
Forced Outage (%)	2.67	2.09	1.59
Unscheduled Maintenance (%)	1.78	0.55	0.79
SF (%)	78.7	62.4	60.9
HRSG Subsystem			
Forced Outage (%)	0.11	0.28	0.21
Unscheduled Maintenance (%)	0.93	0.26	0.26
Steam Turbine Subsystem			
Forced Outage (%)	0.36	0.39	0.81
Unscheduled Maintenance (%)	0.59	0.24	0.26

2.4 percent, whereas for the entire plant (i.e., gas turbine, steam turbine, and HRSG combined), the unexpected outage rate increases to 3.9 percent (2005–2009 timeframe). What this means is that, on average, one should expect annually ~300 hours of lost power generation opportunity due to unforeseen events.

The gas turbine forced outage data in Table 16.6 also illustrate the "learning" or "maturity" effect. In particular, as a technology matures, OEMs and operators become more adept in its design, operation, and maintenance. Even then, effects of equipment aging and changes in the duty cycle of a plant can be detrimental to reliability (and availability). This was indeed supported by data presented in [4], which showed that the mean time between failures for base loaded F-class units decreased by 7 percent between the 1996–2000 and 2001–2005 periods when their SH per start dropped from 137 to 60.

16.2 Failure Mechanisms

Similar to the aircraft engines from which they derive, gas turbines are not as susceptible as other large pieces of equipment in a combined-cycle power plant to the vagaries of heavily cyclic operation with many start–stop cycles. Due to their relatively "lighter" construction, unlike an HRSG or steam turbine with thick-walled metal components, gas turbines suffer less from thermal and pressure stresses leading to low-cycle fatigue (LCF). Nevertheless, this does not mean that excessive numbers of starts and stops are not detrimental to part lives, especially for heavy-duty, industrial gas turbines. (Smaller aeroderivative gas turbines are relatively immune.) The impact of start–stop cycles is accounted for by assigning to each of them an OEM-set "equivalent operating hours" (EOH).

The failure mechanisms associated with base load duty (i.e., long hours of uninterrupted operation at high loads) are the following:

- Creep rupture
- Hot corrosion
- Oxidation
- Erosion

The failure mechanisms associated with cyclic duty (i.e., relatively short hours of operation with frequent start–stop cycles and load up–down ramps) are the following:

- Isothermal LCF
- Thermomechanical fatigue (TMF)

Like in many cases in the power generation jargon, there is a terminology confusion here as well. TMF is defined as "the process of fatigue damage *under simultaneous changes in temperature and mechanical strain*."[1] LCF is a result of plastic deformation in each

[1] Fatigue is the weakening of a material caused by repeatedly applied loads.

repetition (i.e., one cycle) of a certain loading process where the number of cycles is relatively low (e.g., 200 starts of a combined-cycle gas turbine power plant in 1 year). The "load" in question is mechanical stress applied to a material. In other words, strictly speaking, the LCF definition does not include changes in temperature, whereas the TMF definition does not include the number of cycles. Consequently, TMF can be LCF or high-cycle fatigue (HCF). In the latter, the number of cycles is relatively high (e.g., 3,600 revolutions of a turbine or compressor rotor blade). This is why LCF is usually used as a "blanket term" for all TMF phenomena caused by the cyclic operation rhythm of a gas turbine power plant. HCF (e.g., vibration of individual blades or the shaft) and *foreign object damage* (FOD) are failure mechanisms associated with any type of operation.

Risk of failure considerations, especially for the new models of super-heavy, advanced-class gas turbines, are extremely important. Consider that a 300-MWe heavy-duty industrial gas turbine can cost about $65 million with its accessories and field services. The price of a combined-cycle power plant based on this gas turbine is upward of $300 million, not including myriad owner's costs. There are very few entities who can make half a billion dollars of investment out of their own coffers. A good chunk of the requisite money is borrowed from banks. This immediately brings up the question of "insurability." A bank or consortium of lenders funding a new power plant project want to ensure that their money is safe in case the new, "super-duper" gas turbine, which makes the project so attractive in terms of projected cash flows, goes up in a ball of fire and smoke only days after it starts commercial operation.

One might ask: How could that happen? In a study drawing upon 150 operational insurance claims since 2005, one insurer found that losses attributed to the following three events accounted for 95 percent of those 150 losses (totaling more than $1.8 billion in settled losses and reserves for losses that remain ongoing) [6]:

- Weather events (12 percent)
- Fire (8 percent)
- Machine breakdown (76 percent)

In terms of dollar value of losses, machine breakdown accounted for more than 50 percent of the total number or *more than 1 billion dollars*. In terms of specific failure mechanisms, machine breakdown was broken into the following categories (in terms of instances):

- Compressor failure (4 percent)
- Compressor blade failure (7 percent)
- Turbine failure (10 percent)
- Turbine blade failure (26 percent)
- Bearing failure (6 percent)
- Generator failure (21 percent)
- Transformer failure (24 percent)
- Boiler failure (2 percent)

Even though the study does not specify the type of "turbine" (i.e., gas or steam), it is hoped that the reader, based on what has been covered in the book so far, is not surprised at all by this picture. It is no wonder, then, that insurers pay very close attention to gas turbines for risk management.

In order to zoom into the gas turbine, let us look at the data summarized below for the top 20 events contributing to unplanned outage time in 2005–2009 [5].

- For E-class gas turbines, 13 out of 20 were in the generator and one in the main step-up transformer (it burned). Generator problems included a grounded/shorted circuit in rotor and stator windings, vibration, burned wiring, etc. The remaining few causes included broken compressor first-stage stator vanes, installation error in turbine first-stage ring supports, generic control and protection errors, and rotor vibration. There were no combustor events in the top 20.
- For F–class gas turbines, 4 out of 20 events were in the main step-up transformer and one in the generator (grounded circuit in field coils). The others included cracked first-stage turbine disk/wheel, loose combustor fuel nozzles, cracked "zeroth"-stage compressor rotor blades,[2] cracked turbine first-stage rotor blades, and vibration.

Crack propagation leading to blade "liberation" (i.e., breaking off its platform and destroying everything else in its path downstream) is probably the worst thing that can happen to a gas turbine. The most dramatic example of such a catastrophic event is the "corn-cobbing"[3] of the compressor when one of the front-end blades or vanes is liberated. Inspection of blades, especially at the root–platform connection and "dove-tail" parts, which are prone to stress corrosion cracking, using non-destructive testing technology such as eddy current technology during outages is crucial for prevention of this costly and ugly event.

Risk of failure has two facets: likelihood of failure and consequences of failure. The latter includes the following:

- Cost of lost generation (i.e., lost revenue)
- Cost of repairs and parts replacement
- Loss of life or limbs
- Loss of property
- Damage to the environment

Consequently, safe operation practices in gas turbine power plants are extremely important, and the salient aspects thereof are covered in industry standards. In particular, the reader is referred to the International Standard ISO-21789, "Gas Turbine Applications – Safety," first published in 2009. This standard covers the safety requirements and anticipated hazards associated with all types of gas turbine applications (simple- or combined-cycle power generation, pipeline compression, etc.). It also specifies the appropriate preventive measures and processes for reduction

[2] A zero or zeroth stage is the stage added to the front of a compressor (i.e., upstream of the first stage) in an uprate to increase mass flow rate and compression ratio.

[3] The term refers to the fact that the rotor looks like an eaten ear of corn (i.e., a corncob).

Table 16.7 Risk evaluation criteria

	Units	Moderate	Catastrophic
Average Property Damage	MM$/year	Base	25 × base
Average Forced Outage Duration	days/year	39	270
Likelihood	event/unit/year	0.0034	0.00023

or elimination of these hazards. Its references include many ISO and IEC[4] standards on the subject, as well as US NFPA[5] codes. Annex A of ISO-21789 contains a very detailed list of significant hazards (FOD, falling or ejection of objects, contact of persons with "live" parts, etc.). By far the largest part of the list is dedicated to fire or explosion hazards. This is not surprising because fire is one of the major and most common risks. Indeed, the presence of an adequate fire protection system is imperative for risk reduction.

Two papers by Orme and Venturi provide excellent overviews of the risk assessment process in gas turbine power plants [8,9]. The process is based on the following two indices:

- Probable maximum loss (PML)
- Maximum foreseeable loss (MFL)

Both indices are evaluated on an individual equipment (e.g., gas turbine) or entire facility (e.g., combined-cycle power plant) basis. Calculation of PML includes the estimated replacement value of the equipment in question (including installation), damage resulting from the loss event, expediting expenses, and lost revenue. MFL is calculated in a similar fashion, but also considers a catastrophic event that damages not only the particular equipment, but also key balance of plant, neighboring units, and buildings. A typical classification of moderate- and catastrophic-severity events for risk assessment is provided in Table 16.7 [7].

Based on the risk evaluation, a particular facility is rated as follows [8]:

- Excellent: significantly reduced loss potential
- Good: average loss potential
- Fair: somewhat increased loss potential
- Poor: significantly increased loss potential

In order to be rated excellent or good, the facility in question must have taken measures per industry standards and best practices. Such standards and practices prescribe, first and foremost, that the following systems are in place:

- Fire protection system (e.g., automatic water spray system for transformers)
- Fire alarm system (e.g., central alarm panel in the control room)
- Comprehensive maintenance system (e.g., LTSA with the OEM)

[4] International Electrotechnical Commission.

[5] The National Fire Protection Association is a US trade association, but also has some international members.

16.3 Maintenance Considerations

16.3.1 Maintenance Intervals

OEMs account for the different "wear-and-tear" mechanisms resulting from long runs and frequent start–stop cycles in different ways. One method is the aforementioned EOH concept, which translates each cycle into an equivalent number of steady operation hours. Major inspection intervals are based on the total number of EOH. Another method is to base the major inspection intervals on independent counts of starts and hours. In this case, whichever criteria limit is reached first determines the maintenance interval. This is graphically illustrated in Figure 16.1.

As an example, consider that the OEM not using the EOH method has an inspection interval of 24,000 hours or 1,200 starts, whichever is reached first. Let us assume the following two scenarios:

- Scenario 1: annual 4,000 hours and 300 starts (i.e., cyclic or midrange)
- Scenario 2: annual 8,000 hours and 160 starts (i.e., base load)

In the first case, the major inspection would take place after 4 years, determined by the number of starts (i.e., 4 × 300 = 1,200). In the second case, the major inspection would take place after 3 years, determined by the number of hours (i.e., 3 × 8,000 = 24,000).

An OEM using 24,000 EOH as its major inspection interval would have major inspection much sooner, depending on the EOH/start assumption, i.e.,

Scenario 1

Starts/year	300	300	300
Hours/year	4,000	4,000	4,000
EOH/Start	15	20	25
Total EOH	8,500	10,000	11,500
Years	2.8	2.4	2.1

Scenario 2

Starts/year	160	160	160
Hours/year	8,000	8,000	8,000
EOH/Start	15	20	25
Total EOH	10,400	11,200	12,000
Years	2.3	2.1	2.0

Note that the "start" in question in the foregoing discussion is an "equivalent start," which is a combination of the following:

- Normal starts while operating with natural gas fuel;
- Normal starts while operating with backup fuel (usually number 2 fuel oil);

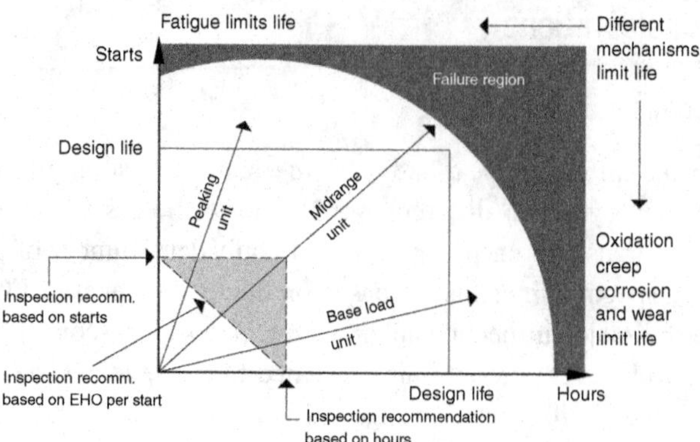

Figure 16.1 Determination of gas maintenance requirements.

- Starts after load rejections;
- Starts after trips;
- Number of rapid load changes.

The exact formula to transfer those events into a total number of equivalent starts varies from OEM to OEM. For example, starting the gas turbine after a trip at full load can be equivalent to as much as 10 normal starts. Restarting after load rejection can be equivalent to five normal starts. Equivalent starts are referred to as "factored starts" by GE, which also defines "factored fired hours" (FFH), which is similar to the EOH concept. The ratio of FFH to actual hours (AH) of operation is the "maintenance factor," which is a number greater than 1.0 because FFH > AH.

There is not much point in debating the advantages and disadvantages of different methods of calculating equipment life expenditure. If you are the owner and operator of a gas turbine manufactured by OEM X, especially one of the state-of-the-art, advanced-class machines, your best bet is to go with the major inspection intervals recommended by the OEM. Typically, the control system that comes with the gas turbine has a counter in it that keeps track of the operation characteristics of the machine and lets the operator know when it is time to schedule an inspection.

There are three types of gas turbine maintenance inspection:

1. Combustion inspection (Type A)
2. Hot gas path inspection (Type B)
3. Major inspection (Type C)

In a major inspection, all internal rotating and stationary components from the inlet of the gas turbine through to its exhaust are examined. For a well-rounded discussion of these inspections and intervals, the reader is referred to GE document GER-3620N (issued in 2017), "Heavy-Duty Gas Turbine Operating and Maintenance Considerations," which is available online for free download.[6] The revision indicator "N" is a

[6] The authors of version "N" are J. Eggart, C.E. Thompson, J. Sasser, and M. Merine. Many of the earlier versions, some of which are also widely available online, were authored by R. Hoeft.

testament to the value and industry acceptance of this 60-page document, which is by far the best source of information on gas turbine maintenance practices. It is expected that readers of these lines in the coming years will have to download revisions "O," "P," and so on.

The goal of all OEMs is to extend the inspection intervals as much as possible in order to offer the customers the most economical solution. In general, these intervals are based on the particular gas turbine technology (or class). Typical numbers are as follows:

- 8,000–12,000 EOH or FFH for Dry-Low-NOx combustors
- 24,000 EOH or FFH for the hot gas path
- 48,000 or FFH for major inspection

A major industry initiative is to eliminate the almost-annual combustor inspection (in terms of actual hours for base loaded units) and combine it with the hot gas path inspection (i.e., roughly every 2 years in AH). The goal is to increase the AF by about 1.5 percent. The means to achieve this goal are provided by the increased adoption of digital technologies by the OEMs to exploit the available sensor data via data analytics. One example of this is "model-based control," which uses a rigorous model of the gas turbine, which is embedded in the distributed control system (DCS), to control the machine instead of the conventional "control curve" method. (Such models are frequently referred to as "digital twins" by marketing people.) Such models enable combination of controls with online, real-time diagnostics such that operators can intervene before an event to fix a pending problem. In other words, a shift from "forensic investigations" to "wellness checks" (e.g., via online "blade health" monitoring) is instituted into the gas turbine operation.

New technologies such as Highway Addressable Remote Transducer (HART) and FOUNDATION™ Fieldbus also contribute to that shift via the following:

- Fully digital, serial, two-way, and multidrop communication;
- Remote access to status and diagnostic information from smart field devices;
- A method to communicate to field instruments over Fieldbus;
- Bus power and intrinsic safety features;
- Faster verification and validation of control loop configurations.

16.3.2 Maintenance Providers

Owner/operators of gas turbines have the following three options available for taking good care of their equipment for a long and trouble-free operational life with maximum revenue:

1. In-house engineering and maintenance staff
2. Third-party service providers
3. OEMs

Once again, one can use the example of automobiles. If you own a top-of-the-line 2017 model passenger car, you take it to the dealership for maintenance whenever the onboard computer sends you a signal that it is time to do so. If you own a, say, >10–year-old model of a widely available brand, in all likelihood, you take it to your neighborhood gas station or favorite mechanic for periodic maintenance and most repairs and to a local franchise of a national chain of specialized service providers (e.g., transmission replacement). Finally, if your car is one of those 1960s or 1970s relics with a simple carbureted engine, you take care of most of the routine maintenance in your own garage (possibly with the help of your neighbor who is handier in such tasks).

This is indeed very much analogous to the situation with industrial gas turbines. Utilities, refineries, and similar organizations who are still operating old Frame 3 or Frame 5 gas turbines have their in-house staff take care of their maintenance. Replacement parts are widely available from third-party suppliers. Older E- and F-class machines representing mature technologies are also amenable to the "do-it-yourself" type of regular maintenance. The operator of a lone power plant (e.g., an independent power producer) may not have the means to employ the requisite personnel to do this. In that case, an *independent service provider* (ISP) is hired to take care of Type A, B, or C inspections and major overhauls. Major OEMs also provide similar services with differing scopes (parts only, operation and maintenance, etc.). Advanced-class gas turbines are almost always covered by OEM-provided LTSAs (it may be a requirement by the insurer). A very good discussion of the advantages and disadvantages of different options can be found in Dagy and Savic [9].

LTSAs are very expensive. As a rough rule of thumb, the price of an LTSA contract is equal to the gas turbine price for base load operation (about 8,000 EOH or FFH and 50 factored starts per year). For a 300-MWe advanced gas turbine, this is nearly \$70–80 million. For cyclic duty (about 4,000 EOH/FFH and 250 factored starts per year), the LTSA price can be up to 50 percent higher. The duration of the LTSA varies from contract to contract. It can be as short as 11–12 years or as long as 20 years.

In terms of mils/kWh-year for a J–class gas turbine, the LTSA price is about twice as much as an F–class gas turbine. The annual payment schedule is also contract-specific with a hefty up-front payment (20–30 percent of the total LTSA price) for capital spare parts and mobilization. Equally divided over the contract duration, the annual LTSA cost is about 5 percent of the gas turbine price.

Consequently, LTSA scope, duration, parts repair and replacement clauses, bonus and penalty, indemnification allowances, and inclusion of ancillary services are extremely important considerations when negotiating the contract with the OEM. Typically, an OEM's *contribution margin* (CM) in the LTSA price is much higher than its CM in the gas turbine price (e.g., perhaps three times as high). (Note: CM is roughly equal to "profit"; i.e., price = cost/[1 – CM].) In addition, the OEM has much more information in its possession to determine reliability and availability risks. Therefore, while undertaking an LTSA is almost unavoidable when procuring an advanced gas turbine with the latest technology, for the older variants, the ISP option should be seriously evaluated. In any event, utmost attention to detail is vital. Even the most (seemingly) mundane items such as storage location of the spares and ownership of the replaced parts must be examined and negotiated.

16.4 Performance Deterioration

Gas turbines are large, air-breathing engines (e.g., 1,500 lb/s or more for most heavy-duty industrial units). In a 24-hour period of operation at full load, such a machine would ingest (and expel, of course)

$$1{,}500 \times 3{,}600 \times 24/2{,}000 = 64{,}800 \text{ st of air,}$$

which is roughly the weight of an M1 Abrams main battle tank. Air ingested by the gas turbine is not pure. Ambient air contains myriad contaminants including solids, liquids, and gases. Typical contaminant concentrations are as follows:

- Farmland and coastal areas, 0.01–0.1 ppm by weight
- Industrial zones, 0.1–10 ppm by weight
- Desert, 0.1–700 ppm by weight

The exact amount of airborne particulate matter (PM) at a specific site can only be determined by detailed air sampling. For general information, the US Environmental Protection Agency (EPA) provides average ambient PM trends by region, county, and city in the USA. Similar data are likely to be found online for other regions in the world as published by their respective government and/or scientific agencies.

If the ambient air contained, say, a total of 10 ppm of PM of any size and type, this would correspond to about 600 kg of solid matter passing through the gas turbine internals (i.e., compressor, combustor, and turbine). Let us pause and consider this for a bit: *3 metric tons* of solids passing through the machine at speeds reaching several hundreds of meters per second (e.g., see Figure 3.1 in Chapter 3) in only a week's worth of operation (5 days). It is clear that, without efficient air filters, a gas turbine could not last a very long time in the field.

Even with the most advanced, self-cleaning air filters, compressor fouling is a major maintenance problem. Fouling is caused by the adherence of PM, typically smaller than 10 μm in diameter (i.e., PM10), to compressor stator vanes, rotor blades, and the inner surface of the casing. Examples of such particles include smoke, oil mists, carbon, and sea salts. In addition, the impact of hard particles larger than 10 μm in diameter on the components can and will lead to erosion of the metal and coatings. These particles include water droplets, which enter the compressor via inlet fogging. (Until recently, for example, GE did not allow the use of inlet foggers with their gas turbines. Operators ignoring this forfeited their warranty.)

Each gas turbine is affected differently by fouling. This is explored in great detail in a paper by Meher-Homji et al. [10]. If the air filter does a good job and prevents ingestion of PM larger than 10 μm, erosion of compressor blades can be precluded. (In fact, air filters are in general very effective for particle diameters of about 5 μm and above.) The other problem is corrosion of the blades, leading to surface "pitting" and thus to increased surface roughness and increased susceptibility to fatigue crack initiation. Even without corrosion, PM sticking on the blades will increase surface roughness, change the blade contour, and thus decrease compressor airflow and efficiency. As a rule of thumb, fouling to the extent of a 1 percent reduction in airflow will

- Decrease output by 2–3 percent;
- Increase heat rate by about 1 percent.

How much time is needed for the airflow to decrease by 1 percent is dependent on many factors, including the severity of the air pollution in the surroundings of the installation and the quality of the air filter, as well as the maintenance practices of the operators. As a rough rule of thumb, at sites with high PM10 concentrations, filters with a high dust-holding capacity are recommended to avoid frequent change-outs. At sites with high PM2.5 (i.e., 2.5 μm diameter or less) concentrations, high-efficiency second-stage filters are necessary to minimize compressor fouling. In other words, the first stage removes the large, erosion-causing particles and reduces the loading on the second stage to extend its life. The second stage removes the remaining large particles and a majority of the small particles that contribute to compressor fouling.

The minimum efficiency reporting value (MERV) rating is a measurement scale designed in 1987 by the American Society of Heating, Refrigerating and Air-Conditioning Engineers (ASHRAE) to rate the effectiveness of air filters. In Europe, EN 1822, "High Efficiency Air Filters (EPA, HEPA and ULPA)," is the standard that is applied to high- and very-high-efficiency air filters with ultra-low penetration (EPA, HEPA, and ULPA) used in the field of ventilation and air-conditioning, as well as in technological processes such as clean-room technology or the pharmaceutical industry. EN 1822 establishes a procedure to determine efficiency based on a method that counts particles with a test liquid aerosol (or alternatively a solid one) and can classify these filters in a normalized way depending on their efficiency. The resulting scale provides the minimum efficiency at the smallest particle size – most penetrating particle size (MPPS) – which is typically about 0.2–0.4 μm for most filters. A filter selection chart devised by EPRI is given in Table 16.8. The majority of filters with a MERV rating of 16, which is the highest ASHRAE rating, or a EN 1822 rating of E10–E12 filter out more than 99 percent of all particles larger than 1 μm.

Progression of compressor fouling can be easily detected by monitoring trends of compressor pressure ratio and efficiency (as indicated by compressor discharge pressure and temperature sensors). Almost all modern gas turbines come with *online* compressor wash systems. Operators can activate the online wash system when the trends indicate appreciable compressor fouling. The arresting effect of periodic online washes (i.e., spraying water into the compressor inlet while the gas turbine is running) on fouling is well established. What is not so clear-cut is whether using a "detergent" (i.e., a chemical agent) helps or not (e.g., see Gülen et al. [12]). The answer depends strongly on the type of fouling agent and it must be evaluated on a case-by-case basis. Even with periodic online washes, which are typically more effective on the front stages of the compressor, after a certain amount of time, a thorough *offline* wash with detergents is requisite to remove the significant buildup on compressor blades in all stages.

There is an interesting aspect to compressor fouling. Starting from the definition of gas turbine efficiency:

$$\eta = \frac{\dot{W}_{GT}}{\dot{m}_f \cdot LHV}.$$

Table 16.8 Gas turbine inlet air filter selection chart [11]

Best Use	ASHRAE Filter Class MERV	ASHRAE 52.2: 2007 Collection Efficiency (%) by Particle Size (μm) E1 0.3–1.0	E2 1.0–3.0	E3 3.0–10.0	EN Filter Class	EN 1822: 2009 Total Filtration Separation Efficiency (%)
Pre-Filter	1			<20	G1	
	2			<20	G2	
	3			<20		
	4			<20		
	5			20–35	G3	
	6			35–50		
	7			50–70	G4	
	8			>70		
Pre-Filter (Very Dirty Environment)	9		<50	>85	M6	
	10		50–65	>85		
	11		65–80	>85	M7	
	12		>80	>90		
Final Filter with Regular Washing	13	<75	>90	>90	F7	
	14	75–85	>90	>90	F8	
	15	85–95	>90	>90	F9	<85
Final Filter without Washing	16	>95	>95	>95	E10	85
					E11	95
					E12	99.5

With some algebra, one can show that

$$\frac{\Delta \dot{m}_f}{\dot{m}_{f,0}} = \frac{1+\dfrac{\Delta \dot{W}_{GT}}{\dot{W}_{GT,0}}}{1+\dfrac{\Delta \eta}{\eta_0}} - 1, \tag{16.1}$$

where subscript "0" denotes the new and clean state of the gas turbine and Δ denotes a change in the particular quantity. Compressor degradation results in a decrease in power output and efficiency. Consequently, both the denominator and the numerator of the first term on the right-hand side of Equation 16.1 are always less than unity. However, since the reduction in power output, *in relative terms*, is always greater than the reduction in efficiency, the fraction itself is less than unity. Therefore, the right-hand side of Equation 16.1 is less than zero. In other words, *fuel mass flow rate at full load (i.e., inlet guide vanes [IGVs] fully open and firing temperature according to the control curve) decreases with increasing compressor fouling*.

This conclusion can also be reached qualitatively starting from the fact that, since a fouled compressor has a lower efficiency and higher discharge temperature, a *smaller amount of fuel* is required to reach a set firing temperature.

What does that mean? Is compressor fouling beneficial after all? The answer, although somewhat counterintuitive, is that "it depends." In monetary terms, the total cost of gas turbine performance degradation due to compressor fouling over a time period of Δt is

$$C_{tot} = \left(c_f \cdot \Delta \dot{m}_f \cdot LHV - e \cdot \Delta \dot{W}_{GT}\right) \cdot \Delta t, \tag{16.2}$$

where c_f is the cost of fuel (i.e., what the plant operator pays for the fuel he or she burns) and e is the rate at which the plant operator is paid for the electric power he or she generates. According to Equation 16.2, the total cost can be negative (i.e., one could actually even save money due to compressor fouling). This would happen in the possible – albeit unlikely – event that

$$\frac{c_f}{e} > \frac{\Delta \dot{W}_{GT}}{\Delta \dot{m}_f \cdot LHV}. \tag{16.3}$$

In order to give a numerical example, consider a gas turbine with 200 MWe output and 39 percent efficiency (i.e., 8,750 Btu/kWh). When compressor fouling leads to a pressure ratio decrease of 2 percent, gas turbine output will drop by 5 percent, whereas the heat rate will increase by 2 percent (e.g., see figure 32 in GER-3620L). This implies a 3.1 percent decrease in fuel mass flow rate (i.e., about 54 MMBtu/h in lower heating value [LHV]). Thus, if c_f in \$/MMBtu (in higher heating value [HHV] with HHV/LHV = 1.109 for natural gas) is greater than about 200 times e in \$/kWh, there will be a saving due to this degradation.

In other words, if the wholesale power sale price is 2 cents per kWh (i.e., \$20 per MWh), the owner will save money because of compressor fouling if the fuel purchase price is higher than \$4.09 per MMBtu (in HHV). The same will hold true if the power sale price is less than \$14.7 per MWh and if the fuel purchase price is 3 \$/MMBtu (in HHV).

Note that these statements hold only for a gas turbine operating at full load and only if the operator does not get penalized by not providing 200 MWe at full load. The same amount of kilowatts as in a new and clean engine can always be produced at the expense of extra fuel consumption in a degraded engine running at part load. In that case, $\Delta \dot{W}_{GT}$ in Equation 16.1 is zero, such that

$$\frac{\Delta \dot{m}_f}{\dot{m}_{f,0}} = \frac{1}{1 + \dfrac{\Delta \eta}{\eta_0}} - 1. \tag{16.4}$$

Since $\Delta\eta/\eta$ is *always negative and less than unity* in magnitude, $\Delta \dot{m}_f$ *is always greater than zero.*

Compressor online and offline wash procedures, water quality and chemicals to be used, and gas turbine operation set-point (for online washes) and/or cooldown period (for offline washes) are prescribed in minute detail by the OEMs. Two terms that frequently appear within the context of compressor wash are *recoverable* and *unrecoverable* degradation. The distinction is usually applied to the recovery of the performance-degrading effects of compressor fouling via online and offline washes

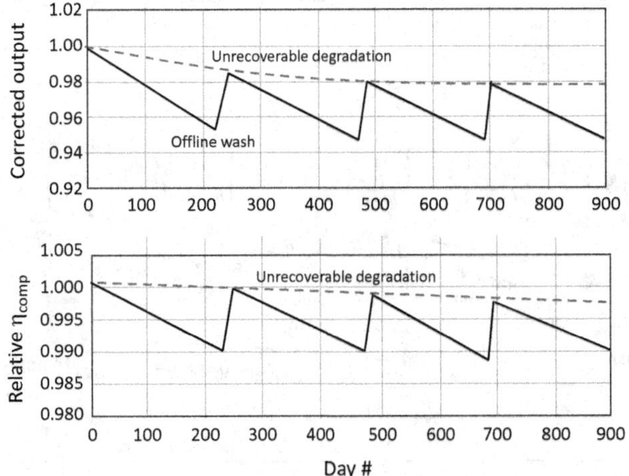

Figure 16.2 Gas turbine and compressor performance recovery via offline washes.

between two major overhauls. This is illustrated by the trends in Figure 16.2 (they are "cartoonized" versions of actual field data).

The data in Figure 16.2 cover a period of about 2.5 years with three offline washes. It clearly shows the significant performance recovery after each offline wash. Nevertheless, at the end of the period, on day 900, the gas turbine is *not* at the same condition as it was on day 0. Its output, corrected to ISO conditions, is 2 percent lower, whereas the compressor efficiency is lower by 0.3 percent (i.e., if it were, say, 90 percent [isentropic] on day 0, it would be 89.7 percent on day 900). Obviously, it is not possible to gauge from the trends in Figure 16.2 what the other contributing factors are to the observed gas turbine output loss. Going by performance derivatives, only about 0.5 percent output loss can be attributed to the observed compressor efficiency degradation.

In any event, in a major overhaul, most of the "unrecoverable" degradation seen in Figure 16.2 will be recovered via repairs, parts replacement, retuning, etc. The bottom line, however, is that as time goes on, no matter how well maintained it is, a gas turbine will never regain its "new and clean" state on the day of its commissioning. In other words, some *truly unrecoverable* degradation is inevitable. On a conceptual basis, the situation is as depicted in Figure 16.3. The bottom dash-dot curve is the truly unrecoverable "wear and tear" (e.g., casing deformation). The dash-dot-dot curve corresponds to what is labeled as "unrecoverable" in Figure 16.2. It really identifies degradation that is not recoverable via offline washes (stator nozzle vanes, rotor blades, and seals, which get damaged via erosion, abrasion, corrosion, spallation, etc., but can be returned to "as new" condition via repair and/or replacement in a major overhaul). Depending on the duty cycle of the power plant in question, the ambient environment, and other factors, a plant may do an offline wash from one to four times annually. Frequent online washing (typically every day when in operation) restores the performance to the dashed line. (In order to make the figure readable, only selected washes are shown.) The exact value of the unrecoverable degradation is an OEM-specific number. Typically, after 25,000 EOH (which can be less than that in real "clock" hours as discussed earlier),

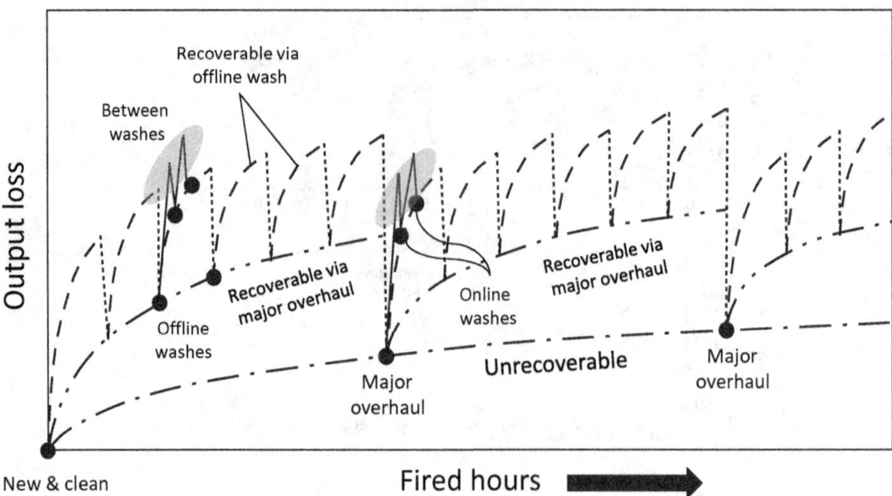

Figure 16.3 Recoverable and unrecoverable gas turbine degradation.

- Output loss can be as high as 5 percent;
- Heat rate increase can be as high as 3 percent.

16.5 Condition Monitoring

Modern gas turbines operating in the field have quite advanced instrumentation, data transmission, and acquisition systems, which are typically built into the controller (e.g., GE's Mark or Siemens' T3000 series turbine controllers). Past and present data are available to the operators for instantaneous observation and trending through the human–machine interface (HMI)[7] in the control room. In addition to the systems, which are provided as part of the OEM's controller, there are also third-party software providers that provide data storage, retrieval, and plotting/analysis platforms (e.g., OSIsoft's PI Historian).

The "condition" of a gas turbine operating in the field is characterized by the following information distilled from the data displayed in real time on HMI monitors and/or from the data stored in the plant historian:

- Thermal performance
  - Generator output, fuel flow rate, efficiency (calculated)
- Mechanical integrity
  - Bearing and rotor vibration, bearing lube oil temperatures
- Operational statistics
  - Cumulative fired hours, number of starts, number of trips, etc.
- Alarms

[7] HMI, also known as MMI (man–machine interface), refers to the user interface of the power plant DCS, which provides a graphics-based visualization of the system components and controls.

According to the American National Standards Institute (ANSI)/International Society of Automation (ISA) standard ISA 18.2, "Alarm Management Standard," an alarm is "an audible and/or visible means of indicating to the operator an equipment malfunction, process deviation, or abnormal condition requiring a response." At any given time during normal operation, the DCS of a gas turbine power plant generates a large number of alarm signals, messages, etc. (which are typically logged by the system historian). A great majority of those alarms are not important and thus "shelved" (i.e., manually suppressed, or simply ignored by the operators [as many as 95 percent per one estimate]). In industry jargon, they are known as "false positives." The goal of an effective alarm system design is to minimize the number of such "nuisance alarms" and generate relevant (and actionable) warning signals to the plant operators without overwhelming them. Recognition of this fact by power and processing industries led to the development of ISA 18.2 standard, which was released in 2009. Combined with data analytics software built into the DCS, the number of nuisance alarms has been reduced significantly in modern power systems.

A typical gas turbine DCS alarm is similar to the engine light coming up on your car's dashboard while driving. Typically, 9 out of 10 times, it is related to the gas tank cap being loose. Whatever it is, you know that it is not something that will require you to pull over immediately and call repair services. However, prudence requires you to take the car to the dealership as soon as possible to prevent a severe engine malfunction that could leave you stranded in the middle of the night in a desolate area. The same principles govern gas turbine operation and maintenance. DCS alarms should be more informative than a simple engine icon turning yellow or red. The operators should be well trained in the system they are overseeing, with detailed work processes in place (as prescribed in ISA 18.2 and OEM manuals). Thus, while the power plant does not have to be shut down immediately while running at a time of peak wholesale prices, parameters in question should be closely monitored with requisite maintenance scheduled at the earliest possible time.

Key data to be monitored for a gas turbine operating in the field are the following:

- Inlet and exhaust pressure losses
- Compressor inlet (bellmouth) temperature
- Compressor discharge pressure and temperature
- Exhaust temperature
- Power output at generator terminals
- Fuel flow rate and temperature
- Diluent water or steam flow rate, pressure, and temperature (if present)
- Rotor and bearing vibrations
- Lube oil temperatures

For steam-cooled gas turbines, cooling steam mass flow rates, pressures, and temperatures can be added to this list. For gas turbines with cooling-air coolers, key process data pertaining to the particular heat exchanger are also measured and recorded. Mechanical (friction) and casing heat losses and any overboard leaks are estimated based on original OEM documents (heat and mass balances). Shaft speed is typically not a concern – after all, it is either 3,000 or 3,600 rpm for machines connected to the grid (except for a few cases – see Section 3.4). Alas, there are places in the world where underfrequency operation (say, 48.5 Hz in a supposedly 50-Hz grid) is a "normal" occurrence. In such cases, keeping an eye on the shaft speed is not a bad idea.

Fuel heating value is another key parameter. In theory, it can be calculated online through the data obtained from a gas chromatograph. Usually, however, periodic samples are taken and sent to a certified laboratory for LHV and HHV determination.

Although it is the most important cycle parameter, turbine inlet temperature (TIT) is not amenable to reliable and continuous measurement in the field. (It is quite difficult for a thermocouple to survive in a nearly 3,000°F environment.) Even if a sturdy and accurate transducer were to be developed, it would be extremely risky for implementation because, in the event of a loss, broken pieces could severely damage the expensive hot gas path components downstream. For characterization tests of a new gas turbine in the field or in a factory test bed (i.e., for a *limited duration* when data are absolutely needed and monetary expenditure is a secondary consideration), there are techniques such as *pyrometry* (i.e., using radiation emitted by the hot components) and *thermal history paints* (chemical compounds whose crystal structures irreversibly change at high temperatures).

Exhaust temperature measurement is rather complicated as well. In this case, the problem is not the magnitude of the temperature to be measured, but its distribution across a very large flow cross-sectional area with significant turbulence and flow/temperature stratification. An exact measurement is only possible by circumferentially arranged "rakes" with mounted thermocouples covering the length of the rakes from the outside wall to the inside wall. Average flow temperature is found by mass-averaging the readings from the individual thermocouples (depending on the size of the gas turbine, well above 200 might be necessary, distributed on 20 or more rakes).

Once again, this is not a big issue for critically important "experimental" or guarantee performance runs. Maintenance of such a large number of rakes and thermocouples over a long period of field operation under a wide operating envelope with sometimes adverse conditions is, however, a problem. Typically, such temporary rakes and thermocouples are removed after the guarantee performance test. The gas turbine control system typically uses the average of permanent thermocouples with about 75 percent immersion on "production rakes" (there is roughly one per combustor can). In the control system, there is a parameter that translates said average to a mass-averaged value across the entire flow area based on the factory or field "characterization test."

The airflow through the gas turbine is probably the most critical parameter for assessing the thermal condition of the gas turbine (since direct measurement of TIT is impractical). Alas, due to the large bell-shaped compressor inlet cross-sectional area (i.e., the "inlet bellmouth"), direct measurement of inlet airflow is not possible, neither on the factory test bed nor in the field. A reliable measurement can only be made by measuring inlet temperature, total and static pressures, and using the principles of compressible fluid flow (Figure 16.4).

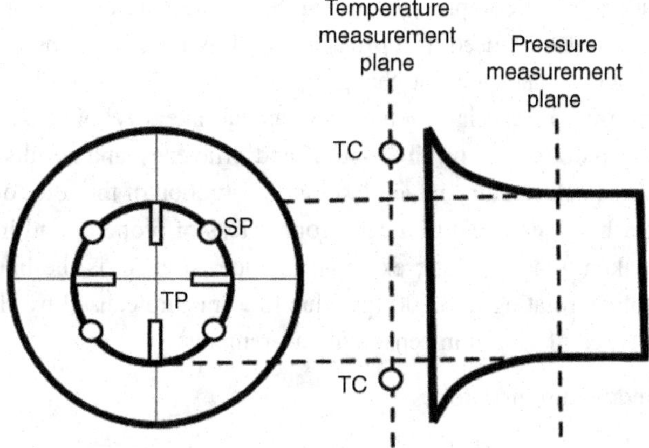

Figure 16.4 Calibrated bellmouth flow measurement instrumentation. TP = total pressure; SP = static pressure; TC = thermocouple.

The formula for airflow calculation is

$$\dot{m}_{air} = C_D \cdot P_t A \sqrt{\frac{\gamma}{RT_t}} Ma \left(1 + \frac{\gamma - 1}{2} Ma^2\right)^{\frac{\gamma+1}{2(1-\gamma)}} \text{ with} \quad (16.5)$$

$$Ma = \left(\frac{2}{\gamma - 1}\left(\left(\frac{P_t}{P_s}\right)^{\frac{\gamma-1}{\gamma}} - 1\right)\right)^{\frac{1}{2}}. \quad (16.6)$$

The "discharge coefficient," C_D, in Equation 16.5 is the calibration constant, which relates the "ideal" airflow from measured P_t, P_s, and T_t to the "actual" airflow. Flow cross-sectional area, A, is evaluated at the pressure measurement plane. The obvious question is: How is the "actual" airflow known? If it is somehow measured, then what is the purpose of all this mathematical rigmarole? In a factory, when the gas turbine is tested at full-speed, no load or, in the most recent facilities adopted by all major OEMs, at full-speed, full-load, the air duct at the exit of the inlet air filter is divided into several "channels," which are small enough to allow flow measurement via ASME-prescribed methods (e.g., nozzle and venturis described in detail ASME PTC 19.5-2004 "Flow Measurement").

In the field, airflow can be "estimated" using P_t, P_s, and T_t from production instrumentation by using Equation 16.5 and the factory value of C_D from the calibrated bellmouth tests. The uncertainty is typically high (i.e., ±2.5 percent or more), but it is useful for trend evaluation (e.g., observing a downward trend) after correcting to a reference ambient set (denoted by the subscript "0") using

$$\dot{m}_{air,corr} = \dot{m}_{air} \frac{P_{amb,0}}{P_{amb}} \sqrt{\frac{T_{amb}}{T_{amb,0}}},$$

which might be indicative of compressor fouling. Static pressure depression at the inlet of the compressor can also be used as a *proxy* for airflow for those time periods when IGVs and variable stator vanes are not changing.

Vibration monitoring is crucial to the mechanical integrity of the gas turbine. Compressor fouling reduces gas turbine output and efficiency and results in revenue loss, but the machine itself remains intact. Excessive vibration of the rotor or individual blades, on the other hand, can lead to a catastrophic loss of property and life if proper measures are not taken. One dramatic example of such an event is the liberation of a turbine last-stage blade rotating at 3,600 rpm due to a fracture caused by HCF.

There are three types of vibration sensors/measurements:

- Proximity (eddy-current) probes
- Velocimeters
- Accelerometers

The last two are "seismic" vibration transducers (i.e., engineering counterparts of putting your hand on the machine to feel it shaking). They are typically mounted on the bearing housing (or the machine casing in aeroderivative gas turbines), either temporarily or permanently. In any event, they are relatively easy to install and remove. Proximity probes are permanently installed transducers that are difficult to install and/or remove (basically, a hole in the bearing housing wall is required). Note that proximity probes are only applicable to fluid-film bearings. Aeroderivative gas turbines typically have rolling element or "ball" bearings, which are monitored with velocimeters or accelerometers mounted on the machine casing.

In order to understand what each sensor measures and its significance, consider a harmonic vibration described by a *sine wave* as

$$x = A \sin(\omega t),$$

where x is the *displacement* with respect to a reference point (x = 0), A is the amplitude of the vibration (i.e., the maximum and minimum displacement described by the sine wave), and ω is the angular speed given by $\omega = 2\pi f$, with f being the frequency of the sine wave describing the vibration. The first derivative of displacement gives the *velocity*, v, i.e.,

$$v = \omega A \cos(\omega t),$$

whereas the second derivative of the displacement gives the *acceleration*, *a*, i.e.,

$$a = -\omega^2 A \sin(\omega t).$$

Proximity probes measure peak-to-peak displacement (i.e., 2A) in mils (one-thousandth of an inch) or micrometers. Velocimeters measure the peak or *root mean square* (rms) value of the velocity (i.e., ωA or $0.707\omega A$, respectively), in in/s (or inches per second [ips]) or mm/s. Accelerometers measure the peak or rms value of the acceleration (i.e., $|\omega^2 A|$ or $|0.707\omega^2 A|$, respectively), in g's (gravitational constant, $g = 9.81$ m/s^2).

Detailed vibration analysis is quite complicated and requires special expertise. The crucial aspect of vibration monitoring is ensuring that moving/rotating parts do not

Table 16.9 Typical gas turbine vibrations

	$X_{pk\text{-}pk}$ (mils)	V_{pk} (ips)
Normal	0.5–3.0	0.1–0.3
Alarm (HI)	3.0–5.0	0.5–0.7
Trip (HI-HI)	6.0–10.0	0.8–1.0

contact stationary parts. For a bearing, this means that the rotating shaft (i.e., the gas turbine rotor) does not contact the bearing journal. In mathematical terms, then, peak-to-peak displacement measured by the proximity probes should be less than the diametral clearance of the bearing by a sufficiently large safety margin. For example, for a bearing with 10 mils of diametral clearance, one should be concerned if the proximity probe reading is 6 mils or larger. Whether a particular vibration reading is acceptable or unacceptable (i.e., cause for a trip or shutdown) is determined by OEM specifications, bearing clearance, and industry standards (e.g., API 616, "Gas Turbines for the Petroleum, Chemical and Gas Industry Services"). For example, for shaft vibration measured with proximity probes, API 616 provides an allowable or normal value of $(12{,}000/N)^{0.5}$ mils, where N is the maximum continuous speed in rpm. Thus, for 3,600 rpm, the allowable or normal peak-to-peak displacement is 1.8 mils. In terms of peak velocity, typically, 0.01–0.02 ips is considered good, whereas 0.32–0.64 ips is considered "rough." In the case of multi-shaft gas turbines such as the aeroderivative gas turbines with balanced "gas generator" (GG) shafts, GG speed (higher than the load turbine or grid-imposed speed) and gas turbine load must be considered as well.

Representative heavy-duty industrial gas turbine vibration levels during normal operation are listed in Table 16.9 [15]. During shutdown, when the unit coasts down from the running speed (e.g., 3,000 or 3,600 rpm) to the turning gear speed, the shaft goes through its critical speed, which leads to a sudden rise in vibration due to resonance. (The same goes for startup when the unit accelerates from the turning gear speed to the synchronous speed.) The spike in vibration can be large enough to cause a rub between the shaft and the journal (the babbitted surface of the bearing pads). Usually, there is a multiplier for the peak-to-peak displacement to indicate the allowable limit during startup and shutdown (typically 2–3 for large gas turbines). For a comprehensive discussion of gas turbine vibration monitoring, the reader is referred to [15].

The primary purpose of condition monitoring is to detect significant changes in sensor readings in order to identify imminent or encroaching faults in the monitored piece of equipment and take corrective actions. Some of those actions are automatically taken by the DCS (e.g., a trip caused by a "HI-HI" reading from a bearing vibration sensor for a certain duration). Almost all critical measurements that might lead to drastic action by the DCS are typically "triple-redundant" with two-out-of-three voting. Other sensor data might require validation, reconciliation, and correction (see Section 16.6) before becoming useful for analysis. A corollary of condition monitoring is *predictive forecasting*. For example, corrected airflow data recorded over a period of, say, several hundred hours of operation are indicators of the rate of compressor fouling and, via extrapolation, can be used to schedule the next offline compressor washing. Unexplained step changes in any key parameter are more often than not flags raised for serious underlying problems.

A comprehensive analysis of the gas turbine sensor data can be used to institute a "condition-based" maintenance schedule in lieu of the predetermined schedules described earlier. This requires specialty software with advanced data analysis capabilities built into the power plant DCS. Combination of the aforementioned three functionalities is referred to as "asset performance management" (APM). The Wikipedia definition of APM is "[combined] capabilities of data capture, integration, visualization and analytics tied together for the explicit purpose of improving the reliability and availability of physical assets." Due to the rapid progress in "information technology" in the first two decades of the twenty-first century, APM has become a very important part of gas turbine products offered by all major OEMs. However, it is hoped that the foregoing discussion made it clear that operator training and a thorough understanding of the machine in question are the first steps before investing in multimillion-dollar software.

16.6 Data Analysis

In gas turbine performance monitoring, there are three critical steps:

- Sensor validation
- Data reconciliation
- Correction to standard conditions

Sensor validation has the following two objectives:

- Making sure that the sensor (transducer) is working (i.e., it is not reading "–999");
- Making sure that the sensor is not failing and/or out of calibration (e.g., compressor discharge temperature unrealistically low or high).

The former is easy enough to check and verify (but, surprisingly, is quite often neglected). The latter requires (1) the existence of an accurate system model running online and comparing sensor data with expected values and (2) utilization of sophisticated statistical analysis tools and software. It is easy to recognize that a compressor discharge temperature reading of 1,300°F is due to a working but seriously "drifting" and/or malfunctioning thermocouple. However, it requires a significant amount of analysis to recognize a failing thermocouple reading, say, 780°F, whereas the "true" value should be closer to, say, 820°F. There is a large body of literature going back to the early 1970s, especially for aircraft engine "health monitoring," which can be found in archival journals and textbooks.[8]

[8] In particular, look for papers by Urban and Volponi.

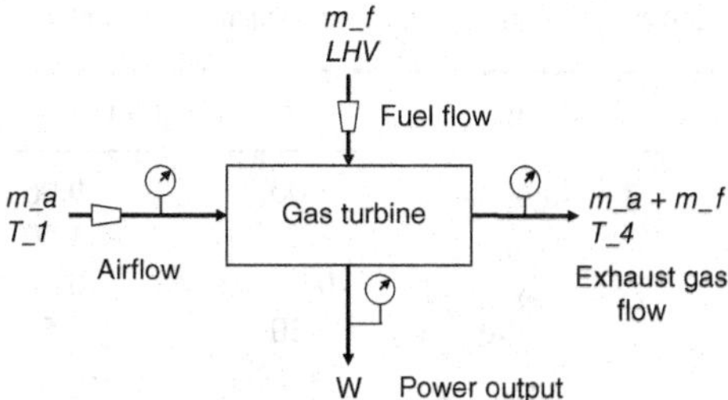

Figure 16.5 Simple gas turbine control volume for the data reconciliation example.

Data reconciliation refers to the adjustment of raw measurements from the data acquisition system using heat and mass conservation principles. This results in

- Reduced measurement uncertainty;
- A thermodynamically consistent set of data.

The mathematical principle underlying data reconciliation and its application to a single-shaft gas turbine combined cycle is described by Gülen and Smith [12]. Its application in an online monitoring system is described in Gülen et al. [13]. The origin of the underlying idea can be traced back to the concept of the "Kalman filter." In very simple terms, when there are, say, two separate measurements of the same physical quantity, one very accurate and the other less so, the best information is *not* obtained from the sole use of the more accurate one (i.e., discarding the other, less accurate measurement).[9] The best and most reliable information is obtained from a *weighted average* of the two measurements, where the weighting factors are determined from their relative uncertainties. In other words, overall uncertainty is minimized by using *all available information*. In a gas turbine framework, application of heat and mass balances to the gas turbine control volume increases the available information for a particular parameter – say, airflow – by other measurements going into those balance equations (which are *not* present in Equation 16.5).

Consider a very simple gas turbine instrumented as shown in Figure 16.5. Small losses and minor energy/mass streams crossing the gas turbine control volume are ignored for clarity. Data obtained during a performance test are summarized in Table 16.10. Substituting the data into the first law (steady-state, steady-flow [SSSF]) equation, it is found that the heat balance error is –1,493 Btu/s (–1,575 kW). In other words, measured power output is 1.6 MW *less* than that indicated by a perfect heat and mass balance.

Closure of the heat balance requires allocation of a portion of the heat balance error to a particular measurement. This allocation is determined by a "weight," which is a function of the "sensitivity" of the heat balance error to the measurement in question and the measurement uncertainty associated with it. In mathematical terms, using the formulae in Gülen and Smith [12], one has the following system to solve:

[9] The accuracy in question is quantified by the measurement uncertainty, which, in problems of interest to gas turbine data analysis, is defined by ASME PTC 19.1-2005 "Test Uncertainty." (See Section 16.7 for more on this subject.)

Table 16.10 Measurements and their uncertainties for the data reconciliation example

i		Measured	Units		$U_M(\%)$
1	m_a	1,498.0	lb/s	2.00	29.96
2	T_1	59.5	R		1.00
3	m_f	28.70	lb/s	0.50	0.14
4	LHV	21,513	Btu/lb	0.30	64.54
5	T_4	1,157.0	R		10.00
6	W	147,500	Btu/s	0.25	368.75

$$w_i \varepsilon = S_i \left(x_i - x_i^*\right), \tag{16.7}$$

$$w_i = \frac{\left(S_i U_{M,i}\right)^2}{\sum_{i=1}^{n} \left(S_i U_{M,i}\right)^2}, \tag{16.8a}$$

$$\sum_{i=1}^{n} w_i = 1, \tag{16.8b}$$

$$\varepsilon = f(x_i), \tag{16.9}$$

$$x_i^* = x_i - \frac{w_i \varepsilon}{S_i}, \tag{16.10}$$

where $\varepsilon = f(x_i)$ is the SSSF energy balance for the gas turbine control volume, with ε representing the heat balance error (0 in theory and for an ideal system). $S_i = \partial f(x_i)/\partial x_i$ is the sensitivity and w_i is the weight described above, with x_i representing the "raw" value of measurement i (there are n total measurements; n = 6 in this example) and x_i^* is the "reconciled" value of it. In other words,

$$\varepsilon = f\left(x_i^*\right) = 0. \tag{16.11}$$

Sensitivities can be calculated using a simple relationship, $dh/dT = c_p$, with c_p values of 0.28 and 0.26 Btu/lb-R for exhaust gas and air, respectively. (This is sufficiently accurate for demonstration purposes. In a "real" application, a bona fide equation of state can be used to evaluate derivatives numerically.) The results are listed below:

Sensitivities

S_1	dε/dm_a	–308.49
S_2	dε/dT_1	389.48
S_3	dε/dm_f	21,189
S_4	dε/dT_4	–427.48
S_5	dε/dW	–1.0
S_6	dε/dLHV	28.7

Table 16.11 Reconciled measurements and their uncertainties

i		Measured	Units	S	w	Reconciled	U_R
1	m_a	1,498.0	lb/s	–308.49	73.2%	1,494.5	15.50
2	T_1	59.5	R	389.48	0.1%	59.5	1.00
3	m_f	28.70	lb/s	21189	7.9%	28.71	0.14
4	LHV	21,513	Btu/lb	28.7	2.9%	21,515	63.58
5	T_4	1,157.0	R	–427.48	15.7%	1,156.5	9.18
6	W	147,500	Btu/s	–1	0.1%	147,498	368.54

Calculated values of reconciled measurements from Equations 16.7–16.10 are listed in Table 16.11. Substitution into Equation 16.11 shows that the heat balance error is reduced to 0.5 Btu/s (i.e., essentially to zero). The reconciled value of the airflow is 1,494.5 lb/s with an uncertainty of ±15.5 lb/s (about 1 percent).

16.7 Uncertainty

Performance analysis, performance monitoring, and data analysis require a good grasp of the basic concepts of statistics, especially error propagation. This is a good place for a short introduction to the subject. The discussion below is not a scholarly treatment; this would be well beyond the scope of the book and the abilities of the author. The objective is to instill in the reader an appreciation of the *probabilistic* nature of key parameters such as output and heat rate, which are covered in an exclusively *deterministic* manner in this book and elsewhere, once they make the transition from the designer's desk to operating in the field.

As the famous proverb goes, the proof of the pudding is in the eating. The "pudding" in question herein is the performance of a gas turbine power plant, in simple- or combined-cycle configuration, as quantified by generator output and thermal efficiency. The proverbial "eating of the pudding" is measurement of said performance either in a carefully conducted performance test (with properly calibrated and highly accurate – thus, quite expensive – test instrumentation) or during normal operation (with standard or "station" instrumentation).

For the sake of simplicity, let us ignore the "net" versus "gross" dichotomy and focus on the directly measured quantities, i.e.,

- Generator output measured at the generator low-voltage terminals using a watt-meter (e.g., refer to section 4.6.3 of ASME test code PTC-46 "Overall Plant Performance");
- Fuel flow measured by an orifice or turbine-type flow meter (e.g., refer to section 4.4.3 of ASME test code PTC-46 "Overall Plant Performance");
- Heating value of the fuel gas either by online chromatography or from laboratory samples, in accordance with ASTM D1945 ("Standard Test Method for Analysis of Natural Gas by Gas Chromatography").

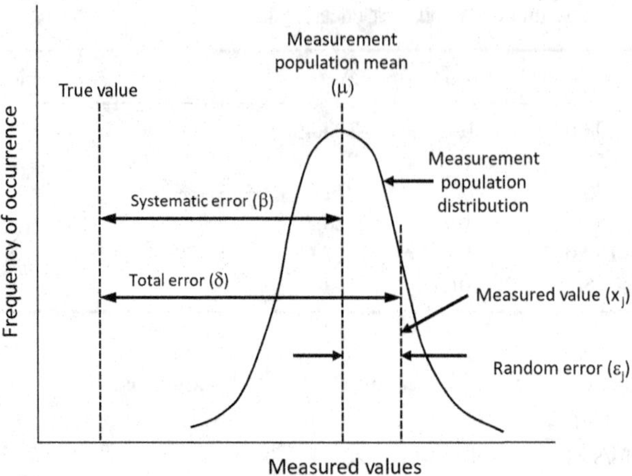

Figure 16.6 Definition of measurement errors (per ASME PTC 19.1-2005).

Each measurement is an "estimate" of the "true" value of the parameter in question. In the context of the discussion herein, the true value is what the designer (say, the OEM) calculates using in-house engineering models and expects (or perhaps hopes) to measure in the field. Alas, once the product (say, the gas turbine) is manufactured, installed, and commissioned in the field, what is measured will deviate from the true value by an amount equal to the sum of *systematic* and *random* errors (see Figure 16.6).

Random errors are caused by difficult-to-control boundary conditions and non-repeatabilities in the measurement activity (including human error). Systematic errors are caused by imperfect calibration, faulty instruments, and/or data reduction methods/tools. Here is the vicious circle one finds oneself in: no matter how rigorous, accurate, proven, etc., one's design tools, manufacturing facilities, and installation tools, procedures, and personnel might be, once the end product is in the field and chugging along, the one and only way to prove that the true (or as designed or expected or hoped for) value of parameter A is *exactly* X_T is to measure it. Since, however, measurements are subject to the aforementioned errors, there is no way to know the true value X_T *without error*. Consequently, the *true* value of A is *unknown* and will remain so (i.e., unknown *before* and unknown *after* the measurement). In other words, one is *uncertain* whether the *measured* value of A, X_M, is indeed X_T or not.

What *is* known with *some certainty*, however, is that the true value X_T lies *somewhere in between* $X_M - U_{XM}$ and $X_M + U_{XM}$, where $|U_{XM}|$ is the total uncertainty of the measurement X_M at a defined level of confidence (see Figure 16.7).

According to the ASME test code PTC 19.1-2005, X_M is the measurement *sample* mean and $|U_{XM}|$ is the *expanded* uncertainty with a 95 percent confidence level, which is given by multiplying the *combined* standard deviation of the sample mean (i.e., including uncertainties caused by systematic *and* random errors) by 2. Note that the *sample* mean is only an estimate of the *population* mean. The dichotomy results from the difficulty of obtaining the average of a parameter for the *entire* population. Therefore, one considers a representative sample and its mean/average as an *estimate* of the

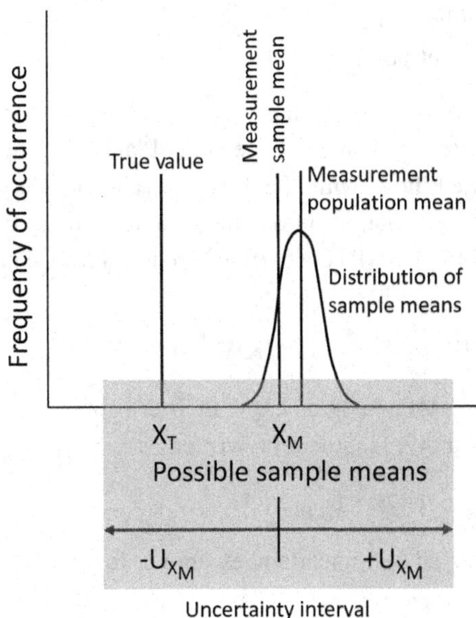

Figure 16.7 Definition of the uncertainty interval (per ASME PTC 19.1-2005).

mean/average of the population. The higher the sample size, the better the estimate. In a performance test, the sample size can be increased via the following approaches:

- Increasing sampling frequency (e.g., every 10 seconds instead of every minute);
- Longer test duration (e.g., 1 hour instead of 30 minutes);
- Multiple consecutive tests (e.g., three tests instead of one).

Thus, for example, a single test of 30 minutes' duration with a sampling frequency of 1 minute will result in 30 data points. Three tests of 1 hour's duration, each with a 30-second sampling frequency, will result in 3 × (3,600/30) = 360 data points. Obviously, there is a limited time window and budget to conduct a performance test so that a compromise is reached at the point of diminishing returns.

A key objective in any performance test is to *minimize* the magnitude of measurement uncertainty, $|U_{XM}|$, which can be accomplished by minimizing systematic and random errors. How to go about it is prescribed in minute detail in performance test codes, which also identify specific uncertainty limits that should be met for each test measurement. For example, according to the ASME PTC 22-2005 ("Gas Turbines"), table 4-1,

- Maximum allowable alternating current (ac) power uncertainty (root sum square [rss] of voltage/current transformers and meters) is ±0.25 percent.
- Maximum allowable fuel gas flow uncertainty with a turbine-meter is ±0.5 percent.

Similarly, according to ASME PTC 46-1996 ("Overall Plant Performance"), table 1.1, for combined-cycle gas turbine power plants with or without supplemental firing, the largest expected test uncertainty is

- ±1.5 percent for *corrected* heat rate;
- ±1.0 percent for *corrected* net power output.

According to PTC 46-1996 (section 4.4.3), gaseous fuel mass flow must be determined with a total uncertainty of no greater than ±0.8 percent. Plant heat input is found by multiplying the measured fuel flow with the heating value of the fuel. For a code-compliant performance test, variation in heating value should be less than ±1.0 percent for the duration of the test (PTC 46-1996, section 4.5.2). Since heat rate (HR) is given by

$$\mathrm{HR} = \mathrm{MF} \times \mathrm{LHV}/\mathrm{KW}$$

(HR in Btu/kWh with fuel flow, MF, in lb/h, LHV in Btu/lb, and output, KW, in kilowatts), *normalized* HR uncertainty, U_{HR}, can be written as

$$U_{HR}^2 = U_{MF}^2 + U_{LHV}^2 + U_{KW}^2.$$

Therefore, using the largest expected test uncertainties above, for the HR, via the rss method, one obtains

$$U_{HR} = \left(U_{MF}^2 + U_{LHV}^2 + U_{KW}^2\right)^{1/2} = \left(0.8^2 + 1.0^2 + 1.0^2\right)^{1/2} = 1.625$$

(i.e., largest expected heat rate uncertainty is ±1.625 percent, which is essentially the upper limit specified in PTC 46-1996). For conceptual study purposes, the best achievable combined-cycle performance test uncertainties on a *corrected* basis can be estimated as

- About ±0.4 percent for output;
- About ±0.6 percent for HR.

What is the point of this *mini-exercise* in basic statistics and error propagation? There are several key points to be made here:

- Whatever one calculates for gas turbine performance, no matter how accurate the methods, the tools, the input data, etc., employed in that endeavor are, one should always be cognizant of the uncertainty involved when the time comes for presenting the proverbial "proof of the pudding."
- This is especially true when different technologies are compared using economic metrics such as cost of electricity, net present value, etc. (see Section 20.5 for more on this).

As an example, consider that you improve the gas turbine performance by making a change to a design feature, which, according to your design tools, results in an output increase of a few hundred kilowatts. If this design tweak costs pretty much nothing in terms of manufacturing and/or installation materials and labor, you should not think twice about implementing it; by all means, go ahead and do it. (For example, think about reducing the chargeable cooling flow slightly by adding "swirl" to the cooling air entering the receiver holes of the turbine wheels and reducing its *total [stagnation]* temperature.)

Table 16.12 Sample design improvement roll-up

Design Feature	Impact (kWe)	Uncertainty (kWe)
1	100	20
2	200	40
3	500	100
4	250	50
5	225	45
6	350	70
7	400	80
8	175	35
9	275	55
10	125	25
Total	2,600	26.4

If, however, the design tweak in question requires significant retooling, etc., and adds to the price of the end product, you must pause and reconsider. If the best measurement resolution (i.e., about ±0.25 percent) is inadequate to prove the proposed improvement unequivocally, it is probably a good idea to either drop it or work on it further to increase its impact (or, of course, to improve the measurement accuracy). If, for instance, the gas turbine in question is a 300-MWe machine, the measurement uncertainty bracket is unlikely to be "tighter" than {−750, +750} in kWe. If you are proposing a design change to increase the machine rating by, say, +500 kWe, it will be quite difficult to declare victory. This is especially true if you consider that the projected +500 kWe is *on paper* and carries its own uncertainty (e.g., see the discussion in Section 20.5).

In reality, though, what happens is this: when a new version of an existing "frame" is planned for introduction to the market, *many* such design tweaks, not just one or two, are piled into the end product. (Typically, each one is assigned a "realization factor" to account for calculation, manufacturing, and operational errors and tolerances.) For a combined-cycle power plant, the list is longer by the design improvements and/or changes applied to the bottoming cycle (steam turbine, HRSG, etc.). Thus, as part of a long list of design improvements, the particular design improvement discussed above, which is worth +500 kWe, becomes very important because it adds to a "bottom line" number, which is much higher than the measurement uncertainty. In addition, even if each individual item is subject to a relatively high uncertainty (e.g., ±20 percent or higher), the final number they roll up to is much more robust. As an example, consider a list of 10 new product design features, each with ±20 percent uncertainty. As shown in Table 16.12, the final improved rating is higher than the base rating by a total of 2,600 kWe with an rss uncertainty of only ±26.4 kWe (i.e., ±10 percent). Thus, going with the example above, quoting the new rating as 2,600 – 25 – 750 = 1,825 kWe higher (an overall realization factor of 0.7; i.e., as 301.8 MWe) would be a very prudent move on a purely technical basis (one should also consider the cost impact).

16.8 Postscript

In the beginning of the chapter, the German World War II heavy battle tank Tiger was mentioned as an example of a "system" superior to its competitors in every measure of "performance," but ultimately suffered due to significant RAM issues. In order to give an idea of this, consider that a Tiger could not be rail-transported with its operational tracks (it was too wide) and therefore had two sets of tracks: one for transportation (*Verladekette*), which was narrower, and one for operations (*Marschkette*) [16]. Today, super-heavy advanced gas turbines pushing 500 MWe in 50-Hz variants are almost in a similar quandary. They will be extremely difficult to transport as a single flange-to-flange unit, which will significantly add to their total installed cost.

In its final production version, Ausführung E, a Tiger weighed close to 60 tons. It was propelled by a 690-hp (515–kW) Maybach V-12 gasoline engine for a power to weight ratio of about 12 hp/ton (cf. about 27 hp/ton for a modern, gas turbine-propelled M1 Abrams tank) [1]. Its top speed was about 25 mph with an operational range of 70–120 miles. Only 1,347 Tigers (Tiger I) were produced. The overall cost of one Tiger was more than that of four Sturmgeschütz III assault guns (these were very successful, especially in defensive operations, which was essentially the situation Germany was in after 1943). At the end, even its massive, more than 10 to 1 "kill ratio" was not enough to compensate for the numerical superiority of the Allied forces. It should be mentioned that Tiger II (also known as Königstiger, the "king" Tiger), the successor to Tiger I, was even heavier with the same Maybach engine [1]. Less than 500 Tiger II variants were built. In comparison, between 1941 and 1945, more than 60,000 T-34s were built, which was more than enough to make up for their huge losses against Tigers and then some.

The low power to weight ratio of the Tiger tanks brings up the obvious question: What if they were equipped with gas turbines as the propulsion unit? In fact, during 1943, there were considerations to equip German heavy tanks with gas turbines, which could use much lower-grade fuel than gasoline-fired Maybachs and generate more power (i.e., 1,150 hp vis-à-vis 690 hp). The gas turbine envisioned for installation on the Panther tank was GT-101 (later versions were GT-102 and GT-103), which was a derivative of the aviation gas turbine BMW 003, which was an axial compressor turbojet engine very similar to Jumo-004 (see Chapter 4). In the GT-102 variant, the original turbojet was used as a gas generator with a free or power turbine driving the drive shaft. It was somewhat different from typical aeroderivatives such as GE's LM gas turbines. Compressed air from the core engine compressor – 30 percent of the inlet airflow – was bled off through a pipe to the two-stage power turbine with its own combustion chamber.

Eventually, no gas turbine engines were deployed on a Panther tank. They would have brought its power-to-weight ratio to about 27 hp/ton, about the same as a modern M1 Abrams tank. Interestingly, Anselm Franz, who oversaw the development of Jumo-004 in wartime Germany, would lead the initial development of Textron Lycoming AGT-1500 in the 1960s, which eventually became the propulsion unit of the M1.

References

1. Forty, G., *German Tanks of World War Two in Action* (London, UK: Blanford Press, 1988).
2. Chivers, C. J., *The Gun* (New York: Simon & Schuster, 2011).
3. Della Villa, Jr., S. A., Koeneke, C., "A Historical and Current Perspective of the Availability and Reliability Performance of Heavy-Duty Gas Turbines: Benchmarks and Expectations," GT2010-23182, ASME Turbo Expo 2010, June 14–18, 2010, Glasgow, UK.
4. Steele, Jr., R. F., Paul, D. C., Rui, T., "Expectations and Recent Experience for Gas Turbine Reliability, Availability, and Maintainability (RAM)," GT2007–27655, ASME Turbo Expo 2007, May 14–17, 2007, Montreal, QC, Canada.
5. Grace, D., Christiansen, T., "Risk Based Assessment of Unplanned Outage Events and Costs for Combined Cycle Plants," GT2012-68435, ASME Turbo Expo 2012, June 11–15, 2012, Copenhagen, Denmark.
6. "Common Causes of Large Losses in Global Power Industry," A Report by Marsh Risk Management Research, Marsh & McLennan Companies, New York, 2013.
7. Orme, G. J., Venturi, M., "Property Risk Assessment for Combined Cycle Power Plants," ASME Turbo Expo 2008, June 9–13, 2008, Berlin, Germany.
8. Orme, G. J., Venturi, M., "Prediction of Power Plant Exposure to Economic Losses through A Property Risk Assessment Methodology," GT2009-59018, ASME Turbo Expo 2009, June 8–12, 2009, Orlando, FL.
9. Nagy, D., Savic, S., "Alternative Gas Turbine Maintenance Concepts by Independent Service Providers," Powergen International 2015, December 8–10, 2015, Las Vegas, NV.
10. Meher-Homji, C. B., Chaker, M., Bromley, A. W., "The Fouling of Axial Compressors – Causes, Effects, Susceptibility and Sensitivity," GT2009-59239, ASME Turbo Expo 2009, June 8–12, 2009, Orlando, FL.
11. Grace, D., Perullo, C., Lieuwen, T., How to Select the Optimal Inlet Air Filters for Your Engine, *Combined Cycle Journal*, **53** (2017).
12. Gülen, S. C., Smith, R. W., A Simple Mathematical Approach to Data Reconciliation in a Single-Shaft Combined Cycle System, *Journal of Engineering for Gas Turbines and Power*, **131** (2009), 021601.
13. Gülen, S. C., Griffin, P. R., Paolucci, S., Real-Time On-Line Performance Diagnostics of Heavy Duty Industrial Gas Turbines, *Journal of Engineering for Gas Turbines and Power*, **124** (2002), 910–921.
14. Lifson, A., Quentin, G. H., Smalley, A. J., Knauf, C. L., Assessment of Gas Turbine Vibration Monitoring, *Journal of Engineering for Gas Turbines and Power*, **111** (1989), 257–263.
15. Fletcher, D., Willey, D., Hayton, M., *Tiger Tank – Owner's Workshop Manual* (Minneapolis, MN: Zenith Press, 2011).

17 Combined Cycle

The basic thermodynamic concepts leading to the gas turbine combined cycle (GTCC) have been established in Chapter 5. They are repeated below for convenience:

- Carnot cycle operating between
  - A hot-temperature *reservoir* (theoretical proxy for maximum cycle temperature), and
  - A low-temperature reservoir (theoretical proxy for minimum cycle temperature)
- Mean effective heat addition temperature (METH)
- Mean effective heat rejection temperature (METL)

In essence, a gas–steam turbine combined-cycle plant is an attempt to "mimic" a Carnot cycle operating between two temperatures (see Figure 17.1):

- Hot reservoir temperature (i.e., gas turbine inlet temperature [TIT])
- Cold reservoir temperature (i.e., the ambient temperature)

It is practically impossible (so far) to devise a machine with isothermal heat addition at a meaningfully high temperature. Thus, a gas turbine with a given cycle/compression pressure ratio (PR) and TIT cannot do better than a "Carnot-like" cycle with a hot reservoir temperature of METH (see Figure 17.1). The deviation from the Carnot ideal on the hot end is thus quantified by the upper-left triangular area on the temperature–enthalpy (T–s) diagram in Figure 17.1.

The situation is much better for the cold end, where the deviation from the Carnot ideal is quantified by the lower-right triangle in Figure 17.1. Even a single-pressure, non-reheat steam Rankine cycle takes care of a big chunk of that problem. The key to that is constant pressure (and temperature) heat rejection via steam condensation (denoted by METL in Figure 17.1), which does a very good job of approximating the "true" cold-reservoir temperature.

The name "combined cycle" is a generic term; it can refer to *any* two or three (or more – at least in theory) thermodynamic cycles connected to each other through their heat rejection and heat addition processes. A GTCC is a specific type of combined cycle comprising

- A gas turbine Brayton cycle; and
- A steam turbine Rankine cycle.

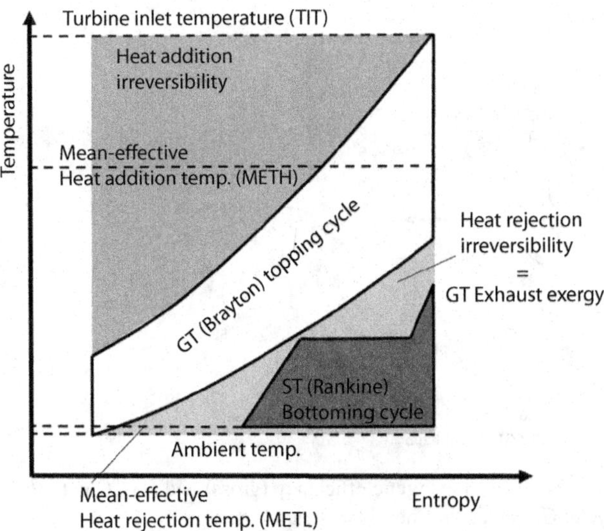

Figure 17.1 T–s diagram of a GTCC. GT = gas turbine; ST = steam turbine.

Due to their respective positions in the T–s diagram, the former is referred to as the "topping cycle"; the latter is referred to as the "bottoming cycle." Strictly speaking, the bottoming-cycle working fluid can be any pure fluid (or a *mixture* of pure fluids) other than steam (e.g., H_2O). Examples are as follows:

- Organic fluids such as pentane (organic Rankine cycle [ORC]);
- Carbon dioxide (i.e., supercritical CO_2 cycle);
- Ammonia–water mixture (the Kalina cycle).

ORCs are available for "low-grade" waste heat recovery applications. Supercritical CO_2 Rankine cycles are also viable candidates for such applications. As will become clear from the discussion below (and in Chapter 22 on closed-cycle gas turbines), neither are viable candidates for utility-scale electric power generation in combination with heavy-duty industrial gas turbines. The Kalina cycle went through a period of intensive research and development in the 1990s, but it did not present a serious challenge to the steam Rankine bottoming cycle due to myriad reasons. Therefore, the focus in this chapter in particular and in the entire book in general is specifically on the steam Rankine bottoming cycle. Without an explanatory moniker, the terms "combined cycle" and "bottoming cycle" refer to the gas–steam turbine combined cycle and steam Rankine cycle, respectively.

Quantitatively, the lower-right triangle in Figure 17.1 is *exactly* equal to the gas turbine exhaust *exergy*, which is a thermodynamic property and, as such, can be calculated *exactly* with known gas composition, pressure, temperature, and mass flow rate using an appropriate equation of state. A *perfect* bottoming cycle (i.e., a Carnot-equivalent engine in and of itself) would convert that exergy into useful work fully and have an exergetic efficiency of 100 percent.

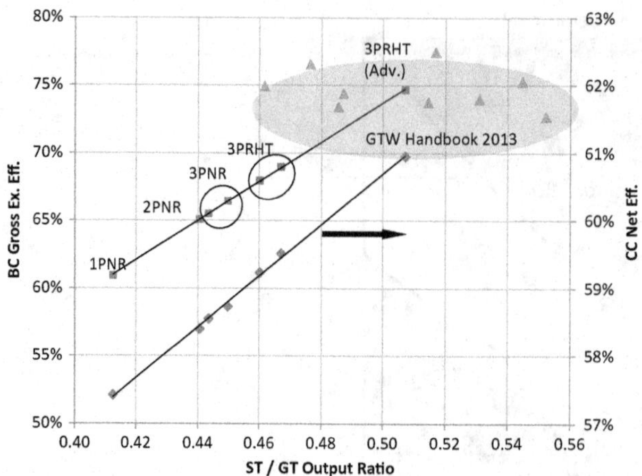

Figure 17.2 Bottoming-cycle (BC) exergetic efficiency (gross) and GTCC net efficiency. CC = combined cycle; GT = gas turbine; ST = steam turbine.

A "real" steam bottoming cycle (obviously) cannot do that – one would require an infinitely large heat recovery steam generator (HRSG; usually pronounced as "her-sig") with infinite pressure levels, an isentropic steam turbine, an infinitely large condenser, and zero losses in the balance of plant (BOP). That is the *bad* news. The *good* news is that, as illustrated in Figure 17.2, a state-of-the-art "three-pressure reheat" (3PRH) bottoming cycle does an impressive job of approaching that ideal. It is compared with the following three lower-rank bottoming cycles:

- 1PNR: one pressure (no reheat)
- 2PNR: two pressure (no reheat)
- 3PNR: three pressure (no reheat)

The triangles in Figure 17.2 are calculated using original equipment manufacturer (OEM) rating numbers for gas turbine exhaust flow and temperature and steam turbine generator (STG) output published in *Gas Turbine World* (GTW) *2013 Handbook*. Other data points are calculated using Thermoflow's GTPRO heat balance simulation tool using typical design inputs from previously published studies and information (see the paper by Gülen [1]). The gas turbine is the same for all cases (typical J class).

As quantified by the ISO base load rating numbers (admittedly a tad too optimistic for commercial reasons), gross exergetic efficiency of a 3PRH bottoming cycle (i.e., based on STG output *excluding* plant auxiliary load) is about 75 percent![1] The interesting takeaway from this figure is the diminishing returns beyond the second pressure level in the HRSG. (Sometimes a *fourth* pressure level in the HRSG is contemplated as a performance improvement; the futility of that should be obvious.) The jump in performance in the last step, from 3PRH to the "advanced" 3PRH, is achieved at maximum possible steam

[1] To convert to of net bottoming-cycle efficiency, subtract 2 percentage points. Gas turbine exhaust gas exergy is calculated for a "dead state" of 59°F and 14.7 psia, per Equation 5.60a.

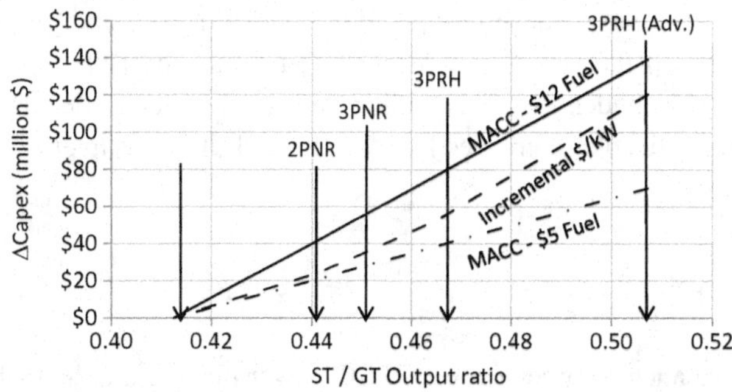

Figure 17.3 Incremental cost of improved bottoming-cycle output. GT = gas turbine; ST = steam turbine.

conditions (160 bar and 600°C high-pressure steam) with very tight pinches in the HRSG (i.e., more than 50 percent larger HRSG heat transfer surface area) and extremely low condenser pressure.

Needless to say, the impressive combined-cycle performance in Figure 17.2, which is driven by the steam bottoming-cycle performance, comes at a steep cost. (Note that each percentage point increase/decrease in bottoming-cycle exergetic efficiency translates to an approximately 0.25-percentage point increase/decrease in net combined-cycle efficiency.) This is illustrated by the chart in Figure 17.3 [1]. Using the PEACE (Plant Engineering and Cost Estimator) add-in of Thermoflow's GTPRO, the total owner's cost for each variant in Figure 17.3 is calculated. Subsequently, the *maximum acceptable increase in capital cost* (MACC) is found by equating before and after levelized cost of electricity values (see Chapter 20) for each bottoming-cycle improvement in Figure 17.3. Incremental capital cost ($/kW) of each bottoming-cycle variant with respect to the base 1PNR design is compared with the corresponding MACC value (per kW) for two different fuel prices, $5 and $12 per million Btu (higher heating value).

According to the curves in Figure 17.3, at a natural gas price of $5/MMBtu (which is about the maximum gas price one can expect in the USA any time soon) or lower, it does *not* pay to have a bottoming cycle better than 3PNR – which is roughly equivalent to a 2PRH bottoming cycle in performance! (Note that a super-heavy gas turbine of H or J class has a net combined-cycle efficiency of ~59 percent, even with a 2PNR bottoming cycle – see Figure 17.2.) This is *not* to suggest that the industry should revert back to >20-year-old technology. For one thing, in many places in the world, natural gas is still very expensive and the most advanced technology is cost-effective (e.g., see the $12/MMBtu fuel price curve, which is *above* the incremental cost curve). Moreover, cheap gas is not a given even in the USA at the time of writing; in the winter of 2014, in some places in the northeast, gas prices hit the $100/MMBtu mark due to pipeline bottlenecks (~$8/MMBtu US average). The main takeaway here is the fact that bottoming-cycle technology development is primarily a matter of cost increase (driven by size *and* materials). With today's state of the art, there is simply very little room left for improvement in a cost-effective manner.

In the limited space of this chapter, the focus will be on the calculation of combined-cycle performance from a limited amount of inputs using fundamental relationships.

There is a huge amount of design, operation, performance, and other technical information pertinent to the HRSG, the steam turbine, and the BOP, specifically the water- and air-cooled condensers and cooling towers. For those details, the reader should consult the applicable references listed at the end of the chapter [2–6].

17.1 First Law Approach

The cardinal performance parameter for a GTCC power plant is the thermal efficiency. It can be calculated fairly accurately by following the thermodynamic process. In essence,

1. A fraction of the chemical energy of the gas turbine fuel, η_{GT}, is converted to shaft work by the gas turbine.
2. The remainder, $1 - \eta_{GT}$, is supplied to the HRSG.
3. In the HRSG, a fraction of the gas turbine exhaust energy, η_{HRSG}, is used to convert water to steam.
4. A fraction of the total energy of steam generated in the HRSG, η_{ST}, is converted to shaft work by the steam turbine.
5. Shaft works of the gas turbine and steam turbine is converted into electric power by their respective synchronous alternating current (ac) generators.

This process is graphically described in a *Sankey diagram* shown in Figure 17.4.[2] It is based on a vintage F–class gas turbine (e.g., General Electric's [GE] 9FA from the late 1990s). However, it is still adequate to paint a reasonably accurate picture for the modern units.

Following the thermodynamic process described above and graphically represented by the Sankey diagram in Figure 17.4, combined-cycle shaft output can be written as

$$\eta_{CC,\,shaft} = \eta_{GT} + (1 - \eta_{GT})\eta_{HRSG}\eta_{ST}. \tag{17.1}$$

Equation 17.1 is the foundation of combined-cycle analysis. Including plant auxiliary power consumption and generator losses, gross and net thermal efficiencies for a combined cycle are given by

$$\eta_{CC,\,gross} = (\eta_{GT} + (1 - \eta_{GT})\eta_{HRSG}\eta_{ST})\eta_{Gen}, \tag{17.2a}$$

$$\eta_{CC,\,net} = \eta_{CC,\,gross}(1 - \alpha), \tag{17.2b}$$

where η_{Gen} is the generator efficiency. For most practical purposes, assuming 99 percent for η_{Gen} is sufficient; for more information on generator efficiency, see Chapter 15.

[2] Named after Irish Captain Matthew Henry Phineas Riall Sankey, who used this type of diagram in 1898 to show the energy efficiency of a steam engine, this is a diagram with arrows proportional in size to the physical quantity they are representing. Probably the most famous diagram of this type is Charles Minard's map of Napoleon's Russian campaign of 1812, which was created in 1869 (it possibly inspired Captain Sankey). It can be easily found on the Internet.

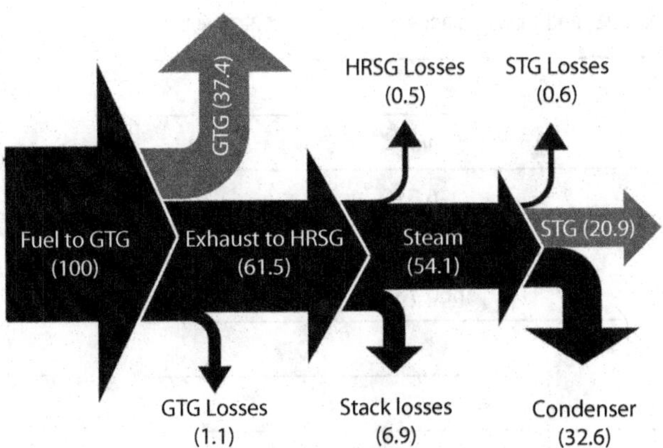

Figure 17.4 Typical energy distribution in a combined-cycle power plant [7]. GTG = gas turbine generator.

Gross power output is plant output at generator *low-voltage* terminals; it does not include step-up transformer losses. Plant auxiliary power consumption as a fraction of gross power output is given by α. For rating performance (e.g., at ISO base load as listed in the GTW Handbook), α can be assumed to be 1.6 percent.

Using Equation 17.2, combined-cycle net output is simply

$$\dot{W}_{CC,\,gross} = \eta_{CC,\,gross}HC, \tag{17.3a}$$

$$\dot{W}_{CC,\,net} = \eta_{CC,\,net}HC, \tag{17.3b}$$

where HC is the total heat consumption of the gas turbine(s). Note that, so far, the assumption is that the combined-cycle power plant in question has no supplementary or duct firing in the HRSG and plant heat input is equal to gas turbine heat consumption.

In Equation 17.1, gas turbine mechanical and heat losses as well as minor contributions due to inlet air enthalpy and sensible heat of the gas turbine fuel are ignored. A more exact representation would be

$$\eta_{CC,\,shaft} = \eta_{GT} + (1 - \eta_{GT} - \lambda_{GT})\eta_{HRSG}\eta_{ST}. \tag{17.4a}$$

Gas turbine losses and minor heat input terms are lumped into the correction term λ_{GT}, which is typically about 1 or 2 percent (depending on the zero-enthalpy reference used in calculations). A complete breakdown of λ_{GT} for an air-cooled gas turbine is given in Equation 17.5:

$$\lambda_{GT} = \lambda_{air} + \lambda_{fuel} + \lambda_{loss} + \lambda_{mecl}, \tag{17.5}$$

where

λ_{air} = energy supplied by compressor inlet air (+)
λ_{fuel} = sensible heat of heated gas turbine fuel (+)
λ_{loss} = turbine heat loss (–)
λ_{mecl} = turbine mechanical losses (–)

Table 17.1 Gas turbine heat and mass balance table. CV = control volume

	Q	
Into the CV	Btu/s	%
Air	–6,310	–0.95%
Fuel LHV at 77°F	661,451	100%
Fuel Sensible Heat	5,648	0.85%
Total In	660,788	99.90%
	Q	
Out of the CV	Btu/s	%
Exhaust Gas	393,928	59.56%
Heat Loss	3,335	0.50%
Mechanical Loss	2,744	0.41%
Total Out	400,007	60.47%
Shaft Power (In – Out)	260,781	39.43%
Heat Balance Error	0	
Generator Loss	2,976	0.45%
Generator Power	257,805	38.98%

The signs in the parentheses indicate the *direction* of the particular heat/energy term (i.e., into the cycle [+] or out of the cycle [–]), but not necessarily whether it is a positive or negative value. If this sounds like a contradictory statement, consider compressor inlet air at 59°F, which *enters* the gas turbine control volume (i.e., its direction is *into* the cycle). The enthalpy of compressor inlet air at 59°F is, however, a *negative* number if it is calculated from an equation of state with a zero-enthalpy reference of 77°F. In order to illustrate this and calculate loss terms, consider the gas turbine heat balance in Table 7.5, which is repeated in Table 17.1 for convenience in a shortened form.

Using the data in Table 17.1,

λ_{air} = –6,310/661,451 = –0.00954 or about –1 percent
λ_{fuel} = 5,648/661,451 = +0.00854 or about +0.9 percent
λ_{loss} = 3,335/661,451 = 0.005 or 0.5 percent
λ_{mecl} = 2,744/661,451 = 0.0041 or 0.4 percent

Thus, from Equation 17.5,

$$\lambda_{GT} = -1\% + 0.9\% + 0.5\% + 0.4\% = +0.8\%.$$

If the heat balance were done with enthalpies with a 59°F zero-enthalpy reference, λ_{air} would be zero, which would result in λ_{GT} = 1.8 percent from Equation 17.5. The exhaust gas energy as a fraction of gas turbine heat consumption can be written as

$$\theta_{exh} = 1 - \eta_{GT} - \lambda_{GT}, \tag{17.6}$$

so that

$$\eta_{CC,shaft} = \eta_{GT} + \theta_{exh} \cdot \eta_{HRSG} \cdot \eta_{ST}. \tag{17.4b}$$

HRSG effectiveness is the ratio of the actual heat transfer from the gas turbine exhaust gas (at the transition duct inlet) to the boiler feed water in the HRSG to the heat energy available in the gas turbine exhaust gas, i.e.,

$$\eta_{HRSG} = 1 - \frac{h_{stck}}{h_{exh}}. \tag{17.7a}$$

Strictly speaking, the definition in Equation 17.7a is

$$\eta_{HRSG} = \frac{h_{exh} - h_{stck}}{h_{exh} - h(T_{ref})}. \tag{17.7b}$$

In some cases, the zero-enthalpy reference can be the same as the ambient temperature used in the calculations (e.g., 59°F). This may lead to the erroneous belief that the HRSG effectiveness is *always* referenced to the ambient temperature. This, however, is not the case. Reference selection for the definition of HRSG effectiveness is quite arbitrary. In theory, without violating the second law, the minimum achievable stack gas temperature is about the same as the feed water inlet temperature,[3] which would be only a few degrees above the steam turbine condenser temperature. (This, of course, would require an astronomically large HRSG.) Indeed, in some references, HRSG effectiveness is so defined, i.e.,

$$\eta'_{HRSG} = \frac{h_{exh} - h_{stck}}{h_{exh} - h(T_{fw})}. \tag{17.8}$$

While there is nothing wrong per se with the definition in Equation 17.8, it is inconvenient for use in, say, Equation 17.4 because gas turbine enthalpies are typically calculated with 59°F or 77°F as T_{ref}. Which temperature is used has an impact on the HRSG effectiveness defined in Equation 17.7. For example, for a gas turbine exhaust temperature of 1,150°F and an HRSG stack temperature of 190°F, HRSG effectiveness is (using Equations 5.61 and 5.62 for calculating enthalpies)

$$\begin{aligned}
\eta_{HRSG} &= 1 - (0.2443 \times 190 - 13.571)/(0.3003 \times 1{,}150 - 55.576) \\
&= 0.8867 \text{ with } T_{ref} = 59^{\circ}F \\
\eta_{HRSG} &= 1 - (0.2443 \times 190 - 17.892)/(0.3003 \times 1{,}150 - 59.897) \\
&= 0.9001 \text{ with } T_{ref} = 77^{\circ}F
\end{aligned}$$

Thus, the difference in HRSG effectiveness due to the selected zero-enthalpy reference is 0.0134, or 1.34 percentage points. For back-of-the-envelope estimates, specifying 0.90 for HRSG effectiveness is adequate for a 3PRH bottoming cycle with advanced F- or H/J-class gas turbines. The preferred convention in this chapter is to assume that T_{ref} = 77°F (since the gas turbine fuel gas LHV is readily available at that reference temperature). Note that, to a very good approximation, Equation 17.7a can be used with temperature values (without evaluating enthalpies), i.e.,

[3] To be precise, feed water temperature at the inlet of the low-pressure economizer section.

$$\eta_{HRSG} \propto 1 - \frac{T_{stck}}{T_{exh}}.$$

The proportionality factor is 1.4915 for T_{ref} = 77°F.

Note that *not* all of the heat extracted from the exhaust gas contributes to the steam turbine heat input. There are several loss mechanisms, e.g.,

- Transition duct heat loss, λ_{duct}, which is typically worth two degrees (~0.12% of HC).
- HRSG heat loss defined by the heat utilization factor (HUF), λ_{HUFt}, which is 99.5 percent for modern designs (~0.25 percent of HC).
- Steam pipe heat losses, λ_{pipe} (~0.05 percent of HC).
- Heat utilized to heat natural gas fuel in a shell-and-tube heat exchanger using intermediate-pressure (IP) feed water bleed from an IP economizer outlet, λ_{FHTR} (~0.8 percent of HC for an F-class gas turbine).

There is additional energy input to the HRSG control volume in the form of feed pump work, w_{pump}, which is about 0.3 percent of gas turbine heat consumption. Thus, the total miscellaneous losses enumerated above can be lumped into a single, non-dimensional steam cycle loss term (as a fraction of gas turbine heat consumption), i.e.,

$$\theta_{misc} = \lambda_{duct} + \lambda_{HUF} + \lambda_{pipe} + \lambda_{FHTR} + w_{pump}. \tag{17.9}$$

Substituting Equation 17.9 into Equation 17.4b, we obtain

$$\eta_{CC,shaft} = \eta_{GT} + (\theta_{exh} \cdot \eta_{HRSG} - \theta_{misc}) \cdot \eta_{ST}, \tag{17.10}$$

and combining with Equation 17.2a, combined-cycle gross output becomes

$$\eta_{CC,gross} = (\eta_{GT} + (\theta_{exh} \cdot \eta_{HRSG} - \theta_{misc}) \cdot \eta_{ST}) \cdot \eta_{Gen}. \tag{17.11}$$

The term in parentheses multiplied by η_{ST} in Equation 17.11 is the steam turbine heat input as a fraction of combined-cycle heat input, which is gas turbine HC in the absence of duct firing (i.e., $\theta = \frac{\dot{Q}}{HC}$). Consequently,

$$\theta_{ST} = \theta_{exh} \cdot \eta_{HRSG} - \theta_{misc}. \tag{17.12a}$$

We can now define another bottoming-cycle loss term as follows:

$$\lambda_{BC} = \frac{\theta_{misc}}{\eta_{HRSG} \cdot \theta_{exh}}, \tag{17.13}$$

so that steam turbine heat input becomes

$$\theta_{ST} = \theta_{exh} \cdot \eta_{HRSG} \cdot (1 - \lambda_{BC}). \tag{17.12b}$$

Similarly, Equation 17.11 can also be rewritten as

$$\eta_{CC,gross} = (\eta_{GT} + \theta_{exh} \cdot \eta_{HRSG} \cdot (1 - \lambda_{BC}) \cdot \eta_{ST}) \cdot \eta_{Gen}. \tag{17.14a}$$

Substituting Equation 17.6, Equation 17.14a becomes

$$\eta_{CC,gross} = (\eta_{GT} + (1 - \eta_{GT} - \lambda_{GT}) \cdot \eta_{HRSG} \cdot (1 - \lambda_{BC}) \cdot \eta_{ST}) \cdot \eta_{Gen}. \tag{17.14b}$$

Obviously, for GTCC net output, we have

$$\eta_{CC,net} = (\eta_{GT} + (1 - \eta_{GT} - \lambda_{GT}) \cdot \eta_{HRSG} \cdot (1 - \lambda_{BC}) \cdot \eta_{ST}) \cdot \eta_{Gen} \cdot (1 - \alpha). \quad (17.15)$$

For modern, 3PRH bottoming cycles with advanced F-, H-, and J-class gas turbines, to a very good approximation, λ_{BC} is about 2 percent. The simplicity of Equation 17.15 is inversely proportional to its predictive power. Consider that modern, heavy-duty industrial gas turbines have cycle efficiencies of about 40 percent (±1 percent). This is about the same as the bottoming-cycle steam turbine efficiency. Thus, from Equation 17.15a, the relationship between the gas turbine and combined-cycle efficiencies becomes

$$\frac{\eta_{CC,net}}{\eta_{GT}} \approx (1 + \theta_{exh} \cdot \eta_{HRSG} \cdot (1 - \lambda_{BC})) \cdot \eta_{Gen} \cdot (1 - \alpha) \quad (17.16)$$

using the following technology factors:

- $\theta_{exh} \approx 60$ percent
- $\eta_{HRSG} \approx 90$ percent
- $\lambda_{BC} \approx 2$ percent
- $\eta_{Gen} \approx 99$ percent
- $\alpha \approx 1.6$ percent

The term on the right-hand side of Equation 17.16 is

$$K = (1 + 0.6 \times 0.9 \times (1\text{-}0.02) \times 0.99 \times (1\text{-}.016) = 1.49.$$

Rating performances of selected air-cooled gas turbines of three major OEMs (from the GTW 2016 Handbook) are summarized in Table 17.2. The average value of the K factor is 1.50 with a standard deviation of 0.034. In passing, this also confirms a well-known "rule of thumb"; namely, that the steam turbine output is roughly 50 percent of the gas turbine output. In other words, for each 100 MWe of gas turbine output in a *simple* cycle, 50 MWe is added to it by adding a steam turbine cycle and making a *combined* cycle.

Using Equation 17.15 along with the data in Table 17.2 and the technology factors discussed above, one can back-calculate the steam turbine efficiency implied by the combined-cycle ratings.[4] This is illustrated in Table 17.3, which includes gas turbine exhaust temperatures. From the data in Table 17.3, steam turbine efficiency is, on average, 42.5 percent (±1.7 percent) – cf. 40.6 percent (±1.4 percent) for the gas turbines in Table 17.2.

Before saying more on steam turbine efficiency, let us look at its impact on combined-cycle efficiency. This can be easily evaluated by the derivative of η_{CC} with respect to η_{ST} from Equation 17.15a:

[4] To be precise, η_{ST} is the "steam-cycle" or "steam turbine-cycle" efficiency. It is influenced by parameters *outside* the flange-to-flange steam turbine itself (e.g., the condenser pressure or main and hot reheat steam temperatures).

Table 17.2 Simple- and combined-cycle efficiencies. CC = combined cycle; GT = gas turbine; MHPS = Mitsubishi Hitachi Power Systems

OEM	Model	GT Efficiency(%)	CC Efficiency(%)	K Factor
MHPS	M701F4	39.9	60.10	1.51
MHPS	M701F5	40.0	61.10	1.53
GE	9F-5	38.7	60.65	1.57
GE	9F-7	41.1	61.25	1.49
GE	9HA.01	42.4	62.65	1.48
GE	9HA.02	42.7	62.75	1.47
Siemens	SGT5-4000F	40.0	58.70	1.47
Siemens	SGT5-8000H	40.0	60.50	1.51

Table 17.3 Combined-cycle steam turbine efficiency. MHPS = Mitsubishi Hitachi Power Systems; ST = steam turbine

OEM	Model	Exhaust Temp. (°F)	ST Efficiency
MHPS	M701F4	1,097	41.7%
MHPS	M701F5	1,131	43.5%
GE	9F-5	1,187	44.1%
GE	9F-7	1,144	42.5%
GE	9HA.01	1,171	43.7%
GE	9HA.02	1,177	43.6%
Siemens	SGT5-4000F	1,074	38.8%
Siemens	SGT5-8000H	1,161	42.3%

$$\frac{\partial \eta_{CC,net}}{\partial \eta_{ST}} = \theta_{exh}\eta_{HRSG}(1-\lambda_{BC})\eta_{Gen}(1-\alpha). \tag{17.17}$$

Using the technology factors,

$$\frac{\partial \eta_{CC,net}}{\partial \eta_{ST}} = 0.516.$$

In other words, each percentage point in steam turbine efficiency is worth about 0.5 percentage points in combined-cycle efficiency. In relative terms, assuming $\eta_{GT} \approx \eta_{ST}$,

$$\frac{\partial \eta_{CC,net}}{\eta_{CC,net}} \cdot \frac{\eta_{ST}}{\partial \eta_{ST}} = \frac{1}{\left(\frac{1}{\theta_{exh} \cdot \eta_{HRSG} \cdot (1-\lambda_{BC})} + 1\right)}. \tag{17.18}$$

Once again, with typical technology factors,

$$\frac{\partial \eta_{CC,net}}{\eta_{CC,net}} \cdot \frac{\eta_{ST}}{\partial \eta_{ST}} = 0.346.$$

This is one of the most widely used "rules of thumb" in GTCC engineering. Each 1 percent (*not* percentage point) change in steam turbine efficiency (or heat rate) is worth about a third of a percent change in combined-cycle efficiency.

Key technologies driving the flange-to-flange steam *turbine* efficiency are as follows:

- Three-dimensional aerodynamic design of buckets and nozzles (for better isentropic efficiency);
- Advanced packing and gland seals to minimize leakages;
- Low-pressure (LP) section last-stage bucket (LSB) size to exploit low condenser pressures;
- Exhaust diffuser design.

Key cycle design parameters driving the overall steam *cycle* efficiency are as follows:

- High-pressure (HP) or *main* steam temperature;
- Hot reheat or IP admission steam temperature;
- HP steam pressure;
- Condenser pressure;
- Volumetric flow rate of steam,

High steam pressures and temperatures as well as steam generation rates (leading to high volumetric flow) are driven by the gas turbine exhaust gas energy, primarily the gas turbine exhaust gas temperature. The relationship between the steam cycle efficiency and gas turbine exhaust gas temperature is linear until the maximum steam temperature (dictated by available turbine casing, rotor and steam piping/valve materials, and cost–performance trade-off) is reached. Up until the early 2000s, the steam temperature limit was 1,050°F. Steam cycle efficiency was between 38 and 40 percent (average ~39 percent).

The data in Table 17.3 reflecting the 2016 state of the art display a jump in advanced F-, H-, and J-class steam turbine cycle efficiencies by three percentage points (on average) between 2006 and 2016 (between 42 and 44 percent for an average of ~43 percent). The main driver of that is the new standard in advanced steam cycle main (i.e., HP) and hot reheat steam temperatures, which is 1,112°F (600°C).

It was shown earlier that each percentage point in steam turbine cycle efficiency was worth 0.5 percentage points in combined-cycle efficiency. Thus, a quite impressive 1.5–percentage point rise in combined-cycle efficiency (at least as represented by OEM-supplied ISO base load ratings in trade publications) between 2006 and 2016 is directly attributable to the bottoming cycle. In order to delve into the veracity of this finding, let us first evaluate the derivative of η_{CC} with respect to η_{GT} from Equation 17.15b:

$$\frac{\partial \eta_{CC,net}}{\partial \eta_{GT}} = (1 - \eta_{HRSG} \cdot (1 - \lambda_{BC}) \cdot \eta_{ST}) \cdot \eta_{Gen} \cdot (1 - \alpha). \tag{17.19}$$

With the technology factors used earlier,

$$\frac{\partial \eta_{CC,net}}{\partial \eta_{GT}} = 0.639.$$

In other words, each percentage point in gas turbine efficiency is worth two-thirds of a percentage point in combined-cycle efficiency.

According to the historical trends displayed in Figure 4.9 (for simple-cycle gas turbines) and Figure 4.10 (for combined-cycle gas turbines), between 2006 and 2016,

- The rise in gas turbine efficiency is 1.5–2 percentage points; whereas
- The rise in combined-cycle efficiency is about 2–3 percentage points.

According to the derivative in Equation 17.19, the combined-cycle efficiency impact of 1.5–2 percentage points in gas turbine efficiency is about 1–1.3 percentage points. This leaves about 1–1.7 percentage points attributable to the improvement in the bottoming (i.e., the steam turbine) cycle efficiency, which was discussed earlier.

Knowing the exhaust and stack gas temperatures is sufficient to calculate the HRSG effectiveness given by Equation 17.7. The former is known from the gas turbine heat balance or from the rating data. Thus, the parameter that defines the HRSG effectiveness is the stack temperature. In general, for a given gas turbine exhaust gas temperature, the lower the stack gas temperature, the higher the HRSG effectiveness. Similarly, the higher the gas turbine exhaust gas temperature, the higher the HRSG effectiveness.

For typical HRSG designs with 2PRH or 3PRH, HRSG effectiveness as a function of gas turbine exhaust gas temperature can be found using the following linear correlations:

$$\begin{aligned} &\text{3PRH}: \quad \eta_{\text{HRSG}}[\%] = 61.2 + 0.025\,T_{\text{exh}}[\text{F}]\left(1,000 \text{ to } 1,200^{\circ}\text{F}\right) \\ &\text{2PRH}: \quad \eta_{\text{HRSG}}[\%] = 58.6 + 0.027\,T_{\text{exh}}[\text{F}]\left(1,000 \text{ to } 1,125^{\circ}\text{F}\right) \end{aligned}.$$

The correlations are obtained from detailed heat and mass balance simulations with the following assumptions:

- 1.2 inches of mercury (about 41 mbar) condenser pressure (i.e., the condensate reaches the HRSG at less than 90°F);
- Gas enthalpies are evaluated with a 77°F zero-enthalpy reference;
- 100 percent methane-fired gas turbine.

Deviations from these assumptions (especially the condenser pressure and liquid fuel-firing) would result in different HRSG performance. However, typical natural gas composition variation should not result in a big error. (The impact of using a different zero-enthalpy reference was discussed earlier.)

The product of HRSG effectiveness, steam turbine/cycle efficiency, and the bottoming-cycle loss term is the *overall* bottoming-cycle efficiency, i.e.,

$$\eta_{\text{BC}} = \eta_{\text{HRSG}} \cdot \eta_{\text{ST}} \cdot (1 - \lambda_{\text{BC}}). \tag{17.20}$$

Substituting into Equation 17.15, we have

$$\eta_{\text{CC,net}} = (\eta_{\text{GT}} + (1 - \eta_{\text{GT}} - \lambda_{\text{GT}}) \cdot \eta_{\text{BC}}) \cdot \eta_{\text{Gen}} \cdot (1 - \alpha). \tag{17.21}$$

To simplify the discussion, let us use the combined-cycle gross efficiency and rewrite Equation 17.21 (with Equation 17.6) as

$$\frac{\eta_{CC,gross}}{\eta_{GT}} = \left(1 + \theta_{exh} \cdot \frac{\eta_{BC}}{\eta_{GT}}\right). \tag{17.22}$$

From Equation 17.22, it is obvious that, for a given gas turbine, maximizing combined-cycle efficiency is equivalent to maximizing bottoming-cycle efficiency. Maximizing bottoming-cycle efficiency, as evidenced by Equation 17.20, requires us to strike a balance between HRSG effectiveness and steam turbine efficiency. It can be easily shown, graphically as well as numerically, that, everything else being equal, the maximum combined-cycle thermal efficiency is achieved at maximum heat recovery, at which point the bottoming-cycle exergetic efficiency is also a maximum – as it should be (but the bottoming-cycle *thermal* efficiency is not!). The reader is referred to the excellent treatment of the trade-off between stack temperature and bottoming-cycle performance by Smith in [7].

17.2 Second Law Approach

If you are not familiar with the second law concepts and terminology used in this section, please read Chapter 5 first or revisit your favorite thermodynamics textbook (as stated earlier, the recommended text by the author is Moran and Shapiro). Otherwise, the discussion below will be very difficult follow.

French sculptor Auguste Rodin (1840–1917) was once asked how he made his remarkable statues. His reply was, "I choose a block of marble and chop off whatever I don't need." Whether the story is apocryphal or not, the idea is a perfect analogy to the second law perspective into the analysis of a thermal power plant.

A thermal power plant, in its bare essence, is a heat engine. A heat engine receives heat from an energy source (a high-temperature reservoir) and rejects a portion of it to an energy sink (a low-temperature reservoir). The balance is converted into useful shaft work. The analogy between the sculpting process and the heat engine operation is simple:

- The potential (theoretically possible maximum) work provided by the energy source is the block of marble. This is referred to as *exergy*.
- The heat engine is the sculptor with his or her chisel and hammer.
- The actual work that the heat engine can produce is the sculpture hidden in the marble block.
- The marble chips that the sculptor had to remove are analogous to that part of exergy that cannot be converted to work in the heat engine. These are exergy losses via heat transfer and/or other mechanisms (e.g., friction, pressure drop), which are collectively referred to as *irreversibilities*.

Thus, one way to look at the combined-cycle power plant analysis is to identify and "hack away" the irreversibilities so that we can find the sculpture hidden in the system: *net power output*.

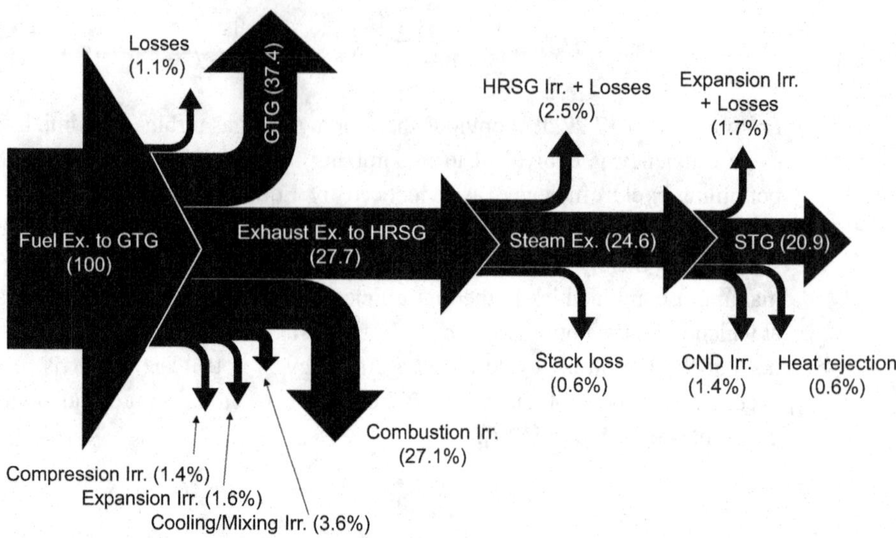

Figure 17.5 Typical exergy distribution in a combined-cycle power plant [7]. CND = condenser; GTG = gas turbine generator.

Note that in the coverage below, the emphasis is largely on the steam (Rankine) bottoming cycle of the GTCC. The reason for this is simple enough.

A Sankey diagram illustrating the distribution of gas turbine fuel exergy, which corresponds to the first law-based Sankey diagram in Figure 17.4, is shown in Figure 17.5. The difference between the "messages" of the two diagrams is strikingly clear. To paraphrase the immortal conclusion of *Animal Farm*,[5] all kilowatts are equal (according to the first law), but some kilowatts are more equal than others (according to the second law). This is best illustrated by the energy lost in the condenser (mainly via heat rejection to the cooling water), which is a whopping one-third of the fuel energy (in LHV), but, from a useful work generation potential perspective, it is worth only a measly 2 percent of the fuel exergy. In layman's terms, then, it is pretty much worthless as a source of power generation.

For a rigorous analysis and quantification of the individual loss mechanisms summarized in the Sankey diagram in Figure 17.5, the reader is referred to the paper by Gülen and Smith [8]. The starting point is the exergy balance around the bottoming-cycle control volume, i.e.,

$$0 = \dot{m}_{exh}(a_{exh} - a_{stck}) + \sum_j \left(1 - \frac{T_{DS}}{\bar{T}_j}\right)\dot{Q}_j - \dot{W}_{BC} - \dot{I}. \tag{17.23}$$

The four terms on the right-hand side of Equation 17.23, from left to right, are as follows:

[5] The allegorical novella by George Orwell, first published in the UK on August 17, 1945.

1. Net exergy transfer into the bottoming cycle via the gas turbine exhaust gas.
2. Net exergy transfer into ($\dot{Q}_j$ >0) or out of ($\dot{Q}_j$ <0) the bottoming cycle via heat transfer, e.g.,
   a. HRSG casing heat loss
   b. Heat rejection from the condenser
   c. Heat supplied to the fuel heater (via feed water extraction from the HRSG)
3. Bottoming-cycle *net* work output (STG output *minus* power consumed by the feed water and condensate pumps[6])
4. Exergy destruction in the bottoming cycle, e.g.,
   a. Heat transfer irreversibility in the HRSG
   b. Heat transfer irreversibility in the condenser
   c. Steam turbine irreversibility

In Equation 17.23, T_{DS} is the dead state temperature (usually the ambient temperature) and $\bar{T}_j$ is the mean effective (average) temperature at which $\dot{Q}_j$ takes place. The reader (especially after going through Chapter 5) can verify that the exergy lost (or added) via $\dot{Q}_j$ is equal to Carnot-cycle work between temperature reservoirs at $\bar{T}_j$ and T_{DS}.

In per-unit-mass terms, Equation 17.23 becomes

$$0 = a_{exh} - a_{stck} + \sum_j \left(1 - \frac{T_{DS}}{\bar{T}_j}\right) q_j - w_{BC} - i.$$

Subtracting the first law equation (i.e., the per-unit-mass energy balance),

$$0 = h_{exh} - h_{stck} + \sum_j q_j - w_{BC}.$$

From the second law equation (i.e., the per-unit-mass exergy balance), one obtains

$$T_{DS}(s_{stck} - s_{exh}) = \sum_j \frac{T_{DS}}{\bar{T}_j} q_j + i,$$

which, of course, is identical to Equation 5.44 ($T_{DS} = T_0$), which is the best mathematical expression of the second law of thermodynamics.

Calculations show that the typical values for advanced F–class gas turbines, as a percentage of gas turbine exhaust exergy, are as follows:

- Stack gas exergy of 2–3 percent;
- Condenser heat rejection and heat transfer irreversibility of about 5.5 percent[7] (total);
- HRSG heat loss and heat transfer irreversibility of about 10–11 percent[8] (total);
- Steam turbine irreversibility of about 6 percent (half of it in the LP turbine);
- Heat supplied to the fuel gas performance heater of about 0.8 percent;
- Miscellaneous losses and irreversibilities (in the pipes, valves, etc.) of about 1–1.5 percent.

[6] Note that the condenser circulating cooling water pump (and cooling tower fan if one is present) is *outside* the bottoming-cycle control volume.

[7] Condenser pressure is 1.2 inches of mercury.

[8] 3PRH with 15°F evaporator pinch.

Thus, one should reasonably expect about 75 percent for bottoming-cycle exergetic efficiency (net bottoming-cycle power output as a fraction of gas turbine exhaust exergy) with advanced gas turbines with exhaust temperatures of around 1,150°F and *reasonable* 3PRH design assumptions.

The key parameter is the gas turbine exhaust temperature, which sets the theoretically possible maximum (i.e., the entitlement) and the percentage of that entitlement value that can be converted to net electric power via technically and economically feasible design choices. Thus, based on the premise that there is a well-designed (or optimal) steam bottoming-cycle design for a given gas turbine exhaust, different technology curves in the form of $f(T_{exh})$ can be generated. Such a curve is given in [8]. This optimal curve goes through the data points representing the bottoming-cycle exergetic efficiencies extracted from trade publications in the 2007–2009 timeframe and is adequately described by the quadratic formula in Equation 17.24.

$$\varepsilon_{BC} = 0.2441 + 0.0746 \cdot \left(\frac{T_{exh}}{100}\right) - 0.00279 \cdot \left(\frac{T_{exh}}{100}\right)^2. \tag{17.24}$$

Equation 17.24 is very well suited to combined-cycle analysis of air-cooled gas turbines with the implicit assumption of a "well-designed" bottoming cycle based on 2007–2009-vintage technology and aggressive equipment design specifications.

Equation 17.24 is a reasonably accurate predictive tool for estimating the bottoming-cycle performance for a given gas turbine exhaust temperature. Note that there is a *spectrum* of steam (Rankine) bottoming-cycle configurations that covers the range of possible gas turbine exhaust temperatures.

- Gas turbine exhaust temperatures below ~1,000°F are not enough to support a feasible reheat design because there is not enough energy to heat the steam *twice* to the same temperature.
- At the high end, for sufficiently high exhaust temperatures, there is a transition from three-pressure to two-pressure because the exhaust gas energy dedicated to HP steam production, superheating, and economizing "squeezes" the IP section out.
- At still-higher temperatures, typically ~1,600°F or higher, the temperature pinch transitions from the LP evaporator to the HRSG economizer inlet.
- At a certain T_{exh}, the gas (cooling) and water (heating) lines for the economizer section will become parallel so that further increase will "pinch" them further together at the HRSG inlet. At that point, the LP evaporator is also "squeezed out" and the system reverts to a single-pressure system.

The last condition is referred to as the bottoming cycle becoming "cold end-pinched." It corresponds roughly to $\mu \approx 0.25$ (depending on steam pressure and temperature and whether the cycle is reheat or non-reheat). From equation A7 in [8], the corresponding T_{exh} can be found as ~1,650°F (~900°C). At this point in time, it is inconceivable that such a gas turbine can ever be built. The only possibility is supplementary (duct) firing upstream of the HP superheater. The reader is referred to the detailed exergetic heat recovery analysis by Smith in [7] for an in-depth discussion accompanied by HRSG heat release diagrams.

Equation 17.24 can be used as the "anchor" to derive an equation similar to Equation 17.21 for combined-cycle efficiency from a second law perspective. Combined-cycle net output is the sum of the STG and gas turbine generator (GTG) outputs after subtracting the plant auxiliary power consumption. For a combined-cycle power plant with N gas turbines and a steam turbine,

$$\dot{W}_{CC} = \left(N \cdot \dot{W}_{GT} + \dot{W}_{ST}\right) - \dot{W}_{Aux}. \tag{17.25}$$

The exergetic efficiency given by Equation 17.24 is the fraction of the gas turbine exhaust exergy that is converted into net bottoming-cycle work. For a combined cycle with N gas turbines and a steam turbine, this results in,

$$\varepsilon_{BC} = \frac{\dot{W}_{BC}}{N \cdot \dot{m}_{exh} \cdot e_{exh}} = \frac{w_{BC}}{e_{exh}}. \tag{17.26}$$

The exergetic efficiency in Equation 17.26 is identical to the second law bottoming-cycle effectiveness of Chin and El-Masri [9] or the "rational" efficiency of Horlock et al. [10], with the implicit assumption that the exergy of the exhaust gas at $T_{stck} = T_{amb} = T_0$ is zero. Using the definition in Equation 17.25, combined-cycle net efficiency becomes

$$\eta_{CC,net} = \frac{N\left(\dot{W}_{GT} + \varepsilon_{BC}\dot{m}_{exh}e_{exh}\right) - \dot{W}_{Aux}}{N \cdot HC}. \tag{17.27a}$$

Eliminating number of gas turbines, N, for simplicity and using the symbol α' (to differentiate it from α in the first law formulation), Equation 17.27a becomes

$$\eta_{CC,net} = \eta_{GT} + \varepsilon_{BC}\frac{\dot{E}_{exh}}{HC}(1 - \alpha'), \tag{17.27b}$$

where the auxiliary load fraction is $\alpha' \approx 1.5 \cdot \alpha$. (Note that the bottoming-cycle exergetic efficiency is on a *net* basis; i.e., it accounts for feed water and condensate pump power consumption. It also includes the generator losses.) Taking the derivative of η_{CC} with respect to ε_{BC}, we obtain

$$\frac{\partial \eta_{CC,net}}{\partial \varepsilon_{BC}} = \frac{\dot{E}_{exh}}{HC}(1 - \alpha'). \tag{17.28}$$

For modern gas turbines with >1,100°F exhaust temperatures, the exhaust *exergy* is about 50 percent of the exhaust *energy*, which in turn is about 60 percent of the gas turbine heat consumption. With about 3 percent for α', from Equation 17.28, it is found that each additional percentage point in bottoming-cycle exergetic efficiency is worth approximately 0.25 combined-cycle net efficiency points.

17.3 Auxiliary Power Consumption

A power plant consumes power to generate power. Gross power is the sum total of generator outputs of prime movers. Net power is nominally what the grid gets (but not the end customer!). Auxiliary power is the difference between net and gross. The combined-cycle power plant auxiliary load is a combination of the following:

- Power consumption of major BOP pumps and fans;
- Power consumption of gas turbine and steam turbine auxiliary equipment (lube and hydraulic oil pumps, blowers, etc.);
- Power consumption of minor BOP pumps and compressors;
- Miscellaneous plant power consumption (lighting, heating, ventilating, and air-conditioning [HVAC], etc.).

Auxiliary power consumption for 39 combined-cycle power plants (65–860 MW) is extracted from data published in a trade publication. The data indicate that the auxiliary load ranges between 1.2 and 2.1 percent of the plant gross output. The average is 1.57 ± 0.24 percent. The linear curve-fit has a slope of 1.67 percent.

A breakdown of typical combined-cycle plant auxiliary power consumption reveals that the boiler feed water pumps account for 30–40 percent of the total. The second biggest contributor is the heat rejection system, which accounts for 15–25 percent of the total. Thus, the two major plant subsystems account for 45–65 percent of the total.

The most common cooling system configurations include:

1. Once-through, open-loop system if a natural cooling water source is available (e.g., a lake, river, or ocean), or
2. A closed-loop system with a mechanical cooling tower.
3. Dry cooling system (i.e., an air-cooled or "A Frame" condenser).

The cooling system power consumers are the cooling water circulating pump and – if applicable – the cooling tower or air-cooled condenser forced draft fans. An estimation methodology for the power consumption of those items can be found in Gülen (see [2] in Chapter 8). Other BOP equipment that consumes power includes myriad small pumps, compressors, miscellaneous plant HVAC, and lighting. These items and generating equipment skids add up to a power consumption that is a small fraction of the combined-cycle gross output (i.e., less than 0.5 percent). The exact value is very much dependent on site and customer requirements.

For quick and quite reliable estimation of combined-cycle auxiliary power consumption, the reader is referred to the information in Table 17.4. (Note that the data in Table 17.4 are derived from actual plant bid data and reflect the nameplate performance quoted by vendors of procured equipment such as pumps, fans, compressors, etc.) For most practical purposes, using the total (as a fraction of combined-cycle gross output; i.e., α) contribution is sufficiently accurate.

In some cases, especially for gas turbines with cycle PRs of about 20 or higher (but definitely for very high PR units such as "sequential combustion" turbines with PR >30), the plant should incorporate a fuel gas "booster" compressor to increase the pipeline gas pressure to the required pressure levels. In those applications where the pipeline gas pressure and the required fuel skid inlet pressure dictate a booster compressor, the power consumption of that compressor should be properly accounted for in

Table 17.4 Standardized combined-cycle (CC) power plant auxiliary load consumption (from [2] in Chapter 8). OT-OL = once-through, open-loop; CW = cooling water; ACC = air-cooled condenser; GT = gas turbine; ST = steam turbine

	OT-OL (Cold)	OT-OL (Normal)	Closed Loop	ACC
Ambient Temperature	ISO			
Condenser Pressure	1.00	1.20	1.50	2.50
Direct				
GT Equipment Auxiliaries		0.25% of GTG output		
ST Equipment Auxiliaries		0.10% of STG output		
Boiler Feed Pumps (% of STG)	1.90	1.95	1.90	2.00
Cooling Tower Fans (% of STG)	NA	NA	0.70	NA
CW Circulating Pump (% of STG)	0.95	1.10	1.00	NA
ACC Fans (% of STG)	NA	NA	NA	1.50
Fuel Compressor (PR >20)		Equation 17.29		
Indirect				
Auxiliary CW Pump (% of STG)		0.10		
Other Pumps		Number of GTGs × 175 kW		
Step-Up Transformer (% of CC Gross)		0.50		
Miscellaneous (% of CC Gross)		0.10		
Total (% of CC Gross)	1.90	2.00	2.20	2.10

the plant auxiliary load. The following formula is suggested for that purpose for gas turbines with cycle PRs >20 (with fuel flow rate in lb/s):

$$\dot{W}_{FC}[\text{kW}] = \dot{m}_{fuel}(80 \ln(\text{PR}) - 225). \tag{17.29}$$

17.4 Importance of the Bottoming Cycle

In a paper presented at the 1983 Tokyo International Gas Turbine Congress, authors from a major OEM declared that "... the future of heavy-duty gas turbine technological development is very promising" [11]. In particular, the authors stated that the water-cooled turbine in development at the time would be the enabler of very high TITs "beyond those possible with air cooling." This potential offered the promise of "major improvements" in combined-cycle efficiency. Alas, at well below 55 percent (from figure 6 in [11]), the severe underestimation of future GTCC performance by the authors was remarkable.[9] (At the time, the upper limit for air-cooled gas turbines was set at ~1,260°C.) This is the same as today's most advanced H/J-class heavy-duty gas turbines, which are rated well above 60 percent net combined-cycle efficiency. Not

[9] Even with water-cooled hot gas path components (a prominent feature of the US Department of Energy's High-Temperature Turbine Technology [HTTT] program of the 1970s) that were shown to be capable of achieving 1,600°C TIT, the authors predicted barely 52 percent net combined-cycle efficiency by 2000 [11].

only that, the projected efficiencies are even below those obtained in field conditions in the top 20 US combined-cycle power plants (some even supplementary-fired) in the first two decades of the twenty-first century (about two decades after the publication date of the cited reference).

At around the same time, in a landmark 1982 paper, Rice described in minute detail a steam-cooled, reheat gas turbine with a cycle efficiency of 42.5 percent and an exhaust temperature of ~700°C [12]. Thus, going by the bottoming-cycle exergetic efficiency in Equation 17.24, [12] should have predicted the commensurate GTCC efficiency as well above 60 percent – more than 30 years ago, when others could foresee barely above 50 percent. Alas, the "possible" combined-cycle efficiency stated in the conclusions section of [12] was a paltry 56 percent. The reason for this is the thoroughly obsolete bottoming cycle with feed water heating and 150°C stack temperature (see figure 11 in [12]), which implied a very poor exergetic efficiency. (It is difficult to estimate a value due to the complex interaction between the topping and bottoming cycles via gas turbine cooling steam.)

These two historical examples provide dramatic illustrations of ignoring the contribution of the bottoming cycle to GTCC performance; there is only so much one can achieve just by burning more fuel.

The other takeaway from this short section is the importance of fundamental thermodynamic principles, without which one cannot hope to have a good handle on future technology developments. In that sense, the second law of thermodynamics and its embodiment in the concept of exergy is an indispensable tool. This is so because, no matter how good an engineer one is (and the authors of the 1983 Tokyo paper above and Ivan Rice were the crème de la crème in their field), the state of the art of technology at the time, combined with the prevailing economic climate (i.e., fuel prices, availability, commodity and labor prices, inflation, income levels, etc.), forces one into a proverbial "box." A good example of that is the bottoming cycle used by Rice, which had feed water heating and, consequently, severely limited the heat recovery in the HRSG to a stack temperature of 150°C (~300°F).

To be fair, another limit on the stack temperature and heat recovery effectiveness was imposed by the most commonly used fuels in gas turbines at the time; namely, number 2 or heavy fuel oil.[10] In order to prevent sulfuric acid condensation on the economizer tubes having a drastic impact on HRSG component life, stack gas temperatures had to be kept above the sulfuric acid dew point. Even without feed water heating, this limited the HRSG design to two or even one pressure level.

Thus, either in order to escape from the limitations imposed by the state-of-the-art box or to avoid being lulled into false expectations (e.g., bottoming-cycle exergetic efficiencies going beyond 80 percent), one must stick with the fundamentals, which have been explored in detail in the preceding sections of this chapter.

[10] The objective of the US Department of Energy's HTTT program was to bring to "technology readiness" an efficient, high-firing temperature gas turbine for application in an integrated gasification combined-cycle power plant.

References

1. Gülen, S. C., Étude on Gas Turbine Combined Cycle Power Plant – Next 20 Years, *Journal of Engineering for Gas Turbines and Power*, **138** (2016), 051701.
2. Kehlhofer, R., Rukes, B., Hannemann, F., Stirnimann, F., *Combined-Cycle Gas and Steam Turbine Power Plants*, 3rd edition (Tulsa, OK: PennWell Corp., 2009).
3. Horlock, J. H., *Combined Power Plants: Including Combined Cycle Gas Turbine (CCGT) Plants* (Oxford, UK: Pergamon Press Ltd., 1992).
4. *HRSG Users Handbook: Design, Operation, Maintenance* (Bozeman, MT: HRSG User's Group, 2006).
5. Sanders, W. P., *Turbine Steam Path Maintenance and Repair*, Volumes 1–3 (Tulsa, OK: PennWell Corp., 2001).
6. Leyzerovich, A., *Large Power Steam Turbines: Design and Operation*, Volumes 1–2 (Tulsa, OK: PennWell Corp., 1997).
7. Smith, R. W., "Steam Turbine Cycles and Cycle Design Optimization: Combined Cycle Power Plants," in *Advances in Turbines for Modern Power Plants*, Ed. T. Tanuma (Duxford, UK: Woodhead Publishing, 2017).
8. Gülen, S. C., Smith, R. W., Second Law Efficiency of the Rankine Bottoming Cycle of a Combined Cycle Power Plant, *Journal of Engineering for Gas Turbines and Power*, **132** (2010), 011801.
9. Chin, W. W., El-Masri, M. A., Exergy Analysis of Combined Cycles: Part 2 – Analysis and Optimization of Two-Pressure Steam Bottoming Cycles, *Journal of Engineering for Gas Turbines and Power*, **109** (1987), 237–243.
10. Horlock, J. H. et al., Exergy Analysis of Modern Fossil-Fuel Power Plants, *Journal of Engineering for Gas Turbines and Power*, **122** (2000), 1–7.
11. Patterson, J. R., Walsh, E. J., "A Manufacturer's Role in Heavy-Duty Gas Turbine Future Technology," ASME Paper 83-TOKYO-IGTC-118, 1983 Tokyo International Gas Turbine Congress, October 23–29, 1983, Tokyo, Japan.
12. Rice, I., The Reheat Gas Turbine with Steam-Blade Cooling – A Means of Increasing Reheat Pressure, Output, and Combined Cycle Efficiency, *Journal of Engineering for Gas Turbines and Power*, **104**: 1 (1982), 9–22.

18 Off-Design Operation

Gas turbine "sticker" or "rating" performances in simple or combined cycles are typically quoted at the following ISO base load conditions:

- Sea level (0 ft/m altitude), 60 percent relative humidity, and 59°F (15°C) ambient;
- Rated fuel flow (i.e., "firing") per the nominal control curve with inlet guide vanes (IGVs) fully open.

(Nowadays, state-of-the-art gas turbines with multiple stages of compressor variable stator vanes [VSVs] do not use "control curves"; they are operated with the "digital twin" of the machine in question built into the controller, which is referred to as *model-based* or *model-predictive* control – see Section 18.3 below.)

Once a gas turbine is installed and commissioned, it operates across a wide envelope of site ambient and loading conditions. Output and efficiency vary accordingly from ISO base load or site-specific design conditions, which may be quite different from ISO specifications. Consequently, a gas turbine engineer must be familiar with the following two types of calculation:

- Correction to design conditions
- Off-design performance estimation

The former is a (mostly) commercial activity associated with performance guarantees. Once the engineering, procurement, and construction (EPC) contractor finishes with the installation and startup work (see Section 14.4), he or she has to prove to the owner that the plant performs in full compliance with the contractual requirements. Running the gas turbine at "full load" is no problem. Obviously, the altitude of the plant site from sea level does not change in the period that passes since the time the design specifications were first laid out. However, waiting for the prevailing weather to be exactly coincident with the contract values of ambient temperature and humidity could be financially prohibitive (e.g., schedule liquidated damages and added construction interest). Other variations from contract specifications may include the fuel gas composition from the pipeline at the time of the performance test and the generator power factor. It may turn out that inlet and exhaust pressure losses recorded during the test are somewhat different from those specified in the contract as well. Ultimately, once the performance test is done, "as-measured" values of output and heat rate must be "corrected" to the contract/design values using mutually agreed-upon "correction curves."

The second type of calculation is associated with myriad engineering activities such as system optimization for given operating scenarios, financial "pro forma" calculations, performance monitoring, etc. In theory, for either type of calculation, one has to use the same tools, methods, and principles. In practice, the situation is slightly different. Correction of performance test measurements to contractually specified site ambient and loading conditions is a strictly commercial activity. Failure to meet contractual performance requirements can lead to significant "liquidated damages" to be paid by the EPC contractor to the owner. On the flip side of the coin, exceeding those requirements can lead to hefty bonuses to be paid by the owner to the EPC contractor. Consequently, it is imperative that the parties involved in this transaction fully understand and agree upon the methods and tools to be used in the measurement and correction of plant performance data. Such methods and tools (i.e., the "correction curves") in question are precisely specified, in minute detail, in performance test standards published by professional organizations such as the ASME. The most widely used test codes (in the USA and many other countries as well) are:

- PTC 46 – "Overall Plant Performance"
- PTC 22 – "Gas Turbines"
- PTC 6 – "Steam Turbines (Rankine Cycle)"
- PTC 6.2 – "Steam Turbines (Combined Cycle)"
- PTC 4 (or 4.1) – "Fired Boilers"
- PTC 4.4 – "HRSGs"
- PTC 19.1 – "Test Uncertainty"

For a specific gas turbine, one should refer to the correction curves provided by the original equipment manufacturer (OEM). This is imperative for accurate representation of the unit performance for commercial purposes. Such proprietary information is clearly not released for publication in technical books and articles. Consequently, one needs physics-based "proxies" for "true" correction curves to be used in optimization studies and similar endeavors. The obvious choice is, of course, high-fidelity models with stage-by-stage simulation of compressor and turbine sections. Developing such models is a significant undertaking requiring thousands of manhours and computational resources or very expensive commercial software. Nevertheless, for most practical purposes, adequate results can be obtained by using the fundamental principles of gas dynamics.

For older-generation E- and F–class gas turbines, on a normalized basis, generic curves (mostly linear) and "rules of thumb" are adequate for most practical purposes. The reason for that can be found by examining the basic aerothermodynamics of a single-shaft gas turbine for electric power generation running at constant rotational speed (i.e., 3,000 or 3,600 rpm for 50- or 60-Hz grids, respectively). (Some smaller and/or multi-shaft industrial or aeroderivative gas turbines have higher speeds for their gas generator shafts or even their power turbines.)

18.1 Performance

The following two off-design scenarios are of prime interest for stationary gas turbines for electric power generation (i.e., constant-speed machines):

- Variation of gas turbine performance (output and efficiency) with ambient temperature;
- Variation of gas turbine efficiency with gas turbine load.

For combined-cycle applications, variation in exhaust gas flow and temperature is also important. For most single-shaft gas turbines with IGVs, as will be demonstrated below, the governing equations are quite simple and differ very little from one OEM to another. State-of-the-art, advanced-class gas turbines with IGVs and typically three VSVs have more degrees of freedom to adjust compressor inlet airflow and hence performance on hot (or cold) days.

A typical E–class gas turbine performance summary (sea level, 60 percent relative humidity, full load) is provided in Table 18.1. Corresponding normalized characteristic parameters are presented in Table 18.2. Similar data for a typical F–class gas turbine are provided in Tables 18.3 and 18.4. Both data sets are obtained from model runs made in Thermoflow's GT PRO software (version 26; engine #85 from the built-in gas turbine model library for the E class and engine #99 for the F class). Characteristic machine parameters were defined in Chapter 11 within the context of compressor off-design performance. The asterisks in Table 18.2 and 18.4 indicate that they are normalized (and non-dimensionalized) by using the values corresponding to a reference point. (The reference point in this case is the ISO base load.) The parameters are summarized here for the reader's convenience:

- Non-dimensional turbine inlet temperature (TIT), τ_3 (T_3/T_1);
- Reduced or corrected speed, ν (Equation 11.41);
- Reduced or corrected flow, μ (Equation 11.43);
- Cycle pressure ratio (PR), π.

A comparison of the data in Tables 18.1–18.4 affirms that, on a normalized basis, E- and F-class gas turbines are practically identical in terms of their performance at varying ambient temperature. This should not come as a big surprise because a gas

Table 18.1 Gas turbine performance variation with ambient temperature (E class)

Ambient Temperature (°F)	20.0	30.0	40.0	50.0	59.0	70.0	80.0	90.0	100.0	110.0	120.0
Normalized Output	1.13	1.10	1.07	1.03	1.00	0.96	0.93	0.89	0.86	0.83	0.79
Efficiency Delta (%)	0.82	0.69	0.54	0.27	0.00	−0.34	−0.67	−1.03	−1.43	−1.88	−2.40
Normalized Efficiency	1.02	1.02	1.02	1.01	1.00	0.99	0.98	0.97	0.96	0.95	0.93
Normalized Exhaust Flow	1.09	1.07	1.04	1.02	1.00	0.97	0.95	0.92	0.90	0.88	0.85
TIT Delta (°F)	−11.6	−6.3	−1.6	−0.2	0	−0.3	−1.6	−4.0	−7.5	−12.3	−18.7
TEXH Delta (°F)	−35.1	−25.8	−16.7	−7.8	0	10.5	19.8	28.9	37.9	46.8	55.7
PR (at Turbine Inlet)	14.7	14.4	14.1	13.8	13.5	13.1	12.8	12.5	12.2	11.9	11.6

Table 18.2 Normalized gas turbine characteristics for the cases in Table 18.1 (E class)

τ_3*	0.996	0.998	0.999	1.000	1.000	1.000	0.999	0.999	0.997	0.995	0.993
μ*	1.05	1.04	1.02	1.01	1.00	0.98	0.97	0.95	0.93	0.92	0.90
ν*	1.04	1.03	1.02	1.01	1.00	0.99	0.98	0.97	0.96	0.95	0.95
π*	1.09	1.07	1.04	1.02	1.00	0.97	0.95	0.93	0.90	0.88	0.86

Table 18.3 Gas turbine performance variation with ambient temperature (F class)

Ambient Temperature (°F)	20.0	30.0	40.0	50.0	59.0	70.0	80.0	90.0	100.0	110.0	120.0
Normalized Output	1.12	1.09	1.06	1.03	1.00	0.96	0.93	0.90	0.87	0.83	0.78
Efficiency Delta (%)	0.87	0.68	0.48	0.23	0.00	−0.35	−0.68	−1.08	−1.50	−2.08	−2.76
Normalized Efficiency	1.02	1.02	1.01	1.01	1.00	0.99	0.98	0.97	0.96	0.94	0.93
Normalized Exhaust Flow	1.06	1.05	1.03	1.02	1.00	0.98	0.96	0.94	0.91	0.89	0.86
TIT Delta (°F)	13.2	9.4	6.1	3.1	0	−2.5	−5.3	−7.2	−9.4	−22.6	−49.1
TEXH Delta (°F)	−19.2	−15.0	−10.6	−5.0	0	8.1	15.6	25.0	34.9	40.8	40.8
PR (at Turbine Inlet)	17.2	16.9	16.7	16.4	16.1	15.8	15.5	15.1	14.8	14.4	14.0

Table 18.4 Normalized gas turbine characteristics for the cases in table 18.3 (F class)

τ_3*	1.005	1.003	1.002	1.001	1.000	0.999	0.998	0.997	0.997	0.992	0.983
μ*	1.02	1.02	1.01	1.01	1.00	0.99	0.98	0.96	0.95	0.93	0.91
ν*	1.04	1.03	1.02	1.01	1.00	0.99	0.98	0.97	0.96	0.95	0.95
π*	1.06	1.05	1.03	1.02	1.00	0.98	0.96	0.94	0.92	0.89	0.87

turbine, after all, is a pretty simple device in terms of its operating principles. The underlying thermodynamic cycle is characterized by two parameters (i.e., cycle PR and TIT). Consequently, differentiation between classes of gas turbines is solely determined by these two parameters. The magnitude of shaft/generator power output is determined by airflow (i.e., size) of the unit. Once these three design parameters are fixed, the design of the actual components (compressor, combustor, and turbine/expander plus the alternating current [ac] generator) follows well-established principles of aerodynamics, fluid mechanics, heat transfer, and electromechanics. (These principles have been covered in detail in the preceding chapters of the book.)

Aerothermodynamic cycle design dictates the selection of materials, coatings, and cooling arrangement that go into the final hardware design. Equally important is the availability of suitable manufacturing technologies for the final product that rolls out of the factory. Differentiation between individual products from different OEMs within each class of gas turbine rests in variations on these aspects of gas turbine design, development, and manufacturing.

When it comes to operation in the field under different "boundary conditions" and loads, however, the "knobs" available to the designer are limited to the following two key flow parameters:

- Compressor inlet airflow
- Fuel flow

These control shaft/generator output and efficiency (heat rate) via TIT and cycle PR. (Interactions between those parameters will be discussed later in the chapter.) The main control mechanism for the inlet airflow is by opening/closing of the IGVs, which are essentially rotating (along a vertical axis via hydraulic or electric actuators) stator vanes upstream of the first stage of the compressors. As such, IGVs increase or decrease the effective flow area and thus the volumetric flow capacity of the compressor (which otherwise is essentially constant at constant rotational speed). For a given airflow, fuel flow determines the TIT. Since TIT cannot be measured directly, the exhaust temperature measurement and cycle PR (from compressor inlet and discharge pressure measurements) are used to calculate an "inferred" TIT to adjust the fuel flow.

In state-of-the-art, advanced-class gas turbines, in addition to the IGVs, stator vanes of the first three compressor stages are also adjustable (or "variable," hence "variable stator vanes" or VSVs). At low mass flow rates (i.e., low absolute velocity, V_1 in Figure 11.1), *aerodynamic stall* is caused by increasing incidence angle (β_1 in Figure 11.1) and resulting flow separation in affected stages, which ultimately leads to compressor surge. IGVs and VSVs alleviate the problem by extending the stable operating regime of the compressor via adjusting the incidence angles of the front stages and maintaining the velocity triangles close to their design shapes. This is the reason why advanced gas turbines have better hot-day performance characteristics. In order to illustrate this, consider the ambient performance data for an HA-class gas turbine in Tables 18.5 and 18.6 (Thermoflow's GT PRO, engine #602 from the model library). ISO base load performances of the E- and F-class gas turbines discussed earlier and the HA-class gas turbine are summarized in Table 18.7.

When the ambient temperature (output and efficiency) data for E-, F-, and HA-class gas turbines are plotted, the difference between the advanced HA-class and the older-

Table 18.5 Gas turbine performance variation with ambient temperature (HA class)

Ambient Temperature (°F)	20.0	30.0	40.0	50.0	59.0	70.0	80.0	90.0	100.0	110.0	120.0
Normalized Output	1.04	1.04	1.03	1.01	1.00	0.98	0.97	0.94	0.91	0.85	0.78
Efficiency Delta (%)	0.31	0.37	0.29	0.14	0.00	−0.24	−0.47	−0.78	−1.11	−1.86	−2.77
Normalized Efficiency	1.01	1.01	1.01	1.00	1.00	0.99	0.99	0.98	0.97	0.96	0.93
Normalized Exhaust Flow	1.00	1.01	1.00	1.00	1.00	1.00	1.00	0.97	0.95	0.90	0.85
TIT Delta (°F)	−4.2	−3.2	4.2	2.1	0	−3.4	−7.1	−9.0	−12.0	−14.6	−47.1
TEXH Delta (°F)	−7.7	−7.9	−1.6	−0.8	0	0.7	1.4	9.8	18.4	40.2	47.7
PR (at Turbine Inlet)	21.5	21.6	21.6	21.5	21.5	21.4	21.4	21.0	20.5	19.4	18.2

Table 18.6 Normalized gas turbine characteristics for the cases in Table 18.5 (HA class)

τ_3*	0.999	0.999	1.001	1.001	1.000	0.999	0.998	0.997	0.996	0.995	0.985	0.971
μ*	0.96	0.98	0.99	0.99	1.00	1.01	1.02	1.00	0.99	0.94	0.90	0.90
ν*	1.04	1.03	1.02	1.01	1.00	0.99	0.98	0.97	0.96	0.95	0.95	0.95
π*	1.00	1.01	1.00	1.00	1.00	1.00	1.00	0.98	0.96	0.90	0.85	0.84

Table 18.7 ISO base load performances

Class	E	F	HA
Vintage	1997	1998	2015
Gas Turbine Gross Power (kW)	114,828	163,546	344,923
Gas Turbine gross Lower Heating Value Efficiency (%)	34.9	37.4	41.6
Gas Turbine Exhaust Mass Flow (lb/s)	782.9	980.3	1,524.4
Gas Turbine Exhaust Temperature (°F)	1,036.2	1,052.9	1,171.3
Cycle Pressure Ratio	14.1	16.6	22.7
Nominal TIT (°F)	2,372	2,552	2,912
Turbine Inlet Mass Flow (lb/s)	717.9	846.0	1,364.0
Turbine Flow Parameter	36.1	36.7	47.1

generation E- and F-class technologies (which, as pointed out earlier, are practically identical) is immediately noticeable. In particular,

- There is reduced lapse in output and efficiency at hot ambients (except at very high ambients; i.e., 120°F).
- There is reduced improvement in output and efficiency at cold ambients.

(It should be emphasized that the actual performance of machines with "model-based control" [MBC] is significantly different from that implied by the data in Tables 18.5 and 18.6 [i.e., "smooth" curves]. The presented data reflect fixed "schedules," which are approximations of the "true" curves. This will be covered in detail in Section 18.3.)

Let us examine the reduced improvement in cold-day performance. As indicated by the data in Table 18.7, the HA–class gas turbine is rated as 345 MW at ISO, which is more than twice that of the F–class gas turbine. If it had the same increase in airflow and output as the F–class unit at 20°F ambient, its output would be nearly 390 MW (instead of 357 MW as indicated by the data in Table 18.5). This may be indicative of a maximum output limit imposed by the generator size and/or shaft torque limit.

The reduced lapse in the hot-day performance (i.e., better output *and* efficiency vis-à-vis E/F–class gas turbines at the same hot-day ambient temperature) can be attributed to a large compressor, which compensates for reduced density by opening the IGVs. This, of course, raises the question of why they were not at their "fully open" position at the ISO point to begin with. The answer, once again, boils down to the maximum output limit imposed by the generator size or, more likely, the shaft torque limit. It is also possible that the OEM might hold back some capacity for future upgrades. (This, in fact, was confirmed by the 2017 rating of the same gas turbine, General Electric's [GE] 7HA.02, published in the *Gas Turbine* 2016–2017 Handbook: 372 MW at ISO base load.)

These points are illustrated by the normalized inlet airflow temperature data where m_1* and T_1* are $\dot{m}_1$ and T_1 divided by their ISO values, respectively.[1] (Note that $\dot{m}_1$ is found by subtracting the fuel flow from the exhaust flow and fuel flow can be found from the output and heat rate by using the lower heating value [LHV] of 100 percent

[1] Caution: temperature ratios are always calculated on an absolute basis (i.e., degrees Rankine or Kelvin).

CH_4 fuel, 21,515 Btu/lb, or 50,044 kJ/kg.) For a very crude (proverbial *back of the envelope*) estimate, one can just assume that

$$m_1^* = \frac{1}{T_1^*}. \tag{18.1a}$$

A better estimate for hot ambients (i.e., $T_1 > 59°F$) is

$$m_1^* = \frac{1}{T_1{}^{*1.4}}. \tag{18.1b}$$

For advanced gas turbines like the HA class, m_1* can be assumed to be unity up to about 75–80°F. For higher ambient temperatures, it can be reduced according to a schedule similar to that shown in Table 18.5. For cold ambients, generalization is difficult; one has to be thoroughly familiar with the design specifications of the particular gas turbine in question. As a first approximation, one can use Equation 18.1a, but it is probably wise to keep the increase in airflow below ~5 percent at cold ambients for more recent vintage, larger gas turbines.

As discussed earlier, the compressor discharge PR is proportional to inlet airflow. In fact, on a normalized basis,

$$\pi^* = m_1^*, \tag{18.2}$$

as verified by the model data in Table 18.5. Consequently, using the turbine flow parameter in Equation 11.45,

$$\frac{\dot{m}_3\sqrt{\frac{R_{unv}}{MW}T_3}}{p_3} = \text{const.},$$

and the data in Table 18.7, one can estimate the TIT. For turbine inlet flow in Equation 11.45, one can use

$$\dot{m}_3 = (1 + f - nch - ch)\,\dot{m}_1, \tag{18.3}$$

where f, nch, and ch are fuel flow, nonchargeable cooling flow, and chargeable cooling flow, respectively, as a fraction of compressor inlet airflow. Fuel fraction can be found from the data in Table 18.7 and it is practically constant across the ambient range. In fact, judging from the fuel flow and cooling flow data in Table 8.5 in Chapter 8, as a first approximation, it is sufficiently adequate to assume

$$\dot{m}_3 \approx 0.8\,\dot{m}_1.$$

As an example, consider the following advanced F-class gas turbine performance at ISO full load with 100 percent methane fuel at 155°F (after the fuel compressor):

- 235 MWe generator output
- 39.1 percent efficiency
- 1,140 lb/s air flow
- 1,187°F exhaust temperature
- 18.21 compressor/cycle PR

Fuel flow is calculated as

$$(235,000/1.05506)/0.391/21,515 = 26.48 \text{ lb}/s.$$

Assuming the same material/coating/cooling technology as the HA class, substituting the data into the input file of the VBA function GasTurbine (see Table 8.4 in Chapter 8), it is found that the TIT for this particular gas turbine is 2,757°F (1,514°C), which is indeed about right for the advanced F-class technology. Using (simplified) Equation 11.45, the turbine flow parameter is estimated as

$$\frac{\dot{m}_3\sqrt{T_3}}{p_3} = \frac{0.8\cdot(1{,}140 - 26.5)\sqrt{2{,}757 + 460}}{18.21\cdot 14.7} = 188.9.$$

Let us estimate the output and efficiency of this gas turbine at 80°F. Firstly,

$$T_1^* = (80 + 460)/(59 + 460) = 1.0405.$$

If this gas turbine is controlled similar to the HA class, as discussed earlier, $m_1^* = 1$. If it is controlled similar to a conventional F-class gas turbine, from Equation 18.1b,

$$m_1^* = (1/T_1{}^*)^{1.4} = 0.946.$$

Assuming that we have no knowledge on this machine's design/control particulars, let us assume that $m_1^* = 0.97$ (i.e., better than old F class but not as good as HA class), so that

$$\dot{m}_1 = 0.97 \times (1,140 - 26.5) = 1,080 \text{ lb/s}.$$

Thus, from Equation 18.2, at 80°F, the cycle/compressor PR is

$$\text{PR} = 0.97 \times 18.21 = 17.66.$$

Turbine PR is about 10 percent lower than compressor PR (CPR), thus

$$p_3 = 0.9 \times 17.66 \times 14.7 = 233.6 \text{ psia}.$$

Since turbine inlet flow and pressure were changed in proportion, for the same turbine flow parameter, T_3 or TIT does not change. We substitute calculated inlet airflow, cycle PR, and specified ambient temperature into the inputs of GasTurbine and vary the fuel flow rate until T_3 is 2,757°F (using the Goal Seek function of Excel). The results are listed in Table 18.8 side by side with the 80°F columns from Tables 18.3 and 18.5.

As the proverb goes, "there is more than one way to skin the cat." For many purposes, representative curves similar to those presented herein can be used for gas turbines with similar technology vintages. Comparable results can also be obtained, with slightly more effort, from more involved heat and mass simulation calculations, as demonstrated by a simple example.

For "exact" calculations (to the extent one can use the term "exact"), one has to resort to rigorous calculations using component performance maps. (One example of such a map is shown in Figure 11.10.) In terms of normalized parameters, maps for similar compressors (i.e., radial or axial) are quite similar to each other. In the absence of the

Table 18.8 Hot-day performance estimation example

	Example	Table 18.3	Table 18.5
Ambient Temperature (°F)	80.0	80.0	80.0
Normalized Output	0.93	0.93	0.97
Efficiency Delta (%)	–0.68	–0.68	–0.47
Normalized Efficiency	0.98	0.98	0.99
Normalized Exhaust Flow	0.97	0.96	1.00
TIT Delta (°F)	0	–5.3	–3.4
Exhaust Temperature Delta (°F)	9.0	15.6	0.7

Table 18.9 Impact of altitude on gas turbine performance

Altitude (ft)	0	500.0	1,000.0	5,000.0
Normalized Output	1.00	0.98	0.96	0.83
Efficiency Delta (%)	0.00	–0.01	–0.03	–0.15
Normalized Exhaust Flow	1.00	0.98	0.96	0.83
TIT Delta (°F)	0	0	–0.1	–0.3
TEXH Delta (°F)	0	0.3	0.6	3.1
PR (at Turbine Inlet)	21.5	21.5	21.5	21.4

"actual" performance map of the gas turbine compressor being investigated (pretty much impossible to obtain unless you happen to be working for the OEM of the gas turbine itself[2]), using a suitable map obtained from the open literature[3] might serve quite well.

For most of its normal operating range, the turbine section is "choked" (i.e., the flat portion of the conceptual operating line in Figure 11.11). While there are performance maps for turbines or turbine stages (e.g., see the book by Münzberg and Kurzke in the references of Chapter 2), using the constant turbine flow parameter as defined by Equation 11.45 is sufficient for industrial gas turbines running at synchronous speed.

Ambient temperature is the most important "boundary condition" parameter. Ambient pressure or, alternatively, site elevation also has a significant impact on gas turbine performance via reduced density at higher altitudes; once the site is specified, there is no variability over the lifetime of the power plant. The data in Table 18.9 illustrate the gas turbine performance variation with altitude and can be used for quick estimates for a given gas turbine.

Ambient relative humidity has very little impact on performance and can be ignored for most practical purposes (the exception to this simplification will be discussed in conjunction with MBC in Section 18.3). A notable exception is applications with inlet evaporative coolers, whose performance is strongly affected by the humidity of

2 Even then, it might be a real chore to obtain it from the compressor aero team unless you provide a good reason as to why you need it.

3 *ASME Journal of Turbomachinery* can be a good source.

Table 18.10 Impact of ambient humidity on "evap-cooled" gas turbine performance

Ambient Temperature (°F)	59.0	90.0	90.0	90.0	100.0	100.0	100.0
Ambient Relative Humidity (%)	60.0	60.0	70.0	80.0	60.0	70.0	80.0
Ambient Wet Bulb Temperature (°F)	51.5	78.4	81.6	84.5	87.1	90.7	94.0
Normalized Output	1.00	0.918	0.909	0.901	0.890	0.881	0.872
Efficiency Delta (%)	0.00	−0.86	−0.98	−1.09	−1.23	−1.37	−1.50
Normalized Efficiency	1.00	0.977	0.974	0.971	0.967	0.964	0.960
Normalized Exhaust Flow	1.00	0.95	0.94	0.93	0.93	0.92	0.91
TIT Delta (°F)	0	−9.7	−11.0	−11.7	−12.3	−13.5	−14.8
Exhaust Temperature Delta (°F)	0	19.6	22.3	25.1	28.3	31.8	35.0

Table 18.11 Impact of inlet and exit pressure losses on gas turbine performance

Inlet Pressure Loss (in. H_2O)	4.0	2.0	6.0	8.0	4.0	4.0	4.0
Exit Pressure Loss (in. H_2O)	12.0	12.0	12.0	12.0	4.0	8.0	16.0
Normalized Output	1.00	1.006	0.994	0.989	1.003	1.001	0.999
Efficiency Delta (%)	0.00	0.06	−0.06	−0.13	0.24	0.12	−0.12
Normalized Exhaust Flow	1.00	1.005	0.995	0.990	1.000	1.000	1.000
Normalized Air Flow	1.00	1.005	0.995	0.990	1.000	1.000	1.000
Normalized Turbine Inlet Flow	1.00	1.005	0.995	0.990	1.000	1.000	1.000
TIT Delta (°F)	0	−0.3	0.3	0.8	−1.5	−0.7	0.8
Exhaust Temperature Delta (°F)	0	−1.8	1.8	3.8	−7.2	−3.6	3.6

incoming air. If the ambient air has high humidity, cooling potential stemming from latent heat absorbed by the water evaporating into the air stream is lower. This is illustrated by the data for a typical F-class gas turbine in Table 18.10. Note that the ambient temperature–relative humidity combinations in Table 18.10 are for illustration purposes only. In general, wet bulb temperatures above 85°F are quite rare. (As an example, the highest recorded wet bulb temperature in Dhahran, Saudi Arabia, was about 92°F.)

The impacts of inlet (i.e., inlet filter, evaporative cooler, etc.) and exit (i.e., transition duct, HRSG, etc.) pressure losses on gas turbine performance are illustrated in Table 18.11 (same firing temperature schedule and IGV position).

The impact of fuel heating value on gas turbine performance is a tricky subject. In theory (as well as in practice), a gas turbine can burn any fuel *as long as its fuel supply and conditioning system and its combustor are designed for that particular fuel*. Once a specific hardware is deployed in the field, only small variations in fuel gas composition and heating value can be tolerated due to combustion stability and emissions considerations. The parameter quantifying the suitability of a given fuel gas composition and heating value for the specific fuel and combustion hardware is the *modified Wobbe index* (MWI). Modern Dry-Low-NOx (DLN) combustors can tolerate only a limited range in MWI variation (due to fuel gas composition, temperature, etc.) due to considerations discussed in detail in Chapter 12. Beyond those limits, operation of a gas turbine is possible only with different combustion hardware.

Table 18.12 Impact of fuel gas heating value on gas turbine performance

Fuel LHV (Btu/lb)	21,518	21,145	20,777	20,415	12,307	9,942	4,240
Fuel LHV (Btu/scf)	881	872	864	855	617	529	264
WI	1,184	1,168	1,152	1,136	749	623	288
MWI	51.1	50.4	49.7	49.0	32.3	26.9	12.4
MWI (Relative)	1.0	0.99	0.97	0.96	0.6	0.5	0.2
Normalized Output	1.00	1.001	1.001	1.002	1.03	1.04	1.04
Efficiency Delta (%)	0.00	0.013	0.026	0.039	0.55	0.75	1.90
Normalized Exhaust Flow	1.00	1.000	1.001	1.001	1.02	1.02	1.04
Normalized Air Flow	1.00	1.000	1.000	1.000	1.00	0.99	0.94
Normalized Turbine Inlet Flow	1.00	1.000	1.001	1.001	1.02	1.03	1.05
TIT Delta (°F)	0	–0.1	–0.2	–0.3	–5.0	–17.0	–104.7
Exhaust Temperature Delta (°F)	0	–0.2	–0.4	–0.5	–7.7	–15.4	–64.4
PR (at Turbine Inlet)	21.5	21.5	21.5	21.5	21.9	22.0	22.2

If the combustor is "happy" with a given fuel heating value, accommodation of variations in turbine mass flow rate via higher or lower fuel gas flow rates is possible via IGV and/or firing temperature controls. This is illustrated by the data in Table 18.12. The first column is the ISO base load performance with 100 percent methane fuel. Down to about 20,400 Btu/lb LHV, increases in fuel flow at the same inlet airflow can be handled by the original combustor, resulting in better gas turbine performance. Fuels with much lower LHVs, provided that the combustor can handle them – which, as the discussion in Chapter 12 highlighted, is not possible – can be accommodated by the gas turbine via IGV closure (lower airflow) and TIT reduction.

Even if the combustor stability aspect is ignored, it is questionable whether the last three columns of Table 18.12 (especially the last two) represent realistic cases. The reason for this is the compressor operability. Since the normalized speed, ν^*, is the same, operating points for those cases in μ^* and π^* represent a shift along the speed line to the left and closer to the surge line (e.g., see Figure 11.12). Unfortunately, the feasibility of such operating points cannot be assessed with heat and mass balance simulation software calculations only. One has to be well-informed about the combustor hardware and the compressor map.

The other important aspect of off-design gas turbine performance is part load operation at specified boundary conditions. A gas turbine's power output (i.e., its "load") can be varied as follows:

- Via inlet airflow change (i.e., IGV opening or closing);
- Via firing temperature change (i.e., increasing or reducing fuel flow);
- A combination of both.

Gas turbine output control using the aforementioned two "knobs" can be via the following three routes:

- Constant airflow (TIT is modulated to set the load);
- Constant turbine inlet or firing temperature (inlet airflow is modulated – via IGVs and VSVs – to set the load);

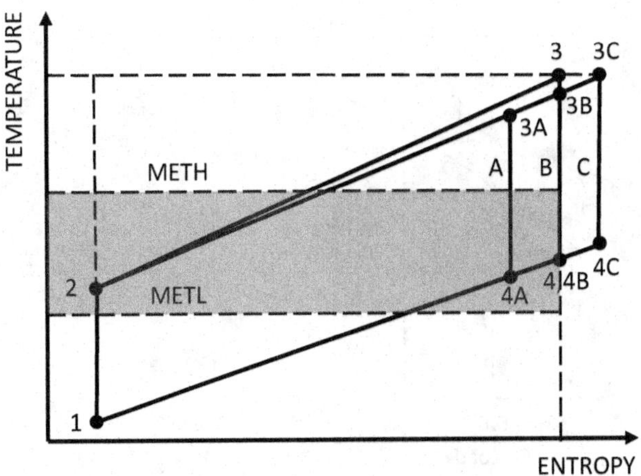

Figure 18.1 Brayton cycle temperature–entropy diagrams for three part load control approaches.

- Constant turbine exhaust temperature (TEXH; inlet airflow and TIT modulated to set the load).

Comparison of the three routes via their respective cycle diagrams on a temperature–entropy surface can be found in Figure 18.1. In particular,

- The base load cycle is {1–2–3–4–1}.
- The Carnot equivalent of the base load cycle is marked by the shaded area, which defines mean effective heat addition and heat rejection temperatures (METH and METL, respectively; see Chapter 5).
- The constant airflow (part load) cycle is denoted by A.
- The constant exhaust temperature (part load) cycle is denoted by B.
- The constant TIT (part load) cycle is denoted by C.

Visual examination of the cycle diagrams in Figure 18.1 clearly shows that the constant airflow method is ranked last in terms of combined-cycle efficiency and gas turbine-specific power (i.e., lowest METH, smallest Brayton-cycle area, and smallest exhaust exergy, which is quantified by the triangular area under the $p_1 = p_4$ isobar – see Chapter 5). The second best is the constant TEXH method, whereas the constant TIT method offers the best combined-cycle efficiency and gas turbine-specific power (i.e., highest METH, largest Brayton-cycle area, and largest exhaust exergy).

In the combined-cycle configuration, for the maximum output from the bottoming cycle, it is desirable to keep the exhaust gas temperature as high as possible. Consequently, load (output) reduction starts first by closing the IGVs and reducing the airflow. Maintaining TIT leads to increasing exhaust gas temperature because cycle PR decreases with decreasing turbine mass flow rate. This continues until the exhaust gas temperature reaches its upper limit (commonly denoted as the "exhaust isotherm"). From that point on, TIT is adjusted to maintain the exhaust isotherm. Once the minimum IGV setting and minimum airflow are reached, the only knob available is the TIT reduction accompanied by a drop in exhaust gas temperature.

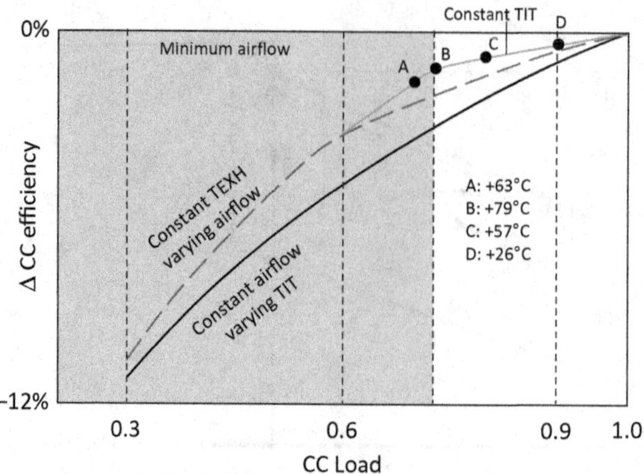

Figure 18.2 Combined-cycle (CC) efficiency impact of three part load control philosophies.

This is illustrated by a numerical example using an ideal (cold) air-standard Brayton-cycle model. For the calculations, the following assumptions are made:

- T_1 of 15°C (59°F)
- T_3 of 1,371°C (2,500°F)
- Cycle PR of 16
- Mass flow of 453.6 kg/s (1,000 lb/s)

Cycle PR is scaled using the choked nozzle assumption (i.e., Equation 11.45). Combined-cycle efficiency is calculated via Equation 5.17. The results are summarized in Figure 18.2. The numbers indicate the delta between TEXH values of constant TIT control and constant TEXH control. For example, at 80 percent load, TEXH with constant TIT control is 57°C (103°F) higher than that with constant TEXH control. When the IGVs are at their fully closed position, inlet airflow is at 70 percent of its base load value.

The main takeaways from Figure 18.2 are the following:

1. Constant airflow control is detrimental to combined-cycle efficiency at a specified part load.
2. Constant TIT control is best until minimum airflow (i.e., maximum IGV/VSV closure) at the base TEXH is reached (at 60% CC load).
3. Thereafter, constant TEXH and constant TIT controls are equivalent.

Clearly, constant airflow is not a feasible approach to controlling combined-cycle output. The extent of the advantage of constant TIT control vis-à-vis constant TEXH control (i.e., the area between the constant TIT and TEXH curves in Figure 18.2) is a function of the maximum allowable TEXH (typically ~1,200°F[4]). The lower that

[4] However, this is very likely to be much higher in near future as evidenced by the cycle data from the major OEMs for their upcoming, advanced gas turbines. See Chapter 21.

Table 18.13 E-class gas turbine ISO part load operation

Ambient Temperature (°F)	59.0	59.0	59.0	59.0	59.0	59.0	59.0	59.0	59.0	59.0
Normalized Output	1.00	0.90	0.80	0.70	0.60	0.50	0.40	0.30	0.20	0.10
Normalized Efficiency	1.00	0.97	0.94	0.88	0.82	0.76	0.68	0.61	0.50	0.33
Normalized Exhaust Flow	1.00	0.90	0.82	0.77	0.71	0.66	0.61	0.60	0.60	0.59
TIT Delta (°F)	0	5.7	4.2	–0.8	–9.1	–49.9	–99.4	–290.3	–518.6	–762.4
TEXH Delta (°F)	0	40.2	68.8	90.5	112.8	114.1	114.1	4.8	–131.4	–276.0
PR (at Turbine Inlet)	13.5	12.1	11.1	10.4	9.6	8.9	8.1	7.6	7.2	6.8

limiting value (i.e., the exhaust isotherm), the smaller the potential benefit of constant TIT control over constant TEXH control.

The comparison in Figure 18.2 and the conclusions drawn from that are with respect to combined-cycle efficiency at the same combined-cycle load only. It provides no information regarding emissions (NOx, CO, etc.), which might render the constant TIT control infeasible at part or all of the load range where it shows an efficiency advantage. Emissions characteristics are highly dependent on the particular gas turbine combustion technology and cannot be investigated using simple models. Furthermore, limitations imposed by gas turbine emissions on the operating envelope are site dependent via existing laws and regulations and not subject to a purely technical study.

Ideal-cycle analysis qualitatively explains the differences between different part load control philosophies. It is now time to look at actual part load performances of heavy-duty industrial gas turbines. We start with ISO part load data for an E–class gas turbine in Table 18.13.

The control scheme in Table 18.13 is characteristic of older-generation gas turbines with low exhaust gas temperatures and an appreciable "gap" between the full load value and the exhaust isotherm. This is also, qualitatively, the TEXH control method adopted by GE and Alstom. Maintaining the exhaust temperature from the get-go is another control strategy as illustrated by the ISO part load data for an F–class gas turbine in Table 18.14. This is the TEXH control method adopted by Siemens (e.g., see the paper by Jansen et al. [1]).

Typically, part load IGV–TIT control schedules or curves are built into the gas turbine controller. Since direct measurement of TIT is not possible, actual control curves are based on CPR and TEXH. There are several built-in CPR–TEXH curves for different TIT values with an "IGV bias" to account for IGV opening (IGV angle) at part load operation. State-of-the-art gas turbines with IGVs and several stages with VSVs use MBC, where the optimal operating point at given boundary/ambient conditions and operator-selected load/output is found by running a high-fidelity gas turbine model inside the controller. As an example, consider the ISO part load data in Table 18.15 for an HA–class gas turbine. In terms of the exhaust temperature variation with load, the E- and HA-class gas turbines in Tables 18.13 and 18.15 are similar. Note, however, the limited gap between full load and isotherm (~50°F) for the advanced gas turbine vis-à-vis the older-vintage gas turbine (~115°F).

Table 18.14 F-class gas turbine ISO part load operation

Ambient Temperature (°F)	59.0	59.0	59.0	59.0	59.0	59.0	59.0	59.0	59.0	59.0
Normalized Output	1.00	0.90	0.80	0.70	0.60	0.50	0.40	0.30	0.20	0.10
Normalized Efficiency	1.00	0.97	0.94	0.90	0.86	0.80	0.74	0.65	0.51	0.32
Normalized Exhaust Flow	1.00	0.94	0.88	0.82	0.76	0.70	0.70	0.70	0.70	0.69
TIT Delta (°F)	0	–43.3	–91.2	–144.4	–204.1	–271.6	–457.3	–647.6	–826.5	–1008.2
TEXH Delta (°F)	0	0	0	0	0	0	–87.9	–174.8	–253.0	–329.3
PR (at Turbine Inlet)	16.1	15.0	14.0	12.9	11.8	10.8	10.3	9.9	9.4	9.0

Table 18.15 HA-class gas turbine ISO part load operation

Ambient Temperature (°F)	59.0	59.0	59.0	59.0	59.0	59.0	59.0	59.0	59.0	59.0
Normalized Output	1.00	0.90	0.80	0.70	0.60	0.50	0.40	0.30	0.20	0.10
Normalized Efficiency	1.00	0.97	0.94	0.90	0.85	0.80	0.73	0.64	0.52	0.34
Normalized Exhaust Flow	1.00	0.94	0.88	0.81	0.75	0.68	0.62	0.56	0.49	0.42
TIT Delta (°F)	0	–25.6	–49.8	–80.9	–120.5	–175.9	–256.7	–356.1	–482.2	–651.7
TEXH Delta (°F)	0	7.4	20.0	32.6	45.2	54.1	54.1	54.1	54.1	52.6
PR (at Turbine Inlet)	21.5	20.1	18.7	17.2	15.7	14.3	12.8	11.3	9.8	8.1

The second law approach to combined-cycle performance analysis, outlined in Chapter 17, is equally handy for performance calculation in off-design conditions. Once the gas turbine performance is available, its exhaust exergy can be used to calculate the bottoming-cycle power output with the exergetic efficiency. Rigorous second law analysis by Gülen and Joseph illustrated that simple corrections to the base exergetic efficiency given by Equation 17.24 are sufficient for this purpose [2]. As shown in Figure 18.3, for part load operation above ~60 percent, using Equation 17.24 "as is" is sufficiently accurate. (The data in Figure 18.3 are for a 1 × 1 combined-cycle system with 50- and 60-Hz frame gas turbines and a once-through, open-loop, water-cooled condenser-based steam turbine heat rejection system. The design point is ISO base load.)

In order to illustrate this with an example, first recall the second law variant of net combined-cycle efficiency from Chapter 17, i.e.,

$$\eta_{CC,net} = \frac{N \cdot \left(\dot{W}_{GT} + \varepsilon_{BC} \cdot \dot{m}_{exh} \cdot a_{exh}\right) - \dot{W}_{Aux}}{N \cdot HC}.$$

For the combined-cycle gross efficiency, we obtain

$$\eta_{CC,gross} = \dot{W}_{GT} + \varepsilon_{BC} \cdot \dot{m}_{exh} \cdot a_{exh}. \tag{18.4}$$

Gas turbine exhaust exergy, a_{exh}, can be obtained from Equation 5.60a.

For an HA-class gas turbine (specifically, GE's 7HA.02, engine #602 from Thermoflow GT PRO's gas turbine library) in single-shaft combined-cycle configuration, ISO part load data from GT MASTER (the off-design companion of GT PRO) are summarized in Table 18.16.

Table 18.16 HA-class combined-cycle ISO part load performance

Gas Turbine Load (%)	100	90	80	70	60	50	40	30	20
Plant Load (%)	100	92	83	75	66	57	48	39	30
Plant Output (MWe)	503.2	461.5	419.6	376.9	333.4	288.7	242.9	196.3	148.6
Plant Net Efficiency (%)	60.41	59.71	58.88	57.89	56.70	55.21	53.30	50.72	47.02
Fuel Input (LHV)	2,843	2,637	2,432	2,221	2,007	1,784	1,555	1,321	1,078

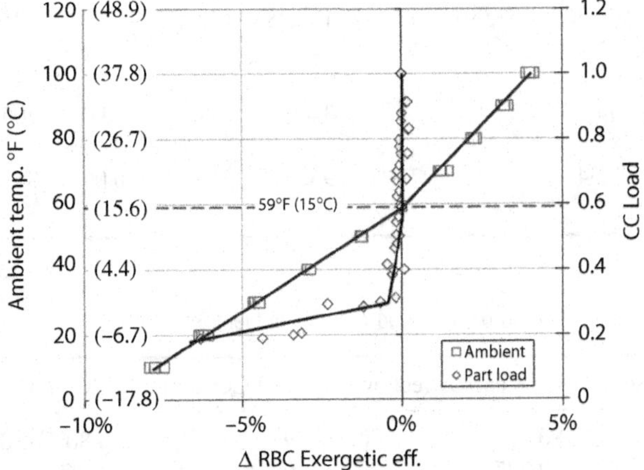

Figure 18.3 Ambient and part load variation of bottoming-cycle exergetic efficiency. CC = combined cycle; RBC = Rankine bottoming cycle.

The data in Table 18.16, on a normalized basis, can be used for any state-of-the-art GTCC as a first estimate. They can also be used as a handy reference to relate gas turbine simple- and combined-cycle part loads.

Part load bottoming-cycle performance (i.e., steam turbine output) estimated from Equation 18.4 is compared to the GT MASTER calculation in Table 18.17. Clearly, for most practical purposes, the simple exergy approach is perfectly acceptable.

Strictly speaking, there is a specific part load curve for a given ambient temperature. As a first approximation, however, combining the ambient and load factors is sufficiently accurate for most purposes. For example, at 80°F and 80 percent load for an E-class gas turbine, combining the data from Tables 18.1 and 18.13, we obtain the estimated correction in Table 18.18. Correction factors obtained from the GT PRO model run are summarized in Table 18.19.

18.1.1 Power Augmentation

Gas turbine power output is a strong function of air mass flow rate. On the other hand, a gas turbine is a constant volumetric flow machine (due to constant rotational speed when connected to the grid). When operating at high ambient temperatures, due to the reduction in air density, gas turbine power output suffers significantly via reduced air

Table 18.17 HA-class bottoming-cycle ISO part load performance

Bottoming-Cycle Exergetic Efficiency (%)	77.40	77.45	77.55	77.65	77.73	77.77	77.77	77.77	77.77
Gas Turbine Exhaust Exergy (Btu/lb)	134.63	135.88	138.13	140.41	142.72	143.85	143.85	143.85	143.85
Steam Turbine Power Output (Exergy Model) (kW)	167,615	159,173	151,075	142,499	133,390	123,110	111,843	100,068	87,540
Steam Turbine Shaft Power at Generator (GT PRO) (kW)	167,793	159,409	151,091	142,262	132,884	122,452	111,173	99,276	86,498
Plant Shaft Output (MW)	516.4	474.1	431.7	388.5	344.6	299.3	253.0	205.9	157.8
Implied Generator Efficiency (%)	97.45	97.34	97.19	97.01	96.77	96.45	96.00	95.30	94.16

Table 18.18 Estimated performance (E class) at 80 percent load at 80°F ambient temperature

	Base	Ambient Correction	Load Correction	Combined Correction
Ambient Temperature (°F)	59.0	80.0	59.0	80°F + 80%
Normalized Output	1.00	0.927	0.80	0.742
Normalized Efficiency	1.00	0.981	0.935	0.917
Normalized Exhaust Flow	1.00	0.948	0.823	0.780

Table 18.19 "Actual" performance (E class) at 80 percent load at 80°F ambient temperature

	Base	80°F + 80%	Combined Correction
Output (kWe)	114,828	85,148	0.742
Efficiency (%)	34.89	31.69	0.908
Exhaust Flow (lb/s)	782.9	625.2	0.799

mass flow rate. In a combined cycle, this loss is amplified by the steam turbine power output loss via reduced steam production in the HRSG (via lower gas turbine exhaust gas mass flow rate).

There are two methods to make up for the power output loss:

- Supplementary firing (also referred to as *duct firing*) in the HRSG to make up for lost steam production due to low gas flow via increased gas temperature;
- Cooling of compressor inlet air.

Since hot summer days, when air conditioners in residences and office buildings (especially in the USA) are going full-blast, are exactly the times when a GTCC power plant has the best opportunity to generate income (typically, wholesale prices are very high), either method or a combination of the two is deployed regardless of the negative impact on fuel efficiency.

From the perspective of the gas turbine only, hot-day power augmentation via inlet air cooling is of interest. There are three widely accepted ways to achieve this ("novel" schemes are ignored because they have not been adopted by plant owners/operators on a meaningful scale):

- Evaporative cooling
- Inlet fogging (with or without "overspray")
- Inlet chiller

One other method, mainly adopted by European OEMs, is "wet compression," which is similar to inlet fogging with overspray (also referred to as "high fogging").

In terms of field deployment, evaporative coolers (commonly referred to as "evap-coolers") rank first by a large margin. Their operating principle is latent heat transfer from hot air into which colder water evaporates. A fundamental thermodynamic analysis can be found in any textbook (e.g., see section 12.10.3 in Moran and Shapiro). Theoretically, their performance limit is the wet bulb temperature (see Tables 12.4 and 12.5), which is a function of ambient relative humidity. Thus, evap-cooling is most effective in dry and hot climates. How close the cooled and humidified air can get to the wet bulb temperature is measured by the evap-cooler effectiveness. One-hundred percent effectiveness means that $T_1 = T_{wb}$. The typical effectiveness of the state-of-the-art evap-coolers used in the field is 90 percent. The positive impact of evap-cooling is somewhat dampened by the additional inlet pressure loss because of the extra hardware in the duct (about 0.5–1.0 in. of water column).

Inlet fogging without overspray makes use of the same thermodynamic principle as the evap-cooler. Since water and air are in direct contact, inlet fogger effectiveness is very close to 100 percent. By the same token, additional inlet pressure loss is smaller, too. On the debit side, evap-coolers are less stringent in their water quality requirements (although they require more water for blow-down); they can use potable water (if equipped with a moisture separator or "drift eliminator"), whereas demineralized ("demin") water is required by inlet fogging systems. Demineralized water is necessary to prevent compressor blade fouling and hot gas path corrosion, which can be caused by the minerals that are naturally present in untreated water. Inlet foggers require high-pressure pumps for atomization of water spray into fine droplets (<50 μm in diameter). There is always the potential for unevaporated droplets to impinge on first-stage compressor blades at very high speeds, leading to parts erosion.

Inlet fogging with *overspray* or *high fogging* works through a combination of two separate physical mechanisms: evaporative cooling and compressor intercooling. The second mechanism is activated by spraying more water into the inlet air stream than is requisite for 100 percent relative humidity (hence "overspray"). Unevaporated water droplets at the inlet evaporate inside the compressor as they absorb latent heat from the compressed air, which becomes hotter as it goes through the compressor stages. Absorption of latent heat creates a continuous intercooling effect and reduces the power consumption of the compressor. (In some wet compression schemes, water is introduced into the compressor downstream of the first or second stage [i.e., not at the inlet]. This is also referred to as *inter-stage water injection*.)

Details of evaporative and wet compression gas turbine "inlet conditioning" technologies for power augmentation, including thermodynamic, operating, and many other relevant aspects, can be found in [3–7].

Inlet chilling refers to compressor inlet air cooling by heat transfer to "refrigerated" water (typically water plus glycol) in a cooling coil in the inlet duct. The refrigeration process can utilize standard package chillers or an absorption chiller. These systems can cool the inlet air well below the wet bulb temperature. Their performance is independent of ambient humidity. Their capital cost is much higher than those of evap-coolers or inlet foggers. In the case of packaged chillers, net performance is severely impacted by the power consumption of the refrigerant compressor. Although the parasitic power consumption of the absorption chiller is much lower (it does not have a compressor), it impacts the combined-cycle performance via heating steam or water extraction from the bottoming cycle. Another disadvantage is that associated piping and HRSG modifications add to the initial capital outlay.

The thermodynamic principles governing chiller operation can also be found in any textbook (e.g., see chapter 10 in Moran and Shapiro).[5] Sample evaluations of gas turbine inlet conditioning options including compressed refrigerant and absorption chillers are reported in [8,9]. There are many similar studies published in the academic and trade literature (for a recent example of the latter, see [10]). The bottom line is that the cost–performance trade-off in the case of mechanical chillers is almost always unfavorable for gas turbine power plants. The author is not aware of a large GTCC power plant in commercial operation equipped with mechanical chillers of any kind (straight refrigerating or absorption).

A comparison of evap-cooling and inlet fogging (with 1 percent overspray; i.e., in effect, wet compression) on GTCC performance for rule-of-thumb estimates can be found in Table 18.20. The amount of overspray is referenced to the water flow rate for achieving 100 percent relative humidity (i.e., wet bulb temperature) for compressor inlet air. This is typically limited to 2 percent, depending on the OEM and the particular machine. (Some OEMs do not allow any fogging due to the detrimental impact on the compressor R0/R1 blades – at least until recently.[6]) The GTCC is a typical 1 × 1 × 1 single-shaft design with an advanced-class gas turbine and three-pressure reheat bottoming cycle. Performances are calculated in Thermoflow's GT MASTER software. The "fine" droplet option was selected for the inlet fogger (see below).

Items of interest in Table 18.20 are as follows:

- In either case, overall increase in combined-cycle output is compensated by increased fuel flow rate (higher airflow into the combustor).

[5] In the movie *The Mosquito Coast*, a somewhat crazy scientist (played by Harrison Ford) constructs a tin shack in the jungle that is a wood-fired freezer operating on the same principle as an ammonia–water absorption chiller (recall that ammonia can be made from urine).

[6] Refer to the two versions of GER-3620, J (2003) and L (2010), to see how the GE view on the detrimental effect of inlet fogging on compressor parts changed (albeit slightly) over time.

Table 18.20 GTCC performance with evap-cooling and inlet fogging. CC = combined cycle; CDT = compressor discharge temperature; GT = gas turbine; ST = steam turbine

	No Inlet Condition	Evap-Cooler	Inlet Fogger
Ambient Conditions	90°F/60% relative humidity		
Overspray	NA	NA	1%
CC Output	1.0	1.039	1.062
CC Heat Rate	1.0	0.996	1.009
CC Efficiency Delta (%)	0	0.23	–0.51
Compressor Inlet Temperature (°F)	90	79	78
GT Fuel Flow	1.0	1.035	1.071
GT Exhaust Flow	1.0	1.029	1.019
CDT Delta (°F)	0	–10.0	–103.0
TIT Delta (°F)	0	0	–20.0
GT Exhaust Temperature Delta (°F)	0	–10.5	–14.3
GT Output	1.0	1.043	1.087
GT Efficiency Delta (%)	0	0.32	0.61
ST Output	1.0	1.028	1.008
HRSG Effectiveness Delta (%)	0	–0.22	–0.45
Steam Cycle Efficiency Delta (%)	0	0.37	–0.19

- In the case of evap-cooling, combined-cycle heat rate is slightly better than the base case.
- In the case of inlet fogging with overspray, combined-cycle heat rate is slightly worse.
- Fuel flow increase in the latter case is higher because of lower compressor discharge temperature (intercooling effect of evaporating overspray droplets).

Note that wet compression (i.e., compression of a two-phase mixture of gaseous air – itself a mixture of several gases including water vapor) and liquid H_2O in the form of microscopic droplets in a multi-stage axial compressor is not amenable to a straightforward calculation model. There are many papers in the literature describing rigorous analytical models to calculate the wet compression process. The reader is referred to the paper by Bhargava et al. [6] and the references therein for further study. Wet compression tests conducted by OEMs on their gas turbines are also available in the literature [11–14]. Results change significantly from OEM to OEM and from product to product. The performance impact closest to that shown in Table 18.20 was reported in [14] (see figure 6 on p. 8 of the cited work).

Caveat emptor: power augmentation via compressor inlet air-conditioning – using any technology – is subject to large deviations from OEM to OEM, from product to product for the same OEM, and from machine to machine for the same product *in addition* to particular site and ambient loading conditions. As an example, the reader is referred to figure 4.11 of [11] (written by W. Kappis) cited in Chapter 2 of the current book. Three gas turbines of the same type by the same OEM show variation in power output increase from ~5.5 to 7.0 percent for the same amount of inlet fogger water spray.

18.2 Controls

There are two types of controls in a heavy-duty industrial gas turbine:

1. Speed control (also known as "governor")
2. Temperature control

In combined-cycle applications, temperature control is combined with IGV control to maintain the highest possible TEXH for maximum combined-cycle efficiency at part load.

There are two types of governor:

1. Droop or "straight proportional" speed control
2. Isochronous or "integrated" (proportional *plus* reset) speed control

In general, droop governor is applicable to operations when connected to a grid (i.e., fixed system frequency and shaft rotational speed). Isochronous governor is applicable when the gas turbine is *not* connected to the grid and operates in an isolated or "islanded" mode. It is essentially a "zero-droop" governor.

The droop is defined as the ratio of the relative change in system frequency to the relative change in generator power output, i.e.,

$$d = \frac{\frac{\Delta f}{f_0}}{\frac{\Delta \dot{W}}{\dot{W}}} \times 100, \tag{18.5}$$

where

Δf = change in system frequency in Hz
f_0 = rated system frequency in Hz
$\Delta \dot{W}$ = change in generator output in kW or MW
$\dot{W}$ = rated generator output in kW or MW
d = droop in percent

For example, if a 1 percent change in frequency (i.e., 0.5 Hz for a 50–Hz grid) causes a 25 percent change in output, the droop is

$$d = 0.01/0.25 \times 100 = 4, \text{i.e.}, 4\%.$$

The droop governor operates as follows:

1. It compares the actual speed to a reference (set-point) value, which is also known as the "speed changer."
2. The difference between the two is referred to as the "speed error."
3. In order to maintain the actual speed at its set-point value, the droop governor changes the fuel flow (i.e., the generator output) in proportion to the speed error.
4. The proportionality factor is d.

Thus, for a 4 percent droop governor, in order to run the gas turbine at 100 percent output/load, the operator sets the "speed changer" to 4 percent because, per Equation 18.5,

$$\frac{\Delta \dot{W}}{\dot{W}} = \frac{\frac{\Delta f}{f_0}}{d} \times 100 = \frac{0.04}{4} \times 100 = 1 \text{ or } 100\%.$$

Similarly, for a 4 percent droop governor, in order to run the gas turbine at 50 percent output/load, the operator sets the "speed changer" to 2 percent because, per Equation 18.5,

$$\frac{\Delta \dot{W}}{\dot{W}} = \frac{\frac{\Delta f}{f_0}}{d} \times 100 = \frac{0.02}{4} \times 100 = 0.5 \text{ or } 50\%.$$

The droop governor curve for 50 percent output at 100 percent speed is shown in Figure 18.4. From the preceding discussion, it should be obvious that 4 percent droop operating lines for 75 and 100 percent loads are constructed by shifting the operating line in Figure 18.4 to the right (i.e., 3 and 4 percent speed error, respectively).

Typically, after synchronizing to the grid and generator breaker closure, a gas turbine is loaded from full-speed, no load (FSNL; i.e., 0 percent) to full-speed, full-load (FSFL; i.e., 100 percent) in 12 minutes (i.e., at a rate of 8.33 percent per minute). With a 4 percent droop governor, speed error change is 4%/12 = 0.333% per minute. This is the rate chosen to minimize transient thermal stresses to the hot gas path components of the turbine. For "fast-start" machines with a double-shell construction, the time is typically halved by doubling the speed error change rate.

What happens when the speed error changes due to a change in the grid frequency? For example, a drop in the grid frequency at a given speed changer will result in a proportional increase in turbine output. In other words, at a set-point of, say, 2 percent speed error (i.e., 50 percent load at 100 percent speed/frequency as shown in

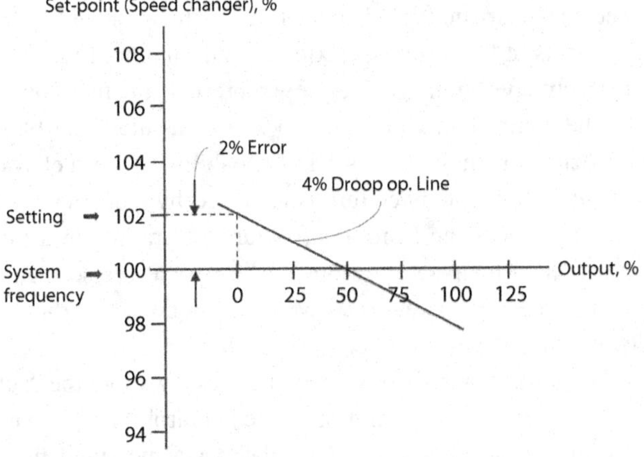

Figure 18.4 Droop governor – 50 percent load at 100 percent speed/frequency.

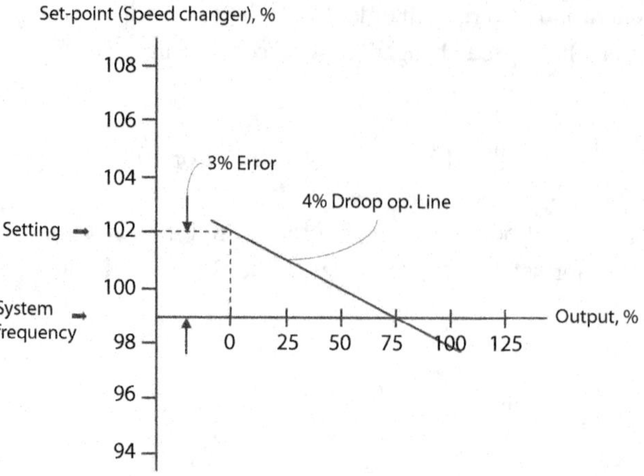

Figure 18.5 Droop governor response to a 1 percent drop in system frequency.

Figure 18.4), a drop in system frequency from 100 to 99 percent will cause the speed error to increase to 3 percent and the turbine output to increase to 75 percent (see Figure 18.5). This action by the droop governor is beneficial to restoring the system frequency back to its rated value because the cause of the frequency drop is a mismatch between system load and generation (i.e., the latter is inadequate to meet the load). Increased turbine generator output is in the direction of reducing this mismatch (imbalance).

There is no inherent problem in gas turbine control responding to a system frequency drop of 1 percent while operating at 50 percent load. The question is: What would happen if the gas turbine was running at 100 percent (i.e., with a 4 percent speed error set-point)? Could it raise its output to 125 percent in accordance with a 4 percent droop governor characteristic? The answer requires a closer look at the operating map of a gas turbine, which is qualitatively depicted in Figure 18.6. Figure 18.6 shows the part load characteristics at hot-day, ISO, and cold-day conditions with IGVs fully open (simple-cycle operation). The characteristic (speed control) lines are bounded by *base* and *peak* exhaust temperature limits at the top and maximum and minimum fuel flow limits on the right and left, respectively. Minimum fuel flow represents the fuel flow necessary to maintain a flame in the combustor, which is also known as the "lean blowout" limit. Maximum fuel is associated with the "full stroke" capability of the fuel system. Lower than full load at a given ambient temperature is achieved by reducing the fuel flow at constant airflow, which reduces the firing temperature and results in a lower exhaust temperature. As the ambient temperature drops, airflow increases so that, at the same firing temperature, the cycle PR increases as well (i.e., "choked" turbine nozzle) and thus results in a lower exhaust temperature.

Coming back to the question posed above, the answer is: without the "auto-peaking" capability, "no." The gas turbine will operate on speed control (droop governor) in part load until the allowable firing temperature (calculated via measured turbine exhaust temperature at the existing compressor discharge pressure or PR) is reached. At that

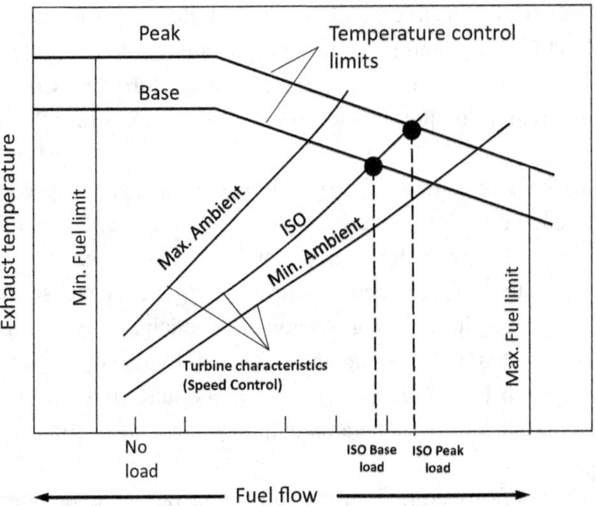

Figure 18.6 Gas turbine operation via speed and exhaust temperature control [15].

point, temperature control becomes active and caps the fuel flow by capping the fuel command signal (known as "fuel stroke reference" [FSR]) from further increase in response to the demands of the droop governor.

The next question is: What about the peak firing? According to the operation map in Figure 18.6, there *is* reserve capacity in the unit via increasing the firing temperature from "base" to "peak." This, however, is typically a *manual* control setting, which must be activated by the operator. The reason for this is the detrimental impact of operation at higher-than-rated gas temperatures on hot gas path component life. Peak-fired operation should be avoided unless absolutely necessary, and then only for the shortest duration possible.

In order to provide reserve capacity while operating on base temperature control, the gas turbine controller might contain an "auto-peaking" feature. This control feature utilizes a speed/frequency detector to automatically ramp the exhaust temperature control set-point from the base curve to the peak curve if the system frequency drops below the detector setting. Consequently, fuel flow control returns from temperature control to the droop governor, which increases the fuel flow in response to the drop in grid frequency. How much reserve capacity this feature adds to the gas turbine is a function of the peak rating of the gas turbine in question (*if* it has a peak firing capability in the first place). In combined cycles, another factor to take into consideration is the allowable exhaust gas temperature limit imposed by the HRSG design and materials. In any event, 25 percent (e.g., 75 MW for a gas turbine rated at 300 MW at ISO full load) is not physically possible. Even if the firing temperature were allowed to get there, shaft torque and/or generator limits would preclude it. While it ultimately depends on the particular gas turbine, a 10 percent boost over base load is probably the upper limit. (As a rule of thumb, increasing TIT 30–50°C – 54–90°F – results in 5–10 percent higher output.)

Once the grid frequency is restored and detected by the frequency detector, the gas turbine is returned to base temperature control after an adjustable time interval. Amount of overfiring (via measured exhaust temperature) and its duration is recorded by the equivalent operating hours counter in the control system to adjust the maintenance interval.

On the flip side of the coin, a jump in the grid frequency at a given speed changer will result in a proportional decrease in turbine output. For example, at a set-point of 2 percent error, a jump in system frequency from 100 to 101 percent will cause the speed error to decrease to 1 percent and the turbine output to decrease to 25 percent. Once again, this action by the droop governor is beneficial to restoring the system frequency back to its rated value since the cause of frequency increase is also a mismatch between system load and generation (i.e., the latter is *more* than needed to meet the load). Decreased turbine generator output is, once again, in the direction of reducing this mismatch.

The concept of a droop governor is not intuitive; in fact, it is counterintuitive. In contrast, an isochronous governor is perfectly intuitive. In the example above, when the system/grid frequency goes up, an isochronous governor would simply decrease the fuel flow and generator output to maintain the rated frequency. Why is it not chosen? The reason is that without some form of droop, turbine speed regulation would always be unstable in a networked system (i.e., the electric grid supplied by many generators). This can be best understood by an analogy to a simple physical system shown in Figure 18.7.

In Figure 18.7, four men pull a 100–unit load, which is analogous to a 50-Hz electric grid. Consequently, the four men are analogous to four gas turbine generators. Each man (generator) provides a certain fraction of the load. All of a sudden, a 20–unit load block falls off. The mismatch between the force exerted by the four men (100 units) and

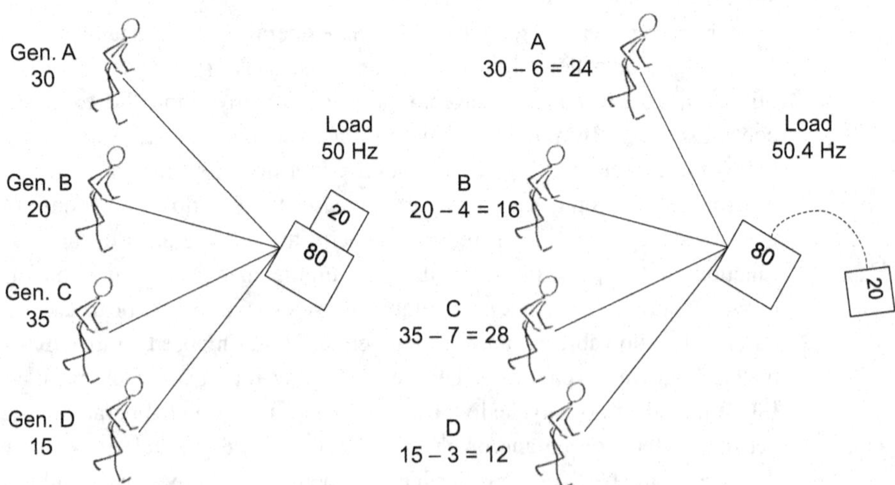

Figure 18.7 Physical analogy to generators with a droop governor responding to a loss in load.

the load (80 units) would increase the speed of the system, which, in the grid analogy, corresponds to a frequency increase of 0.4 Hz. Here is the catch: without droop, generators would not know how to divide the new load (80 units) amongst themselves!

If the generators/men were all operating on an isochronous governor, they would all try to correct the frequency/speed on their own, which would lead to utter chaos and system breakdown.

With a 4 percent droop governor, however, each generator knows exactly how much output correction it needs to make. For generator A, for example, using Equation 18.5,

$$\Delta\dot{W} = \frac{\frac{\Delta f}{f_0}}{d} \times 100 \times \dot{W} = \frac{\frac{0.4}{50}}{4} \times 100 \times 30 = 6 \text{ units.}$$

Similar corrections are made by generators B, C, and D *independently*, and their total output matches the new system load (80 units) and the system goes back to its stable running speed/frequency.

If a particular gas turbine generator is the only unit on the grid, an isochronous governor is indeed the logical choice because the generator can set its speed and thus the grid frequency. But for a generator connected to a large electrical grid of tens of gigawatts, it is the grid that determines the frequency and speed of the particular electric generator. For almost all cases, the grid disturbance and frequency variation are too big to be influenced by a single generator. A droop governor, however, does *not* try to change the speed/frequency of the system. It simply decreases or increases the governor speed reference as the load increases or decreases. This allows the governor to vary the gas turbine load since the speed, which is determined by the grid frequency, cannot change.

The droop setting, d, can be adjusted in the field to between 2 and 10 (i.e., 2 and 10 percent droop). Most national grids in the world operate at 4 or 5 percent droop. A low value of d means a large load change for a small change in system frequency. In the same vein, a high value of d means a small load change for a large change in system frequency. In the former case, load control stability is not very good (i.e., very large load swings in response to a small disturbance in grid frequency). In the latter case, the grid frequency-restoring or -regulating contribution from the particular generator is marginal. A simulation of the speed governor response to a unit load change (i.e., from 0 to 100 percent) for different droop settings is conceptually shown in Figure 18.8. The situation shown in Figure 18.8 is typical of a smaller gas turbine (e.g., GE's Frame 5 or 6); for larger units (e.g., Frame 9) with much higher inertia, the oscillations are damped quickly. The difference between the "stiff" (d = 2 percent) and stable cases (d = 10 percent) is striking.[7]

How the actual fuel flow command is generated is described in a simplified block diagram in Figure 18.9. Note that, for simplicity, only two major control modes (i.e.,

[7] In mathematics, a stiff equation is a differential equation for which the numerical solution is unstable (i.e., it oscillates), unless the step size is taken to be extremely small. In analogy, with d = 10 percent, the full load range is divided into 10 "steps" vis-à-vis only two "steps" with d = 2 percent.

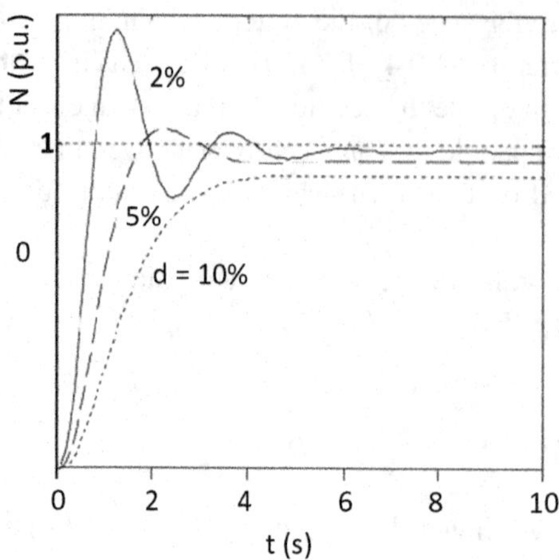

Figure 18.8 Gas turbine speed governor responses with different droop settings.

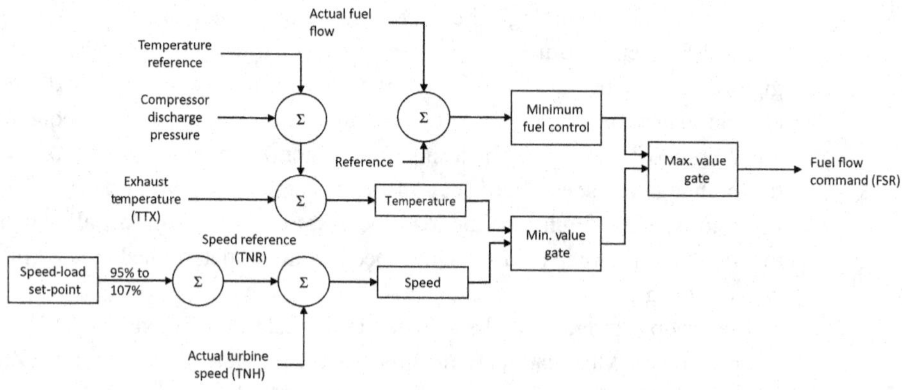

Figure 18.9 Simplified fuel control block diagram [16].

temperature and speed) are shown going into the minimum value gate. Other modes (i.e., startup, acceleration, manual entry, and synchronization) are omitted. The command signal for fuel flow, in GE control parlance, is the FSR. The input to the system is the operator command for speed (when disconnected from the grid) or load (when connected). The output is fuel flow command (liquid or gaseous – herein only the latter will be covered). Note that between FSNL and FSFL, the gas turbine fuel flow changes between about 0.25 and 1.0 in normalized terms. This can be verified from the data in Tables 18.13–18.15. (Fuel flow at FSNL was given by Rowen as 0.23 for vintage frame units [4].)

In essence, FSR is changed by the turbine controller in proportion to the difference between the actual turbine–generator speed (TNH) and the requested speed reference (TNR). The reason for the acronym "TNH" goes back to the days of old, two-shaft Frame 5 gas turbines and refers to the speed of the high-speed shaft (hence "H"). In modern single-shaft turbines (E, F, and H class), of course, it is the actual shaft speed.

The reference speed, TNR, determines the load of the turbine as discussed earlier. The range for heavy-duty industrial gas turbines is from 95 percent (minimum) to 107 percent (maximum) speed. In a 4 percent droop governor, it is 104 percent (see the discussion above). The controller also ensures that the FSR does not go below the value that would cause flame-out during a transient condition (minimum fuel control block).

As shown in the simplified block diagram in Figure 18.9, the temperature controller function has the following three components:

1. Measured exhaust temperature signal TTX (or TTXM; i.e., mean TTX, which is the average of exhaust thermocouple readings)
2. Compressor discharge pressure (CDP) bias
3. Temperature reference (TTXR)

In simple, conceptual terms, fuel flow command, FSR, is generated such that TTX does not exceed the "base" or "peak" (whichever is selected) allowable value, TTXR, corresponding to the firing temperature found from CDP and TTX. As long as the FSR from speed–load control is lower than that, it is picked by the minimum value gate. Fuel flow to the combustor cans is changed in response to the FSR command, which is the percentage of the maximum fuel flow (when peak-fired at minimum ambient temperature) required by the control system. A schematic description of the gas fuel control system is provided in Figure 18.10.

The fuel gas control system comprises the following three valves:

1. ANSI Class VI shutoff valve (SOV)
2. Stop/speed ratio valve (SRV)
3. Gas control valve (GCV)

If fuel supply conditions (pressure and temperature) do not exceed the Class VI shutoff limitation[8] on the SRV, the functionality of a separate SOV can be provided by the SRV. In DLN combustors, there are four GCVs (i.e., GCV-1, -2, -3, and -4) for each fuel passage, PM1, PM2, PM3, and D5 (see Figure 12.6).

The SRV controls the gas supply pressure P2 upstream of the GCV, which is a function of the gas turbine speed (TNH), and P3, which is a function of the compressor PR and the fuel nozzle size. The P2 set-point is lower at low speeds in order to provide better flow control at low fuel flow and increases linearly to the 100 percent rated speed point. Since the GCV does not have to reduce the fuel line pressure, it can open further to provide better low-flow control.

The rated P2 set-point must be sufficiently high in order to maintain the critical pressure drop, PR_CRIT, across the GCV at maximum flow. The SRV-controlled P2 and the GCV's design PR_CRIT ensure that the valve stroke is proportional to the fuel

[8] There are six different seat leakage classifications defined by ANSI/FCI 70-2 2006 (European equivalent standard IEC 60534-4). Class VI is known as soft seat classification. Soft seat valves are those where either the plug or seat or both are made from some kind of composition material such as polytetrafluoroethylene (PTFE) or similar. Maximum seat leakage when closed is 0.01 percent of rated valve capacity.

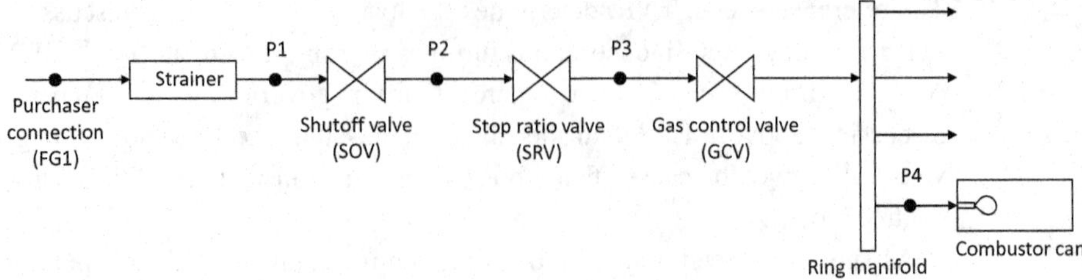

Figure 18.10 Gas fuel control system.

flow. In other words, the FSR command is approximately linear with the megawatt output of the gas turbine. As discussed in Chapter 12, the percentage of fuel to each passage (and associated GCV) is a function of the combustor reference temperature (CRT) and the DLN operating mode. The two-valve control system (SRV and GCV) provides a turndown ratio in excess of 100:1 and thus ensures stable control in an operating range extending from startup to maximum flow.

18.3 Model-Based Control

Gas turbine controls are based on "schedules," i.e.,

- Exhaust temperature as a function of compressor PR;
- DLN combustor fuel splits as a function of combustion reference temperature;
- Compressor inlet bleed heat (IBH) flow as a function of compressor inlet temperature (CIT; to prevent icing);
- Compressor operating line limit (OLL) as a function of corrected speed and IGV position (see Chapter 11).

These schedules are defined in order to protect the gas turbine against trespassing the "boundaries" of its safe operating envelope. Such boundaries (i.e., limit values of particular parameters) include the following:

- Emission limits (e.g., NOx and CO; see Chapter 12);
- Combustor dynamics (i.e., pressure pulsations leading to hardware vibration and "humming"; see Chapter 12);
- Combustor lean blowout limit;
- Compressor surge;
- Compressor inlet icing;
- Compressor and turbine hardware limits (vibration, clearances, thermal stress, etc.).

Such boundaries are determined from data obtained from factory tests and field operation. Inputs to the control system include (i) measured operating parameters, (ii) calculated operating parameters, and (iii) operator settings, some of which in turn become inputs to the schedules. Measured operating parameters are primarily the following:

- Ambient conditions (pressure, temperature and humidity);
- Fuel flow;
- Turbine exhaust temperature (from the exhaust rakes);
- CIT;
- CPR from compressor inlet and discharge pressures;
- Compressor discharge temperature;
- Rotational speed;
- Generator output.

Note that "data reconciliation" as described in Chapter 16 (see Section 16.6) can be applied to measured parameters in order to improve their fidelity. Calculated operating parameters of high importance are as follows:

- Compressor airflow;
- Combustor fuel/air ratio;
- TIT;
- Combustor flame temperature.

The outputs of the schedules are used to adjust the turbine control system outputs, i.e.,

- Fuel flow
- Combustor fuel split
- IGV position
- IBH flow

which are known as "effectors." In biology, the term "effector" refers to a *molecule* that activates, controls, or inactivates a *process* or *action*. In other words, it describes an "agent." In the context herein, for example, IGV position is an "agent" that controls the "air-sucking action" of the compressor.

In conventional gas turbine controls, schedules are rigidly predetermined to the point that they cannot *actively* accommodate changing operational boundary conditions such as component deterioration, fuel gas composition, and site ambient conditions in an effective manner. The reason for this is to keep a safe margin from the "boundaries" even in the worst-case scenario. Since components foul and degrade during operation, one or more boundaries may change (e.g., the compressor OLL). Continuous, automated tuning is not a feature of the conventional system and frequent retuning is impractical. Consequently, the only practical remedy is to set the boundaries artificially low (e.g., large OLL margin) to *preemptively* accommodate hardware deterioration and other effects. Not to do so would increase the risk of costly failure.

Furthermore, each schedule sets a single boundary with multiple schedules and multiple boundaries active in a given gas turbine at any given time. The *coupling* between them may lead to suboptimal operation and/or tuning difficulties. For example,

- Schedule A determines exhaust temperature (TEXH) from CPR.
- CRT (which is TTRF in GE controls), a proxy for the firing temperature, is determined from the exhaust temperature (via built-in transfer function).
- Fuel flow (FSR – see the preceding section) is determined based on CRT.

- Schedule B determines the combustor fuel split from CRT.
- Fuel flow and fuel split in turn determine the exhaust temperature via the turbine expansion physics.

In effect, a process comprising simultaneous interaction of multiple operating parameters in a single, seamless "loop" is handled by two independent and predetermined schedules (based on a *worst-case* scenario at the outset), which lose their "optimality" quite quickly as the machine operates away from its "new and clean" condition and is subjected to varying boundary conditions. There is no flexibility to adapt to the changing conditions, and if one schedule is retuned, eventually the other has to be retuned as well.

Before moving on, it should be noted that Schedule A in the example above actually refers to a "collection" of schedules, which are commonly known as *temperature control curves* (TCCs). The collection includes different TCCs for

- Gas and liquid (usually number 2 fuel oil as backup) fuels
- Simple- or combined-cycle operation
- Base, part, or peak loads
- Water/steam injection for NOx control or power augmentation (not applicable with modern DLN combustors except when running on liquid fuel)

In reality, the TCCs are composites of line segments or "pieces." There are two, three, or more pieces/segments depending on the particular type of TCC. The base load TCC is a "three-piece" curve as shown in Figure 18.11.

The base load TCC is constructed based on the philosophy of maintaining the nominal TIT. As the ambient temperature and, consequently, CIT decrease, airflow increases, leading to a rise in CPR. The controller adjusts the fuel flow to arrive at the TEXH indicated by the first segment of the TCC, which implies constant TIT (which cannot be measured directly). Similarly, as the ambient temperature and CIT increase, airflow decreases, leading to a decrease in CPR. The controller adjusts the fuel flow to arrive at the TEXH indicated by the second segment of the TCC. At a sufficiently high ambient temperature, the exhaust temperature limit is reached and the control adjusts the fuel flow to maintain it.

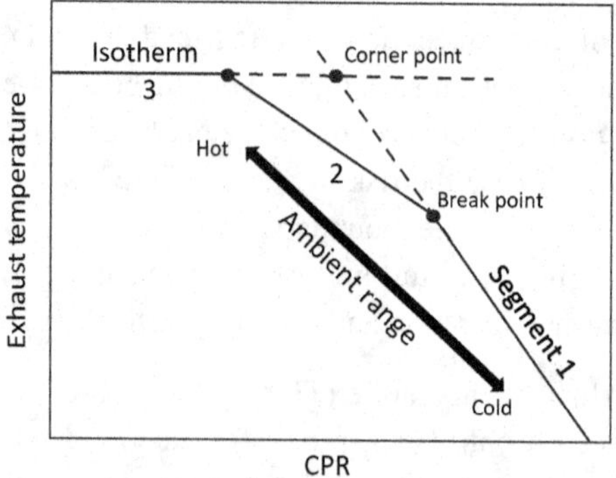

Figure 18.11 Base load TCC.

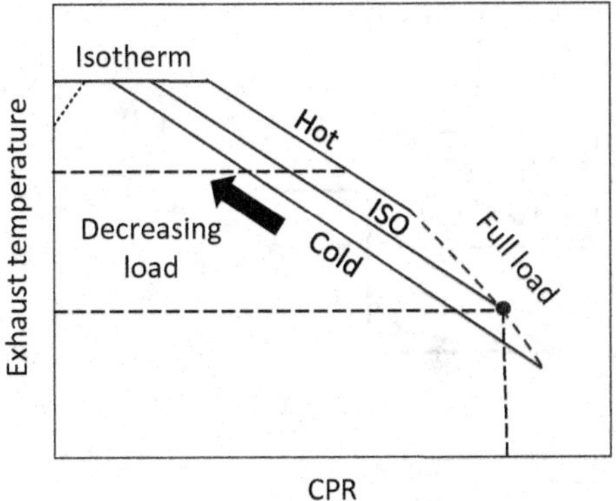

Figure 18.12 Part load TCCs.

Also known as the "isotherm," the exhaust temperature limit is dictated by the materials used in the exhaust end of the gas turbine (i.e., the exhaust frame, the exhaust diffusers, the struts, etc.; e.g., see Section 10.7). Its value is set to protect the turbine hardware from damage caused by excessive temperature. Its typical value for the F-class gas turbines is around 1,200°F (650°C). In modern H/J-class gas turbines with TITs exceeding 1,600°C, it is higher. In some cases, though, especially in combined-cycle power plants built in the late 1990s and early 2000s, HRSG part life concerns, mainly due to extended low-load operation and frequent cycling, can require a reduction in the isotherm value. This requires construction of new TCCs to be installed in the controller. For more details, the reader is referred to the paper by Mendoza et al. [17].

Part load control curves are typically "two-piece" curves as shown in Figure 18.12. In combined-cycle applications, load is decreased by closing the IGVs and reducing the airflow. This decreases the CPR and increases the TEXH at nominal TIT. There is a unique part load curve for a specified ambient temperature. A third linear segment (dotted line) is on the left when the IGVs reach their fully closed position. Further reduction in load is achieved by reducing the TIT, which reduces TEXH as well because of limited reduction in CPR (caused by lower T_3). This third linear segment is typically very short or not present at all.

MBC eschews the problems in conventional gas turbine controls by replacing the PID character of the gas turbine control function by a dynamic model of the gas turbine system. The dynamic model in question is also known as *adaptive real-time engine simulation* (ARES). The generic control process is shown in Figure 18.13.[9] (SP in Figure 18.13 is the "set-point" and PV is the "process variable.") The ARES concept is illustrated in Figure 18.14 in simple block diagram format. As is shown in Figure 18.14, the "error" between the measured PV and model-generated PV', denoted as PV_Err, is used to create corrections to the model parameters in a continuous "online tuning"

[9] How the gas turbine speed PI control works will be discussed in detail in Section 19.1 in the context of gas turbine dynamic simulation.

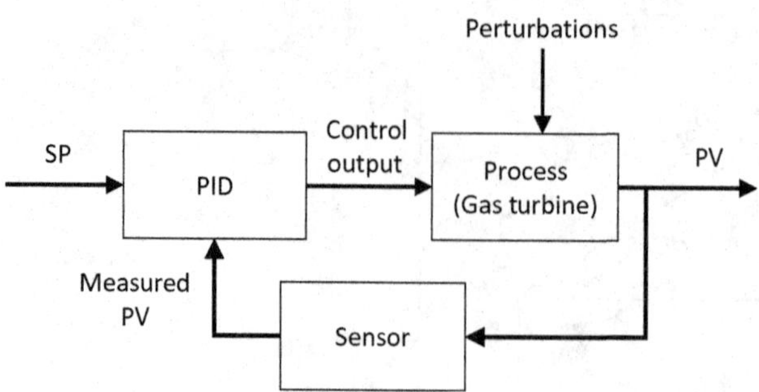

Figure 18.13 Generic block diagram of a PID control process.

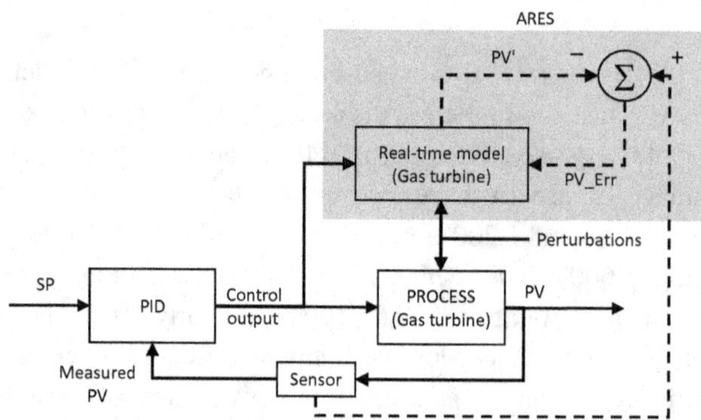

Figure 18.14 ARES running in parallel with the actual system.

process. The correction is typically done via a "gain matrix" or "Kalman filter." The underlying mathematical principle is analogous to the data reconciliation process described in Section 16.6 on data analysis. Note that the process of data reconciliation includes replacement of missing values from broken sensors as well as correction of badly erroneous values resulting from calibration drift or another malfunction.

How the MBC works with ARES is illustrated in Figure 18.15. "Virtual parameters" created by the ARES include "reconciled" measured parameters as well as non-measured parameters such as the TITs. Therefore, a complete and accurate "snapshot" of the gas turbine is generated continuously for the control logic to act upon and deliver the output requested by the operator with maximum efficiency while ensuring all machine limits are observed for safe operation.

The MBC process in Figure 18.15 can be summarized as follows:

1. Operating parameters are picked up by sensors and transmitted to ARES.
2. Using sensor data, ARES generates a set of tuned and reconciled parameters supplemented by another set of non-measured but model-calculated parameters that describe the gas turbine operating status as accurately as possible.
3. The "virtual" parameter set generated by ARES is used by the controller to calculate/estimate

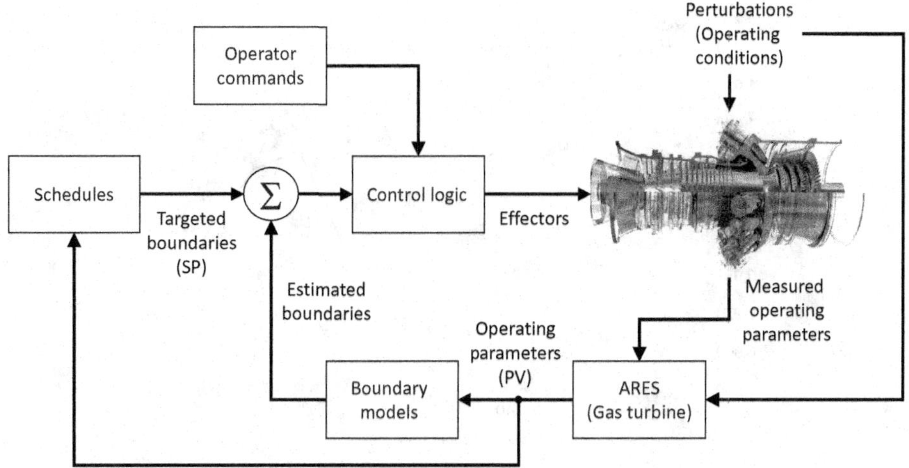

Figure 18.15 Block diagram of MBC.

   a. Existing values for the gas turbine operating boundaries, and
   b. Desired values of said boundaries.

4. Comparison of desired and estimated boundary values for multiple schedules results in "error" signals.
5. The control logic comprising simultaneous PI loops uses the error signals to generate operational commands ("effectors") to the gas turbine to minimize said error signals.
6. This process is repeated continuously (i.e., as many as 20–30 times in a second).

Note that some of the ARES-generated parameters are themselves "boundaries" (exhaust temperature [measured and reconciled], firing temperature [non-measured but calculated], CPR, etc.). Boundary models calculate emission limits, combustor dynamics-related limits, aeromechanical limits, etc.

The simple, mostly linear, and/or smoothly curved continuous variation of gas turbine performance with site ambient conditions and loading we have seen in Section 18.1 is not the case with MBC. This is illustrated by the *actual* combined-cycle net output correction "curve" depicted in Figure 18.16 (the gas turbine is equipped with inlet evap-coolers).

There are *seven* distinct segments of the correction curve in Figure 18.16:

1. From 15°F to 0°F, IGVs are closed to maintain turbine exit (axial) Mach number limit.
2. From 45°F to about 15°F, IGVs are opened while the controller maintains the combustor temperature rise limit by controlling the firing temperature (i.e., fuel flow).
3. From about 59°F to 45°F, IGVs are still opening and firing temperature is rising.
4. From 59°F to 70°F, the inlet evap-cooler is on; firing temperature increases up to its *nominal* value (IGVs are still opening).

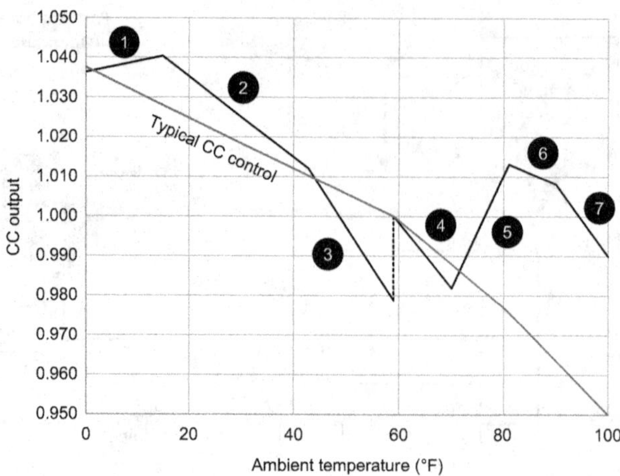

Figure 18.16 Combined-cycle (CC) output versus ambient temperature with MBC.

5. From 70°F to about 80°F, firing temperature increases towards its *maximum* value (IGVs are still opening and the evap-cooler is on).
6. From about 80°F to 90°F, IGVs are fully open; firing temperature further increases towards its maximum value (the evap-cooler is on).
7. From 90°F to 100°F, IGVs are fully open and firing temperature is at its maximum value (the evap-cooler is on).

Also shown in Figure 18.16 is the conventional combined-cycle control with IGVs fully open and the TEXH-CPR schedule active at high ambient temperatures. The dramatic improvement in performance, especially with the firing temperature pushed to its maximum, is striking.

At ambient temperatures of 42°F or below, the "anti-icing" system is active. In most gas turbines, this system uses air "bled" from the compressor discharge to keep the compressor inlet warm and thus prevent the formation of ice on the surfaces. The reason for this is that such sheets of ice are prone to breaking off in chunks and flowing into the compressor at high speed. Impact of ice chunks on the stator and rotor blades of the inlet stages can lead to severely shortened part lives and deteriorating performance (even culminating in catastrophic loss). The icing prevention system described above is known as IBH.

In some modern gas turbine combined cycles, the anti-icing system uses intermediate-pressure steam from the HRSG to heat a glycol–water mix that goes to a heating coil in the gas turbine inlet. This system activates whenever ambient temperature drops below 42°F. When the anti-icing system is active, it monitors the ambient dew-point temperature. The IBH or heating coil works in the following two different ways:

- When the ambient temperature is below 32°F, the inlet air is heated to a temperature of 10°F above the ambient dew-point temperature.
- When the ambient temperature is between 32°F and 42°F, the inlet air is heated to 10°F above the ambient dew-point temperature or to 42°F, whichever is lower.

Note that, in the presence of an inlet evap-cooler or mechanical chiller, compressor inlet conditions are different from the ambient conditions (i.e., lower temperature and close to 100 percent relative humidity). For example, at an ambient relative humidity of 50 percent, the ambient temperature range of 59–70°F in Figure 18.16 corresponds to a CIT range of ~50°F to ~59°F. In other words, the MBC operation in segment 4 is the same as in segment 3.

For most ambient conditions, the impact of ambient of relative humidity on combined-cycle output is quite small (i.e., within the 0.995–1.005 band; ±2.5 MWe for a 500-MWe combined-cycle power plant). Even within that band, it becomes significant at either "bone-dry" conditions (relative humidity of about 30 percent or less) or very humid conditions (relative humidity of about 70 percent or more).

At humid (i.e., relative humidity about 70 percent or more) *and* cold ambients (i.e., 30°F or lower), with the compressor inlet anti-icing system active, significant output loss can be experienced, especially in systems that use intermediate-pressure steam from the bottoming cycle as the inlet heating source.

References

1. Jansen, M., Schulenberg, T., Waldinger, D., Shop Test Result of the V64.3 Gas Turbine, *ASME Journal of Engineering for Power*, **114** (1992), 676–681.
2. Gülen, S. C., Joseph, J., Combined Cycle Off-Design Performance Estimation: A Second-Law Perspective, *ASME Journal of Engineering for Gas Turbines and Power*, **134** (2011), 011801.
3. Chaker, M., Meher-Homji, C. M., Evaporative Cooling of Gas Turbine Engines, *ASME Journal of Engineering for Gas Turbines and Power*, **135**: 8 (2013), 081901.
4. Chaker, M., Meher-Homji, C. M., Selection of Climatic Design Points for Gas Turbine Power Augmentation, *ASME Journal of Engineering for Gas Turbines and Power*, **134**: 4 (2012), 042001.
5. Bhargava, R. K., Meher-Homji, C. B., Chaker, M. A., Bianchi, M., Melino, F., Peretto, A., Ingistov, S., Gas Turbine Fogging Technology: A State-of-the-Art Review – Part I: Inlet Evaporative Fogging – Analytical and Experimental Aspects, *ASME Journal of Engineering for Gas Turbines and Power*, **129** (2006), 443–453.
6. Bhargava, R. K., Meher-Homji, C. B., Chaker, M. A., Bianchi, M., Melino, F., Peretto, A., Ingistov, S., Gas Turbine Fogging Technology: A State-of-the-Art Review – Part II: Overspray Fogging – Analytical and Experimental Aspects, *ASME Journal of Engineering for Gas Turbines and Power*, **129** (2006), 454–460.
7. Bhargava, R. K., Meher-Homji, C. B., Chaker, M. A., Bianchi, M., Melino, F., Peretto, A., Ingistov, S., Gas Turbine Fogging Technology: A State-of-the-Art Review – Part III: Practical Considerations and Operational Experience, *ASME Journal of Engineering for Gas Turbines and Power*, **129** (2006), 461–472.
8. Ondryas, I. S., Wilson, D. A., Kawamoto, M., Haub, G. L., Options in Gas Turbine Power Augmentation Using Inlet Air Chilling, *ASME Journal of Engineering for Gas Turbines and Power*, **113**: 2 (1991), 203–211.
9. Cortes, C., Willems, D., "Gas Turbine Inlet Air Cooling Techniques: An Overview of Current Technologies," POWERGEN 2003, December 9–11, 2003, Las Vegas, NV.

10. McGuigan, P., Why Keeping Cool Keeps Output High, *Power Engineering*, **122**: 2 (2018), 38–44.
11. Nuding, J.-R., Johnke, T., "Increasing GT Power and Efficiency through Wet Compression," POWERGEN 2002, 11–13 June, 2002, Milan, Italy.
12. Hoffmann, J., Agostinelli, G.-L., "High Fogging Commissioning Test in the Alstom Test Center Birr (Switzerland)," POWERGEN 2002, 11–13 June, 2002, Milan, Italy.
13. Lecheler, S., Hoffmann, J., "The Power of Water in Gas Turbines: Alstom's Experience with Air Inlet Cooling," POWERGEN Latin America 2003, November 11–13, 2003, Sao Paolo, Brazil.
14. Jolly, S., Cloyd, S., "Performance Enhancement of GT24 with Wet Compression," POWER-GEN International 2003, December 9–11, 2003, Las Vegas, NV.
15. Johnson, D., Miller, R. W., Ashley, T., "Speedtronic™ Mark V Gas Turbine Control System," GER-3658E, 39th GE Turbine State-of-the-Art Technology Seminar, 1994.
16. Rowen, W. I., "Operating Characteristics of Heavy-Duty Gas Turbines in Utility Service," ASME Paper No. 88-GT-150, ASME Gas Turbine and Aeroengine Congress, June 6–9, 1988, Amsterdam, The Netherlands.
17. Mendoza, E., Lin, T., Jiang, X., "Reduction of Gas Turbine Exhaust Temperature Limit Due to HRSG Limitations by Change on Control Curve to Optimize Plant Operation and Performance," ASME Paper GT2015-43784, ASME Turbo Expo 2015, June 15–19, 2015, Montreal, QC, Canada.

19 Transient Operation

There are several basic operating regimes of the gas turbine, i.e.,

- Startup from standstill to synchronization and full-speed, no-load (FSNL);
- Load ramp from FSNL to full-speed, full-load (FSFL) or operator-selected load level;
- Steady-state operation at varying loads while connected to the grid;
- Load up/down ramps at operator-selected rate;
- Emergency modes (e.g., response to over- or under-frequency events as prescribed by the applicable grid codes, load rejection, and trip);
- Normal shutdown.

With the exception of steady-state operation to generate electric power and deliver it to the grid, all operating regimes are *transient* (i.e., lasting only a few seconds or minutes) with varying speed and/or output along with varying cycle pressures, temperatures, and air/fuel flow rates. Steady-state operation characteristics are adequately covered in Chapter 18 within the context of off-design performance. This chapter will go through some very basic concepts pertaining to the analytical study of unsteady mass/energy transport and look into the details of key power plant transients.

Steady-state analysis of gas turbines and gas turbine components is simple and straightforward.[1] For most practical purposes, it can be reduced to a single equation describing the energy (enthalpy) balance around the particular control volume. For more mundane components such as pipes and ducts, one does not even bother with an equation; a coefficient to quantify pressure loss (maybe supplemented by a temperature loss) suffices. This simplicity completely disappears when it comes to the analysis of time-dependent phenomena, when thermal and mechanical parameters of interest change with time. In addition to the control volume energy/enthalpy balance, which is not simple anymore, one also has to consider the momentum balance (e.g., to solve for the shaft speed) and the full version of the continuity equation. Except for the most basic problems supplemented with many simplifying assumptions, closed algebraic solutions are not available and one has to resort to tedious and time-consuming numerical methods.

For the gas turbine in particular and other power plant equipment in general (heat exchangers, pumps, compressors, turbines, pipes, valves, etc.), transient analysis

[1] This is not the case at all for the actual design process, of course, which requires highly sophisticated 3D computational fluid dynamics and finite-element analysis tools.

involves three major types of calculations: (i) shaft speed, (ii) "heat soak," and (iii) "volume packing."

For gas turbine jet engines used in aircraft, marine, and land propulsion, shaft speed change is part of normal operation. In electric power generation applications, wherein the gas turbine generator runs at constant speed in synchronous connection to the electric grid, it is of importance only during startup and shutdown (with the exception of the "gas generator" spool of the aeroderivative gas turbines). The same is also true for the volume packing effect, which refers to filling and emptying of large volumes between components. In a gas turbine, the only large volume of note is the combustor or, as is the case in most advanced machines, the total volume occupied by the circumferentially arranged combustor cans. In a combined-cycle configuration, of course, there are several such volumes (e.g., large steam pipes).

The term "heat soak" refers to heating or cooling of the metal mass of machine components (e.g., turbine casings, shafts, stator vanes, and rotor blades) during transient events. In the combined-cycle power plant, one has to also include HRSG drums, heat exchanger tubes, large steam valves, and steam pipes. The importance of heat soak events lies in the magnitude of thermal stresses generated in large metal parts of plant equipment via steep temperature gradients. In general, unless a heavy-duty industrial gas turbine is rolled to FSNL and then ramped to FSFL very fast (see Section 19.1 for more on "fast" versus "slow" or conventional starts), thermal stresses are not of big concern. The same is true for much "lighter" aeroderivative gas turbines, which can go from "cold iron" to full load in 10 minutes or less. In comparison, the gas turbine jet engines of a civil aircraft accelerate from "flight idle" to 95 percent thrust in less than 10 seconds.[2]

Thermal stresses are of extreme importance in thick-walled components such as the high-pressure (HP) drum of the heat recovery steam generator (HRSG) and the steam turbine rotor and casings. One can add HP steam pipes and steam valves to the mix as well. Consequently, thermal stress control during transient events such as startup is of prime importance in gas turbine combined-cycle practice.

It is patently impossible to provide a reasonably well-rounded but simplified coverage of simple- or combined-cycle gas turbine *unsteady*-state performance à la, say, Chapter 17 in this book. (In fact, the usage of the term "performance" is of dubious value in this context.) Strictly speaking, in terms of quantitative analysis of transient events encountered in gas turbine or other thermal power plant operation, one moves from the realm of "calculation" to the realm of "simulation" or, more specifically, "dynamic simulation." There are no treatises known to the author that provide a comprehensive guide to someone who is interested in developing such dynamic simulation. The closest example is the paper by Gülen and Kim that describes a simplified, physics-based method derived from fundamental relationships to accurately predict the dynamic response of the steam bottoming cycle of a combined-cycle power plant to changes in gas turbine exhaust temperature and flow rate [1].

[2] See *Gas Turbine Performance* by Walsh and Fletcher (referenced in Chapter 2), especially chapter 8.

The method presented in the aforementioned paper enables rapid calculation of various modes of combined-cycle transient performance such as startup, shutdown, and load ramps for conceptual design and optimization studies. The method in question can be best described as a tool kit rather than a single model per se; it is based on one fundamental heat transfer formula (i.e., the *lumped capacitance method* [LCM]) and one simplifying principle (i.e., the *distributed thermal mass system* [DTMS]). Thus, it ensures transparency of governing principles and solution methods for ease of use by a wider range of practitioners. The reader is referred to the cited work for a full list of formulae and methods that are readily amenable to implementation on a computational platform of one's choice.

Some of the salient items from [1] and a few basic considerations pertaining to thermal stresses are covered below in Section 19.2. They should be sufficient to make some rough estimates about the thermal response of major power plant equipment within common transient event scenarios. The rest of the chapter focuses on three of the most important transients in gas turbine operation:

- Startup
- Shutdown
- Response to grid events

The reader may wonder about the relative dearth of gas turbine-specific calculations in Section 19.2 below. The reason for this is quite simple. With the exception of the rotational inertia of its shaft, which does not come into play in electric power generation applications except during startup and shutdown, the gas turbine is an essentially linear, nondynamic device. There is a small transport delay associated with the combustion reaction time (when the fuel flow is increased or decreased for load control) and a small time lag associated with the compressor discharge (and the combustor) volume. They are all much less than *one second* in duration and can be safely ignored in comparison to the thermal and transport time lags associated with the HRSG and the steam turbine, which can run into *minutes*, as will be shown below.

As a result, the biggest need for rigorous gas turbine dynamic simulation is for control system analysis. In such applications, simplified block diagram models suffice for most purposes. (Matlab-Simulink developed by MathWorks® is excellent for this type of modeling and analysis.) In the design, development, and tuning of commercial gas turbine control system software, original equipment manufacturers (OEMs) typically use the actual controller software itself, which interfaces with simulated transducer inputs that mimic the real gas turbine behavior using simple mathematical relationships. This will be the subject of Section 19.1, which should be considered only a brief introduction to the subject matter in order to enable the reader to delve deeper into the existing literature.

19.1 Gas Turbine Dynamic Simulation

Transient simulation of a gas turbine is mostly limited to functional representation of the machine and its controls (see the discussion in Section 18.2) using block diagrams.

The seminal 1983 paper by Rowen is probably still the best single source for the simulation of a heavy-duty gas turbine and its controls [2]. A similar 1992 paper by Rowen for mechanical drive gas turbines, which included inlet guide vanes (IGVs), completes the coverage of the subject for most practical purposes [3]. Figure 1 in [2] is a simulation block diagram of a single-shaft gas turbine. Its heart is the control block diagram in Figure 18.9, which is supplemented by the block diagrams for the fuel system, combustor, and shaft (rotor). It is pretty much the beginning point for all gas turbine transient simulation work done in the last three decades (e.g., see Yee et al. [4]). Components are represented by "blocks" encapsulating their time-dependent characteristics ("transfer functions") in Laplace transforms. In other words, the transfer function of a component is its actual behavior in *time domain*, f(t), which is translated into *s-domain*, F(s), by application of a Laplace transform.

Arguably the most important transient event in a heavy-duty gas turbine for electric power generation is startup (i.e., roll-up and loading from "cold iron" at turning gear [TG] to 100 percent load at 100 percent speed). Consequently, modeling of the turbine-generator shaft is of prime importance. In simple terms, when a torque, τ, is applied to a body with moment of inertia, I, the angular acceleration, α, is given by

$$\alpha = \tau/I, \tag{19.1}$$

$$\alpha = d\omega/dt, \tag{19.2}$$

$$\omega = d\theta/dt, \tag{19.3}$$

where θ is the angular displacement (in radians; e.g., 180° is equal to π radians). For a gas turbine with rotational speed, N, in rpm,

$$\omega[\text{rad/s}] = 2\pi N/60.$$

The torque generated by the gas turbine is a function of the fuel mass flow rate (i.e., system energy input) and rotational speed. The transfer function for a typical gas turbine is given by Rowen [2] as

$$\frac{\tau}{\tau_0} = 1.3\left(\frac{\dot{m}_f}{\dot{m}_{f,0}} - 0.23\right) + 0.5\left(1 - \frac{N}{N_0}\right), \tag{19.4a}$$

where the subscript "0" denotes the FSFL values of respective parameters. Thus, in "per-unit" or p.u. terms,

$$t = 1.3\,(m - 0.23) + 0.5\,(1 - n), \tag{19.4b}$$

where t, m, and n are normalized (i.e., per-unit) values of torque, fuel flow rate, and shaft speed, respectively. Substituting Equation 19.4b into Equation 19.1 and using Equation 19.2, we end up with

$$\frac{d\omega}{dt} = \frac{2\pi}{60}\frac{dN}{dt} = \frac{\tau_0}{I}[1.3\,(m - 0.23) + 0.5\,(1 - n)] \text{ or} \tag{19.5a}$$

$$\frac{dn}{dt} = \frac{30}{\pi}\frac{\tau_0}{N_0 I}[1.3\,(m - 0.23) + 0.5\,(1 - n)]. \tag{19.5b}$$

Table 19.1 Selected turbine-generator characteristics

	Speed (rpm)	Rating (MWe)	Torque (N-m)	Moment of Inertia (kg-m^2)	τ_I (s)
Frame 6B	5,100	35.9	67,140	1,892	15.1
Frame 7E	3,600	75.0	198,966	6,447	12.2
Frame 9E	3,000	106.7	339,612	18,521	17.1
Frame 9F	3,000	275.0	875,451	59,078	21.2

Equation 19.5b gives the acceleration of the gas turbine-generator shaft as a function of the *net* torque applied to it in "per-unit" terms (with the implicit assumption that the *load* torque is zero). Let us call the term in the square brackets on the right-hand side of Equation 19.5b the "turbine torque function," f_t, so that

$$\frac{dn}{dt} = \frac{1}{\tau_I} f_t \text{ with} \tag{19.6}$$

$$\tau_I = \left(\frac{30}{\pi} \frac{\tau_O}{N_O I} \right)^{-1}, \tag{19.7}$$

where τ_I is the turbine-generator shaft's (rotor) inertia time constant in seconds. Selected data for several General Electric (GE) frames (in metric units to save the reader lbf-to-lbm conversion aggravation) are listed in Table 19.1. Rotor time constants are from Equation 19.7. Torque values are calculated from rated output and speed. Moment of inertia values, except for 9F, are from Rowen [2].

The Laplace transform of a function f(t) is defined as

$$F(s) = \int_0^\infty e^{-st} f(t) dt.$$

Therefore, it can be shown that the Laplace transform of the first derivative of a function df(t)/dt or f'(t) is given by

$$sF(s) - f(0) = sF(s),$$

if the initial value of the function in question is zero (i.e., f[t = 0] = 0). If the function f(t) is the speed of the gas turbine-generator shaft as a function of time (i.e., f[t] = n[t]), from Equation 19.6, we obtain

$$sF(s) = \frac{1}{\tau_I} f_t, \tag{19.8a}$$

$$F(s) = \frac{1}{\tau_I s} f_t. \tag{19.8b}$$

Equation 19.8b is the transfer function for the normalized (i.e., per-unit) gas turbine-generator shaft (rotor) speed, n. As such, it constitutes the "rotor or shaft block" in the gas turbine simulation model block diagram. Before delving deeper into the gas turbine block diagram, let us go through a small exercise that will illustrate the proverbial "heart" of almost all dynamic/transient simulation problems.

For the initial cranking of the turbine-generator from the TG speed (a few rpm for GE units) to the ignition speed (about 15 percent of the synchronous speed), Equation 19.5b, with substitution of Equation 19.7, can be written as

$$\frac{dn}{dt} = \frac{0.5(1-n)}{\tau_{II}}. \tag{19.9}$$

The solution of the differential equation in Equation 19.9 can be found by defining a new variable, y = 1 – n, which results in

$$\frac{d}{dt}\ln y = -\frac{1}{2\tau_{II}}, \tag{19.10}$$

which, upon integration, gives

$$\ln y = -\frac{t}{2\tau_{II}} \rightarrow y = \exp\left(-\frac{t}{2\tau_{II}}\right) \rightarrow 1 - n = \exp\left(-\frac{t}{2\tau_{II}}\right),$$

$$n(t) = 1 - \exp\left(-\frac{t}{2\tau_{II}}\right), \tag{19.11}$$

for the initial condition n(t) = 0 at t = 0. Note that the rotor time constants in Table 19.1 are based on the *full torque* generated by the particular gas turbine at full speed. Thus, they are only applicable for normal operation at different loads and/or speeds. During the initial cranking of modern gas turbines, the generator is used as a "motor" by the static starter (see Section 19.3 below for more on this), which can apply a maximum torque whose magnitude is only a fraction of the full system torque. Let us assume that said fraction is 5 percent; for the Frame 9 gas turbine in Table 19.1, then, the time constant becomes 424 seconds. In passing, note that the static starter torque for the Frame 9 at 5 percent of the full value is about 44,000 N-m, which is nearly equal to the *full torque* of the Frame 6 gas turbine in Table 19.1! In other words, in order to crank a very large industrial gas turbine to its ignition speed in a reasonably short time period, without the static starter, one would need a small industrial gas turbine. In any event, integrating Equation 19.11, we can easily find that n = 0.15 at t = 138 seconds or slightly above 2 minutes.

Let us now concentrate on Equation 19.11 itself and plot it for τ_I = 12 seconds (roughly Frame 7E in Table 19.1) as shown in Figure 19.1. The second term on the right-hand side of Equation 19.11 is the "exponential decay" function, which is common in many physical processes. The factor multiplying time t in the argument of the exponential function ($1/2\tau_I$ in this case) is the "decay constant." Its inverse ($2\tau_I$ in this case) is the "characteristic time" of the physical process in question. (It is also referred to as "mean lifetime" in some problems.) In power plant transient problems,

- Time-dependent cooling or slowing processes are characterized by the *exponential decay function* (EDF).
- Time-dependent heating and acceleration processes are characterized by (1 – EDF).

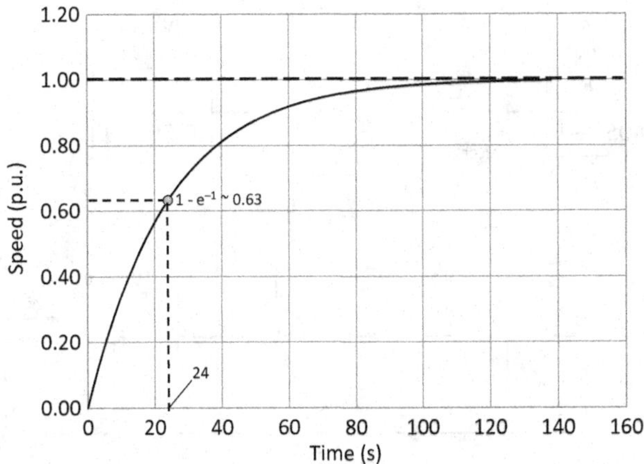

Figure 19.1 Gas turbine-generator shaft acceleration.

In Figure 19.1, it is shown that a 7E power train *could* be accelerated from zero to full speed in about 2 minutes (with no assistance from the gas turbine) with the application of an external torque equivalent to the full system torque at FSFL operation. The characteristic time of the acceleration process is about 24 seconds.

Coming back to the gas turbine simulation, we start with figure 1 in [2] (mentioned earlier as the *genesis* of all similar work) and simplify it by eliminating acceleration and temperature controller blocks. For typical applications (except startup) where frequency deviations are within a limited band (e.g., ± 1 percent), acceleration control does not enter the picture except in a loss-of-load situation (see Section 19.5 below for more on this subject). Temperature control is approximated by setting the upper bound of the limiter block to the rated load and takes the exhaust temperature out of the picture. These simplifications essentially reduce the gas turbine simulation to the simulation of shaft torque, which can be expressed as a simple transfer function of fuel flow and rpm (e.g., Equation 19.4). One can, of course, replace that simple transfer function with a full-blown gas turbine model, but the gain in information would be minuscule in comparison to the computational effort spent to achieve it.

The aforementioned changes are equivalent to eliminating the "minimum-value gate" (also known as the "low-value selector") in Figure 18.9. A speed controller with a droop governor is the only active controller. The simplified block diagram is shown in Figure 19.2. The lower part of the diagram is essentially the s-domain representation of Equation 19.6. The "summing block" determines the *net* or *accelerating* torque acting on the shaft as the difference of the gas turbine-generated torque and the load torque, and the "rotor block" integrates the result to find the p.u. shaft/rotor speed, n(t).

As shown in Figure 19.2, the speed (droop) governor block acts on its input, which is the speed error generated by the summing block on the far left in the image, and generates an output proportional to it. The proportionality factor (the "gain" in control theory) is determined by the (adjustable) droop setting d (e.g., 4 percent in most cases). The time constant associated with the governor, 0.05 seconds, is typical of digital

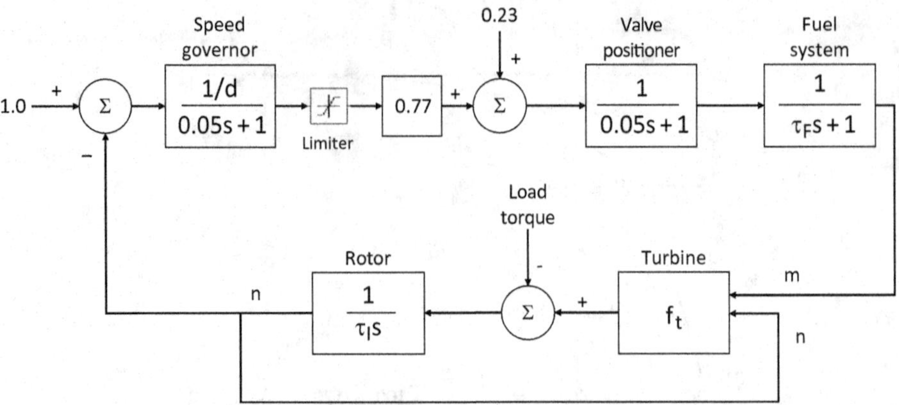

Figure 19.2 Simplified block diagram of a natural gas-fired heavy-duty industrial gas turbine with a droop governor.

logic-based governors (instead of analog governors of yore). Strictly speaking, the fuel system's output command is in proportion to the product of the speed governor output signal times the unit speed, but this can be ignored for small variations in speed (as is the case in Figure 19.2).

Let us look at the speed governor a bit more closely. It is actually a simplification of the full block diagram of a proportional-integral (PI) controller with a gain of 1/d as shown in Figure 19.3. After the requisite algebra, the block diagram in Figure 19.3 reduces to the single block shown in Figure 19.2. Thus, for a 4 percent droop, the controller gain is 1/0.04 = 25. The time constant of the controller, 0.05, suggests that the integrator gain in Figure 19.3 is K = 25/0.05 = 500.

The summing block between the governor output (after the limiter) and the valve positioner is a result of the fact that even at no load, the gas turbine requires nearly a quarter of its FSFL fuel input for self-sustaining, full-speed rotation (0.23 in this example). Since the governor operates for an active load range of 0–100 percent, its output is corrected to a fuel flow range of 0.23–1.00 in order to be compatible with the gas turbine operation.

The fuel gas control system in Figure 18.10 is simplified to account for only two effects: (i) the time constant associated with the gas control valve (GCV) positioning system and (ii) the volumetric time constant associated with the downstream piping and fuel gas distribution manifold, τ_F.

The simulation described by the block diagram in Figure 19.2 is simplified to a point that it can be easily solved in Excel. Its more complicated variants are typically implemented in a software tool such as Simulink by MathWorks, which is a graphical programming environment for modeling, simulating, and analyzing multidomain dynamical systems.[3] Similar block diagrams can also be developed for the Rankine

[3] www.mathworks.com, Corporate Headquarters: 1 Apple Hill Drive, Natick, MA 01760-2098, USA (as of November 2017).

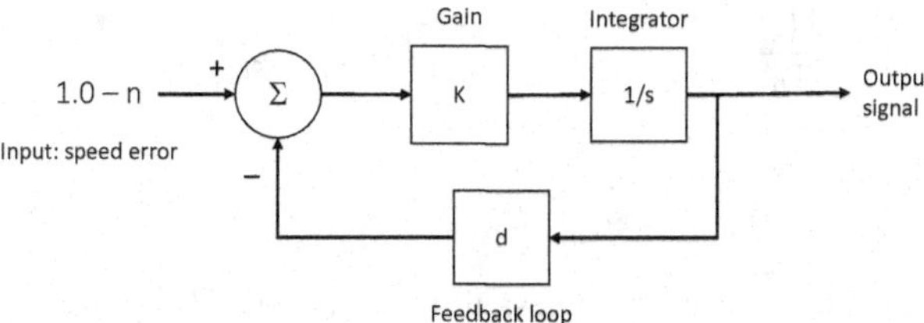

Figure 19.3 Block diagram of a PI controller with steady-state feedback.

steam bottoming cycle of a combined-cycle power plant [5]. Application examples can be found in the literature [6,7]. For comprehensive coverage of the subject matter, the reader is referred to the superb book by Kundur [8].

19.2 Thermal Stress Management

Mathematically, what happens in a transient event is quite simple: some parameter changes and creates a "perturbation" in the system and some other parameters respond to it with a certain "time lag." Simplified mathematical treatment of perturbation–response problems is based on several concepts, i.e.,

- Thermal capacitance
- Thermal resistance
- Characteristic time or time constant

As mentioned in the introduction to the chapter, thermal stress management is one of the key considerations in power plant transients, which requires evaluation of heat transfer in plant equipment. (Other important considerations are controlling compressor and turbine clearances, combustion dynamics, and compressor surge, which are taken care of by the gas turbine controller.) More specifically, one has to be able to obtain a rough quantitative idea about the heating or cooling of large metal components (e.g., how hot or how cold and how fast).

If you have not already done so, it is highly recommended that you read the section on thermal stress in Chapter 13 (Section 13.6.1). In order to analyze the heating or cooling of components with large metal masses and/or thick walls, one has to find a solution of the governing differential equation for T(x,t) for the component in question. For a solid body of a non-uniform geometry, this is a very difficult task and can only be handled by numerical methods such as finite elements. Exact and approximate solutions for simple geometries (infinite cylinders, plane walls, etc.) are available in most textbooks on heat transfer [9]. Even the approximate solution is tedious and requires tabulated values such as the Bessel functions. Nevertheless, they are relatively easy to implement in an Excel spreadsheet. For example, the solution relevant to the most important practical problem (i.e., steam turbine rotor thermal stress management) is the one for the infinite cylinder. Thus, for Biot numbers greater than 1, at the surface of a turbine rotor modeled as a cylinder, the temperature difference as a function of time is

$$\frac{T(t) - T_\infty}{T_0 - T_\infty} = C_1 \exp\left(-\zeta_1^2 Fo\right) J_0(-\zeta_1), \tag{19.12}$$

$$Fo = \frac{\delta t}{(D_0/2)^2}. \tag{19.13}$$

The *Fourier* number, Fo, is a dimensionless time with *thermal diffusity*, δ, which was defined in conjunction with Equation 13.7 earlier. Equation 19.12, which is an *approximate* solution for Fo >0.2, gives the surface temperature of an infinite cylinder with diameter D_0, which is initially (i.e., at time t = 0) at a uniform temperature, T_0, and suddenly subjected to a fluid flow at temperature T_∞.

The *exact* solution of the infinite cylinder problem is similar to Equation 19.12, where the right-hand side of the equation is an infinite series of the same form. In dimensionless terms,

$$\theta^* = \sum_{n=1}^{\infty} C_n \exp\left(-\zeta_n^2 Fo\right) J_0(-\zeta_n r^*), \tag{19.14}$$

$$\theta^* = \frac{T(t) - T_\infty}{T_0 - T_\infty}, \tag{19.15}$$

$$r^* = \frac{r}{(D_0/2)}, \tag{19.16}$$

$$C_n = \frac{2}{\zeta_n} \frac{J_1(\zeta_n)}{J_0^2(\zeta_n) + J_1^2(\zeta_n)}, \tag{19.17}$$

where the discrete values of ζ_n are positive roots of the *transcendental* equation

$$Bi = \zeta_n \frac{J_1(\zeta_n)}{J_0(\zeta_n)}. \tag{19.18}$$

The quantities J_1 and J_0 are *Bessel* functions of the first kind, and their values can be found using the Microsoft Excel worksheet function BESSELJ. The first four roots of the transcendental equation are tabulated in Table 19.2.

Table 19.2 Roots of the transcendental equation [10]

Bi	ζ_1	ζ_2	ζ_3	ζ_4
0.01	0.1412	3.8343	7.017	10.1745
0.1	0.4417	3.8577	7.0298	10.1833
0.5	0.9408	3.9594	7.0864	10.2225
1	1.2558	4.0795	7.1558	10.271
2	1.5995	4.291	7.2884	10.3658
5	1.9898	4.7131	7.6177	10.6223
10	2.1795	5.0332	7.9569	10.9363
20	2.2881	5.2568	8.2534	11.2677
50	2.3572	5.4112	8.484	11.5621

19.2.1 Generic Combined-Cycle Problem

For the evaluation of overall plant transient performance, the gas turbine is the driver or, in terms of the mathematical representation of the power plant, the *perturbation* to which the bottoming cycle *responds* dynamically with a certain *time lag*. Thus, the key endeavor in the combined-cycle transient performance evaluation is the estimation of the bottoming-cycle dynamic response to the changes in the gas turbine exhaust flow and temperature. The basic philosophy underlying said dynamic response is the determination of the following items in a chronologically coherent manner:

1. Time-dependent warm-up of the HRSG heat exchanger metal (tubes, headers, drums, etc.).
2. Time-dependent steam generation in HRSG evaporators.
3. Steam pressure and temperature at steam turbine admission points (HP, intermediate pressure [IP], or hot reheat [HRH], and low pressure [LP]) as a function of time.
4. Achievement of steam turbine steam admission start permissives.
5. Steam turbine ramp to FSNL (in multishaft systems or in single-shaft systems with prime movers separated by a clutch) via steam admission into the HP or IP (HRH steam) section.
6. Steam turbine rotor and HP evaporator drum wall thermal stress evaluation.
7. Steam turbine steam flow, pressure, and temperature and load ramps dictated by allowable steam turbine thermal stresses.

The reader is referred to the paper by Gülen and Kim, which explains in detail how this philosophy can be put into action using physics-based, relatively simple relationships that are amenable to implementation in an Excel spreadsheet [1]. The governing physical principles are explained in detail in an article by Gülen that appeared in *Power Engineering*.[4] It is available online and is strongly recommended as a companion to the present discussion. There are several types of combined-cycle transient operation modes (startup, shutdown, load ramps, trips, primary and secondary responses to a drop in grid frequency, etc.). All of them are important, but both publications focus on the combined-cycle startup, which includes all of the salient aspects of the bottoming-cycle dynamic response. The same is true for the coverage herein, too. The salient aspects of thermal stress management during gas turbine combined-cycle startup will be demonstrated by a snapshot of the entire event in the Section 19.2.2. High-level details will be covered in section 19.3.2.

19.2.2 A Sample Problem

Let us look at a hot-start problem in a combined-cycle power plant with the cascaded steam bypass system (also known as a "wet reheater" system), which is illustrated in Figure 19.4. HP steam generated in the HRSG during startup is bypassed around the HP turbine to the cold-reheat line, where it mixes with the IP steam and is sent to the

[4] "Gas Turbine Combined Cycle Fast Start: The Physics behind the Concept," *Power Engineering*, www.power-eng.com, June 2013, pp. 40–49.

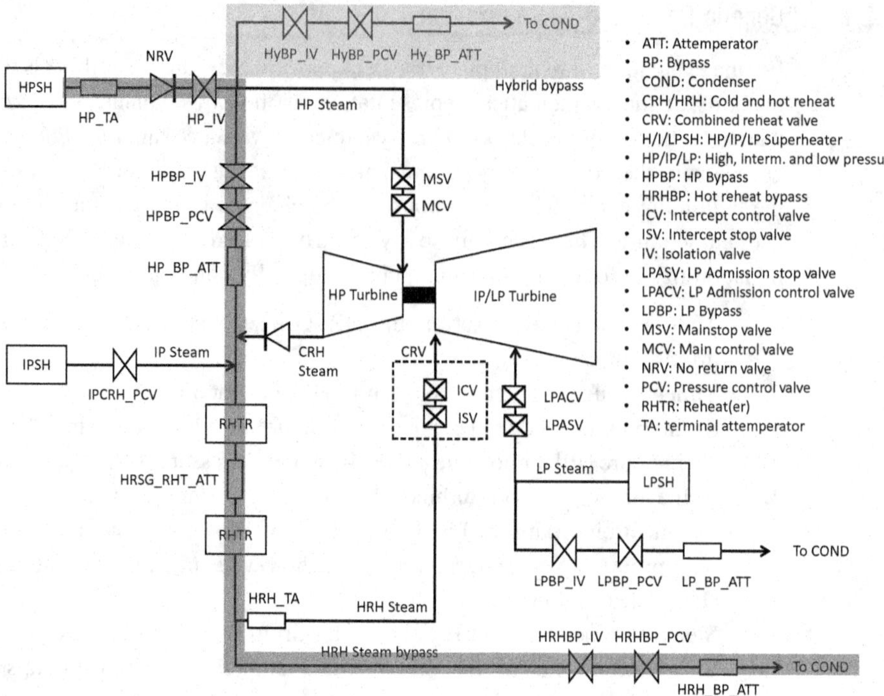

Figure 19.4 Cascaded steam bypass diagram.

reheater. The HP bypass line is equipped with a pressure control valve and an attemperator (desuperheater) utilizing spray water from the boiler feed water pump discharge. The HRH steam bypasses the IP turbine and is sent to the condenser. The HRH bypass line also includes a pressure reduction and attemperation station similar to that in the HP bypass line. The HRH bypass attemperators utilize spray water from the condensate pump discharge. This system maintains a continuous steam flow through the reheater to keep the reheater tube metal at a safe temperature. Otherwise, as in a parallel steam bypass system with a "dry reheater," higher (and thus more costly) alloy tube materials would be required.

Selected results from a transient simulation model (to be specific, in Aspen HYSYS) are shown in Figure 19.5 in a sanitized form. The HP steam production rate in the HRSG (curve C in Figure 19.5) is driven by the gas turbine exhaust flow and temperature. At the 28-minute mark, the main control valve (MCV; i.e., the HP throttle valve) opens and steam is admitted into the HP turbine (curve E). Until that point, all HP steam bypassed the steam turbine (curve D). In coordination with the MCV, the HP bypass valve closes until all HP steam is admitted into the HP turbine (i.e., none is bypassed), which happens at about the 33-minute mark. Well before the 28-minute mark, steam temperature reaches 1,000°F and is maintained at that level via terminal attemperators (TAs; curve A). Controlled opening and closing of the steam valves ensures that the HP steam pressure is maintained at 1,200 psia (curve B). At the time of steam admission start, the steam turbine rotor temperature is 900°F. The goal of the current exercise is to

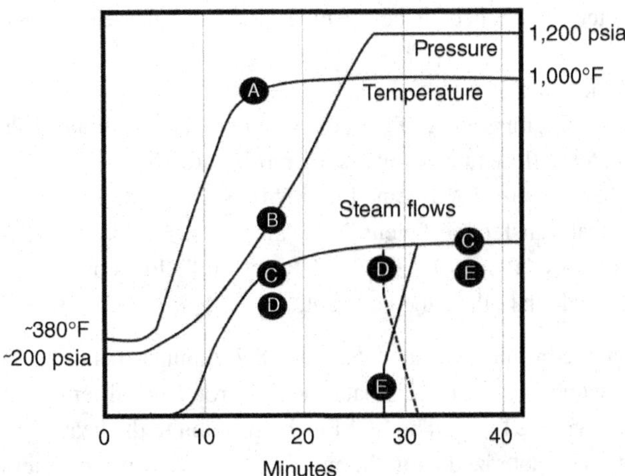

Figure 19.5 HP turbine steam admission during hot start.

find out whether the described HP steam admission process introduces unacceptable levels of thermal stress on the steam turbine rotor with an adverse impact on the component's cyclic life expenditure.

A few remarks are in order before proceeding:

1. The HP rotor surface stress during this hot-start steam admission process is *compressive* (see Section 13.6.1 for details).
2. Assuming that the steam turbine rotor is a long, smooth cylinder heated or cooled uniformly along its surface (which is the case herein) is a highly idealized model. Nevertheless, comparison of results with those obtained for a number of actual rotor geometries indicates that this approximation is satisfactory. Thermal stresses in an actual rotor are about 10 percent higher on the surface and 10 percent lower at the bore than those calculated for a smooth, infinitely long cylinder.
3. The magnitude of the surface centrifugal stresses in high-temperature turbine rotors is low compared to that of the thermal stresses. Thus, their impact on cyclic life expenditure is neglected. In fact, during heating, tensile centrifugal stresses are even beneficial in that they counter the compressive thermal stresses. (The situation is different at the *rotor bore*.)

The calculations are done using the following assumptions:

1. Rotor material properties are approximate for CrMoV (pronounced "chromoly-vee"), which is a low alloy. In particular,
   a. Young's modulus as a function of bulk of temperature is found from the data in Table 13.4.
   b. Other pertinent material properties such as $\rho = 475$ lb/ft^3 (7.61 ton/m^3), $\alpha =$ 1.2E-05 K^{-1}, Poisson's ratio = 0.3, and k = 30 W/m-K are typical average values for alloy steels from Table 13.3. The thermal diffusivity, δ, obtained from these parameters is 8.30E–06 m^2/s.

2. Rotor diameter is assumed to be 22 in. Rotor bore diameter is assumed to be 0.01 in.
3. Initial rotor metal temperature, T_i, is 900°F.
4. Steam admission temperature, T_∞, and pressure are 1,000°F and 1,200 psia.
5. Normalized steam flow rate is obtained from Figure 19.5.
6. Allowable thermal stress is assumed to be 50 ksi.
7. Convection heat transfer coefficient, h, for the free stream (i.e., steam) is assumed to be 450 W/m^2-K (about 79.3 Btu/h-ft^2-R) at "full" flow conditions. This is the value used to calculate the time constant.

The thermal time constant is calculated as 18.7 minutes using the data in the assumptions. This value is used to calculate the bulk rotor metal temperature, T_{BULK}, as a function of time via the LCM. The LCM is the solution to the heat transfer problem where a solid body, characterized as a "lump of mass" with a uniform temperature, is subjected to a sudden change in its environment (e.g., flow of a colder or hotter fluid around it). The implication is that the "lump of mass" has *infinite* conductivity so that no temperature gradients form in it. The solution is yet another manifestation of the EDF, i.e.,

$$\frac{T(t) - T_\infty}{T_i - T_\infty} = e^{-\frac{t}{\tau}}, \tag{19.19}$$

where T_i is the initial temperature of the solid body, T_∞ is the free stream temperature, and τ is the time constant. The solution of Equation 19.19, T(t), is T_{BULK} as a function of time.

For the outer surface stress, we have to consider the outer layer of the rotor, which can be thought of as a "pipe" with a wall thickness, t_w. Assuming that $t_w = 0.025 \times D_O$, the thermal time constant for that imaginary pipe is calculated as 55 seconds. This value is used in Equation 19.19 to calculate the rotor surface metal temperature, T_{SURF}, as a function of time (see Section 13.6.1).

The bore temperature, T_{BORE}, at the radius r = 0 as a function of time is calculated using the "exact" solution for an infinite cylinder as described above (i.e., Equations 19.14–19.17). (D_i = 0.01 in. is too small and can be safely ignored.) Fourier and Biot numbers are obtained from Equations 19.13 and 19.18, respectively. The latter requires a reliable calculation of the convective heat transfer coefficient, h (see Appendix D).

The thermal stress at the rotor surface is calculated using Equation 13.1, i.e.,

$$\sigma = E\frac{\alpha}{1-\nu}K_T(T_{SURF} - T_{BULK}), \tag{19.20}$$

where the thermal stress concentration factor, K_T, is assumed to be 1.5. The thermal stress at the rotor bore is also calculated using Equation 19.20 with a K_T of 1. Instead of T_{SURF}, T_{BORE} is used to calculate the temperature difference.

Results of the calculation for the sample problem are summarized in Figure 19.6. The inputs are the steam admission temperature and flow rate from Figure 19.5, which are approximated as linear line segments (the dashed lines in Figure 19.6). Thermal stresses are normalized by the maximum allowable value, which is assumed to be 50 ksi.

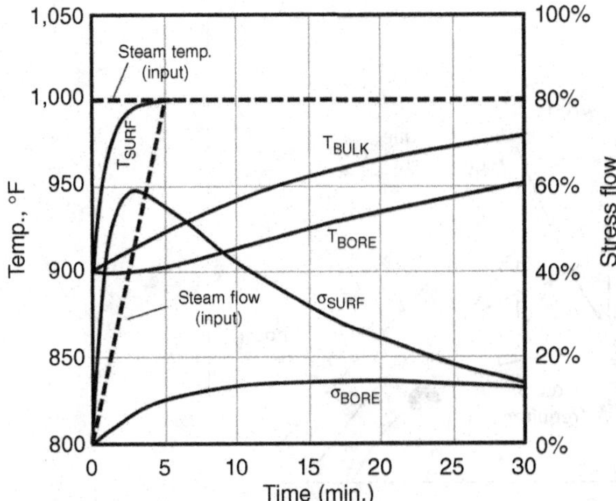

Figure 19.6 Steam turbine HP rotor metal temperature and thermal stress during a hot start.

As shown in Figure 19.6, the surface temperature comes rapidly to a thermal equilibrium with the steam flow, at which point the maximum temperature difference and, as suggested by Equation 19.20, maximum thermal stress are reached. Note that the driver of the surface thermal stress is the difference between T_{SURF} and T_{BULK}, with the latter responding much more slowly to the step change in the temperature at the boundary between the steam and the rotor. Thereafter, the temperature difference and the thermal stress gradually come down as the rotor body slowly heats up.

19.3 Startup

19.3.1 Simple-Cycle Startup

Like other internal combustion engines, a gas turbine cannot produce torque at zero speed. A starting system must be used to supply the torque required for startup. In the past, when gas turbines were relatively small compared to the state-of-the-art behemoths rated anywhere from 300 MWe to more than 500 MWe, electric motors, small diesel engines, torque converters, and expansion turbines were used for this purpose. With the advent of larger gas turbines, high-power, solid-state frequency converters (i.e., static starters) have become a viable source of gas turbine starting power.

The graph in Figure 19.7 illustrates the torque requirement for starting a typical gas turbine. As shown in Figure 19.7, the starting system is assumed to comprise an electric motor and torque converter. Actual values of torque are omitted since they vary widely between different units, but the basic shapes of the curves are consistent. However, for numerical estimates, consider that (from Equations 19.1–19.3)

$$\tau = I\frac{d\omega}{dt},$$

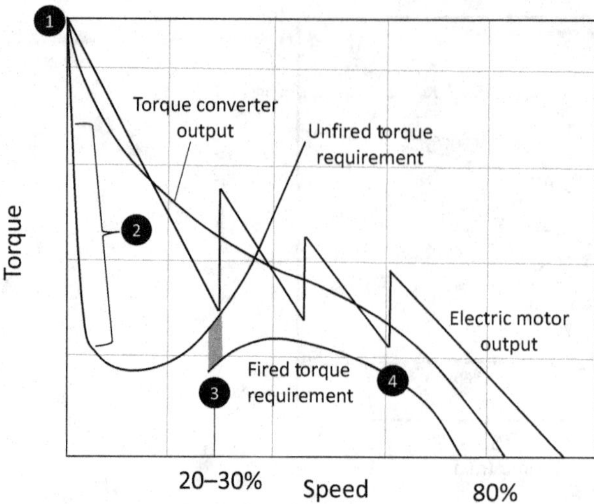

Figure 19.7 Gas turbine starting torque requirement.

where τ is the torque in lbf-ft (~1.356 N-m in SI units), ω is the angular velocity in rad/s, I is the moment of rotating inertia in lb-ft^2 (kg-m^2 in SI units), and t is time in seconds. Physically, I is the mechanical property of a body with mass m, which characterizes the "flywheel effect" of the mass when said body is rotating. The relationship between torque and shaft power is given by

$$\dot{W}\,[\text{in hp}] = \tau \frac{N\,[\text{in rpm}]}{5252}.$$

(Note that 1 hp is 0.7457 kW.) Torque is somewhat of an "odd duck" in turbomachine dynamics. In particular, torque has the units of energy or work (i.e., 1 N-m = 1 Joule), but it is *neither* energy *nor* work. In the US customary system, this distinction is emphasized by the units used for torque (i.e., lbf-ft), as opposed to the units used for work (i.e., ft-lbf).

By definition, torque is associated with *angular acceleration* (i.e., α = dω/dt). Since a gas turbine generator connected to the electric grid is rotating at a constant speed (i.e., 3,000 or 3,600 rpm for 50- and 60-Hz grids, respectively), would this mean that its shaft torque must be zero because α is zero? The answer is, of course, no. There *is* a shaft torque generated by the gas turbine, but it is opposed exactly by a *counter-torque* exerted by the grid (i.e., the "load") so that the *net* force/torque acting on the shaft, and thus angular acceleration, is zero.

One rarely needs torque calculations in practice unless one is directly involved in rotordynamic design of the gas turbine and generator components. Nevertheless, it is useful to have an idea about the magnitude of torque encountered in typical transient problems (e.g., gas turbine startup).

For a 50-Hz (3,000–rpm) gas turbine rated 250 MWe, shaft torque at base load operation is

$$\tau = 250{,}000/0.7457 \times 5{,}252/3{,}000 = 586{,}921 \text{ lbf-ft, or in SI units}$$
$$\tau = 586{,}921 \times 1.356 = 795{,}757 \text{ N-m}$$

The total rotor mass of such a gas turbine (turbine plus generator) is about 80 metric tons. Assuming a radius of gyration of 0.75 m, the moment of inertia is

$$I = 80{,}000 \times 0.75 \times 0.75 = 45{,}000 \text{ kg-m}^2,$$

which is about 1,068 kp-ft^2 (1 kp = 1,000 lb). In a typical *fast* start, the turbine is accelerated from TG to FSFL in 6 minutes, i.e.,

$$\alpha = d\omega/dt = 2\pi \times (3{,}000/6)/3{,}600 = 0.87 \text{ rad/s}^2.$$

Thus, the requisite torque for this acceleration rate is

$$\tau = 45{,}000 \times 0.87 = 39{,}270 \text{ N-m},$$

which is about 5 percent of the base load torque. In other words, a starter system to accomplish this feat should be rated at about 5 percent of the base load rating of the gas turbine generator in question.

Coming back to Figure 19.7, at zero speed, as shown in the figure, (1) the torque required for breakaway is very high due to the static friction between the turbine rotor and the bearings. Note that large industrial gas turbines always start from the TG and utilize lift oil to separate the bearing and the journal. Therefore, the breakaway torque is somewhat obsolete and may apply only to the old units in the field. It is included herein for completeness.

The starting torque first decreases quickly (2) as bearing lubrication films are established, but then the need for torque increases with increasing speed due to the power required to drive the axial-flow compressor.

As shown in Figure 19.7, at about 20–25 percent of the rated speed, the airflow is sufficient to fire the gas turbine (3). (This is typical for an older-generation machine. In modern frame machines, firing takes place at about 15 percent speed.) After firing, the turbine section of the unit produces some torque of its own. This reduces the amount of torque required from the starting system. Nevertheless, the gas turbine still needs additional torque to reach a self-sustaining speed (4), which is about 80 percent for state-of-the-art large gas turbines. Once the turbine achieves a self-sustaining speed, no torque is required from the starting system.

A static starter provides variable-frequency power directly to the generator terminals (i.e., using it as a synchronous motor) to start the gas turbine and bring it to a self-sustaining speed. There are numerous advantages to a static starter. For one, its price is less than that of a separate electric motor and its accessories to provide the requisite power and torque. Furthermore, the maintenance requirements of the generator are significantly less than those of the electric motor. Since most sites are already connected to a power grid, alternating current (ac) power to operate the static starter is readily available. (A large diesel engine might still be required for "black start" capability.) On the other hand, a static starter requires more sophisticated electronic controls than other types of starting systems.

The static start system consists of the following major components:

1. Load commutated inverter (LCI)
2. Isolation transformer
3. LCI disconnect switch
4. Slow roll motor (turning gear)

The LCI is a static, adjustable-frequency drive system that controls a synchronous machine from zero to rated speed (see Figure 19.8). The *isolation transformer* isolates the static starter from the ac line and provides the correct voltage at the input terminals of the *rectifier* (a thyristor bridge circuit). The gating of the thyristors is controlled to produce a variable direct current (dc) output. The output of the rectifier is fed to a *dc link reactor*, which is basically an inductor. The dc link reactor smooths out the current flow and keeps it continuous over the entire operating envelope of the system. The output of the dc link reactor is fed to the *inverter*, which is also a thyristor bridge circuit and converts the dc input into a variable-frequency ac output. The ac output of the inverter feeds the stator terminals of the generator. The thyristors in the rectifier and the inverter are controlled by electronic microprocessors in the static starter circuits. (Note that *commutation* refers to the process of turning off a conducting thyristor.)

The TG assembly contains the following elements:

1. Electric (induction) motor
2. Worm gear reducer
3. Synchro-self-shifting (SSS) clutch
4. Flexible coupling

The TG provides the power necessary to "break the turbine away" and rotate it prior to the start. It also rotates the turbine after shutdown to avoid deformation of the rotor. After the breakaway, the TG "slow rolls" the turbine at about 5–7 rpm. In the event of a power failure, the TG is equipped with a feature for manual turning of the rotor. Lubricating oil for the reduction gear is self-contained, but lubrication of the SSS clutch and the output shaft bearings requires a continuous oil supply from the main lube oil system.

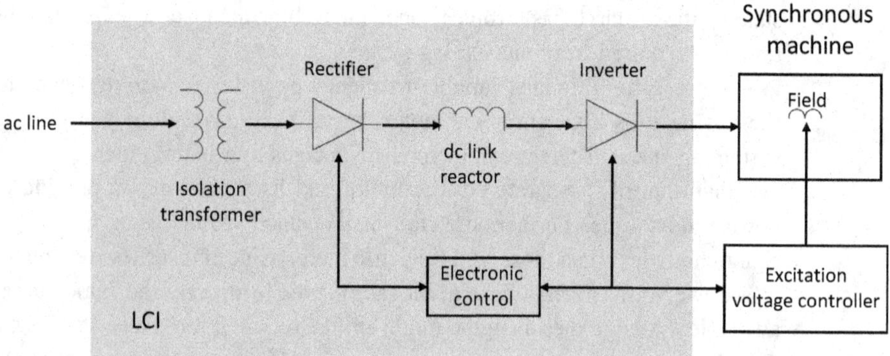

Figure 19.8 Load commutated inverter.

The TG is connected to the turbine/generator shaft via the SSS clutch and the flexible coupling. The SSS clutch is a positive tooth-type (i.e., running in a clockwise direction) overrunning clutch, which is self-engaging in breakaway and slow roll modes. It overruns whenever the turbine/generator shaft speed exceeds the TG drive speed. The insulated flexible coupling allows for angular and parallel misalignment, as well as for the axial expansion of the generator shaft.

Once the gas turbine startup sequence is initiated, the control system activates the *bearing lift oil pump* (BLOP). The BLOP provides HP lube oil to hydrostatically raise the turbine shaft. This reduces the static friction between the gas turbine rotor and the bearings and thus reduces the amount of breakaway torque required for starting the turbine. It also minimizes bearing damage via excessive friction during startup.

The control system then activates the motor of the TG assembly. The SSS clutch engages as soon as the motor begins driving the worm gear, applying the requisite breakaway torque. Once the TG has the gas turbine shaft rotating at a slow roll, the LCI delivers a "pulse-commutated" output to the stator of the ac generator. As such, the ac generator is used as an ac *motor* to accelerate the gas turbine. When the turbine-generator rotor driven by the ac machine overruns the TG, the SSS clutch disengages and the TG shuts down.

After starting in the "pulse-commutated" mode, the LCI changes to the "load-commutated" mode at about 10 percent speed when there is sufficient emf in the ac machine (see Chapter 15, Equations 15.1 and 15.2 and the associated text). During startup, the generator field (excitation) current is controlled so that the generator stator flux does not exceed the allowable limits. The LCI output voltage increases linearly from 0 to its rated value (about 4 kV) as the shaft speed increases from near 0 to 30 percent of the rated speed (1,080 rpm for a 60-Hz unit), providing constant volts/Hertz (flux) on the generator. Above 30 percent speed, the LCI output voltage is held constant and the flux falls off inversely with speed by reducing the excitation current (known as "field weakening").

The power of the LCI is primarily determined by the required startup time of the turbogenerator from standstill until synchronization at 3,000 or 3600 rpm. As the LCI must only be connected until ~60 percent speed is reached (i.e., 1,800–2,160 rpm depending on grid frequency, which is referred to as the *self-sustaining* speed), the LCI active time is somewhat shorter. As the sample calculation above indicated, for state-of-the-art "fast-start" or "rapid-response" gas turbines, the LCI rating should be about 5 percent of the gas turbine base load rating.

During gas turbine startup, there is a preset schedule of acceleration rates and speed set-points, which are programmed into the turbine controller (e.g., see Figure 19.9). Once the operator pushes the "start" button, the turbine controller relays this information to the LCI controller. By controlling generator field voltage and stator current, the LCI controller adjusts the torque produced by the generator (which now runs as a *motor*) and thus controls the acceleration and speed of the turbine-generator set.

Prior to the "start" command, the gas turbine-generator set is spinning at TG speed in order to prevent a bow in the gas turbine rotor. For GE gas turbines, the TG speed is typically less than 10 rpm. In contrast, the TG speed for Siemens gas turbines is more

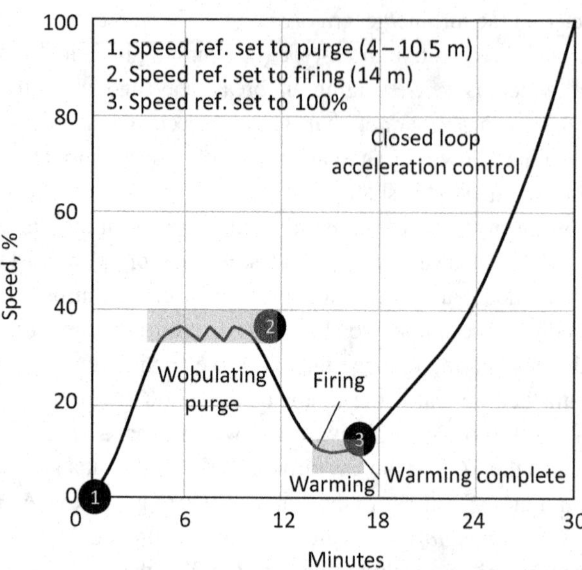

Figure 19.9 Typical gas turbine startup sequence (GE gas turbine).

than 100 rpm. The reason for this high TG speed (as claimed by the OEM) is to eliminate the turbine rotor blade *chatter* that is (claimed to be) common to other gas turbines when running on the TG with a much lower rpm. In other words, by eliminating the chatter via the centrifugal force generated by the higher rpm (i.e., extending the blades radially from the center of the rotor and into a position much closer to their normal operating position), any wear that would result from the blades chattering and rubbing would be negated.

At startup, the LCI connects to the generator stator and assumes control of the exciter field voltage reference. The LCI then accelerates the turbine to the purge speed set-point (typically 25 percent of synchronous speed). The turbine is held at purge speed for approximately 6 minutes, during which time combustible gases are expelled from the non-combustion portions of the system. Note that the purge sequence in Figure 19.9 is identified as "wobulating purge." Wobulation refers to the variation of the turbine speed by the controller in a "sawtooth" pattern. This method is adopted in single-shaft combined cycles without a SSS clutch separating the steam turbine from the generator. (In other words, the steam turbine rotates at the same speed as the gas turbine.) The goal is to avoid excitation of a natural frequency of the long last-stage buckets of the steam turbine (or the bearings) at a low speed by not staying for long at a resonance condition. (See Section 13.6.2 for the underlying principles.)

Upon completion of the purge cycle, the LCI is turned off and the turbine coasts down to the ignition speed of 15 percent. Once at the firing speed, the LCI is turned on again and the ignition commences. The turbine is briefly held at constant speed to allow for warming. Thereafter, the LCI accelerates the turbine to its self-sustaining speed, at which point the LCI is turned off and disconnected from the generator. From that point on, the controller accelerates the machine to FSNL and synchronization to the grid.

Upon turbine shutdown, as the turbine decelerates to below the TG speed, the SSS clutch engages if the TG motor is energized to provide slow-roll rotor cooldown. The cooldown continues until gas turbine wheel-space temperatures drop to the ambient level.

19.3.2 Combined-Cycle Startup

In a combined-cycle startup, there are many considerations in addition to the basic task of bringing the gas turbine to FSNL, synchronizing it to the grid, and loading it to FSFL at OEM-prescribed rates. Most important of all, there is a *second* turbogenerator that needs to be rolled to FSNL, synchronized, and loaded to FSFL: the steam turbine generator. In addition, correct steam chemistry, establishment of steam seals, maintaining proper vibration, and overspeed and thrust limits are all vital for acceptable component life and reliability, availability, and maintainability. When all is said and done, however, the single most important issue from a fast-start perspective is steam turbine thermal stress management. Furthermore, if the HRSG is a drum-type design, HP drum thermal stress management becomes an integral part of the problem.

The combined-cycle startup optimization problem can be formulated as needing to minimize the time required to reach the dispatch power (e.g., full load or a specific part load) without "breaking anything" in the process – literally. The failure mode to avoid is crack initiation and propagation. Failure to control thermal stresses results in cracks via *low/high-cycle fatigue* (LCF and HCF) and brittle fracture. In fact, LCF is found to account for roughly two-thirds of steam turbine rotor life, with the remainder attributable mainly to creep. In particular, thick-walled components such as the HP drum, steam turbine valves, casings, and rotor are exposed to LCF due to thermal cycling (start–stop sequence or load up–down ramps) and the associated thermal stress–strain loop.

In principle, the solution is simple enough: *thermal decoupling* of gas turbine and steam turbine start processes. Thus, the gas turbine is started and rolled to FSNL at the maximum rate dictated by the size of the LCI, shaft torque limit, and particular Dry-Low-NOx (DLN) combustion system limits (availability of heated fuel gas, minimum fuel requirement by the lean blowout margin, Wobbe index variation, etc.). Following synchronization, the gas turbine is loaded as fast as possible first to minimum emissions-compliant load (MECL) and then to FSFL.

There are two basic combined-cycle startup methods (see Figure 19.10):

1. Conventional or standard (i.e., "slow")
2. Accelerated (i.e., "fast" or "rapid")

Before going into the details of the two startup methods, the gas turbine combined-cycle start time definition must be clarified. The definition hinges on when to start the chronometer. Unless specified unambiguously, it is impossible to be sure when time $t = 0$ is, and the difference can be significant. For a conventional start with HRSG purge and normal loading rate (i.e., no holds for HRSG warming), the difference between the "start" command and ignition is 20 minutes (see Figure 19.10). Thus, the same start

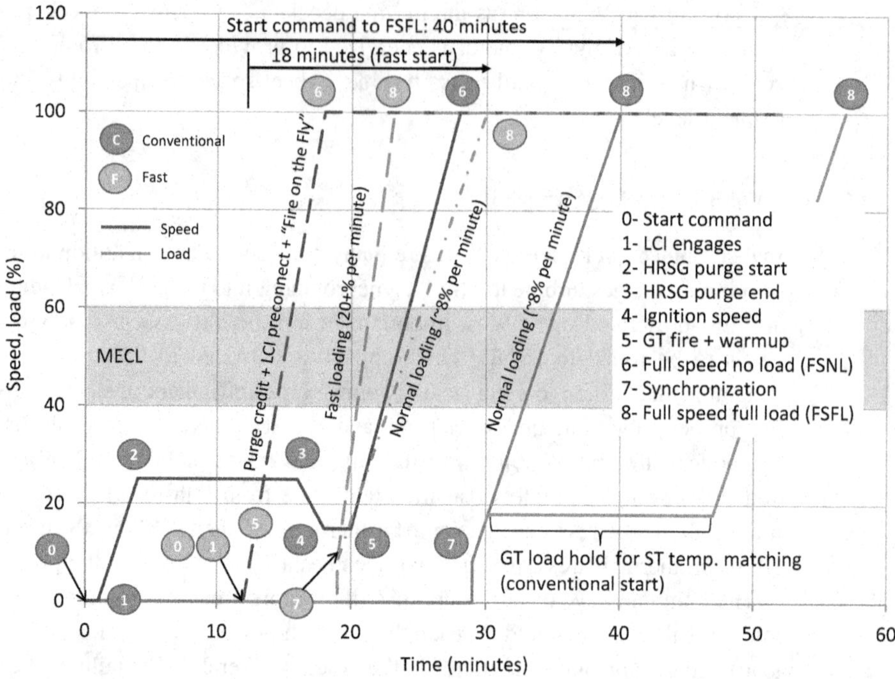

Figure 19.10 Typical gas turbine (GT) startup diagram (conventional and fast versions). ST = steam turbine.

time (40 minutes to be exact) can be quoted as 20 minutes if one sets the time t = 0 at ignition. State-of-the-art, "fast" gas turbines with features like "purge credit," LCI preconnect, and "fire on the fly" can reach FSFL in 18 minutes or less from the "start" command (depending on the loading rate). Herein, the assumption is that the clock starts running when the operator pushes the "start" button.

In a "conventional" start, the following sequence of events happen when the "start" button is pushed (Step 0):

1. The LCI engages (see Section 19.3.1).
2. The gas turbine is rolled to 25 percent speed (e.g., 900 rpm for a 60-Hz machine) and purging of the HRSG starts.
3. HRSG purging ends after 12 minutes and the LCI is turned off.
4. The gas turbine rolls down to 15 percent speed, at which point the ignition starts.
5. After a warm-up period when all combustor cans are "cross-fired," the gas turbine starts rolling again.
6. With the help of LCI, FSNL is reached at the 28-minute mark.
7. Synchronization is complete and the breakers are closed 1 minute later; the gas turbine load is about 10 percent.

Once the generator breakers are closed, the gas turbine is rolled to a very low load – less than 20 percent – and held there by the distributed control system (DCS) for a prescribed amount of time, which is a function of the time elapsed since the last shutdown. The longer that time, the "colder" the bottoming-cycle metal (i.e., the HRSG; heat exchanger tubes, headers, steam drums, etc.), steam pipes, and valves between the HRSG and the steam turbine, steam turbine casing, and rotor. The colder the metal, the longer the hold time at low load so that it warms up gradually to prevent excessive thermal stresses, especially in thick-walled components (e.g., the HP steam drum of the HRSG or the steam turbine casing).

This hold time is requisite for steam temperature "matching." The reason for this can be seen from the typical gas turbine part load data (e.g., Table 18.13 or Table 18.14). If the gas turbine is loaded to its MECL (typically 50 percent for older E- and F-class gas turbines), its exhaust temperature will be either very high (e.g., at the *isotherm* for GE F–class gas turbines; i.e., 1,200°F) or at its base value (e.g., for Siemens gas turbines). In either case, steam generated in the HRSG will be too high in temperature for admission into the steam turbine. Interstage steam *attemperators* (also referred to as *desuperheaters*) between superheater rows are not suitable for accurate steam temperature control. Thus, the HRSG steam temperature is controlled at the requisite level (as dictated by the steam turbine controller) by controlling the gas turbine exhaust temperature. Steam admitted at that low temperature and OEM-prescribed low flow (and controlled by the DCS via *steam bypass* to the condenser) into the steam turbine gradually warms up the rotor and casing metal to allowable levels. Obviously, this warmup process takes a longer time for colder startups. We can now continue with the startup event list.

8. Once the controller deems that the steam turbine is "warm enough," the gas turbine is loaded to FSFL at an OEM-prescribed rate, which is about 8 percent per minute.

Shortening the time to MECL is critical for a reduction of startup emissions. The reason for this lies in the basic design philosophy of modern DLN combustors with fuel–air premixing, which are designed to run near the lean limit for low emissions. As discussed at great length in Chapter 12, this is accomplished by piloted, multi-nozzle fuel injectors via sequential activation of fuel flow through individual nozzles (known as staging) to prevent lean blowout and combustion dynamics while staying within the narrow equivalence ratio band to control NOx and CO emissions. For older E- and F-class gas turbines, MECL is 60 percent; for modern gas turbines, the low load limit is around 40–50 percent (maybe 35 percent for the most advanced systems). The exception to the rule is sequential combustion (reheat) gas turbines (i.e., GE/Alstom GT24 and GT26), which can turn off their second combustors (low-pressure or SEV combustor in OEM parlance) to operate at 20 percent or lower load while remaining emissions-compliant. Advanced-class gas turbines with fuel staging in their can-type combustors are also expected to have this ability in the near future (see Chapter 21).

The following two steps are instrumental in reducing gas turbine and combined-cycle start times:

1. Elimination of the HRSG purge sequence (by performing it right after shutdown in compliance with NFPA® 85).

2. Elimination of hold time at low load with reduced exhaust energy (flow and temperature) to control the HRSG steam production rate and steam temperatures (at the HP drum and HP superheater exit).

Elimination of direct HRSG steam temperature control via gas turbine load and exhaust energy is the "thermal decoupling," which is the key enabler of fast start. It can be accomplished via a *bypass stack* and modulated damper controlling the exhaust flow to the HRSG. A recently proposed technique is "air attemperation" of the gas turbine exhaust gas flow via air injection into the transition duct. Ignoring the obvious but wasteful practice of "sky venting," the currently accepted method is a "cascaded" steam bypass system with TAs. Steam generation and temperature–pressure ramp rates in the HP drum are dictated by gas turbine exhaust energy, whereas final steam temperature control is accomplished by the TAs, which are located at the exit of the HRSG (in addition to the conventional attemperators located between superheater sections). Until steam temperatures reach acceptable levels for admission into the steam turbine, steam is bypassed via a route including the reheat superheater so that the latter is pressurized and "wet" (i.e., cooled by steam flow, obviating the need for expensive alloys). This is commonly known as "cascaded bypass."

The "fast" start in Figure 19.10 has the following steps after the "start" button is pushed (Step 0):

1. The LCI is pre-connected and engages immediately (no time lag).
2. Eliminated.
3. Eliminated.
4. Eliminated.
5. "Fire-on-the-fly" ignition.[5]
6. With the help of LCI, FSNL is reached in 6 minutes.
7. Synchronization is complete and the breakers are closed 1 minute later; the gas turbine load is about 10 percent.
8. Loading to FSFL
   a. At a normal rate (~8 percent per minute)
   b. At a faster rate (~20 percent per minute)

In comparison to the conventional start, which took 40 minutes from start to FSFL, the fast mode takes only 18 minutes, which can be even shorter (12 minutes) if the load ramp rate is increased to 20 percent per minute. Fast-start benefits vis-à-vis the conventional start are as follows:

1. Faster time to "full load" with about 15 percent less fuel consumption;
2. Reduced startup emissions;
3. Improved dispatch ranking.

[5] The warm-up hold period immediately after ignition shown in Figure 19.9 is eliminated. Ignition occurs at the standard rotor speed and airflow. Acceleration is paused until flame is detected and resumes thereafter.

There is a third startup option that is in between the conventional and fast startups. Note that the fast-startup capability requires additional hardware to facilitate the elimination of the steam temperature matching hold: TAs and "parallel" bypass pipes.

In conventional cascaded bypass, excess HP steam is first routed to the reheater and thereafter to the condenser. In a combined cycle capable of fast start, extra steam bypass capacity is needed, which is accommodated by a "parallel" bypass line directly from the HP superheater exit to the condenser (the shaded section in Figure 19.4). This is known as a *hybrid bypass system*. It requires significantly longer and more expensive piping with additional desuperheating station.

By loading the gas turbine to MECL and holding it there for steam temperature matching (albeit for a much shorter duration vis-à-vis the conventional method), parallel bypass piping can be eliminated. This is accomplished by reduced steam production in the HRSG, which can be handled by the standard cascaded bypass piping.

Furthermore, even a "cheaper" variant is possible for highly cycled power plants with daily shutdown and hot starts. Due to the limited cooling of the steam turbine during an overnight shutdown, the terminal attemperation system can be eliminated as well for even more reduction in capital cost. This, of course, requires an advanced gas turbine with low MECL capability (i.e., 40 percent or lower). A combined-cycle "hot" start of this type is shown in Figure 19.11.

The description of the combined-cycle start in Figure 19.11 is provided below.

1. Steam and gas turbines are on the TG; the plant is ready to start; operator pushes "start" button.
2. The LCI accelerates the gas turbine to the purge speed (the steam turbine is still on the TG; in a single-shaft configuration, it is separated from the rest of the power train via the SSS clutch).
3. Purge completed; the gas turbine slows down to ignition speed; light off and acceleration to FSNL.
4. The gas turbine synchronizes to the grid and is ramped to MECL (45 percent load as shown) and held there while the HRSG is warmed up.

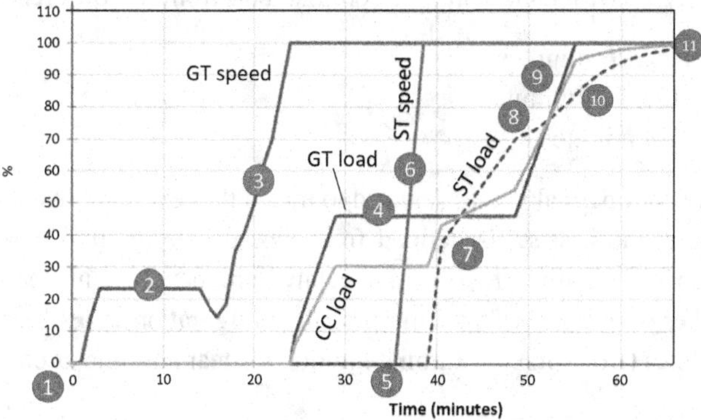

Figure 19.11 Combined-cycle (CC) "hot" startup with gas turbine (GT) hold at MECL for temperature matching. ST = steam turbine.

5. HRSG warm-up is complete; the steam turbine bypass valves are open; TAs are in "temperature match" mode.
6. Steam turbine roll-off to FSNL starts by admitting steam into the IP section (via the intercept control valves); the SSS clutch engages.
7. The steam turbine is loaded to the *forward flow transfer* point (where HP steam is admitted into the HP turbine via the MCVs).
8. Steam turbine loading continues to the *inlet pressure control* mode, at which point the bypass valves are closed and gas turbine loading to full load starts.
9. The plant (i.e., the steam turbine – the gas turbine is already at full load) is ramped to full load.
10. TAs are turned off; the steam turbine is in full steam admission mode.
11. The plant is at 100 percent (full) load.

Figure 19.11 illustrates another startup definition issue. The first one discussed earlier pertained to the definition of time t = 0 (i.e., when to *start* the chronometer). This one pertains to when to *stop* it. Due to the inverse exponential decay nature of the bottoming-cycle "heat soak" process, the last part of reaching "true" 100 percent combined-cycle load takes a long time. A commonly used definition for quantifying the combined-cycle start time is to stop the chronometer when

1. Bypass valves are closed,
2. TAs are off, and
3. The steam turbine is in full admission mode.

This is a quite logical choice because, beyond that point, there is no control knob left available to the DCS to impact the combined-cycle output. The gas turbine is at full load with IGVs fully open and the temperature control active. All bypass valves are fully closed and all admission valves are fully open. In other words, there is nothing left except to wait until all components reach their fully heat-soaked, steady-state operating level. In any event, based on the definition used for beginning and ending points, the start time can change by 10–20 minutes for a hot start.

In several recent combined-cycle projects (2 × 2 × 1 multi-shaft configuration) with a conventional drum-type HRSG, the start time commitments from two different OEMs were

- Cold start: 226/220 minutes
- Warm start: 128/115 minutes
- Hot start: 38/58 minutes

The listed times are commercial values selected to match the requirements of the plant air permit. For example, the first set is defined from gas turbine ignition to HRSG stack emissions compliance (including the selective catalytic reduction) with the gas turbines at MECL. In another project (3 × 3 × 1 multi-shaft configuration), OEM-estimated (*not* guaranteed) start times (from first gas turbine ignition to steam turbine at base load) were

- Cold start: 197 minutes
- Warm start: 136 minutes (72 hours of downtime)
- Hot start: 85 minutes (16 hours of downtime)

The sample start times listed above are significantly longer (especially for warm and cold starts) than advertised fast-start capabilities by OEMs in the trade literature, e.g.,

- Cold start: 60 minutes
- Warm start: 45 minutes
- Hot start: 30 minutes

In fact, they are more in line with the conventional combined-cycle start times that can be found in papers and articles from 10–20 years ago, i.e.,

- Cold start: 3 hours
- Warm start: 2 hour
- Hot start: 1 hour

The reason for the discrepancy is not that the new technologies are not capable of achieving what they promise. There is simply very little commercial need for fast start after an extended shutdown (e.g., for maintenance), which are events that are planned in advance and take place as scheduled. Therefore, there is little or no financial incentive to put the plant equipment through unnecessary thermal stresses. Consequently, after an extended shutdown (several days or longer), operators choose to start their equipment at a slow pace with ample time for gradual warming *as long as they are in compliance with their air permits*. A typical permit (for a 3 × 3 × 1 advanced F-class gas turbine combined-cycle power plant) reads something like this:

Pursuant to the best available technology requirements of [applicable code sections], the total emissions from the combined-cycle gas turbines combined (including HRSG duct burners) shall not exceed the following totals in any 12 consecutive month period. (These emissions limits also include those during startup and shutdown events.)

- Nitrogen oxides – 230.16 tons
- Carbon monoxide – 388.9 tons
- Volatile organic compounds – 71.6 tons
- Sulfur oxides – 54.3 tons
- Total particulate matter (including PM10 and PM 2.5) – 197.0 tons
- Sulfuric acid mist – 26.6 tons
- Ammonia – 213.7 tons

For a gas turbine with a duct-fired HRSG (including an SCR), the typical NOx emission guarantee is 2 ppmvd at 15 percent O_2. This corresponds to about 75 tons per year of NOx as NO_2 emissions (8,760 hours at full load with duct-firing) for an F–class gas turbine with ~1,200 lb/s exhaust flow. For three gas turbines, one is looking at 225 tons/year. Prima facie, this leaves 5 tons/year for emissions during startups and shutdowns. Obviously, the actual plant is going to be run much less severely than the nonstop, round-the-clock duct-fired assumption above. Thus, operators will have a good idea about their emissions budget when deciding how to start the plant at a given time in a given year.

Economic Implications

One can easily calculate the total energy generation, E, for a given startup profile similar to that given in Figure 19.12, i.e.,

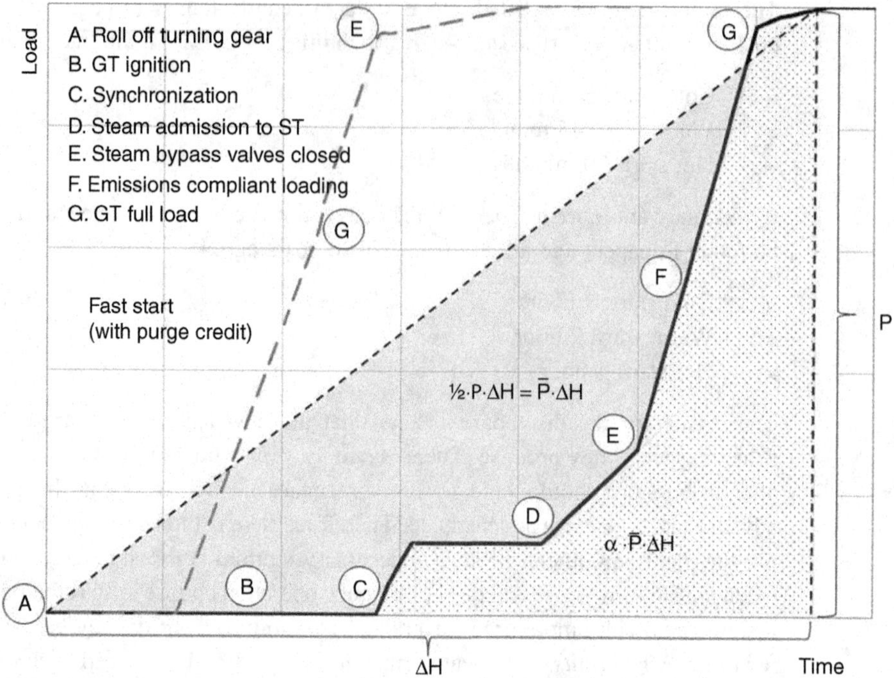

Figure 19.12 Conventional combined-cycle start (per figure 3 of [11]). GT = gas turbine; ST = steam turbine.

$$E = \alpha \cdot \bar{P} \cdot \Delta H, \tag{19.21}$$

for $\Delta H = 1$ hour. The parameter α quantifies the true MWh generated (the area below the start curve) as a fraction of the simple estimate, which is the triangular area below the straight line between $t = 0$ and $t = \Delta H = 1$ hour, i.e.,

$$\frac{P \cdot \Delta H}{2} = \bar{P} \cdot \Delta H.$$

For the startup period, then, the "load factor," λ, is found as

$$\lambda = \frac{\alpha \cdot \bar{P} \cdot \Delta H}{P \cdot \Delta H} = \frac{\alpha \cdot \bar{P}}{P}. \tag{19.22}$$

If the startup profile were a simple straight line, λ would be equal to 0.5. Fuel consumption can be estimated in the following two steps:

1. Read the "mean effective" heat rate or efficiency for the startup period at λ from the applicable part load curve (see Section 18.1).
2. Use it to calculate the fuel consumption from

$$\dot{Q}_f = \frac{\bar{P}}{\bar{\eta}}.$$

The value of α depends on the type of start (e.g., hot, warm, or cold) and the plant technology, i.e.,

- For the conventional start in Figure 19.12, α is about 0.33;
- For the fast start in the same chart (dashed line) with *purge credit*, α is about 0.5.

Reemphasizing a point made earlier, definition of startup time is critical to a proper evaluation of its impact on energy generation. Herein, the period between the gas turbine rolling off the TG (point A in Figure 19.12) and the point with (i) the gas turbine at full load and (ii) all steam bypasses closed (point G for the conventional or point E for the fast start in Figure 19.12) is defined as the start time. However, alternative definitions setting time as t = 0 at gas turbine ignition or synchronization are also available in the literature. One should be cognizant of such differences when making comparisons.

The operational flexibility benefits of a combined-cycle power plant with an advanced F-class gas turbine are explained in detail in articles in trade publications (e.g., see Gülen and Jones [11]). A key feature is the capability to reach full load in less than 30 minutes in a *hot* start following overnight shutdown. This is less than half the time required by a *conventional* hot start. A comparison of the fast and conventional hot-start technologies is shown in Figure 19.12 (cf. figure 3 of [11]). The benefit of the faster start is primarily due to the lower amount of time spent at low loads, which translates into a higher load factor and average efficiency. For a 425 MW-58% (net ISO baseload) single-shaft combined cycle with a state-of-the-art F–class gas turbine, the benefits are quantified and summarized in Table 19.3.

In particular, fast start with purge credit results in ~30 percent less fuel consumption and CO_2 emissions (planned or scheduled start). Furthermore, in the same time period with simultaneous roll off of the TG, the fast-start technology results in more than four times the power production during conventional start (unplanned or unscheduled start). There are two operating scenarios for assessing startup generation:

- *Unplanned (unscheduled) starts.* The fast plant starts at the same time as the slow plant and reaches base load quicker, allowing it to be dispatched for a longer

Table 19.3 Comparison of conventional and fast starts

	Conventional	Fast	
	Planned/Unplanned	Planned	Unplanned
Start Time, t (Min)	60	28	60
Normalized Generation (%-h)	15.4	12.9	65.7
Load Factor, λ	15.4%	27.7%	65.7%
Effective Efficiency, η	36.3%	43.1%	53.1%
α	0.31	0.55	
CO_2 Generation (Tons)	39	28	113
Energy Generation, E (MWh)	65.5	55.0	279.3
Heat Consumption (MWh-th)	180.5	127.5	525.7

period and therefore generating more revenue, but at the expense of higher total fuel consumption and emissions (see Table 19.3). This is most likely to occur in an emergency situation, when immediate dispatch is required to exploit an opportunity or to make up for a sudden or imminent loss in generation.

- *Planned (scheduled) starts.* The fast plant starts later than the slow plant and reaches base load at the same time. The benefit to operating in this manner is lower fuel consumption and emissions, but at the expense of less generation (i.e., less revenue). This is the norm for regular plant starts. (Note that the less generation and revenue argument is somewhat murky when electricity sale prices are taken into account. The price before the prescribed time when the market price becomes effective is probably too low, to the point that it is below the variable cost of generation.)

It is customary to express the total energy production given by Equation 19.21 in *normalized* terms over the period of H hours (in %-hours or %-h; i.e., e = E/P. In other words, if P is 500 MW, for running 1 hour at P output, instead of 500 MWh (megawatt-hours), one can use 100%-h (i.e., 100 percent-hours). Similarly, for only 30 minutes at P output, one ends up with 250 MWh or 50%-h.

19.4 Shutdown

In a combined-cycle power plant, normal plant shutdown is accomplished by ramping down the load of the gas turbines and disconnecting them from the grid. The steam turbine is taken offline by the DCS at a predetermined load as the gas turbines are ramping down. The steam bypass valves are modulated by the controller to maintain the allowable main and HRH steam pressures. After all turbine-generators are disconnected from the grid and coasted down to the turning speed, the units are placed on the TG. The HRSG stack damper is closed and the HRSG is "bottled up" to reduce heat losses during short outages (e.g., overnight) and thus to reduce the restart time (i.e., to enable "hot" start).

In addition to a normal (also called "fired") shutdown, which is initiated by the operator pushing the "stop" button, the power plant can be stopped by the DCS in response to plant malfunctions or disturbances in the grid as necessary for personnel safety and/or equipment protection. This is commonly known as a "trip." When a gas turbine or steam turbine trips, it coasts down to TG speed and is available for restart when the restart interlocks have cleared.

A combined-cycle power plant can be designed with the capability to shut down or trip the steam turbine without causing the gas turbines to trip. This is accomplished by rapid opening of the steam bypass valves to prevent overpressure in the HRSG. In a single-shaft combined-cycle power plant, this is made possible by the SSS clutch. How long this operation mode can be continued depends on the gas turbine operating conditions. A multi-shaft combined-cycle plant with two or more gas turbines typically has the capability to shut down or trip one of the gas turbines without causing the entire plant to trip.

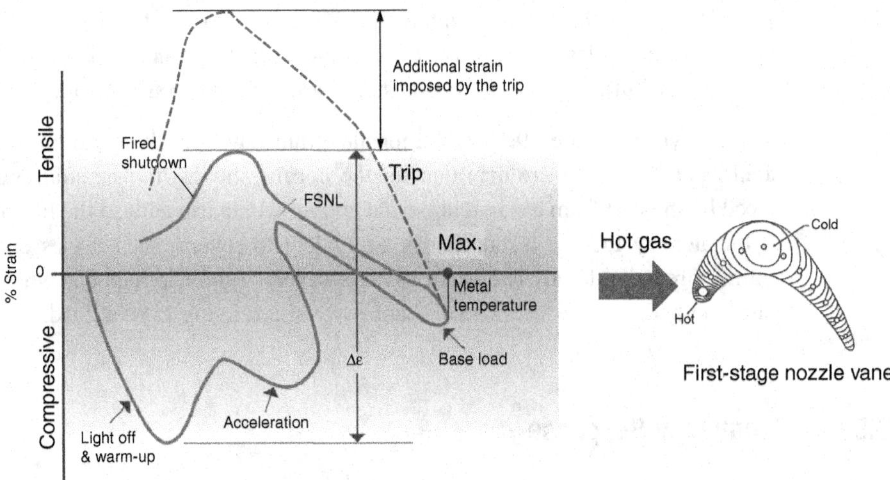

Figure 19.13 Start–stop cycle (for LCF considerations).

In case of a plant full load rejection (i.e., the entire plant output is suddenly disconnected from the grid), all turbine-generators will trip and proceed to a safe shutdown condition. When the gas turbine trip is initiated by the DCS, fuel system GCVs are slammed shut ("trip-closed") and the IGVs close based on the "trip schedule." At the same time, steam turbine MCVs and intercept control valves are also trip-closed and the bypass valves are opened to minimize shaft overspeed and prevent HRSG overpressure. Once the generator breakers open, the turbines coast down to the TG speed.

Trip from full load is detrimental to gas turbine part lives. This is illustrated by the metal temperature–strain diagram in Figure 19.13, which covers a full start–stop cycle with steady-state full load operation in the middle. Also shown in Figure 19.13 is the temperature distribution in a stage 1 nozzle vane.

The leading edges of the airfoils (i.e., stator vanes and rotor buckets) respond to the hot gas temperature change more quickly than the thicker middle section, which creates temperature gradients in the airfoil cross-section. These gradients lead to transient thermal stresses during startup and shutdown. In particular,

1. Light-off and acceleration produce transient *compressive* strains in the airfoil as the fast responding leading edge *heats up* more quickly than the thicker bulk section of the airfoil.
2. As the gas turbine is rolled to FSNL, the temperature gradients diminish as the inner regions of the airfoil warm up, which reduces the temperature gradient and the thermal stress.
3. During the loading of the gas turbine from FSNL to FSFL in temperature control mode, the temperature gradient between the leading edge and the center of the airfoil and hence the thermal stress increase again.
4. At full load, the airfoil reaches its maximum metal temperature and a *compressive* strain is produced from the normal steady-state temperature gradients that exist in the cooled part.

5. When the shutdown is initiated, the reverse process takes place (i.e., the fast-responding leading edge *cools down* more quickly than the thicker bulk section of the airfoil), which leads to transient *tensile* strains in the airfoil.

Also shown in Figure 19.13 is the tensile strain range resulting from a trip, which is significantly more severe than that for the normal shutdown. The strain range experienced by the part from a start–stop cycle ending with a trip adds to the life consumption as dictated by LCF at a rate that is equivalent to several such cycles ending with a normal fired shutdown. The impact is less severe if the trip happens when running at part load and more severe when it happens while running at peak load.

19.5 Frequency Response

A gas turbine in electric power generation duty very rarely operates on its own. Whether in a simple- or combined-cycle configuration, the gas turbine, via its ac generator, is connected to an interconnected network, which is commonly referred to as an electrical or power grid or, simply, "the grid." The ultimate duty of the grid is to transmit electric power generated by a multiplicity of generators (fossil fuel-fired, nuclear, renewable) to an order-of-magnitude larger number of end users (industrial and residential) via a complex network of transmission lines at the highest possible reliability and the lowest possible cost. Since electric power must be consumed when generated, this requires a continuous balancing act of matching demand from the end users (which fluctuates from minute to minute, hour to hour, and day to day) with the supply from generators with very different performance and operability characteristics.

Consequently, operation and maintenance of the grid is subject to rigorous rules and regulations, which constitute the "grid code." A grid code specifies technical criteria pertaining to the design and operation of electric power generation plants such as

- Quality of supply (e.g., voltage and frequency)
- Protection (e.g., backup in case of system failure)
- Generating unit specifications (power factor, frequency response, etc.)
- Metering and monitoring requirements

There are different grid codes developed and enforced by different countries and/or group of countries (e.g., the EU). In the USA, for example, there are three main power grids (i.e., the Eastern, Western, and Texas Interconnects) and 10 regional reliability councils of the North American Energy Reliability Council (NERC). Members of those councils come from all segments of the electric industry (investor-owned utilities, rural electric cooperatives, independent power producers, etc.).

Frequency stability and control are the most important grid code requirements. Grid frequency is the single most important measure of the balance between electric power supply and demand. If power generation and consumption (the "load") in the grid are exactly matched, the system frequency is exactly equal to the rated frequency (i.e., 50 or 60 Hz depending on the region). Unexpected malfunctions in any given part of the grid such as an emergency trip of a large generation station create an imbalance between supply and demand, which is reflected by a change in system frequency.

In fact, the grid frequency is a continuously changing variable. If demand or load is greater than generation, the system frequency falls (i.e., generators connected to the grid slow down). If generation is greater than demand, the system frequency goes up (i.e., generators connected to the grid speed up). A mechanical analogy to this was presented earlier in Chapter 18 (e.g., see Figure 18.7).

In general, three types of event cause a change in grid frequency:

- Loss of generation (supply)
- Loss of load (demand)
- Normal variations in load and generator output

For stable operation, grid frequency should be held within narrow limits defined by the applicable grid code. Minor deviations such as 200 mHz or the absence thereof indicate that there is a balance between generation and load. Unexpected, sudden events such as plant trips and load losses result in system faults and lead to larger deviations, which, unless corrected, can destabilize and collapse the grid.

There are two types of faults:

- Faults *within* a controllable range
- Faults *outside* the controllable range

Faults in the former category lead to fluctuations within a small band (e.g., ±200 mHz) and can be ridden out by adjustments on the generation side. In principle, it must be possible to ride out the loss of the largest generator in the system without a frequency excursion outside the controllable band. Faults in the latter category cannot be alleviated by increasing or decreasing power generation only. In severe cases of underfrequency (i.e., loss of a significant amount of generation), the system survives by "load shedding" (i.e., throwing certain segments of users into "darkness"). Even if the underfrequency event takes place very rapidly (i.e., within a few seconds), as long as it remains above a grid code-specified limit, generators must continue operating stably as required by the grid code.

In the UK, for example, according to the National Grid Code (NGC), each generating unit (e.g., a gas turbine) must satisfy several minimum requirements for frequency control. One of those requirements stipulates that the generating unit, if operating at full load, must be capable of maintaining power output if the grid frequency drops to 49.5 Hz (i.e., 500 mHz below the normal 50 Hz). Thereafter, the generating unit should be able to stay online, albeit with a reduction in power output (no more than pro rata with grid frequency), down to 47 Hz.

In other words, a gas turbine at 100–MWe output (full load) at 3,000 rpm must be able to run at 100 MWe at 2,970 rpm. Furthermore, it should be able to stay synchronized to the grid down to 2,820 rpm, at which point it should be able to generate 95 MWe. Since reduced speed reduces the "suction" capacity of the compressor, compliance with this requirement means that the gas turbine must have the capability to *overfire* in order to compensate for the loss in mass flow rate.

On the other side, the gas turbine should be able to stay synchronized to the grid up to 3,120 rpm (4 percent overspeed), at which point it should be able to generate 68 MWe. Compliance with this requirement means that the gas turbine must have the capability to reduce the airflow via IGV closure and reducing the firing per the built-in control algorithm.

The NGC underfrequency response requirement described above is of the "ramp" type. There are other grid codes with different types of underfrequency response requirements, e.g.,

- "Step" type (e.g., in Germany, 50–Hz grid)
- "Intrinsic" type (e.g., in Brazil, 60–Hz grid, and in Australia, Greece, Italy, etc., all 50–Hz grids)

It is easy to appreciate that designing a gas turbine to meet *all* possible grid code requirements *anywhere in the world* is a monumental task. OEMs approach this problem by collecting data from different countries showing frequency distributions over a long period of time to identify a "normal range" and "outliers beyond six sigma" (i.e., six standard deviations or 6σ). Consequently, they strive to design their gas turbines to meet the requirements for the most likely events from a representative set covering the major market segments.

The response of a power plant to a grid frequency deviation caused by an event in the grid is handled by the frequency control. Using the NGC as an example, let us look at the key components of the frequency control problem. The response of a generating unit (e.g., a gas turbine power plant in simple or combined cycle) to a drop in grid frequency is divided into two steps as shown in Figure 19.14.

The primary response capability (P in Figure 19.14) of a generating unit is the minimum increase in active power output between 10 and 30 seconds after the start of the frequency ramp.

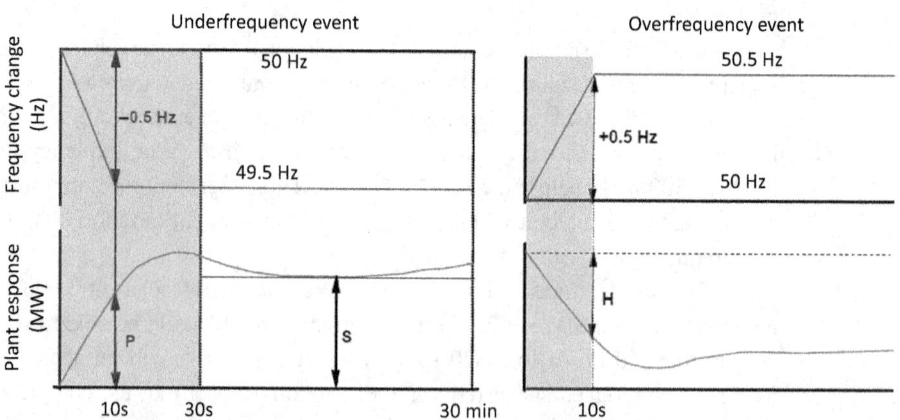

Figure 19.14 Primary and secondary responses to frequency excursions (UK NGC).

The secondary response capability (S in Figure 19.14) of a generating unit is the minimum increase in active power output between 30 seconds and 30 minutes after the start of the frequency ramp.

While the specific requirements of each response vary from code to code, this two-step structure is the basic code compliance framework for gas turbine power plants.

Similarly, according to the UK NGC, the high-frequency response capability of a generating unit (H in Figure 19.14) is defined as the decrease in active power output provided 10 seconds after the start of the frequency ramp and sustained thereafter.

From a simple- or combined-cycle gas turbine power plant perspective, the key operability issue is the primary and secondary response to a frequency drop event. Obviously, due to its large thermal inertia, the bottoming cycle of a combined-cycle power plant (i.e., the steam turbine generator) cannot contribute to the primary response. The gas turbine, on the other hand, can respond to an underfrequency event to provide the primary response by opening the IGVs (i.e., more inlet airflow) and/or by increasing the firing temperature.

At this point, if you have not already done so, it is suggested that you go over the discussion of "droop" in Chapter 18 (specifically, Section 18.2). If you are already familiar with the concept, let us continue. For a 50-Hz system, 4 percent droop at ±200 mHz (i.e., ±0.2 Hz) *controllable* band means that the generator power should increase by (per Equation 18.5)

$$\frac{\Delta \dot{W}}{\dot{W}} = \left| \frac{\frac{-0.2}{50}}{4} \times 100 \right| = 0.1 \text{ or } 10\%.$$

Different grid codes require this response in different times.

The contribution of a generator (i.e., a gas turbine) to the correction of a disturbance in the grid depends mainly upon the droop and the *primary control reserve* (PCR) of the generator in question. Figure 19.15 shows a diagram of variations in the generating

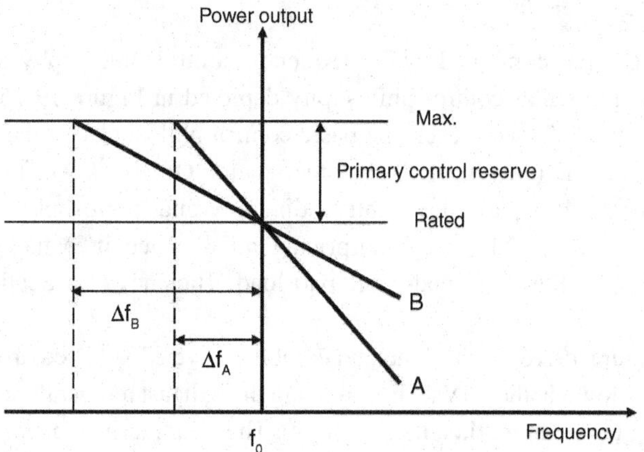

Figure 19.15 PCR illustration.

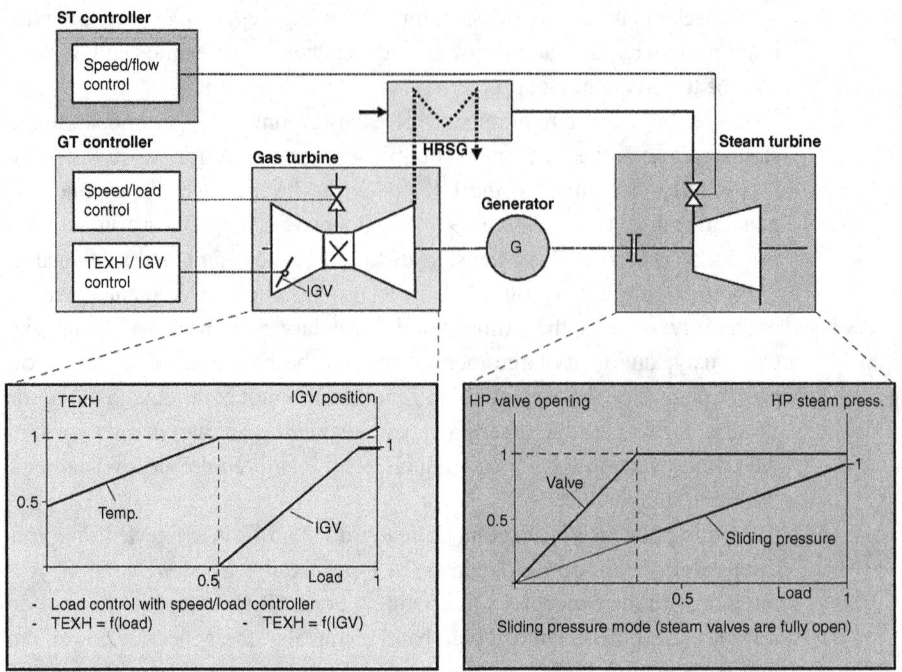

Figure 19.16 Combined-cycle control for grid response. GT = gas turbine; ST = steam turbine. [12]

output of two generators (e.g., gas turbines) A and B with different droop characteristics (i.e., A has a smaller droop than that of B), but with identical PCRs.

Let us assume that there is a minor grid disturbance of magnitude $\Delta f < \Delta f_B$ such that the frequency drops from the set value of f_0 to $f_0 - \Delta f$. The contribution of generator A with the smaller droop to the correction of the disturbance will be greater than that of generator B. However, the PCR of A will be exhausted well before that of B (i.e., corresponding to a disturbance of Δf_A), even though both A and B have the same PCR. The contributions of both generators to the primary control will be equal (since they have the same PCR).

A typical gas turbine combined-cycle frequency control philosophy is shown in Figure 19.16. The particular control philosophy depicted in Figure 19.16 is adopted from a Siemens paper [12]. However, the basic control philosophy of the gas turbine is similar to that of GE (and most likely other major OEMs as well). In order to verify this assertion, compare the control diagram and philosophy with those covered in Section 18.2. This is not surprising at all since it is the gas turbine control that maximizes the combined-cycle part load efficiency (see Section 18.1 for more details).

As shown in Figure 19.16, gas turbine and combined-cycle load is controlled mainly by changing the airflow via the IGVs while keeping the exhaust temperature constant by adjusting the firing temperature (i.e., the fuel flow). The steam turbine operates at valves wide open (VWO) or "sliding pressure" mode until the throttle pressure reaches the "floor" value. (Reduction in the throttle pressure is driven by the reduction in gas turbine exhaust energy, which reduces the HRSG steam production.) Beyond that point, the throttle valve (MCV) is closed to maintain the floor pressure.

With the steam turbine in sliding pressure mode, the gas turbine is the primary control provider in the combined-cycle power plant. This places a significant onus on the load-changing capability and/or flexibility of the gas turbine. Typically, in a combined-cycle power plant, primary and secondary control can be provided in the range of 65–100 percent of the plant output (as dictated mainly by the IGV range). The control is accomplished via the gas turbine speed-load controller as discussed earlier. In particular,

- Peak fire to >70–100°F;
- Open IGVs to the maximum position (i.e., maximum airflow);
- Turn on the online wash system (additional mass flow).

These control actions increase gas turbine output and exhaust energy, which is bound to increase the steam turbine power output and contribute to the overall power plant output. However, the time it takes for the steam turbine to reach its new output level is determined by the thermal inertia of the HRSG and the plant balance of plant, which has a time constant of several minutes. Consequently, in VWO or sliding pressure mode, the steam turbine is not suitable for primary response/control (which must take place within seconds). However, it will be able to engage in secondary response/control.

Enhancements to the steam turbine response or contribution to the grid code requirements – specifically secondary control/response – are possible. Examples are as follows:

- Throttle valve closure to maximum HRSG pressure (HP or IP admission valves);
- Reduced attemperation for limited-time steam temperature excursion as prescribed by the OEM (e.g., from 1,050°F to 1,085°F for up to 15 minutes);
- Reduced gas turbine fuel temperature (more IP steam production).

In certain cases, controller, combustion, and/or component life limits indicate that the primary control cannot be provided by the gas turbine only. Therefore, the ability to meet a certain grid code requirement rests in the ability of the steam turbine to contribute to the primary control as well. By running in a valve-throttled mode with a fast-acting valve, discharge of stored energy in the HRSG during a grid event can achieve that. Typically, this feature is combined with running the combined-cycle power plant at part load (say, at 90 percent load) with the gas turbine IGVs partly closed. This provides a readily available "reserve capacity" to be deployed when the grid frequency drops suddenly. Running the combined cycle in such a "grid control" mode comes with a plant heat rate penalty in comparison to running at full load. This loss should be balanced against the bonus payments for the ancillary service provided (for grid frequency control) and hot gas path life saving due to reduced overfiring.

An alternative method for increasing the power output capacity of a steam turbine is the inclusion of an "overload valve." This valve is essentially a "bypass" valve that redirects a portion of the HP steam (up to about 20 percent) to an admission point downstream of the HP turbine inlet. The overload valve is closed during normal operation. In a grid event, when HP steam generation in the HRSG increases via higher gas turbine exhaust energy, it opens and lets out the extra steam admitted at the downstream admission point. This ensures that the plant is run at full load and the extra steam flow does not push the throttle pressure beyond its maximum rating.

19.5.1 Example 1

Consider a heavy-duty industrial gas turbine rated at nominal 250 MW and operating in a 50-Hz grid whose total generating capacity is 8,000 MW. Let us assume that all turbogenerators on the grid, including the aforementioned gas turbine, are operating in 4 percent droop-control mode. Thus, the speed governor of the gas turbine will take

$$250/8{,}000 = 3.125\%$$

of any load demand change. For this example, assume that the gas turbine is generating 250 MWe with firing up to 2,420°F and IGVs in their fully open position. Here is what happens: grid frequency drops by 5 mHz to 49.995, which is 5/50,000 = 0.0001 or 0.01 percent in magnitude.

Using the droop equation, Equation 18.2, requisite load change is

$$0.0001 \times 100/4 = 0.0025 \text{ or } 0.25\%.$$

For the whole grid, this amounts to 0.25% × 8,000 = 20 MW.

For the sample gas turbine, this amounts to 0.25% × 250 = 0.625 MW.

In other words, the share of the gas turbine in restoring the grid frequency is 0.625 MW. The remaining 20 – 0.625 = 19.375 MW is handled by the remaining generators on the grid.

At 49.995 Hz, the gas turbine speed would be (for a synchronous ac generator with two poles)

$$N = 49.995 \times (120/2) = 2{,}999.7 \text{ rpm}.$$

In order to generate 250 + 0.625 = 250.625 MW, which amounts to an increase of 0.625/250 = 3.125%, at the same IGV opening and 2,999.7 rpm, the firing temperature of the gas turbine should increase by 3.6°F.

As seen in this example, droop control ensures that each and every generator on the grid does its share to restore the grid frequency without having to know what the others are doing.

Without droop, there would be chaos. In order to prevent the chaos, one generator could be designated as the "grid stabilizer" while operating in isochronous control. This is a possibility in small power grids (e.g., on an island in the Caribbean) where the largest generator operates in isochronous mode. Note that, in any system of connected prime movers, only one can operate in isochronous mode, with the remainder controlled via a droop governor. In a scenario like this, a sudden increase in grid load and the commensurate drop in grid frequency would be handled by the generator operating with an isochronous governor. The isochronous governor detects the drop in speed/frequency and increases the generator output by increasing the fuel flow in order to pick up the new load demand.

Would this be possible in our example? If the sample gas turbine is asked to generate 250 + 20 = 270 MW and thus to meet the entire load demand increase of the grid, with no room for extra airflow via opening the IGVs, it would have to be fired to 2,529°F. The required overfiring of more than 100°F is quite unrealistic.

19.5.2 Example 2

According to the UK NGC, at 95 percent frequency (i.e., 47.5 Hz or 2,850 rpm for the synchronous ac generator), a power plant should be able to deliver 96 percent of the registered load (e.g., ISO base load). How onerous is this requirement? Let us examine this.

For the gas turbine similar to that in Example 1 above, in a 1 × 1 × 1 configuration, the registered output is 386.8 MW. At constant firing temperature and IGV opening, gas turbine and combined-cycle output drops with decreasing rotational speed, as shown in Figure 19.17 (via lower airflow). The deficiency of about 5 percent in combined-cycle output should be made up by a combination of gas turbine overfiring and IGV opening. Cycle calculation shows that, when a steady-state condition is reached about 10 minutes after the event – based on the thermal inertia and time constant of the bottoming cycle – 386.8 × 0.96 = 371.3 MW is achieved at a 2,488°F firing temperature (IGVs in the fully open position). At that operating condition, gas turbine output is 235.8 MW and steam turbine output is 143.1 MW.

However, the steam turbine contribution is not available during the first minute or two and then picks up slowly per the exponential decay formula controlled by the HRSG time constant. Thus, during the first 10–30 seconds (primary response), the steam turbine output will essentially be the same as it was just prior to the grid event (i.e., 141.3 MW). Thus, within 10 seconds, instead of 235.8 MW indicated by the steady-state performance, the gas turbine should reach 237.6 MW in order to cover the

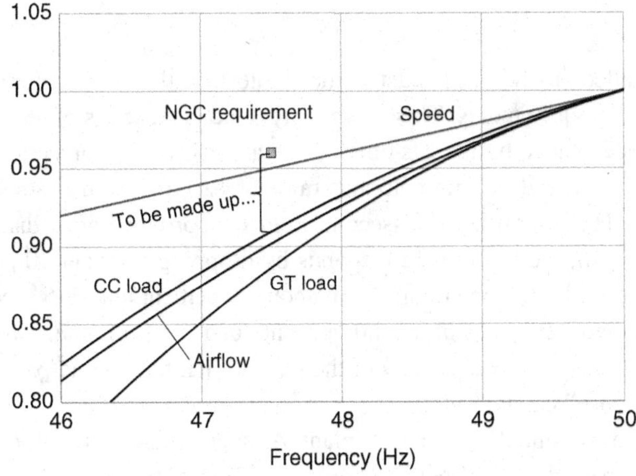

Figure 19.17 Gas turbine (GT) combined-cycle (CC) output as a function of speed (ISO).

steam turbine deficiency of 143.1 – 141.3 = 1.8 MW. This requires overfiring to 2,499°F instead of 2,488°F.

In such an event, the frequency ends up outside the band for normal operation (i.e., ±500 mHz per the NGC). In other words, the grid cannot be stabilized by the actions of the generating side only. Consequently, load shedding to 96 percent takes place. If nothing is done, the plant can run only at about 90 percent load. Since the steam turbine response has a time constant of 4 minutes, the gas turbine responds in 10 seconds by overfiring to 2,499°F and covers for the steam turbine. As the steam turbine output picks up, the controller reduces the gas turbine in lockstep until steady-state is reached and the combined-cycle power plant runs at 96 percent load with a 2,488°F firing temperature.

Note that, in this example, at ISO conditions, we assumed that there was no assistance from the IGVs (i.e., more airflow). Even then, with 70–80°F overfiring, the gas turbine is still able to satisfy the grid code primary and secondary response requirements. Overfiring can be reduced by turning on the online water wash skid (i.e., increasing mass flow through the machine) and opening the IGVs below their normal "fully open" position (if there is room for doing that). This would significantly alleviate the negative impact of overfiring on hot gas path part lives.

The situation is more precarious on hot days when the gas turbine airflow and output are reduced significantly via lower ambient air density. The gap between the normal part load operation curve and the NGC requirement is more than 10 percent without power augmentation on a 90°F day with 40 percent relative humidity. Overfiring cannot close the gap by itself anymore. Additional knobs are required, including power augmentation via inlet cooling (evaporative cooler or mechanical chiller), which compensates for the reduction in ambient air density. Due to the variations in compressor operating limit line, exhaust isotherm, and other hardware/material capabilities of the heavy-duty industrial gas turbines, appropriate control strategies to satisfy myriad grid code requirements should be determined on a case-by-case basis.

19.5.3 Example 3

Gas turbines installed in the field must demonstrate that they can comply with the applicable grid code requirements. This is typically done via a series of tests conducted prior to the final acceptance by the customer. Such a test involves "injection" of >200 mHz in 10 seconds. Instead of a monotonous ramp at +20 mHz/s, the simulated event includes a +100 mHz/s ramp and a 2-second hold to reproduce an actual event. As shown in Figure 19.18, the gas turbine responds by dropping to about 50 percent load (plant basis) in 4 seconds and then ramps up to about 70 percent load in 5 seconds (most likely to MECL) and settles at about 80 percent load in 1 minute. The first part determines the "fast-response" capability of the gas turbine; the second part determines its "sustained-response" capability.

Note that this is a combined-cycle power plant. As such, in the first few seconds, only the gas turbine can participate in frequency response. Therefore, in terms of megawatts, all of the 50 percent drop in plant output comes solely from the gas turbine. This means

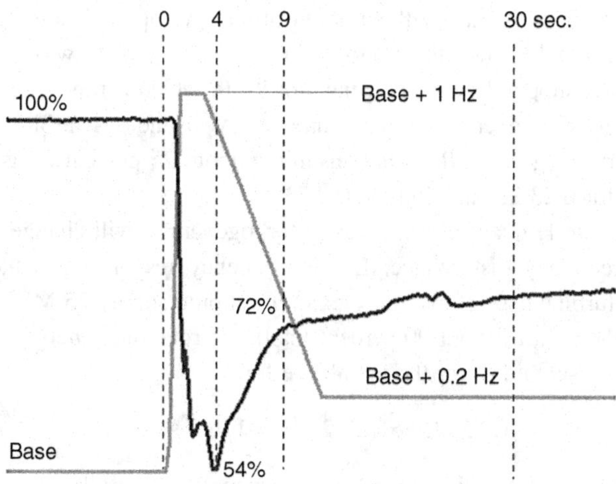

Figure 19.18 Gas turbine response to >200–mHz frequency injection.

that the gas turbine drops to about 35 percent load, at which point its fuel consumption is about 50 percent. Clearly, the DLN combustor of the gas turbine goes through several mode changes.

Mini test: Can you figure out the droop of the governor of this turbine if the grid in question is 50 Hz?[6]

19.5.4 Rotational Inertia

The inertia constant H, which is defined as the kinetic energy at rated speed divided by the rated power, quantifies the amount of energy stored in a turbogenerator. For a generic gas turbine generator, we have shown that the turbine-generator shaft's (rotor) inertia time constant in seconds is given by Equation 19.7 as

$$\tau_I = \left(\frac{30}{\pi}\frac{\tau_0}{N_0 I}\right)^{-1}.$$

Going through the simple algebra, we can verify that

$$H = \tau_I/2.$$

From the data in Table 19.1, H is around 5–10 seconds for typical gas turbines. Using the 275–MW 9F gas turbine in Table 19.1 as an example, while running synchronized to the grid at 3,000 rpm, this gas turbine has

$$275 \text{ MW} \times (21.2/2) = 2{,}915 \text{ MJ}$$

[6] A 1–Hz change in frequency is 2 percent (1/50 = 0.02), which resulted in a 50 percent output change (i.e., a 4 percent change in frequency would cause a 100 percent output change – in other words, 4 percent droop [plant]).

of rotational energy. This is exactly the amount of energy supplied to the gas turbine-generator when it is cranked from the TG to FSNL during startup. As was shown earlier, part of this energy is supplied by the ac generator itself, which is run as a motor by the LCI, and part of it is supplied by burning natural gas in the gas turbine combustor. When the unit is running at FSNL, it is consuming about 25 percent of its FSFL heat input (i.e., the factor 0.23 in Equation 19.4).

The inertia constant H quantifies how fast a turbogenerator will change its speed in case of a grid frequency disturbance. If the frequency response demands that the 275–MW 9F gas turbine used above as an example should supply 25 MW more while running at 250 MW (i.e., at about 90 percent load), the rotational energy stored in the gas turbine-generator would cover the unbalance for

$$(275/25) \times (21.2/2) = 116.7 \text{ s}$$

(i.e., for about 2 minutes). Needless to say, the frequency will rapidly excur beyond the grid code limit well before then. How can we estimate that? Let us write the equation for the kinetic energy of the turbogenerator:

$$W_{kin} = \frac{1}{2} I\omega^2 = \frac{1}{2} I(2\pi f)^2. \tag{19.23}$$

Taking the derivative with respect to time, we obtain

$$\frac{dW_{kin}}{dt} = 4\pi^2 If \frac{df}{dt}, \tag{19.24a}$$

which gives us the power imbalance between the generator and the grid, i.e.,

$$\Delta\dot{W} = 4\pi^2 If \frac{df}{dt}. \tag{19.24b}$$

Combining Equation 19.7 with the correlation between torque and power, we find that

$$\tau_I = \frac{4\pi^2 I f_0^2}{\dot{W}_0}.$$

Substituting into Equation 19.24, we can write that

$$f \frac{df}{dt} = \frac{\Delta\dot{W}}{\dot{W}_0} f_0^2 \frac{1}{\tau_I}. \tag{19.25}$$

Integrating Equation 19.25 (noting that $f = f_0$ at $t = 0$), the rate of change in rotational frequency is found as

$$f = f_0 \left(2 \frac{\Delta\dot{W}}{\dot{W}_0} \frac{t}{\tau_I} + 1 \right)^{\frac{1}{2}}. \tag{19.26}$$

Using our example again, Equation 19.26 comes up with 4.6 seconds before the frequency drops to 49 Hz if this were an "islanded" grid with only one generator (i.e., our 9F gas turbine) on it and nothing was done to restore the power balance.

Equation 19.26 is valid for a constant imbalance between power supply (by the turbogenerator) and demand (by the grid). In reality, as we have already seen, the self-regulating nature of the power grid will ensure that the imbalance is alleviated as the frequency changes.[7] Thus, Equation 19.25 should be modified as

$$f\frac{df}{dt} = \frac{\Delta\dot{W} - \kappa\dot{W}_i(f_0 - f)}{\dot{W}_0} f_0^2 \frac{1}{\tau_I}, \tag{19.27}$$

where $\dot{W}_i \leq \dot{W}_0$ is the turbogenerator load at time t = 0 and κ is the load change per unit frequency as a fraction. As we have seen earlier, in the case of a frequency disturbance, the primary control (PC) reserve (as prescribed by the applicable grid code), $\dot{W}_{pc}$, will kick in to further reduce the imbalance. In that case, Equation 19.27 will become

$$f\frac{df}{dt} = \frac{\Delta\dot{W} - \kappa\dot{W}_i(f_0 - f) - \dot{W}_{pc}}{\dot{W}_0} f_0^2 \frac{1}{\tau_I}. \tag{19.28}$$

19.5.5 Example 4

Let us assume that 10 Frame 9F gas turbines are running at about 91 percent load (i.e., at 250 MW each) to satisfy a grid load of 2,500 MW (at 50 Hz). Suddenly, one generator trips. Without self-regulation, from Equation 19.28, we find that the grid frequency would drop to 0 Hz in about 2 minutes. With self-regulation (κ = –0.01), Equation 19.27 can be numerically integrated using fourth-order Runge–Kutta methods (with a time step of 0.1 seconds) to find that the frequency settles at 40 Hz in about 2 minutes. In other words, if the remaining nine gas turbines keep running at 250 MW, generation and demand are matched at 2,250 MW at a frequency of 40 Hz. This, of course, is neither acceptable (from a grid perspective) nor physically possible (from a turbogenerator perspective; i.e., the units will not be able to generate 250 MW at 2,400 rpm).

As mentioned earlier, grid code requirements set the maximum deviation from the nominal frequency during disturbances such as the loss of a major generation asset or load. Clearly, self-regulation is not sufficient to restore the system frequency. The remaining nine gas turbines operating at 91 percent load (250 MW) with 4 percent droop control have about 9 percent primary reserve or 25 MW each. Thus, loading up to 275 MW will restore the frequency at 49.82 Hz (see Equation 18.5) and 2,475 MW. To restore the original load will require slight overfiring of the gas turbine generators to 250 + 27.8 = 277.8 MW or 101 percent load each. At that point, the grid frequency is 49.8 Hz (*not* 50 Hz!).

This is a very important finding: primary reserves operating at droop control cannot restore the grid frequency by themselves. In this example, the difference of 200 mHz (a

[7] For example, if the grid frequency drops, electricity demand also drops slightly because synchronous ac motors in the grid demand less power at lower running speed. However, they constitute only a fraction of the total load and, as such, in large power grids in developed nations, κ is typically only –0.01. In smaller grids, it would be practically zero.

so-called quasi-steady-state deviation) will remain. This is when the secondary reserves (also known as *frequency restoration* reserves) and, if necessary, tertiary reserves (also known as *replacement* reserves) come online. In this example, once the secondary reserves engage and restore the load and frequency, the nine gas turbines turn back down to 250 MW on 4 percent droop control.

References

1. Gülen, S. C., Kim, K., Gas Turbine Combined Cycle Dynamic Simulation: A Physics Based Simple Approach, *Journal of Engineering for Gas Turbines and Power*, **136** (2014), 011601.
2. Rowen, W. I., Simplified Mathematical Representations of Heavy-Duty Gas Turbines, *ASME Journal of Engineering for Power*, **105** (1983), 865–869.
3. Rowen, W. I., "Simplified Mathematical Representations of Single-Shaft Gas Turbines in Mechanical Drive Service," ASME Paper 92-GT-22, ASME International Gas Turbine and Aeroengine Congress and Exposition, June 1–4, 1992, Köln, Germany.
4. Yee, S. K., Milanovic, J. V., Hughes, F. M., Overview and Comparative Analysis of Gas Turbine Models for System Stability Studies, *IEEE Transactions on Power Systems*, **23**: 1 (2008), 108–118.
5. De Mello, F. P., Ahner, D. J., Dynamic Models for Combined Cycle Plants in Power System Studies, *IEEE Transactions on Power Systems*, **9**: 3 (1994), 1698–1708.
6. Hannett, L. N., Khan, A., Combustion Turbine Dynamic Model Validation from Tests, *IEEE Transactions on Power Systems*, **98**: 1 (1993), 152–158.
7. Hannett, L. N., Feltes, J. W., "Testing and Model Validation for Combined Cycle Power Plants," IEEE Power Engineering Society Winter Meeting, January 28–February 1, 2001, Columbus, OH.
8. Kundur, P., *Power System Stability and Control* (New York: McGraw-Hill, Inc., 1994).
9. Incropera, F. P., Dewitt, D. P., *Introduction to Heat Transfer*, 4th edition (New York: John Wiley & Sons, Inc., 2002).
10. Özışık, N., *Conduction Heat Transfer* (New York: John Wiley & Sons, Inc., 1988).
11. Gülen, S. C., Jones, C. M., "GE's Next Generation CCGT Plants: Operational Flexibility Is the Key," *Modern Power Systems*, June 2011, pp. 16–18.
12. Diegel, D. et al., "Fullfillment of Grid Code Requirements in the Area Served by UCTE by Combined Cycle Power Plants," POWER-GEN Europe 2004, May 25–27, 2004, Barcelona, Spain.

20 Economics

Designing a "machine" without an eye on cost and price is just wishful thinking. Every part, every feature, every improvement that goes into the machine and, ultimately, the machine itself has a "maximum acceptable" cost. No matter how good the machine is, if it costs more than that, it is not going to be an economically viable product. Then again, there will be times when, even if it costs *less* to develop and manufacture than it did before, it will not command the price it did before and it will still not be economically viable. (Recall the bursting of the combined-cycle power plant-building "bubble" of the 1990s in 2000.)

The dichotomy between cost and price is not always well understood. Engineers can control the cost of the gas turbine that they are working on, but they cannot control its selling price. The latter is dictated primarily by the prevailing market conditions. By the same token, however, engineers must be aware of the market conditions and their impact on the price *and* availability of the materials and labor that go into the gas turbine they are designing. Not only that, engineers should also be aware of the economic principles governing the price of the end product of the gas turbine they conceived (i.e., electricity supplied to the grid and ultimately to the end users). There is a significant body of scholarship and an accumulated knowledge base on all these subjects. It is impossible to cover them in a meaningful manner in any single book, let alone a chapter in a book. Nevertheless, there are certain concepts that can be distilled into relatively simple parameters and provide a good handle on the cost–performance trade-off (inherent in *any* engineering endeavor) specific to electric power generation. The most important of those is the *levelized cost of electricity* (LCOE).

The discussion in this chapter is within the framework of a *gas turbine combined cycle* (GTCC), which is by far the most widely used plant configuration for natural gas-fired electric power generation. Originally intended for mid-range load duties with a limited number of start–stop cycles, GTCC power plants are widely used in cyclic duty (about 4,000–5,000 hours per year with 200 or more starts) in the USA and in Europe. In Japan, South Korea, and other industrialized (or industrializing) countries in Southeast Asia, they are frequently used for liquefied natural gas (LNG)-fired base load power generation.

20.1 How Much Does It Cost?

The total plant cost (TPC) is by far the most difficult number to come up with in economic evaluation of a power plant. The problems start with the exact definition of the "cost scope" (i.e., what exactly to include in that number). Since there is not an industry standard on that subject, the literature is full of widely varying numbers based on myriad assumptions, estimation models, escalation factors, etc. To make matters worse, even the terminology is a total mess.

"Capex" is the short form of capital expenditure, which is defined as "funds used by a company to acquire or upgrade physical assets such as property, industrial buildings or equipment."[1] It is commonly used for the "total installed cost" of a power generation "asset" (i.e., a gas turbine simple- or combined-cycle power plant). Strictly speaking, the *total installed cost* (TIC) incorporates many more cost buckets than capex. Why perpetuate a misuse of a carefully defined term? Because, somewhat similar to using a photocopier *brand name* (i.e., "Xerox") for the *action* of photocopying a document (or the physical photocopy of it) using a machine by any other manufacturer, the term "capex" has permeated the power generation industry as a proxy of *any type* of capital outlay. In the absence of a published standard by an organization such as the American Society of Mechanical Engineers (ASME), the American National Standards Institute (ANSI), or the International Standards Organization (ISO) setting the rules for the terminology and definition of power plant costing, the closest documents to doing that job are the National Energy Technology Laboratory (NETL) Quality Guidelines [1] and the Technical Assessment Guide (TAG®) by the Electric Power Research Institute (EPRI).

As defined in [1], TPC[2] is the engineering, procurement, and construction (EPC) cost plus contingencies. With the addition of myriad owner's costs, TPC becomes total overnight capital (TOC). Financing costs, especially interest during construction – also known as *allowance for funds during construction* (AFUDC), can be very significant for major capital projects, which take several years from conception to commissioning. Adding those to the TOC gives the total "as spent" capital (TASC).

In a similar vein, EPRI TAG defines the *total plant investment* (TPI), which is the TPC plus AFUDC. This does *not* include myriad owner/developer costs such as royalties, permits, preproduction (startup) costs, working capital, land, etc., which are added to the TPI to find the *total capital requirement* (TCR), which is equivalent to US Department of Energy (DOE)/NETL TASC. TASC or TCR is essentially the capex for a power plant.

On top of all this, there is the price–cost dichotomy. Depending on market demand–supply dynamics (a good example to remember is the natural gas-fired power "bubble" in the late 1990s), commodity prices, craft labor availability, and many other factors, what is estimated to "cost" $100 today can easily be $300 or $500 tomorrow – well beyond what is suggested by simple inflation. Even projects with mature technologies (e.g., GTCC) are subject to changes in TIC estimates depending on site- and/or project-specific complexities. For example, adding a *zero-liquid discharge* system to a GTCC project, which in other aspects is pretty much identical to earlier ones, can result in substantial cost increases. The only remedy is learning gained during early

[1] Per www.investopedia.com (as of May 18, 2017)

[2] Also known as "overnight" construction cost.

Table 20.1 AACE process plant cost estimate classification

Class	Project Definition(%)	Cost Estimating Purpose	Cost Estimating Methodology	Cost Estimating Accuracy Range(%)	Preparation Effort
5	0–2	Concept screening	Capacity factored or parametric models; engineering judgment; analogy	Low: –20 to –50 High: +30 to +100	1 (base)
4	1–15	Study or feasibility	Equipment factored or parametric models	Low: –15 to –30 High: +20 to +50	2–4
3	10–40	Budget authorization or control	Semi-detailed unit costs with assembly-level line items	Low: –10 to –20 High: +10 to +30	3–10
2	30–75	Control or bid/ tender	Detailed unit cost with forced detailed takeoff	Low: –5 to –15 High: +5 to +20	4–20
1	65–100	Check estimate or bid/tender	Detailed unit cost with detailed takeoff	Low: –3 to –10 High: +3 to +15	5–100

Table 20.2 EPRI power plant cost estimate classification

Class	AACE Equivalent	Preparation Effort	Project Contingency(%)
I	Class 5/4	Simplified	30–50
II	Class 3	Preliminary	15–30
III	Class 3/2	Detailed	10–20
IV	Class 1	Finalized	5–10

implementation so that potential risks and pitfalls are better understood and quantified at different stages of the subsequent projects.

There are different cost classifications that try to make sense of the myriad difficulties encountered in process and/or power plant installed cost estimation. Two of them must be intimately known by the gas turbine engineer:

1. American Association of Cost Engineering (AACE) International Recommended Practice No. 18R-97, Cost Estimate Classification System – As Applied in Engineering, Procurement, and Construction for the Process Industries.
2. EPRI Cost Classification.

The two classifications are summarized in Tables 20.1 and 20.2.

There are two types of contingencies: the first one is the *project* contingency, which is an allocation for anticipated but a priori unknown problems that are expected to crop up in any large engineering design, procurement, and construction project. The second one is *process* contingency, which is a cost adder to inflate the estimated capital cost in order to quantify the uncertainty in the technical performance and cost of the commercial-scale equipment requisite for a new technology. Per EPRI guidelines, process contingency for a new concept with limited data is 40 percent or more. In other words, if one estimates \$1,000/kW for a new technology, it is likely to end up \$1,400/kW or even more.

The other component that makes capital cost estimation difficult is the "lead time" (i.e., the total time spent on project definition, environmental permitting, licensing, conceptual and final engineering design, and finally construction followed by startup and commissioning). Some activities can overlap at certain times. The lead time can have a substantial impact on the total cost through financing requirements during that time (e.g., AFUDC). Typical values for lead times are 4 years for GTCC power plants (actual construction time is about 2 years). It can be as short as 1 year for a simple-cycle peaker plant comprising one or more packaged aeroderivative gas turbine generator sets. Even slight variations and/or unforeseen problems (mainly due to a lack of prior experience) can substantially add to the lead time, especially to the construction phase, so that the final bill becomes even more inflated than what was suggested by the original estimates.

A GTCC power plant, due to its amenability to standardized "reference plant" design and relatively short construction period (roughly 2 years), has the lowest cost estimation uncertainty. This is the reason why gas turbine simple- and combined-cycle *budgetary price estimates* can be found in trade publications in tabular form (e.g., the *Gas Turbine World* [GTW] *Handbook*). When using GTW Handbook budgetary price estimates, a few things should be kept in mind. First and foremost, plant total cost going into the LCOE formula should include the cost of equipment, materials, labor, engineering and construction management, contingencies related to the construction of a facility, and owner's costs (land acquisition, licenses, and administrative costs). If the interest used during construction (typically 2 years for GTCC) for funds borrowed to pay for them is excluded, this gives the total overnight construction (TOC) cost (TPC per the EPRI definition above).

For example, the 2015 budgetary price for a typical GTCC is \$675 (GTW Handbook scope) with ±15 percent uncertainty. At least 30 percent should be added to this for TOC (interest during construction excluded). This gives a range of \$750–\$1,000 per kW. (Note that a 2010 survey of construction costs for GTCC power plants indicated a range of costs from about \$670 to slightly more than \$1,400 per kW of installed capacity.)

The capacity-cost scaling exponent for GTCC plants is about 0.75 (similar to the famous *six-tenths rule* with 0.60 as the exponent) for plants rated ~500 MWe or less. Here is what this means: assume that you know the cost of a recent P_0 MWe GTCC, which cost C_0 dollars. You are now contemplating a GTCC of similar scope, which is rated at P MWe. The cost of the new power plant can be estimated as

$$C = C_0 \times (P/P_0)^{\alpha},$$

where α is the *capacity scale factor*. For larger plants, α is 0.93 (from the chart on p. 44 of the GTW 2014–15 Handbook). In other words, there is little economy of scale for large GTCC plants. Doubling the capacity from, say, 500 to 1,000 MWe will reduce the specific capital cost (\$ per kW) by only 5 percent. (The corresponding value for α in the GTW 2016–17 Handbook was 0.915.)

According to the GTW 2016–17 Handbook, α is 0.818 from pricing data for simple-cycle gas turbines rated above 100 MW and 0.735 for those rated below 100 MW.

Similar data are not available for steam turbine generators (STGs). Cost of materials and manufacturing the steam turbine equipment is a function of the following:

- Number of casings (high pressure [HP], intermediate pressure [IP], and low pressure [LP], combined HP/IP and LP, etc.);
- LP turbine configuration (single-flow axial, double-flow down exhaust, etc.);
- Last-stage bucket length;
- Steam conditions (pressure and temperature);
- Steam flow rate.

Price, of course, is dependent on the market conditions, which determine the *contribution margin* (CM) charged by the original equipment manufacturer (OEM), i.e.,

$$Price = \frac{Cost}{1 - \mathrm{CM}}.$$

In general, STG prices do not vary significantly within a class (e.g., 100–150 MWe single-flow axial LP turbine unit with [nominal] steam pressures up to 1,815 psia and steam temperatures up to 1,050°F). Such units (turbine *and* generator) come at $25–$30 million (in 2015 dollars). In order to estimate the price of the remaining "power island" (including the heat recovery steam generator [HRSG], duct firing system [if any], selective catalytic reduction [if any], distributed control system, and continuous emissions monitoring system), one can add 30–40 percent to the STG price. For conceptual studies, it is easier to use a bottoming-cycle price, which can be extracted from the GTW Handbook simple- and combined-cycle budgetary price data. For example, from the GTW 2014–15 Handbook,

$$\mathrm{BC\_Price} = 1,238 \times \exp(33.4/\mathrm{STG\_MW})\ \$/\mathrm{kW}.$$

For more detailed information and methodologies regarding power plant capital cost estimation and LCOE calculations, the reader is encouraged to consult [2,3] and the works cited therein. A good source for large power plants of all types is the recent report by the US Energy Information Administration (EIA) [4] (or its more recent editions when/if available). A pricey but highly useful source for technical and cost information is the TAG published by the EPRI (the latest 2013 edition has a price tag of $75,000). Readers with access to it (through their organizations or a library) are encouraged to consult it as the first source. It must be emphasized that any *generic* cost information from the cited and similar sources must be verified and/or adjusted by real project information (if/when available, of course).

20.2 Levelized Cost of Electricity

In evaluating electric power generation technologies using dollars and cents, the most widely used metric is the LCOE, which combines plant output, efficiency, and operations and maintenance (O&M) with capital investment and fuel expenditure in a simple formula. This metric is useful when comparing power generation alternatives that use *similar technologies*. The standard formulation of the *first-year* COE is the sum of capital, fuel, and O&M costs of plant ownership [5], i.e.,

$$COE = \frac{\beta \cdot C}{P \cdot H} + \frac{f}{\eta} + \left\{ \frac{OM_f}{P \cdot H} + \mu \cdot OM_{v,b} \right\}, \tag{20.1}$$

where

β = capital charge factor (i.e., cost of money)
C = TPC ($)
H = annual operating hours
P = net rated output (kW)
f = fuel cost in $ per kWh (in lower heating value [LHV])
η = net rated LHV efficiency
OM_f = fixed O&M costs ($ or $/kW-yr)
$OM_{v,b}$ = variable O&M costs for base load operation ($/kWh)
μ = maintenance cost escalation factor (1.0 for base load)

The cost of generation as provided by this COE formula can be interpreted as the price at which electricity must be sold in order to cover all fixed and variable generating expenses and to match the return on a company's equity implicit in the assumed *capital charge factor* (β). In other words, it is the price of electricity that would make the *net present value* (NPV) of the power project in question zero. The COE is limited to a single operating condition, typically new and clean rated performance, and is usually calculated at ISO base load.

The fuel cost used in Equation 20.1 should be in $/kWh (LHV) and is given by

$$f = \frac{f_0 \cdot h}{293.071},$$

where

h = ratio of fuel higher heating value (HHV) to LHV (1.109 for natural gas as 100 percent CH_4)
f_0 = base fuel cost in $ per MMBtu (HHV)

The LCOE can be calculated as

$$LCOE = LF \times COE, \tag{20.2}$$

where LF is the *levelization factor*.

Typical values for the financial parameters that can be used in the LCOE equation are as follows [1]:

- For *investor-owned utilities* (IOUs), β is 10–13 percent.
- For *independent power producers* (IPPs), β is 15 to >20 percent.
- For IOUs, LF is 1.268; for IPPs, LF is 1.169.

Financial parameters are subject to change from year to year (prevailing economic conditions) and from developer to developer. At the time of writing, for the IPPs, a more appropriate value of β is in the range of 11–15 percent. The reader should always strive to use the most reliable, up-to-date (and pertinent) information in his or her studies.

A good rule of thumb for the fixed O&M cost for GTCC is \$10–\$15/kW-yr (higher values for advanced H- and J-class machines). For the variable O&M, one can use \$0.50–\$1.50/MWh. The lower value is appropriate for base loaded plants, the higher value for cyclic operation. Somewhat higher values may be adopted for H- and J-class machines comprising advanced single-crystal alloys and coatings. The reader is encouraged to refer to [4] for more information on generic O&M costs for gas turbine and other fossil fuel technologies. In a GTCC power plant, the gas turbine is the largest O&M cost contributor. Gas turbine maintenance comprises three major activities: combustor inspection, hot gas path inspection, and major inspection [6]. Periodicity of these maintenance activities is a function of fired operating hours and start–stop cycles. For the state-of-the-art advanced machines comprising high-technology parts and operating at high firing temperatures, the prudent but costly approach is to enter into *long-term service agreements* with the OEM.[3] For the more mature technologies such as the older F- or E-class machines, independent (third-party) service providers can be the less costly choice with minimal risk [7].

Prevailing and historic US natural gas spot prices can be easily found on the Internet. In the past, before the shale gas glut hit the market, this price was as high as \$13 per MMBtu (HHV). Depending on unexpected weather conditions (pushing up the demand for home heating), it can still show significant spikes (e.g., winter of 2014). For the time being, \$3 natural gas seems to be a reality in the USA, which makes widespread acceptance of advanced but proven technologies such as integrated gasification gas turbines (IGCCs; especially with their cost and schedule creep problems) extremely difficult. Overseas, especially in countries dependent on imported natural gas or LNG, this price is around \$7–\$8 at the time of writing (2016–2017), but it went up to as high as \$15 in the recent past (*landed* price; i.e., as received at the terminal). This is one reason why coal-burning technologies such as advanced ultra-supercritical and carbon capture are expected to make significant inroads in Europe and Japan well before the USA.

Strictly speaking, the basic LCOE formula has very limited applicability; it is only appropriate for comparing *similar technologies of the same vintage* (i.e., an F-class, three-pressure with reheat GTCC power plant with another GTCC of the same configuration with the same features). The technologies must have similar rated performance, part load and ambient efficiency lapse, degradation, reliability, availability, and maintainability (RAM; see Chapter 16), emissions, etc. In other words, for a given operating scenario, they must deliver approximately the same annual megawatt-hour generation. Unfortunately, the LCOE formula has been used for comparing different technologies (e.g., IGCC with GTCC) with total disregard to the aforementioned technology characteristics. The most glaring omission is RAM, whose importance can be demonstrated by a simple example.

[3] Also known as contractual service agreements.

Suppose that two technologies, A and B, are considered for base load generation at an 800-MW nominal rating for 7,000 hours a year. Technology A is proven and has a reliability of 99 percent; that is, one has to account for an unexpected failure probability of only 1 percent. Technology B is relatively new (and more efficient) with only a few operating units in the field; its reliability is estimated at 80 percent (i.e., there is a 20 percent probability that it will be down when it is expected to run and generate power, due to an unforeseen component failure). Thus, while both technologies have nominally the same annual generation capability (i.e., 800 × 7,000 = 5,600,000 MWh), in reality,

- Technology A has an *expected* annual generation capability of 0.99 × 800 × 7,000 = 5,544,000 MWh, whereas
- Technology B has an *expected* annual generation capability of 0.80 × 800 × 7,000 = 4,480,000 MWh.

If one goes ahead with technology B, allowance must be made for the purchase of nearly 1,000,000 MWh from other generators on the grid. The simple LCOE formula in Equation 20.1 does not and cannot account for that huge handicap.

While Equation 20.1 is still reasonably adequate as a *figure of merit* for base loaded power plants (as was the case for plants installed in the mid-1990s to the early 2000s), the fact is that at the time of writing this book (2016–2017), gas turbine power plants are mostly intended for *cyclic* operation with frequent starts and stops as well as large load swings. Recent rapid expansion in renewable generation capacity (e.g., solar and wind) in the USA and abroad and the concomitant requirement for spinning reserve and backup power has been a strong factor in this regard. Thus, even though the natural gas prices are at the $3/MMBtu level (due to the "shale gas boom" in the USA), gas-fired combined-cycle plants are expected to continue to be operated as intermediate and cyclic duty plants [8–12].

Typically, the following four general types of plant operating regimes are possible:

- Peaking
- Cyclic
- Intermediate
- Base loaded

These operating regimes can be characterized by the number of annual fired starts in conjunction with operating hours. A comparison based on generic definitions is shown in Table 20.3.

From Table 20.3, it is clear that simply plugging in H (i.e., 4,250 hours per year), along with requisite plant P and η, into Equation 20.1 is a very simplistic and totally inaccurate method to estimate the COE of this plant. This is especially true when one has to evaluate offers from multiple bidders for a given project, who might be comparable in terms of product performance and price, but offer significantly different off-design and operability features. Examples for the latter are (but are not limited to) the following:

Table 20.3 Definition of plant operating regimes used in this analysis [13]

	Cold Starts	Warm Starts	Hot Starts	H	μ
Downtime	≥72 hours	≤48 hours	≤2 hours		
Start time	1 hour	2 hours	3 hours		
Peaking	40	100	10	600	10×
Cyclic	5	45	225	4,800	2.5×
Base Load	9	10	2	7,500	Baseline
Continuous	3	5	10	8,200	1.1×

- Better hot-day heat rate;
- Better part load performance;
- Better emissions performance;
- Faster start and stop times.

Obviously, Equation 20.1 is unable to take those features into consideration. Admittedly, there are ways to evaluate all of these features (and much more) diligently and accurately via rigorous simulation utilizing detailed power plant models. Examples are as follows:

- Cash flow pro forma;
- Dispatch modeling;
- Network modeling (including forecasting).

These methods incorporate detailed thermal system models, price duration curves, electricity sale prices, and forecasts of grid/dispatch changes and exercise them over the projected life of the plant per specific operating scenarios to calculate revenue, NPV, return on investment (ROI), return on equity, etc. There are even techniques to perform these calculations using stochastic models (e.g., "real" options based on financial option theory) to account for the inherent uncertainty in making future predictions. However, this kind of rigor comes at the cost of significant resources in terms of algebraic and/or numerical modeling, programming, and computation time. Thus, there is always a need for relatively simple and reliable methods to accomplish certain tasks before deploying such expensive resources.

In particular, basic comparison and feasibility study stage evaluation of power plant options and verification of detailed simulation models are necessary tasks and can be easily accomplished by an expanded variant of Equation 20.1, which uses annual load and ambient-weighted averages (i.e., *mean effective* values) of plant output and efficiency, $\bar{P}$ and $\bar{\eta}$, respectively, i.e.,

$$\mathrm{COE} = \frac{\beta \cdot C + \Omega_f}{\bar{P} \cdot H'} + f\bar{\eta} + \mu \cdot \Omega_{v,b}. \tag{20.3}$$

Note that the terms in Equation 20.1 are regrouped such that the first term on the right-hand side of Equation 20.3 represents the *fixed* costs and the remaining two represent the *variable* costs. Total operating hours, H', in Equation 20.3 are larger than the nominal operating hours, H, by the total time spent during plant starts (shutdowns are ignored), i.e.,

$$H' = H + \sum_i N_i \cdot t_i, \tag{20.4}$$

where N_i is the number of a particular start (i.e., hot, warm, or cold, which are referred to by the indices i = 1, 2, and 3, respectively) and t_i is the corresponding start time. Using the numbers in Table 20.3 and the sample cyclic duty considered above with H = 4,250 hours, we find the total effective operating hours as

$$H' = 4{,}250 + 200 \cdot 1 + 48 \cdot 2 + 2 \cdot 3 = 4{,}552 \text{ hours.}$$

Consequently, the denominator of the first term on the right-hand side of Equation 20.3 is the total energy generation, E, which includes plant starts and load ramps (see below).

For a generic application, the beginning point would be an *annual load profile* in terms of actual megawatts generated, which can be represented in a tabular form. Entries in each cell of such a table would be the operating hours at the corresponding ambient temperature with the corresponding power output. If a GTCC product is unable to satisfy a particular load requirement (e.g., a high load point on a very hot day), it is left out when calculating its total MWh production. The balance is made up by other generating resources, which would enter the COE via capacity/energy replacement terms. For individual projects, such tables can be constructed on a case-by-case basis. For conceptual studies, master tables can be established for product size classes (say, 500–MW class, 1,000–MW class, etc.) and ambient-load profiles reflecting major global markets (say, southern Europe, North America, etc.).

Total annual operating hours and energy generation would be obtained from a summation of the H_i and $P_i{\cdot}H_i$ terms across the entire table (the subscript "i" corresponds to each individual cell in the table).

Dividing the total energy production by the total operating hours would result in the mean effective plant output, $\bar{P}$, to be used in Equation 20.3.

Each cell in the annual load profile table corresponds to a particular ambient-load point for the power plant in question (e.g., 60 percent load at 90°F ambient, and so on). Some would be unattainable without power augmentation via inlet chilling, duct firing, etc. (e.g., 550 MW on a 90°F day from a product rated 500 MW at ISO baseload). When comparing alternative offerings, those can contain maximum possible values if no power augmentation is available. If this is the case, in the COE analysis, as will be shown below, missing megawatt-hours are going to be accounted for via *capacity* and *energy replacement* terms.

For each cell (i.e., operating point) in the annual load profile table, a heat consumption (HC) or heat rate term is calculated by running a system heat balance model or using correction curves provided by the OEM (whichever is available and/or more convenient).

Alternatively, the table could be provided to the OEMs bidding for the particular project to populate the cells with the appropriate heat rate data.

Care must be taken that, especially for the hot-day points, any applicable power augmentation feature (inlet chilling, evaporative cooling, duct firing, etc.) is accounted for in the heat rate or HC calculations.

Once the complete table is available, total fuel consumption in thermal megawatt-hours is calculated from a summation of the HC_i or $P_i \cdot HR_i$ terms across the entire table. Thus, the mean effective plant efficiency, $\bar{\eta}$, to be used in Equation 20.3 is obtained by dividing the total electricity production by the total fuel consumption.

This is quite straightforward and is easy enough to accomplish if system models and/or correction curves (or, even better, HC or heat rate tables) are readily available. Whenever possible, it should be the preferred approach because it returns the *exact* fuel consumption value. An alternative *approximate* approach, which is perfectly adequate for preliminary evaluations and also highly useful to illustrate the underlying principles, is to calculate $\bar{\eta}$ from the baseload value, η_0, via part load and ambient correction factors using the *load factor*, λ, which is defined as

$$\lambda = \frac{E}{(P_0 \cdot \kappa_0 \cdot (1 - d_p)) \cdot H'} = \frac{\bar{P}}{P_{corr}}, \tag{20.5}$$

where κ_0 is the ambient correction factor for the power plant net output (see below) and d_p is the output degradation factor (see Equation 20.9 below).

Note that the load factor, λ, is essentially the same as the *output factor* (OF), defined by industry standards such as Institute of Electrical and Electronics Engineers (IEEE) 762 (see Chapter 16 for a detailed discussion). It is the average (or *mean effective*) load at which the plant would have to run nonstop for H' hours to generate the same total kilowatt-hours as it would generate with the actual duty profile.

Continuing with the earlier sample calculation at a nominal 500–MW base load power requirement and 4,250 hours of cyclic operation with 1,700,000 MWh total generation, ignoring the seasonal ambient temperature variation and the time spent during startups and load ramps (i.e., assuming them to take place instantaneously), λ is calculated as $1{,}700{,}000/(4{,}250 \times 500) = 0.80$.

One more parameter is required for the simple approach to calculating $\bar{\eta}$: *average* annual ambient temperature, $\bar{T}_{amb}$, which can be calculated using the exact same method for calculating $\bar{P}$. For example, for a typical site in southern Europe,

- A straight average over a 1-year period (8,760 hours) gives 63.3 ± 12.4°F.
- A load-weighted average for a typical cyclic duty is calculated as 64.4°F by stepwise (hour-by-hour) integration.

Using a straight average of the pertinent meteorological data (if available) is sufficient if the plant operation is evenly distributed across the entire year. Otherwise, one has to do the more "exact" (but quite straightforward) averaging, which would return a different answer. For example, if the plant were more frequently dispatched for longer durations during the hot summer period, the load-weighted average would be higher, with the exact number being dependent on the particular plant duty.

Using λ and $\bar{T}_{amb}$, the requisite correction factors can be obtained from OEM-supplied curves. There are three of them:

- In addition to κ_0, the ambient correction factor for the power plant net output, one also needs the following:

- The ambient correction factor for net efficiency or heat rate (from the correction curve at $\bar{T}_{amb}$), κ_1, and
- The part load correction factor (from the correction curve at λ), κ_2, for the plant net efficiency.

The mean effective power plant efficiency to be used in Equation 20.3 is then calculated as

$$\bar{\eta} = \eta_0 \cdot \kappa_1(\bar{T}_{amb}) \cdot \kappa_2(\lambda) \cdot (1 - d_e), \qquad (20.6)$$

where d_e is the efficiency degradation factor (see Section 20.2.2 below).

If one is interested in evaluating the benefits of dynamic plant features such as fast-start and rapid-load ramps using the COE method, power generation and fuel consumption during those transient events as well as the time spent during the plant starts (shutdowns can be ignored with little loss of accuracy) have to be estimated. They should then be added to the total operating hours, the total energy generation, and total fuel consumption. While this might be somewhat more tedious than simple summation of numbers in the cells of a table, the extra effort is warranted, as will be shown in the example below.

For example, for the 1-hour hot startup in Figure 19.12, using Equation 19.21, e = E/P (E normalized with rated power) can be calculated as 22.7%-h (with $\Delta H = N = 1$ and $\alpha = 0.33$). Similarly, for the warm and cold starts, E can be calculated as 50.0 and 75.0%-h, respectively, with $\alpha = 0.5$. Thus, the total energy production during plant starts within the calendar year can be found as

$$e_{start} = 200 \cdot 22.7 + 48 \cdot 50 + 2 \cdot 75 = 5,883\%\text{-h}.$$

For the 500-MW rating used as an example, this is equivalent to 29,417 MWh of additional generation. Load ramps can be factored in a similar fashion. Note that symmetrical load–unload pairs can be ignored to simplify the estimation. For example, instantaneous ramping *down* from 100 to 40 percent load overestimates total generation by exactly the same amount as instantaneous ramping *up* from 40 to 100 percent load underestimates it.

Ramping down from 100 percent load to 50 percent five times a week and 50 weeks per year for a total of 250 times at a rate of 5 percent load per minute has *reduced* electric generation (vis-à-vis instantaneous unloading assumption) given by

$$\Delta e_{ramp} = 250 \cdot \tfrac{1}{2} \cdot (100 - 50)^2/(5 \cdot 60) = 1,042\%\text{-h}.$$

For the 500-MW rating used as an example, this is equivalent to 5,208 MWh (i.e., *missing* generation during the 10 minutes between the load ramp-down command and achievement of the desired 50 percent load at the prescribed time). Adding these two unsteady-state generation quantities to the nominal generation of 1,700,000 calculated above, one arrives at

$$E = E_{nominal} + E_{start} + \Delta E_{ramp} = 1,700,000 + 29,417 - 5,208 = 1,724,209 \text{ MWh}.$$

Thus, the correction due to the dynamic effects results in the *effective* load factor of

$$\lambda = \frac{1{,}724{,}209\,\text{MWh}}{500\,\text{MW} \cdot 4{,}552\text{ hrs}} = 0.758.$$

The reduction in load factor from 0.80 to 0.758 when plant performance during starts is accounted for is a significant factor in COE calculation.

20.2.1 Availability and Reliability

A key consideration in the evaluation of a power generation system is quantified via two probabilities: *reliability* and *availability* (see Chapter 16). The former is the probability of not being forced out of service when the unit is needed and the latter is the probability of being available, independent of whether the unit is needed or not. As probabilities, they can only be estimated using data from existing units (same or similar technology) in the field. When used in formulae, they result in *expected* values. For proven technologies, such as E- and F–class gas turbines, reliability is very high (i.e., 97 percent or even higher) [14]. In other words, if a particular unit is intended for 1,000 hours of operation, it will be expected to run for 970 hours and to lose 30 hours of generation due to unforeseen events. (Note that this is an *expected* value; the unit may very well run for the entire 1,000 hours without a mishap or just run for 200 hours before being destroyed in a freak accident caused by human error or a manufacturing defect.)

The availability of combined-cycle power plants based on advanced gas turbines is quite high (i.e., more than 90 percent), especially for F-class units with large fleet sizes [14]. For the cyclic and/or medium-load duties, which are characteristic for most fossil fuel-fired plants nowadays (i.e., the second decade of the twenty-first century), this is an extremely robust number and does not require specific attention.

The reliability can be taken into account by multiplying the total energy generation, E, by the reliability factor, R; in other words, using R = 99 percent for the example above,

$$E' = R \cdot E = 99\% \cdot 1{,}724{,}209\text{ MWh} = 1{,}706{,}967\text{ MWh}.$$

The expected reduction in the unit service hours due to a less-than-perfect reliability (i.e., R <100 percent) is thus reflected in a reduction of the unit's total energy generation and, per Equation 20.5 above, the load factor, λ. The reduction in the latter will also manifest itself in effective plant efficiency and, as will be shown below, in costs associated with capacity and energy replacement. Note that this is not an arbitrary correction. By recognizing that $R = 1 - P_f$, where P_f is the *probability of failure*, the proposed correction (ignoring the effect of λ) is equivalent to rewriting the COE as

$$COE' = \frac{COE_{fix}}{1 - P_f} + COE_{var},$$

$$COE' \approx COE_{fix} \cdot (1 + P_f) + COE_{var},$$

$$COE' = COE + P_f \cdot COE_{fix}. \quad (20.7)$$

In other words, the base COE is increased by an amount equal to the product of the probability of failure and the *cost of failure*, which is the same as the fixed part of the base COE. At a minimum, cost of failure is the sum total of cost of repairs (parts, materials, and labor) and downtime (lost opportunity of power generation) and can be much higher in certain cases such as major accidents. Its exact value is not that important from a conceptual analysis perspective. If it can be quantified reasonably accurately, it can be accounted for explicitly on a case-by-case basis. Herein, the proposed correction via reduced energy generation is adopted. However, the downtime cost via lost power generation is accounted for via *replacement* costs, as will be described later in the chapter.

20.2.2 Degradation

The economic life of a combined-cycle power plant is measured in decades, with fired generation hours reaching tens of thousands of hours. Even assuming ideal running scenarios and normal maintenance including online and offline compressor washes and major component inspection at OEM-prescribed intervals, the plant is expected to experience an *unrecoverable* performance loss [15]. Typical trends of unrecoverable combined-cycle performance loss in terms of plant output and efficiency were depicted conceptually Figure 16.3 (the lower curve labeled "unrecoverable"). A similar curve is usually provided by the OEMs for heat rate degradation as well, which can be translated into efficiency degradation.

The time-dependent nature of the performance degradation in Figure 16.3 is easy enough to account for in rigorous simulation studies, which calculate hourly values of output and efficiency using system models with detailed inputs such as duty profiles, seasonal ambient variation, and other boundary conditions. In order to account for degradation in a simple formula such as the LCOE in Equation 20.3, the approximation described below should be sufficient. Using OEM-provided curves similar to those shown in Figure 16.3, the representative or *averaged* output degradation factor, d_p, can be determined via integration as

$$d_p = 1 - \frac{\int_0^{H_{LIFE}} (1 - d_p(t)) \cdot dt}{H_{LIFE}}. \quad (20.8)$$

This is the *mean effective* (i.e., constant) degradation factor for output, which does not change with time and, when applied to the rated "new and clean" performance over the same lifetime operating hours, H_{LIFE}, results in the same total energy generation. The mean effective degradation factor for efficiency, d_e, can be looked up from the heat rate degradation curve at the same location corresponding to d_p. They are factored into the COE formula via λ and $\bar{\eta}$.

A typical degradation curve for GTCC output is given by

$$d_p(t) = 6.145 \left(1 - \frac{1}{1 + \left(\frac{t}{21{,}695} \right)^{0.6172}} \right), \quad (20.9)$$

where t is in hours from the time point when the unit is declared as "new and clean" and d_p is the percentage degradation (i.e., $d_p = 2$ means 2 percent degradation). Degradation in efficiency is usually specified as a fraction of output degradation (i.e., $d_e = a \cdot d_p$). A typical value for a is 0.67.

20.2.3 Emissions

The COE formula in Equation 20.3 can be expanded to include an emission cost component, e.g.,

$$\mathrm{COE} = \frac{\beta \cdot C + \Omega_f}{\bar{P} \cdot H'} + f\bar{\eta} + \mu \cdot \Omega_{v,b} + \sum_i c_i \cdot m_{p,i}, \tag{20.10}$$

where c_i is the price/cost of pollutant i in terms of \$/ton and $m_{p,i}$ is the plant generation of pollutant i in tons/kWh. The suggested formulation in Equation 20.10 should be considered as a reasonable placeholder. At the time of writing (and presumably for the foreseeable future), there is no industry-wide accepted method to convert specific plant emissions (i.e., pounds of pollutants such as NOx, SOx, CO_2, unburned hydrocarbons, particulate matter, etc., per generated MWh) into a COE contributor.

For the natural gas-fired GTCC plant, the primary pollutants in the stack gas are NOx and CO_2, the latter of which is a notorious *greenhouse gas* (GHG). While not a concern at baseload, CO emissions can be a problem when the unit is turned down, especially during plant starts when the unit spends a considerable amount of time at low loads. A modern (unfired) combined cycle generates about 750–800 lb/MWh of CO_2 (less than 1 lb/MWh of NOx with advanced Dry-Low-NOx combustors) at base load. For the sample 500-MW plant and the operation discussed earlier, the total generation of pollutants would add up to ~680,000 tons/year and ~500 tons/year for CO_2 and NOx, respectively.

Obviously, the heart of the problem is determining the price or cost, c_i, for the particular pollutant. In theory, a value for c_i of the GHGs can be derived from an economic pricing model such as the *marginal abatement cost curve*, which is utilized by carbon traders. Another approach would be to use the applicable carbon tax and/or emission permitting fees. These, however, are strongly site dependent (country, region, municipality, etc.) based on existing rules and regulations and can only be evaluated by someone who is well versed in the details of the particular project at hand.

For other pollutants such as NOx, SOx, CO, etc., emission allowances (if applicable) can be substituted for c_i. The particular numerical value will vary from case to case based on many factors.

20.2.4 System Considerations

Note that a power plant does not operate in a vacuum. It is part of a large, complex, and interconnected system in which generating units are dispatched in an optimal manner to minimize generating costs. Many individual unit and system characteristics are factored into the determination of the order of commitment (generation demand, electricity price, incremental heat rate, [start] reliability, availability, sunk costs, variable [fuel and maintenance] costs, etc.). These can only be handled via sophisticated dispatch or network simulation models designed with the rigor and depth commensurate with the complexities of the task. For a simple approach such as the COE discussed herein to be minimally acceptable for a comparison of power plant alternatives in a larger system context, the requirement of the same energy generation (in kWh or MWh) should be met [16]. This would require the same or very closely rated output, generation hours, and reliability.

The analogy to system planning or life cycle production cost analysis is appropriate to facilitate an apples-to-apples comparison of plant alternatives using COE analysis on an *equal energy output* and *equal reliability* basis. (Recall that R was factored into calculation of $\bar{P}$ via E and λ, as described earlier.) This is done by adding the following two more terms to the COE equation to account for the *two* key system dispatch planning considerations:

- A deficiency in *energy* (kWh or MWh) should be made up by bringing another unit in the system online.
- A deficiency in *capacity* (kW or MW) should be made up by purchasing firm capacity from the neighboring systems.

This will bring the alternative generating systems being compared to each other via an LCOE analysis onto a *common* annual total energy generation and reliability basis. The proposed modification, when applied to Equation 20.10, results in the following COE formula:

$$\mathrm{COE} = \frac{\beta \cdot C + \Omega_f}{\bar{P} \cdot H'} + f\bar{\eta} + \mu \cdot \Omega_{v,b} + \sum_i c_i \cdot m_{p,i} + \frac{S_c \cdot \Delta P + S_e \cdot \Delta E}{\bar{P} \cdot H'}, \tag{20.11}$$

where capacity to be replaced (in kW) is given by

$$\Delta P = \bar{P}_b - \bar{P}, \tag{20.12}$$

and energy to be replaced (in kWh) is given by

$$\Delta E = \bar{P}_b \cdot H'_b - \bar{P} \cdot H'. \tag{20.13}$$

In Equation 20.13,

- H'_b is the annual operating hours to be used as a *basis* in COE comparison. For multiple unit comparisons, it can be taken as the highest H' among the alternatives.
- $\bar{P}_b$ is the annual load and ambient-weighted average unit output to be used as a *basis* in COE comparison. For multiple unit comparisons, it can be set to the highest $\bar{P}$ among the alternatives.

The energy replacement cost, S_e, is dependent on the makeup of the particular generating system, for which the alternatives are evaluated using COE. Note, however, that the economic dispatch principle dictates that the replacing unit's variable generating costs

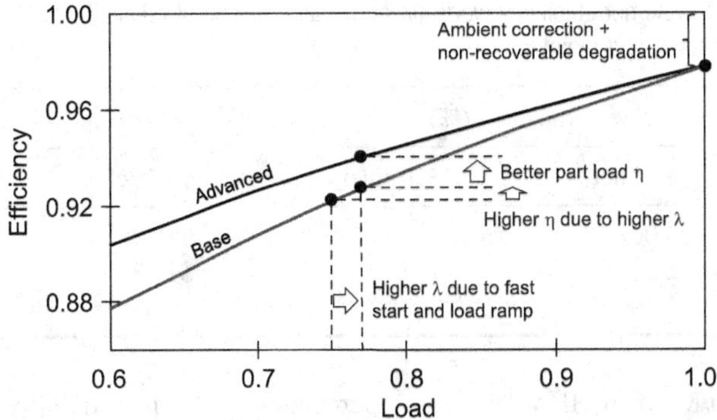

Figure 20.1 Impact of better part load heat rate, fast start, and load ramps on plant efficiency (normalized basis).

must be at least equal to or higher than the units under consideration. Otherwise, that unit itself would already be online and generating electric power. (The variable costs are *inclusive* of emissions costs, at least within the framework of the present discussion.) Only a system operator can provide the "right" number to be used in a "real" analysis.

The same goes for the capacity replacement cost, S_c, which requires even more theory to estimate. A recent method, introduced by PJM (a regional transmission organization [RTO]), is the *reliability pricing model* (RPM), which was first implemented in 2007. The capacity-weighted average price for different regions as determined by the RPM between 2007 and 2010 ranged from ~$100/MW-day to ~$175/MW-day.

Detailed examples of the application of Equation 20.11 can be found in [5] and [17]. More important than numbers are the fundamental "drivers" of economic advantage (as measured by the COE concept), which are captured by Equation 20.11, but not by Equation 20.1. A graphical illustration of the impact of said drivers is provided in Figure 20.1. The impact of better part load efficiency or heat rate is pretty straightforward to evaluate. Improved "agility" of the power plant (i.e., fast startup, shutdown, and ramps) is represented by the improvement in the load factor, λ (i.e., it is higher). A higher load factor, of course, translates into a higher efficiency or lower heat rate (everything else being constant).

20.2.5 Limitations of LCOE Analysis

First and foremost, life-cycle COE is *not* the correct metric for evaluating specific operability decisions and costs. It is more suitable to assessing the long-term impact of plant performance and operability *characteristics*. This can be done via *value* analysis, which calculates the capital cost equivalence of COE variables. Using Equation 20.11, for example, it can be shown that only a 1−percentage point reduction in reliability (e.g., R changing from 99 to 98 percent) is more than enough to wipe out all of the value created by a fast-start plant. The sensitivity analysis in Table 20.4 shows the net combined-cycle efficiency and capital cost equivalence (i.e., value) of reliability, performance degradation, and O&M ramification of cyclic duty. (For example, for the

Table 20.4 Combined-cycle net efficiency worth (in percentage points) of reliability, mean effective unrecoverable degradation, and maintenance factor

	Efficiency (%)		Capex (million $)	
	$5 Fuel	$10 Fuel	$5 Fuel	$10 Fuel
1% in R	2.00	1.10	$15.9	$22.8
1% in d_p	2.20	1.30	$17.7	$19.7
10% in μ	1.40	0.80	$10.0	$10.0

same COE, at $5/MMBtu[HHV] fuel, a 1 percent decrease in reliability, R, can be compensated by a 2 percent increase in net combined-cycle efficiency or a $15.9 million reduction in total capital cost.)

Even with the expanded version presented herein and paying heed to the caveats outlined earlier (i.e., using LCOE comparison only for similar technologies of similar capacity and maturity), the fact still remains that the LCOE concept is a "relic" from the days when the electricity markets were dominated by regulated utilities. When it first came out, LCOE represented the average lifetime cost for providing a kilowatt-hour of electricity for given full-load hours. In other words, as mentioned earlier, it gave the utility planners the constant price of electricity for which the NPV of the investment they were planning (say, a 750-MWe coal-fired power plant for base load generation) would be equal to zero. This of course helped the utility planners to assess the level of electricity tariffs.

Since the 1990s when the electricity markets were liberalized in the developed world, the LCOE metric has not been a reliable gauge to assess the competitiveness of a given power generation investment. This is so because competitive markets establish prices that reflect the *marginal costs* rather than the *average costs* in the LCOE formula, which are independent from system requirements. In the USA, the following three major market developments can be cited in that regard:

- Passed in 1996, Federal Energy Regulatory Commission (FERC) Order 888 forced utilities to provide nondiscriminatory market access to merchant generators. As a result, IPPs were able to sell power into different markets for the first time. In conjunction with FERC Order 2000, Order 888 also established the framework for independent system operators/RTOs to open up investment access to many new investors.
- Around 2000, independent system operators introduced *locational marginal pricing* (LMP), which allows more certain pricing data to be known across many different points in a marketplace, as opposed to a single zonal price. These better price signals help developers to site new generation in locations that are most in need and to offer the highest returns.
- First introduced in 2006, *capacity markets* provide an incentive for new plant investment by providing a more predictable revenue stream. Capacity markets are functioning in independent system operators across the USA such as the New York ISO (Independent System Operator) and PJM.

Once electricity prices are inputs into investors' profitability calculations (rather than outputs of them), LCOE loses its predictive value. Investors use detailed pro forma spreadsheets to calculate the NPV in order to assess whether the cash flow of a new power project is sufficient to reimburse their investment in financing it. Those calculations are based on expected external electricity prices as well as their variation and uncertainty over time.

In pro forma calculations, usually two types of revenues are considered: capacity and energy. The former is a simple product of the power generation asset's capacity and the capacity price. Energy revenue is a product of three factors: capacity, dispatch hours, and the electricity price during dispatch hours. The level of dispatch depends on the bid made by the generating asset operator. In a competitive market, the bid price reflects the variable costs of the generation asset (i.e., fuel price, O&M costs, and any environmental allowance costs plus any *uplift* payments[4] required by the operator). (In a competitive market, the hourly dispatch of a plant will be based on simple arithmetic. If the plant's variable costs are lower than the hourly market price, the plant will be dispatched, otherwise not.)

Competitive wholesale or spot electric energy prices are determined on an hourly basis by matching supply (i.e., available generating resources) with demand. In each hour, the prevailing spot price of electric energy will be approximated by the short-run marginal cost of production of the most expensive unit operating in that hour. Reasonably accurate projections of spot and capacity prices and combining them with detailed performance data of the generating asset to come up with a reliable cash flow stream and NPV are requisite for making investment decisions that can run into a billion dollars or even more. For a detailed look into such modeling endeavors, the reader is referred to a 2010 report issued by the US DOE's NETL, "Investment Decisions for Baseload Power Plants" (NETL Report 402/012910, January 29, 2010), which is available online free of charge.

20.3 Maximum Acceptable Capital Cost

For quick evaluation of technology improvements for a given system, a useful tool is *maximum acceptable increase in capital cost* (MACC), which is the capital cost equivalent of a given plant performance improvement (output and/or heat rate) with no change in COE. Using the basic COE formula and ignoring the change in O&M costs, MACC is calculated by equating before and after values of COE for a given plant improvement, i.e.,

$$\mathrm{MACC} = \frac{\mathrm{H} \cdot \mathrm{f}}{\beta} \cdot \left(\frac{\Delta \mathrm{P}}{\eta_0} - \Delta \mathrm{HC} \right) + \mathrm{k}_0 \cdot \Delta \mathrm{P}, \tag{20.14}$$

[4] "Uplift" is the gap between the revenues collected in market settlements and the compensation that generators require (based on their bids). The exact magnitude of the uplift depends on the methodology used by the independent system operator to calculate market prices.

where f is the levelized fuel cost (in \$ per kWh in LHV), ΔP is the change in plant net output (kW), ΔHC is the change in plant HC (in kW), η_0 is the base plant net efficiency, and k_0 is the base plant-specific capital cost (\$/kW).

The two terms on the right-hand side of the MACC formula in Equation 20.14 give the value of plant heat rate, HR (or, its equivalent, efficiency, η) and output, P. In general, an improvement in combined-cycle heat rate (efficiency) comes with a change in output and heat (i.e., fuel) consumption, HC. If the improvement is limited to the bottoming-cycle or gas turbine hot gas path section, however, there is no change in gas turbine HC (i.e., $\Delta HC = 0$). Using the relationship between efficiency and heat rate (i.e., $\eta = P/HC$ and $HR = 3{,}412/\eta$ Btu/kWh in US customary system), it can be shown that the value of a 1–Btu/kWh reduction in heat rate is given by

$$VHR = 10^{-6} \cdot \frac{H \cdot f}{\beta} \cdot P_0 \cdot \left(1 + \frac{\Delta P}{P_0}\right) \approx 10^{-6} \cdot \frac{H \cdot f}{\beta} \cdot P_0, \tag{20.15}$$

where P is in kW with f in \$/MMBtu (LHV) and P_0 is the base value of plant output. Alternatively, one can also use the value of one *basis point* improvement in net efficiency[5] as a yardstick, i.e.,

$$VEFF = 10^{-4} \cdot \frac{H \cdot f}{\beta} \cdot \frac{HC_0}{\eta_0} \cdot \left(1 + \frac{\Delta HC}{HC_0}\right) \approx 10^{-4} \cdot \frac{H \cdot f}{\beta} \cdot \frac{HC_0}{\eta_0}. \tag{20.16}$$

Note that HC in Equation 20.16 is in MMBtu/h with f in \$/MMBtu, both in LHV. HC_0 is the base value of the gas turbine (i.e., plant) HC. Note the difference between VHR and VEFF. The former is independent of the base technology, which is typically represented by net efficiency or heat rate. In other words, whether the base technology is a low-efficiency E class or advanced H or J class, for the same base rating P_0, VHR is the same. The same is not true for VEFF, which *is* a function of the base technology (i.e., η_0). If you combine Equations 20.15 and 20.16 and go through the simple algebra, you can verify that VHR and VEFF are equal to each other only for

$$\eta_0 = \sqrt{3{,}412.14} = 58.41$$

(as a percentage), where 3,412.14 is the unit conversion factor between kW and Btu/h (i.e., 1 kW = 3,412.14 Btu/h). Thus, in US customary units,

- For "old" technology with $\eta < 58.41$, VEFF > VHR;
- For advanced technology with $\eta > 58.41$, VEFF < VHR.

In SI units, of course, you can verify that VHR and VEFF are equal to each other only for

$$\eta_0 = \sqrt{3{,}600} = 60$$

(as a percentage), where 3,600 is the unit conversion factor between kW and kJ/h.

[5] A basis point is one-hundredth of one percentage point or 1/10,000.

This is the implicit reason why, in commercial evaluations, VHR is used to put a dollar value on power plant technology (i.e., "efficiency"). For specified economic and financial criteria (i.e., β and f) and generating capacity (i.e., P_0H in megawatt- or kilowatt-hours), VHR is *invariable*.

Equations 20.15 and 20.16 are frequently used to evaluate the value of 1 Btu of heat rate or 1 basis point of efficiency. Unfortunately, for the special cases where ΔHC is zero, they are incorrect and overstate the value of heat rate by a very large margin. This was shown via detailed mathematical analysis in [18]. The inherent fallacy can be easily grasped by considering the following two special cases of combined-cycle efficiency improvement:

- Via an improvement in the *bottoming* cycle (e.g., a better steam turbine; i.e., ΔHC is zero);
- Via an improvement in the *topping* cycle (e.g., a better compressor efficiency; i.e., ΔHC is non-zero – in fact, negative [i.e., less fuel consumed for the same power output]).

Note that

- In the first case (i.e., when ΔHC is zero), the plant owner's fuel bill does *not* change at all.
- In the second case, the plant owner pays *less* for fuel for the *same* power output.
- In either case, however, Equation 20.15 returns a positive value.

The "intrinsic" value of heat rate (or efficiency) is fully independent of any change in power output and can be evaluated by applying a *realization factor* (RF), which is rigorously evaluated in [18] and is represented by the curve in Figure 20.2. The independent variable, E, is the *efficiency improvement factor*, which is defined as percentage change in η per percentage change in P.

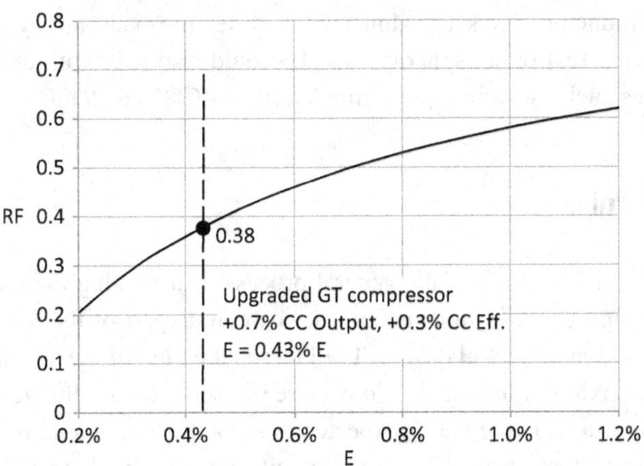

Figure 20.2 RFs to be applied to VHR in Equation 20.15 as a function of E, which is the efficiency improvement factor, defined as percentage change in η per percentage change in P. CC = combined cycle; GT = gas turbine.

As an example, consider a combined-cycle power plant with a 500–MWe net output and 58 percent net LHV efficiency. The gas turbine compressor is upgraded with new technology to give 0.7 percent higher combined-cycle output and 0.3 percent higher combined-cycle efficiency (i.e., the new combined-cycle performance is 503.5 MWe and 58.17 percent). Therefore,

- The efficiency improvement factor E is 0.3/0.7/100 $\approx$ 0.0043 (or 0.43 percent for each 1 percent in P).
- Using this value, from Figure 20.2, RF is read as 0.38. At \$3 fuel, using Equation 20.15, VHR is found to be \$50,000.
- Thus, using the RF of 0.38, the intrinsic value of 1 Btu in heat rate at \$3 fuel is 0.38 $\times$ \$50,000 = \$19,000.

In addition to the significant difficulty and uncertainty involved in capital investment estimation, the uncertainty in predicting the future price of fuel, load or demand growth, inflation, and other economic and/or financial fluctuations (e.g., interest rates) resulting from social and political turmoil make LCOE a very tricky tool to use. As such, to the extent possible, probabilistic methods such as *Monte Carlo simulation* should be preferred over deterministic comparisons. By assigning probabilities to the key parameters in the LCOE formula (by no means an easy task), the end result from any LCOE comparison should be the *probability* of "option A being lower/higher than option B," and *not* whether "option A *is* lower/higher than option B."

Application of *real options theory* to the valuation of power generation assets is a powerful technique that should be superior to *deterministic* methods such as LCOE evaluation. This is especially true for the currently (and most likely in the future as well) prevalent deployment mode of fossil generation assets – especially the simple- and combined-cycle gas turbines. While real options theory provides a powerful *stochastic* tool for economic analysis, its application requires sophisticated mathematical modeling and computer programming expertise. Nevertheless, the reader is strongly encouraged to consult the introductory book by Mun [19] in order to obtain an idea on the key principles underlying real options theory, which should also help with applying probabilistic techniques such as Monte Carlo simulation to LCOE modeling.

20.4 Commercial Margins

Commercial margin is the basis of the general process of managing the *uncertainty* in the gas turbine design process. In essence, it is the counterpart of the safety factor or *margin of safety* used in structural design. The latter is the ratio of the actual strength of a structure to the maximum load it should ever see in service (i.e., the design requirement). If one designs a structure that is expected to see a maximum load of 100 units in service such that it is able to carry a load of 150 units, the margin of safety built into the structure is 50 percent.

Analogously, if an OEM guarantees a gas turbine performance of 100 MWe (to be verified in a performance test at the end of the commissioning period – see

Section 14.5), the OEM makes sure that, as designed, the gas turbine is capable of delivering, say, 110 MWe at the guarantee conditions. The difference of 10 MWe is the *commercial margin* and is intended to minimize the probability of *not* making the guarantee performance and being hit with liquidated damages. How the commercial margins for output and heat rate are determined is explained below in general terms. The terminology and the rough outline are borrowed from General Electric practices. Nevertheless, the underlying principles are based in fundamental statistics and it is expected that each OEM has a process similar to this in its general features.

The OEM obtains the design performance from a proprietary, in-house computer program, which solves the aerothermodynamic relationships governing the compression, combustion, and expansion processes in a gas turbine. Fundamental physics-based formulae from the diverse branches of engineering (fluid dynamics, heat transfer, thermodynamics) are fortified by myriad empirical "corrections" encapsulating the engineering know-how gathered over more than half a century of gas turbine design. In General Electric, this program is known as the "cycle deck" (the terminology goes back to the early days when computer programs were punched cards stored as a *card deck* in boxes and fed into a reader mechanically), which is a FORTRAN code. Similar "decks" are employed by all major OEMs.

The performance calculated from this "engineering deck" is the starting point and is commonly known as the *P50* performance. This relates to the fact that each input parameter that went into the calculation of that performance has an inherent uncertainty. In other words, an input parameter x is really $x \pm \Delta x$, where Δx is the half-width of the uncertainty band around x. Thus, when the product is actually operating in the field, x can be anywhere between $x + \Delta x$ and $x - \Delta x$. Typically, Δx is assumed to be the same as the *standard deviation*, σ, which reflects the typical *machine-to-machine* (MTM) variation observed in the field in hundreds of similar units (due to manufacturing tolerances and other similar factors).

There are N such inputs of x_i to the engineering deck (i.e., x_i with $i = 1, 2, \ldots, N$), such that if one could run a Monte Carlo simulation with those inputs to the engineering deck (e.g., using Crystal Ball software), the deck outputs would be $y_j \pm \Delta y_j$, with $j = 1, 2, \ldots, M$ denoting M output parameters (e.g., y_1 is power output, y_2 is efficiency, and so on). Note that each output parameter y_j is a unique function f_j of inputs x_i, i.e.,

$$y_j = f_j(x_i).$$

Thus, a particular value such as y_1 would have a 50–50 probability reflecting the mean of the distribution (hence, P50) with a standard deviation, $\sigma = \Delta y_1$.

The designer who wants a higher certainty than 50–50 (say, 85–15; i.e., 85 percent chance that his design target y_j is obtained at its specified value or better) would quote the performance obtained from his engineering cycle deck at a value *one standard deviation lower* (i.e., not y_j, but $y_j - \Delta y_j$). This value is referred to as the *P85* value. (This is so because, according to the Gaussian normal distribution, $\pm 1\sigma$ around the mean covers 68.2 percent of the area under the bell curve or 15.9 percent under each tail.)

During the early stages of design, when the product is maybe several years away from the first field unit in reliable operation, there is an additional uncertainty, which is known

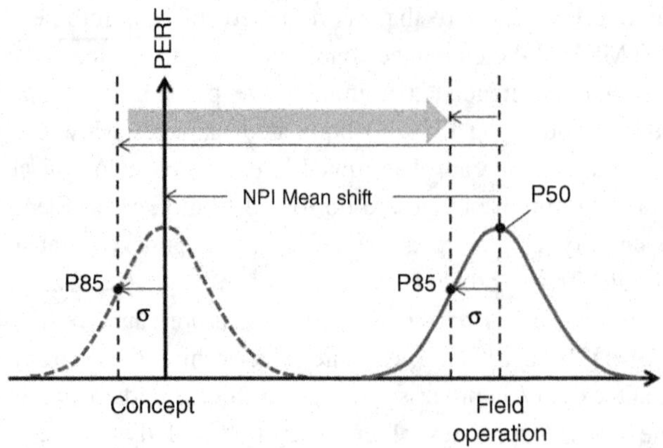

Figure 20.3 Graphical illustration of MTM variation (σ) and NPI mean shift. PERF = performance.

as the "mean shift." The mean shift reflects the fact that there is always a gap between the "design intent" in the beginning of a *new product initiation* (NPI) process and the final measured *and* validated performance achieved in the field. Once the product is field-validated (typically, three units with more than 8,000 cumulative hours), the mean shift becomes zero. These considerations are graphically summarized in Figure 20.3.

Thus, the gas turbine designer sets his or her engineering deck outputs to

$$y_j - \Delta y_j - \delta y_j,$$

where

y_j = engineering deck output (P50)
Δy_j = MTM variation (1σ)
δy_j = NPI mean shift

A cycle deck containing these *uncertainty reduction corrections* is given to the engineers who are involved in commercial transactions. This deck is known as the "commercial cycle deck" and it reports a P85 performance with mean shift (which ideally becomes progressively smaller as the NPI program approaches the finish line).

It is important to understand that the commercial cycle deck does *not* calculate a commercial performance. It is a tool that informs the commercial team that the design team has 85 percent confidence that the performance generated by the cycle deck can actually be achieved in the field. It is up to the commercial team to decide to apply *commercial margins* to the P85 performance so that they can ensure the maximum ROI for the OEM by (i) avoiding liquidated damages, (ii) ensuring bonus payments (if any), and (iii) still quoting a price with an ample CM commensurate with the quoted performance. As such, it is not unknown that sometimes a commercial performance guarantee is made with a *negative* commercial margin (i.e., with a performance *above* P50 and, as such, with a higher risk).

At this point, it should be pointed out that the process described so far predates the modern-day situation of full-speed, full-load factory testing capability being possessed by the major OEMs. This capability, *in theory*, eliminates the need for having at least three units operating in the field with more than 8,000 accumulated full load hours (although there are strong counterarguments). Unless the OEM starts "selling" a particular new gas turbine very early in the NPI process, the mean shift essentially disappears.

The symbiotic relationship between P50 and P85 performances can be broken from the engineering side as well. In terms of thermal performance, the gas turbine is represented by the engineering cycle deck. The engineering cycle deck changes when the gas turbine it represents changes. A change in the gas turbine can mean one of the following two outcomes: either the hardware is modified or replaced (aerodynamic profiles of the compressor or turbine airfoils, the airfoils themselves, cooling flows, coatings, etc.) or its performance is found to be different from what was originally predicted (component efficiency, cooling air flow orifice coefficient, pressure drop, etc.).

The cycle deck typically changes in one of the following two ways:

- Something in the code itself changes – more specifically, hardwired data such as compressor and turbine maps, scalers, coefficients, fudge factors, etc.;
- One or more of the default hardware and/or software (control) inputs change.

Thus, once the cycle deck (i.e., the machine and/or its intended control scheme) is changed, with the exact same inputs as used in the previous version, one is bound to get different outputs. Obviously, even with the *same* deck, one can get different outputs if one changes the inputs, especially those pertaining to the boundary conditions (e.g., changing the inlet loss by 0.5 in. of H_2O). There are, however, several other inputs that can be classified as "tuners," which are usually only accessible to the design engineers. The most common examples are inlet guide vane settings and firing temperature controls.

This latter type of input change via tuning knobs is not available to the users of the commercial cycle deck and cannot be accessed by them. When the machine is "tuned" to a better performance via these knobs, at the same site ambient and loading conditions, the P50–P85 performance gap can open up even further than indicated by the fundamental relationship shown in Figure 20.3.

In order to demonstrate the "economic, risk-based thinking" that goes into determining commercial margins, let us work out an example problem. Note that this example is overly simplistic, although it is adequate for illustration purposes. Actual analysis done by OEMs to determine commercial margins and pricing is much more complicated, with many factors taken into account simultaneously.

The nomenclature is as follows:

LD = total liquidated damages ($)
PE = engineering (expected) power output (kWe)
PG = guaranteed power output (kWe)
PM = measured power output (kWe)
TOL = test tolerance (kWe)
MRG = commercial output margin (kWe)

The commercial contract stipulates that the OEM is liable to pay the owner \$1,000 for each missed kilowatt, which is the difference between the measured power output (corrected to the guarantee boundary conditions) and the guaranteed power output. Thus, with zero test tolerance allowed (TOL = 0),

$$\text{PG} > \text{PM} \rightarrow \text{LD} = 1000 \times (\text{PG} - \text{PM}).$$
$$\text{PG} < \text{PM} \rightarrow \text{LD} = 0.$$

If the contract allows for test tolerance (which is rare these days),

$$\text{PG} > \{\text{PM} + \text{TOL}\} \rightarrow \text{LD} = 1000 \times (\text{PG} - \text{PM} - \text{TOL}).$$
$$\text{PG} < \{\text{PM} + \text{TOL}\} \rightarrow \text{LD} = 0.$$

Let us assume that the OEM's engineering deck, calibrated to the factory test-bed findings, calculates 250 MWe at the guarantee conditions. The test tolerance allowed by the contract is ± 0.5 percent of the instrument reading (95 percent confidence interval per ASME PTC 19.1). This corresponds to 1.25 MWe or 1,250 kWe. The OEM has full confidence in the calculated performance, but field experience shows that one should expect ± 0.5 percent MTM performance variation (1σ) due to manufacturing and installation tolerances, etc. Since the 95 percent confidence interval for a normal distribution is roughly equivalent to $\pm 2\sigma$, it translates into ± 1.0 percent (i.e., $\pm 2{,}500$ kWe). The OEM is extremely risk-averse, so it guarantees (assuming that the contract stipulates zero test tolerance)

$$\text{PG} = \text{PE} - \text{TOL} - \text{MRG}$$
$$\text{PG} = 250{,}000 - 1{,}250 - 2{,}500 = 246{,}250 \text{ kWe}.$$

We intuitively know that, unless something goes very badly awry (e.g., a "black swan" event), the probability of measuring an output lower than PG is very low. In fact, a Crystal Ball Monte Carlo simulation shows that it is only ~0.4 percent. The question is this: How much money is left on the table by being so risk-averse? For example, what if the guaranteed output were PG = 248 MWe? The simulation shows that the probability of PM <248 MWe is ~8 percent and the probability that the OEM would pay LD of more than \$100,000 is 6.75 percent. Thus, if the price delta that the OEM can ask for a 248-MWe guarantee output vis-à-vis 246.25 MWe *for the exact same machine* is much more than \$100,000, then being too risk-averse certainly leaves a lot of money on the table. (It can also lead to losing a competitive bid, which means a much bigger loss.)

It is practically impossible to "reverse engineer" commercial margins. There is simply not enough information. Heat and mass balance analysis of the guarantee performance can provide some clues, but separating the margins in generator output and heat rate is not possible. The situation is further complicated if the OEM bids the "power island" (i.e., steam and gas turbines, as well as the HRSGs). In an "equipment-only" bid, there might be margins applied to exhaust flow and temperature as well.

20.5 Dangers of Deterministic Thinking

The technical literature on economic evaluation of gas turbine alternatives (or any power generation technology, fossil-fired or not) using LCOE, NPV, ROI, or any other metric is full of "cut-and-dried" conclusions, e.g.,

- Option A has an LCOE of $5.675/MWh, Option B has an LCOE of $5.665/MWh; ergo, Option A is better than Option B.
- Option A has an ROI of 15.77 percent, Option B has an ROI of 15.91 percent; ergo, Option B is better than Option A,
- Et cetera.

Without a proper statistical analysis of the calculated metric, such deterministic proclamations of success (or failure) are highly error-prone.

In statistical hypothesis testing, a *type I error* is the incorrect rejection of a true null hypothesis (also known as a "false-positive" finding), while a *type II error* is incorrectly retaining a false null hypothesis (also known as a "false-negative" finding). In other words, a type I error is to falsely infer the existence of something that is not there, while a type II error is to falsely infer the absence of something that is.

In economic evaluation of gas turbine alternatives A and B, we have a quite similar situation, i.e.,

- A type 1 error is rejecting A when B is the inferior alternative.
- A type 2 error is accepting A when B is the superior alternative.

In order not to fall into this trap, one should use a Monte Carlo simulation approach to the calculated metrics. Let us use an example to illustrate this. Two GTCC alternatives are selected from the GTW 2016–17 Handbook with the price–performance information summarized below.

	A	B
Output (MWe)	701	717
Efficiency (%)	62.3	63.1
Heat Rate (Btu/kWh)	5,477	5,408
Budgetary Price (MM$)	455.5	463.0

LCOEs for A and B are calculated using Equation 20.1 with the following assumptions:

- 5,000 hours/year operation
- $3/MMBtu (HHV) fuel
- 15 percent capital charge factor
- Levelization factor of 1.169
- Owner's costs $135 million (same for both)
- O&M costs ignored (for simplicity)

The results are shown below, which indicate that alternative B is the better one.

	A	B
Capex ($/MWh)	25.27	25.02
Fuel ($/MWh)	18.22	17.99
LCOE ($/MWh)	50.84	50.28

Typical LCOE analyses usually stop at this point. And why not? B has 16 MWe more output and a 0.8-percentage point better efficiency. The question is this: Is it worth the $7.5 million premium in price? In order to investigate this, we run a Monte Carlo simulation with the following inputs:

- Normal distribution for outputs with $\sigma = \pm 1{,}500$ kWe
- Normal distribution for heat rate with $\sigma = \pm 0.6$ percent
- Uniform distribution for budgetary prices {–15%,+25%}, which is in line with AACE Class 4 (see Table 20.1)
- Maximum extreme distribution for fuel price (see Figure 20.4)
- 5,000 trials

The "forecast" of the Monte Carlo simulation, FC, is the LCOE delta between B and A, i.e.,

$$FC = LCOE_B - LCOE_A,$$

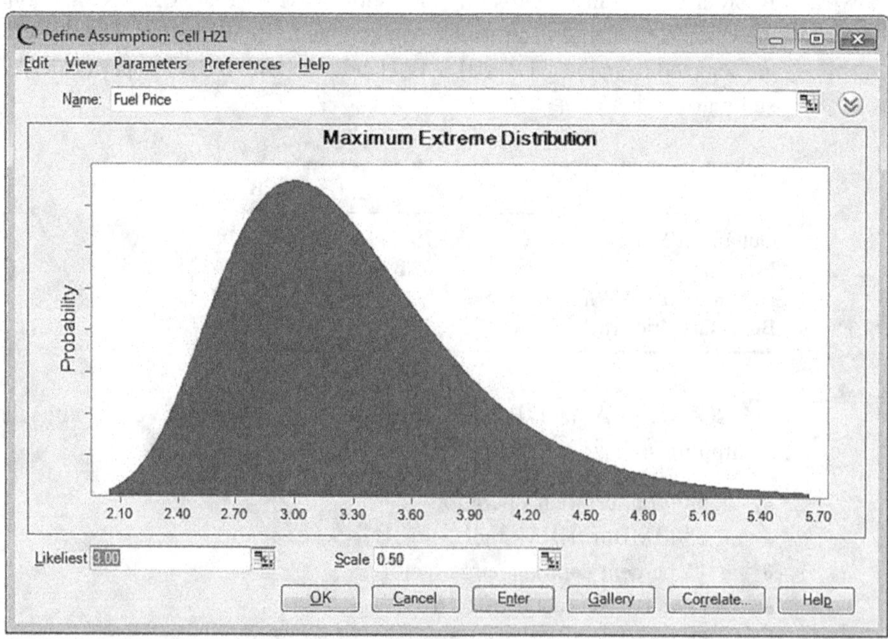

Figure 20.4 Maximum extreme distribution for fuel price.

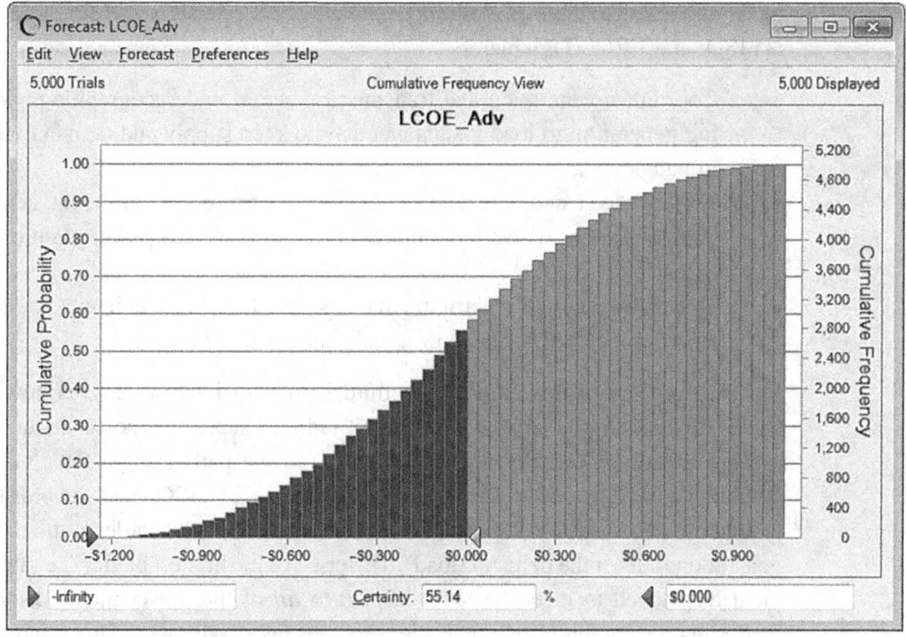

Figure 20.5 Cumulative frequency chart.

which is \$0.56/MWh as calculated earlier. After the 5,000-trial simulation is completed, the cumulative frequency chart (see Figure 20.5) shows that the cumulative probability that FC <0 (i.e., plant B having an LCOE lower than that for plant A) is 55.14 percent. In other words, in terms of LCOE, when accounting for the uncertainty in capital cost estimation (AACE Class 4) and in performance test measurement, whether to choose A or B is pretty much a "coin toss." Plant B is indeed better than plant A, but only marginally so.

20.6 Market Considerations

One frequently hears or reads about the "deregulation" of the electricity markets in the USA, which is essentially *shorthand* for the two orders issued by the Federal Energy Regulatory Commission (FERC) in April 1996 (FERC Orders 888 and 889). Prior to that, IOUs owned and controlled the entirety of the generation, transmission, and distribution assets. Consequently, consumers could not freely choose the provider of the electricity they were consuming (i.e., the electricity market was not competitive). Thus, with Orders 888 and 889, the FERC aimed to remove the barriers to competition and "liberate" the wholesale electricity markets. This led to the formation of two quite similar organizations:

- RTOs
- Independent system operators

In North America, there are 10 independent system operators and RTOs (e.g., CAISO, NYISO, and ERCOT), which

- Coordinate generation and transmission across wide geographic regions, matching generation to load instantaneously to keep supply and demand for electricity in balance;
- Forecast load and schedule generation to ensure that sufficient generation and backup power are available in case demand rises or a power plant or power line is lost;
- Operate wholesale electricity markets that enable participants to buy and sell electricity on a day-ahead or a real-time spot market basis.

For the purposes of this chapter, the third item is of interest. Here is how the system works (in very simple terms): IPPs submit bids for generating power to an independent system operator. Typically, the bid is based on the particular power plant's marginal operating costs (i.e., cost of fuel and variable O&M). The ISO ranks the bids from lowest to highest and dispatches the bidding power plants in that order until it has enough generation to meet the demand (load). The price of the highest bidder (i.e., the *last* power plant dispatched to meet the load) is paid to *all* of the power plants dispatched. The increasing generation from renewable resources (especially wind farms) has a significant impact on this dynamic because they are always dispatched first. In the summer of 2017, for instance, in the ERCOT area (in Texas), wind generation frequently surpassed 10 GWe and, as a result, even modern combined-cycle power plants with respectable efficiencies and burning less than $3 natural gas had to sit idle – generating no income to pay their fixed expenses. The situation is also somewhat complicated by a new structure called LMP, according to which an independent system operator pays a generator higher prices for power generated near areas with the largest demand.

Wholesale power prices fluctuate with the time of day as well as the day of the week. Typically, during weekdays, prices peak in the late afternoon and early evening and fall to their minimum level in the early morning hours. In some cases, prices fall to zero if the supply exceeds the demand, or even to *negative* values (i.e., operators *pay money* in order to continue generating). The reason for this is two-fold. In particular, older-generation fossil-fired power plants take a long time to restart, with excessive fuel burn and negative impact on part lives. Operators also want to be in "spinning reserve" mode and be ready when the opportunity to generate at significant profit reappears. On the flip side of the coin, prices can also jump to astronomic levels when the load demand increases significantly due to extreme weather conditions and the supply is barely enough (or is insufficient) to meet it.

As an example, consider the "price duration" data in Table 20.5 (actual data for ERCOT in 2016). The left-most column is the number of hours in a 1-year period, when the wholesale electricity price was *higher* than the value listed in the middle column. The third column is the gross income per each megawatt of electricity generated by a given power plant during that price bracket. For example, there was 3-hour period (7 – 4) when wholesale prices were between $400 and $500/MWh. A power plant running in that period would make $1,324 per each MWe of output.

Table 20.5 ERCOT price duration data (2016)

Hours	Price ($/MWh)	Income ($/MW)
2	600	$1,300
4	500	$1,034
7	400	$1,324
14	300	$2,369
23	200	$2,286
30	180	$1,319
37	160	$1,175
43	140	$903
57	120	$1,801
76	100	$2,042
111	80	$3,150
160	60	$3,377
298	40	$6,568
839	30	$18,062
3758	20	$67,386
8379	10	$78,329
8651	0	$0

Similarly, for the *entire* 7-hour period when wholesale prices were above $400, a power plant running in that period would make

$$\$1{,}300 + \$1{,}034 + \$1,324 = \$3,658$$

per each MWe of output.

Note that, according to the data, there was a 109-hour time period in 2016 (i.e., 8,760 – 8,651) when wholesale prices were actually negative.

Based on the example presented above, at $3 fuel, fuel cost is $18/MWh – more like $20/MWh for a more realistic efficiency of around 58 percent for state-of-the-art power plants operating in the field. Variable O&M costs for a typical combined cycle are around $2–$3/MWh. Thus, a realistic "break-even" wholesale price for a modern combined-cycle power plant is $22–$23/MWh. Even at $20/MWh, there were only 3,758 hours in 2016 when a combined-cycle power could be "in the money" (~43 percent of period hours). Assuming that a power plant ran whenever it was in the money (i.e., no forced or planned outages), its gross profit per megawatt of output for different break-even prices is summarized in Table 20.6. Marginal cost is simply the product of total variable costs (fuel plus O&M) with in-the-money hours. The importance of efficiency (heat rate) is clearly visible, i.e.,

- In the $20–$30 range, typical of combined-cycle gas turbines, variation in gross profit is nearly $18,000, whereas
- In the $30–$40 range, typical of simple-cycle gas turbines, it is less than $5,000.

Returning back to the original example, with $3/MWh variable O&M and $15,000/MW-yr fixed operating costs, the 1-year financial picture for sample plants A and B is shown in Table 20.7. The price duration data in Table 20.5 are used to determine in-the-

Table 20.6 Gross profit as a function of break-even price

Break-Even Price ($)	Marginal Cost	Gross Profit
100	$7,600	$7,954
80	$8,880	$9,824
60	$9,600	$12,481
40	$11,920	$16,729
30	$25,170	$21,541
20	$75,160	$38,938
10	$83,790	$108,637

Table 20.7 Sample plant comparison

	A	B
Variable O&M, per MWh	$20.22	$19.99
Marginal Cost, per MW	$75,994	$75,125
Gross Profit, per MW	$38,020	$38,644
Fixed O&M, per MW	$15,000	$15,000
Debt Payment, per MW	$42,118	$41,702
Capital Charge, per MW	$126,355	$125,105

money hours (3,758 hours in 2016). Cost of debt is assumed to be 5 percent. The takeaway from Table 20.7 is quite sobering. Either plant, if only operated when they are in the money, simply does not generate sufficient gross profit to cover their fixed operating costs and debt payments, never mind providing the ROI implied by the capital charge factor. Actually, in order to generate income to pay their fixed expenses, they would have to run when they were *out of the money* (i.e., when wholesale prices are *less* than their marginal costs).[6] This bleak situation is a direct outcome of increased wind power generation in the ERCOT region, which exceeded 15,000 MW in November 2016.

References

1. "Cost Estimation Methodology for NETL Assessments of Power Plant Performance," US DOE Report DOE/NETL-2011/1455, April 2011.
2. Bejan, A., Tsatsaronis, M. M., *Thermal Design & Optimization* (New York: John Wiley & Sons, 1996).

[6] This example using ERCOT, which does not have a "capacity market" mechanism, is quite pessimistic. The situation is better in an independent system operator/RTO like PJM, which has a capacity market, called the *reliability pricing model*, to ensure long-term grid reliability by procuring the appropriate amount of power supply resources needed to meet predicted energy demand 3 years in the future. Under the "pay-for-performance" model, power plants that receive a "capacity payment" even if they do not run must deliver on demand during system emergencies or owe a significant payment for non-performance.

3. Peters, M. S., Timmerhaus, K. D., West, R. E., *Plant Design and Economics for Chemical Engineers*, 5th edition (New York: McGraw Hill, Inc., 2004).
4. "Updated Capital Cost Estimates for Utility Scale Electricity Generating Plants," US Energy Information Administration, April 2013, Washington, DC.
5. Gülen, S. C., Mazumder, I., An Expanded Cost of Electricity Model for Highly Flexible Power Plants, *Journal of Engineering for Gas Turbines and Power*, **135** (2013), 011801.
6. Balevic, D., Burger, R., Forry, D., "Heavy-Duty Gas Turbine Operating and Maintenance Considerations," GER-3620K, GE Energy, 2004.
7. Nagy, D., Savic, S., "Alternative Gas Turbine Maintenance Concepts by Independent Service Providers," Powergen International 2015, December 8–10, 2015, Las Vegas, NV.
8. US Energy Information Administration, "Annual Energy Outlook 2010," December 2009, DOE/EIA-0383, 2009.
9. Maize, K., Peltier, R., "The U.S. Gas Rebound," *POWER*, January 2010, pp. 20–31.
10. Carrino, A. J., Jones, R. B., "Coal Plants Challenged as Gas Plants Surge," *POWER*, January 2011, pp. 47–49.
11. Cox, J., "Implications of Intermittency," *Modern Power Systems*, January 2010, pp. 22–23.
12. Peltier, R., "Flexible Turbine Operation Is Vital for a Robust Grid," *POWER*, September 2010, pp. 50–54.
13. Husak, M., Jones, C., Tegel, D., "Combined Cycle Plant Operational Flexibility," POWER-GEN International 2006, November 28–30, 2006, Orlando, FL.
14. DellaVilla, S., Koeneke, C., "A Historical and Current Perspective of the Availability and Reliability Performance of Heavy Duty Gas Turbines: Benchmarks and Expectations," GT2010-23182, ASME Turbo Expo 2010, June 14–18, 2010, Glasgow, UK.
15. Diakunchak, I. S., Performance Deterioration in Industrial Gas Turbines, *Journal of Engineering for Gas Turbines and Power*, **114** (1992), 161–168.
16. Marsh, W. D., *Economics of Electric Power Utility Power Generation* (New York: Oxford University Press, 1980).
17. Gülen, S. C., "A More Accurate Way to Calculate the Cost of Electricity," *POWER*, June 2011, pp. 62–65.
18. Gülen, S. C., "What Is the Worth of 1 Btu/kWh of Heat Rate?" *POWER*, June 2013, pp. 60–63.
19. Mun, J., *Real Options Analysis: Tools and Techniques for Valuing Strategic Investment and Decisions*, 2nd edition (Hoboken, NJ: John Wiley & Sons, Inc., 2005).

21 The Hall of Fame

There is a wide range of industrial gas turbines for electric power generation with ratings from a few megawatts (e.g., small aeroderivatives) to >500 MWe for state-of-the-art behemoths (in 50 Hz) such as General Electric's (GE) 9HA.02. In that entire range, there are many original equipment manufacturers (OEMs) with diverse products and licensed packagers or manufacturers. However, at the pinnacle of the gas turbine technology are four "families" of products by four major OEMs:

- Siemens' H/HL class
- GE's HA class
- Mitsubishi Hitachi Power Systems' (MHPS) J/JAC class
- Ansaldo Energia's GT36

Those four product families incorporate each OEM's best technologies in compressor, combustor, and turbine design as well as packaging and auxiliary systems. Each product family is covered in detail in the paragraphs below. Furthermore, the "genesis" of each product family is given ample attention as well. In other words, it is demonstrated how each state-of-the-art technology finds its roots in its respective forerunner (e.g., the *fully* steam-cooled H-System [forerunner of the HA class], the steam-cooled G class [forerunner of the J/JAC class], and sequential combustion GT24/26 [forerunner of GT36]).

It should be noted that these gas turbines, especially the 50-Hz variants, are sometimes referred to as the "jumbo frames." The aptness of this nickname is evidenced by the data summarized in Table 21.1.

21.1 Siemens HL Class

This technology is an evolution of the Siemens' air-cooled H class. The "HL" designation stands for "H technology on the way to 'L' technology for 65 percent combined-cycle efficiency." It has the same major technology features as in the H class, i.e.,

- Four-stage air-cooled turbine with 3D buckets; first three stages with thermal barrier coating (TBC);
- Twelve-stage compressor with two variable stator vanes (VSVs) and 3D blades;
- Single tie-bolt, steel rotor with Hirth serration (air-cooled);

Table 21.1 "Jumbo frame" gas turbines (50 Hz, 3,000 rpm). (?) = "guesstimate"

OEM	SGT5-9000HL	9HA.01	9HA.02	GT36-S5	M701J	M701JAC
	Siemens	GE		Ansaldo	MHPS	MHPS
Output, MW	545	446	544	500	478	493
Efficiency	42.0%	43.1%	43.9%	41.5%	42.3%	42.9%
Cycle Pressure Ratio	24.0	23.5	23.8	25.0	25 (?)	25 (?)
Exhaust Temperature, °F (°C)	1,256 (680)	1,164 (629)	1,177 (636)	1,155 (624)	1,166 (630)	1,186 (641)
Exhaust Flow, lb/s (kg/s)	2,205 (1,000)	1,862 (845)	2,168 (983)	2,227 (1,010)	1,975 (896)	1,975 (896)

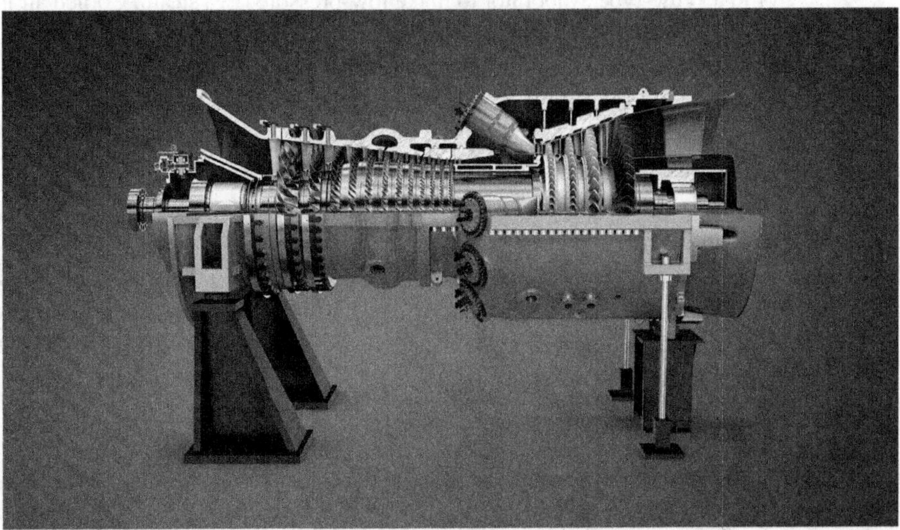

Figure 21.1 Siemens HL class gas turbine. (courtesy: Siemens Energy, Inc. © 2018. All Rights Reserved)

- Hydraulic Clearance Optimization (HCO);
- Can-annular combustor (12 and 16 cans for 60-Hz and 50-Hz units, respectively) (Figure 21.1).

HCO is a patented Siemens "active clearance control" technology that is based on axially shifting the rotor *against* the airflow direction to optimize turbine clearances during steady-state operation. The rotor shift is performed automatically by hydraulic pistons behind the compressor thrust bearing. During startup, clearances are at a level to prevent the blades from rubbing against the casing wall. The gain in power and efficiency on the turbine side (i.e., the clearances between the rotating blades and the casing inner wall becoming smaller) is expected to be higher than the losses on the compressor side (i.e., the clearances becoming larger). Despite being an integral part of the new Siemens gas turbines, HCO is also available as an upgrade to existing gas turbines.

Like the other three state-of-the-art, heavy-duty industrial gas turbines covered in this section, the HL–class units are of cold-end generator drive, two-bearing construction. Assuming a polytropic efficiency of 92.5 percent, first the compressor discharge temperature (CDT) is found as

$$\text{CDT} = \text{T}_{\text{in}}\text{PR}^{\frac{\gamma-1}{\eta_c\gamma}},$$

$$\text{CDT} = (59 + 460)\cdot 24^{\frac{1.4-1}{0.925\cdot 1.4}} - 460 = 925^{\circ}\text{F}.$$

Using CrMoV steel for compressor rotor construction (and still being able to handle such high air temperatures) is made possible by cooling air cooling, which has been used in Siemens (and Westinghouse before that) F–class gas turbines. Typically, cooling air extracted from the combustor shell is sent to an air-to-air (fin-fan) cooler or to a kettle-type evaporator to make low-pressure (LP) steam. Cooled air is returned to the "torque tube" (the stub-shaft between the turbine and compressor shafts) to provide seal air and cooling air for the turbine wheels and first-, second-, and third-stage rotor blades. Using cooled air for rotor blades limits corrosion and oxidation of the internal cooling passages. Due to the very high exhaust gas temperature, the large fourth-stage turbine blade is internally cooled as well.

Simple-cycle performances of 50- and 60-Hz units, SGT5-9000HL and SGT6-9000HL, respectively, are summarized in Table 21.2. Siemens claims that the HL class is capable of "more than 63 percent" combined-cycle efficiency. This is put to the test using what we learned in this book, specifically in Chapter 5 and Chapter 6.

Using the data in Table 21.2, the exhaust exergy of the 50-Hz SGT5-9000HL from Equation 5.60a is

$$a_{\text{exh}} = 0.1961\cdot 1256 - 86.918 = 159.4\,\text{Btu/lb}.$$

Multiplying by the exhaust mass flow rate, *total* exhaust exergy is found as

$$A_{\text{exh}} = 2205\ \text{lb/s} \times 159.4\,\text{Btu/lb} = 351,440\,\text{Btu/s},$$

$$A_{\text{exh}} = 1.05506 \times 351,440\,\text{Btu/s} = 370,790\,\text{kW}.$$

In other words, if the bottoming cycle of this gas turbine *were* a Carnot-equivalent cycle, its output would be about 371 MWe. From Figure 6.4, however, the expected technology

Table 21.2 Simple-cycle performance of SGT-9000HL. SCR = selective catalytic reduction

	50 Hz	60 Hz
Net Power Output	545 MW	374 MW
Gas Turbine Ramp-up	85 MW/min	85 MW/min
Net Efficiency	42%	41.70%
Pressure Ratio	24	24
Exhaust Mass Flow	1,000 kg/s (2,205 lb/s)	700 kg/s (1,543 lb/s)
Exhaust Temperature	680°C (1,256°F)	680°C (1,256°F)
NOx Emissions	2 ppm with SCR	2 ppm with SCR
CO Emissions	10 ppm	10 ppm

factor can be as high as 0.80; that is, only a fraction, albeit a very high one, of the theoretical maximum is available to the design engineer. Using Equation 17.24 for the technology factor, the expected *net* bottoming-cycle output (high-end estimate) is

$$0.74 \times 371 \sim 280.3 \text{ MWe}.$$

According to the data in Table 17.4, 1.95 percent of the steam turbine generator (STG) output is spent by feed water and condensate pumps. Therefore, the expected STG output is

$$280.3/(1 - 1.95\%) \sim 285.9 \text{ MWe},$$

so that combined-cycle *gross* output (in a 1 × 1 × 1 configuration) becomes

$$545 + 285.9 = 830.9 \text{ MWe}.$$

Gas turbine fuel consumption is

$$545/42\% = 1,298 \text{ MWth}.$$

Thus, combined-cycle gross efficiency becomes

$$830.9/1,298 = 64\%.$$

Assuming a plant auxiliary load of 1.6 percent (about the bare minimum for *Gas Turbine World* handbook ratings), net combined-cycle efficiency is

$$64\% \times (1 - 1.6\%) \sim 63\%.$$

In conclusion, Siemens HL technology is indeed capable of ~63 percent net combined-cycle efficiency at ISO base load with minimal auxiliary power consumption (highly dependent on the STG heat rejection system and site regulations concerning emissions and water conservation).

What makes this performance possible? Using the technology formula developed in Chapter 6 (i.e., Equation 6.1), we can estimate the firing temperature for the HL class, i.e.,

$$\text{TFIRE} = \frac{\text{TEXH} + 460}{0.882\left(\text{PR}^{-0.25} - 1\right) + 1} - 410,$$

$$\text{TFIRE} = \frac{1256 + 460}{0.882\left(24^{-0.25} - 1\right) + 1} - 410 = 2912°\text{F}.$$

Using the generic correlation between turbine inlet and firing temperatures in Figure 6.2, the turbine inlet temperature (TIT) for the HL class is estimated as 3,195°F or 1,757°C. As expected, within the uncertainty of the technology factors used, HL is clearly a 1,700°C-class gas turbine, which is widely advertised as the next evolutionary step in heavy-duty industrial gas turbine technology.

21.2 GE HA Class

In a nutshell, GE's HA (i.e., H class with air cooling) class is a revival of the company's technologically successful but commercially disappointing steam-cooled H-System [1,2]. The background of the steam-cooled H–class technology has already been covered in Chapter 4. (Note that GE always referred to an "H-System" rather than an H–class *gas turbine*.) A venerable industry veteran, Ivan Rice, once proposed that steam-cooled gas turbine-based "combined-cycle" systems should have been renamed "integrated-cycle" systems. This is due to the fact that the gas turbine Brayton and steam turbine Rankine cycles of the H-System are intertwined via cooling steam exchange.[1] In contrast, in standard air-cooled gas turbine-based systems, the two cycles are merely "attached" to each other. (Performance fuel heating utilizing intermediate-pressure [IP] economizer feed water extraction can be ignored without introducing significant error.) The symbiotic correlation between the two cycles is even more pronounced in the 107H with fuel moisturizing.

Although the six H-System combined cycles (two 60-Hz 107H and four 50-Hz 109H) were successful in terms of field performance, the technology was a flop commercially. Eventually, GE gave up on steam cooling, but utilized the field-proven key enablers (e.g., four-stage turbine, single-crystal [SX] materials, and advanced DVC[2] coatings) in the air-cooled HA class (Figure 21.2).

The compressor in the HA class is an aerodynamically scaled-up version of the compressor in the advanced F class (designated as FB or FA.05) to accommodate ~12 percent higher airflow. A higher pressure ratio (PR) with the same stage count (14 stages; i.e., ~22 for the HA vis-à-vis ~18 for the FA.05) indicates higher-loaded stages. The FA.05 compressor has been extensively tested (both as a stand-alone component and within the FA.05 engine) in GE's Greenville, SC, factory test facility. It contains 3D aero, field-replaceable blades with "super finish." The inlet guide vane and the three stages of VSVs are electrically actuated.

The Dry-Low-NOx (DLN) 2.6+ combustor in the HA class differs from its forebears in a very significant aspect: *axial fuel staging* (AFS). AFS technology is requisite for a high TIT (i.e., 1,600°C or 2,912°F) with reasonable NOx emissions (see the in-depth discussion in Chapter 12) and low turndown while remaining emissions compliant (i.e., about 30 percent minimum emissions-compliant load, but the tested capability is claimed to be even lower). The technology is very simple *in principle*: a mixture of fuel and air is injected into the main flow path through the combustor or the transition piece downstream of the main combustion zone as shown in Figure 21.3. This enables the achievement of high combustor exit temperatures without excessively high flame zone temperatures. The latter, as already discussed in detail in Chapter 12, is the primary cause of NOx production via the thermal (or *extended* Zel'dovich) mechanism.

[1] Indeed, GE has never published simple-cycle performance for either of its H-class units, 7H and 9H.

[2] Dense vertically cracked TBC. It is a GE proprietary plasma spray process that aims to achieve a durable, high-performance coating in terms of strain tolerance, spallation resistance, and component life.

Figure 21.2 GE HA–class gas turbine. (courtesy: General Electric)

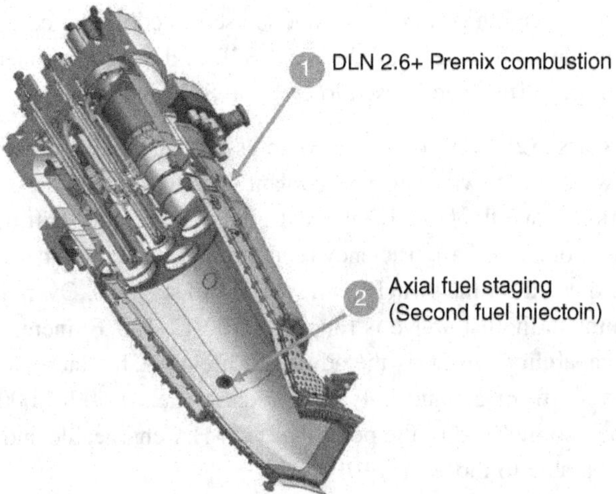

Figure 21.3 Dry-Low-NOx (DLN) 2.6+ combustor with axial fuel staging.

GE has been active in AFS technology development. There are many GE patents covering the technology (under the name *late lean injection*) with contributors from different organizations (i.e., corporate research and development, power generation, and aircraft engines). GE states that the operability of the new combustor design has been verified via extensive test-bench runs in the combustion laboratory (42 shifts of lab testing with 350 fired hours plus demonstration of AFS in a 7EA gas turbine in Texas).

The four-stage turbine (hot gas path [HGP]) section of the HA–class gas turbine is the key to achieving the advertised >41 percent simple-cycle efficiency. This mainly comes from the following three sources:

1. Lower average stage loading (higher turbine efficiency – see Section 10.2)
2. Improved turbine stage 1 materials (e.g., SX first-stage buckets with thin wall casting)

3. Reduced secondary flows
   a. Optimal use of compressor extractions for HGP cooling
   b. Other improvements to "tweak" secondary flows (see Chapter 9):
      i. Improved platform and end-wall cooling
      ii. Short-shank buckets (reduces wheel-space volume to be purged)
      iii. Near flow path seals

The first turbine stage (S1) of GE's advanced F-class gas turbines comprised conventionally cast (also known as *equiaxed* investment casting) GTD-111 vanes/nozzles (S1N) and first-generation SX René N4 buckets/blades (S1B). The S1N components of the HA class are manufactured from equiaxed René 108 (same as in the steam-cooled H-System S3N), whereas S1B components are made from the rhenium-free René N500 superalloy,[3] which is similar to the second-generation SX René N5 (same as that used in H-System).

The secondary flow system of the HA class is described below:

1. Air bled from the eighth compressor stage is used to cool S3N components.
2. Air bled from the 10th stage is used to cool S2B and S3B components.
3. Air bled from the 11th stage is used to cool the S2N components.

In the advanced F class, S2B was cooled by compressor stage 13 extraction. Modeling in Thermoflex shows that the switching S2B coolant source from compressor discharge to stage 11 extraction is worth about 1.5 percent more gas turbine shaft power output and +0.3 percentage points in shaft efficiency (everything else being the same). A similar benefit can be expected for the switch from stage 13 to stage 10. Overall, the impact of the last three items in the list above is roughly a 100°C (180°F) increase in turbine HGP temperature capability[4] vis-à-vis the advanced F class. This increase in lockstep with the TIT increase from advanced F to HA class (i.e., 1,500–1,600°C) should alleviate the turbine cooling load to the point that total HA chargeable and nonchargeable flows are comparable to those for FB/FA.05.

Performances of 60-Hz and 50-Hz HA-class gas turbine products are listed in Tables 21.3 and 21.4, respectively. The data are taken from the *Gas Turbine World 2016–2017 Handbook*, which did not have the exhaust flow rate numbers. The missing flow information is estimated using a heat and mass balance analysis as described in Chapter 7 (assuming 77°F fuel).

As shown in Tables 21.3 and 21.4, each HA product has two variants, denoted by the suffixes 0.01 ("dot oh one") and 0.02 ("dot oh two"), respectively. The 0.02 variants, which have about 20 percent more airflow than their 0.01 counterparts, are "flow-

3 Rhenium-free N500 and low-rhenium N515 have been developed by GE in response to dwindling supplies and increasing prices for the rhenium used in N5.

4 "Material temperature capability" is admittedly a vague term. There are several definitions used in the literature. For example, the numbers in Figure 13.2 are roughly in line with the GE definition, which is "rupture temperature under 20 ksi of stress at 100,000 hours" [3]. Another definition is "temperature for 1000-hour creep life at 20 ksi" and is about 100°C higher than that shown in Figure 13.2.

Table 21.3 GE 60-Hz HA–class gas turbine performance. LHV = lower heating value

	7HA.01	7HA.02
Speed (rpm)	3,600	3,600
Output (MWe)	289	372
Cycle PR	21.6	23.1
LHV Efficiency	41.9%	42.5%
Exhaust Flow (lb/s)	1,272	1,565
Exhaust Temperature (°F)	1,161	1,181

Table 21.4 GE 50-Hz HA-class gas turbine performance. LHV = lower heating value

	9HA.01	9HA.02
Speed (rpm)	3,000	3,000
Output (MWe)	446	544
Cycle PR	23.5	23.8
LHV Efficiency	43.1%	43.9%
Exhaust Flow (lb/s)	1,862	2,168
Exhaust Temperature (°F)	1,164	1,177

scaled" versions of the latter. As usual, the 50-Hz "frame 9" is a "speed-scaled" version of the 60-Hz "frame 7" (see Section 3.2 and Figure 3.3 in Chapter 3).

In order to get more output out of the same gas turbine architecture at the same rotational speed (3,600 or 3,000 rpm), more mass flow should be squeezed through the compressor and turbine sections without significant changes to the major dimensions and airfoil aerodynamics. According to GE [4], this has been achieved by "[opening] the first five stages of the compressor blading, nominally four inches, and in the back of the turbine [by radially moving] the blade tips a shade less than four inches to hold the exit Mach [number]." Opening or "re-staggering" the compressor blading refers to changing the stagger or blade chord angle of the airfoils in a stage row in such a manner that the flow area is increased (e.g., see Figure 21.4, which describes the staggering of turbine stage 1 nozzle vanes to increase the flow area). Furthermore, the blades were redesigned for optimal aerothermodynamic performance (due to the significant flow increase). For example, the *titanium* R1 blade of 7HA.02 is 4 inches taller and several pounds heavier than that of 7HA.01.

In order to prevent flow breakdown and excessive total pressure loss in the exhaust diffuser, the final stage (i.e., the fourth stage in an HA turbine) exit Mach number (M*a*) of a gas turbine is targeted to be around 0.5 (initial design).[5] The allowable

[5] Total pressure loss is proportional to $(1 - \mathrm{RF})\cdot\mathrm{Ma}^2$, where RF is the diffuser recovery factor. A higher RF and a longer (i.e., more expensive) diffuser section are required to balance excessive exit Mach number. A high RF and an insufficient diffuser length might result in flow breakdown. See the in-depth discussion in Section 10.6 in Chapter 10.

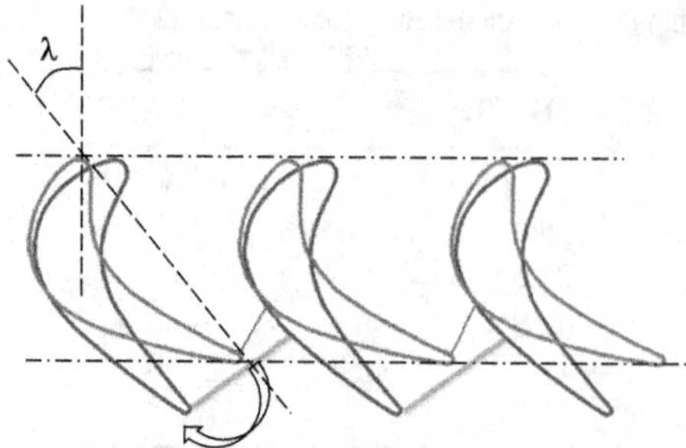

Figure 21.4 Conceptual description of re-staggering airfoils to accommodate larger flow (λ is the stagger angle).

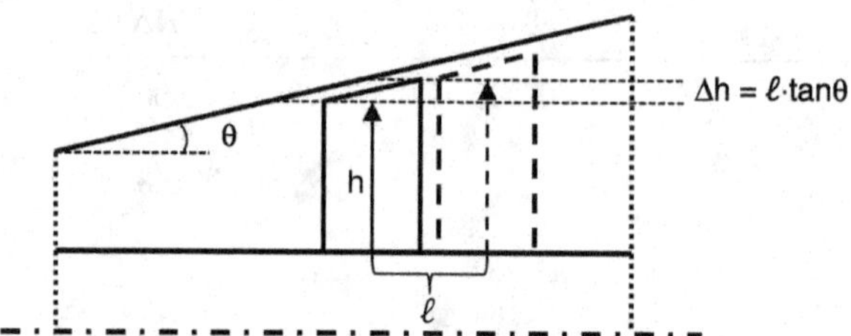

Figure 21.5 Conceptual stage 4 redesign to accommodate 7HA.01 to 7HA.02 growth.

upper limit is typically around 0.8 to provide some room for future growth (and cold-day performance) with an acceptable value of AN^2, which is proportional to the centrifugal stress acting on the long last-stage blade. In order to limit the exit Mach number increase (or to keep it constant), one has to increase the exit annulus area, which is equivalent to increasing the last-stage blade height. Within the same casing architecture, this can be accomplished by pushing the last-stage blade row out as shown in Figure 21.5.

It is difficult to estimate the turbine exit Mach number accurately. Nevertheless, by modeling the 7HA HGP in Thermoflex to match the published 7HA.01 and 7HA.02 performance in [4], the stage 4 geometry is estimated as shown in Table 21.5.

The Mach number in Table 21.5 is calculated using the formula below (gas molecular weight [MW] = 28.33 lb/lbmol, gas constant $\gamma = 1.33$):

$$A_{an} = \frac{\dot{m}_{gas}\sqrt{\frac{R_{unv}}{MW}\frac{T_{gas}}{\gamma}\cdot\left[1+\frac{\gamma-1}{2}Ma^2\right]^{\frac{\gamma+1}{\gamma-1}}}}{P_{gas}Ma\cos(\alpha)}.$$

The inner casing slope is assumed to be around 15°. Thus, l in Figure 21.5 is calculated as about 15 inches. (Note that l can [and probably is] distributed over the four stages of the turbine to minimize gap friction losses and increase nozzle inlet areas of the upstream stages – see below for more on this.) This is in agreement with the GE statement in [4] that the 7HA.02 engine is 19 inches longer than the 7HA.01.

Table 21.5 Stage 4 geometry

	HA.01	HA.02	Δ
Airflow (pps)	1,230	1,477	20%
Output (MW)	275	330	20%
Exhaust Flow (pps)	1,259	1,512	20%
Exhaust Temperature (°F)	1,142	1,142	
Exhaust Pressure (psia)	14.92	14.92	
Annulus Area, A_{an} (sqft)	54	61	13.2%
AN^2 (10^9 sqin)	101	114	13.2%
Exit Swirl Angle, α (°)	5	5	
Mach Number (*Ma*)	0.61	0.68	11.3%
Blade Length, *h* (in.)	28.1	32.1	4.0
Pitch Diameter, D_p (ft)	6.2	5.9	
Hub–Tip Ratio[a]	0.52	0.46	
Tip Speed (fps)[b]	1,825	1,874	

Notes:
[a] Ideal value is 0.6
[b] Should be kept below 1,900 fps

In order to maintain compressor surge margin, fuel system design and turbine exhaust temperature (for optimal heat recovery), cycle PR must be kept constant (or the increase in PR kept to a minimum). Note that the first-stage nozzle vanes of the turbine (as well as the other stages, too, most likely) operate in a choked condition. Thus, from the equation above with $Ma = 1$ and $\alpha = 0°$,

$$\dot{m}_{in} = P_{in}A_{in}\sqrt{\frac{MW}{R_{unv}}\frac{\gamma}{TIT}\left(\frac{2}{\gamma+1}\right)^{\frac{\gamma+1}{\gamma-1}}},$$

with the gas conditions and the annulus area at the nozzle inlet. Accordingly, with a 20 percent higher gas flow rate and the same TIT, the nozzle inlet annulus area should be increased by the same amount as shown in Figure 21.4. From a practical perspective, opening or re-staggering nozzle guide vanes to the tune of 20 percent seems excessive. Typical experience from converting frame units to low- or medium-Btu syngas operation is around 10 percent or less. Indeed, the higher cycle PR of 7HA.02 vis-à-vis 7HA.01 implies that the entire flow increase could not be accommodated by the S1N area increase. Conceivably, some of the stage 1 nozzle inlet area might come from slightly longer blades (similar to the redesign for the last stage discussed above). In order to maintain the stage loading distribution, similar nozzle inlet area increases for stages 2 and 3 are requisite as well (the stage 4 nozzle inlet area increase happens "by default" via accommodation of the exit Mach number limit as explained above).

Long turbine buckets such as those in 7HA.02 (see Table 21.5 for the estimated length) have a pair of circumferentially extending shrouds on the airfoil, one projecting from the pressure surface and one projecting from the suction surface. The shrouds are located at a radial location between the blade dovetail and the blade tip. In some other designs, the shrouds may be located at the tip of the blade airfoil. During normal operation of the turbine, the blades twist and the shrouds on adjacent blades contact each other, forming a shroud ring that provides support to the blades. As such, the shroud ring resists vibration and twisting of the long blades.

Going through the same calculation sequence in Section 21.1 for the Siemens HL class, the following data are calculated for 9HA.02:

- Exhaust exergy 329.1 MW
- Net bottoming-cycle output 246.7 MWe (technology factor 0.74)
- STG output 255.6 MWe (1.95 percent for boiler feed water and condensate pumps)
- Combined-cycle gross output 799.6 MWe
- Gas turbine fuel consumption 1,239.2 MWth
- Combined-cycle gross efficiency 64.5 percent
- Combined-cycle net efficiency 63.5 percent (1.6 percent auxiliary power consumption)

The *Gas Turbine World 2016–2017 Handbook* combined-cycle performance for 9HA.02 is 804 MWe and 63.5 percent (net lower heating value [LHV]), which is very close to the estimate herein.

The CDT is estimated as 922°F with 92.5 percent polytropic efficiency and a PR of 23.8.

The firing temperature is found as 2,754°F from Equation 6.1. The corresponding TIT is estimated as 2,991°F (1,644°C) using the transfer function in Figure 6.2. These numbers are probably on the low side. Assuming 10 percent for nonchargeable and 15 percent for chargeable flows, detailed heat and mass balance calculations (using the VBA function GasTurbine – see Section 8.3 in Chapter 8) return the following:

- 3,016°F (1,658°C) for TIT;
- 2,808°F (1,542°C) for rotor inlet temperature (RIT)/TFIRE;
- 1,443°C for TIT per ISO-2314.

It is expected that the "true" numbers are somewhere in between.

21.3 MHPS J/JAC Class

The Mitsubishi J class started as an evolutionary version of their steam-cooled G class, both with features adopted from their F–class gas turbines, including the basic architecture. It was the first gas turbine with a 1,600°C TIT. As already explained in Chapter 4, some of the design features can be traced back to the Westinghouse gas turbine technology (e.g., the compressor rotor with bolt-connected disks and torque pins and the turbine rotor discs with Gleason Curvic coupling [similar to Siemens' Hirth serration technology, which goes back to Jumo-004 – see Chapter 4]).

The 15-stage compressor is based on the technology used in Mitsubishi's "fully steam-cooled" (similar to GE's H-System) H–class gas turbine compressor with a PR of 25.

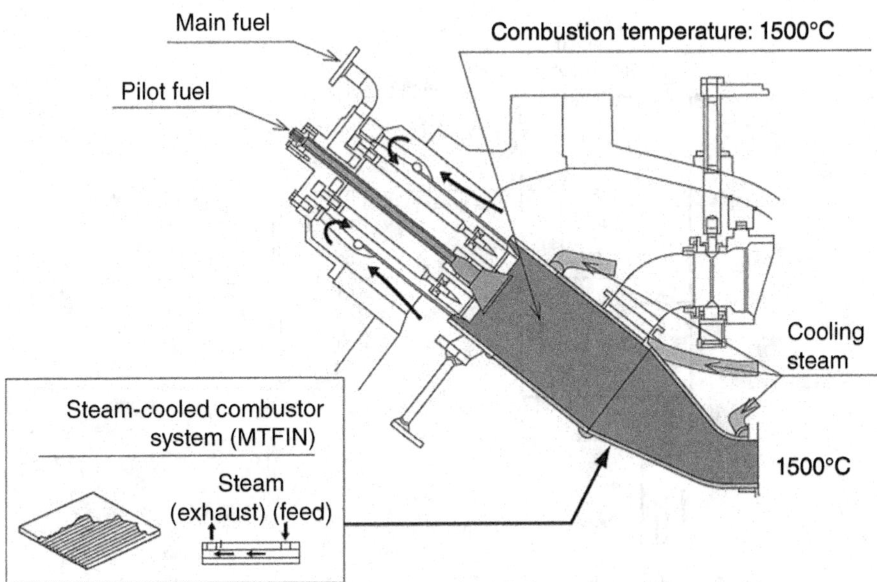

Figure 21.6 Mitsubishi G-class steam-cooled combustor.

The J-class compressor has a PR of 23. It has three stages of VSVs and 3D blades (which is practically de rigueur for all OEMs).

The J-class combustor is based on the steam cooling system used in the G class. Use of low-NOx technologies, such as reducing the local flame temperature in the combustion area by improving the fuel nozzle, ensured that G-class emissions were not exceeded at the 100°C (180°F) higher TIT. The Mitsubishi steam-cooled combustor and the geometry of the cooling passages, which is called MT-FIN (Mitsubishi Takasago FIN), are shown in Figure 21.6 [5]. As can be seen in Figure 21.6, the combustor basket and the transition piece form a single component.

Note the combustor air bypass valve in Figure 21.6 (on the right, attached to the L-shaped piping coming out of the combustor casing). Air bypass at low load ensures high combustion efficiency and low CO emissions. A similar system is also available in some Siemens engines. When the bypass valve is open, airflow to the flame zone decreases and the flame temperature is maintained at an optimum level. The GAC version does not have the bypass valve.

The gas turbine combustor and turbine stage 1 blade ring steam cooling system are shown in Figure 21.7. Steam from the IP superheater at about 600–650°F is sent to the gas turbine components to be cooled. The MHPS G-class gas turbine uses about 50 t/h (115,000 lb/h) steam for cooling. The estimated cooling duty is of the order of 7,000–8,000 Btu/s (about 7 MWth). Hot steam from the gas turbine is returned to the bottoming cycle at the hot reheat (HRH) steam line upstream of the intercept control valve. It is attemperated to match the HRH steam temperature (typically 1,050°F) via cold reheat steam extracted from the HP turbine exhaust (control valve CV1).

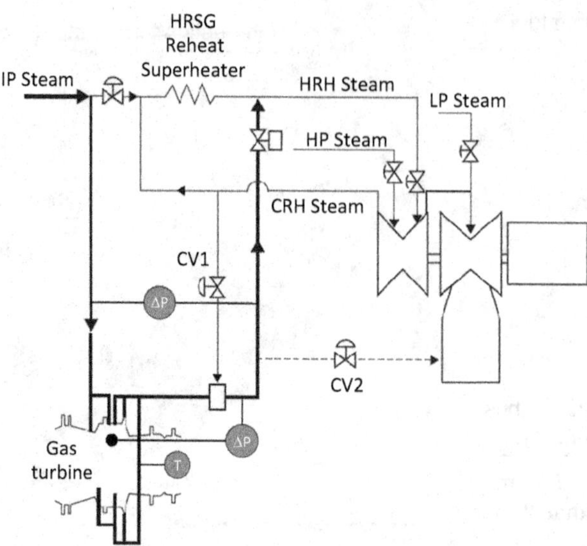

Figure 21.7 Mitsubishi steam cooling flow diagram and controls. HRSG = heat recovery steam generator.

The return steam temperature is monitored to detect possible overheating, which leads to an alarm and gas turbine runback [6].

The pressure delta between cooling steam return and the combustor shell is also monitored to detect any cooling steam leak (at higher pressure) into the combustor. (Such leakage can result in a gas turbine trip.) The pressure delta between the cooling steam supply and return lines is monitored as a proxy for cooling steam flow. (Insufficient flow can result in a gas turbine trip.) Excessive pressure in the return line is relieved via bypass to the condenser (control valve CV2).

Steam cooling is a proven technology, but comes at a cost in terms of extra equipment, control complexity, and diligent monitoring and maintenance requirements [6]. The steam and feed water purity requirements are similar to those requisite for the heat recovery steam generator (HRSG) steam drums at comparable pressures (e.g., feed water pH value of about 9.5, 7 parts per billion [ppb] dissolved oxygen in the feed water, 20 ppb silica in cooling steam, and 0.3 μS/cm cation conductivity). There is a redundant steam supply from the HRSG HP boiler as a backup. Steam from an auxiliary boiler is utilized during startup. (In GE's fully steam-cooled H-System, the gas turbine is started on air cooling and switched to steam cooling at about 15 percent load.) Once the gas turbine is synchronized and the HP steam pipe warm-up is complete, cooling steam supply transfer from the auxiliary boiler to the HRSG HP evaporator takes place. At about 25 percent load, cooling steam supply transfer from the HP evaporator to the IP superheater is accomplished (the former is available as a backup).

Steam cooling of the turbine stage 1 ring optimizes the blade tip clearance (space between the blade tip and the inner surface of the ring segment) during startup and normal operation. Cooling steam flowing through the circumferential channel formed

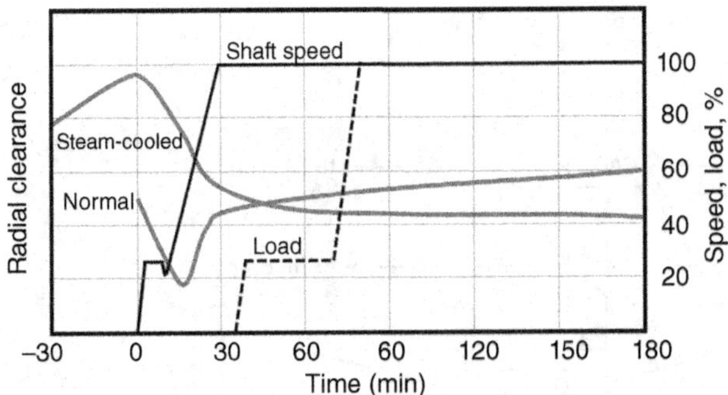

Figure 21.8 Turbine stage 1 blade tip clearance with and without steam cooling.

by the turbine inner casing and the ring segments, which is hotter than the metal initially, warms the ring up and expands it to increase the space separating it from the blade tip. As the gas turbine rolls up to full-speed, full-load and synchronizes and loads to its full load, cooling steam, now colder than the metal, contracts the ring to decrease the space separating it from the blade tip. This is shown in Figure 21.8 [6].

In 2012 and 2013, MHPS converted four steam-cooled M501J (60-Hz) units in Korea to air cooling because the country had to stop its nuclear plants due to the incident in Fukushima. They needed to operate in simple cycle while the combined-cycle construction was in progress. Those units were converted to "JAC" (for "J Air Cooled") in a speedy manner and a bypass stack was installed for them to operate in simple cycle. The conversion was accomplished via modifications to the combustor and enhanced TBC in the turbine section to allow for higher firing temperatures. The units were operated in cyclic duty for close to a year. Once the combined-cycle plant construction was completed, they were converted back into steam-cooled M501J (a relatively simple conversion because of commonality).

In order to improve the efficiency and operability of the J class, and encouraged by the experience in Korea, MHPS decided to add JAC into its product lineup. In 2014, the M501J in the T-Point facility of Mitsubishi's Takasago Machinery Works in Japan was also converted to air cooling. In 2015, the same M501J was retrofitted with *enhanced air cooling* (EAC; see below) to raise the TIT first to 1,620°C and later to 1,650°C. The JAC-class gas turbine was officially unveiled at the end of 2016 after a long testing and verification period.[6]

The key enabler of the next-generation JAC with a 1,650°C TIT is the EAC technology described in Figure 21.9. In the EAC system, compressor discharge air extracted from the combustor casing is cooled by two cooling air coolers (CACs) and pressurized by a booster compressor (single-stage radial compressor with PR of 1.3). After the pressurized air is used for cooling the combustor, it is returned back into to the combustor casing. The MT-FIN cooling structure of the combustor in the EAC system

[6] Note that MHPS 60-Hz JAC gas turbines are being built and shipped from the Savannah Machinery Works in Georgia, as well as the Takasago Works in Japan.

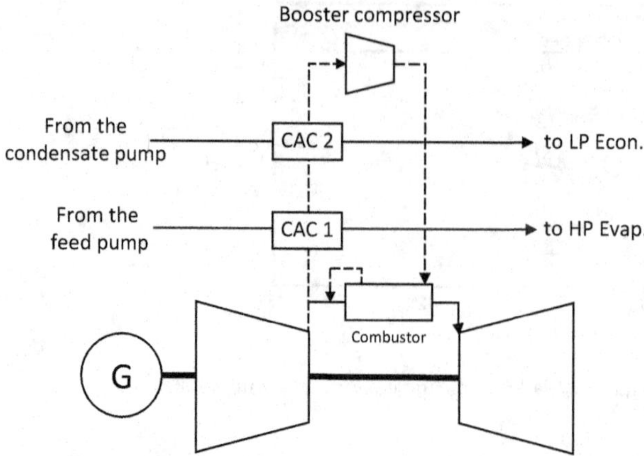

Figure 21.9 Schematic description of MHPS EAC technology. CAC = cooling air cooler.

is the same as that in the steam-cooled combustors. The upstream side of the combustor is cooled by compressor discharge air and the downstream side is cooled by the EAC air in order to minimize the parasitic load of the booster compressor. A detailed description of this can be found in Yuri et al. [7].

The benefit of the EAC is achieving a high TIT without excessive dilution of combustion products via conventional wall cooling and preventing a high flame zone temperature, which is the main driver of NOx emissions. In addition, enhanced cooling air is also used to optimize the turbine stage 1 and stage 2 blade tip clearances. The air supply line is provided with a three-way valve that allows the diversion of cooling air through the blade rings prior to entering the transition piece and combustor. Thus, during startup, when large clearances are required, the cooling air is routed directly to the transition piece without any effect on the turbine blade tip clearances. When the gas turbine reaches a certain load, the three-way valve is used to redirect the cooling air flow through the blade rings to cool and contract the ring metal and reduce the blade tip clearance for improved efficiency. The key principle and the end effect are similar to those described above for the steam-cooled version.

MHPS simple-cycle performances at ISO base load are listed in Tables 21.6 and 21.7 (from MHPS product brochure PSB0-01 GT08E1-E-0). No cycle PR was listed in the brochure. All four gas turbine variants are listed with a cycle PR of 23 in *Gas Turbine World 2016–2017 Handbook.* According to figure 10 in [7], "[s]imilar to M501H gas turbine, a pressure ratio of 25 is adopted to suppress the exhaust gas temperature rise." Detailed heat and mass balance analysis with the GasTurbine VBA function indicated that the latter is closer to the "true" value for the data listed in Tables 21.6 and 21.7.

Going through the same calculation sequence in Section 21.1 for the Siemens HL class, the following data are calculated for M701JAC:

- Exhaust exergy 303.5 MW
- Net bottoming-cycle output 227.5 MWe (technology factor 0.74)
- STG output 232 MWe (1.95 percent for boiler feed water and condensate pumps)

Table 21.6 MHPS 50-Hz J/JAC-class performance

	M701J	M701JAC
Speed (rpm)	3,000	
Output (MWe)	478	493
LHV Efficiency	42.3%	42.9%
Exhaust Flow (lb/s)	1,975	1,975
Exhaust Temperature (°F)	1,166	1,186

Table 21.7 MHPS 60-Hz J/JAC-class performance

	M501J	M501JAC
Speed (rpm)	3,600	
Output (MWe)	330	370
LHV Efficiency	42.1%	42.6%
Exhaust Flow (lb/s)	1,367	1,477
Exhaust Temperature (°F)	1,175	1,211

- Combined-cycle gross output 725 MWe
- Gas turbine fuel consumption 1,149.2 MWth
- Combined-cycle gross efficiency 63.1 percent
- Combined-cycle net efficiency 62.1 percent (1.6 percent auxiliary power consumption)

The MHPS brochure number for M701JAC is 717 MWe and 63.1 percent (net LHV).

The CDT is estimated as 907°F with 92.5 percent polytropic efficiency and a PR of 23 (943°F with a cycle PR of 25).

Using a cycle PR of 23, the firing temperature is found to be 2,750°F from Equation 6.1. The corresponding TIT is estimated as 2,986°F (1,641°C) using the transfer function in Figure 6.2. With a cycle PR of 25, the firing temperature and TIT values are 2,802°F and 3,052°F (1,678°C), respectively.

Assuming 10 percent for nonchargeable and 15 percent for chargeable flows, detailed heat and mass balance calculations (using the VBA function GasTurbine – see Section 8.3 in Chapter 8) return the following:

- Cycle PR of 25 (it is impossible to match 493 MWe with PR = 23);
- 3,112°F (1,711°C) for TIT;
- 2,899°F (1,593°C) for RIT/TFIRE;
- 1,453°C for TIT per ISO-2314.

It is expected that the "true" TIT is somewhere between 1,650°C and 1,700°C.

Heat rejection from the CAC is estimated as 15,000 Btu/s (about 15.8 MWth). EAC booster compressor power consumption is estimated as 400 kWe.

The heat and mass balance of M701JAC is summarized in Table 21.8. Note the handling of the EAC booster compressor; it is an energy stream entering the gas turbine

Table 21.8 M701JAC heat and mass balance

			m (lb/s)	p (psia)	T (°F)	h (Btu/lb)	Q Btu/s	Q kWth
M1	+	Air In	1,914.8	14.7	59.0	–0.2	–462	–487
M2	+	Fuel In	50.63	558.4	77.0	8.9	1,089,753	1,149,754
M4	–	Overboard Leak	1.9	349.0		215.6	413	436
M5	+	Exhaust Frame Blower	11.5		110.0	12.07	139	146
M6	–	Turbine Exhaust	1,975.0	14.9	1,186.0	302.1	596,591	629,439
E1	–	Casing Heat Loss					2,723	2,873
E3	–	CAC					15,000	15,826
E4	+	EAC Booster Comp					400	
E5	–	Mech. Loss					2,250	1,952
		Shaft Output (kW)						498,888
		Generator Output (kWe)						493,400
		Gas Turbine Auxiliary (kWe)						400
		Gas Turbine Net Output (kWe)						493,000
		Gas Turbine Net Efficiency						42.90%
		Gas Turbine Net Heat Rate (Btu/kWh)						7,954

control volume (+), but it is subtracted from the generator output as an auxiliary power consumer to arrive at the "net" output. Enthalpy and combustion calculations (to find the exhaust gas composition) are done using the property data and methods in ASME PTC 4.4 – 2008 (see [5] in Chapter 12). The underlying VBA code is included in the function COMBUST in Appendix C.

21.4 Ansaldo Energia GT36

Strictly speaking, GT36 is an Alstom gas turbine, which is a derivative of Alstom's sequential combustion GT24/26 gas turbines. In 2015, Ansaldo Energia acquired Alstom's GT26 (50-Hz) and GT36 (50- and 60-Hz GT36-S5 and GT36-S6, respectively) gas turbine assets and technology, as required by European regulators. The acquisition followed GE's completion of a deal to acquire the remainder of Alstom's vast power and grid business for $10.6 billion. The European Commission (EC) had granted its approval to GE's merger with Alstom on the condition that Alstom divested "central parts" of its heavy-duty gas turbines business and key personnel to Ansaldo. According to the EC, the divestiture was intended to avoid the possibility of higher prices being imposed by a quasi-monopoly resulting from the merger.

Future development, design, and manufacturing of GT26 and GT36 are going to be undertaken by Ansaldo Gas Turbine Technology Co., Ltd., which is a joint venture between Ansaldo Energia SpA and Shanghai Electric Group Company, Ltd., established in Shanghai in 2014. Ansaldo Energia SpA is 59.9 percent owned by Cdp Equity in the Cassa Depositi e Prestiti Group, an Italian state-owned entity, and 40 percent by Shanghai Electric Group of China.

Under an agreement that ran from 1991 through to October 2004, Siemens had licensed its "V"-class turbine technology to Ansaldo Energia. Those gas turbines (i.e., V64.3, V94.2, and V94.3) were further developed, manufactured, and sold by Ansaldo as AE64.3, AE94.2, and AE94.3, respectively. Following the acquisition of Alstom technology by Ansaldo, Siemens filed an arbitration action to block the agreements and terminate the license with Ansaldo Energia. Siemens claimed the agreements were outside the original license and allowing them to go forward would damage its gas turbine business and its position in the market. In 2016, however, Ansaldo Energia successfully defended the case in the International Chamber of Commerce Arbitral Tribunal regarding its right to continue to use the "V" technology under the license agreement.

Ansaldo GT36 is essentially a 1,600°C TIT-class gas turbine in direct competition with the other three major OEMs' H/HA- and J/JAC-class gas turbines. Instead of GT24/26's two combustors in series design with a single-stage HP turbine in between (PR of 2), GT36 uses the *constant-pressure sequential combustor* (CPSC) with 12/16 circumferentially arranged cans (60/50-Hz versions, respectively). It has a 15-stage compressor with three VSVs and 3D aero blades. The one-piece rotor is of welded-disk type (proven Alstom technology going back to the Brown Boveri Corp. [BBC] days – see Chapter 14). The air-cooled, four-stage turbine is similar to the LP turbine of GT24/26 and uses the same materials. Although the GT36 has a lower cycle PR vis-à-vis the sequential combustion machines (i.e., 24/25 for 60/50-Hz versions, respectively, vis-à-vis >30) turbine end stages have similar temperatures due to highly loaded front stages (see Chapter 10 for the underlying principles). The first three stages have TBC to compensate for the temperature increase. The first-stage rotor blades have a 3D core design to improve the cooling effectiveness. The third- and fourth-stage rotor blades are shrouded. Further design and development details can be found in the paper by Ruedel et al. [8]. The simple-cycle performance of GT36 is summarized in Table 21.9.

The design and development of the CPSC in particular and the entire gas turbine had already started in the Alstom days. The validation campaign included HP combustion

Table 21.9 Ansaldo Energia GT36 simple-cycle performance

	GT36-S5	GT36-S6
Speed (rpm)	3,000	3,600
Output (MWe)	500	340
Cycle PR	25	24
LHV Efficiency	41.5%	41.0%
Exhaust Flow (lb/s)	2,227	1,543
Exhaust Temperature (°F)	1,155	1,166

tests at the German Aerospace Center in Köln and at the newly built test plant in Birr, Switzerland. Full engine testing started in May 2016 with the first fire and ended in April 2017, under Ansaldo Energia ownership, with the assistance of the Swiss Federal Office of Energy. A detailed report was published in December 2017 [9].

The machine in the test stand was the 60-Hz GT36-S6, which was connected to the 50-Hz Swiss grid via an "active generator." The active generator is an Alstom technology that is now owned by GE. (Ansaldo is using it under a license.) It is a four-pole, 120–Hz generator with associated electronics. The generator itself is similar in size to a typical two-pole, 50/60–Hz advanced gas turbine generator. It generates four phases, which go through a two-step static frequency conversion process. In the first step, 27 phases are generated from the original four from the generator. The electronics for this are stored in a large box, which is about three-quarters of the size of the generator. In the second step, a large, three-story bank of switches actively select the three best matches (out of 27) for the three-phase grid.

As a sidenote, it is worth mentioning that the "active generator" concept was once thought of as a "disruptive technology." It allows one to use a large 50-Hz gas turbine in a 60-Hz grid. On a 2 × 2 × 1 gas turbine combined-cycle basis, one goes from a 1,000–MWe power plant to an approximately 1,450–MWe power plant in one fell swoop. Unfortunately, in comparison to a standard 50- or 60-Hz generator, the entire setup occupies a larger footprint and costs three times as much. On top of that, the three-story bank of switches requires cooling, and the significant cooling load results in an approximately 1 percentage point hit to the overall machine efficiency (i.e., say a 39 percent instead of a 40 percent efficient gas turbine). Although the system proved quite reliable, without a big cost reduction, it is difficult to see a transition occurring from the factory test bed to commercial application.

Detailed validation test findings are reported in [9]. Performance measurements made in Birr, after correction for the aforementioned active generator losses, indicated a gas turbine efficiency above 41 percent. As stated by Ansaldo, this level of performance requires "firing temperatures of over 1800 K,"[7] which translates into approximately >1,500°C (~2,800°F). We will take a closer look at this later in the section.

The operating principle of the CPSC and its benefits are similar to those for the original reheat gas turbine's EV and SEV combustors (see Figure 21.10).[8] The first stage is always on from 10 percent to full load with almost constant exit temperature and low NOx/CO emissions. Load control is achieved by adjusting the fuel flow to the second stage, which operates in an "Ultra-Low-NOx" regime with complete air–fuel mixing and a short residence time. At any given load, stage 2 exit temperatures determine the TIT.

Several problems encountered in the tests (e.g., torsional rotor vibrations at full-speed, no load and low loads, inability to meet CO emissions requirements at 40 percent load, etc.) remain to be fixed. According to Ansaldo, as of December 2017, memoranda of understanding for three 50-Hz units have been signed. The first unit is expected to be

[7] "Feuerungstemperaturen von über 1800 Kelvin" in the original German.

[8] EV stands for "environmental"; SEV stands for "sequential EV."

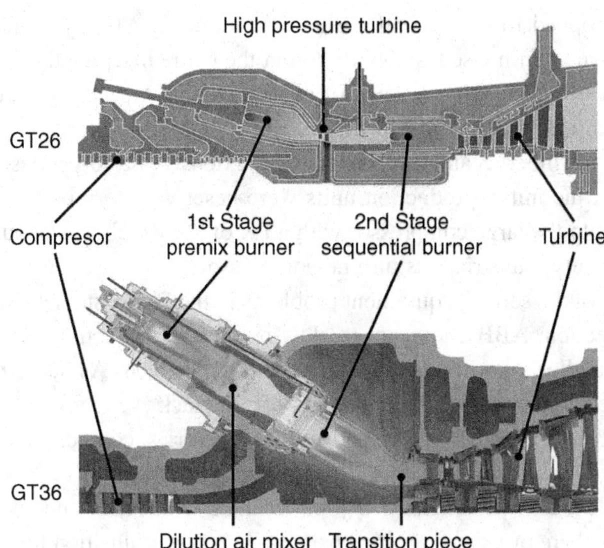

Figure 21.10 Comparison of GT26 and GT36 architectures. (courtesy: Ansaldo Energia)

delivered to the customer (a power plant in China) at the beginning of 2019. The gas turbines will be manufactured in Genoa, Italy.

21.4.1 Reheat Gas Turbine Background

Reheat or sequential combustion has been around for quite a long time. Stodola explicitly referred to it as a means to increase efficiency in an article he wrote right after he oversaw the performance testing of the world's first industrial gas turbine in 1939 (see the book by Eckardt, [4] in Chapter 4). Soon after, in 1948, BBC built and tested two such gas turbines in Beznau, Switzerland. These machines were quite different from the compact "jet engines on steroids" one tends to associate with the term "gas turbine" these days. They were plants in their own right with intercooled two-shaft configurations comprising separate LP and HP compressor-turbine trains and single-can combustors. (See Section 14.2 for more on these unique machines.) In the 1950s, BBC supplied these "tailor-made" units all over the world (e.g., four 25-MW units to Port Mann Station in Vancouver, BC, one unit to Lima, Peru). A more recent and well-known example is the Huntorf compressed air energy storage (CAES) plant's single-shaft HP–IP turbine with two silo combustors [10,11]. The same reheat gas turbine architecture is also to be found in the 110-MWe McIntosh CAES power plant in Alabama, USA [12].

Thus, Asea Brown Boveri (ABB; descendant of the venerable BBC) took the obvious evolutionary path when they introduced the more compact GT24/26 with two annular combustors comprising their proprietary EV and SEV burners in 1993. (The initial designs included a compressor intercooler, which significantly added to the engine length and was dropped from the final design.) At the time, ABB, just like the other OEMs, did not have an in-house test facility to put the entire machine through its paces before shipping it to the customers. This, of course, put the first few field units in the uncomfortable position of being *guinea pigs*.

The first GT24 at Gilbert Station, NJ, underwent extensive prototype testing. Unfortunately, however, the initial production units were beset with serious technical problems – largely due to the large compressor with a PR of 30, which was about twice that of existing heavy-duty industrial gas turbine compressors.

At first, in spite of the serious equipment problems in the combustors as well as in the HP compressor section, ABB chose not to slow down its sales activities, and did not publicize (to the author's knowledge) its poor field experience. As such, rumors and half-truths replaced cold facts and the technology got a black eye that would remain for years to come. Ultimately, ABB terminated further deliveries, compensated clients for damages, and devoted significant resources to fixing the problems.

In 1999–2000, Alstom first formed a joint venture with ABB and then acquired ABB's 50 percent share of its gas turbine business. Until its acquisition by GE in 2015, Alstom had been the OEM of reheat combustion GT24/26 gas turbines. In a 2000 press release, Alstom acknowledged the severity of the design flaws and field problems associated with GT24/26, stating that it was setting aside close to €1 billion to address those issues. Since then, though, it is fair to state that GT24/26 gas turbines have established themselves as reliably performing, efficient power generation systems.

Reheat or sequential combustion is a modest approximation of isothermal heat addition, which can be found in any undergraduate textbook. It has been covered in Section 8.5 using basic thermodynamic arguments (see Figure 8.13). The underlying concept is to realize an increase in the cycle mean effective heat addition temperature without increasing TIT, which is the maximum cycle temperature.

On a "real" cycle basis, unfortunately, the textbook advantage of the ideal reheat cycle turns out to be a mirage. As illustrated in Figure 21.11, the realistically achievable

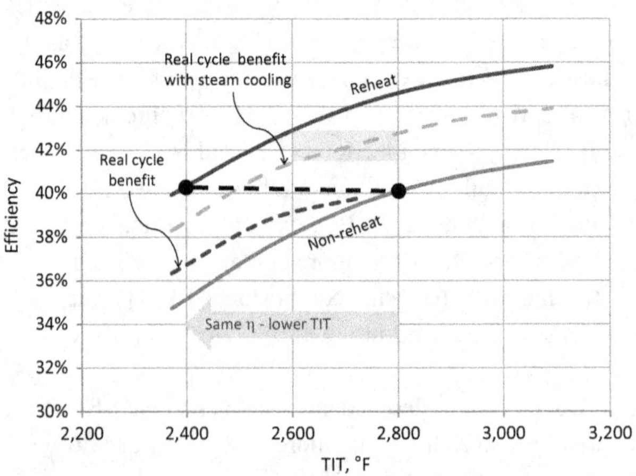

Figure 21.11 Advantage of a "real" reheat Brayton cycle over a non-reheat variant.

performance advantage is much more modest than that predicted by the ideal-cycle comparison. The primary drivers for this are increased HGP component cooling load and combustor design requirements. In fact, for TIT values of ~1,450°C (~2,642°F) or above, the reheat cycle efficiency advantage disappears due to significant increases in cooling losses (see [2] in Chapter 5).

In a recent article, Gülen has shown that this disadvantage can be alleviated by using closed-loop steam cooling for the HP and LP turbine stage 1 stators [13]. Using a stage-by-stage model in Thermoflex, it was shown that the reheat Brayton-cycle efficiency could be brought close to its entitlement value. The shaded rectangle in Figure 21.11 designates the ideal application range of a closed-loop, steam-cooled reheat gas turbine. In other words, with the highest cycle temperature at around 1,500°C, combined-cycle efficiencies well above 60 percent can be achieved (i) without resorting to the most expensive alloys and coatings and (ii) with mature DLN combustion technology and single-digit NOx emissions.

Note that the reheat gas turbine with open-loop steam cooling was proposed by Rice in his 1982 paper [14]. The gas turbine envisioned by Rice had an efficiency of 42.5 percent and a 1,299°F exhaust temperature. It was a bona fide >61 percent net combined-cycle efficiency enabler. Unfortunately, as already mentioned in Section 17.4, Rice was not as visionary with his choice of bottoming cycle (he had a two-pressure cycle with a 300°F HRSG stack and feed water heating) and ended up projecting well below 60 percent efficiency.

As far as a potential steam-cooled reheat gas turbine is concerned, the most likely approach is to keep the current GT24/26 architecture and integrate the proven cooling steam delivery system into the welded rotor design. Even though the performance entitlement offered by a "fully steam-cooled" turbine is highly tempting, in all likelihood, cost and complexity issues will preclude it.

However, steam-cooled HP and LP turbine inlet nozzle vanes provide most of the proverbial "bang for the buck" and should be eminently achievable with reasonable cost and engineering effort. Conceivably, the first (EV) annular combustor can be replaced by a can-annular GE design with axial fuel staging to get the highest possible HP TIT. In that case, the second (SEV) annular combustor would be retained for the most compact final configuration.

There is no doubt that a steam-cooled, reheat combustion, integrated-cycle power plant would be quite expensive. However, the plant would not be as inflexible as existing opinion suggests; it would retain the low-load capability of the existing reheat machines and would not be too sluggish in terms of warm/cold starts and load ramping. Obviously, it would not be as nimble as an air-cooled "fast-start" unit readily amenable to daily, two-cycled load following and/or standby operations. Then again, this is not the intended application for a highly efficient and pricey system most suitable to base load duty.

21.4.2 Reheat in a Single Can

The direction chosen by the major OEMs is quite clear from the state-of-the-art machines covered in this section. Retaining the simple cycle/machine architecture and achieving high simple- and combined-cycle performance via ever-higher TITs and exhaust temperatures (well above 1,200°F) seems to be the path taken toward 65 percent net combined-cycle efficiency.

This can be seen from the design philosophy of GT36 (started by Alstom and now continued by Ansaldo) covered above. Using the simple-cycle data in Table 21.9 and going through the same calculation sequence in the preceding sections, the following data are calculated for GT36-S5:

- Exhaust exergy 328 MW
- Net bottoming-cycle output 245.7 MWe (technology factor 0.74)
- STG output 250.6 MWe (1.95 percent for boiler feed water and condensate pumps)
- Combined-cycle gross output 750.6 MWe
- Gas turbine fuel consumption 1,204.8 MWth
- Combined-cycle gross efficiency 62.3 percent
- Combined-cycle net efficiency 61.3 percent (1.6 percent auxiliary power consumption)

The Ansaldo brochure number for a 1 × 1 × 1 GT36-S5 combined cycle is 720 MWe and 61.5 percent (net LHV).

The CDT is estimated as 943°F with 92.5 percent polytropic efficiency and a PR of 25.

The firing temperature is found to be 2,742°F (1,506°C) from Equation 6.1. The corresponding TIT is estimated as 2,975°F (1,635°C) using the transfer function in Figure 6.2. Assuming 10 percent for nonchargeable and 15 percent for chargeable flows and using detailed heat and mass balance calculations (using the VBA function GasTurbine – see Section 8.3 in Chapter 8), the TIT and RIT (firing temperature) were found to be 2,999°F (1,648°C) and 2,765°F (1,535°C), respectively.

References

1. Matta, R. K. et al., "Power Systems for the 21st Century – 'H' Gas Turbine Combined-Cycles," GER-3935B, 2000.
2. Pritchard, J. E., "H-System™ Technology Update," ASME Paper GT2003-38711, ASME Turbo Expo 2003, June 16–19, 2003, Atlanta, GA.
3. Schilke, P. W., "Advanced Gas Turbine Materials and Coatings," GER-3569G, 2004.
4. Probert, T., "Exelon Will Be First to Debut GE's New 7HA.02 Gas Turbine," *Gas Turbine World*, September–October 2014, pp. 14–19.
5. Tsutsumi, A. et al., Description of the Latest Combined Cycle Power Plant with G type Gas Turbine Technology in the Philippines, *Mitsubishi Heavy Industries Technical Review*, **40**: 4 (2003), 1–5.
6. Koeneke, C., Steam Cooling of Large Frame Gas Turbines One Decade in Operation, *VDI-Berichte*, **1965** (2006), 33–42.
7. Yuri, M. et al., Operating Results of J-series Gas Turbine and Development of JAC, *Mitsubishi Heavy Industries Technical Review* **54**: 3 (2017), 16–22.

8. Ruedel, U. et al., "Development of the New Ansaldo Energia Gas Turbine Technology Generation," ASME Paper GT2017-64893, ASME Turbo Expo 2017, June 26–30, 2017, Charlotte, NC.
9. Meier, P., "Validierung einer neuartigen, sauberen, und hocheffizienten Verbrennungstechnologie," Final Report, BFE-Vertragsnummer SI/501154-01, Ansaldo Energia, 2017.
10. Brown-Boveri Company, BBC Publication No. D GK 1274 86 E, "Operating Experience with the Huntorf Air Storage Gas Turbine Power Station," 1979.
11. Brown-Boveri Company, BBC Publication No. D GK 90202 E, "Huntorf Air Storage Gas Turbine Power Plant," 1979.
12. Nakhamkin, M. et al., "Compressed Air Energy Storage: Plant Integration, Turbomachinery Development," ASME Paper 85-IGT-4, 1985.
13. Gülen, S. C., "General Electric – Alstom merger brings visions of the Überturbine," *Gas Turbine World*, July/August 2014, pp. 28–35.
14. Rice, I. G., The Reheat Gas Turbine with Steam-Blade Cooling – A Means of Increasing Reheat Pressure, Output, and Combined Cycle Efficiency, *Journal of Engineering for Gas Turbines and Power*, **104**: 1 (1982), 9–22.

Part IV

Special Topics

22 Closed-Cycle Gas Turbine

In Figure A.1, the gas turbine (Brayton) cycle heat rejection process between state-points 4 and 1 and the associated heat exchanger are depicted with dashed lines (i.e., they are "imaginary"). This is why the standard gas turbine cycle is an "open" cycle (to be precise, an oxymoron); the working fluid, which itself varies in composition via combustion, does not complete a full loop. In that sense, using the term "cycle" for an actual gas turbine is erroneous in the first place.

A "true" gas turbine cycle with a single working fluid that stays constant in composition would be a "closed" cycle. Obviously, in this case, the term "cycle" is scientifically correct, but the moniker "closed" is superfluous. Nevertheless, this is the widely accepted terminology one has to live with.

Readers interested in the theory and history of closed-cycle gas turbines are referred to the one and only monograph on the subject matter; namely, the excellent book by Hans Ulrich Frutschi [1]. Hans Frutschi was assisted in writing the book and with its translation into English by several venerable industry veterans (i.e., Peter Rufli, Hans Wettstein [who wrote the foreword], Septimus van der Linden, and Axel von Rappard). He had actually led and/or assisted in the development of almost all closed-cycle gas turbine power plants in Europe by Escher Wyss AG.

The basic patent for the cycle was registered in Switzerland in 1935 by Ackeret (of ETH Zurich) and Keller (of Escher Wyss [EW] in Zurich). This is why the cycle is frequently referred to as the Ackeret–Keller or AK cycle. The first test installation, AK36, with air as the working fluid was in the EW factory in Zurich in 1939. In the 1950s and 1960s, under licenses provided by EW, several AK-cycle power plants were built in the UK, France, Germany, Austria, and Japan. The largest, commissioned in 1972 in Vienna, was rated at 30 MWe. The range of the remaining AK-cycle power plants with air as the working fluid was 2–17 MWe [1].

Several closed-cycle concepts with helium as the working fluid were developed primarily with nuclear power plant applications in mind. Helium, having the lowest neutron capture cross-section (it is chemically and neutronically inert or *non-fertile*[1]), was considered as an ideal coolant for high-temperature nuclear reactors and fast breeders. Supercritical carbon dioxide and its mixture with helium were also considered

[1] A fertile material can be converted into a fissile material by neutron absorption and subsequent nuclei conversions. An example is U-234, which can be converted into U-235.

as working fluids for similar applications, and many studies were done on "direct" cycles with either working fluid [2].

Interestingly, the only closed-cycle gas turbine that was actually built and operated had nitrogen as the working fluid, and it was part of a mobile 400-kWe unit built for the US Army, the ML-1 [1]. The nitrogen turbine had a pre-cooler and recuperator, but the compressor was not intercooled. Nitrogen was "charged" to 9 bara at the compressor inlet. At the turbine inlet, the nitrogen temperature was about 650°C. The unit actually worked, but generated only about 200 kWe. It was beset by many problems, including excessive pressure losses. The recuperator was probably overkill for the type of application envisioned by the Army. The concept was eventually dropped in 1965, at a time when the Vietnam War was rapidly escalating and money for research and development was becoming scarce.

In 1974, as an offshoot of the German–Swiss high-temperature reactor project with a large helium turbine (1,000 MWe), a 50-MWe coke oven gas-fired helium turbine was built in Oberhausen, Germany. Like all AK cycles, this was an intercooled-recuperated cycle. Hot gas at the high-pressure turbine inlet was at 750°C (~1,400°F) and 27 bara (~400 psia) and the helium flow rate was 85 kg/s (about 190 lb/s). Although designed for 34.5 percent cycle efficiency, the plant achieved 30 MWe at an efficiency of only 23 percent. A Thermoflex model of the Oberhausen II helium turbine is shown in Figure 22.1 (after figure 44 in [1]). The reasons for the unit falling far short of the design performance are explained in detail in [1].

Although envisioned for nuclear power applications, the first helium turbine was developed in the USA (by James La Fleur of Los Angeles) to be used in a cryogenic air separation plant. It was an ingenious design, with a conventional helium turbine comprising a natural gas-fired heater. It was equipped with a bleed flow (about 50 percent) from the compressor discharge, which formed a second, "mirror-image" cycle. In this second cycle, helium went through the following processes in the order listed:

1. First cooling by circulating cooling water;
2. Second cooling in the regenerator of the second, mirror-image cycle;

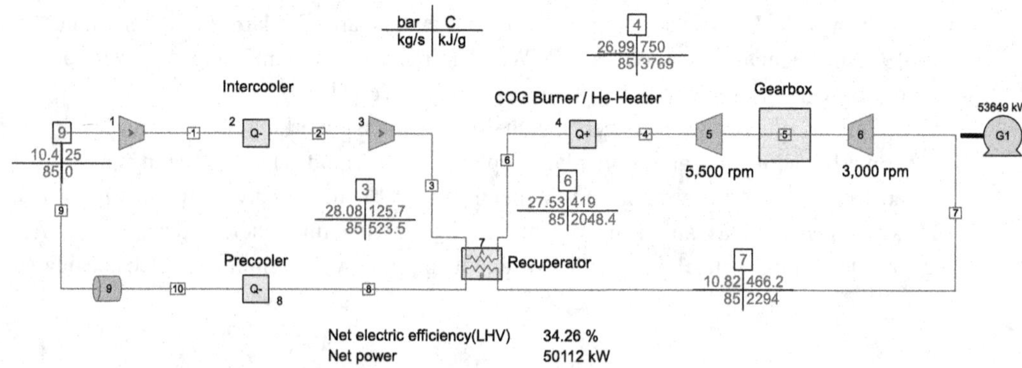

Figure 22.1 Oberhausen II helium turbine heat and mass balance. COG = coke oven gas.

3. Heating in the cryogenic air separation plant's air condenser;
4. Expansion through the turbine of the second cycle.

The described helium turbine, rated at about 6 MWe, was deployed in the Dye Oxygen plant in Phoenix, Arizona, in 1966 [1].

In the 1950s and 1960s, with their ability to utilize "difficult" fuels such as coal, steel mill off-gases, and peat, closed-cycle gas turbines had some success. Their inherent size disadvantage (just imagine the size of a modern H–class gas turbine power plant with an air mass flow rate approaching 1,000 kg/s operating in a closed cycle with near-atmospheric pressure at the compressor inlet) was minimized by charging the working fluid to high pressures. Charging not only helped to reduce the size of the piping and heat exchangers from a volumetric flow perspective, but also enhanced the heat transfer coefficients as well. (See Table 13.6 to get a feel for the dramatic impact of fluid pressure on the convective heat transfer coefficient.) Furthermore, since the velocity triangles in the rotating components changed very little with charging, part load control of closed-cycle turbines with nearly constant efficiency via pressure control (i.e., letting working fluid into or out of the cycle) was a clear advantage.

Nevertheless, the inability of closed-cycle machines to match the ever-increasing turbine inlet temperatures, in addition to the simplicity and agility of industrial gas turbines, which are similar to jet engines in construction and thermal cycle, ended their run in the early 1970s. In a nutshell, a cost-effective closed-cycle gas turbine design is out of question at the high temperatures requisite for acceptable efficiencies.

Although there are certain aerothermodynamic design challenges associated with closed-cycle helium turbines (mainly stemming from the unique properties of helium), there is a good chance of them making a comeback in the event of a long-awaited nuclear "renaissance." There is a considerable body of work, both past and present, including field experience gained from Oberhausen II and a few other installations, which could make the timely deployment of an actual helium turbine feasible. The reader is referred to [2–6] for more on this subject.

This is pretty much all that can be said about the closed-cycle gas turbines in this book, and if it had been written, say, 5 or 10 years ago, this chapter would have ended here. But something happened to change things in the last few years.

22.1 Supercritical CO_2 Cycles

Sometime around 2010 and thereafter, trade publications and archival journals were inundated with articles and papers filled with hyperbole and lofty claims about closed-cycle *supercritical* CO_2 turbines and their merits with little regard to reality being dictated by fundamental thermodynamic principles. Some even suggested – not so subtly either – that the technology was the next big thing in power generation, to the point of creating excitement and enthusiasm comparable to that experienced in a landmark engineering program of the twentieth century. Apparently, there is a not insignificant pie of research dollars out there and, clearly, very respectable investigators and institutions, including government laboratories and original equipment manufacturer research and development arms, are pursuing credible paths with modest and achievable goals (i.e., in incremental steps). Nevertheless, there is also a significant amount of "junk science" being peddled around. This makes separating seriously worthy projects with down-to-earth goals from the proverbial snake oil a very difficult task. For more information on the realistic description and assessment of supercritical CO_2 technology, the reader is referred to [7–12].

In principle, a closed-cycle power generation system with supercritical CO_2 as its working fluid (henceforth SCO2) is an excellent *bottoming cycle* to complement a *topping cycle* with low-grade exhaust energy (e.g., an aeroderivative gas turbine or reciprocating gas-fired engine). It is also a good fit for advanced nuclear reactors (e.g., generation IV reactors such as liquid metal reactors [LMRs]). In particular, a sodium-cooled fast-breeder LMR is ideally suitable to the SCO2 Brayton cycle with ~500°C at the exit of the intermediate heat exchanger.[2] Similarly, an SCO2 turbine that is small enough to be put at the top of a solar tower along with a solar receiver is an intriguing option for *concentrated solar power*. However, even in those "good-fit" scenarios, things are not as simple as the proponents of the technology claim. But there is at least a conceivable and realistic path to feasible, cost-effective performance, which would be difficult and/or costly to achieve with, say, a steam cycle.

While some variants of the SCO2 cycle hold promise (such as the semi-closed Allam cycle – see Section 22.2), as demonstrated via fundamental thermodynamics in an article by Gülen [13], closed-cycle SCO2 technology *cannot* be a viable alternative (regardless of how much money is spent on developing the technology for any length of time) to large, utility-scale power generation systems where the current performance (and economic) benchmark is set by gas turbine combined cycles (GTCCs) with advanced H- or J-class heavy-duty industrial gas turbines (>300 MW, >40 percent efficiency). Complemented by a three-pressure reheat (3PRH) steam bottoming cycle utilizing advanced steam turbine technology (>2,500 psia, 1,100°F steam), a modern GTCC can easily deliver field-proven >60 percent efficiency at >500 MW block sizes for about $1,000/kW total installed cost.

This is so because, due to their inherent thermodynamic cycle (i.e., low cycle pressure ratio (PR) and need for extensive recuperation to maintain a respectable thermal efficiency), SCO2 cycles (Brayton or Rankine) do *not* lend themselves to

1. Heat recovery comparable to a 3PRH steam cycle, so that they cannot be viable *bottoming-cycle* alternatives to advanced H/J-class gas turbines.
2. Heat rejection comparable to an advanced heavy-duty gas turbine with >40 percent efficiency and ~1,200°F exhaust temperature, so that they cannot be viable *topping-cycle* alternatives.

[2] See the *Gas Turbine World* March–April 2007 article "Closed Cycle Nuclear Plant Rated at 165 MW and 41% Efficiency" by S. van der Linden for a recuperated helium Brayton-cycle turbine envisioned for a pebble-bed modular reactor.

There is little doubt that the SCO2 technology, whether in recuperated Brayton or Rankine closed-cycle forms, is quite promising (see [13]). On its own, with a modest turbine inlet temperature that is similar to those of advanced steam cycles for coal-fired power plants, it can match the efficiency of an ultra-supercritical, pulverized coal system – as long as a sufficiently cold heat sink is available. It also seems to be a very suitable candidate for low-grade waste heat recovery, which until recently had been the playing field of organic Rankine cycles with heavy hydrocarbon working fluids.

Unfortunately, the "tug-of-war" between heat recovery potential and thermal efficiency on either end of the SCO2 cycle (Rankine or Brayton; i.e., heat addition and heat rejection) ensures that it is a "lone wolf" technology. In other words, although the cycle is very efficient on its own, only limited room is left for waste heat recovery, and very little room is left for another cycle on the heat rejection end. Thus, it is of limited use as a bottoming cycle, and it is no good as a topping cycle.

As a consequence, when it comes to being a player in the big leagues – say, as an enabler of 65 percent net GTCC efficiency with an advanced heavy-duty gas turbine – SCO2 technology is simply not that player.

As already covered in minute detail in Chapter 17, when it comes to bottoming-cycle design, thermal efficiency is only a part of the puzzle. Equally important – maybe more so – is the extent of heat recovery as quantified by the *heat recovery effectiveness*. In [13], it was shown that, in terms of first law (thermal) efficiency, a SCO2 Rankine cycle with split-flow recompression leaves even the crème de la crème of GTCC steam turbines in its dust! However, since that performance comes at the expense of heat recovery effectiveness, it is not a good candidate for the GTCC bottoming cycle. For a detailed analysis, the reader is referred to [13].

In conclusion, in the realm of utility-scale, advanced GTCCs with super-heavy-duty H/J–class gas turbines, the SCO2 closed-cycle technology is light-years behind the established state of the art (e.g., exergetic efficiency in the mid-40s under the most aggressive assumptions). Significant improvement is precluded by the basic nature of the thermodynamic cycle, which, limited by its low cycle PR, requires recuperation for meaningful thermal efficiency. This, however, limits the heat recovery potential and condemns the cycle to low-grade exhaust energy applications. It is an either-or situation (i.e., either low stack temperature or high thermal efficiency, but not both!). The trade-off determines the optimal bottoming-cycle efficiency, which is slightly above 21 percent (i.e., about 15 percentage points below that of the state of the art) [13].

Closed-cycle gas turbine technology with SCO2 working fluid has its rightful place in the arsenal of power generation (e.g., low-grade waste heat recovery, concentrated solar power, and nuclear power, especially LMRs or small modular reactors). Advanced GTCCs, alas, are beyond its thermodynamic grasp.

22.1.1 How Small Is Small?

One of the big selling points of the SCO2 power plant is the miniature size of the turbomachinery. A 10-MWe SCO2 Brayton-cycle power plant has a 15-MW turbine. Thermodynamic data of the turbine from the Thermoflex model are shown in

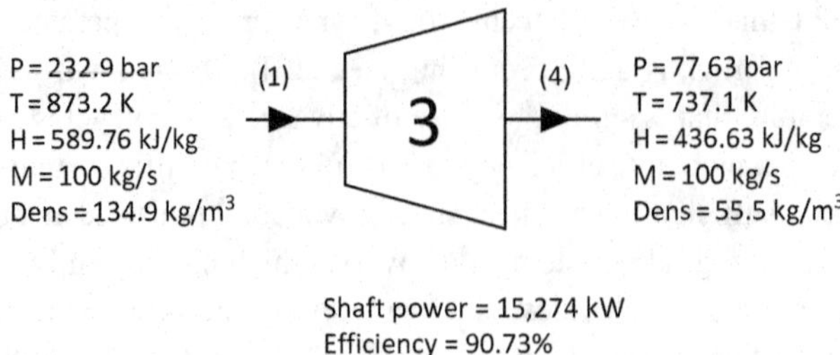

Figure 22.2 15-MW expander section for the 10-MWe SCO2 turbine.

Figure 22.2. For the turbine flow path calculation in TURB_STG, the following approximate data are used (molecular weight of CO_2 is 44.01):

- c_p = 0.2687 Btu/lb-s (1,126 J/kg-K)
- h = c_pT(°F) – 177.43 Btu/lb (multiply by 2,326 to convert to J/kg)
- γ = 1.205

It should be noted that supercritical CO_2 is a dense fluid – neither liquid nor gas in the usual sense – and as such it cannot be described by the ideal gas formulae. The simple model described above is only applicable to the states shown in Figure 22.2. It is derived from a linear curve-fit to turbine inlet and exit conditions. The specific heat ratio is found to be 1.2723, but the value used in TURB_STG is 1.205 to ensure that the simple isentropic p–T correlation (Equation 3.13) used to calculate stage exit temperatures leads to the final turbine output that agrees with the value in Figure 22.2.

In order to determine the shaft size and its rotational speed, we run the VBA function Num_Stages (see Chapter 10) for a range of speeds from 3,600 to 24,000 rpm for (average) ψ = 1.5 and ϕ = 0.5. Based on the results summarized in Table 22.1, a three- to five-stage design with a 24,000-rpm shaft speed seems to be a reasonable choice. A three-stage design is chosen for illustrative purposes.

The function TURB_STG is run with the inputs shown in Table 22.2. Note that stage i – 1 exit pressure, temperature, and absolute flow angle are inputs to stage i. TURB_STG outputs are listed in Table 22.3. Stage-by-stage geometry is summarized in Table 22.4.

As the numbers in Table 22.4 illustrate, the gas flow path of this 15-MW turbine/expander with 100 kg/s (~220 lb/s) working fluid flow is indeed very compact. Assuming an aspect ratio of 1.5 (blade height divided by blade width), the same width for the nozzle vanes, and a spacing between the vanes and nozzles set to a quarter of the blade width, we end up with an axial length of 129 mm (stage 1 stator inlet to stage 3 rotor exit). Thus, the entire flow path would fit into a 200-mm (~8-inch) diameter and a 150-mm (about 6-inch) long cylinder.

For a larger variant (in the distant future at the time of writing), it would be desirable to connect the SCO2 turbine directly to the generator. Thus, the shaft speed has to be 3,000 or 3,600 rpm. For a 100-MWe rating, the turbine output (and the mass flow rate) would be 10 times that shown in Figure 22.2. With ψ = 1.5, ϕ = 0.4, and α_3 = 15°, the outputs from Num_Stages are shown in Table 22.5.

Table 22.1 Num_Stages outputs for a 10-MWe SCO2 turbine

N (rpm)	3,600	10,800	18,000	24,000
# of stages	53	12	6	4
r_m (mm)	116	81	68	62
A (m^2)	0.003	0.014	0.010	0.008
u_m (m/s)	21.9	45.6	64.1	77.6
Ma	0.05	0.10	0.15	0.18
AN_2 ($in.^2 - rpm^2 \times 10^{-9}$)	0.6	2.6	5.1	7.5
σ (ksi)	0.8	3.3	6.6	9.6

Table 22.2 TURB_STG inputs for a 10-MWe SCO2 turbine

	Inputs	Stage 1	Stage 2	Stage 3
1	Shaft rpm	24,000	24,000	24,000
2	ψ	1.546	1.370	1.188
3	ϕ	0.5	0.5	0.5
4	α_3	20	15	5
5	Mass flow rate (kg/s)	100	100	100
6	T_1 (total) (K)	873.2	826	781
7	P_1 (total) (bar)	232.9	161.5	112.0
8	Stage PR	1.44	1.44	1.44
21	α_1	5	20	15

Table 22.3 TURB_STG outputs for a 10-MWe SCO2 turbine

	Outputs	Stage 1	Stage 2	Stage 3
1	α_2	69.9	68.0	66.4
3	β_2	36.1	25.3	16.1
4	β_3	67.1	66.2	64.4
5	V_2	269.3	256.8	252.1
6	V_3	98.6	99.7	101.3
7	V_{R2}	114.7	106.5	105.1
8	Rotor-relative T_2 (total) (K)	847	802	757
9	T_2 (total) (K)	873	826	781
12	Stage Δh (J/kg)	53,134	50,809	48,430
13	T_2 (static) (K)	841	797	753
14	T_3 (static) (K)	822	776	733
16	T_3 (total) (K)	826	781	738
21	Uncooled stage η	89.5%	90.5%	91.2%
24	Ma_u	0.42	0.45	0.49

The function TURB_STG is run again with the inputs shown in Table 22.6 (using five stages). Stage-by-stage geometry is summarized in Table 22.7. As the numbers in Table 22.7 illustrate, while not "tiny" anymore, the gas path of this 153-MW turbine/ expander with 1,000 kg/s (~2,200 lb/s) working fluid flow is also quite compact (vis-à-vis,

Table 22.4 Stage geometry for a 10-MWe SCO2 turbine

	Stage 1	Stage 2	Stage 3
Stage Work, kW	5,313	5,081	4,843
Hub Radius, r_h	62	62	62
Tip Radius, r_t	85	91	98
Pitch-Line (Mean) Radius, r_m, mm	74	77	80
Blade Height, h, mm (in)	23 (0.91)	29 (1.15)	36 (1.43)
υ (Hub–Tip Ratio)	0.73	0.68	0.63
ε (Total Deflection Angle)	103.14	91.52	80.49
Degree of Reaction, $R(r_m)$	0.41	0.45	0.45
$R(r_h)$	0.17	0.16	0.08
$R(r_t)$	0.56	0.61	0.63

Table 22.5 Num_Stages outputs for a 100-MWe SCO2 turbine

Parameter	Value
N (rpm)	3,600
# of stages	6
r_m (mm)	340
A (m^2)	0.13
u_m (m/s)	51.0
Ma	0.12
AN_2 ($in^2 - rpm^2 \times 10^{-9}$)	2.6
σ (ksi)	3.3

Table 22.6 TURB_STG inputs for a 100-MWe SCO2 turbine

	Inputs	Stage 1	Stage 2	Stage 3	Stage 4	Stage 5
1	Shaft rpm	3,600	3,600	3,600	3,600	3,600
2	ψ	1.443	1.445	1.373	1.301	1.220
3	ϕ	0.4	0.6	0.6	0.6	0.6
4	α_3	15	15	15	15	5
5	Mass flow rate (kg/s)	1,000	1,000	1,000	1,000	1,000
6	T_1 (total) (K)	873.2	845	817	789	763
7	P_1 (total) (bar)	232.9	187.0	150.1	120.5	96.7
8	Stage PR	1.25	1.25	1.25	1.25	1.25
21	α_1	5	15	15	15	15

say, a steam turbine of the same rating). Assuming an aspect ratio of 2.0 (blade height divided by blade width), the same width for the nozzle vanes, and a spacing between the vanes and nozzles set to a quarter of the blade width, we end up with an axial length of 450 mm (stage 1 stator inlet to stage 5 rotor exit). Thus, the entire hot gas path would fit into a 950-mm (~38-inch) diameter and a 500-mm (~20-inch) long cylinder. (For detailed turbine flow path layout calculations, one should use a special software package, such as SoftInWay, Inc.'s AxStream.)

Table 22.7 Stage geometry for a 100-MWe SCO2 turbine

	Stage 1	Stage 2	Stage 3	Stage 2	Stage 3
Stage Work, kW	32,205	31,375	30,552	29,756	28,804
Hub Radius, r_h	367	368	367	367	367
Tip Radius, r_t	425	436	447	460	475
Pitch-Line (Mean) Radius, r_m, mm	396	402	407	414	421
Blade Height, h, mm (in)	58 (2.28)	68 (2.67)	80 (3.14)	93 (3.66)	108 (4.25)
υ (Hub–Tip Ratio)	0.86	0.84	0.82	0.80	0.77
ε (Total Deflection Angle)	110.2	103.1	95.3	86.1	83.7
Degree of Reaction, $R(r_m)$	0.39	0.42	0.46	0.50	0.46
$R(r_h)$	0.28	0.31	0.34	0.36	0.30
$R(r_t)$	0.47	0.51	0.55	0.59	0.58

Before concluding this section, the reader is cautioned that the foregoing discussion was strictly limited to the SCO2 turbine flow path design. Considering that the working fluid pressure is well above 3,000 psia (more than 200 bara), it is not difficult to imagine the size and weight of the double-shell turbine casing. Even with the casing thrown in, the turbomachinery is dwarfed by the huge size of the cycle heat exchangers. A very dramatic visual illustration of these two facts (oft neglected in the hype surrounding the technology) can be found in the paper by Johnson et al. [14] (specifically, figures 6 and 7 in the cited work).

22.2 The Allam Cycle

Unlike the closed-loop SCO2 cycles analyzed in this chapter, the Allam cycle is a semi-closed cycle with oxy-combustion wherein CO_2 constitutes 95 percent of the fluid flow in the combustor (by mass) with the rest (5 percent) made up by oxygen and fuel (see Figure 22.3). Oxygen for combustion is generated by a cryogenic air separating unit (ASU). Carbon dioxide generated by the combustion is taken away from the cycle at the CO_2 pump discharge to maintain the cycle mass balance (hence *semi-closed*). The resulting combustion product is 90 percent CO_2 and the ASU parasitic power consumption is minimized by the lower O_2 requirement. The claimed net lower heating value efficiency of the Allam cycle is nearly 59 percent [15,16].

Construction of a 50-MWth demonstration plant in Texas started in early 2016 (at a total projected program cost of $140 million undertaken partly by major players in the power industry). The author visited the construction site in La Porte, TX, in September 2017. The plant construction was essentially complete (note that O_2 is delivered to the demonstration plant from a nearby chemical process facility and captured CO_2 is discharged to the atmosphere). A separate combustor test bench was being constructed next to the turbine building (to test the combustor without running the entire plant). At the time, the projection was to finish combustor testing by the end of 2017 and to commission the pilot plant in early 2018.

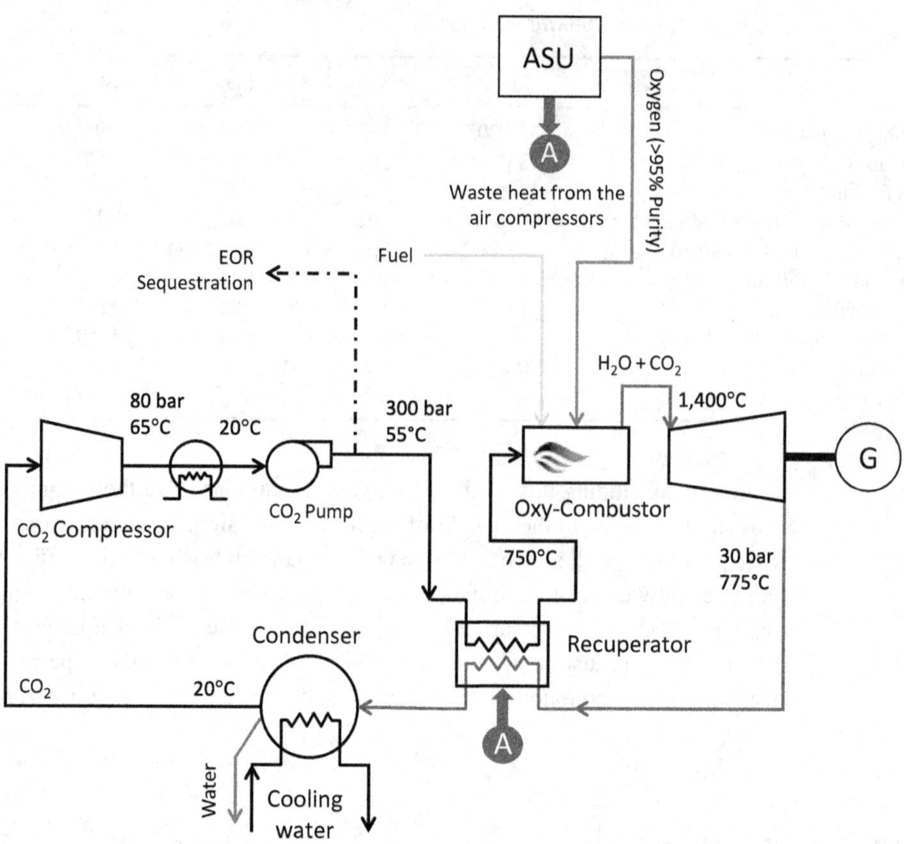

Figure 22.3 Semi-closed oxy-combustion Allam cycle. EOR = enhanced oil recovery.

This demonstration plant is the initial step on the road to the first 295-MWe commercial-scale power plant based on this technology. Just like all the other oxy-combustion cycles, the heart of the cycle is the combustor and turbine (a hybrid of steam and gas turbine technologies), which are under development by Toshiba. Design details and other information can be found in [15]. The turbine in the La Porte demonstration plant is a six-stage design with a PR of 10. The turbine inlet temperature and pressure are 1,150°C and 300 bar, respectively.

The Allam cycle with 59 percent net efficiency has a technology factor of 0.758, which is well within the realm of the current state of the art in GTCC technology (see Chapter 6 for a thorough discussion of technology factors). Although quite reasonable at first glance, this technology factor might, however, be somewhat high for a system with first-of-a-kind equipment (cf. 0.82 for the cascaded humidified air turbine cycle – which never materialized – in Section 8.5.2). Even so, considering that the cited performance is inclusive of O_2 generation as well as CO_2 capture and compression (presumably), even coming up a few percentage points short would result in an impressive performance.

For more information on oxy-combustion technologies from a high-level perspective, please refer to [11] in Chapter 8. For a comparative evaluation of oxy-combustion cycles, please see the paper by Bolland et al. [17]. For another oxy-combustion technology that has had test bench-scale demonstration, please refer to the papers by Anderson et al. [18,19].

References

1. Frutschi, H. U., *Closed-Cycle Gas Turbines: Operating Experience and Future Potential* (New York: ASME Press, 2005).
2. Lee, J. C., Campbell, Jr., J., Wright, D. E., "Closed-Cycle Gas Turbine Working Fuels," ASME Paper 80-GT-135, Gas Turbine Conference & Products Show, March 10–13, 1980, New Orleans, LA.
3. McDonald, C. F., Orlando, R. J., Cotzas, G. M., "Helium Turbomachine Design for GT-MHR Power Plant," General Atomics Report GA-A21720, 1994.
4. McDonald, C. F., "Helium and Combustion Gas Turbine Power Conversion Systems Comparison," ASME Paper 95-GT-263, International Gas Turbine and Aeroengine Congress and Exposition, June 5–8, 1995, Houston, TX.
5. No, H. C., Kim, J. H., Kim, H. M., A Review of Helium Gas Turbine Technology for High-Temperature Gas-Cooled Reactors, *Nuclear Engineering and Technology*, **39**: 1 (2007), 21–30.
6. Kodochigov, N. D. et al., "Development of the GT-MHR Vertical Turbomachine Design," ASME Paper HTR2008-58309, the 4th International Topical Meeting on High Temperature Reactor Technology, September 28–October 1, 2008, Washington, DC.
7. Conboy, T. et al., Performance Characteristics of an Operating Supercritical CO_2 Brayton Cycle, *Journal of Engineering for Gas Turbines and Power*, **134** (2012), 111703.
8. Fleming, D. et al., "Scaling Considerations for a Multi-Megawatt Class Supercritical CO_2 Brayton Cycle and Path Forward for Commercialization," GT2012-68484, ASME Turbo Expo 2012, June 11–15, 2012, Copenhagen, Denmark.
9. Held, T. J., "Initial Test Results of A Megawatt Class Supercritical CO2 Heat Engine," The 4th International Symposium – Supercritical CO_2 Power Cycles, September 9–10, 2014, Pittsburgh, PA,
10. Hejzlar, P. et al., Assessment of Gas Cooled Fast Reactor with Indirect Supercritical CO_2 Cycle, *Nuclear Engineering and Technology*, **38**: 2 (2006), 109–118.
11. Johnson, G. A. et al., "Supercritical CO_2 Cycle Development at Pratt & Whitney Rocketdyne," GT2012-70105, ASME Turbo Expo 2012, June 11–15, 2012, Copenhagen, Denmark.
12. Utamura, M., Thermodynamic Analysis of Part-Flow Cycle Supercritical CO_2 Gas Turbines, *Journal of Engineering for Gas Turbines and Power*, **132** (2010), 111701.
13. Gülen, S. C., Supercritical CO_2 – What Is It Good For? *Gas Turbine World*, September–October 2016, pp. 26–34.
14. Johnson, G. A. et al., "Supercritical CO_2 Cycle Development at Pratt & Whitney Rocketdyne," ASME Paper GT2012-70105, ASME Turbo Expo 2012, June 11–15, 2012, Copenhagen, Denmark.
15. Isles, J., Gearing Up for a New Supercritical CO_2 Power Cycle System, *Gas Turbine World*, November–December 2014, pp. 14–18.

16. Allam, R. J. et al., "The Oxy-Fuel, Supercritical CO_2 Allam Cycle," ASME Paper GT2014-26952, ASME Turbo Expo 2014, June 16–20, 2014, Düsseldorf, Germany.
17. Bolland, O., Kvamsdal, H. M., Boden, J. C., "A Thermodynamic Comparison of the Oxy-fuel Power Cycles," International Conference on Power Generation and Sustainable Development, October 8–9, 2001, Liège, Belgium.
18. Anderson, R. et al., "Oxy-fuel Gas Turbine, Gas Generator and Reheat Combustor Technology Development and Demonstration," ASME Paper GT2010-23001, ASME Turbo Expo 2010, June 14–18, 2010, Glasgow, UK.
19. Anderson, R. E. et al., "Adapting Gas Turbines to Zero-Emission Oxy-fuel Power Plants," ASME Paper GT2008-51377, ASME Turbo Expo 2008, June 9–13, 2008, Berlin, Germany.

23 Aeroderivative Gas Turbine

The aircraft gas turbine engine, specifically the *turbojet* engine, differs from its land-based cousin in its output, which is *net thrust* instead of mechanical shaft output, which, of course, is translated into electrical output via the synchronous alternating current (ac) generator. If a turbojet engine is fixed on a pedestal and run (i.e., the air velocity of the aircraft is zero), the thrust being produced is referred to as the static thrust and given by

$$F_T = \dot{m}_g \cdot V_J$$

(i.e., the product of the jet exhaust mass flow rate and velocity [at the jet nozzle exit]). In SI units, it is given by Newton (N), and in US customary units by lbf (pound-force) via division by the factor g. Jumo-004 had 1,980 lbf (~8,800 N) thrust with about 46.6 lb/s airflow and 0.77 lb/s fuel consumption (i.e., fuel–air ratio $f \approx 0.016$). The Messerschmitt ME-262, propelled by Jumo-004 jet engines, had a maximum speed of 524 mph at 20,000 feet altitude (about 234 m/s). Converting it to thrust power, we obtain

$$P = 8,800 \text{ N} \times 234 \text{ m/s} \sim 2,060 \text{ kW}.$$

This is a rough estimate of the power that could be generated if a "power turbine" were attached to the end of a stationary Jumo-004 engine. In a nutshell, this is the idea behind an aeroderivative gas turbine for mechanical drive and electric power applications. There are three major aircraft engine original equipment manufacturers (OEMs) whose jet engines are used as "gas generators" in industrial gas turbines:

- General Electric (GE)
- Pratt & Whitney (P&W)[1]
- Rolls-Royce (offered by Siemens)

Typically, the parent aircraft engine is in a single- or two-shaft arrangement. In the former, conversion from the aircraft engine to an industrial shaft or generator drive variant is accomplished simply by replacing the nozzle with a power turbine (sometimes also referred to as a "free turbine") [1]. The examples include Rolls-Royce Avon and GE LM2500. In the latter, a high-pressure (HP) "spool" (compressor and turbine on a common shaft) is arranged concentrically with a low-pressure (LP) spool. Conversion includes the following:

[1] The industrial engines division of P&W is now owned by Mitsubishi Hitachi Power Systems.

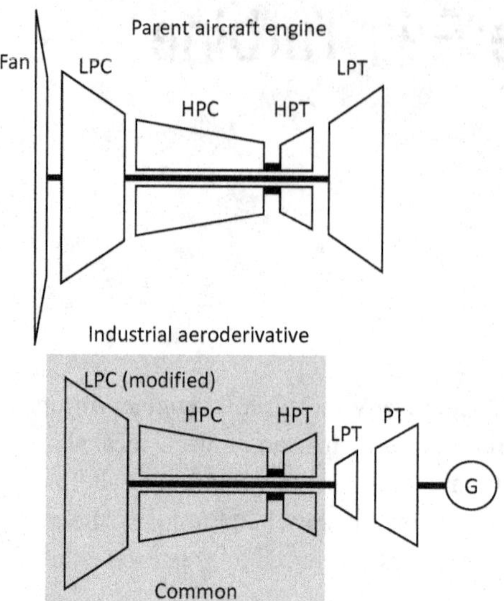

Figure 23.1 Commonality between the parent aircraft engine and the aeroderivative gas turbine. HPC = high-pressure compressor; HPT = high-pressure turbine; LPC = low-pressure compressor; LPT = low-pressure turbine; PT = power turbine.

- Removal of the fan;
- Modification of the LP compressor (to overtake the duty of the removed fan);
- New LP turbine to drive the modified LP compressor;
- New power turbine.

This is illustrated in Figure 23.1 [2]. The three-shaft Rolls-Royce RB211 is converted in a similar fashion. The fan and its turbine are removed to result in a two-shaft gas generator. GE's LM6000 is based on a different philosophy. There is no separate power turbine. The LP compressor and the LP turbine are redesigned so that the LP shaft is the "driven equipment" driver as well via direct coupling of the gas turbine to the load. Both hot- and cold-end drive arrangements are possible (with a free power turbine, obviously, the only possibility is hot-end drive). This is similar to the approach taken by Rolls-Royce in converting the *aircraft* Trent to the *industrial* Trent. The two-shaft intermediate-pressure (IP) and HP cores are retained. The fan turbine is replaced by a new turbine, which also drives a modified LP compressor. The gas turbine is connected to the load via the third shaft.

A representative list of aeroderivative gas turbines for electric power generation by major OEMs is provided in Table 23.1. Note that each gas turbine has multiple variants (with Dry-Low-Emissions [DLE] or *single annular combustors* [SACs in GE parlance, using water injection for NOx controls], with 3,000- or 3,600-rpm shaft speeds for 50-Hz or 60-Hz applications, respectively, etc.). GE's LM6000 has a SPRINT™ (Spray-Intercooled Turbine) variant with atomized water injection at both LP compressor (LPC) and HP compressor (HPC) inlet plenums to reduce compressor discharge temperature via "wet compression." This system provides a nearly 30 percent

Table 23.1 Selected aeroderivative gas turbine data. PR = pressure ratio; WI = water-injected

	Model	Combustor	Output (MWe)	Efficiency (%)	PR	MEXH (lb/s)	TEXH (°F)	rpm
GE	LM2500+ G4	DLE	32.7	39.7	23.0	197.6	951	6,100
	LM6000PF+	DLE	52.0	41.5	33.0	298.2	926	3,911
	LMS100PA+	WI	114.0	43.3	42.5	509.8	792	3,000
P&W	FT4000	WI	68.7	41.1	36.7	388.0	799	3,000/3,600
Siemens	RB211-GT30	DLE	32.5	38.3	22.3	215.8	932	3,600
	Trent 60	DLE	54.0	42.5	33.6	347.4	808	3,600

increase in power output at 90°F ambient temperature. Similarly, Siemens' Trent 60 has an inlet spray intercooling variant, which accomplishes the same performance enhancement as SPRINT.

Note that, in 2017, Siemens adopted a unique naming convention for the aeroderivative gas turbines acquired with the Rolls-Royce deal. The naming rule is as follows:

SGT-A##XX,

where SGT stands for "Siemens Gas Turbine" and A stands for "Aeroderivative." The numbers following A are indicative of the power output in megawatts (nominal rating, can be ±5 MW). The two letters at the end (XX) indicate the aero-engine origin, i.e.,

- RB: derived from Rolls-Royce RB211 family
- TR: derived from Rolls-Royce Trent family
- AV: derived from Rolls-Royce Avon family
- AE: Allison engine family[2]

Thus, the first Siemens engine in Table 23.1 is now referred to as SGT-A35 RB and the second one as SGT-A65 TR.

GE has by far the largest share of the aeroderivative market. The flagship unit is the LM6000, which originates from GE's CF6 jet engine found in the B747, B767, and Airbus A330. As of 2017, LM6000 has had more than 1,270 units delivered and 37 million operating hours. Its output ranges from 44 to 57 MW in simple-cycle and 117 to 149 MW in 2 × 1 combined-cycle operation. LM2500 is also based on the CF6 aircraft engine. Introduced in 1971, it has three models with multiple configurations and output ranging from 22 to 37 MW. Water-injected and dry combustor options are available. With more than 90 million operating hours across 2,300 units, LM2500 is the most widely used aeroderivative gas turbine. The newest addition to the GE family of aeroderivative gas turbines is LM9000, which is based on GE90-115B, and has been the engine of the Boeing 777 since 2004. LM9000 is a 66–75 MW unit rated at 43 percent simple-cycle efficiency. It has dual-fuel capability with a DLE combustor

[2] Allison Engine Company was acquired by Rolls-Royce in 1995.

and maintenance intervals of up to 36,000 hours for the hot section and 72,000 hours for overhaul. The first LM9000 is slated for shipment in 2019.

Modern aeroderivative gas turbines are very efficient due to their high cycle pressure ratios (PRs). Their lightweight construction (i.e., low rotational and thermal inertia) enables them to start and reach 100 percent load in less than 10 minutes from a cold state with no adverse impact on cyclic life. They can operate on a very wide range of fuels with low NOx emissions using DLE combustors or water injection. They can usually achieve at or below 25 ppm NOx without selective catalytic reduction. As such, they are uniquely suitable to peaking duties, especially in support of renewable generation resources with fluctuations and variability. As a result of their low exhaust temperatures (due to high cycle PR), they are not particularly suitable to combined-cycle applications. However, they are frequently used for industrial cogeneration applications. They are usually offered as packaged units with prefabricated accessory modules for rapid installation and commissioning. (In a simple-cycle configuration, the time from the order to commissioning is about 1 year.) Their modular design also enhances their operability and maintainability.

Aeroderivative gas turbines are usually started by a pneumatic starter using compressed air or a direct-drive ac start system comprising an induction motor and variable-frequency drive. The electric motor drives the HP spool via a clutch, gearbox, and bevel gear. Pneumatic starters are ideal for "black start capability" (i.e., the ability to bring the gas turbine online when there is no external electric power available). Even with a diesel engine-driven hydraulic starter, low shaft weight and inertia are advantageous for keeping the size and cost of the starter system low. As an example, GE's LM2500 with a starting torque of less than 750 ft-lbs (1,017 N-m) requires about 2,000–2,600 scfm compressed air to start. Electric and hydraulic starters can provide black start capability as well if equipped with a backup battery (to drive the hydraulic pump in the latter case). In that case, the gas turbine has to run at regular intervals to charge the direct current (dc) batteries.

The fuel gas delivery system of an aeroderivative gas turbine is similar to those for their larger, heavy-duty industrial counterparts. Fuel gas heating above the hydrocarbon dew point is accomplished in an electric heater. Performance fuel heating is usually not possible (because there is no heat recovery steam generator [HRSG] to supply, say, IP feed water as a heat source). Other auxiliary systems include the following:

- Inlet air filter
- Lube oil system
- Control system (on-skid or off-skid)
- Vibration monitoring system
- Water-wash skid
- Fire protection skid

Most aeroderivatives have an engine-mounted, accessory-drive gearbox for starting the unit and supplying power for critical accessories. The accessory gearbox provides mounting for the fuel pump (for liquid fuel), lube oil pump, air/oil separator, and pneumatic starter. The accessory drive assembly (see Figure 23.2) is driven through the compressor rotor shaft via the inlet gearbox, radial drive shaft, and transfer gearbox.

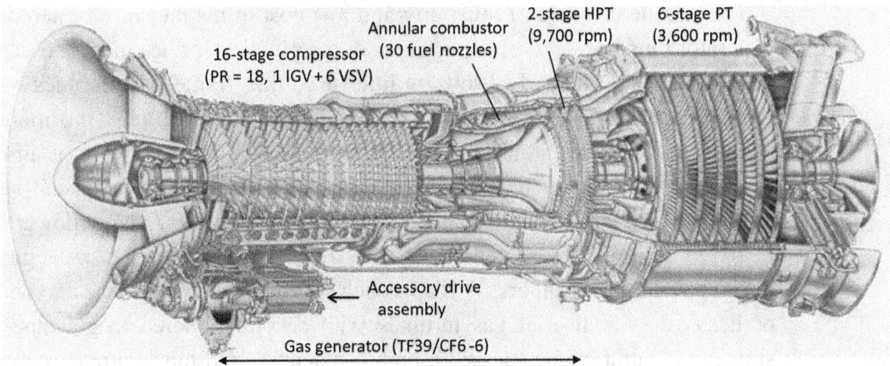

Figure 23.2 GE's LM2500 aeroderivative gas turbine. (courtesy: General Electric) HPT = high-pressure turbine; IGV = inlet guide vane; PT = power turbine; VSV = variable stator vane.

The rapid startup capability of aeroderivative gas turbines, combined with their low capacity factor, is advantageous for making additional use of the synchronous ac generator as a "synchronous condenser." The gas turbine has a synchronous, self-shifting (SSS) clutch between the turbine and generator that allows the turbine to be shut down after the generator is brought up to speed. The generator then acts as a large motor, or *synchronous condenser*, and adds *reactive power* (VAR) to the grid (or absorbs it from the grid). This is discussed in more detail in Chapter 15.

As a result of their low-weight rotors, aeroderivative gas turbines such as GE's LM2500 utilize roller bearings, which consume less lubricating oil (hence simpler/smaller lube oil systems) than journal bearings in larger industrial gas turbines. They are also easier and cheaper to maintain and replace.

23.1 History of Aeroderivatives

At the time of writing (2016–2017), aeroderivative gas turbines are typically installed as self-contained units, ranging from a single unit up to four or five units per site. In the past, however, especially in the 1960s, plant designers came up with very unique configurations combining "jet engines" with large power turbines.

In 1958, Columbia Gulf installed a P&W GG4 aeroderivative unit (based on the P&W J75/JT4 jet engine[3]) with a Cooper-Bessemer power turbine in a transcontinental pipeline. (The "GG" stands for "gas generator.") The unit ran fairly well and more orders followed. Subsequently, major pipeline companies started using aeroderivative packages in increasing numbers from a variety of OEMs (e.g., GE LM2500 and the Rolls-Royce RB-211). Power turbines were typically provided by independent manufacturers (Cooper-Bessemer, Worthington, etc.), but also by the engine OEMs.

[3] J75 is the military version, while JT4 is the commercial one. The former was used in the Republic F-105 Thunderchief and Convair F-106 Delta Dart. The latter was used in the McDonnell Douglas DC-8 and Boeing 707. Some 1,005 out of a total 2,579 were commercial engines.

The excellent record of reliability and low cost of maintenance of aeroderivative gas generator plus power turbine packages in the pipeline service did not escape the notice of utilities. Until the early 1960s, utilities used the "frame" gas turbines with increased firing temperatures for peaking duties of limited durations (e.g., no more than 1,000 hours per year [with number 2 fuel oil as the fuel]). Increased use of air-conditioning resulted in unanticipated power shortages, which led utilities to install aeroderivative gas turbines in larger numbers and capacities. In those days, a large power station could be based on multiple gas generators (i.e., 4, 8, 10, or even more "jet engines") feeding one or two large expanders. This approach was dropped a long time ago with the advent of heavy-duty industrial gas turbines with several hundred megawatts of output in simple and combined cycles with high efficiencies and much simpler configurations.

The breakthrough came after the Great Northeast Blackout of November 9, 1965, which was actually caused by a single faulty relay at a transmission station in Ontario, Canada. Nevertheless, one end result of the event was that the electric utilities throughout the USA were mandated by their regional "Reliability Councils" (e.g., North American Energy Reliability Council [NERC] for the northeast) to increase their system reserve margins by installing a certain percentage of their overall capacity in the form of smaller, localized fast-start generating units, the majority of them with "black start" capability to ensure that large plants and grids could be restarted even in the event of another major outage.[4] This mandate was further fortified by the fact that the summers of 1966 and 1968 saw major heat waves and record peak demands. The result was a wave of aeroderivative gas turbine generator installations, chosen as the fastest and most economical way to meet this mandate.

Some of the stations were truly unique with ingenious arrangements of jet engine gas generators. One example is the Cincinnati Gas and Electric Company's 100-MWe station, where 10 GE J79 aircraft jet engines are clustered circumferentially in a vertical structure to feed a single-stage 1,200-rpm expander built by GE's large steam turbine division [3]. (The unit was nicknamed "bird roaster.") Gas generators could run both on aircraft kerosene or on natural gas. Plant efficiency at rated output was 24 percent with kerosene and 22.7 percent with natural gas. In 1965, it cost $9.5 million or $95/kW to build, which is about $735/kW in 2017 dollars. In comparison, a single LMS100, which can deliver the same output with more than 40 percent efficiency, has a budgetary price of about $400/kW.

In the UK and the USA, such power stations with multiple gas generator-free (power) turbine modules or the clustered gas generator-single free turbine arrangements were exclusively used for peaking duties in the 1960s and 1970s. Multiple-module power plants retained the impressive agility of individual aeroderivative gas turbines (i.e., 8 minutes from "cold iron" to full load). They also had the schedule/cost advantage from one size to another by simply adding or subtracting modules. The larger power turbine of the clustered gas generator arrangement increased the plant startup time due

[4] It was a GE Frame 5 "black start" unit in Southampton, NY, that initiated the restoration of power in Long Island and New York City after the November 1965 blackout (as recalled by Dave Lucier in a *Combined Cycle Journal* article, "The Venerable Frame 5," in the 3Q/2012 issue).

to higher thermal stresses. The relatively poor access for maintenance and severely reduced modularity could offset the economies of scale gained from one large power turbine vis-à-vis multiple smaller ones as well.

The last unconventional aeroderivative power plant arrangement is P&W's (now under Mitsubishi Hitachi Power Systems) FT8 TwinPac™. The FT8 is a 30-MW aeroderivative gas turbine with a P&W JT8D-219 core (deployed in the Boeing 727 and 737). In TwinPac arrangement, a single generator is connected to two FT8 gas turbines from both ends for a total rating of 60 MWe. The new designation for the system is FT8 SWIFTPAC. In some applications, the turbine–generator connection can be via SSS overrunning clutches for improved flexibility (including synchronous condenser application – see Chapter 15).

23.2 Aeroderivative Combined Cycle

Although aeroderivative gas turbines are deployed predominantly in simple-cycle configurations, there are combined-cycle applications as well. This route is usually taken by industrial customers who need their own electric power supply with high reliability and availability (especially in countries with unreliable voltage and frequency stability). A very illustrative example is provided in a paper by Kolp and Levey, which describes a LM2500+ single-shaft combined cycle in Turkey, which entered service 1998 [4]. The plant is capable of generating more than 35 MW at 48.5 percent efficiency. The bottoming cycle is two-pressure, non-reheat with a 9.5-MW (maximum rated power) steam turbine generator and air-cooled condenser. (It is a single-pressure steam turbine with 675 psia-743°F main steam; the LP evaporator is for deaerating the feed water.) The bypass stack between the LM2500+ and the HRSG and the SSS clutch between the steam turbine and the generator (which is "in the middle" between the gas and steam turbines) enables the plant to run in a simple-cycle mode as well. Primary gas turbine fuel is natural gas with naphtha as backup. The SAC combustor of the LM2500+ is water-injected for NOx control (see Section 23.7 for more on aeroderivative gas turbine combustors).

The interesting feature of this small combined-cycle power plant is its ability to stabilize generation frequency via a 12-MW *load bank*. When the national grid frequency and voltage fluctuations reach a level to threaten the operation of the customer facility (a textile plant), the plant automatically separates from the grid. The rejected load is picked up by the load bank so that the plant runs normally while being protected from frequency and voltage fluctuations. The load bank is necessary because it was difficult for this gas turbine (when running in simple-cycle mode) to shed nearly half of its load (i.e., 12 MWe out of about 26 MWe for the LM2500+) and maintain the frequency deviation within the requisite band of ±1 percent (±0.5 Hz for this site) with stable combustion.[5]

[5] Modern H-class gas turbines can do that. See Section 19.5.3 and Figure 19.18.

The exhaust gas flow rate of the LM2500+ is 180 lb/s at 954°F, whose exergy is (from Equation 5.60a)

$$a_{exh} = 0.1961 \cdot 954 - 86.918 = 100.2 \text{ Btu/lb},$$

$$A_{exh} = 180 \text{ lb/s} \times 100.2 \text{ Btu/lb} \times 1.05506 \text{ kW-s/Btu} \sim 19{,}022 \text{ kW}.$$

Assuming 1.5 percent for feed water and condensate pumps, net bottoming-cycle output is (using the steam turbine design point output of 8,763 kWe)

$$\dot{W}_{BC} = 8763 \cdot (1 - 1.5\%) = 8{,}632 \text{ kWe},$$

which translates into a bottoming cycle *exergetic efficiency* of 8,632/19,022 = 45.4%. This is well below the state-of-the-art 3PRH bottoming cycles with advanced-class heavy-duty industrial gas turbines (about 75 percent – see Chapter 17). The famous rule of thumb (i.e., the bottoming-cycle steam turbine output is roughly half of that of the topping-cycle gas turbine) can be modified for larger aeroderivatives (i.e., 25 MWe or higher) by replacing "half" by "a third."

The combined cycle in question cost $23 million in 1998 or about $635/kW installed, which is equivalent to $950/kW in 2017 dollars. Some of the added cost is associated with the load bank and the air-cooled condenser (which is more expensive than water-cooled condensers). Nevertheless, due to their smaller size and low exhaust exergy (requiring more expensive equipment to "squeeze" each kilowatt of electric power out of it), combined cycles with aeroderivative gas turbines are typically not economic for electric power generation. They are more amenable to unique applications, especially those with cogeneration opportunities similar to that in the example covered above.

LM2500+ has a PR of 22.9. Using the isentropic relationship with 93 percent polytropic efficiency and $\gamma = 1.4$, from Equation 5.1,

$$\frac{T_2}{519} = 22.9^{\frac{0.2857}{.93}},$$

we find that T_2 is about 900°F. The nominal turbine inlet temperature (TIT) of LM2500+ is estimated to be 2,250°F so that from Equation 5.3b

$$\bar{T} = \text{METH} = \frac{(2250 - 900)}{\ln\left(\dfrac{2250 + 460}{900 + 460}\right)}$$

(i.e., METH is about 1,500°F). Similarly, with 954°F exhaust temperature, from Equation 5.4,

$$\bar{T} = \text{METL} = \frac{(954 - 59)}{\ln\left(\dfrac{954 + 460}{519}\right)}$$

(i.e., METL is about 435°F). Consequently, the *ideal* Brayton-cycle efficiency for LM2500+ is

$$\eta = 1 - \frac{\text{METL}}{\text{METH}} = 1 - \frac{435 + 460}{1500 + 460} = 54.3\%$$

Since the lower heating value efficiency of LM2500+ is 37 percent, the estimated Carnot or technology factor for this engine is 37/54.3 = 0.68, which is in excellent agreement with the state of the art in heavy-duty industrial gas turbines (about 0.7 on average – see Chapter 6).

23.3 Gas Turbine–Battery Hybrid

In terms of installed capacity, not surprisingly, aeroderivative gas turbines larger than 10 MWe typically constitute less than 20 percent of the worldwide market for *all* gas turbines. In the more focused segment of the market, however (i.e., small gas turbines rated between 18 and 65 MW), aeroderivatives account for two-thirds of the installed capacity.[6] Note that, even outside the USA, GE's LM2500 and LM6000 products hold nearly a 70 percent market share (in 2012).[7] The cited numbers include *all* aeroderivatives with applications ranging from marine propulsion to the oil and gas and pipeline industry to electric power generation (about a third of the total). (In passing, naval applications for ship propulsion are dominated by GE's LM2500 gas turbines.) There is no reason to expect that the future role of aeroderivative gas turbines in electric power generation will differ significantly from their current role: peaking (emergency) power, renewable support, and reactive power in addition to focused industrial power generation with cogeneration (combined with battery storage).

In conjunction with the last item, note GE's "LM6000 Hybrid Electric Gas Turbine (EGT)," which combines an aeroderivative gas turbine with battery storage. The new system was introduced in 2016 to provide 50 MW of spinning reserve (without burning fuel), flexible capacity, and peaking energy, 25 MW of high-quality regulation, and 10 MVAR of reactive voltage support and primary frequency response when not online (i.e., as synchronous condensers). The way the hybrid system works is as follows: a standard LM6000 can start in less than 5 minutes. The battery included with the LM6000 provides electricity for the first 4.5 minutes. So, the combined LM6000 plus battery hybrid EGT will function as one asset and bid into spinning reserve while the gas turbine is switched off. If the request for spinning reserve comes in, the battery will provide electricity for the first 4.5 minutes and the gas turbine will start up in that time and provide seamless electricity after the 4.5 minutes. This is how it delivers "green" 50 MW of spinning reserve without burning fuel. The battery can be charged at a later time when the gas turbine has taken over or, preferably, via excess wind or solar energy, which otherwise has to be curtailed.[8] The first two LM6000 hybrid EGT units went commercial in California in March 2017. Each unit integrates a 10-MW/4.3-MWh

[6] As presented by Mark Axford in Western Turbine Users, Inc.'s (WTUI) 2013 Conference (March 10–13, San Diego, CA).

[7] Ibid.

[8] Provided, of course, regulatory and/or pricing mechanisms are favorable, which may or may not be the case.

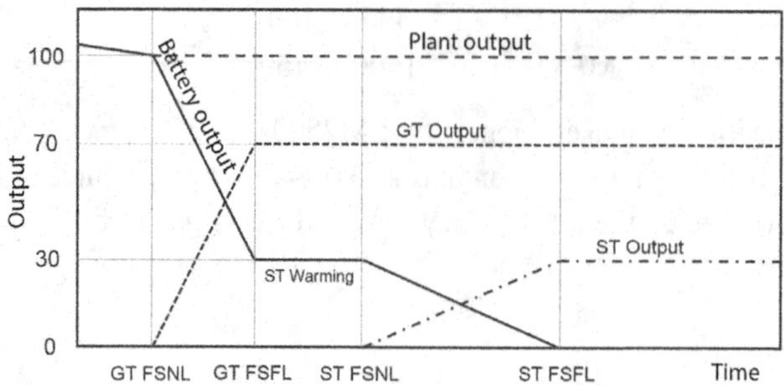

Figure 23.3 Gas turbine (GT) and battery hybrid combined-cycle power plant startup. FSFL = full-speed, full-load; ST: steam turbine.

battery energy storage system with the 50-MW aeroderivative gas turbine. They are owned and operated by Southern California Edison (SCE).

Siemens offers the Hybrid EGT capability under the trade names SIESTART and SIESTORAGE [6]. Unlike GE, Siemens does not limit the concept to simple-cycle aeroderivative gas turbines (at least not on a product basis). How the system operates is conceptually demonstrated in Figure 23.3. When the power plant is dispatched from "cold iron" in the event of an unpredicted generation loss, the battery engages almost instantaneously and provides full power to the grid while powering the load commutated inverter for the gas turbine start. Once the gas turbine rolls up to full-speed, no load (FSNL) and starts its load ramp, the battery contribution goes down in lockstep so that full power is still delivered to the grid. After a warming period, during which the battery contributes to the plant output steadily, the steam turbine rolls up to FSNL and starts its load ramp. This time, the battery contribution goes down in lockstep with the steam turbine load ramp until the steam turbine and, consequently, plant full load are reached. The sequence of events shown in Figure 23.3, minus the steam turbine, would be conceptually the same for the GE Hybrid EGT.

23.4 Not "Derivative" but Close ...

In addition to the "true" aeroderivatives by three major OEMs discussed above, one should also mention two manufacturers of small industrial gas turbines similar in basic architecture to the aeroderivatives. The first one is Solar Turbines (owned by Caterpillar), whose one- and two-shaft small industrial gas turbines, while not deriving literally from a "parent" aircraft engine, can be considered as aeroderivative gas turbines *in spirit*. (The company was founded in 1927 as Prudden-San Diego Airplane Company, but its airplane saga was cut short by the Great Depression. Renamed Solar Aircraft Company in 1929, the company eventually found its niche in small industrial gas turbines.)

Solar's largest electric power generation product is Titan 250, which is a two-shaft gas turbine rated at 21.745 MWe and 38.9 percent efficiency at ISO base load. It comprises a 16-stage axial compressor with a PR of 24, one inlet guide vane (IGV) and five variable stator vanes (VSVs) driven by a 10,500–rpm two-stage HP turbine. It has an annular DLE combustion chamber (using proprietary SoLoNOx™ technology) with 14 injectors. The power turbine is a three-stage design with a maximum speed of 7,000 rpm. Note that Titan, unlike comparable aeroderivatives such as GE's LM2500, has tilting-pad journal bearings. Exhaust gas conditions are 150.4 lb/s and 865°F.

European Gas Turbines (EGT, Ltd.) was created in 1989 from Ruston Gas Turbines, Ltd., and it was the manufacturer of one- and two-shaft small industrial gas turbines similar to those of Solar Turbines. Ruston was the subsidiary of the General Electric Company (GEC) of the UK (from 1968). In 1998, EGT changed its name to Alstom Gas Turbines, Ltd., when GEC-Alsthom changed its name to Alstom (GEC had become part of GEC-Alsthom in 1989). In 2003, Siemens acquired the company from Alstom, and the product line (except the *Tornado*), with new names, is currently offered by Siemens Industrial Turbomachinery, Ltd.:

- Typhoon (SGT-100)
- Tempest (SGT-300)
- Cyclone (SGT-400)

Selected Solar and Siemens Industrial (formerly EGT) products are summarized in Tables 23.2 and 23.3.

In addition to the former EGT product line, Siemens also acquired the former Asea Brown-Boveri small and mid-size gas turbines, i.e.,

- GT10 (SGT-600)
- GTX-100 (SGT-800)

Table 23.2 Selected Solar gas turbines. GG = gas generator; PT = power turbine

		Output (kW)	Efficiency (%)	PR	MEXH (lb/s)	TEXH (°F)	GG rpm	PT rpm
Mercury 50	Single-shaft	4,600	38.5	9.9	39.3	690	NA	15,000
Taurus 70	Single-shaft	7,965	34.3	17.6	59.3	945	NA	11,000
Mars 100	Two-shaft	11,350	32.9	17.7	93.8	905	10,780	8,625/8,570
Titan 250	Two-shaft	21,745	38.9	24.1	150.4	865	10,500	7,000

Table 23.3 Siemens small gas turbines (formerly EGT)

	Output (MWe)	Efficiency (%)	rpm	PR	MEXH (lb/s)	TEXH (°F)
SGT-400	14.3	35.4	9,500	18.9	97.7	1,004
SGT-400	12.9	34.8	9,500	16.8	86.8	1,031
SGT-300	7.9	30.6	14,010	13.7	66.6	1,008
SGT-100	5.4	31.0	17,384	15.6	45.4	988

Siemens later came up with a derivative of SGT-600 – SGT-700 – which has higher output and better efficiency. SGT-600/700 is a two-shaft gas turbine with a gas generator spool comprising an 11-stage compressor (10-stage for SGT-600) and two-stage turbine thermally connected to a free power turbine (two uncooled stages) rotating at 6,500 rpm (SGT-700) or 7,700 rpm (SGT-600). Both gas turbines are equipped with DLE combustors.

A new design deriving from SGT-600/700 is the 40-MWe (40 percent efficient) SGT-750 with 115 kg/s exhaust flow and 468°C (875°F) exhaust temperature. Its key design features are as follows:

- 13-stage compressor (PR of 24) with IGVs and two VSVs (controlled diffusion airfoils – see Chapter 11);
- Can-annular DLE combustion system with eight cans;
- Two-stage compressor turbine (both stages cooled; stage 1 has thermal barrier coating);
- Two-stage (free) *counter-rotating* power turbine (uncooled) – 6,100 rpm in power generation mode (variable speed in mechanical drive application) – with shrouded blades.

The question that arises naturally is: Why the two-shaft configuration? Mid-size gas turbines like SGT-750 (see Figure 23.4) are widely used in large mechanical drive applications such as pipeline compressor drivers. This requires variable-speed capability, which is very difficult to achieve efficiently with a single-shaft design. Consider that the power turbine of SGT-750, which becomes the drive turbine in mechanical applications, rotates at varying speeds between 3,050 and 6,405 rpm as dictated by the driven component (typically a large gas compressor).

Figure 23.4 Siemens SGT-750 industrial gas turbine. (courtesy: Siemens Energy, Inc. © 2018. All Rights Reserved)

23.5 Intercooled Aeroderivative Gas Turbine

It is (or should be) by now clear that the "secret sauce" of aeroderivative gas turbines is their high cycle PR, which leads to high Brayton-cycle efficiency, but limits their size. Obviously, one is tempted to ask: How much higher can one go in cycle PR before hitting the proverbial "brick wall?" The answer depends on many things; of primary concern are available materials and HP combustors with acceptable emissions. There is also the question of designing a machine (which will definitely be quite expensive) with a meaningful output rating to keep the installed cost at an economically justifiable level (in dollars per kilowatt). With that in mind, designers in GE came up with LMS100, a 100–MWe, three-shaft aeroderivative (nearly 35–40 MWe larger than Siemens/Rolls-Royce Trent and LM6000) with a cycle PR of 42 (see Figure 23.5). Based on the key technologies contributing to it, LMS100 can more aptly be called an aeroderivative (LM in its designation; i.e., "Land and Marine") and "frame" hybrid. Its key characteristic feature is the intercooler between the LP and HP compressors.

Compressed air from the LP compressor (LPC) is sent to the intercooler, cooled to a temperature close to that at the inlet of the LPC, and then redelivered to the HP compressor (HPC). The result is a reduction in compressor power consumption for a higher PR and increased mass flow (about 460 lb/s at ISO). Note that fuel pipeline pressure has to be set high enough to obviate the need for a fuel gas booster compressor. This may not always be the case, with a concomitant detrimental impact on net plant output.

LMS100 contains two "free spools" or balanced shafts: one comprising the LPC and intermediate-power turbine (IPT) at 5,500 rpm and the other comprising the HPC and HP turbine (HPT) at 9,500 rpm. The five-stage power turbine (low-pressure turbine [LPT] or power turbine [PT]), which is based on LM6000, drives the synchronous ac generator (at 3,000 or 3,600 rpm). There is no gearbox. For each speed, there is a different PT stage 1 stator nozzle design. Cooling tower heat rejection is about 30,000 Btu/s (about 31.1 MWth) with a fan power consumption of about 150 kWe. Condensate knockout from the (cooled) return air stream is about 1 kg/s or 15 gpm. The TIT is

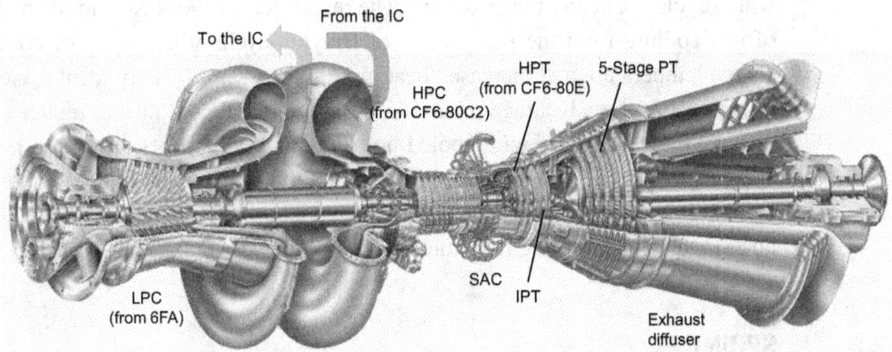

Figure 23.5 GE's LMS100 aeroderivative gas turbine. (courtesy: General Electric) HPC = high-pressure compressor; HPT = high-pressure turbine; IC = intercooler; IPT = intermediate-power turbine; LPC = low-pressure compressor; PT = power turbine; Single annular combustor.

1,400°C (F class). A very detailed description can be found in Reale [5], including plant layout, part load curves, startup times, etc.

Let us calculate the technology factor of LMS100. (A representative, hot ambient [90°F] heat and mass balance (the heat and mass balance [HMB] of LMS100 from Thermoflex is utilized to obtain key cycle data.) This requires a slight modification to the calculation of METL for cycle heat rejection, which takes places in *two different paths*:

1. Cooling of compressed air from 362°F to 100°F in the intercooler;
2. Cooling of exhaust gas (hypothetically, that is, this is an *open* cycle) from 806°F to 90°F.

Thus, expanding on the original Equation 5.4, we obtain

$$\bar{T} = \text{METL} = \frac{(362 - 100) + (806 - 90)}{\ln\left(\dfrac{362 + 460}{100 + 460}\right) + \ln\left(\dfrac{806 + 460}{90 + 460}\right)}.$$

The denominator of the equation above is the sum total of *entropy deltas* for paths 1 and 2; the numerator is the sum total of *enthalpy deltas* for paths 1 and 2. The result is METL = 343°F. Cycle METH is calculated as 1,495°F using the Thermoflex HMB data so that the *ideal* Brayton-cycle efficiency for LMS100 is

$$\eta = 1 - \frac{\text{METL}}{\text{METH}} = 1 - \frac{343 + 460}{1495 + 460} = 58.9\%,$$

Using the net output of 96,825 – 250 = 96,575 kWe (accounting for cooling tower fan and circulator pump power), net efficiency is 42.4 percent and the technology factor for LMS100 is 42.4/58.9 = 0.72. In most cases, due to its very high cycle PR, LMS100 would require a fuel gas booster compressor, which would consume about 1,000–1,200 kWe. This would bring the net efficiency down to 41.9 percent and the technology factor down to 0.71.

Note that the METH calculated for LMS100 is about the same as the METH calculated for LM2500+ within 5°F, even though the TIT for the former is 300°F higher with a cycle PR nearly twice as high. The reason for this goes back to the main drawback of intercooling from the perspective of Brayton-cycle efficiency: low compressor discharge temperature and increased heat input for the same TIT. In ideal cycles, this tends to cancel out the advantage gained in higher net cycle output via reduced compressor power. In actual cycles with cooled hot gas path components, however, a reduction in chargeable air consumption further adds to the benefits of intercooling, which also include higher specific power output via increased airflow. Combined with the 300°F push in TIT, intercooling turns out to be a good cycle feature for LMS100 after all.

23.6 SPRINT™

SPRINT (Spray Inter-Cooled Turbine) is the GE trademark for the patented gas turbine "inlet fogging" technology that is offered with the twin-shaft LM6000 aeroderivative

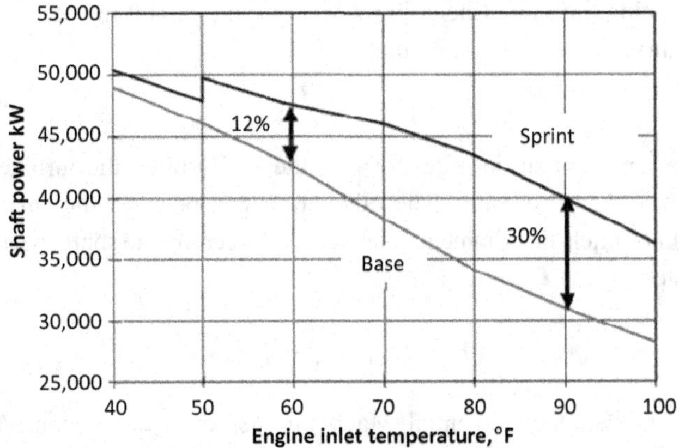

Figure 23.6 SPRINT impact on LM6000 performance [2].

gas turbine. Atomized water is injected at both LPC and HPC inlet plenums. Injection water is atomized by utilizing air bled from the eighth stage of the HPC, which is further compressed by an HP booster compressor. Atomized air is injected into the compressor inlet plenums via spray nozzles arrayed on water manifolds. (Note that additional water injection in the combustor for NOx abatement is still needed.) The performance impact is illustrated in Figure 23.6 (ISO base load, natural gas with water injection to 25 ppm NOx; 5/10 in. H_2O inlet/exit losses).

23.7 Aeroderivative vs. Industrial Gas Turbine

The biggest visible difference between the two gas turbine variants is obviously in size and weight. In the same rating class (say, about 25–50 MWe), industrial gas turbines (i.e., GE's Frame 5 and 6) have a specific weight of about 4–5 short tons (st) per MW output.[9] The cited number is for a flange-to-flange gas turbine excluding the generator (whose size, of course, is independent of the type of its prime mover). For aeroderivative gas turbines (i.e., GE's LM2500 and LM6000), the specific weight is about 0.3–0.5 st per MW output (i.e., about a tenth of their land-based cousins). Up to about 30 MWe, they can be shipped in trailers[10] to a site and erected rapidly (in about 10 days for a pre-commissioned package) for emergency power generation purposes.

This is not surprising, of course, because the main building block of an aeroderivative gas turbine is an aircraft jet engine, which is deliberately designed to be small in size and light in weight. The contributing factors are a high cycle PR and a high shaft speed.

[9] One short ton is 2,000 lb or 0.907185 metric tons (hence the reason for the moniker "short").

[10] At 30 MW, typically in two trailers (e.g., P&W MOBILEPAC), one containing the gas turbine, electric generator, exhaust collector, diffuser, and engine lube oil system, and the other containing the switchgear, control system, operation panel, protective relays, batteries and charger, motor control center, and the hydraulic start package.

Everything else being the same (including working fluid mass flow rate), starting from the mass continuity, it is easy to show that

$$D_m \propto PR^{-2},$$

where D_m is the mean (pitch-line) diameter of the HP turbine (the turbine of the gas generator, which is the jet engine). The other trade-off, between the shaft speed and compressor mean or pitch-line diameter, can be easily seen by the shaft speed formula at the mean diameter

$$U = \omega r_m = \frac{\pi N}{30} r_m.$$

In order to ensure that the tangential Mach number does not become supersonic, increasing speed should be balanced by decreasing r_m.

In terms of off-design performance, there is not a big difference between the two types of gas turbine. They both suffer similarly at high ambient temperatures and at high altitudes (i.e., reduced mass flow rate and output). They are also similar in heat rate lapse with decreasing load. Although the same basic principle of component matching can be applied to the analysis of what is happening inside the "guts" of the machine, due to the presence of one or even two balanced "spools" with a power turbine or power spool, this becomes very tedious to do even on a simplified basis. The reader can refer to Saravanamuttoo et al. for an excellent treatment of this subject (especially the section on matching procedures for twin-spool engines).

Cold and hot ambient temperature performance and part-load heat rate lapse (at ISO conditions) of three gas turbine types are compared. The selected gas turbines are as follows:

- A typical multi-shaft aeroderivative (GE's LM-6000);
- A typical single-shaft E–class industrial gas turbine;
- GE's intercooled LMS100.

The part-load heat rate curves are shown in Figure 23.7. Output and heat rate curves at varying ambient temperatures are shown in Figures 23.8 and 23.9, respectively. (No inlet evaporative cooling or chilling is used for hot-day power augmentation.) Thermoflow's GTPRO software is used for the calculations. The dotted line in Figure 23.8 is the effect of inlet evaporative cooling on LMS100.

There is no big difference in part-load performance of the three gas turbines. The intercooled, three-spool LMS100 shows a pronounced difference from the other two in terms of power output at low ambient temperatures. The reason for this is not provided in myriad papers and articles published by GE in the open literature. The most likely driver of the small dip in power output below ISO is perhaps the shaft speed limit of the free HP spool (the gas generator), which is GE's aircraft jet engine CF6-80C2/80E. Typically, the speed of the free HP spool in aeroderivative gas turbines such as LM6000 goes up at low ambient temperatures and goes down at high ambient temperatures. The derivative is a roughly 0.3 percent drop in shaft speed with each 10°F rise in ambient temperature.

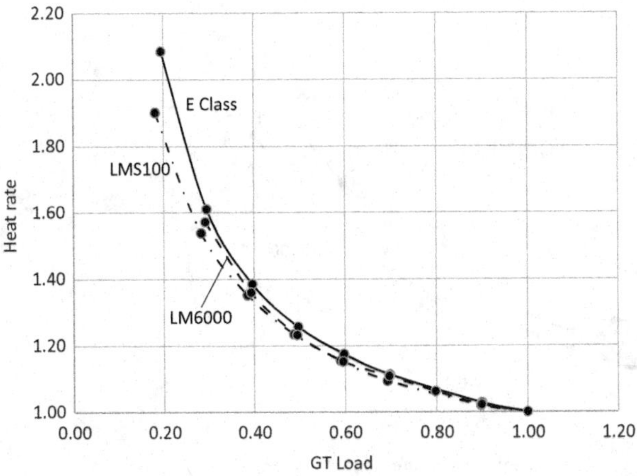

Figure 23.7 Part-load heat rate of aeroderivative and industrial gas turbines (GTs).

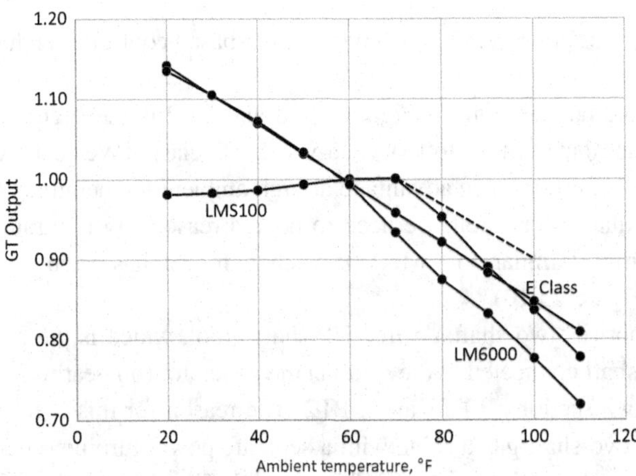

Figure 23.8 Output of aeroderivative and industrial gas turbines (GTs) as a function of ambient temperature.

At high ambient temperatures, on the other hand, at least until 90°F, LMS100 has a better output lapse than the E class and LM6000. The reason for this should be tied to intercooling, which helps maintain low HPC discharge temperature (the source of HP turbine cooling air). Consequently, the LMS100 firing temperature can be held at ISO levels at high-ambient temperature operation. This is in contrast to other gas turbines *without* intercoolers, especially the aeroderivatives with uncooled LPT or PT with maximum allowable inlet temperature limits, which typically reduce their firing temperatures in lockstep with increasing ambient temperature. In other words, if the ISO firing temperature is T°F, at, say, 90°F ambient, it is (T – 31)°F. The reduction factor is the difference between ambient temperatures (i.e., 90 – 59 = 31). Note that this should be considered to be a reasonable "rule of thumb." Different OEMs and/or different

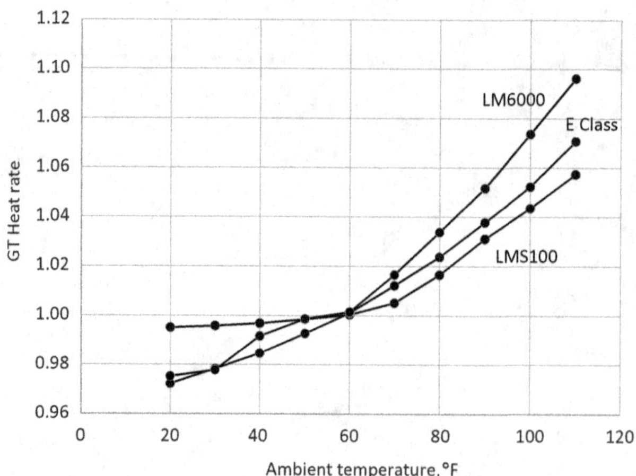

Figure 23.9 Heat rate of aeroderivative and industrial gas turbines (GTs) as a function of ambient temperature.

models, especially those of recent vintage with model-based controls, can have different characteristics.

The other notable takeaway from Figures 23.8 and 23.9 is the higher hot ambient output and heat rate lapse of LM6000 vis-à-vis the E class. (We just explained the source of LMS100's performance advantage at high ambient temperatures.) In order to explain this particular observation, we need to have a reasonably accurate model of a two-spool gas turbine similar to LM6000. Such a model has been constructed in Thermoflex (see Figure 23.10).

The model comprises two shafts: a free HP shaft/spool, which is the gas generator, and a power (LP) shaft connected to the grid via the generator (no gearbox). The LPT of the LP shaft has two sections: LPT1 and LPT2. The reason for this is to maintain the ability to model a two-shaft gas turbine with a separate power turbine (i.e., LPT2). For simplicity, hot gas path cooling of the HPT is ignored. The model is *loosely* calibrated to GE's LM6000 aeroderivative gas turbine in the "design" mode and run in "off-design" mode. Grid frequency is assumed to be 60 Hz (i.e., the LP shaft speed is 3,600 rpm; HP shaft speed is assumed to be 9,500 rpm). Each compressor's operation is controlled by the default axial compressor map in Thermoflex. Component matching is automatically achieved by the fixed swallowing capacities of the HPT and LPT (i.e., fixed nozzle areas). The LPC sets the airflow based on the pressure balance, whereas the HPC uses the airflow to set its inlet pressure (i.e., its PR) based on the downstream pressure signals created by the HPT and LPT inlet nozzle areas and gas flows. Combustor exit temperature (HPT inlet temperature; i.e., the TIT) is user-specified. Shaft balance for the free HP shaft is achieved by varying the shaft speed.

An ambient sweep done with the model showed excellent agreement with the "real" machine in terms of power output. The LM6000 output lapse displayed in Figure 23.8 was almost exactly replicated. In terms of heat rate, the agreement was excellent below ISO, but above it, the model essentially replicated the heat rate lapse displayed by the

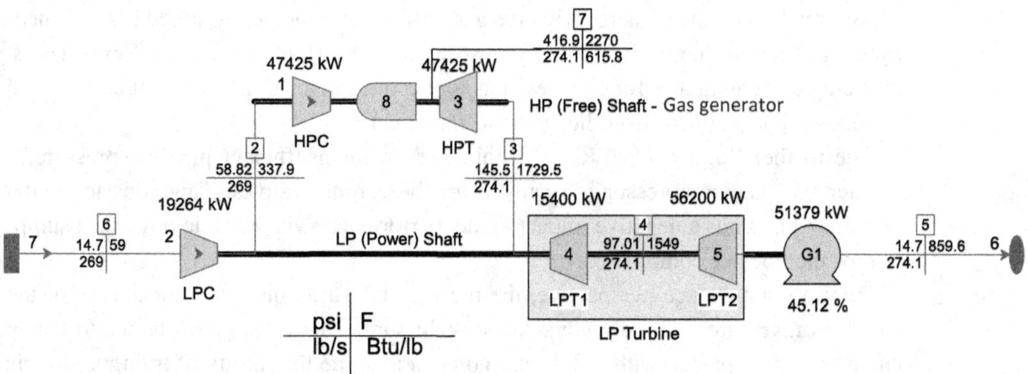

Figure 23.10 Thermoflex model of a two-shaft aeroderivative gas turbine.

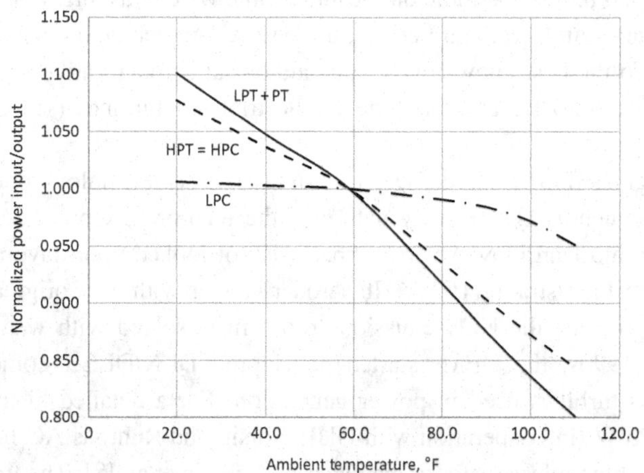

Figure 23.11 HPT and LPT power outputs of the turbine in Figure 23.10.

E-class gas turbine in Figure 23.9. In other words, it was too optimistic vis-à-vis the "real" LM6000 heat rate lapse at hot ambient temperatures. Nevertheless, for our purposes herein, the model is deemed acceptable.

Power output and consumption (input) of the HPT and LPT and compressors are shown in Figure 23.11. They explain why the output lapse of a two-shaft aeroderivative gas turbine is more pronounced vis-à-vis an industrial gas turbine. The LPT, which is the turbine of the power shaft, loses output much faster than the reduction in power consumption of the LPC on the same shaft. At 110°F, the reduction in LPT power output is 20 percent (referenced to ISO) vis-à-vis only a 5 percent reduction in the LPC power consumption.

Several other aeroderivative versus industrial gas turbine comparison items are touched upon briefly below. (For a quite exhaustive comparison, the reader is referred to the paper by Perkavec [7] – unfortunately, only available in German.)

As already mentioned, aeroderivative gas turbines are not ideally suited to combined-cycle application due to their low exhaust temperature (high cycle PR). Nevertheless, they are good candidates for cogeneration applications in a varying range of commercial and industrial facilities (from hospitals to factories).

Due to their high cycle PR, in certain sites with insufficient pipeline pressure, a booster fuel gas compressor is requisite for the aeroderivatives. This adds to capital investment and has a negative impact on net performance via parasitic power consumption of the booster compressor.

From a maintenance perspective, the modular construction and compactness of the aeroderivatives are definitely advantageous. In some cases, the entire flange-to-flange engine can be replaced within 2 days. (You can imagine the futility of trying to do this with an H–class behemoth at about 400 tons.) In some cases, only a part of the engine – say, the gas generator – can be removed and replaced in the same amount of time.

Like their industrial brethren, aeroderivatives are also prone to fouling, more in some sites than others, and require periodic online and offline water washing. For specific hot gas path maintenance and overhaul periods, the best recommendation is to consult the OEM literature. To the best knowledge of the author, an aeroderivative counterpart of the venerable GER-3620 for the heavy-duty industrial gas turbines (see Chapter 16) does not exist.

Aircraft jet engines do not use the Dry-Low-NOx (DLN) technology due to safety-based reliability concerns. The finicky nature of the lean-premix process, which can easily degenerate into lean blowout or flashback, is not looked upon favorably by the Federal Aviation Administration (FAA). In aeroderivatives with their original diffusion combustion components, the NOx emissions problem is solved with water or steam injection. In the 1990s, the OEMs started developing DLN/DLE[11] combustors for aeroderivative gas turbines used in power generation. For a detailed history of such an endeavor by P&W (in cooperation with GHH Borsig and Ruhrgas AG in Germany) for the FT8 gas turbine, please refer to the paper by Schlein et al. [8]. The first six lean-premix DLE combustor-equipped LM6000 gas turbines were operational by the end of 1995 when DLE combustor development for LM2500 started. (Due to the limited room available in the compact gas generator, simple replacement of existing cans with DLN combustor cans developed for the heavy-duty industrial gas turbines and squeezing them into the annular combustor space was not an option.) The reason for choosing LM6000 as the first engine to be fitted with DLE combustors was its high cycle PR (about 30) and resulting high pressure and temperature (about 1,000°F) at the combustor inlet. Detailed information on GE's DLE combustor development for aeroderivative gas turbines can be found in a series of excellent papers published in the 1990s [9–11].

Firstly, Leonard and Stegmaier demonstrated that the conventional thinking of high pressure-ergo-high NOx did not apply to well-premixed systems below 1,900K (~3,000°F) [9]. Furthermore, unlike in a diffusion combustor, high combustor inlet temperature did not automatically lead to high flame temperature (thus, high NOx)

[11] DLE is the GE name for DLN in aeroderivative gas turbines.

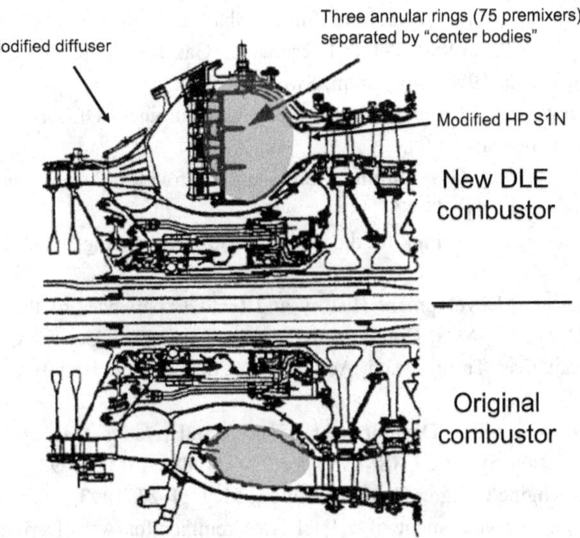

Figure 23.12 Comparison of old (diffusion) and new (DLE) combustors. Figure reprinted from [9] with permission from the American Society of Mechanical Engineers

because in a premixed combustor the latter can be controlled independently of the former [9]. Thus, the high PR and compressor discharge temperature of the LM6000 (or any aeroderivative, for that matter) was not an inherent disadvantage. One key requirement was to increase the combustor volume in order to increase the residence time to achieve low CO and unburned hydrocarbon emissions without adversely affecting NOx emissions (as long as the flame temperature was below 1,900K). Due to the limited axial space inside the existing frame, this required a "bulbous" combustor (see Figure 23.12). Accommodation of the large-volume DLE combustor required modifications to the combustor back-frame (i.e., a new diffuser) and HP S1N, which are described in detail in the cited references. One challenge to overcome was cooling the heat shields and center bodies with a limited cooling air budget, which was solved by cooling those parts from the backside convectively. Similar large-volume combustors were eventually developed for the other "LM" products.

References

1. Horlock, J. H., Aero-Engine Derivative Gas Turbines for Power Generation: Thermodynamic and Economic Perspectives, *Journal of Engineering for Gas Turbines and Power*, **119** (1997), 119–123.
2. Badeer, G. H., "GE Aeroderivative Gas Turbines – Design and Operating Features," GER-3695E, GE Power Systems, 2000.
3. Yeager, B. J., Clark, G. W., "A 100,000-kW Reserve Power Plant for the Cincinnatti Gas & Electric Company," Proceedings of the American Power Conference, March 26–28, 1963, Chicago, IL.

4. Kolp, D. A., Levey, C. E., "LM2500+ Single Shaft Combined Cycle with Frequency Stabilization," ASME Paper 98-GT-418, International Gas Turbine & Aeroengine Congress & Exhibition, June 2–5, 1998, Stockholm, Sweden.
5. Reale, M. J., "New High Efficiency Simple Cycle Gas Turbine – GE's LMS100™," GER-4222A, GE Power Systems, 2004.
6. Mieckowski, C., "Battery-Gas Turbine Combination Provides Power Plant Flexibility," *POWER*, February 2018, pp. 44–45.
7. Perkavec, M., "Gasturbinentechnik für die stationäre Stromerzeugung," BWK Bd. 53 (2001), pp. 54–58.
8. Schlein, B. C. et al., "Development History and Field Experiences of the First FT8 Gas Turbine with Dry Low NOx Combustion System," ASME Paper 99-GT-241, ASME 1999 International Gas Turbine and Aeroengine Congress and Exhibition, June 7–10, 1999, Indianapolis, IN.
9. Leonard, G., Stegmaier, J., "Development of an Aeroderivative Gas Turbine Dry Low Emissions Combustion System," ASME Paper 93-GT-288, ASME 1993 International Gas Turbine and Aeroengine Congress and Exposition, May 24–27, 1993, Cincinnatti, OH.
10. Joshi, N. D. et al., "Development of a Fuel Air Premixer for Aero-Derivative Dry Low Emissions Combustors," ASME Paper 94-GT-253, ASME 1994 International Gas Turbine and Aeroengine Congress and Exposition, June 13–16, 1994, The Hague, the Netherlands.
11. Joshi, N. D. et al., "Dry Low Emissions Combustor Development," ASME Paper 98-GT-310, ASME 1998 International Gas Turbine and Aeroengine Congress and Exhibition, June 2–5, 1998, Stockholm, Sweden.

24 Epilogue

Every book is a reflection of its times (i.e., the exact point in history when its author penned it). This is even more apt for non-fiction books dealing with highly technical subjects. There is a tendency of referring to certain (indisputably great) works as "timeless," which, at least in this author's opinion, is a misnomer. Every book, no matter how great, has its time of practical usefulness, beyond which it is merely of historical importance at best. This book is no different either. It reflects the state of the art in stationary gas turbine technology for electric power generation at the end of the first two decades of the twenty-first century. It is the author's hope that it will be of *some* value to practitioners in the field and students in graduate school until the end of the first half of the present century. This would be a significant feat in and of itself.

In order to improve the likelihood of this hoped-for outcome, the approach undertaken in the first 23 chapters has been as follows:

- Providing an exhaustive look at the history of gas turbines, with the guiding light of the second law of thermodynamics;
- Establishing the link between historical precedents and modern technology features as much as possible;
- Using the second law of thermodynamics to evaluate the state of the art in order to
    - Pinpoint the level of technology precisely;
    - Establish the "room" left for further improvement unequivocally;
- Introducing simple quantitative methods to critically analyze and evaluate most performance and operability problems to be encountered for decades to come.

For example, as long as gas turbines are used for electric power generation, they will have to be started and stopped quite frequently. It does not matter whether the year is 2017 or 2071. Basic principles, rules of thumb, and other quantitative/qualitative information and/or tools introduced in Chapter 19 will still be helpful to gain a handle on such transient events.

Even more importantly, as long as natural gas is available and gas turbines are allowed by regulatory agencies to burn it to generate electric power, there will be no escape from the "fangs" of the second law. It does not matter whether the year is 2017 or 2071 or 2171. The limits set by the combustion stoichiometry, Carnot engine, and cost-effectively achievable technology factor are indeed "timeless."

Talk has been, is, and will be cheap. One will continue to hear bogus claims about this or that "paradigm-shifting" technology and its magical outcomes. The analytical tools provided in this book (along with "real-life," quantitatively elaborate examples) are hopefully adequate to enable the reader to separate marketing hyperbole from what is *cost-effectively and reliably* possible, today and tomorrow.

To the reader who has read most of the preceding chapters, it should be clear that further improvement lies in two key areas: namely, hot gas path (HGP) materials and coatings. But this has been the case from the dawn of the "jet age." Engineers of all nations involved in jet engine development were fully aware of the need for materials that can withstand higher and higher turbine inlet temperatures to achieve higher and higher thermal efficiencies. This is exactly why a modern gas turbine jet engine is not too different from a Jumo-004 in major design aspects. By far the greatest share of the technology evolution (described in detail in Chapter 4) is attributable to the development of superalloys (including casting techniques) and thermal barrier coatings.

The concomitant hurdle to overcome is related to low-emission, stable combustion at ever-higher temperatures made achievable by advanced materials and/or coatings (say, ceramic matrix composites). It is certainly true that, when push comes to shove, alleviation of NOx and CO emissions can be shifted to post-combustion cleanup technologies (e.g., selective catalytic reduction in the heat recovery steam generator [HRSG]). Another possibility (already under investigation) is exhaust gas recirculation. But these technologies add significantly to the system complexity (with ramifications on reliability, availability, and maintainability) and cost so that they may not be feasible beyond a certain point. In any event, requisite tools to evaluate any such future prospects, both technically and economically, are provided in this book. It is hoped that the reader will make good use of them.

At the time of writing, state-of-the-art gas turbine simple-cycle efficiencies are comfortably above 40 percent net (see Figure 24.1). How much better they can get (and how quickly they can get there) is a question whose answer has been hinted at in Chapter 4 (e.g., Figure 4.9) and Chapter 6 (e.g., Figure 6.1). It is left to the reader to draw his or her conclusions.

At the time of writing, state-of-the-art gas turbine combined-cycle efficiencies are comfortably above 60 percent net *under favorable site ambient and loading conditions with unfired HRSGs*. How much better they can get (and how quickly they can get there) is a question whose answer has also been hinted at in Chapter 4 (e.g., Figure 4.10) and Chapter 6 (e.g., Figure 6.3). Once again, it is left to the reader to draw his or her conclusions. However, one last nugget of information is provided to assist the reader in this endeavor.

The chart in Figure 24.2 illustrates the evolution of the gas turbine combined-cycle (GTCC) technology as represented by the technology factor (in case you have forgotten, it is the ratio of published GTCC efficiencies to the theoretical maximum for the same turbine inlet temperature [TIT] and pressure ratio [PR]). Data points are from trade publications such as the annual *Turbomachinery International Handbook* (diamonds). The sole triangle represents original equipment manufacturer (OEM) data culled from several published sources.

The ideal combined-cycle efficiency can be calculated using the following formula (for convenience – otherwise it is not so difficult to calculate it "exactly" – see Chapters 5 and 6):

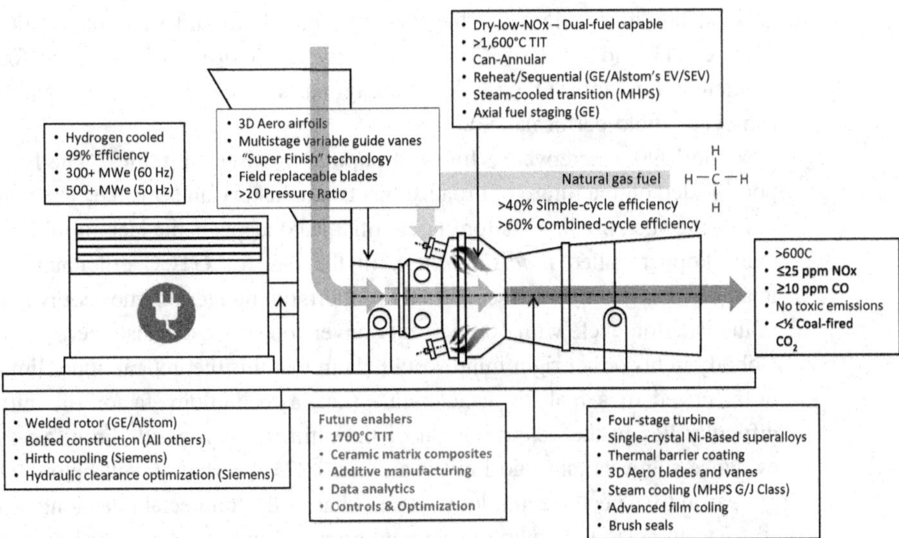

Figure 24.1 Gas turbine state of the art and future enablers. GE = General Electric; MHPS = Mitsubishi Hitachi Power Systems; TIT = turbine inlet temperature.

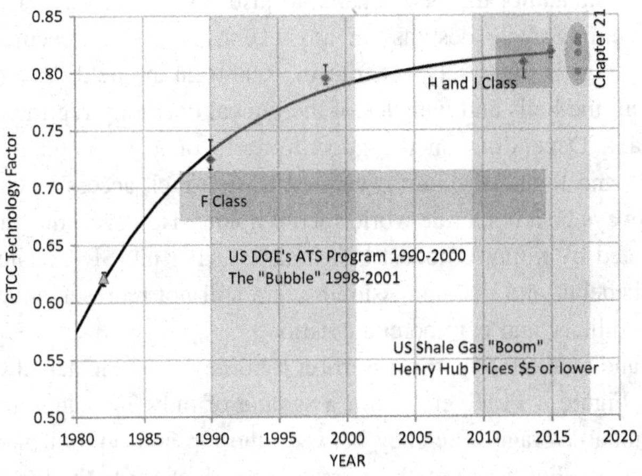

Figure 24.2 Evolution of GTCC technology. ATS = Advanced Turbine Systems; DOE = Department of Energy.

$$\eta_{CC,\,max} = 1 - PR^{-0.227} \frac{\ln\left(\dfrac{\tau_x}{PR^{0.0561}}\right)}{\tau_x - PR^{0.0561}}.$$

In the equation above, τ_x is the gas turbine exhaust temperature (the most readily available piece of information, used as a proxy for TIT) normalized by dividing by T_1 (59°F for ISO levels). The trend shown in Figure 24.2 speaks for itself and the sudden step change in OEM-claimed combined-cycle efficiencies (in 2017) for their latest-and-greatest products is inexplicable to the author. It should be pointed out that, with a 1700°C TIT and a PR of 25, a technology factor of 0.85 is required for 65 percent efficiency (on an ISO base load rating basis – *not* consistently and comfortably achievable field performance).

In the 1990s, there were a flurry of popular treatises with titles "The End of . . ." that pontificated on the future of broad subjects like history and science. The main weakness in making such predictions for unquantifiable concepts is the lack of inherent testability (Karl Popper called it *falsifiability*). In the case of GTCC efficiency, "the end" is unambiguously set by the second law of thermodynamics: Carnot-equivalent efficiency of the Brayton cycle with 100 percent conversion of the exhaust exergy to useful work. Nobody in his or her right mind would claim that this theoretical upper limit can indeed be achieved in a realistic heat engine (i.e., a technology factor of unity). The real difficulty lies in the determination of the ultimate, *asymptotic* limit. So far, available evidence suggests that said limit is around 0.825 – certainly well below 0.85. There is no reason to expect a technology leap to change this appreciably as long as the industry sticks to the classic Brayton-cycle configuration and the brute force approach of ever-higher TITs. Incremental improvements are unlikely to be more successful than Achilles trying to catch up with Zeno's turtle.

Yet, all of a sudden, in 2017, the OEMs seem to have found a way to come very close to the 0.85 bogey. The author has been unable to justify those claims with reasonable, cost-effective bottoming-cycle designs. In any event, excessive speculation on the future of sticker ratings is ultimately not fruitful. Reality in the field demands a metric that accounts for all the bells and whistles of the typical operating regime of a modern GTCC power plant. Depending on the good fortunes of a given plant site and an owner's economic and financial criteria (very expensive fuel, access to cheap capital, etc.), there will always be a bona fide world-record holder. However, the "true" state of the art is represented by a more modest benchmark. (After all, SR-71 and Concorde, decades after their debut and decommissioning, are still not representative of today's state of the art in military and commercial aviation.)

Quantitatively and qualitatively, this assertion is borne out by the actual GTCC plant data presented in Figure 4.11. Even within a sample of only 20 plants, the spread in field-recorded, annual-average efficiency between the top and bottom plants is about three percentage points. Furthermore, the average is much closer to the lower end of the band (i.e., only a few plants are truly the "best of the best"). Nevertheless, however one might look at it, the bottom line is unmistakable: over the last decade, in the largest 60-Hz market in the world, the state of the art in GTCC, as represented by the "best of the best," hovered within 56–57 percent (lower heating value) with no clear upward trend.[1] On the other hand, such an upward trend is indeed clearly present for the larger sample of 20 units (as measured by the sample mean, which only recently hit

[1] While working on this book, in December 2017, the author checked the EIA-923 data for 2017 (until November of that year). The best GTCC (with a brand new H/J–class gas turbine) clocked at 57.46 percent. In the June 2018 issue of *Power Engineering* magazine, the top three GTCC plants in heat rate were listed at 58, 57.33, and 57.13 percent (net lower heating value efficiency) for the year 2017.

55 percent). It is not clear, however, whether this is a direct result of new technology creeping in or higher load factors (or, more likely, a combination of both).

It thus seems that the industry needs a well-defined mean effective load factor and corresponding efficiency, in addition to the traditional ISO base load rating performance, to represent the true capability of GTCC products. ISO base load rating is a "sticker performance" that is rarely (if ever) achieved in a real power plant. Site ambient and loading conditions, recoverable and unrecoverable degradation, start–stop cycles, load ramps, and planned or unplanned outages all contribute to a performance much below that at *new and clean* conditions – let alone the rated performance advertised in trade publications. General Electric (GE) recognized this and came up with a definition to account for those factors, *FlexEfficiency*, which is the ratio of "profitable annual megawatt-hours" to annual fuel consumption. In that sense, it is essentially the same as the EIA-923 efficiencies in Figure 4.14. Using a predefined operating profile (about 200 starts/year and a mix of base load, part load, and minimum turndown hours), GE calculated the typical advanced combined-cycle power plant FlexEfficiency as about 56 percent (or 58.5 percent for the FlexEfficiency 50 GTCC with a base load efficiency of 61 percent). Indeed, this number is in line with the upper end of the EIA-923 data in Figure 4.11.

To estimate the *load factor*, EIA-923 averages are normalized using the technology curve in Figure 4.10, which is described by the formula

$$\eta = 63.2 \cdot \left(1 - 0.315 \cdot e^{-\left(\frac{Y-1980}{16.7}\right)}\right),$$

after subtracting one percentage point to account for the *commercial* performance. The results are shown in Figure 24.3. The EIA-923 efficiencies fall within a band of 0.919–0.943, which implies a load factor range of 58–75 percent. (Note that the GE FlexEfficiency 50 value is 58.5/61 = 0.959, which corresponds to a load factor of 76 percent on the curve labeled "NEWER.") The two curves in Figure 24.3 are representative of the advanced, *newer* F- (or H-) and *older* F-class technologies. Ambient effects are ignored. (Unless a power plant is run exclusively in a certain part of the year [e.g., in summer], the quasi-sinusoidal annual distribution of the ambient temperature ensures that the load-weighted average is not too different from the ISO value – only about 4–5°F higher even for a site in southern Europe.)

A conceptual depiction of the operating regime of a modern GTCC is shown in Figure 24.4. The average load factor is the plant load, at which the plant should run at *steady state* over the same time period to generate the same megawatt-hours (represented by the shaded rectangle in Figure 24.4). In mathematical terms, it is a *mean effective average* value. In a 2011 ASME journal paper, Gülen and Mazumder calculated the average load factor as 74.9–77 percent when taking into account all factors that impact the performance of a GTCC power plant.[2]

The main advantage of mean effective load/efficiency as a measure of the GTCC technology is its realism. It not only corresponds to the actual situation in the field; it

[2] [2] in Chapter 20.

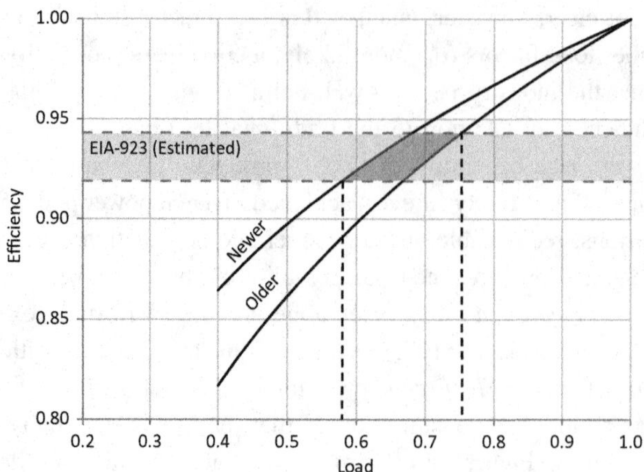

Figure 24.3 Generic GTCC part-load performance (normalized).

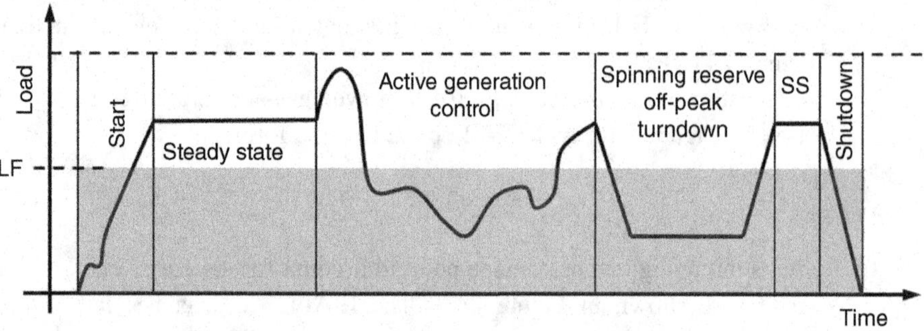

Figure 24.4 Typical operating regime of a modern GTCC. LF = load factor; SS = steady-state.

also lends itself to more meaningful future projections with a lot of room to grow (with abundant engineering innovation) without chasing after (ultimately) stoichiometric combustion and designing quasi-Carnot engines.

24.1 Brayton-Cycle Variations

As discussed at great length in several chapters of the book, there are three variants of the basic Brayton cycle that found (albeit limited) commercial deployment in the realm of electric power generation in the last quarter of the twentieth century:

- Closed-loop steam cooling (H and G/J)
- Reheat (sequential) combustion
- Regeneration and/or intercooling

Closed-loop steam cooling, especially the "H" version (i.e., GE's H-System) with the first two turbine stages fully steam-cooled (apart from bucket platforms and wheel spaces), had a brief commercial life with only six units in the field. Although the technology was quite successful in terms of field performance (all six are operational at the time of writing), it is a very safe bet that it is dead in terms of any future applications. The main reason for this is the excessive complexity of the alloy piping system and single-crystal HGP materials, which adversely impact not only the initial investment, but the lifetime operations and maintenance (O&M) expenditures as well. (A major outage for an H-System gas turbine can take up to a month.)

Even though the H-System was well ahead of its time in terms of firing temperature (due to the much lower temperature drop across the first-stage stator vanes), subsequent advances in metallurgy, coatings, and film-cooling techniques obviated the need for the complex plumbing requisite for closed-loop cooling.

In contrast to GE, Mitsubishi Heavy Industries (MHI)/Mitsubishi Hitachi Power Systems (MHPS) stuck to a limited version of closed-loop steam cooling (combustor liner and transition piece mainly plus the first-stage turbine ring) and had a quite successful run with their G–class gas turbines. The technology is still offered by the OEM, both in G- and J-class gas turbines. Nevertheless, recently, MHPS has been putting a bigger emphasis on their air-cooled GAC and JAC gas turbines with enhanced air cooling replacing closed-loop steam cooling.

After a quite rocky period of field introduction, Asea Brown-Boveri/Alstom GT24/26 sequential combustion gas turbines had a reasonably successful run, especially in the 50-Hz market. After the GE acquisition of Alstom, the 60-Hz version, GT24, ended up with GE, whereas the 50-Hz version, GT26, was handed over to Ansaldo Energia. At the time of writing, to the best knowledge of the author, although there has been no specific acknowledgment by GE, it is almost a foregone conclusion that there will be no new GT24 units on offer.[3] (The technology could never overcome its bad reputation in the 60-Hz market stemming from the early field problems.) Ansaldo Energia is putting its focus on the new GT36, but, at least for the time being, the OEM will continue to offer GT36 to new customers.[4]

The Achilles' heel of GT24/26 is similar to that of GE's H-System (and to a lesser extent to MHPS G/J with steam-cooled parts). New technologies such as GE's and Alstom/Ansaldo's staged fuel injection combustors deliver almost the same effect (granted, at much higher TITs), but within a much simpler machine architecture. Whether the technology will survive deep into the twenty-first century remains in question. The author is of the opinion that, at least from a purely thermodynamic perspective, there is merit to a steam-cooled (limited to the stator vanes of the high-pressure [HP] turbine and the first stage of the low-pressure [LP] turbine) reheat gas turbine.[5] But it is extremely unlikely that this will become a reality.

[3] However, as indicated on their services website, GE offers upgrades for both GT24 and GT26.

[4] On March 11, 2016, Ansaldo Energia Switzerland announced in a news release that it had received two contracts for eight new GT26 gas turbine generator sets and other equipment for combined-cycle duty worth €600 million split for two separate GTCC power plants in the Sultanate of Oman.

[5] See [11] in Chapter 21.

Recuperation is harmful to combined-cycle efficiency (see the discussion in Section 8.5.1). It has no merit from a thermodynamic perspective, nor from an economic one. Its combination with intercooling is even worse. These two technologies are condemned to textbook pages for eternity as far as large gas turbines for electric power generation are concerned.

Intercooling (without recuperation) definitely makes sense for aeroderivative gas turbines with very high cycle pressure ratios. GE's LMS100 has been a notable success (see Section 23.5). It is, however, neither cheap nor simple. Whether there will be other large aeroderivative gas turbines (rated at 100 MWe or higher) from GE or other OEMs (most likely Siemens, the new owner of Rolls-Royce technology) remains to be seen.

A fourth Brayton-cycle variant, which has been around since the early 1950s – on paper that is – is constant-volume combustion. Personally, the author fully recognizes its significant potential to increase cycle thermal efficiency [3]. But practical difficulties of achieving a good approximation of the "true" constant-volume combustion (akin to that taking place in the cylinder of a reciprocating internal combustion engine) inside the framework of a steady-state, steady-flow machine make the prospects of a commercial product very dim. A significant amount of research was done on "pulse(d) detonation combustion" in the early 2000s. It looks like the major researchers in the field, both in industry and in academia, as well as government labs, have given up on it. If it ever comes to fruition, it will most likely be in the form of a "rotating pulse detonation" combustor. The interested reader should refer to the Gülen paper cited above and the references listed therein.

24.2 Other Future Prospects

The preceding section should have made it clear that, in all likelihood, the gas turbine of 2071, *if gas turbines are still around for electric power generation at that time*, will not be much different from the gas turbine of 2017 in basic design and architecture. If you find this hard to believe, please revisit Chapter 4 and compare what we have in 2017 with what we had in the 1940s. There may of course be some "variations on the theme." One example is the "integrated combustor and vane" concept, which proposes combining the (typically rectangular) exit of the combustor can transition piece in can-annular configurations with stage 1 stator nozzle vanes into a seamless whole [1]. The goal is to eliminate the detrimental effect of vortices shed from the combustor transition piece exit walls on the stator vane leading edge film cooling "by periodically removing the coolant flow from the leading edge surface" (the hottest part in the hot gas expansion path). Using numerical analysis, the authors predicted up to a 25 percent saving in cooling flow and also ventured that the simplified design could lead to lower manufacturing costs as well.

There is one gas turbine power plant configuration or application that deserves mentioning: cogeneration (or "combined heat and power" as it is called in Europe). *Cogeneration* is nothing new, of course, and has been widely used in Europe, although not as much in the USA. Consequently, no ink will be wasted herein on the cogeneration basics. The only point to be made is that, if distributed generation becomes the norm in the future in lieu of large central power stations, cogeneration makes a lot of economic sense – especially, with highly efficient and compact aeroderivative gas turbines. As discussed in the book, these gas turbines are not suitable for efficient and cost-effective combined-cycle applications due to their low exhaust temperatures (caused by high cycle pressure ratios). But they are perfectly suitable to the generation of low-pressure saturated steam and/or hot water in a compact HRSG. This would increase the fuel utilization effectiveness tremendously and increase the capacity factor of the gas turbines for low levelized cost of electricity via higher generation hours and reduced O&M costs (i.e., reduced start–stop cycles). The system can be further enhanced by a storage component using battery technology (conventional lithium ion batteries, or one of the emerging technologies).

Hybrids like *fuel cell-gas turbines* are somewhat beyond the scope of this book. Skepticism of the author aside, the hybrid technology may very well be the shining star of power generation technology in the mid-twenty-first century. Alas, the author does not have enough background in fuel cell technology to do justice to a critique of it in a depth comparable to that in purely thermal systems. Fortunately, there are articles, papers, books, and reports galore that the interested reader can consult for more information. A simple Google search must turn up hundreds if not thousands of them.

The *integrated gasification combined cycle* (IGCC) is a subject deserving of a whole book on its own and, unsurprisingly, there are many such books out there. From the perspective of the flange-to-flange gas turbine, there are the following two major design aspects:

- Modification of an existing product to burn syngas generated by the gasifier (after cleanup);
- Integration with the gasification island (e.g., utilization of nitrogen generated in the air separation unit (ASU) as a diluent for NOx reduction).

The ability of gas turbines to burn different gaseous and liquid fuels, including gasification products, has been touched upon briefly in Chapter 12. A high-level review and pertinent references can be found in [11] in Chapter 8. For simplified modeling of an IGCC power plant, the reader should consult the paper by Gülen and Driscoll [2]. The IGCC was once proclaimed to be the savior of coal. The extreme complexity of the system (comprising three bona fide plants; i.e., power generation, gasification, and the ASU), astronomical capital expenditure, and long construction periods (subject to mistakes, schedule lapses, and more costs) practically killed the technology – at least in the USA at the time of writing.

In any event, burning low-Btu fuels such as syngas and blast furnace gas/coke oven gas requires modifications to the combustion system, HGP, and/or compressor of a unit from the particular OEM's standard natural gas product line. (There will *never* be a gas turbine developed from the proverbial blank sheet specifically for a low-Btu fuel. There is not a large enough market to justify the stupendous investment.) The first one includes replacement of the Dry-Low-NOx (DLN) combustor with a diffusion combustor that uses steam or nitrogen as a diluent for NOx abatement. The basic technology ingredients and references to consult for more in-depth study can be found in Chapter 12.

The HGP (i.e., the turbine section) can be modified by "opening" the stage 1 nozzle (S1N) inlet area to increase the swallowing capability of the turbine in order to handle part of the increased hot gas mass flow rate (caused by the large amount of fuel gas mass flow rate to compensate for the lack of heat content). See Figure 21.4 and the accompanying text – these should provide you with an adequate understanding of the basic idea. There is enough information in the book distributed among Chapters 10, 11, and 18 to enable you to do the requisite calculations in Excel (albeit a bit tedious) or in a commercial software tool such as Thermoflex or AxCycle.

The ultimate goal is to maintain the cycle pressure ratio in order to ensure that the compressor keeps operating near its original design point (away from the surge line or the operating limit line). S1N opening (maximum 6 percent or so increase in inlet area) cannot do the job by itself. Typical syngas plus diluent increases the turbine mass flow rate by about 15 percent or more. Reduced TIT helps a bit (see the turbine flow parameter in the aforementioned chapters), but it cannot be used as a full "knob" due to the detrimental impact on cycle efficiency. The reduction in TIT is mainly done in order to compensate for the high H_2O content in the hot gas, which changes the heat transfer characteristics in the HGP and reduces the part lives.

Some OEMs (e.g., GE) use air extraction to compensate for excess fuel flow rate in addition to S1N opening and TIT increase. Closing inlet guide vanes (IGVs) might help as well. Other OEMs, specifically MHI/MHPS, modify the compressor via blade "tip-cut" to shorten them and reduce the airflow (without using the IGVs).

Whatever the final modified package is, the generator output will increase, i.e.,

- Either by higher turbine shaft power generation and about the same compressor shaft power consumption; or
- By about the same turbine shaft power generation and lower compressor shaft power consumption.

The limit on how much output increase can be accommodated will be imposed by the torque capability of the original shaft and/or generator rating. The latter can be handled by using a larger generator. In the first modification option, the last-stage bucket might require a redesign to accommodate the higher HGP gas flow in accordance with the exit Mach number limits.

Burning hydrogen, either 100 percent H_2 fuel (which is the Holy Grail) or a CH_4–H_2 mixture, has been covered in Chapter 12. This will be a goal to be pursued quite aggressively by the OEMs with the support of government research labs and academia. The key technology requisite for widespread commercial deployment (provided, of course, "cheap" H_2 is available – a big hurdle) is a DLN combustor that can handle the large amounts of H_2 without running into stability problems.

24.3 Additive Manufacturing Technologies

Additive manufacturing (AM) is the technologist's term for 3D printing. The latter term also covers *rapid prototyping*. AM refers to the process of building a part – say, a turbine blade, which is a solid three-dimensional object – layer by layer from sliced computer-aided design models. The interested reader can find more information on the process from OEM websites or other online articles, papers, etc.

In 2017, Siemens announced the completion of the first full-load engine tests for gas turbine blades manufactured (the entire blade with a conventional blade design) by 3D printing at full engine conditions. The blades in question were for the HP rotor of an SGT-400 gas turbine (rated at 13 MW) with about a 1,250°C TIT and a shaft speed of 9,500 rpm. Siemens also reported having tested a new blade design with a completely revised and improved internal cooling geometry manufactured using 3D technology.

AM is obviously not going to change the laws of thermodynamics and the Euler turbine equation. In other words, it is not a magic wand to enable the design and development of 50 percent efficient gas turbines and 70 percent efficient combined cycles in direct violation of all the physics-based limits and constraints explored in detail in this book. AM, however, will be a key enabling technology to manufacture cooled HGP parts with intricate internal cooling flow channels and microscopic exit holes to approach the ideal limit of effusion cooling. If it can be pulled off successfully, this could provide a significant boost to turbine efficiency via drastic reduction of parasitic coolant flows. (See Chapter 8 for the sensitivity factor associated with cooling flows and Chapter 9 for the underlying theory.) This could also help with NOx emissions because lower flame zone and combustor exit temperatures will be needed to achieve a given firing temperature.

Another potential advantage of AM is in spare parts production upon demand in order to reduce spare parts inventories. This should reduce the repair and/or replacement costs of expensive HGP parts made from advanced superalloys and intricate cooling flow channels and holes. Presently, except from minor damage, which can be (maybe) blended or recoated, repairing such parts (e.g., weld repair) is practically impossible. Timely replacement requires maintaining a very expensive inventory on site or paying a premium to the OEM (in particular, stage 1 and 2 parts with single-crystal alloys, exotic coatings, and intricate structures cannot be provided by third-party suppliers for a long time to come), which adds significantly to the owner/operator O&M and long-term service agreement costs.

From the OEM perspective, AM significantly reduces the design and development cycle via rapid prototyping and reduces the (usually exorbitant) cost of bringing the new products and upgrades to the market. Combined with the ability of all major OEMs to carry out full-speed, full-load testing in their factory test beds, this could be a big enabler for bringing new technologies onto the market in quick succession with reduced risk. Having said that, considering the maturity of the underlying technology in the three major components, miracles should not be expected. (Consider the polytropic efficiency of modern axial compressors as an example, which is around 93 percent. How much more should one expect, especially with increasing pressure ratios driven by higher firing temperatures and increasing inlet Mach numbers driven by higher airflows?)

24.4 Data Analytics

Data analytics, big data, Industrial Internet of Things (frequently used as an acronym; i.e., IIoT) are the buzzwords (buzz-*terms*?) du jour at the time of writing this book (late 2016 to early 2018). They are frequently used interchangeably or in combination. (One should add "the cloud" to the bunch, too.) It is very unlikely that the readers of this book are not familiar with these terms. Therefore, copy-pasting stock definitions one can easily find in Wikipedia here is pointless.

Just like AM or 3D printing covered in the preceding section, neither of these technologies will make Messrs. Kelvin and Planck "roll over" in their graves and tell Mr. Euler "the news." They won't even budge. For about 90 percent of the subject matter covered in this book, IIoT and co. are and will be irrelevant.

The sole exception is Chapter 16, especially its coverage of condition-based monitoring and data reconciliation. There are hundreds of instruments attached to a gas turbine operating in the field, generating millions of bytes of data every few seconds. During normal operation, operators pay attention to a limited few, and only to the serious alarms (those labeled "HI-HI," typically; many low-level alarm notices generated by the distributed control system [DCS] simply go unnoticed or are simply ignored – to be looked at in the next outage). Trending real-time "raw" data to generate meaningful insights into the condition of a gas turbine running in the field requires filtering, reconciliation, correction (to a "base" state of ambient and load conditions), and reformulation (i.e., combining several parameters into a single representative metric).

Real-time performance monitoring systems have been developed and implemented in the field going back to the 1990s (e.g., see [14] in Chapter 16). The basic governing principles are covered in Section 16.6. State-of-the-art computing resources, IIoT, and complex mathematical and statistical algorithms are undoubtedly capable of enhancing the usefulness of existing and/or future monitoring tools. One example is collecting the time-series data generated by the DCS and transmitting it to a central analysis location run by a particular OEM, whose dedicated staff can monitor their fleet's operating status and act proactively. This is pretty much the model adopted by the major OEMs such as GE, who launched its *Predix* software, a cloud-based platform for creating IIoT applications, in 2015. MHPS introduced their version of a "digital solutions platform," *MHPS-TOMONI*, in 2017. Siemens has the "Digital Factory" to provide their power generation customers (and customers in other industries under the corporate umbrella) with similar capabilities.

Based on the author's long hands-on experience with performance monitoring and data analysis (Refs. [13,14] in Chapter 16), the weakness of conventional performance monitoring software has always been the lack of time, training, and engineering wherewithal of the plant personnel to make use of the information generated by the software. Consequently, a built-in ability of the DCS (say, via GE's Predix) to send the data to an OEM processing center to be looked at by subject matter experts is indeed a very good idea. Furthermore, if the software can carry out the "data cleansing" processes automatically and makes meaningful metrics available for study by OEM engineers, then that is a big plus, too. The icing on the cake is, of course, the software's ability to analyze the metrics automatically ("machine learning" and "artificial intelligence" are the buzzwords here) and come up with actionable solutions to existing or "soon-to-happen" problems.

So far, so good – this is all fine and dandy in theory. But there are several problems. The reader who is mesmerized by the corporate marketing hype generated around data analytics should be cognizant that the $80 million H–class behemoth can keep running like a Swiss chronograph, but a mundane $100,000 pump can go down and bring the power plant to a screeching halt for the next few days. In modern power plants, there are many auxiliary systems that are quite complicated and capricious to operate, but are mandated by the regulatory agencies. (One good example is the zero liquid discharge system, whose major components are extremely finicky.) The "on-paper" solution is simple enough: instrument and monitor *everything* in the power plant down to the last valve. The tools are certainly there: "the cloud," IIoT, supercomputers, and PhDs from Ivy League schools trampling each other to apply their math theories to the "big data." This might just work, of course, if the OEM of the gas turbine were also the OEM of *everything* in the power plant down to that last valve, which, of course, is not, never has been, and never will be the case.

Even if one assumes that the proprietary data-sharing problem is solved, there is the question of who is going to track hundreds of power plants transmitting data to the OEM data center, make note of software-generated reports (including actionable "solutions"), and decide to take action for each individual power plant. Furthermore, will the owner/operators of the power plant be ready to submit to the will of a computer thousands of miles away if it wants to shut down their unit in the middle of summer when power prices are $500/MWh because it thinks that there *might* be a problem with an obscure part? Automating everything and eliminating people is, of course, the ultimate goal of the artificial intelligence proponents. Whether this will happen or not remains to be seen. Whether this is a good thing or not is an altogether different matter. Here is a cautionary real story.

On the night of June 1, 2009, Air France Flight 447 (an Airbus 330), on route from Rio de Janeiro to Paris, crashed mysteriously (at the time) into the Atlantic Ocean. Sadly, 228 people, passengers and crew, perished. Almost 2 years later, in May 2011, the plane's flight data recorders were recovered from a depth of nearly 4 kilometers (a fascinating story in and of itself), which helped solve the mystery. Apparently, the pitot tubes generating the airspeed (by far the most vital information for an airplane since it is the flow of air across the wings generating the lift to keep it in the air to begin with) had malfunctioned in the inclement weather due to icing. This resulted in the disengagement of the autopilot.

Let us pause here for a moment. Modern passenger airplanes, especially wide-bodied jets on intercontinental routes, fly probably 90 percent of their entire journeys from point A to point B on autopilot. This is probably by far the widest use of artificial intelligence in a human endeavor. The ultimate dream of data nerds – full operation without humans – is a near reality in air travel today. Yet, what happened? A piece of equipment, barely worth a few hundred dollars, failed and catastrophe ensued. Was this inevitable? As it turns out, yes and no. If (i) one takes the human factor away and/or (ii) the "human factor" in question is not up to par, catastrophe is 100 percent guaranteed. Whether the mise-en-scène is an Airbus 330 plane or an H/J-class GTCC power plant is completely immaterial.

As it turned out (with no disrespect to the memory of the deceased pilots), once the autopilot disengaged, control of the airplane reverted back to the crew, *who could have easily resolved the problem.* (Remember: *there was absolutely nothing wrong with the airplane, its engines, or any of its electric/hydraulic/mechanic/whatever systems!*) Alas, to make a long story short, according to the final report released, "the crew failed to recognize that the aircraft had stalled and consequently did not make inputs that would have made it possible to recover from the stall."

Several months before the loss of Air France Flight 447, on January 15, 2009, a different kind of mishap had happened to US Airways Flight 1549 right after takeoff from the LaGuardia Airport in New York City. The plane was disabled by striking a large flock of Canada geese and lost power in both engines. If the plane were flown by so-called artificial intelligence, the 155 people on board would have been certainly doomed to a fiery death. Luckily, Captain Chesley Sullenberger was at the helm. A former US Air Force fighter pilot who flew in McDonnell-Douglas F-4 Phantoms, "Captain Sully" immediately determined that they would be unable to reach any airport (as they had no working jet engines!); he piloted the plane to a water landing on the Hudson River. All aboard were rescued by nearby boats.[6]

The moral of the story is clear: there is a very high likelihood – close to a certainty – that if Captain Sully had been in command of Air France Flight 447 on that fateful night in June 2009, the Airbus 330 with 228 people in it would have continued flying on its route to Paris with barely a hiccup.

There is absolutely no replacement for in-depth knowledge of a system, say, a gas turbine, or *any* system for that matter, which comes

- From years of undergraduate and graduate study of the subject matter;
- From decades of hands-on experience in the field, putting your hand on the actual chunk of steel day in and day out; and
- From decades of hands-on experience (concurrent or in succession, preferably field *before* office) in the "shop" designing, optimizing, and analyzing the same chunk of steel.

No IIoT, no artificial intelligence, no "cloud," no fancy-name software can replace an experienced engineer making his or her daily plant walk and listening to the hum of the machinery, looking into its guts and vital signals (on the DCS human–machine interface), and ensuring that the entire machinery is in good working order. This is the venerable and irreplaceable "look, listen, and feel" approach, which is the first step in preventive maintenance of mostly mundane but vital (for safety *and* performance) plant equipment.

Let this be your final takeaway from this book.

[6] In 2016, Clint Eastwood directed the Hollywood movie *Sully* with Tom Hanks in the role of Captain Sully and told the story.

References

1. Rosic, B., Denton, J. D., Horlock, J. H., Uchida, S., Integrated Combustor and Vane Concept in Gas Turbines, *Journal of Turbomachinery*, **134**: 3 (2012), 031005.
2. Gülen, S. C., Driscoll, A. V., Simple Parametric Model for Quick Assessment of IGCC Performance, *Journal of Engineering for Gas Turbines and Power*, **135** (2013), 010802.
3. Gülen, S. C., "Pressure Gain Combustion Advantage in Land-Based Electric Power Generation," GPPF2017-0006, 1st Global Power and Propulsion Forum, GPPF 2017, January 16–18, 2017, Zürich, Switzerland.

Appendix A Nomenclature

Late Walter Traupel's two-volume *Thermische Turbomaschinen* is arguably the best book ever written on turbomachinery fluid and thermodynamics (and much more!). Unfortunately, it still waits to be translated from German into English. However, even if you have a good command of technical German, the book is a real chore to read due to Dr. Traupel's extremely convoluted nomenclature and his idiosyncratic tendency to group several parameters into new dimensionless ones.

In this book, for the reader's convenience and maximum readability, standard, modern US textbook nomenclature is utilized for the most common thermodynamic and other engineering parameters of interest. In other words, p, T, v, h, u, and s are used for pressure, temperature, specific volume, specific enthalpy, specific internal energy (rarely needed in turbomachinery analysis), and specific entropy, respectively. The term "specific" means "per unit mass, m." Total values of volume, enthalpy, internal energy, and entropy are denoted by capital letters (i.e., V, H, U, and S, respectively).

For exergy, *a* and A are used (for *availability*, which is the term used in US textbooks such as Moran and Shapiro). This is one of the few cases where the same letter (i.e., *a*) is used for another parameter – in this case, the *speed of sound*. However, due to the significant difference of the contexts in which they appear, this is not expected to lead to any confusion. The other one involves V, used for total volume (in ft^3 or m^3) and velocity (ft/s or m/s). Once again, significant difference of the contexts in which they appear should preclude any confusion.

Specific heat capacities at constant pressure and volume are c_p and c_v, respectively; their ratio is γ. The term $(\gamma - 1)/\gamma$, which frequently appears in thermodynamic equations, is represented by the symbol *k*. (Note that in certain textbooks and papers, authors use *k* or κ for the specific heat ratio.)

The most common Greek letters used in the nomenclature are ρ (density), ω (angular speed), η (efficiency), and τ (torque or non-dimensional temperature – obvious within the context). Others are defined where they are used. Note that π is reserved for "pi" – *not* for pressure ratio (PR in this book). This should not be an issue for the majority of the readers, but there was a time when the symbol θ was used for temperature in US technical books (mainly in the 1950s and 1960s), which is, luckily, not the case anymore.

The universal gas constant is R_u and its division by the molecular weight, MW, gives the gas constant R for the fluid in question.

Power or work is denoted by $\dot{W}$ on a total basis (kW or MW) or as w on a per unit mass basis (Btu/lb or kJ/kg).

Heat transfer rate is denoted by $\dot{Q}$ on a total basis (Btu/s, kWth, or MWth) or as q on a per unit mass basis (Btu/lb or kJ/kg).

A "dot" over a parameter signifies a "rate of change," i.e.,

- $\dot{m}$ or "m dot" is mass flow rate;
- $\dot{V}$ or "V dot" is volumetric flow rate.

A "bar" over a parameter signifies a "mean" or "average," i.e.,

- $\bar{T}$ or "T bar" is average or mean temperature;
- $\bar{\eta}$ or "eta bar" is average or mean efficiency.

Other parameters are defined where they are used.

Subscripts

Subscripts are logical for most cases, i.e.,

a: Air
amb: Ambient
comb: Combustor
comp: Compressor
exh: Exhaust
f: Fuel
in or i: Inlet
out or ex or x: Outlet (exit)
stck: Stack
stm: Steam
turb: Turbine

Other subscripts (and superscripts, if any) are defined where they are used.

It is thus hoped that a reader with 4 years of undergraduate study in mechanical engineering under his or her belt will be able to follow the narrative without flipping back and forth through hundreds of pages.

State-Point Numbering

Standard numbering is used for *air-standard Brayton cycle* describing the ideal gas turbine, i.e.,

1. Compressor inlet
2. Compressor exit (heat adder inlet)
3. Combustor exit (turbine/expander inlet)
4. Turbine/expander exit

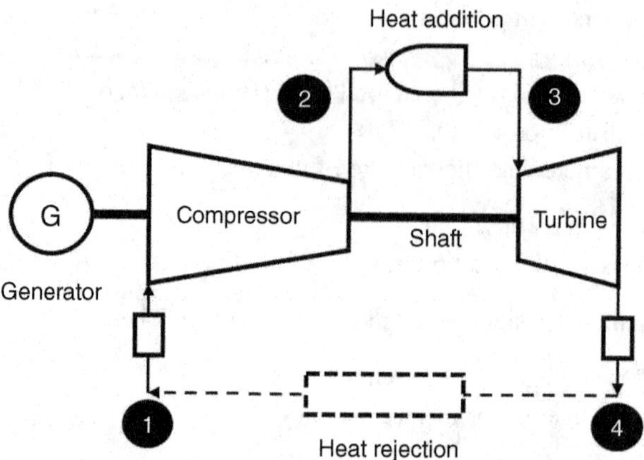

Figure A.1 Basic gas turbine diagram and state-point numbering.

This is illustrated in Figure A.1. Note that in this book the term "turbine" is reserved for the *expander*, whereas the entire unit is referred to as the "gas turbine." Since the coverage is limited to land-based electricity generation units, the more descriptive "gas turbine generator" and "gas turbine" should be understood to be one and the same. Also note the placement of the generator in Figure 3.1: it is attached to the compressor. This is not an arbitrary choice; it represents the configuration of most land-based gas turbines for electric generation and is known as "cold-end drive." Some older units (e.g., General Electric's 7EA) were "hot-end drive" machines (i.e., with the generator connected to the shaft on the turbine side.

Modern gas turbines, heavy-duty industrial or "frame" units, as well as the aeroderivative variants are "open-cycle" (i.e., "internal combustion") engines. There is no heat exchanger between state-points 4 and 1. Compositions of the working fluid at state-points 3 and 4 are different from the compositions of the working fluid (i.e., air) at state-points 1 and 2 due to combustion. The air-standard Brayton cycle assumes a hypothetical "heat adder" between state-points 2 and 3 and a hypothetical "heat ejector" between state-points 4 and 1. Composition of the working fluid does *not* change. The moniker "air-standard" means that the working fluid is air modeled as a *calorically perfect gas*, i.e.,

1. It obeys the ideal gas law, $pv = R_aT$ or $p = \rho R_aT$.
2. Specific heats, c_p and c_v, are constant.

Improvement on the air-standard Brayton cycle in terms of accuracy can be made by considering the working fluid as an *ideal gas* (i.e., by using temperature-dependent functions for c_p, one for state-points 1 and 2 [air] and another one for state-points 3 and 4 [combustion products]). For even better accuracy, pressure losses at the inlet, exhaust, and across the heat adder can be worked into the formulae. (This impacts calculations involving component isentropic or polytropic efficiencies. Otherwise, since ideal gas properties, with the exception of entropy, depend on temperature only, pressure losses do not impact energy balance calculations using enthalpies or specific heats.)

In general discussions, the following shorthand is used:

1. Gas turbine inlet (or ambient) temperature (TAMB) is T_1.
2. Compressor discharge temperature (CDT) is T_2.
3. Turbine inlet temperature (TIT) is T_3.
4. Turbine exhaust temperature (TEXH) is T_4.
5. Cycle pressure ratio is the compressor pressure ratio, PR = P_2/P_1.

Textbook gas turbine cycle improvements of intercooling and recuperation are not used in commercially available products nowadays.[1] There are, however, two exceptions:

1. General Electric's LMS100 aeroderivative gas turbine with intercoolers (Chapter 23);
2. Solar Turbines (owned by Caterpillar) Mercury 50 with a recuperator (Chapter 8).

These particular gas turbines are touched upon briefly in the book. Their requirement for an extended set of state-point numbering should thus not present an undue source of confusion.

[1] In the early days (1950s–1970s), however, recuperative gas turbines were widely used in pipeline and refinery applications (e.g., General Electric's Frame 3 and 5 gas turbines). Some of those old workhorses are still operational. They are discussed within the context of gas turbine development history in several places in the book.

Appendix B Acronyms and Abbreviations

Like any field of specialized expertise, the gas turbine technical literature is full of acronyms. In this book, acronyms are usually defined where they first appear in a chapter. Nevertheless, some acronyms appear so frequently (in this book and/or trade/academic literature) that they are defined in a central location anyway for easy reference. (Note that some acronyms [e.g., PR and CDT] are also used as variables in certain formulae.) Even so, it behooves to repeat the main caveat of the book you are holding once more. This is *not* a book intended for beginners. In fact, 90 percent of the acronyms used in the text are industry standards (HRSG, DLN, BOP, etc.) and, in all likelihood, most readers would require minimal need for a full-blown glossary.

The acronyms used in the book (and some *not* used in the book, but commonly encountered in the trade and academic literature) are listed below. Note that some of the terms are really not acronyms per se, but rather capitalized and "shortened" parameters such as TFIRE (firing temperature) or TAMB (ambient temperature). In order to make life easier for the reader unfamiliar with "industry-speak" (whose first reaction would be to go to the acronyms list when coming across an all-capitals term), they are also included herein.

2-D or 2D: Two-dimensional
2PRH(T): Two-pressure with reheat (steam cycle type)
2PNRH (T): Two-pressure with no reheat
3-D or 3D: Three-dimensional
3PRH(T): Three-pressure with reheat
3PNRH (T): Three-pressure with no reheat
ABB: Asea Brown-Boveri
AC or ac: Alternating current
ACC: Air-cooled condenser
AM: Advanced manufacturing (technical term for "3D printing")
ANSI: American National Standards Institute
AOO: Apache OpenOffice (software suite similar to Microsoft Office)
ASME: American Society of Mechanical Engineers
BBC: Brown-Boveri Company (forerunner of ABB and Alstom)
BC: Bottoming cycle

BOP: Balance of plant
CAC: Cooling air cooling (or cooler)
CAD: Computer-aided design
CAES: Compressed-air energy storage
CC: Combined cycle
CCS: Carbon capture and sequestration
CDA: Controlled diffusion airfoil
CDP: Compressor discharge pressure
CDT: Compressor discharge temperature
CEMS: Continuous emissions monitoring system
CFD: Computational fluid dynamics
CHAT: Cascaded humidified air turbine
CHP: Combined heat and power (European terminology for "cogeneration")
CIT: Compressor inlet temperature
CL-SC: Closed-loop steam cooling (e.g., GE's H-System or MHPS's G/J–class gas turbines)
CMC: Ceramic matrix composite
COD: Commercial operation date
COE: Cost of electricity
COT: Combustor outlet temperature – same as TIT (not used in this book)
CPC: Constant-pressure combustion
CPSC: Constant-pressure sequential combustion
CRH: Cold reheat
CSA: Contractual service agreement (a GE term, same as LTSA)
CT: Cooling tower[1]
CTQ: Critical to quality (a GE acronym, but used by others, too)
CVC: Constant-volume combustion
DC or dc: Direct current
DCS: Distributed control system
DLE: Dry-Low-Emissions (same as DLN)
DLN: Dry-Low-NOx
DOE: Department of Energy
EGC: Exhaust gas recirculation
EGT: Evaporative gas turbine
EOS: Equation of state
EPC: Engineering, procurement, and construction
EPRI: Electric Power Research Institute
FEA: Finite-element analysis
FERC: Federal Energy Regulatory Commission
FOD: Foreign object damage

[1] Not in this book, but in some, especially older, references, CT refers to "combustion turbine" (i.e., the gas turbine).

FSFL: Full-speed, full-load
FSNL: Full-speed, no load
FUA: Fuel Use Act
GE: General Electric (or gas engine)
GG: Gas generator (balanced shaft of a multi-shaft gas turbine)
GTCC: Gas turbine combined cycle (same as CC)
GTG: Gas turbine generator
GTW: *Gas Turbine World* (trade publication)
HAT: Humidified air turbine
HC: Heat (i.e., fuel) consumption
HCF: High-cycle fatigue
HCO: Hydraulic clearance optimization (Siemens technology – see Section 21.1)
HGP: Hot gas path
HHV: Higher (gross) heating value
HMI: Human–machine interface
HPC: High-pressure compressor (in aeroderivatives)
HPT: High-pressure turbine (in aeroderivatives and steam turbines)
HRB: Heat recovery boiler (same as HRSG)
HRH: Hot reheat
HVAC: Heating, ventilation, and air-conditioning
HRSG: Heat recovery steam generator
ICAD: Intercooled aeroderivative gas turbine
ICE: Internal combustion engine
ICV: Intercept control valve (steam turbine)
IGCC: Integrated gasification gas turbine
IGV: Inlet guide vane
IOU: Investor-owned utility
IPP: Independent power producer
ISO: International Standards Organization
KWU: Kraftwerkunion (forerunner of Siemens)
LCF: Low-cycle fatigue
LCI: Load commutated inverter
LCM: Lumped capacitance method (in heat transfer)
LCOE: Levelized cost of electricity
LHV: Lower (net) heating value
LPC: Low-pressure compressor (in aeroderivatives)
LPG: Liquid petroleum gas
LPT: Low-pressure turbine (in aeroderivatives and steam turbines)
LSB: Last-stage (gas or steam turbine) bucket
LTSA: Long-term service agreement (see CSA)
MBC: Model-based control
MCA: Multi-circular arc

MCV: Main (steam) control valve (steam turbine)
MECL: Minimum emissions-compliant load
MHI: Mitsubishi Heavy Industries
MHPS: Mitsubishi Hitachi Power Systems (MHI's new corporate name)
MOU: Memorandum of understanding
MWI: Modified Wobbe index
NDT: Non-destructive testing
NERC: North American Energy Reliability Council
NETL: US DOE's National Energy Technology Laboratory
NFPA: National Fire Protection Association
NGC: National Grid Code (UK)
NGCC: Natural gas-fired gas turbine combined cycle (same as CC or GTCC)
NPV: Net present value
OED: Oxford English Dictionary
O&M: Operations and maintenance
OEM: Original equipment manufacturer
OL-AC: Open-loop, air-cooling (HGP cooling with air extracted from the compressor)
OLL: Operating limit line
OT-CL: Once-through, closed-loop (water-cooled steam turbine condenser with a cooling tower)
OT-OL: Once-through, open-loop (water-cooled steam turbine condenser without a cooling tower)
OV: Open ventilated (gas turbine generator)
PCC: Post-combustion capture
PED: Pressure Equipment Directive (European equivalent of the ASME Boiler and Pressure Vessel Code)
PR: Pressure ratio
PT: Power turbine (or "free" turbine) in aeroderivatives
PURPA: Public Utility Regulatory Policies Act
R&D: Research and development
RAM: Reliability, availability, and maintainability
RH: Relative humidity or reheat (in the steam cycle)
RIT: Rotor inlet temperature[2]
ROI: Return on investment
RPM or rpm: Revolutions per minute
RTO: Regional Transmission Organization (similar to ISO)
SAC: Single annular combustor
SCC: Stress corrosion cracking
SCR: Selective catalytic reduction (emissions control technology)

[2] Applies only to the first-stage rotor of the turbine and is the same as the "firing temperature."

SCR:	Short-circuit ratio (in synchronous ac generators)
SPRINT:	Spray-intercooled gas turbine (GE trademark)
SSS:	Synchronous, self-shifting (overrunning clutch)
SSSF:	Steady-state, steady-flow
STG:	Steam turbine generator
STIG:	Steam-injected gas turbine (GE trademark)
TAMB:	Turbine ambient temperature
TBC:	Thermal barrier coating
TEXH:	Turbine exhaust (exit) temperature
TET:	Turbine exit temperature – same as TEXH (not used in this book)
TEWAC:	Totally enclosed water air-cooled (gas turbine generator)
TIT:	Turbine inlet temperature
TMI:	*Turbomachinery International* (trade publication)
UHC:	Unburned hydrocarbons
UPS:	Uninterruptible power supply
USUF:	Uniform-state, uniform-flow
VB(A):	Visual Basic (for Applications)
VFD:	Variable-frequency drive
VIGV:	Variable inlet guide vane (same as IGV)
VS(G)V:	Variable stator (guide) vane
VWO:	Valves wide open (for steam turbine)
WHR:	Waste heat recovery[3]
ZLD:	Zero liquid discharge

[3] Rarely used for electric power generation applications; typically reserved for small-scale cogeneration cases.

Appendix C Visual Basic Functions

Several methodologies, formulae, utility functions, etc., described in the text are coded in Excel VBA functions and subroutines, which are made available to the reader online on the Cambridge University Press website at www.cambridge.org/gasturbines.

Appendix D Convective Heat Transfer Coefficient

The most confusing and tedious task in heat transfer calculations encountered in steam turbine rotor heating/cooling and thermal stress problems is the estimation of the convective heat transfer coefficient (HTC), h, which governs the temperature profile in the boundary layer between the steam flow and the rotor surface (more on that in Chapter 19). Estimates of the average HTC based on the steam flow, pressure, and temperature properties can be obtained from the famous Dittus–Bölter (D–B) equation for certain geometries (e.g., flow over a flat plate of length L [[9] in Chapter 19]). For "real" problems with complex geometries, the "core" of the D–B formula is of interest, i.e.,

$$\bar{h} \propto \dot{m}^{0.8}.$$

In other words, for the scaling of the convective HTC at the steam flow conditions encountered in startup processes, proportionality to the flow rate raised to a power of 0.8 is the key. A method for the determination of HTCs inside steam turbine cylinders and buckets is given in the paper by Breitkopf [1]. One way to come up with simplified HTC transfer functions is to use the D–B formula to generate a polynomial or power-law function in temperature at a given steam pressure and velocity and modify it via flow rate scaling and a *calibration factor* (CF) for adjustment to known test or literature data (i.e., in US customary units, HTC is in Btu/h-ft^2-R and temperature is in °F):

$$\bar{h} = \mathrm{CF}\left(aT^2 + bT + c\right)\left(\frac{\dot{m}}{\dot{m}_{full}}\right)^{0.8} \tag{D.1}$$

$$\bar{h} = \mathrm{CF}\left(aT^b\right)\left(\frac{\dot{m}}{\dot{m}_{full}}\right)^{0.8} \tag{D.2}$$

where $\dot{m}_{full}$ is the steam flow rate at full load. For 1,200 psia and 500 ft/s, the coefficients in Equation D.1 are

a = 0.000202355
b = –0.471223
c = 382.102

and calculated HTC values are 850 to 700 Btu/h-ft^2-R for temperatures 800 to 1,100°F. (Note that the second-order polynomial fit would not extrapolate well outside its data range.) With CF = 1, these values are within ±50 Btu/h-ft^2-R of those reported in [1].

For lower-pressure steam at 120 psia, the coefficients in Equation D.2 are

$a = 1275.5$
$b = -0.366$

and calculated HTC values are 135 to 95 Btu/h-ft^2-R for temperatures 500 to 1,100°F. For startup steam admission applications, the recommended CF is 0.1, which results in HTC values similar to those used in the turbine stress controller.

Reference

1. Breitkopf, G. E., Determination of Heat Transfer Coefficients in Steam Turbines, *Wärme- und Stoffübertragung*, **13** (1980), 195–204.

Index

Printed in the United States
by Baker & Taylor Publisher Services